Reza Rezaee, Brian J. Evans (Eds.)

Natural Hydrogen Systems

Also of interest

Hydrogen Production, Storage and Utilization.
Thermochemical, Electrochemical, Sonochemical, Biological and Photocatalytic
Processes
Slimane Merouani, Oualid Hamdaoui (Eds.), 2025
ISBN 978-3-11-162375-7, e-ISBN (PDF) 978-3-11-162386-3,
e-ISBN (EPUB) 978-3-11-162396-2

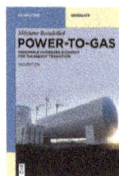

Power-to-Gas.
Renewable Hydrogen Economy for the Energy Transition
2nd Edition
Méziane Boudellal, 2023
ISBN 978-3-11-078180-9, e-ISBN (PDF) 978-3-11-078189-2,
e-ISBN (EPUB) 978-3-11-078200-4

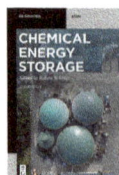

Chemical Energy Storage
2nd Edition
Robert Schlögl (Ed.), 2022
ISBN 978-3-11-060843-4, e-ISBN (PDF) 978-3-11-060845-8,
e-ISBN (EPUB) 978-3-11-060859-5

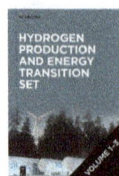

ENERGY, ENVIRONMENT AND NEW MATERIALS (SET)
Vol. 1: Hydrogen Production and Energy Transition
Vol. 2: Hydrogen Storage for Sustainability
Vol. 3: Utilization of Hydrogen for Sustainable Energy and Fuels
Marcel Van de Voorde (Ed.), 2021
ISBN (Set) 978-3-11-075497-1

Natural Hydrogen Systems

Properties, Occurrences, Generation Mechanisms,
Exploration, Storage and Transportation

Edited by
Reza Rezaee and Brian J. Evans

DE GRUYTER

Authors

Prof. Reza Rezaee
Petrophysics and Reservoir Characterization
Faculty of Science and Engineering
WASM: Minerals, Energy and Chemical
Engineering
Curtin University
Kent Street
Perth, Bentley, WA 6102
Australia
r.rezaee@curtin.edu.au

Prof. Brian J. Evans
Geophysics and Reservoir Characterization
Faculty of Science and Engineering
WASM: Minerals, Energy and Chemical
Engineering
Curtin University
Kent Street
Perth, Bentley, WA 6102
Australia
b.evans@curtin.edu.au

ISBN 978-3-11-143703-3
e-ISBN (PDF) 978-3-11-143704-0
e-ISBN (EPUB) 978-3-11-143761-3

Library of Congress Control Number: 2025934926

Bibliographic information published by the Deutsche Nationalbibliothek
The Deutsche Nationalbibliothek lists this publication in the Deutsche Nationalbibliografie;
detailed bibliographic data are available on the Internet at http://dnb.dnb.de.

© 2025 Walter de Gruyter GmbH, Berlin/Boston, Genthiner Straße 13, 10785 Berlin
Cover image: Scharfsinn86/iStock/Getty Images Plus
Typesetting: Integra Software Services Pvt. Ltd.

www.degruyter.com
Questions about General Product Safety Regulation:
productsafety@degruyterbrill.com

Preface

In 1859, Edwin Drake drilled the first successful oil well in Pennsylvania, marking the birth of the modern oil industry. This groundbreaking discovery, prompted by observations of natural oil seeps, laid the foundation for today's energy sector and catalyzed transformations in industrialization, transportation, and the global economy. However, at that time, knowledge of petroleum systems – how oil is generated, migrates through the subsurface, and is trapped in reservoirs – was extremely limited. The industry took decades to fully understand the complexities of these systems, building the knowledge we now take for granted.

Today, we find ourselves at a similar juncture with **natural hydrogen**. Like early oil exploration, our understanding of hydrogen systems – how it forms, accumulates, and can be extracted – is in its infancy. However, unlike the pioneers of the oil industry, we have the benefit of modern scientific techniques and the wealth of experience gained from over a century of oil and gas exploration. Yet, there remain significant gaps in our understanding that must be filled to unlock natural hydrogen's potential as a viable energy resource.

This book is designed to shed light on these critical aspects of natural hydrogen systems. Throughout the chapters, we will explore the geological processes involved in hydrogen generation, migration, and storage, while also examining the tools and methodologies that can help identify and quantify potential hydrogen reserves. Each chapter is dedicated to a key element of the natural hydrogen system, from the geology and geochemistry of hydrogen formation to the technical challenges of reservoir discovery and production.

By the end of this book, readers will have a deeper understanding of how natural hydrogen systems work, the parallels with petroleum systems, and the unique challenges and opportunities that hydrogen presents. With this knowledge, we can build a more robust framework for natural hydrogen exploration, enabling it to play a significant role in the future of sustainable energy.

Reza Rezaee and Brian J. Evans
February 2025

https://doi.org/10.1515/9783111437040-202

Contents

Part I: Geological foundations of natural hydrogen

Part II: Natural hydrogen generation mechanisms

Part III: Hydrogen exploration and detection techniques

Editorial note

Natural hydrogen has recently emerged as a promising clean energy resource with the potential to significantly contribute to the global energy transition. Unlike conventional hydrogen production methods, natural hydrogen occurs geologically in various settings, offering a cost-effective and low-carbon alternative. This book provides a comprehensive exploration of natural hydrogen, including its geological properties, generation mechanisms, exploration methodologies, global case studies, and challenges in storage, transportation, and environmental impact. The chapters in this volume collectively represent the latest advancements and research efforts driving the development of natural hydrogen as a sustainable energy resource.

The first part, **Geological foundations of natural hydrogen**, establishes the geological principles governing natural hydrogen systems. Chapter 1 introduces a comparative framework between natural hydrogen and petroleum systems, highlighting key geological elements such as hydrogen source rock quantitative analysis through serpentinization, migration pathways, and caprock sealing efficiency. Chapter 2 explores subsurface hydrogen accumulations identified from drilling results, analyzing migration dynamics, and the role of Fe(II) oxidation as a primary source. Chapter 3 discusses the similarities and differences between natural hydrogen and hydrocarbon exploration, emphasizing established workflows that may be adapted to hydrogen prospecting. Chapter 4 provides a detailed review of the geological and geochemical pathways responsible for onshore natural hydrogen generation, with a focus on Archean and Neoproterozoic cratons, water-rock interactions, and serpentinization processes.

The second part, **Natural hydrogen generation mechanisms**, investigates the fundamental processes driving hydrogen generation. Chapter 5 investigates the role of shallow peridotites at magma-poor rifted margins, highlighting how serpentinization contributes to hydrogen production, particularly in ocean-continent transition zones. Chapter 6 examines the significance of cratons as stable geological environments conducive to hydrogen generation and storage, presenting case studies from West Africa, Brazil, and Australia. Chapter 7 introduces an alternative hypothesis, suggesting that hydrogen could originate from organic matter decomposition and catalytic reactions in sedimentary basins. Chapter 8 presents the primordially hydridic Earth concept, which theorizes that Earth's interior retains vast hydrogen reserves from planetary formation.

The third part, **Hydrogen exploration and detection techniques**, focuses on methodologies for detecting and quantifying natural hydrogen resources. Chapter 9 investigates how rock physics properties, such as density, elastic moduli, and seismic velocity changes, can be used to detect hydrogen-bearing formations through geophysical techniques. Chapter 10 expands on geophysical responses associated with trapped hydrogen, including seismic, magnetic, and gravity survey methods. Chapter 11 introduces an innovative airborne Raman LIDAR system for hydrogen detection, enabling large-scale, noninvasive hydrogen seep mapping. Chapter 12 explores sur-

https://doi.org/10.1515/9783111437040-204

face gas geochemistry as an exploration tool, highlighting challenges in distinguishing geological hydrogen from biological and anthropogenic sources. Chapter 13 presents natural hydrogen favorability maps (NHFMs) as a systematic approach for integrating geological, geophysical, and geochemical data to improve exploration success rates. Chapter 14 concludes the part by discussing numerical basin modeling techniques, illustrating how these models simulate hydrogen generation, migration, and phase behavior at a regional scale.

The fourth part, **Global case studies and regional insights**, presents case studies from different geological settings, where natural hydrogen has been observed. Chapter 15 examines Brazil's hydrogen exploration efforts, emphasizing fieldwork methodologies and long-term monitoring of seepage zones. Chapter 16 provides an overview of China's hydrogen occurrences, linking them to geological processes such as radiolysis and mantle degassing. Chapter 17 investigates the potential for hydrogen reservoirs in the Korean Peninsula based on regional tectonic history. Chapter 18 highlights Japan's ongoing research efforts, including mechanochemical hydrogen generation and potential hydrogen-bearing formations. Chapter 19 discusses the geological and geophysical factors contributing to Saudi Arabia's hydrogen system potential, integrating data from previous hydrocarbon exploration projects to assess key reservoir-seal pairs and hydrogen migration pathways.

The final part, **Hydrogen storage, transportation, environmental and technological challenges**, addresses the practical challenges associated with storing, transporting, and utilizing natural hydrogen. Chapter 20 draws lessons from underground hydrogen storage (UHS) research, highlighting parallels between UHS and natural hydrogen reservoirs. Chapter 21 evaluates the feasibility of hydrogen storage in salt caverns in Oman, with a detailed assessment of site selection criteria and storage capacity. Chapter 22 discusses the logistical and technological challenges of transporting hydrogen at an industrial scale, covering different transport methods such as pipelines, liquefaction, and solid-state carriers. Chapter 23 examines the environmental impacts of hydrogen production, storage, and utilization, focusing on greenhouse gas emissions, leakage risks, and nitrogen oxide formation. Chapter 24 explores engineering challenges in drilling, well integrity, and completion technologies for hydrogen extraction, with a focus on materials selection and subsurface hydrogen containment. The book concludes with Chapter 25, which analyzes interactions between hydrogen, rock salts, and clays, offering insights into hydrogen retention and contamination risks in geological reservoirs.

This book provides an invaluable resource for researchers, geoscientists, and energy professionals involved in the exploration and development of natural hydrogen. By integrating geological, geochemical, and technological perspectives, this volume underscores the growing significance of natural hydrogen as a viable energy resource in the global transition to clean energy.

Reza Rezaee and Brian J. Evans

Contributing authors

Chapter 1
Prof. Reza Rezaee
WA School of Mines: Minerals, Energy and
Chemical Engineering
Curtin University
Perth, Bentley, WA
Australia
r.rezaee@curtin.edu.au

Chapter 2
Prof. Eric Deville
IFPEN-IFP School
av. Bois-Préau
92852 Rueil-Malmaison
France
eric.deville59@gmail.com; eric.deville@ifpen.fr

Chapter 3
Dr. Gabor C. Tari
OMV Energy
Trabrennstrasse 6-8
1020 Vienna
Austria
gabor.tari@omv.com

Chapter 4
Dr. Kanchana Kularatne
Laboratoire des Fluides Complexes et leurs
Réservoirs
Université de Pau et des pays de l'Adour
Pyrénées-Atlantiques
France
kanchana.kularatne@univ-pau.fr

Dr. Isabelle Moretti
LFCR
UPPA Rue de l'Université
64012 Pau
France
isabelle.moretti@univ-pau.fr

Chapter 5
Dr. Javier García-Pintado
MARUM Center for Marine Environmental
Sciences

Universität Bremen, Bremen
Germany
jgarciapintado@marum.de

Prof. Marta Pérez-Gussinyé
MARUM Center for Marine Environmental
Sciences
Universität Bremen, Bremen
Germany
mpgussinye@marum.de

Chapter 6
Prof. Humberto Reis
HR Consulting Energy and Geosciences Ltda.
Rua Nunes Vieira, 35, 401, Belo Horizonte, Minas
Gerais
Brazil
and
Departamento de Ciência da Computação/
Instituto de Geosciencias
Universidade Federal de Minas Gerais (UFMG)
Campus Pampulha, Belo Horizonte, Minas Gerais
Brazil
hr@hrenergyandgeosciences.com,
reis.humbertols@gmail.com

Dr. Olivier Lhote
ENGIE, Paris La Défense
France
Olivier.LHOTE@storengy.com

Chapter 7
Mr. John Hanson
Independent Correspondence
England, UK
jcph100@gmail.com

Chapter 8
Mr. Vitaly Vidavskiy
Western Australian School of Mines
Curtin University
Kensington, 6151 WA
Australia
and

https://doi.org/10.1515/9783111437040-205

AVALIO
West Perth, WA 6005
Australia
Vv@avalio.net

Mr. Nikolay Larin
Natural Hydrogen Energy (NH2E)
Greeley, CO, USA
nl@avalio.net

Chapter 9
Dr. Yashee Mathur
Department of Energy Science and Engineering
Stanford University
Stanford, CA, USA
yashee@stanford.edu

Prof. Tapan Mukerji
Department of Energy Science and Engineering
Stanford University
Stanford, CA, USA
mukerji@stanford.edu

Chapter 10
Prof. Brian J. Evans
Curtin University
Perth, WA
Australia
b.evans@curtin.edu.au

Chapter 11
Prof. Charlie Ironside
Department of Physics and Astronomy
Curtin University
Perth, WA
Australia
Charlie.Ironside@curtin.edu.au

Prof. Mervyn Lynch
Department of Physics and Astronomy
Curtin University
Perth, WA
Australia
m.lynch@curtin.edu.au

Dr. Jacob Martin
Department of Physics and Astronomy
Curtin University

Perth, WA
Australia
jacob.w.martin@curtin.edu.au

Prof. Mark Paskevicius
Department of Physics and Astronomy
Curtin University
Perth, WA
Australia
M.Paskevicius@curtin.edu.au

Dr. Mauricio Di Lorenzo
Department of Physics and Astronomy
Curtin University
Perth, WA
Australia
Mauricio.Dilorenzoruggeri@curtin.edu.au

Prof. Craig E. Buckley
Department of Physics and Astronomy
Curtin University
Perth, WA
Australia
C.Buckley@curtin.edu.au

Mr. Andrew Lockwood
Xcalibur Multiphysics
Perth, WA
Australia
andrew.lockwood@curtin.edu.au

Chapter 12
Dr. Giuseppe Etiope
Istituto Nazionale di Geofisica e Vulcanologia
Sezione Roma 2
Rome
Italy
giuseppe.etiope@ingv.it

Alexandra Orbán
Faculty of Environmental Science and Engineering
Babes-Bolyai University
Cluj-Napoca
Romania
alexandra.orban@ubbcluj.ro

Chapter 13
Assoc. Prof. Mahmoud Leila
School of Mining and Geosciences
Nazarbayev University
Astana
Kazakhstan
Mahmoud.leila@nu.edu.kz

Dr. Fiammetta Mondino
Terra-A AG
Wuhrstrasse 14, 9490 Vaduz
Liechtenstein
Mahmoud.leila@nu.edu.kz

Dr. Aya Yasser
Faculty of Science
Mansoura University
Mansoura
Egypt
Mahmoud.leila@nu.edu.kz

Prof. Randy Hazlett
School of Mining and Geosciences
Nazarbayev University
Astana
Kazakhstan
randy.hazlett@nu.edu.kz

Chapter 14
Dr. Nicolas Ferrando
IFPEN
1-4 av. de Bois-Préau
92500 Rueil-Malmaison
France
nicolas.ferrando@ifpen.fr

Dr. Marie-Christine Cacas-Stentz
IFPEN
1-4 av. de Bois-Préau
92500 Rueil-Malmaison
France
marie-christine.cacas-stentz@ifpen.fr

Dr. Francesco Patacchini
IFPEN
1-4 av. de Bois-Préau
92500 Rueil-Malmaison
France
francesco.patacchini@ifpen.fr

Dr. Françoise Willien
IFPEN
1-4 av. de Bois-Préau
92500 Rueil-Malmaison
France
françoise.willien@ifpen.fr

Dr. Benjamin Braconnier
IFPEN
1-4 av. de Bois-Préau
92500 Rueil-Malmaison
France
benjamin.braconnier@ifpen.fr

Chapter 15
Assoc. Prof. Corinne Arrouvel
UFRJ–IMQ/CM Campus Macaé
Rio de Janeiro, Brazil
corinne.arrouvel@imq.macae.ufrj.br

Chapter 16
Dr. Xueying Yin
H2Terra Ltd Co.
Beijing 102600
China
xyin@h2terra.com

Dr. Bingchuan Yin
H2Terra Ltd Co.
Beijing 102600
China
byin@h2terra.com

Chapter 17
Prof. Hyeong Soo Kim
Department of Earth and Environmental
Sciences
Korea University
Seoul 02841
Republic of Korea
haskim2@korea.ac.kr

Chapter 18
Assoc. Prof. Yunfeng Liang
Department of Systems Innovation
Graduate School of Engineering
The University of Tokyo
Tokyo 113-8656, Japan
liang@sys.t.u-tokyo.ac.jp

Dr. Wuge Cui
Department of Systems Innovation
Graduate School of Engineering
The University of Tokyo
Tokyo 113-8656
Japan
wuge.cui@hotmail.com

Dr. Arata Kioka
Department of Systems Innovation
Graduate School of Engineering
The University of Tokyo
Tokyo 113-8656
Japan
kioka@sys.t.u-tokyo.ac.jp

Prof. Takeshi Tsuji
Department of Systems Innovation
Graduate School of Engineering
The University of Tokyo
Tokyo 113-8656
Japan
tsuji@sys.t.u-tokyo.ac.jp

Chapter 19
Dr. Manzar Fawad
Center for Integrative Petroleum Research
King Fahd University of Petroleum and Minerals
Dhahran 31261
Saudi Arabia
manzar.fawad@kfupm.edu.sa

Prof. Scott Andrew Whattam
Department of Geosciences
King Fahd University of Petroleum and Minerals
Dhahran 31261
Saudi Arabia
scott.whattam@kfupm.edu.sa

Dr. Abdullah Alqubalee
Center for Integrative Petroleum Research
King Fahd University of Petroleum and Minerals
Dhahran 31261
Saudi Arabia
abdullah.alqubalee@kfupm.edu.sa

Dr. Ahmed Al-Yaseri
Center for Integrative Petroleum Research
King Fahd University of Petroleum and Minerals

Dhahran 31261
Saudi Arabia
ahmed.yaseri@kfupm.edu.sa

Chapter 20
Dr. Quan Xie
Discipline of Energy Engineering
Curtin University
26 Dick Perry Avenue
6151 Kensington
Australia
quan.xie@curtin.edu.au

Mr. Adnan Aftab
Discipline of Energy Engineering
Curtin University
26 Dick Perry Avenue
6151 Kensington
Australia
adnan.aftab@postgrad.curtin.edu.au

Assoc. Prof. Mohammad Sarmadivaleh
Discipline of Energy Engineering
Curtin University
26 Dick Perry Avenue
6151 Kensington
Australia
mohammad.sarmadivaleh@curtin.edu.au

Dr. Lingping Zeng
CSIRO Energy
Melbourne
VIC 3168
Australia
lingping.zeng@csiro.au

Dr. Alireza Safari
Department of Earth Resources Engineering
Graduate School of Engineering
Kyushu University
744 Motooka, Nishi Ward
Fukuoka
Japan
alireza.safari@curtin.edu.au

Chapter 21
Dr. Mohammed Al Kindi
Earth Sciences Consultancy Centre (ESCC)
Ghala Heights

Muscat
Oman
malkindi@omanescc.com

Mr. Muhannad Al Hinai
Ministry of Energy and Minerals (MEM)
Al-Khuwair, Ministry Streets, Muscat
Oman
malkindi@omanescc.com

Mr. Ahmed Al Abri
Hydrom Oman
Mina Al Fahal, Muscat
Oman
malkindi@omanescc.com

Dr. Zaid Al Siyabi
Vision Advanced Petroleum Solutions L.L.C
(VAPS)
Rusail Industrial City
Al Seeb, Muscat
Oman
malkindi@omanescc.com

Chapter 22
Dr. Christopher Lagat
WA School of Mines: Minerals, Energy and
Chemical Engineering
Curtin University
Perth, WA
Australia
christopher.lagat@curtin.edu.au

Chapter 23
Dr. David A. Wood
DWA Energy Limited
Lincoln
UK
dw@dwasolutions.com

Prof. Reza Rezaee
WA School of Mines: Minerals, Energy and
Chemical Engineering
Curtin University
Perth, WA 6102
Australia
r.rezaee@curtin.edu.au

Chapter 24
Prof. Md Mofazzal Hossain
Discipline of Energy Engineering
School of WASM Minerals, Energy and Chemical
Engineering
Curtin University
Perth, WA
Australia
md.hossain@curtin.edu.au

Chapter 25
Dr. Yongqiang Chen
WA School of Mines: Minerals, Energy and
Chemical Engineering
Curtin University
Kensington, WA 6151
Australia
yongqiang.chen@csiro.au and
chenyongqiang86@foxmail.com

Part I: **Geological foundations of natural hydrogen**

Reza Rezaee
Chapter 1
Evaluating natural hydrogen systems through the lens of petroleum exploration

Abstract: This chapter investigates the potential of natural hydrogen as a sustainable energy source by comparing its system elements with those of conventional petroleum systems. Emphasizing hydrogen generation through the continental serpentinization of ultramafic rocks, this chapter explores key processes shared by both energy systems. It introduces a quantitative framework for assessing natural hydrogen resources, incorporating multiple regression analysis of laboratory data under varying temperature and pressure conditions to estimate hydrogen generation rates. Additionally, it performs volumetric calculations of hydrogen production during serpentinization, evaluates caprock sealing efficiency, and analyses migration pathways to identify conditions favorable for natural hydrogen accumulation. These findings represent an important step toward understanding natural hydrogen's feasibility as a clean energy resource and highlight the need for continued investigation into its commercial potential.

Keywords: natural hydrogen, generation mechanisms, serpentinization, hydrogen generation rate, volumetric calculations of hydrogen, caprock sealing efficiency, migration pathways

1.1 Introduction

There is a growing need for sustainable and environmentally friendly energy sources. Among various clean energy options, natural hydrogen emerges as a promising alternative. Traditionally, there are several methods to generate hydrogen through industrial processes such as steam methane reforming (SMR) [1], where methane reacts with steam at high temperatures to yield hydrogen, carbon monoxide, and carbon dioxide. Alternative methods include partial oxidation (POX) [2], coal gasification [3], and electrolysis [4]. However, there is another type of hydrogen, known as natural or geologic hydrogen, where hydrogen is generated and accumulated in the subsurface by natural processes somehow comparable to fossil fuel generation and accumulation.

While biotic sources of hydrogen in the subsurface are known, hydrogen is also found in recent hot igneous rocks where no microbes can survive, indicating possible

Reza Rezaee, WA School of Mines: Minerals, Energy and Chemical Engineering, Curtin University, Perth, WA, Australia, e-mail: r.rezaee@curtin.edu.au

https://doi.org/10.1515/9783111437040-001

abiotic sources [5, 6]. Exploration of relatively pure hydrogen in Mali [7] and the detection of molecular hydrogen (H_2) in numerous deep and shallow boreholes without a clear source explanation have triggered the concept of natural hydrogen. Several studies from different countries have also reported naturally degassing hydrogen from the surface of the Earth [7–15].

Hydrocarbon has been the basis of the global energy supply for over a century and its system is well-understood, with established methods for exploration, production and utilization. The fundamental elements of a petroleum system including source rock, reservoir, migration pathways, caprock, and trap, have been extensively studied, leading to sophisticated models that guide oil and gas exploration. In contrast, the natural hydrogen system represents a relatively new and under-explored study area. For natural hydrogen to accumulate in significant quantities, similar to hydrocarbons in petroleum systems, the key elements and processes that govern hydrogen generation, migration, and trapping within geological systems must be present.

This study compares the similarities and differences between petroleum and natural hydrogen system elements. The elements that will be discussed include source or generation mechanisms, the quantitative hydrogen generation rate through serpentinization at the continent, reservoir rocks, caprock sealing efficiency, and migration pathways. However, because the first element – source or generation mechanisms and the quantitative hydrogen generation rate through serpentinization – is crucial for establishing the concept of a natural hydrogen system, it will be discussed in greater detail. In contrast, minimal attention will be given to the elements of a petroleum system, as this topic is already well-developed and extensively covered in numerous other publications. Therefore, they will be discussed only briefly, emphasizing the key points. To address these objectives, this chapter is organized into several sections, including the geological processes underlying the formation of natural hydrogen, serpentinization as the primary source of natural hydrogen, quantitative calculations of hydrogen generated through serpentinization, fluid-rock interactions in hydrogen reservoirs, caprock sealing efficiency for hydrogen, and hydrogen migration pathways.

1.2 Source and generation mechanisms of natural hydrogen

Several potential sources have been proposed to explain the presence of natural hydrogen. Natural hydrogen, also known as "white" or "gold" hydrogen, is produced through natural geological processes and differs from "green" hydrogen (produced via water electrolysis) or "gray" and "blue" hydrogen (derived from fossil fuels). Unlike petroleum systems where organic-rich, thermally mature formations are the primary hydrocarbon sources, the origin of natural hydrogen remains a topic of debate.

Proposed sources include abiotic, biotic, and anthropogenic processes, with Milkov [16] identifying 32 mechanisms. Significant sources include hydrogen produced through the serpentinization of ultramafic rocks and mantle-derived hydrogen, supported by evidence from magmatic gases [17–22]. Thermal maturation of organic matter also contributes under specific conditions, with hydrogen generation peaking at a vitrinite reflectance of ~2.0% and declining thereafter [23]. Radiolysis, where ionizing radiation from radioactive elements breaks down water molecules into hydrogen and oxygen, further illustrates the diverse origins of natural hydrogen, particularly in granite or uranium-bearing formations [24, 25].

While numerous geological processes produce hydrogen, only a few, such as serpentinization and deep-seated mantle degassing, potentially generate quantities significant enough for commercial exploitation. Understanding these mechanisms is crucial for advancing the utilization of natural hydrogen as a sustainable energy resource. The following sections focus on serpentinization, the primary mechanism for hydrogen generation.

1.2.1 Serpentinization

Because serpentinization occurs in ultramafic rocks, it is important to provide an overview of their composition and mineralogy. Ultramafic rocks are a class of igneous rocks characterized by low silica content (less than 45%) and high concentrations of iron (Fe) and magnesium (Mg). Ultramafic rocks are commonly associated with tectonic settings such as ophiolites, subduction zones, and mantle peridotites, where serpentinization can occur under specific pressure-temperature conditions. Understanding their mineral composition helps in evaluating their potential for hydrogen generation and other geochemical transformations.

These rocks predominantly consist of minerals such as olivine, orthopyroxene, and clinopyroxene [26, 27], which play a crucial role in serpentinization. Table 1.1 summarizes the primary minerals found in ultramafic rocks, their chemical composi-

Table 1.1: Primary minerals in ultramafic rocks and their role in serpentinization.

Mineral	Chemical formula	Occurrence	Role in serpentinization
Olivine	$(Mg,Fe)_2SiO_4$	Common in peridotite and dunite	Highly reactive, primary precursor to serpentine
Orthopyroxene	$(Mg,Fe)_2Si_2O_6$	Present in peridotites, websterite, and harzburgites	Can undergo alteration but is less reactive than olivine
Clinopyroxene	$(Ca,Mg,Fe)_2Si_2O_6$	Found in wehrlite, websterite, and gabbro	More resistant to serpentinization processes

tion, and their role in serpentinization. Figure 1.1 further classifies ultramafic rocks based on the modal proportions of olivine (Ol), orthopyroxene (Opx), and clinopyroxene (Cpx).

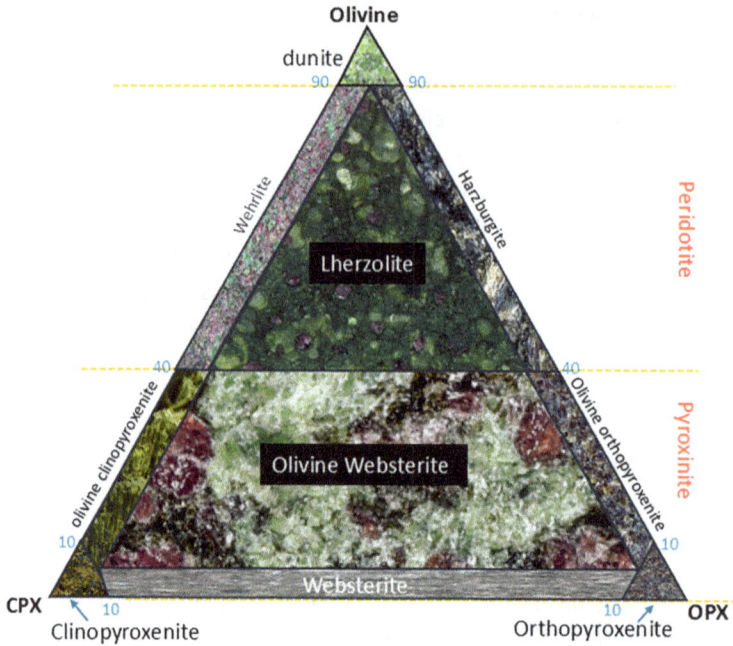

Figure 1.1: Ultramafic rocks are classified based on the relative proportions by volume, of olivine (Ol), clinopyroxene (Cpx), and orthopyroxene (Opx) (modified after [28]).

The chemical stability of ultramafic minerals can be understood through Bowen's reaction series (BRS) (Figure 1.2), which describes the order of mineral crystallization as magma cools. Developed by Norman L. Bowen in the early twentieth century, this model explains how different minerals solidify at specific temperatures, offering insight into the formation and stability of igneous rocks. One key aspect of BRS is that minerals crystallizing at higher temperatures – such as olivine – are less stable at the Earth's surface, whereas minerals that form at lower temperatures tend to be more resistant to alteration. This is because high-temperature minerals are far from their equilibrium state under surface conditions, making them highly reactive. For instance, olivine, which crystallizes early in the series, is particularly prone to alteration, including transformation through serpentinization when exposed to water.

Interaction of ultramafic rocks with water at temperatures below 400 °C, which marks the stability threshold of olivine, initiates serpentinization, leading to the formation of serpentine and other secondary minerals. This reaction often results in the production of hydrogen-rich fluids [30–35]. Olivine, which is abundant in dunite and

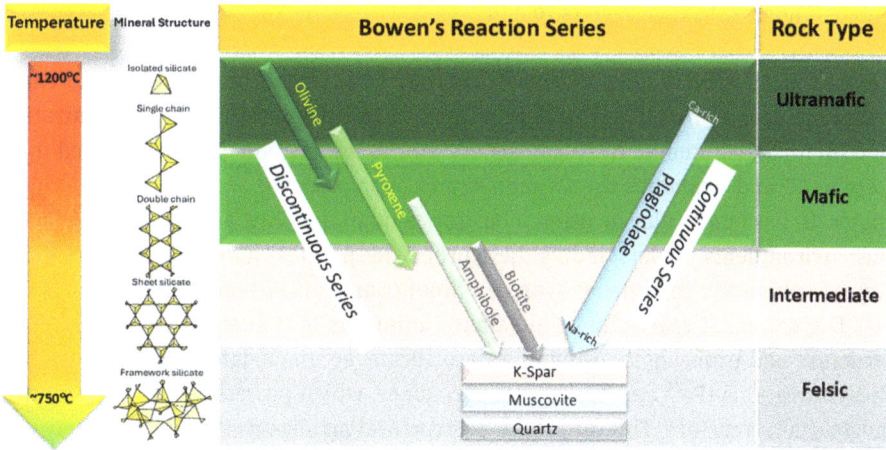

Figure 1.2: Bowen's reaction series (modified after [29]) highlights the mineral composition of ultramafic rocks, which are primarily composed of olivine, with lesser amounts of pyroxene. The figure also depicts the mineralogical structures, showing that olivine, which has a relatively simple atomic arrangement, crystallizes at higher temperatures and is highly susceptible to alteration at lower temperatures.

peridotite (Figure 1.1), is a dominant mineral in the Earth's mantle, oceanic crust, and ophiolitic sequences [36–38]. The alteration of olivine has been widely recognized as a key process in molecular hydrogen generation, following the reaction:

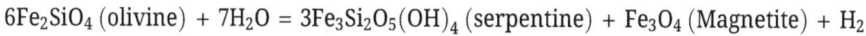

$$6Fe_2SiO_4 \text{ (olivine)} + 7H_2O = 3Fe_3Si_2O_5(OH)_4 \text{ (serpentine)} + Fe_3O_4 \text{ (Magnetite)} + H_2$$

This equation demonstrates how olivine reacts with water to form serpentine and magnetite, with molecular hydrogen (H_2) released as a byproduct [39]. Although other ultramafic minerals, such as orthopyroxene and clinopyroxene, can also undergo serpentinization and contribute to hydrogen production, their lower reactivity and more complex crystalline structures typically result in slower reaction rates and lower hydrogen yields [31].

Hydrogen production in serpentinization is closely linked to the oxidation of ferrous iron (Fe^{2+}) to ferric iron (Fe^{3+}) via its interaction with water, which follows the general reaction:

$$2FeO + H_2O \rightarrow Fe_2O_3 + H_2$$

The overall quantity of H_2 generated during serpentinization is influenced by iron partitioning among the reaction products. The incorporation of ferric iron (Fe^{3+}) into serpentine and magnetite promotes hydrogen production, whereas ferrous iron (Fe^{2+}) retained in brucite or serpentine does not contribute to hydrogen release. Consequently, the distribution of iron from olivine among magnetite, serpentine, and brucite plays a critical role in determining the total hydrogen output as the reaction progresses [30, 35].

1.2.1.1 Mineral stability and its role in serpentinization

The chemical stability and Fe-driven reactivity of olivine, orthopyroxene, and clinopyroxene play a crucial role in determining their behavior during serpentinization and their contribution to hydrogen generation. The discussion below is adapted from Rezaee 40 and examines the structural and compositional characteristics of these minerals, with particular focus on their iron content, stability, and reactivity in aqueous environments. Understanding these properties provides a foundation for estimating the volumetric potential of hydrogen production from ultramafic rocks.

The chemical reactivity of ultramafic minerals is influenced by their crystal structure and iron content. Olivine, a nesosilicate, accommodates higher amounts of divalent cations (Fe^{2+} and Mg^{2+}) than pyroxenes, which possess a more constrained single-chain structure. This difference in structural arrangement influences their susceptibility to oxidation. Additionally, during magma cooling, Fe and Mg are progressively depleted, leading to compositional variations among the ultramafic minerals. As a result, olivine tends to have a higher Fe content than orthopyroxene and clinopyroxene, as shown in Table 1.2. Among these minerals, olivine exhibits the lowest chemical stability, primarily due to its simple structure, which lacks the robust bonding networks found in pyroxenes. This makes olivine highly reactive, especially under low-temperature aqueous conditions, where it readily alters to form serpentine and magnetite, releasing hydrogen in the process. With FeO content ranging between 10% and 35% [41], olivine is particularly susceptible to oxidation-driven alteration. Given its high Fe content and structural simplicity, olivine is assigned a stability index of 1.0 (Table 1.2). Orthopyroxene, in contrast, has greater resistance to alteration due to its single-chain silicate structure, which provides increased thermodynamic stability. Its Fe content (5% to 25% FeO) [41] is lower than that of olivine, making it less prone to rapid oxidation. Furthermore, its chain structure shields Fe^{2+} from direct exposure to reactive environments, reducing its alteration rate. Based on these properties, orthopyroxene is assigned a stability index of 0.7. Clinopyroxene is the most chemically stable among the three minerals. Its more complex chain structure, along with calcium incorporation, reduces its reactivity by lowering Gibbs free energy and enhancing structural resilience. Additionally, its relatively low Fe content (2% to 15% FeO) [41] contributes to its greater resistance to oxidation. For these reasons, clinopyroxene is assigned a stability index of 0.5, reflecting its superior resilience to alteration.

Using olivine as the reference mineral (assigned a stability value of 1.0 or 100%), the relative stability of orthopyroxene and clinopyroxene can be ranked accordingly (Table 1.2). The lower the stability index, greater is the resistance to alteration under serpentinization conditions. These rankings align with BRS, where minerals crystallizing at higher temperatures (such as olivine) are generally less stable than those forming at lower temperatures.

These values are provisional and provide a preliminary estimation based on the general thermodynamic principles of mineral stability, which correlate with BRS: high-temperature minerals (like olivine) are less stable, while lower-temperature minerals are more stable. Future experimental studies may refine these stability rankings and further clarify the interplay between iron oxidation, mineral stability, and hydrogen production in natural systems.

Table 1.2: Chemical stability of ultramafic-mafic minerals.

Mineral	Fe content (%)	Stability rank (C)	Explanation
Olivine	10–35%	1.0	Olivine is the least stable mineral under surface conditions due to its susceptibility to alteration. It has a high Fe content, increasing in more Fe-rich varieties, especially in the fayalite end-member (Fe_2SiO_4).
Orthopyroxene	5–25%	0.7	Orthopyroxene is more stable than olivine, but still reactive under serpentinization conditions. Orthopyroxene (e.g., enstatite) contains moderate Fe content, with Fe^{2+} substituting for Mg^{2+} in its crystal structure. Higher Fe in Fe-rich varieties such as ferrosilite.
Clinopyroxene	2–15%	0.5	Clinopyroxene is slightly more stable than orthopyroxene, making it less reactive to surface alteration. Clinopyroxene (e.g., augite) contains less Fe than olivine and orthopyroxene, with a greater proportion of Mg and Ca in its structure.

A lower stability rank indicates that a mineral is more stable and less prone to alteration in surface environments (from, Rezaee, [40]).

1.2.1.2 Serpentinization environments

Ultramafic rocks are primarily found in the Earth's mantle. They can be brought closer to the surface at mid-ocean ridges, where they form new oceanic crust. Some ultramafic rocks can be found closer to the Earth's surface or even exposed due to geological processes like uplift and erosion. They can intrude into the continental crust and can occur as ophiolite packages, which are sections of oceanic lithosphere thrust onto continents.

Serpentinization typically occurs in mid-ocean ridges (MORs) [42–45], subduction zones, ophiolite complexes [33, 46–48], and any other places where iron-rich, ultramafic rocks of the mantle have been emplaced by geological processes to a shallower depth at the continent. At mid-ocean ridges, fluids rich in hydrogen are often found alongside nitrogen (N_2) and methane (CH_4), while black smokers additionally emit carbon dioxide (CO_2) [49, 50]. Serpentinization is strongly influenced by mid-ocean ridge spreading rates. Periods of super-continental breakup characterized by slow-spreading ridges have resulted in extensive serpentinization [51].

Natural hydrogen (H_2) seepages are also observed onshore, associated with ophiolite complexes such as in Oman, the Philippines, the northwestern Pyrenees, Turkey [52–56], and the Hakuba Happo area in Japan (recent visit by the author), where serpentinization of ultramafic mantle bodies emplaced at shallow depths generate hydrogen. Hydrogen emissions in these locations are considered evidence of ongoing serpentinization processes. Vacquand et al. [57] presented a comparative analysis of reduced gas seepages within ophiolitic complexes across Oman, the Philippines, Turkey, and New Caledonia. The seepages predominantly consist of hydrogen, methane, and nitrogen in varying proportions. The hydrogen-rich gases were found to be associated with ultrabasic spring waters or directly seeping through fracture systems within the ophiolitic rocks. Methane is produced through the reaction of dissolved CO_2 with mafic-ultramafic rocks during serpentinization, or subsequently through interactions involving H_2 and CO_2 [58, 59] through $CO_2 + 4H_2 = CH_4 + 2H_2O$ reaction. The presence of hydrogen in the gas indicates the system has depleted available carbon for further methane generation.

The presence of water is essential for the serpentinization process to occur, and the availability of water at greater depths becomes limited, which can restrict the process of serpentinization. Consequently, serpentinization will only take place where ultramafic rocks come into contact with water, either through deep fluid infiltration in fracture zones and faults or by uplift and exposure to the water circulation depth realm. In sea floors and MORs, rocks are in direct contact with seawater, and the subduction of the oceanic crust carries water down into deeper parts of the Earth. However, other mechanisms must be considered for deep water circulation on continents.

Fault zones act as pathways for fluid migration through the lithosphere due to enhanced permeability from brittle fracturing [60] and ductile flow, which can establish a dynamic granular fluid pump in ductile shear zones [61]. Mcintosh and Ferguson [62] reported meteoric water circulation depths across North America reaching up to 5 km. Prigent et al. [63] suggested that seawater percolation extended to depths of 20–25 km and that serpentinization extended to approximately 11–13 km. Evidence for the passage of surface-derived fluids through retrogressive ductile shear zones at temperatures of 400–450 °C and depths of about 10 km has recently been obtained in the French Pyrenees [64].

Another mechanism that can provide water for serpentinization is through pelitic rock metamorphism. Water contained in pelitic sediments amounts to approximately 10% of the total mass of Earth's oceanic water [65]. During metamorphic processes, these fluids, which are chemically bound within minerals, are released and ascend towards the Earth's surface [66].

1.2.1.3 Hydrogen generation rates in serpentinization

As outlined by Rezaee [40], the estimation of hydrogen generation rates during serpentinization is essential for assessing its viability as a potential natural hydrogen source. Several factors influence hydrogen production efficiency, including mineral composition, temperature, pressure, and the water-to-rock (w/r) ratio. Fe-rich olivine plays a dominant role in hydrogen generation, producing more hydrogen compared to Fe-poor olivine [30]. The optimal temperature range for serpentinization-driven hydrogen production is 200–315 °C, although the reaction can occur over a broader range [35, 67]. The w/r ratio significantly affects redox conditions and, consequently, hydrogen concentrations. Generally, higher w/r conditions generate more hydrogen per unit of rock. Pressure enhances serpentinization by accelerating mineral-fluid interactions and enabling deeper water infiltration into ultramafic rocks, thereby increasing hydrogen production. Beyond merely affecting reaction rates, pressure also drives phase transitions in olivine and influences porosity and permeability, ultimately regulating fluid movement and hydrogen release in subsurface reservoirs [68–72].

1.2.1.3.1 Experimental and field studies on hydrogen generation

Experimental research studies have provided quantitative insights into serpentinization-driven gas production, with several studies investigating hydrogen yields under varying conditions of temperature, pressure, grain size, and pH (Table 1.3). Studies by Berndt et al. [32], McCollom and Seewald [73], and Allen and Seyfried Jr [68] have demonstrated that hydrogen production is significantly influenced by reaction conditions, with yields varying from 0.012 mmol/kg to over 244 mmol/kg depending on mineralogy, fluid composition, and thermal regimes. For instance, Berndt et al. [32] recorded 158 mmol/kg hydrogen generation from powdered olivine at 300 °C and 500 bar, while Grozeva et al. [74] found 65 mmol/kg hydrogen from harzburgite under seawater-rich conditions at 300 °C and 350 bar. At lower temperatures, Mayhew et al. [75] observed hydrogen yields ranging from 0.04 mmol/kg to 0.3 mmol/kg during the serpentinization of peridotite at 55–100 °C. Similarly, McCollom et al. [76] found that higher pH values accelerate olivine alteration, enhancing hydrogen production up to 105 mmol/kg at 230 °C and 350 bar.

Field-based studies provide further insights into serpentinization-associated hydrogen emissions. Donval et al. [77] analyzed hydrothermal fluids from the Rainbow vent field at 360 °C, detecting 13 mmol/kg hydrogen. Charlou et al. [42] reported 16 mmol/kg hydrogen emissions from ultramafic-hosted hydrothermal fluids along the mid-Atlantic Ridge, where hydrogen comprised more than 40% of the total extracted gases.

1.2.1.3.2 Modeling and large-scale estimations

Numerical and geochemical models have also been used to predict hydrogen concentrations under different serpentinization conditions. McCollom and Bach [35] found that hydrogen generation peaks at approximately 360 mmol/kg at 315 °C, but decreases to less than 70 mmol/kg at 400 °C and 7 mmol/kg at 50 °C. Klein et al. [30] reported that peridotite-dominated lithologies generate more hydrogen than pyroxene-rich lithologies, with estimated yields of 200–340 mmol/kg at 250–340 °C, dropping to 50–150 mmol/kg below 200 °C. Further studies by Albers et al. [31] indicated that serpentinization at MORs can yield 200–350 mmol H_2 per kg of rock, while ultraslow-spreading ridges produce 50–150 mmol/kg under temperatures below 200 °C.

According to Liu et al. [78], the theoretical maximum hydrogen yield from serpentinization is 454 mmol/kg, assuming complete olivine conversion. However, iron partitioning into non-reactive phases such as brucite significantly reduces this yield, with practical values estimated at 227 mmol/kg, closely matching experimental results by Huang et al. [79]. Table 1.3 taken from Rezaee [40] provides a collection of quantitative key studies.

Table 1.3: List of available quantitative studies for hydrogen generation during the serpentinization of ultramafic minerals and rocks (modified from Rezaee [40]).

Study	H_2 mmol/kg	Conditions	Mineral/grain size	Method
Berndt et al. [32]	158	300 °C, 500 bar, 69 days	Olivine (30–80 um)	Lab experiment
McCollom and Seewald [73]	74	300 °C, 350 bar, 133 days	Olivine (fine powder)	Lab experiment
Allen and Seyfried Jr [80]	25	400 °C, 500 bar, 60 days	Olivine and pyroxene (60 µm)	Lab experiment
Jones et al. [81]	12	200 °C, 300 bar, 55 days	olivine (100um)	Lab experiment
Neubeck et al. [82]	0.5–1.5	30–70 °C, low pressure, 270 days	Olivine (0.250 and 0.500 mm)	Lab experiment
Marcaillou et al. [83]	14	300 °C, 300 bar, 70 days	Peridotite (~1 µm)	Lab experiment
Mayhew et al. [75]	0.04	55 °C, low pressure, 100 days	Peridotite (53–212 µm)	Lab experiment
Mayhew et al. [75]	0.3	100 °C, low pressure, 100 days	Peridotite (53–212 µm)	Lab experiment
Okland et al. [84]	0.012	25 °C, low pressure, 99 days	Dunite (<63 µm)	Lab experiment

Table 1.3 (continued)

Study	H$_2$ mmol/kg	Conditions	Mineral/grain size	Method
McCollom and Donaldson [85]	0.2	90 °C, near-surface, 213 days	Olivine (53–212 µm)	Lab experiment
Grozeva et al. [74]	65	300 °C, 350 bar, 85 days	Harzburgite (63–125 µm)	Lab experiment
Miller et al. [86]	0.47	100 °C, low pressure, 97 days	Dunite (53–212 µm)	Lab experiment
Huang et al. [79]	244	300 °C, 3,000 bar, alkaline solution, 8 day	Olivine (>250 µm)	Lab experiment
McCollom et al. [76]	105	230 °C, 350 bar, high pH, 375 days	Olivine (80%) and pyroxene (20%) (38–212 um)	Lab experiment
Donval et al. [77]	13	NA	NA	MOR (Rainbow hydrothermal field at mid-Atlantic ridge)
Charlou et al. [42]	16	NA	NA	MOR (Rainbow hydrothermal field at md-Atlantic ridge)
Kelley et al. [50]	15	NA	NA	MOR (Lost City hydrothermal field)
McCollom and Bach [35]	360	NA	Olivine	Modeling
Klein et al. [30]	200–340	NA	Olivine	Modeling
Klein et al. [30]	50–150	NA	Olivine	Modeling
Albers et al. [31]	200–350	NA	Mantle rocks	Modeling
Etiope [48]	0.72	NA	Dunite	Extraction

1.2.1.4 Empirical modeling of hydrogen generation rates

Quantifying hydrogen generation rates during serpentinization is essential for evaluating its potential as a natural hydrogen source. To establish a predictive framework, Rezaee [40] conducted a multiple regression analysis using available experimental data (see Table 1.3), leading to the development of an empirical equation (1.1) that correlates hydrogen generation rate (RH₂) with temperature and pressure. Despite a relatively small dataset, the regression model achieved a coefficient of determination (R^2) of approximately 0.86, indicating a strong relationship between these variables and hydrogen production efficiency (Figure 1.3). The derived equation is as follows:

$$R_{H2}(mmol/kg) = 3.102 + (0.12T) + (0.073P) \tag{1.1}$$

where T is the temperature (°C) and P represents pressure (bar). This equation is valid up to 315 °C, at which point hydrogen production reaches its peak, aligning with findings from McCollom and Bach [35]. Beyond 315 °C, hydrogen generation declines significantly, dropping to approximately 25 mmol/kg at 400 °C [80].

Given this trend, a single equation cannot accurately capture the behavior across the entire temperature range. Instead, two separate models are required: one for temperatures below 315 °C and another for the range between 315 °C and 400 °C, where hydrogen production gradually declines.

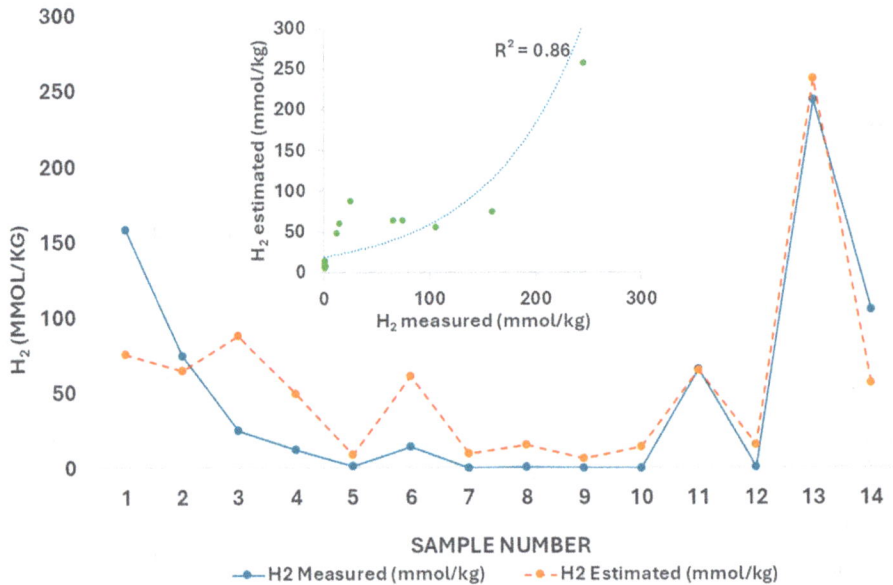

Figure 1.3: Correlation between experimentally acquired hydrogen from serpentinization and the hydrogen estimated using eq. (1.1). In most cases, the data points show good agreement. The sample numbers are assigned based on their order as presented in Table 1.3 (modified from Rezaee, [40]).

1.2.1.4.1 Refining the hydrogen generation model

Rezaee [40] presented a comparative analysis of hydrogen generation rates (Figure 1.4) predicted by eq. (1.1) (dashed line) and a tentative refined model (gray-shaded area) that more closely reflects real-world serpentinization conditions. The refined curve accounts for variations in depth/temperature, and mineral alteration kinetics, particularly as ultramafic minerals destabilize and serpentinization initiates. This pattern is analogous to thermal maturation curves in petroleum systems, which define the transformation of organic matter into hydrocarbons. While petroleum models are well-established due to decades of research, the development of a similarly reliable hydrogen generation curve requires further experimental validation.

To improve accuracy, regression analysis was performed on two distinct temperature ranges, yielding two empirical equations, eqs. (1.2) and (1.3):

$$H_2(mmol/kg) = 0.43T - 9.33 \qquad \text{for temperature less than 315 °C} \qquad (1.2)$$

$$H_2(mmol/kg) = -1.38T + 571.3 \qquad \text{for temperature between 315 °C and 400 °C} \quad (1.3)$$

where T is the temperature in degrees Celsius.

Notably, for $T < 315$ °C, both eqs. (1.1) and (1.2) yield similar results. However, beyond 315 °C, eq. (1.1) tends to overestimate hydrogen production, necessitating the use of eq. (1.3) instead. Since pressure effects were already incorporated in the original regression model, a separate pressure-dependent term was not required for these equations.

In Figure 1.4, the gray-shaded area is interpreted as the total hydrogen generation potential, estimated at approximately 227 mmol/kg of ultramafic rock. Hydrogen production initiates at depths where olivine begins to destabilize (~400 °C) and continues toward shallower environments over geological timescales. The cumulative hydrogen yield can be approximated based on the initial volume of ultramafic rock undergoing alteration. At different depth intervals, variations in temperature and pressure conditions influence the rate of hydrogen production, with peak generation occurring around 315 °C, where serpentinization is most efficient.

1.2.1.5 Volumetric estimation of hydrogen generated through serpentinization

A key aspect of assessing a natural hydrogen system is estimating the total hydrogen yield (GH$_2$) possibly produced by the serpentinization of ultramafic rock masses. This estimation is essential for determining the potential accumulation of hydrogen in subsurface reservoirs with effective caprocks. The volumetric calculation framework introduced in Rezaee [40] consists of five key stages, outlined below:

1. **Determining the gross rock volume (GRV)**
   - The dimensions of the ultramafic rock body are established, including its surface area and thickness, to calculate its gross rock volume (GRV) in cubic meters:

$$GRV(m^3) = Area\ (m^2) \times thickness\ (m) \qquad (1.4)$$

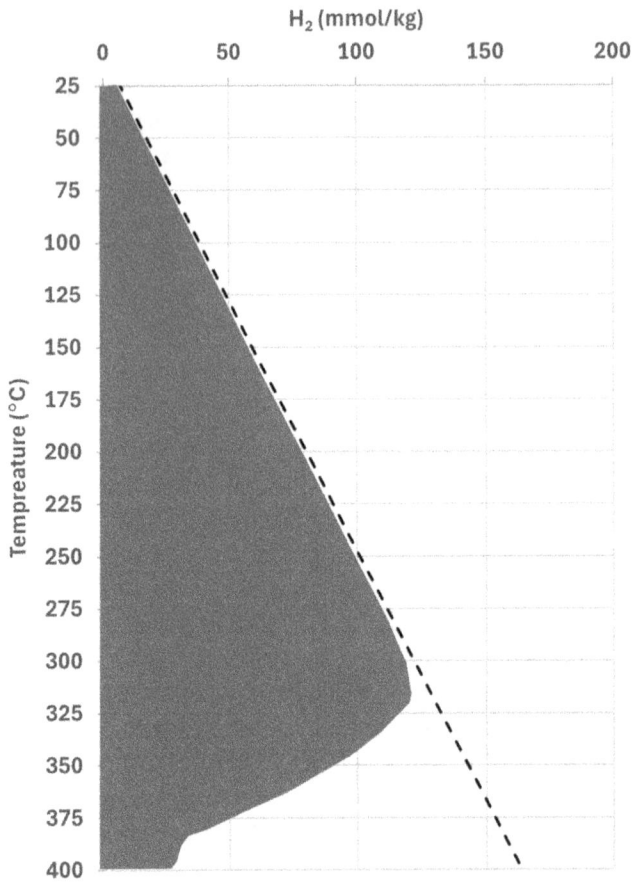

Figure 1.4: Temperature versus hydrogen generation (adapted from Rezaee, [40]). The dashed line represents hydrogen production estimates derived from the empirical equation (1.1), assuming a geothermal gradient of 30 °C/km. The gray-shaded region indicates potential real-world variations in hydrogen output, suggesting that additional thermodynamic and geochemical factors beyond temperature and pressure may influence the process. Hydrogen generation reaches its peak around 315 °C, aligning with the observations of McCollom and Bach [35].

2. **Converting GRV to gross rock mass (GRM)**

   – The GRV is multiplied by the density (ρ_{rock}) of the ultramafic rock to obtain the gross rock mass (GRM) in kilograms:

$$\text{GRM (kg)} = \text{GRV} \left(\text{m}^3\right) \times \rho_{rock}\left(\text{kg/m}^3\right) \tag{1.5}$$

3. **Applying a mineralogical correction factor $(C)^*$**
   – A correction factor $(C)^*$ is introduced to account for the chemical stability and Fe content of the ultramafic rock, as detailed in Table 1.2. If the precise

mineral composition is unknown, a default value of $C = 1$ can be applied, assuming olivine-dominated lithologies, which are prevalent in the upper mantle.

4. **Determining the hydrogen generation rate (RH₂)**
   - The rate of hydrogen production (RH₂) is computed based on temperature and pressure conditions, using the empirical relationships established in eqs. (1.1)–(1.3).
5. **Calculating the total generated hydrogen (GH₂)**
   - The total hydrogen mass generated is obtained by integrating these parameters into the following equation:

$$G_{H2}(kg) = 2 \times 10^{-6} GRM \times C \times R_{H2} \tag{1.6}$$

GH_2 represents the total hydrogen generated (kg), 2×10^{-6} is a conversion factor for computing hydrogen mass from mmol, R_{H2} denotes the hydrogen generation rate (mmol/kg), and C^* accounts for the mineralogical composition of the rock.

This quantitative approach enables a systematic assessment of hydrogen potential in ultramafic terrains, facilitating more accurate reservoir modeling and exploration strategies.

*To refine hydrogen generation estimates during serpentinization, a correction factor (C) is applied based on the chemical stability and Fe content of ultramafic minerals (See Table 1.2). As outlined in Rezaee [40], this factor accounts for variations in mineral composition and their respective contributions to hydrogen production, using olivine as the reference standard ($C = 1.0$). The Fe content of olivine is variable, influenced by magma composition, geological setting, and pressure-temperature conditions. However, since the empirical equation (1.1) for hydrogen generation rate (R_{H2}) was primarily developed from experiments on olivine-rich rocks, eq. (1.6) assumes $C = 1.0$ for olivine, reflecting the natural serpentinization process. For lithologies with lower olivine content or those rich in pyroxenes, a correction is necessary, as serpentinization rates and hydrogen yields differ. For rocks with mixed mineralogy, the correction factor is determined by weighting the contribution of each mineral. For example, in a rock composed of 60% olivine and 40% orthopyroxene (where $C = 1.0$ for olivine and 0.7 for orthopyroxene), the overall correction factor is: $C = (0.6 \times 1.0) + (0.4 \times 0.7) = 0.88$. If non-reactive minerals such as feldspar are present, they are assigned $C = 0$, as they do not undergo serpentinization or contribute to hydrogen generation. This ensures that only reactive minerals are considered in hydrogen yield calculations. While this correction factor provides a logical framework for estimating hydrogen production, it remains an approximation based on existing experimental data. Further laboratory research will be required to refine mineral-specific hydrogen generation rates and improve model accuracy. Nevertheless, Klein's thermodynamic modelling [30] suggests that serpentinization should produce roughly equivalent amounts of H₂ from a given mass of rock across protolith compositions – ranging from dunite to harzburgite and lherzolite – when reacted at the same temperature.

Rezaee [40] attempted to calculate the volume of hydrogen generated (G_{H2}) through serpentinization for the Giles Complex. This complex, a large ultramafic-mafic intrusion in Australia, is a mantle-derived complex, formed approximately 1,070 million years ago. It consists of around 20 layered mafic-ultramafic intrusions with notable variations in composition and emplacement depth. The Giles Complex volume is not explicitly stated, but estimates suggest it is large. The Mantamaru-Cavenagh portion contributes ~20,000 km^3, and the major ultramafic intrusions in the Mamutjarra Zone are estimated at ~82,000 km^3 [87]. Aitken et al. [88] estimate the volume at ~32,000 km^3, with a total likely exceeding 100,000 km^3 when including deeper intrusions and eroded material. The intrusion ranges from 20-km depth to surface exposure.

Rezaee [40] considered a volume of 32,000 km^3, a temperature range of 623 °C to 23 °C (corresponding to depths from 20 km to the surface with a 30 °C/km geothermal gradient), and an assumed composition of 50% olivine and 25% orthopyroxene and 25% clinopyroxene, with a C value of 0.8. He calculated the total rock volume by dividing the intrusion into 100-meter depth intervals. Hydrogen generation was considered negligible for temperatures above 400 °C, while eq. (1.3) was applied for temperatures between 400 °C and 315 °C, and eq. (1.1) was used for temperatures below 315 °C (Rezaee, [40]). He estimated hydrogen yield for the Giles Complex to be around 2.24×10^{15} kg. However, not all ultramafic rock undergoes complete serpentinization. If we consider that only a small portion of the ultramafic mass (e.g., 0.01%) has undergone serpentinization so far due to its complex composition, serpentinization heterogeneity, and considering that some of the generated hydrogen may have been consumed in reactions like CO_2 reduction to methane, etc., the estimated hydrogen production calculated to be approximately 2.24×10^{13} kg, equivalent to 8.88 Pscf (Peta standard cubic feet) of hydrogen gas at standard conditions.

While this calculation provides an approximation of hydrogen potential, it does not account for trapping and accumulation mechanisms, which require further geological evaluation. Similar to petroleum system assessments, where hydrocarbon generation is linked to source rock properties, this estimation offers an initial guide for exploring natural hydrogen reservoirs in ultramafic-mafic complexes.

1.3 Reservoir rock

Like a petroleum system, gas generated from the source rock must be accumulated in porous and permeable reservoir rock. However, unlike hydrocarbons, which have little-to-no impact on the rock's petrophysical properties through chemical alteration of rock-forming minerals, hydrogen may react with these minerals.

1.3.1 Hydrogen interaction with the reservoir rock

Hydrogen-induced redox reactions can alter the rock's mineral composition, affecting processes of mineral dissolution and precipitation [89–91]. These mineralogical changes can impact the physical properties of the rocks, particularly their porosity and permeability.

Two primary types of interactions can occur within hydrogen reservoirs: abiotic interactions, which involve geochemical reactions between fluids and rocks that alter the physical properties of the reservoir rocks and biotic reactions, driven by microbial activity, that modify the composition of the hydrogen.

1.3.1.1 Abiotic interactions

Hydrogen's interaction with reservoir rock has the potential to significantly alter the rock's petrophysical properties by triggering both mineral dissolution and precipitation processes. As hydrogen reacts with various minerals within the rock, some minerals may dissolve, which can increase the porosity and permeability of the rock. On the other hand, the precipitation of new minerals, such as clay or sulfides, can reduce these properties by clogging pore spaces. These changes can affect the overall quality and fluid flow within the reservoir.

Yekta [91] conducted experimental and numerical studies on abiotic reactions in sandstone to assess hydrogen's reactivity with minerals under typical reservoir conditions. Limited reactions were observed, leading to the conclusion that hydrogen behaves largely as an inert gas in sandstone. In contrast, Flesch et al. [92] found that hydrogen can alter pore fluid chemistry, dissolving pore-filling minerals like anhydrite and carbonate cement, which increases porosity and permeability by creating an interconnected pore network, though no silicate alterations were detected. Similar findings were reported by Pudlo et al. [93] in studies on heterogeneous sandstone from Europe and Argentina. Several other studies have concluded the same – that hydrogen has low reactivity with sandstone due to the stability of silicate minerals, leading to only minor changes in rock properties (e.g., [94–97]) This suggests that hydrogen's interaction with sandstone is limited, resulting in minimal alteration to the rock's physical characteristics, such as porosity and permeability.

Carbonate rocks are more reactive to hydrogen than sandstone due to the presence of minerals like calcite and dolomite. At high pressure and temperature, hydrogen can generate gases such as methane and CO_2, which may influence rock pore properties [98] though no significant impact on pore properties was found. Similarly, Veshareh et al. [99] observed no calcite dissolution in Danish North Sea chalk. Al-Yaseri et al. [100] also reported no major changes in porosity or significant hydrogen loss in limestone and dolomite after 125 days of exposure at 75 °C and 1,400 psi.

1.3.1.2 Biotic interactions

Similar to petroleum systems, where bacterial activity in hydraulically active conditions can degrade hydrocarbons, one of the main concerns for hydrogen in reservoir rocks is the effect of biotic activities [101]. Methanogenic archaea, in the presence of CO_2, can consume hydrogen through hydrogenotrophic methanogenesis, as shown by the reaction [102]:

$$CO_2 + 4H_2 = CH_4 + 2H_2O$$

This reaction leads to the consumption of hydrogen and alters the gas composition in the reservoir. Based on an experimental study [103] this interaction stops simultaneously with the depletion of CO_2 from the gas phase when calcite is consumed. Additionally, sulfate-reducing bacteria (SRB) can use hydrogen to reduce sulfate to hydrogen sulfide (H_2S), following the reaction:

$$4H_2 + SO_4{}^{2-} + H^+ \rightarrow HS^- + 4H_2O$$

In the presence of sulfate minerals like anhydrite, hydrogen can be converted into H_2S, resulting in hydrogen loss [104]. These microbial processes acidify pore water, causing both mineral dissolution (e.g., carbonate, gypsum, feldspar, chlorite) and mineral precipitation (e.g., illite, iron sulfide). While dissolution increases porosity, mineral precipitation reduces permeability, impacting the overall reservoir properties [105, 106].

Numerous microbial communities with abundant metabolic functions including sulfate reduction, sulfur oxidation, nitrate and iron reduction, methane oxidation, and methanogenesis exist in a subsurface environment [107]. However, in the long run, when clastic reservoirs are depleted of reactive components such as CO_2, methane, and sulfate minerals, biotic activity may significantly decline or even cease entirely. The availability of these compounds is crucial for microbial metabolisms, particularly for sulfate-reducing bacteria and methanogens, which rely on them for energy production. Without a steady supply of CO_2 and sulfate, these micro-organisms may no longer be able to sustain their metabolic processes, leading to a sharp decrease in biotic activity.

1.4 Caprock

A caprock acts as a flow barrier to hydrogen, either completely preventing its upward migration or significantly reducing the migration rate due to the low permeability of the seal material. Under this definition, many rock types, including both sedimentary and non-sedimentary rocks, can function as caprocks with efficient sealing capacities (e.g., shale, salts and evaporitic layers, crystalline rocks, etc.). In this study, the sealing efficiency of shale/mudstone, one of the most prevalent rock types in sedimentary systems and a critical caprock for petroleum systems, has been specifically evaluated.

The sealing efficiency of a caprock is determined by several parameters including the capillary threshold pressure (P_{th}), which is influenced by pore throat size, the interfacial tension (IFT), fluid's contact angle (θ), and the density difference between hydrogen and brine.

Hydrogen's low density (0.084 kg/m^3 at 20 °C and 0.1 MPa) compared to formation brines creates a significant buoyancy effect. This buoyancy exerts an upward force, pushing hydrogen towards the caprock and increasing the pressure against it. This capillary pressure promotes the potential for hydrogen to escape through the pore networks of the caprock. If the buoyancy pressure of the hydrogen column exceeds the capillary threshold pressure of the seal, the hydrogen can penetrate the sealing stratum, potentially leading to hydrogen leakage.

Capillary pressure can be expressed by Washburn's equation [108]:

$$P_C = -\frac{2\gamma \cos \theta}{r} \times C \tag{1.7}$$

where P_C is the capillary pressure (psi), Y is the IFT (mN/m); θ is the contact angle (degrees), r is the pore radius (µm), and $C = 0.145$ (constant to convert to psi).

Furthermore, capillary pressure can also be related to the height (h), to which the fluid rises above the free water level. This also indicates the buoyancy pressure exerted on the seal due to a column of height, h, of hydrogen under hydrostatic conditions and is given by

$$P_c = 0.434 \left(\rho_{brine} - \rho_{hydrogen} \right) h \tag{1.8}$$

where Pc is capillary pressure equal to buoyancy pressure (psi), ρ_{brine} and $\rho_{hydrogen}$ are specific gravities of brine and hydrogen, respectively (g/cc), and h is hydrogen column height in ft.

A caprock will leak if the buoyancy force (see eq. (1.8)) exceeds the threshold pressure of the caprock, a parameter that can be determined using mercury injection capillary pressure (MICP) data (see explanation in Section 1.4.1.1).

The maximum hydrogen column height (h) in the reservoir beneath the caprock is estimated by the balance between buoyancy and capillary forces:

$$h = \frac{4.6 \ \gamma \ Cos \ \theta}{r\Delta\rho} \tag{1.9}$$

The calculation of maximum column heights of the non-wetting lighter phase from MICP data is a well-established method discussed by several researchers [109–112]. The capillary pressure of the H_2-brine can be equated to the buoyancy pressure of the accumulated hydrogen column. To calculate the maximum column heights of hydrogen, capillary pressure from an air/mercury system has to be converted to an H_2-brine system using the following equation:

$$Pc_{H_2-brine} = Pc_{air-Hg} \frac{\gamma_{H_2-brine} \cdot \cos\theta_{H_2-brine}}{\gamma_{air-Hg} \cdot \cos\theta_{air-Hg}} \qquad (1.10)$$

By combining eqs. (1.8) and (1.10), the following equation is derived, which can be used to calculate the maximum hydrogen column that the caprock can support:

$$h_{H_2} = P_{th_{air-Hg}} \frac{\gamma_{H_2-brine} \cdot \cos\theta_{H_2-brine}}{\gamma_{air-Hg} \cdot \cos\theta_{air-Hg}} \times \frac{1}{0.434\left(\rho_{brine} - \rho_{H_2}\right)} \qquad (1.11)$$

In eqs. (1.9)–(1.11), h_{H2} is the maximum column height of hydrogen, $P_{thair-Hg}$ is threshold pressure from MICP data, σ_{air-Hg} is IFT between air and mercury, $\sigma_{H2-brine}$ is IFT between hydrogen and brine, θ_{air-Hg} and $\theta_{H2-brine}$ are contact angles between air/mercury and H_2/brine, respectively, ρ_{brine} and ρ_{H2} are brine and H_2 densities, respectively. For an air/mercury system, an IFT of 480 mN/m and a contact angle of 140° are considered.

1.4.1 Controlling parameters of maximum column height of hydrogen

To apply eq. (1.11) to quantify the maximum hydrogen column height that a specific shale caprock can support, several critical parameters are required. These include the caprock's threshold pressure, the IFT and the contact angle between brine and hydrogen, as well as the densities of both brine and hydrogen under reservoir conditions. The accuracy of the hydrogen column height calculation depends on the precise measurement or estimation of these parameters, as they govern the capillary forces and buoyancy effects, which ultimately determine the maximum height of the hydrogen column that the caprock can effectively seal. In the following sections, these parameters are discussed in detail to emphasize how they can be obtained and their significance in ensuring reliable predictions.

1.4.1.1 Caprock threshold pressure (P_{th})

MICP curves are used to estimate threshold pressure. The rapid increase in mercury injection rate identifies the threshold pressure, where mercury begins to intrude into the pore structure. Thompson et al. [113] define it as the inflection point where the curve turns convex upward. For hydrogen, this threshold pressure indicates when the non-wetting hydrogen phase displaces water within the pore networks of the caprock.

To assess the variation in the maximum column height of hydrogen in a reservoir with a shale caprock, 67 MICP data points were collected from a diverse range of shale samples across several basins (Figure 1.5). These include 5 shale samples from

South Perth Basin (Harvey Well #3) in the Jurassic Eneabba and Yalgorup formations; 11 samples from the Gippsland Basin (Latrobe Group); 4 samples from the Canning Basin (Theia Well, Goldwyer Formation); 4 samples from the North Perth Basin (Carynginia Formation); 5 samples from the Northern Carnarvon Basin wells (Thebe-2, Gorgon Well-1ST1 (GDW-1), and Saracen-1) from the shale formations of Muderong, Dupuy, and Dingo; 5 samples from the Otway Basin wells (Redman-1, CRC-1, and CRC-2) from the shale formations of the Belfast Mudstone, Paaratte Formation, and Pember Mudstone; 2 samples from the Darling Basin (Mena Murtee-1) and the Cooper Basin (Tindilpie-11, Murteree Shales); and 31 samples provided by Lohr and Hackley [114] from the Gulf Coast Tuscaloosa Marine Shale in Mississippi and Louisiana, USA.

As shown in Figure 1.5, the threshold pressure determined from these MICP curves ranges from approximately 600 psi to 12,000 psi, with a predominant value of around 6,000 psi.

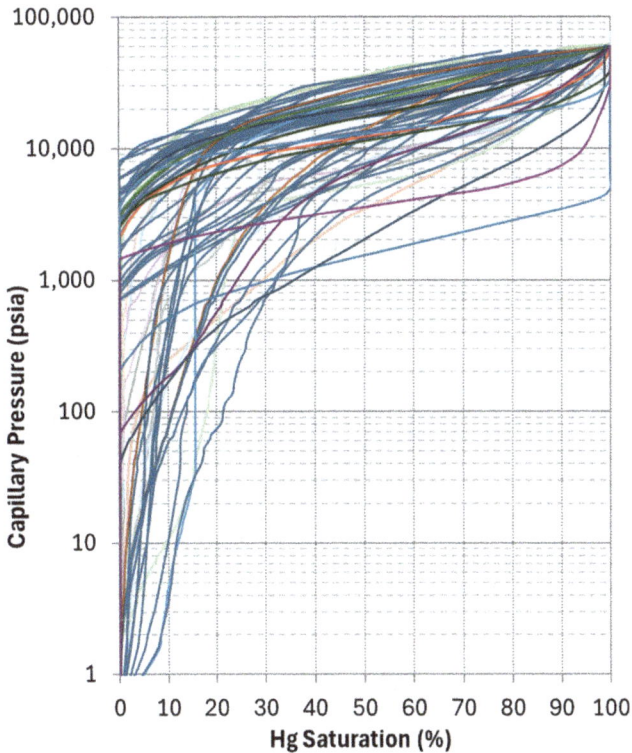

Figure 1.5: Mercury injection capillary pressure curves for 67 shale samples collected from several basins.

1.4.1.2 Interfacial tension (IFT)

A limited number of studies have reported the IFT of the brine-hydrogen system [115–117]. Hosseini et al. [118] measured the IFT (γ) between hydrogen and brine as a function of pressure, temperature, and water salinity. They also introduced an empirical equation to predict the H_2–brine IFT in the following form:

$$\gamma = Am + B \tag{1.12}$$

where γ is IFT (mN/m); A is a linear function of temperature (K) and pressure (MPa); m is brine molality (mol/kg); and B is the product of two linear functions.

Based on the fitting parameters provided in [118], the equation can be written as:

$$\gamma = ((-0.47876 - (0.01004P) + (0.00593T)) \times m) + (135.41479 \tag{1.13}$$
$$- (0.38368P) - (0.2052T) + (0.00084PT))$$

where P and T are pressure and temperature in MPa and kelvin, respectively.

The IFT between hydrogen and brine, generally decreases with increasing temperature and pressure (Figure 1.6). At higher temperatures, the molecular interactions at the interface become less pronounced, reducing the IFT. Similarly, increasing pressure tends to compress hydrogen, altering the fluid properties and lowering the IFT. This relationship is important in the subsurface as it can influence capillary forces and the ability of caprocks to effectively seal hydrogen accumulations.

1.4.1.3 Contact angles

Wettability, the preference of a solid surface to be in contact with one fluid over another in the presence of two immiscible fluids, plays a critical role in determining how hydrogen interacts with the shaly caprock. Shale typically exhibits water-wet characteristics due to its clay-rich composition. The high affinity for water molecules on shale surfaces is due to the presence of hydrophilic minerals such as clays (e.g., kaolinite and illite). In a water-wet system, water tends to coat the rock's surface, making it difficult for hydrogen, as a non-wetting phase, to displace water and migrate through the shale pore networks.

There are limited experimental data available on brine-hydrogen contact angles, primarily measured on quartz and sandstone samples [119–122]. These studies reported contact angles ranging between 10° and 50° under varying pressures and temperatures. Al-Yaseri et al. [123] examined the clay-brine-hydrogen system and concluded that wettability varies with clay type. They also introduced equations to calculate clay-brine-hydrogen contact angles based on hydrogen density (ρ, kg/m^3) for three different clays:

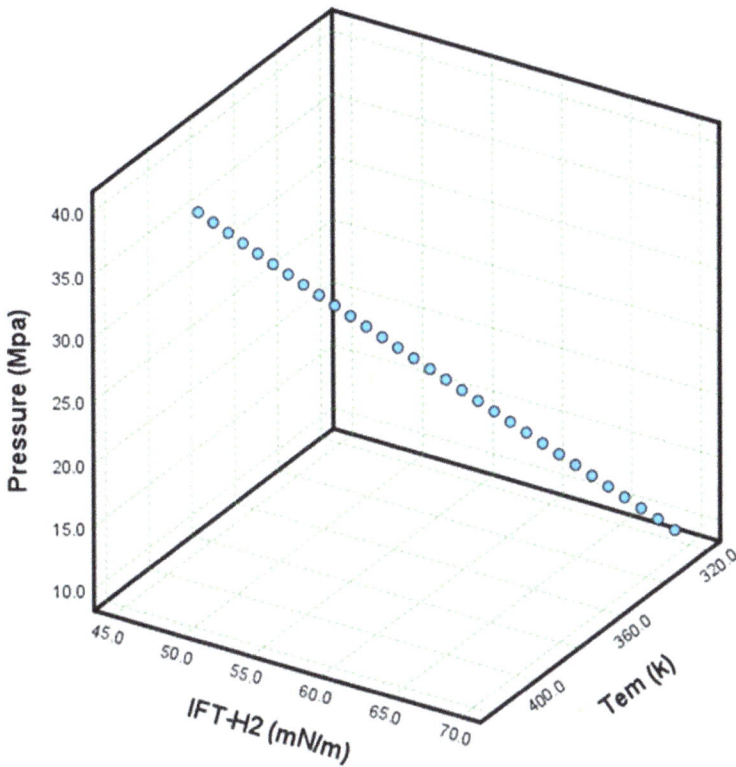

Figure 1.6: The relationship between interfacial tension (IFT), pressure, and temperature shows that IFT decreases as both temperature and pressure rise. For example, IFT can reduce from about 70 mN/m to 50 mN/m as temperature increases from 326 K to 416 K and pressure from 10 MPa to 40 MPa.

$$\cos\theta l = -0.0171\rho + 1 \quad \text{for montmorillonite} \tag{1.14}$$

$$\cos\theta l = -0.0117\rho + 1 \quad \text{for illite} \tag{1.15}$$

$$\cos\theta l = -0.0079\rho + 1 \quad \text{for kaolinite} \tag{1.16}$$

Based on their study, all three clays exhibited strong water-wetting behavior, with brine-hydrogen contact angles consistently below 40°. Figure 1.7 compares brine-hydrogen contact angles for three clay types calculated based on [123] equations eqs. (1.14)–(1.16). Their findings reveal a gradation in the water-wettability of clays with hydrogen: kaolinite is more hydrophilic, exhibiting stronger water adsorption and water-wet behavior, while illite and montmorillonite are less hydrophilic, affecting their capillary forces and sealing efficiency in subsurface environments.

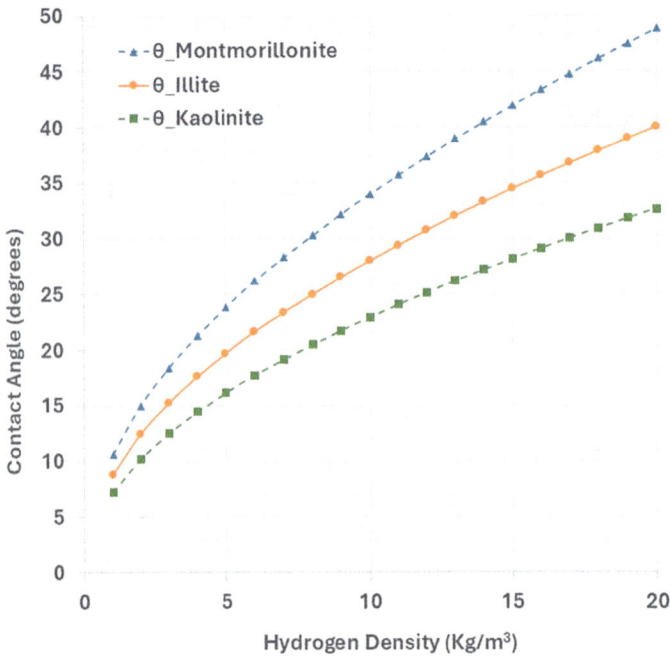

Figure 1.7: Contact angle versus hydrogen density for three types of clays.

1.4.1.4 Fluid density

The density of hydrogen can be calculated using the ideal gas law as an approximation:

$$\rho_h = \frac{P * M}{ZRT} \tag{1.17}$$

$$\rho_h = \frac{P * 0.002016}{1 * 8.314 * T} \tag{1.18}$$

where ρ_h is the density of hydrogen (kg/m³); P is the pressure in Pa; M is the molar mass of hydrogen = 0.002016 kg/mol; Z is the compressibility factor, which is typically around 1.15 at very high pressure; R is the universal gas constant = 8.314 J/(mol · K); T is the temperature in kelvin (K).

The density of brine (ρ_b) can be approximated using the general equation (1.19), considering a negligible change in pore water salinity with depth:

$$\rho_b = 1,000.0 + (A * S) + (B * T) + (C * P) \tag{1.19}$$

where ρ_b is the density of brine (kg/m³); S is the brine salinity (NaCl, ppm/eq); T is the temperature (centigrade); P is the pressure (kPa); A = 0.8 kg/m³ per g/kg (salinity effect); $B = -0.2$ kg/m^3 per °C (temperature effect); $C = 4.5 \times 10^{-5}$ kg/m^3 per kPa (pressure effect, water compressibility).

1.4.2 The maximum hydrogen column height calculation

To calculate the maximum hydrogen column height in a subsurface reservoir, key parameters explained in the previous sections need to be integrated. As an example, for the top of a reservoir rock at a depth of approximately 2,000 m, the following parameters can be estimated for further calculations:

– With a pore pressure gradient of about 0.45 psi/ft, the pressure at this depth would be around 2,967 psi, equivalent to 20.45 MPa.
– Assuming an average geothermal gradient of 30 °C/km and a surface temperature of 23 °C, the estimated temperature at this depth would be approximately 83 °C (356 K and 183.8 °F).
– The estimated density of brine with a moderate salinity of 60,000 ppm NaCl would be around 1.03 g/cc.
– The estimated density of hydrogen at these temperature and pressure conditions would be around 0.014 g/cc or 14 kg/m^3.
– The H$_2$-brine contact angle, calculated using the equation provided by Al-Yaseri et al. [123] for illite-based caprock and calculated hydrogen density, would be approximately 32° (with cos $\theta = 0.84$).
– The IFT of the H$_2$-brine system, calculated using the approximation provided by Hosseini et al. [118], is estimated to be around 62.07 mN/m.
– An average threshold pressure of 6,000 psi was derived from 67 MICP data points of shale samples (Figure 1.5).

Considering all the input parameters specified above and applying eq. (1.11), the maximum height of the hydrogen column is calculated to be approximately 586 m. However, this height may range from about 60 to 1,188 m, depending on the threshold pressures of the caprock, which are assumed to vary between 600 and 12,000 psi (see Figure 1.5).

This calculation assumes a negligible reservoir threshold pressure, meaning any resistance due to capillary forces or pore structure within the reservoir is minimal and does not significantly influence the column height. This estimation represents a theoretical limit for the hydrogen column height under the given conditions, offering valuable insights into the reservoir's capacity to store hydrogen in subsurface environments, whether for natural hydrogen accumulations or underground hydrogen storage (UHS) projects.

1.4.3 Hydrogen-water-caprock interactions

Recent research has brought increased attention to the interactions between hydrogen, water, and rock in subsurface formations, which are critical for understanding the integrity of caprocks when they are in contact with hydrogen (e.g., [97, 124–126]). Zeng et al. [127] explored the impact of hydrogen on caprocks and found that hydrogen penetration is limited, with a minimal effect on caprock integrity. Over a 30-year period, hydrogen diffusion into shale led to less than 1% mineral dissolution, indicating that hydrogen, brine, and shale interactions are unlikely to significantly compromise caprock stability. This suggests that the integrity of the caprock remains robust, with hydrogen infiltration being restricted to only a few meters beneath the surface. However, the research also highlighted potential concerns related to specific minerals within the caprock. Bo et al. [128] demonstrated that the presence of calcite in shale formations could reduce sealing efficiency due to its susceptibility to dissolution. Calcite, being reactive to hydrogen, may dissolve more readily, which could compromise the caprock's ability to maintain its sealing capacity. These reactions could create pathways for hydrogen migration, undermining the effectiveness of the caprock as a barrier.

In another study, Shi et al. [129] conducted experiments in which core samples were injected with hydrogen and natural gas, resulting in a noticeable reduction in caprock permeability. This suggests that hydrogen may alter the physical properties of the caprock, potentially influencing its overall integrity over time. Surprisingly, this alteration could be favorable as it may reduce the risk of sealing failure. The changes induced by hydrogen, such as permeability reduction, can lead to tighter seals, thus improving the long-term containment of hydrogen and reducing the chances of leakage through the caprock.

Additionally, Zeng et al. [125] reviewed the role of reactive minerals, particularly carbonates, sulfates, sulfides (such as pyrite), and Fe^{3+}-bearing minerals in caprock integrity. These minerals are considered "sensitive" because they can react with hydrogen through redox processes, leading to dissolution and precipitation reactions. Such chemical changes are more likely in caprocks rich in these reactive minerals, potentially altering the porosity and permeability of the rock. Over time, these processes could weaken the rock structure, reducing its sealing capacity.

Thus, while hydrogen's direct impact on caprock integrity might appear limited, the composition of the caprock, especially the presence of reactive minerals, plays a crucial role in determining its overall effectiveness as a seal.

1.4.4 Hydrogen diffusivity through the caprock

Hydrogen molecular diffusion is a crucial yet under-explored mechanism in assessing caprock quality. Its small molecular size enhances its potential for diffusion [130]. High hydrogen diffusivity may lead to gas losses through the caprock [131]. Hydrogen's diffusion rate is four times greater than that of natural gas. Research employing experimental and molecular dynamics simulations has yielded significant insights into hydrogen diffusion coefficients. For instance, Mostinsky [132] reported a diffusion coefficient of 3×10^{-11} m^2/s for hydrogen in water-saturated clays at 25 °C, while a broader range of 6×10^{-8}–12.2×10^{-8} m^2/s was found for self-diffusion coefficients in montmorillonite clays under varying conditions [133]. Krooss [134] reports that in clayey, water-saturated rocks at 25 °C, the effective diffusivity constant for hydrogen is 3×10^{-7} cm^2/s. Krooss also estimates that 1.5–2.2% of hydrogen can be lost through diffusion into the caprock. Moreover, Bhimineni et al. [135] estimated hydrogen diffusion into brine with different cations, validating coefficients between 0.46×10^{-8} and 1.31×10^{-8} m^2/s at temperatures from 10 °C to 60 °C. Similarly, Ghasemi et al. [133] assessed diffusion through various water-saturated clays, finding values from 3.4×10^{-9} to 12×10^{-9} m^2/s. Liu et al. [130] reported that with a hydrogen diffusion coefficient of 4×10^{-8} m^2/s at 353 K and 10 MPa, it is estimated that it would take roughly 250 years to achieve steady-state diffusive flux through a 10 meter-thick caprock with a 10% porosity. Once the steady state is reached, hydrogen losses due to diffusion could amount to about 0.656 kg/m^2 annually.

To calculate the amount of hydrogen that may diffuse through a caprock, we would need to consider the following factors:

- Steady-state diffusive flux: This is the rate at which hydrogen is lost through the caprock over time. For example, based on the value from Liu et al. [130], the steady-state hydrogen loss is 0.656 kg/m^2 per year.
- The surface area of the caprock: We need to know the surface area of the caprock through which hydrogen is diffusing.
- Time: Over what duration do you want to calculate the loss?

The equation to calculate the amount of hydrogen that diffuses through the caprock over time is:

$$\text{Amount of hydrogen diffused (kg)} = \text{Steady} - \text{state flux} \left(\text{kg}/\text{m}^2/\text{year} \right)$$
$$\times \text{area} \left(\text{m}^2 \right) \times \text{time (years)} \tag{1.20}$$

For example, if the steady-state flux is 0.656 kg/m^2 per year, the surface area of the caprock is 1,000 m^2, and you want to calculate the amount over 10 years:

Amount of hydrogen diffused = 0.656 kg/m^2/year $\times$ 1,000 m^2 $\times$ 10 years = 6,560 kg

This means 6,560 kg of hydrogen would diffuse through the caprock over 10 years.

1.5 Migration pathway

In a natural hydrogen system, the migration pathway, like in a petroleum system, involves the movement of fluids from the source rock into a reservoir rock through conduits such as fractures and fault zones. The process in petroleum systems is typically divided into primary and secondary migration phases. During primary migration, hydrocarbons are generated in the source rock (e.g., shale), which has low porosity and permeability. The transformation of kerogen into hydrocarbons generates internal pressures that create microfractures, facilitating the movement of hydrocarbons through the source rock as a key mechanism of primary migration. Secondary migration occurs when hydrocarbons move into the reservoir rock, driven by buoyancy forces resulting from the density difference between hydrocarbons and brine.

In a natural hydrogen system, the migration pathway operates somewhat differently from a petroleum system. While we can also define primary and secondary migration phases for natural hydrogen systems, key distinctions exist. Hydrogen generated during the serpentinization process in primary migration must migrate out of the rock, typically serpentinized ultramafic rocks. These rocks or more broadly igneous crystalline rocks, generally exhibit negligible matrix porosity and permeability. Additionally, unlike petroleum systems where hydrocarbons are generated from kerogen with increasing temperature, natural hydrogen systems require the introduction of water into the rock for serpentinization to occur, enabling hydrogen generation. The w/r ratio plays a key role in controlling the redox conditions for H_2 production, with higher ratios producing more hydrogen per unit of rock and fluid.

The natural hydrogen systems in continental regions are often distant from subduction zones and without or very limited access to seawater. Therefore, serpentinization requires the development of fluid pathways through typically impermeable rock. While tectonic processes by generating faults and fracture systems can create these pathways, the serpentinization reaction itself enhances the process by inducing fracturing due to volume expansion facilitating continued fluid flow and forming a pervasive permeability network by creating preferential fluid pathways [136]. This self-enhancing mechanism created by internal stresses from volume changes forms distinct structures like mesh textures and rectangular domains extensively studied by researchers [68–72] and plays a critical role in sustaining serpentinization. These structural features act as conduits, enabling hydrogen to migrate from deeper rocks and accumulate in traps closer to the surface.

The tectonic activity associated with faulting creates zones of increased permeability, which facilitates this upward migration. An example of this is seen in the East African Rift, where rifting and active tectonics create ideal conditions for hydrogen migration, allowing it to escape along faults and fractures. This process is essential for hydrogen's journey from its generation zones to potential storage reservoirs, emphasizing the role of tectonic features in shaping hydrogen migration patterns.

Alternatively, shallower-emplaced ultramafic rocks onshore may interact directly with groundwater without needing significant fault or fracture systems. If these rocks are in direct contact with water, they can generate hydrogen, which may accumulate in a shallow trap if one is present. However, if there is no trap or competent caprock to retain the hydrogen, it may simply escape to the surface. Understanding these dynamics is essential for assessing the potential for hydrogen accumulation in various geological settings.

In the natural hydrogen system, secondary migration follows principles similar to those in petroleum systems. Buoyancy forces, driven by the density difference between hydrogen and formation brines, cause the lighter hydrogen to migrate upwards in a reservoir rock. Capillary pressure influenced by the pore sizes and wetting properties of the reservoir rocks also plays a role in this movement. As hydrogen migrates, it accumulates at the top of the reservoir, typically beneath a low-permeability caprock.

One important consideration is the role of fracture and fault systems in delivering water to rocks for serpentinization. Fracture and fault systems not only facilitate water movement but also serve as conduits for the migration of natural hydrogen, allowing it to escape from deep geological sources to the surface, where the water is sourced. This process can hinder the effective trapping of hydrogen, limiting its accumulation potential. The proximity of surface manifestations of hydrogen degassing, such as circular depressions, to major fault lines supports the hypothesis that hydrogen migrates along the same pathways that initially provided water required for the serpentinization process, highlighting the interconnected roles of geology, hydrology, and hydrogen dynamics.

Furthermore, for example in Western Australia, many circular depressions align with river drainage systems (Figure 1.8). While studies attribute these features to processes such as tectonic activity, climate changes, wind erosion, and paleodrainage evolution – classifying many as playas or salt lakes [137–139] – their alignment with paleodrainage systems may suggest long-term water flow through these regions. Additionally, these paleovalleys or paleodrainage systems may have been aligned parallel to structural trends, such as deep basement faults [13]. This alignment could direct flowing water to the subsurface, facilitating serpentinization.

For hydrogen to be effectively trapped in the subsurface, smaller-scale faults (e.g., synthetic and antithetic faults) and fracture networks must be present. These fracture systems can redirect the generated hydrogen away from primary water-supply pathways, potentially guiding it toward traps at shallower depths. This redirection is crucial for preventing hydrogen from escaping along the same routes that supply water for serpentinization (Figure 1.9).

These dynamics highlight the complex interplay between water flow, fault systems and hydrogen migration. Understanding these interactions is essential for accurately assessing hydrogen accumulation potential in different geological settings, especially when considering areas with active tectonics or pre-existing fault systems. This comprehensive understanding is key to identifying viable subsurface hydrogen reservoirs and designing strategies for exploration and extraction.

Figure 1.8: Example of circular/subcircular depressions in WA that are aligned with paleochannels.

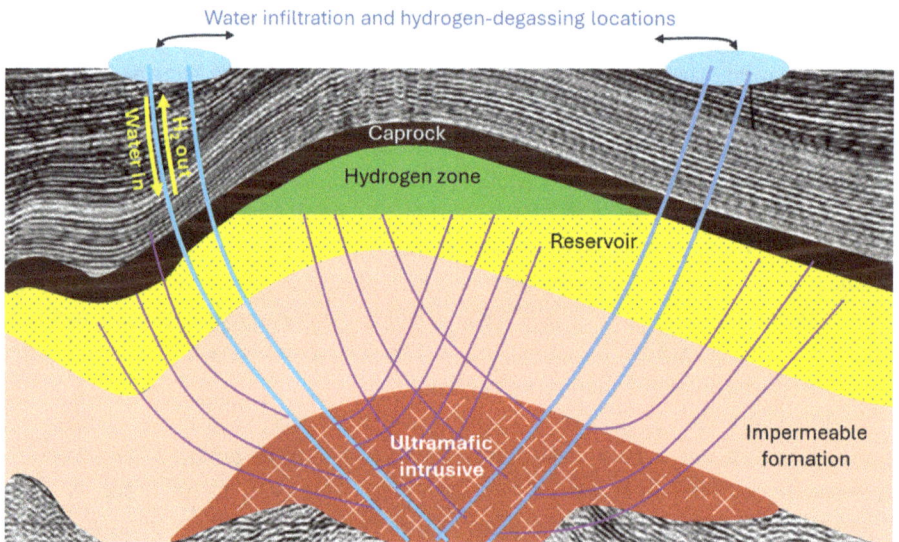

Figure 1.9: A simplified schematic illustration shows major faults (blue) transporting surface or near-surface water into intruded ultramafic rocks (red), facilitating serpentinization. These faults can also serve as pathways for some of the generated hydrogen to escape to the surface. Additionally, smaller-scale faults and fractures (purple) may direct hydrogen toward reservoirs capped by impermeable, competent caprock, enabling accumulation within subsurface traps (green zone).

1.6 Discussion and conclusion

This study provides a comparison between the well-understood petroleum and natural hydrogen systems. It emphasizes the critical need for further exploration of natural hydrogen, an emerging energy resource while outlining the similarities and distinctions between the two systems. The primary focus of this study was on the quantitative hydrogen generation rate through serpentinization, caprock efficiency, and migration pathways.

1.6.1 Source of natural hydrogen

The study highlights the serpentinization of ultramafic rocks as a primary process for natural hydrogen generation, emphasizing the key factors influencing this reaction, including temperature, pressure, mineral composition, and w/r ratio. Temperature is a key factor in determining the efficiency of hydrogen production through serpentinization. Similarly, pressure exerts a complex and multifaceted influence. It enhances serpentinization by accelerating mineral-fluid interactions and enabling deeper water infiltration, thereby increasing hydrogen generation.

Among ultramafic minerals, olivine plays a dominant role due to its high Fe content and reactivity, while pyroxenes contribute at lower rates. The optimal temperature range for hydrogen production falls between 200 °C and 315 °C, beyond which reaction rates slow down due to thermodynamic constraints. A critical component of this work is the empirical modeling of hydrogen generation rates, which demonstrated a strong correlation between hydrogen output, temperature, and pressure. The model provides a framework for estimating hydrogen production potential, though further refinement is necessary for improved accuracy.

Comparisons with petroleum system models suggest that hydrogen generation follows depth-temperature-dependent trends, but unlike hydrocarbons, it occurs as ultramafic bodies are uplifted and exposed to water circulation zones.

Applying the volumetric hydrogen estimation method to the Giles Complex in Australia by Rezaee [40] provides insights into the scale of natural hydrogen generation. Even under conservative assumptions, where only 0.01% of the ultramafic mass undergoes serpentinization, the estimated yield reaches 2.24×10^{13} kg. Similar large-scale hydrogen fluxes have been estimated for other geological settings, reinforcing the global relevance of natural hydrogen exploration. Liu et al. [78] estimated total production of 4.3×10^{18} mol (8.6×10^{12} metric tons) in the North Atlantic margins, while Cannat et al. [140] reported a global hydrogen flux of 1.67×10^{11} mol/year (3.34×10^5 tons annually).

These volumetric estimates quantify the hydrogen generated from the ultramafic intrusion but do not account for trapping and accumulation processes, which require further assessment. Similar to petroleum system analysis, this approach provides valuable insights into hydrogen exploration potential, helping refine exploration strategies and identify high-potential development areas.

This study developed a practical approach for estimating hydrogen generation through serpentinization, but key limitations must be acknowledged. Beyond temperature and pressure, factors such as ultramafic rock composition, w/r ratio, surface area, and fluid chemistry significantly influence hydrogen production. Finely ground ultramafic rocks generate more hydrogen due to higher reactivity, but laboratory experiments using rock powders do not fully replicate natural conditions, where water access is restricted by porosity, permeability, and fracture networks [141].

The w/r ratio controls reaction progress, hydrogen yield, and mineral transformations. In natural systems, limited fluid infiltration leads to diffusion-limited reactions, heterogeneous alteration fronts, and lower hydrogen output than in laboratory settings [141]. Serpentinization is highly heterogeneous, influenced by mineralogy, fluid availability, and tectonic settings [30, 142]. High fracture density enhances hydrogen production by improving water infiltration, while low-permeability zones restrict fluid flow. Variations in olivine-to-pyroxene ratios also affect Fe^{2+} oxidation, a key driver of hydrogen release [34]. Mantle body studies show serpentinization heterogeneity ranging from 3% to 62% [142].

Water availability further constrains serpentinization, especially at depth. While ultramafic rocks at mid-ocean ridges interact directly with seawater, continental settings rely on deep fluid migration through fault zones and shear structures [60]. Meteoric water circulation can reach 5 km [62], seawater percolation may extend to 20–25 km with serpentinization at 11–13 km [63], and surface-derived fluids have been observed at ~10 km depth in the French Pyrenees [64]. Additionally, metamorphic dehydration of pelitic sediments provides deep-seated water, representing ~10% of Earth's oceanic water [66].

Fracture density plays a key role in deep serpentinization, as high fracture density sustains water-rock interactions, while lower fracture densities limit reactions and hydrogen production [143]. Serpentinization textures in peridotites show veins and microfractures act as fluid pathways, with magnetite-rich fractures indicating past fluid flow [68]. Fluid chemistry also affects reaction pathways, with alkaline conditions promoting Fe^{2+} oxidation and hydrogen release [18], while dissolved CO_2, sulfate, or organics may alter hydrogen yields [73, 144, 145].

The empirical equation proposed here must be interpreted within geological timescales, as serpentinization unfolds over millions of years, akin to petroleum system evolution. While this study does not explicitly address permeability, fracture density, and deep fluid flow, these factors are crucial in controlling hydrogen generation and warrant further research into fluid-rock interactions and geochemical constraints.

1.6.2 Reservoir rock

The interactions between hydrogen and reservoir rocks can play a critical role in the natural hydrogen system. Unlike hydrocarbons, hydrogen's ability to chemically alter the mineral composition of reservoir rocks introduces unique challenges and opportunities for a natural hydrogen system. The findings indicate that abiotic interactions, particularly in sandstone, demonstrate limited reactivity, with hydrogen behaving largely as an inert gas. This suggests that sandstone formations may maintain their structural integrity over time, as hydrogen-induced changes to porosity and permeability are minimal. However, hydrogen can enhance fluid flow by dissolving pore-filling minerals like anhydrite.

The role of biotic interactions introduces another layer of complexity to a natural hydrogen system. The consumption of hydrogen by methanogenic archaea and SRB could be significant, as it not only alters the gas composition but also has implications for the overall hydrogen balance within the reservoir. The reactions leading to the formation of methane and hydrogen sulfide highlight potential pathways for hydrogen loss with time.

Moreover, the findings underscore the necessity of considering microbial activity in the assessment of hydrogen reservoirs. As biotic processes are contingent on the availability of reactive compounds such as CO_2 and sulfate, depletion of these resources could lead to a decline in microbial activity, potentially resulting in reduced hydrogen consumption rates over time.

The results highlight the need for a deeper understanding of potential biotic reactions occurring within hydrogen reservoirs and the conditions under which these reactions may lead to the consumption of accumulated hydrogen. It is crucial to identify the specific chemical, temperature, and pressure conditions that are most conducive to the degradation of hydrogen and the depletion of the reservoir. Moreover, it is essential to recognize that reservoirs, similar to petroleum systems, are dynamic environments. They may be subject to continuous replenishment of hydrogen as the ultramafic rock undergoes alteration over geological time. Understanding this dynamic nature will help in developing more accurate models for hydrogen generation and retention, ultimately informing strategies for effective hydrogen resource management. By addressing these factors, future research can enhance our understanding of hydrogen behavior in geological formations and optimize exploration efforts for this promising energy source.

1.6.3 Caprock

The integrity of caprock plays a critical role in the hydrogen system, serving as a barrier that limits upward flow and potential leakage of hydrogen from reservoirs. Its sealing efficiency is determined by several factors, notably capillary threshold pressure (P_{th}), which is influenced by pore throat size, IFT, contact angles, and the density contrast between hydrogen and brine. The buoyancy effect of hydrogen, due to its low density compared to brine, exerts significant pressure against the caprock, potentially leading to hydrogen escape if the buoyancy pressure surpasses the capillary threshold pressure of the seal.

The reliance on Washburn's Equation to relate capillary pressure to fluid migration within the caprock highlights the importance of accurately assessing the maximum column heights of hydrogen that the caprock can sustain. By correlating MICP data with hydrogen-brine interactions, researchers can estimate the threshold pressures necessary to predict shaly caprock performance under various conditions. Notably, a range of threshold pressures (e.g., 600–12,000 psi) derived from extensive shale sample analysis suggests variability in caprock effectiveness across different geological settings.

Furthermore, the relationship between IFT and reservoir conditions is crucial for understanding capillary forces at play. Recent studies indicate that IFT between hydrogen and brine decreases with increasing temperature and pressure, which may enhance the caprock's sealing capability. Conversely, high pressures can compress hydrogen, altering its buoyancy effects and potentially impacting the effectiveness of the seal.

Wettability, defined by contact angles, is another key factor influencing hydrogen migration. Shale's water-wet characteristics, primarily due to its clay-rich composition, hinder hydrogen displacement within the caprock. The findings from studies on brine-hydrogen contact angles suggest that various clay types exhibit different wettability behaviors, with kaolinite displaying a stronger water affinity than illite and montmorillonite. This gradation in hydrophilicity can significantly affect capillary forces, influencing the overall sealing efficiency of the caprock.

Lastly, the densities of both hydrogen and brine under reservoir conditions are pivotal for evaluating the buoyancy and capillary pressures that govern hydrogen retention. Accurate calculations of these densities, alongside empirical data on IFT and contact angles, are essential for reliable predictions of hydrogen column heights and caprock performance.

The calculation of maximum hydrogen column height in a subsurface reservoir is a critical aspect of understanding hydrogen accumulation capabilities beneath caprock. By integrating key parameters highlighted in this study, we can estimate the potential for hydrogen accumulation within a given reservoir system.

The maximum hydrogen column height (m) varies as caprock threshold pressure (P_{th}), IFT, and contact angle change. The maximum hydrogen column height is directly proportional to both the caprock threshold pressure and the IFT. This means that as P_{th} and IFT increase, the height of the hydrogen column that can be retained beneath the caprock also increases. Conversely, there is an inverse relationship between the maximum hydrogen column height and the contact angle. As the contact angle increases, the ability of the caprock to retain a hydrogen column decreases, leading to a shorter column height. This reflects the importance of wettability and rock-fluid inter-actions in controlling the trapping efficiency of the caprock for hydrogen storage.

In summary, the complex interactions between hydrogen, brine, and caprock ma-terials govern the efficient accumulation of hydrogen. A comprehensive understand-ing of these dynamics is essential for assessing the maximum potential hydrogen col-umn height within a reservoir and its caprock.

Recent research has increasingly focused on the interactions between hydrogen, water, and rock in subsurface formations, which are vital for assessing caprock integ-rity when in contact with hydrogen. Studies indicate that hydrogen penetration into caprock is limited, with minimal effects on caprock stability. However, certain miner-als within the caprock, particularly calcite, pose potential risks. For example, calcite's susceptibility to dissolution in the presence of hydrogen could reduce sealing effi-ciency, creating pathways for hydrogen migration and undermining the caprock's barrier function. The role of reactive minerals, such as carbonates and sulfides, is also crucial in determining caprock effectiveness since these minerals may undergo redox reactions with hydrogen, potentially altering the rock's porosity and permeabil-ity over time and compromising sealing capacity.

Hydrogen molecular diffusion is a vital but often-neglected factor in assessing cap-rock integrity. Its small size results in high diffusivity, making hydrogen diffuse four times faster than natural gas, which can lead to significant gas losses through caprocks. Research indicates varying diffusion coefficients, reporting a value of 3×10^{-11} m^2/s in water-saturated clays [132]. It is estimated that 1.5–2.2% of hydrogen could be lost due to diffusion in clayey rocks; however, it would take about 250 years to reach a steady-state diffusive flux through a 10 meter-thick caprock, resulting in annual losses of approxi-mately 0.656 kg/m^2 [130].

Overall, while hydrogen's direct impact on caprock integrity appears limited, the composition and mineralogy of the caprock significantly influence its sealing capa-bility.

1.6.4 Migration pathway

The migration pathways in natural hydrogen systems share similarities with those in petroleum systems but also exhibit distinct differences due to the unique geochemical processes involved. A key consideration in understanding hydrogen production and migration in ultramafic rocks is the relationship between fault and fracture systems. These pathways not only facilitate water infiltration for serpentinization but also pro-vide routes for hydrogen generated during this process to escape to the surface. Loca-tions where hydrogen degasses likely correspond to areas where water has already penetrated the rocks. However, since hydrogen follows the same fracture systems as water, surface degassing points may not accurately indicate subsurface gas accumula-tions. Consequently, relying solely on these surface degassing points to identify gas accumulations could be misleading. Actual gas reservoirs may exist along alternative pathways within the fracture network, far from the observed surface degassing locations (Figure 1.9). Therefore, comprehensive mapping of fracture networks and an understanding of their roles in both water infiltration and gas migration are crucial for effective exploration.

Additionally, it is important to investigate whether some of the circular depressions exhibiting hydrogen degassing result from the degassing process itself or if they are merely morphological features that facilitate water accumulation and infiltration. This water infiltration may trigger hydrogen generation through other minor mechanisms such as ferrolysis and biological activity.

These dynamics emphasize the complex interactions between water flow, fault systems, and hydrogen migration. Understanding these relationships is crucial for assessing hydrogen accumulation potential in various geological settings, particularly in areas with active tectonics or pre-existing faults. This knowledge is vital for identifying viable subsurface hydrogen reservoirs and developing effective exploration and extraction strategies.

References

[1]   Dincer I., Acar C. Review and evaluation of hydrogen production methods for better sustainability. International Journal of Hydrogen Energy. 2015, 40(34): 11094–11111.

[2]   Momirlan M., Veziroglu T. N. The properties of hydrogen as fuel tomorrow in sustainable energy system for a cleaner planet. International Journal of Hydrogen Energy. 2005, 30(7): 795–802.

[3]   IEA (2022), Global Hydrogen Review 2022, IEA, Paris https://www.iea.org/reports/global-hydrogen-review-2022.

[4]   Shoko E., et al. Hydrogen from coal: Production and utilisation technologies. International Journal of Coal Geology. 2006, 65(3–4): 213–222.

[5]   Klein F., Tarnas J. D., Bach W. Abiotic sources of molecular hydrogen on Earth. Elements: An International Magazine of Mineralogy, Geochemistry, and Petrology. 2020, 16(1): 19–24.

[6]   Worman S. L., et al. Abiotic hydrogen (H2) sources and sinks near the Mid-Ocean Ridge (MOR) with implications for the subseafloor biosphere. Proceedings of the National Academy of Sciences. 2020, 117(24): 13283–13293.

[7]   Prinzhofer A., Cissé C. S. T., Diallo A. B. Discovery of a large accumulation of natural hydrogen in Bourakebougou (Mali). International Journal of Hydrogen Energy. 2018, 43(42): 19315–19326.

[8]   Larin N., et al. Natural molecular hydrogen seepage associated with surficial, rounded depressions on the European craton in Russia. Natural Resources Research. 2015, 24(3): 369–383.

[9]   Zgonnik V., et al. Evidence for natural molecular hydrogen seepage associated with Carolina bays (surficial, ovoid depressions on the Atlantic Coastal Plain, Province of the USA). Progress in Earth and Planetary Science. 2015, 2(1): 1–15.

[10]   Prinzhofer A., et al. Natural hydrogen continuous emission from sedimentary basins: The example of a Brazilian H2-emitting structure. International Journal of Hydrogen Energy. 2019, 44(12): 5676–5685.

[11]   Zgonnik V. The occurrence and geoscience of natural hydrogen: A comprehensive review. Earth-Science Reviews. 2020, 203: 103140.

[12] Moretti I., et al. Long-term monitoring of natural hydrogen superficial emissions in a brazilian cratonic environment. Sporadic large pulses versus daily periodic emissions. International Journal of Hydrogen Energy. 2021, 46(5): 3615–3628.

[13] Rezaee R. Assessment of natural hydrogen systems in Western Australia. International Journal of Hydrogen Energy. 2021, 46(66): 33068–33077.

[14] Frery E., et al. Natural hydrogen seeps identified in the North Perth Basin, Western Australia. International Journal of Hydrogen Energy. 2021, 46(61): 31158–31173.

[15] Vidavskiy V., et al. Natural hydrogen in North Perth Basin, WA Australia: Detection in soil gas for early exploration. doi: 10.20944/preprints202404.0532.v1 2024.

[16] Milkov A. V. Molecular hydrogen in surface and subsurface natural gases: Abundance, origins and ideas for deliberate exploration. Earth-Science Reviews. 2022, 230: 104063.

[17] Larin N. Hydridic earth the new geology of our primortidally hydrogen-rich planet translation. Maple Ridge, BC, Canada: Polar Publishing, 1993.

[18] Gilat A., Vol A. Primordial hydrogen-helium degassing, an overlooked major energy source for internal terrestrial processes. HAIT Journal of Science and Engineering B. 2005, 2(1–2): 125–167.

[19] Yang X., Keppler H., Li Y. Molecular hydrogen in mantle minerals. Geochemical Perspectives Letters. 2016, 2(2): 160–168.

[20] Vidavskiy V., Rezaee R. Natural deep-seated hydrogen resources exploration and development: Structural features, governing factors, and controls. Journal of Energy and Natural Resources. 2022, 11(3): 60–81.

[21] Fischer T. P., Chiodini G. Volcanic, magmatic and hydrothermal gas discharges Encyclopaedia of Volcanoes (second ed.) (2015), pp. 779-797, doi: 10.1016/B978-0-12-385938-9.00045-6

[22] Mao H.-K., et al. When water meets iron at Earth's core–mantle boundary. National Science Review. 2017, 4(6): 870–878.

[23] Mahlstedt N., et al. Molecular hydrogen from organic sources in geological systems. Journal of Natural Gas Science and Engineering. 2022, 105: 104704.

[24] Bourdet J., et al. Natural hydrogen in low temperature geofluids in a precambrian granite, South Australia. Implications for hydrogen generation and movement in the upper crust. Chemical Geology. 2023, 638: 121698.

[25] Lin L.-H., et al. The yield and isotopic composition of radiolytic H2, a potential energy source for the deep subsurface biosphere. Geochimica Et Cosmochimica Acta. 2005, 69(4): 893–903.

[26] Winter J. D. An introduction to igneous and metamorphic petrology. Prentice-Hall Inc., New Jersey (2001) 697 p.

[27] Schumann W. Handbook of Rocks, Minerals, and Gemstones. HarperCollins Publishers and Houghton Mifflin Company, New York, NY, 1993.

[28] Streckeisen A. Classification and nomenclature of plutonic rocks recommendations of the IUGS subcommission on the systematics of igneous rocks. Geologische Rundschau. 1974, 63: 773–786.

[29] Bowen N. L. The Evolution of the Igneous Rocks Princeton University Press. Princeton, New Jersey, 1928.

[30] Klein F., Bach W., McCollom T. M. Compositional controls on hydrogen generation during serpentinization of ultramafic rocks. Lithos. 2013, 178: 55–69.

[31] Albers E., et al. Serpentinization-driven H2 production from continental break-up to mid-ocean ridge spreading: Unexpected high rates at the West Iberia margin. Frontiers in Earth Science. 2021, 9: 673063.

[32] Berndt M. E., Allen D. E., W.e. S. Jr Reduction of CO2 during serpentinization of olivine at 300 C and 500 bar. Geology. 1996, 24(4): 351–354.

[33] Sleep N., et al. H2-rich fluids from serpentinization: Geochemical and biotic implications. Proceedings of the National Academy of Sciences. 2004, 101(35): 12818–12823.

[34] Seyfried W. Jr, Foustoukos D., Fu Q. Redox evolution and mass transfer during serpentinization: An experimental and theoretical study at 200 C, 500 bar with implications for ultramafic-hosted hydrothermal systems at Mid-Ocean Ridges. Geochimica Et Cosmochimica Acta. 2007, 71(15): 3872–3886.

[35] McCollom T. M., Bach W. Thermodynamic constraints on hydrogen generation during serpentinization of ultramafic rocks. Geochimica Et Cosmochimica Acta. 2009, 73(3): 856–875.

[36] Rossman D., Castañada G., Bacuta G. Geology of the zambales ophiolite, Luzon, Philippines. Tectonophysics. 1989, 168(1–3): 1–22.

[37] Nicolas A., Boudier F. O. Large mantle upwellings and related variations in crustal thickness in the Oman ophiolite. Special Papers-Geological Society Of America, 2000, 67–74.

[38] Iyer K., et al. Serpentinization of the oceanic lithosphere and some geochemical consequences: Constraints from the Leka Ophiolite Complex, Norway. Chemical Geology. 2008, 249(1–2): 66–90.

[39] Frost B. R. On the stability of sulfides, oxides, and native metals in serpentinite. Journal of Petrology. 1985, 26(1): 31–63.

[40] Rezaee, R., Quantifying Natural Hydrogen Generation Rates and Volumetric Potential in Onshore Serpentinization. Geosciences, 2025. 15(3): p. 112.

[41] Deer W. A., Howie R. A., Zussman J. An introduction to the rock-forming minerals. Mineralogical Society of Great Britain and Ireland, 2013.

[42] Charlou J., et al. Geochemistry of high H2 and CH4 vent fluids issuing from ultramafic rocks at the Rainbow hydrothermal field (36 14′ N, MAR). Chemical Geology. 2002, 191(4): 345–359.

[43] Fouquet Y., et al. others. Geodiversity of hydrothermal processes along the Mid-Atlantic Ridge and ultramafichosted mineralization: A new type of oceanic Cu-Zn-Co-Au volcanogenic massive sulfide deposit. in Rona P.A. Devey C.W. Dyment J. Murton B.J., eds., Diversity of Hydrothermal Systems on Slow Spreading Ocean Ridges : American Geophysical Union Geophysical Monograph 188, p. 321–367.

[44] Proskurowski G., et al. Abiogenic hydrocarbon production at Lost City hydrothermal field. Science. 2008, 319(5863): 604–607.

[45] Konn C., et al. Hydrocarbons and oxidized organic compounds in hydrothermal fluids from Rainbow and Lost City ultramafic-hosted vents. Chemical Geology. 2009, 258(3–4): 299–314.

[46] Chavagnac V., et al. Characterization of hyperalkaline fluids produced by low-temperature serpentinization of mantle peridotites in the Oman and Ligurian ophiolites. Geochemistry, Geophysics, Geosystems. 2013, 14(7): 2496–2522.

[47] Monnin C., et al. Fluid chemistry of the low temperature hyperalkaline hydrothermal system of Prony Bay (New Caledonia). Biogeosciences. 2014, 11(20): 5687–5706.

[48] Etiope G. Natural hydrogen extracted from ophiolitic rocks: A first dataset. International Journal of Hydrogen Energy. 2024, 78: 368–372.

[49] Kumagai H., et al. Geological background of the Kairei and Edmond hydrothermal fields along the Central Indian Ridge: Implications of their vent fluids' distinct chemistry. Geofluids. 2008, 8(4): 239–251.

[50] Kelley D. S., et al. A serpentinite-hosted ecosystem: The Lost City hydrothermal field. Science. 2005, 307(5714): 1428–1434.

[51] Merdith A. S., Atkins S. E., Tetley M. G. Tectonic controls on carbon and serpentinite storage in subducted upper oceanic lithosphere for the past 320 Ma. Frontiers in Earth Science. 2019, 7: 332.

[52] Miller H. M., et al. Modern water/rock reactions in Oman hyperalkaline peridotite aquifers and implications for microbial habitability. Geochimica Et Cosmochimica Acta. 2016, 179: 217–241.

[53] Abrajano T., et al. Geochemistry of reduced gas related to serpentinization of the Zambales ophiolite, Philippines. Applied Geochemistry. 1990, 5(5–6): 625–630.

[54] Etiope G., Schoell M., Hosgörmez H. Abiotic methane flux from the Chimaera seep and Tekirova ophiolites (Turkey): Understanding gas exhalation from low temperature serpentinization and implications for Mars. Earth and Planetary Science Letters. 2011, 310(1–2): 96–104.

[55] Etiope G. Massive release of natural hydrogen from a geological seep (Chimaera, Turkey): Gas advection as a proxy of subsurface gas migration and pressurised accumulations. International Journal of Hydrogen Energy. 2023, 48(25): 9172–9184.

[56] Lefeuvre N., et al. Natural hydrogen migration along thrust faults in foothill basins: The North Pyrenean Frontal Thrust case study. Applied Geochemistry. 2022, 145: 105396.

[57] Vacquand C., et al. Reduced gas seepages in ophiolitic complexes: Evidences for multiple origins of the H2-CH4-N2 gas mixtures. Geochimica Et Cosmochimica Acta. 2018, 223: 437–461.

[58] Fu Q., et al. Abiotic formation of hydrocarbons under hydrothermal conditions: Constraints from chemical and isotope data. Geochimica Et Cosmochimica Acta. 2007, 71(8): 1982–1998.

[59] Foustoukos D. I., W.E. S. Jr Hydrocarbons in hydrothermal vent fluids: The role of chromium-bearing catalysts. Science. 2004, 304(5673): 1002–1005.

[60] Sibson R., Moore J. M. M., Rankin A. Seismic pumping – A hydrothermal fluid transport mechanism. Journal of the Geological Society. 1975, 131(6): 653–659.

[61] Fusseis F., et al. Creep cavitation can establish a dynamic granular fluid pump in ductile shear zones. Nature. 2009, 459(7249): 974–977.

[62] McIntosh J. C., Ferguson G. Deep meteoric water circulation in Earth's crust. Geophysical Research Letters. 2021, 48(5): e2020GL090461.

[63] Prigent C., et al. Fracture-mediated deep seawater flow and mantle hydration on oceanic transform faults. Earth and Planetary Science Letters. 2020, 532: 115988.

[64] McCaig A., Wickham S. Oxygen isotope variations in metasomatically altered shear zones from the Pyrenees. Terra Cognita. 1987, 7: 137.

[65] Walther J. V., Orville P. M. Volatile production and transport in regional metamorphism. Contributions to Mineralogy and Petrology. 1982, 79: 252–257.

[66] Skelton A. Flux rates for water and carbon during greenschist facies metamorphism. Geology. 2011, 39(1): 43–46.

[67] McCollom T. M., et al. Temperature trends for reaction rates, hydrogen generation, and partitioning of iron during experimental serpentinization of olivine. Geochimica Et Cosmochimica Acta. 2016, 181: 175–200.

[68] Plümper O., et al. The interface-scale mechanism of reaction-induced fracturing during serpentinization. Geology. 2012, 40(12): 1103–1106.

[69] Malvoisin B., Brantut N., Kaczmarek M.-A. Control of serpentinisation rate by reaction-induced cracking. Earth and Planetary Science Letters. 2017, 476: 143–152.

[70] Macdonald A., Fyfe W. Rate of serpentinization in seafloor environments. Tectonophysics. 1985, 116 (1–2): 123–135.

[71] Iyer K., et al. Reaction-assisted hierarchical fracturing during serpentinization. Earth and Planetary Science Letters. 2008, 267(3–4): 503–516.

[72] Zhang L., et al. Modeling porosity evolution throughout reaction-induced fracturing in rocks with implications for serpentinization. Journal of Geophysical Research: Solid Earth. 2019, 124(6): 5708–5733.

[73] McCollom T. M., Seewald J. S. A reassessment of the potential for reduction of dissolved CO2 to hydrocarbons during serpentinization of olivine. Geochimica Et Cosmochimica Acta. 2001, 65(21): 3769–3778.

[74] Grozeva N. G., et al. Experimental study of carbonate formation in oceanic peridotite. Geochimica Et Cosmochimica Acta. 2017, 199: 264–286.

[75] Mayhew L. E., et al. Hydrogen generation from low-temperature water–rock reactions. Nature Geoscience. 2013, 6(6): 478–484.

[76]   McCollom T. M., et al. The effect of pH on rates of reaction and hydrogen generation during serpentinization. Philosophical Transactions of the Royal Society A. 2020, 378(2165): 20180428.

[77]   Donval J. P., Charlou J. L., Douville E., Knoery J., Fouquet Y., Poncevera E., Jean-Baptiste P., Stievenard M., German C. High H_2 and CH_4 content in hydrothermal fluids from Rainbow site newly sampled at 36° 14'N on the AMAR Segment, Mid-Atlantic Ridge (Diving FLORES cruise, July 1997): Comparison with other MAR sites, Eos Trans. Am. Geophys. Union 78 (46): 832.

[78]   Liu Z., et al. Mantle serpentinization and associated hydrogen flux at North Atlantic magma-poor rifted margins. Geology. 2023, 51(3): 284–289.

[79]   Huang R., et al. Influence of pH on molecular hydrogen (H2) generation and reaction rates during serpentinization of peridotite and olivine. Minerals. 2019, 9(11): 661.

[80]   Allen D. E., W. S. Jr Compositional controls on vent fluids from ultramafic-hosted hydrothermal systems at mid-ocean ridges: An experimental study at 400 C, 500 bars. Geochimica Et Cosmochimica Acta. 2003, 67(8): 1531–1542.

[81]   Jones L. C., et al. Carbonate control of H2 and CH4 production in serpentinization systems at elevated P-Ts. Geophysical Research Letters. 2010, 37(14) p. L14306.

[82]   Neubeck A., et al. Formation of H 2 and CH 4 by weathering of olivine at temperatures between 30 and 70 C. Geochemical Transactions. 2011, 12: 1–10.

[83]   Marcaillou C., et al. Mineralogical evidence for H2 degassing during serpentinization at 300 C/300 bar. Earth and Planetary Science Letters. 2011, 303(3–4): 281–290.

[84]   Okland I., et al. Formation of H2, CH4 and N-species during low-temperature experimental alteration of ultramafic rocks. Chemical Geology. 2014, 387: 22–34.

[85]   McCollom T. M., Donaldson C. Generation of hydrogen and methane during experimental low-temperature reaction of ultramafic rocks with water. Astrobiology. 2016, 16(6): 389–406.

[86]   Miller H. M., et al. Low temperature hydrogen production during experimental hydration of partially-serpentinized dunite. Geochimica Et Cosmochimica Acta. 2017, 209: 161–183.

[87]   Aitken A. R., et al. Magmatism-dominated intracontinental rifting in the Mesoproterozoic: The Ngaanyatjarra Rift, central Australia. Gondwana Research. 2013, 24(3–4): 886–901.

[88]   Aitken A., Aitken A. Imaging crustal structure in the west Musgrave Province from magnetotelluric and potential field data. Geological Survey of Western Australia, Report, 114 (2013) (81 pp.).

[89]   Truche L., et al. Sulphide mineral reactions in clay-rich rock induced by high hydrogen pressure. Application to disturbed or natural settings up to 250 C and 30 bar. Chemical Geology. 2013, 351: 217–228.

[90]   Ganzer L., et al. The H2STORE project-experimental and numerical simulation approach to investigate processes in underground hydrogen reservoir storage. European association of geoscientists and engineers. EAGE, Netherlands, pp 679–687. https://doi.org/10.2118/164936-ms.

[91]   Yekta E. Characterization of geochemical interactions and migration of hydrogen in sandstone sedimentary formations: Application to geological storage (Doctoral dissertation, Université d'Orléans). Applied Geochemistry. 2017, 95: 182–194.

[92]   Flesch S., et al. Hydrogen underground storage – Petrographic and petrophysical variations in reservoir sandstones from laboratory experiments under simulated reservoir conditions. International Journal of Hydrogen Energy. 2018, 43(45): 20822–20835.

[93]   Pudlo D., et al. The impact of hydrogen on potential underground energy reservoirs. In: EGU general assembly conference abstracts. p. 8606, 2018.

[94]   Ebrahimiyekta A. Characterization of geochemical interactions and migration of hydrogen in sandstone sedimentary formations: Application to geological storage. PhD Dissertation, Université d'Orléans, France. 2017.

[95]   Al-Yaseri A., et al. Pore structure analysis of storage rocks during geological hydrogen storage: Investigation of geochemical interactions, Fuel, 361 (2024), Article 130683, doi: 10.1016/j. fuel.2023.130683.

[96]   Hassanpouryouzband A., et al. Geological hydrogen storage: Geochemical reactivity of hydrogen with sandstone reservoirs. ACS Energy Letters. 2022, 7(7): 2203–2210.

[97]   Labus K., Tarkowski R. Modeling hydrogen–rock–brine interactions for the Jurassic reservoir and cap rocks from Polish Lowlands. International Journal of Hydrogen Energy. 2022, 47(20): 10947–10962.

[98]   Zeng L., et al. Hydrogen storage in Majiagou carbonate reservoir in China: Geochemical modelling on carbonate dissolution and hydrogen loss. International Journal of Hydrogen Energy. 2022, 47(59): 24861–24870.

[99]   Veshareh M. J., Thaysen E. M., Nick H. M. Feasibility of hydrogen storage in depleted hydrocarbon chalk reservoirs: Assessment of biochemical and chemical effects. Applied Energy. 2022, 323: 119575.

[100] Al-Yaseri A., Al-Mukainah H., Yekeen N. Experimental insights into limestone-hydrogen interactions and the resultant effects on underground hydrogen storage. Fuel. 2023, 344: 128000.

[101] Reitenbach V., et al. Influence of added hydrogen on underground gas storage: A review of key issues. Environmental Earth Sciences. 2015, 73: 6927–6937.

[102] Panfilov M., Gravier G., Fillacier S. Underground storage of H2 and H2-CO2-CH4 mixtures. In: ECMOR X-10th European conference on the mathematics of oil recovery. European Association of Geoscientists & Engineers, 2006.

[103] Haddad P., et al. Geological storage of hydrogen in deep aquifers–an experimental multidisciplinary study. Energy & Environmental Science. 2022, 15(8): 3400–3415.

[104] Hemme C., Van Berk W. Hydrogeochemical modeling to identify potential risks of underground hydrogen storage in depleted gas fields. Applied Sciences. 2018, 8(11): 2282.

[105] Allan M. M., Turner A., Yardley B. W. Relation between the dissolution rates of single minerals and reservoir rocks in acidified pore waters. Applied Geochemistry. 2011, 26(8): 1289–1301.

[106] Pudlo D., et al. The impact of diagenetic fluid–rock reactions on Rotliegend sandstone composition and petrophysical properties (Altmark area, central Germany). Environmental Earth Sciences. 2012, 67: 369–384.

[107] Beaver R. C., Neufeld J. D. Microbial ecology of the deep terrestrial subsurface. The ISME Journal. 2024, 18(1): wrae091.

[108] Washburn E. W. The dynamics of capillary flow. Physical Review. 1921, 17(3): 273.

[109] Berg R. R. Capillary pressures in stratigraphic traps. AAPG Bulletin. 1975, 59(6): 939–956.

[110] Schowalter T. T. Mechanics of secondary hydrocarbon migration and entrapment. AAPG Bulletin. 1979, 63(5): 723–760.

[111] Watts N. Theoretical aspects of cap-rock and fault seals for single-and two-phase hydrocarbon columns. Marine and Petroleum Geology. 1987, 4(4): 274–307.

[112] Rezaee R., et al. Shale alteration after exposure to supercritical CO2. International Journal of Greenhouse Gas Control. 2017, 62: 91–99.

[113] Thompson A., Katz A., Raschke R. Estimation of absolute permeability from capillary pressure measurements. In: SPE annual technical conference and exhibition Dallas, 27–30 September, 1987.

[114] Lohr C., Hackley P. Mercury injection capillary pressure data in the US Gulf Coast Tuscaloosa Group in Mississippi and Louisiana collected 2015 to 2017. US Geological Survey Data Release. doi: 10.5066/F7BC3XTK. 2018.

[115] Jander J. H., et al. Hydrogen solubility, interfacial tension, and density of the liquid organic hydrogen carrier system diphenylmethane/dicyclohexylmethane. International Journal of Hydrogen Energy. 2021, 46(37): 19446–19466.

[116] Pan B., Yin X., Iglauer S. Rock-fluid interfacial tension at subsurface conditions: Implications for H2, CO2 and natural gas geo-storage. International Journal of Hydrogen Energy. 2021, 46(50): 25578–25585.

[117] Chow Y. F., Maitland G. C., Trusler J. M. Interfacial tensions of (H2O+ H2) and (H2O+ CO2+ H2) systems at temperatures of (298–448) K and pressures up to 45 MPa. Fluid Phase Equilibria. 2018, 475: 37–44.

[118] Hosseini M., et al. H2– brine interfacial tension as a function of salinity, temperature, and pressure; implications for hydrogen geo-storage. Journal of Petroleum Science and Engineering. 2022, 213: 110441.

[119] Iglauer S., Ali M., Keshavarz A. Hydrogen wettability of sandstone reservoirs: Implications for hydrogen geo-storage. Geophysical Research Letters. 2021, 48(3): e2020GL090814.

[120] Ali M., et al. Hydrogen wettability of quartz substrates exposed to organic acids; Implications for hydrogen geo-storage in sandstone reservoirs. Journal of Petroleum Science and Engineering. 2021, 207: 109081.

[121] Hashemi L., et al. Contact angle measurement for hydrogen/brine/sandstone system using captive-bubble method relevant for underground hydrogen storage. Advances in Water Resources. 2021, 154: 103964.

[122] Yekta A., et al. Determination of hydrogen–water relative permeability and capillary pressure in sandstone: Application to underground hydrogen injection in sedimentary formations. Transport in Porous Media. 2018, 122(2): 333–356.

[123] Al-Yaseri A., et al. Hydrogen wettability of clays: Implications for underground hydrogen storage. International Journal of Hydrogen Energy. 2021, 46(69): 34356–34361.

[124] Mu Y., et al. Hydrogen-water-rock interaction from the perspective of underground hydrogen storage: Micromechanical properties and mineral content of rock. International Journal of Hydrogen Energy. 2024, 70: 79–90.

[125] Zeng L., et al. Storage integrity during underground hydrogen storage in depleted gas reservoirs. Earth-Science Reviews. 2023, article 104625.

[126] Perera M. A review of underground hydrogen storage in depleted gas reservoirs: Insights into various rock-fluid interaction mechanisms and their impact on the process integrity. Fuel. 2023, 334: 126677.

[127] Zeng L., et al. Role of geochemical reactions on caprock integrity during underground hydrogen storage. Journal of Energy Storage. 2023, 65: 107414.

[128] Bo Z., et al. Geochemical reactions-induced hydrogen loss during underground hydrogen storage in sandstone reservoirs. International Journal of Hydrogen Energy. 2021, 46(38): 19998–20009.

[129] Shi Z., Jessen K., Tsotsis T. T. Impacts of the subsurface storage of natural gas and hydrogen mixtures. International Journal of Hydrogen Energy. 2020, 45(15): 8757–8773.

[130] Liu J., et al. Hydrogen diffusion in clay slit: Implications for the geological storage. Energy & Fuels. 2022, 36(14): 7651–7660.

[131] Carden P., Paterson L. Physical, chemical and energy aspects of underground hydrogen storage. International Journal of Hydrogen Energy. 1979, 4(6): 559–569.

[132] Mostinsky I. Diffusion coefficient. Florida, USA: CRC Press, 1996, DOI.

[133] Ghasemi M., et al. Molecular dynamics simulation of hydrogen diffusion in water-saturated clay minerals; implications for Underground Hydrogen Storage (UHS). International Journal of Hydrogen Energy. 2022, 47(59): 24871–24885.

[134] Krooss B. Evaluation of database on gas migration through clayey host rocks. Belgian National Agency for Radioactive Waste and Enriched Fissile Material (ONDRAF-NIRAS). RWTH Aachen: Aachen, Germany, 2008.

[135] Bhimineni S. H., et al. Machine-learning-assisted investigation of the diffusion of hydrogen in brine by performing molecular dynamics simulation. Industrial & Engineering Chemistry Research. 2023, 62(49): 21385–21396.

[136] Roumejon S., et al. Serpentinization and fluid pathways in tectonically exhumed peridotites from the Southwest Indian Ridge (62–65 E). Journal of Petrology. 2015, 56(4): 703–734.

[137] Bettenay E. The salt lake systems and their associated aeolian features in the semi-arid regions of Western Australia. Journal of Soil Science. 1962, 13(1): 10–17.

[138] Van-De-Graaff W. E. Relict early Cainozoic drainages in arid Western Australia. Zeitschrift Für Geomorphologie. 1977, 21(4): 379–400.

[139] English P. Ancient origins of some major Australian salt lakes: Geomorphic and regolith implications. In: Fourth Australian Regolith Geoscientists Association conference, Thredbo, New South Wales. pp.7–10, 2016.

[140] Cannat M., Fontaine F., Escartin J. Serpentinization and associated hydrogen and methane fluxes at slow spreading ridges. Diversity of Hydrothermal Systems on Slow Spreading Ocean Ridges. 2010, 188: 241–264.

[141] Ely T., et al. Huge variation in H2 generation during seawater alteration of ultramafic rocks. Geochemistry, Geophysics, Geosystems. 2023, 24(3): e2022GC010658.

[142] Loiseau K., et al. Hydrogen generation and heterogeneity of the serpentinization process at all scales: Turon de Técouère lherzolite case study, Pyrenees (France). Geoenergy. 2024, 2(1): geoenergy2023–024.

[143] Farough A., et al. Evolution of fracture permeability of ultramafic rocks undergoing serpentinization at hydrothermal conditions: An experimental study. Geochemistry, Geophysics, Geosystems. 2016, 17(1): 44–55.

[144] Alt J. C., W.c. S. III Sulfur in serpentinized oceanic peridotites: Serpentinization processes and microbial sulfate reduction. Journal of Geophysical Research: Solid Earth. 1998, 103(B5): 9917–9929.

[145] Wang X., et al. Serpentinization, abiogenic organic compounds, and deep life. Science China Earth Sciences. 2014, 57: 878–887.

Eric Deville
Chapter 2
Subsurface natural H$_2$ systems: some lessons from drilling results

Abstract: The existence of gas-phase accumulations of natural molecular hydrogen (H$_2$) in the subsurface of continents has been demonstrated for many decades. Over the past 20 years, the discovery of numerous H$_2$ seeps on land has shown that the genesis of native H$_2$ is a widespread process that had been largely underestimated. However, can we consider that natural H$_2$ could represent an energy source capable of playing a major role in the energy mix? To answer such a question, it is necessary to better define hydrogen systems in terms of source, migration, reservoir, trap, and production. Valuable information is provided by the results of drilling operations, in which H$_2$ has been discovered in significant quantities. The largest accumulations have been found in cratonic domains. In these contexts, the latest studies carried out from drilling results suggest that the origin of H$_2$ is mainly linked to Fe(II) oxidation processes. H$_2$ has been discovered in conventional reservoirs (comparable to those in which hydrocarbons are discovered): karstified carbonates, porous sandstones, and fractured granites. The seals of the H$_2$ accumulations are particularly effective, such as very low-permeability clay levels or massive dolerites. Deep aquifers also contribute to H$_2$ retention. Indeed, H$_2$ is fairly soluble in water at high pressure and temperature, that is, at depth, but is very poorly soluble at low pressure and temperature, toward the surface. As a result, H$_2$ becomes less and less likely to diffuse easily to the surface at shallow depths, due to its low solubility. It then tends to form a gas phase in relatively shallow reservoirs (of the order of a few hundred meters, or even a hundred meters in the case of Mali, which is the best documented to date). In Mali, the results show that the main H$_2$ reservoir is the shallowest and corresponds to a dynamic system that is progressively recharged with H$_2$ over the production timescale. This is because H$_2$, which is largely dissolved at depth in deep aquifers (confirmed by mud gas logging vs. logs below 800 m), gradually degasses during production and thus recharges the upper reservoir. This maintains a relatively constant pressure in the gas reservoir. These studies emphasize that H$_2$ production can take place in relatively shallow gas reservoirs, but that H$_2$ resources cannot be estimated without assessing the potential and evolution of dissolved H$_2$ at depth. Also, H$_2$ exploration cannot be

Eric Deville, IFPEN-IFP School, av. Bois-Préau, 92852 Rueil-Malmaison, France,
e-mail: eric.deville59@gmail.com, eric.deville@ifpen.fr

https://doi.org/10.1515/9783111437040-002

based solely on the presence of a H_2 generation process at depth. It must also be based on the presence of a trapping system capable of effectively slowing down H_2 migration in the gas phase.

Keywords: natural hydrogen, subsurface, drilling, reservoir, seal

2.1 Introduction

If there was a stubborn prejudice about natural gas till just a decade ago, it was the unfounded assertion that molecular hydrogen (H_2) did not exist in nature even if long H_2 had indeed been discovered in the subsurface [1]. Despite this widespread prejudice, we now know with certainty that H_2 is very frequently found on the Earth in the form of natural, diffuse, and continuous surface emanations. A number of recent studies have shown that this gas is present in many parts of the planet, notably in the form of bubble seepages or diffusive flows in soils or fractured rocks (see, for instance, the review of [2] and references therein). Natural H_2 emissions (sometimes associated with helium) have been found in many places around the world, notably at sea, along mid-ocean ridges ([3–5] and others), but also on land, in large ophiolitic massifs (notably in Oman [6–8]; in the Philippines [9]; in New Caledonia [10]; and others), in intraplate domains, notably ancient cratons ([11–14] and many others), in volcanic systems [15, 16] and others. Beyond these obvious surface clues to the genesis of H_2 in the subsurface, the question arises: can H_2 accumulate at depth and be exploited in a significant way by drilling in the subsurface, comparable to techniques used to exploit fossil hydrocarbons (HCs), and can this gas be recovered in sufficient abundance to provide a significant source of new energy?

Numerous exploration projects on an industrial scale by various private companies are currently underway on every continent of the planet. A sufficiently abundant source of natural H_2 could be used to meet current industrial needs, which are already in excess of 100 Mt/year and if this source proves substantial, it could be used as a partial substitute for carbon-rich fossil fuels. This would make it possible to avoid consuming fossil HC reserves or using large quantities of electricity to produce it. Natural H_2 would thus become a primary source of energy, a possible sustainable source of energy (production being a continuous phenomenon, linked to the Earth's dynamics), and a clean source of energy (combustion of H_2 produces just water, H_2O). Recovering this gas at the surface from punctual gas bubble outlets, or from zones where diffusion of this gas is known to occur, seems complicated, and only allows us to envisage the recovery of extremely limited flows of H_2. To be able to produce H_2 from the natural environment in industrial quantities, there is a need to drill for this gas in the subsurface, as has been done for decades in the search for HC gases.

In drilling, H$_2$ has indeed been encountered in different geological contexts (Figure 2.1). Some have been discovered in sedimentary basins with layers rich in very mature organic matter (Vaulx-en-Bugey in France [17, 18] and China [19]). Some volcanic systems contain H$_2$ in significant quantities, particularly within the hydrothermal fluids (in Iceland [16]; in the Afar [15]). Ophiolitic massifs with only partially serpentinized peridotites are also characterized by significant seepages of H$_2$, which have been encountered in mines or wells (in Turkey [7], Roumania [20], and Albania [21]). However, H$_2$ has been found most frequently in wells in cratonic domains in the form of seepages (in Canada and Scandinavia [22], and in South Africa, notably in the Witwatersrand Basin [23]), or in the form of accumulations (South Australia [1], European Russia [12], Mid-Rift System in the USA [24, 25], Mali [26–28], and Australia [29]), where Precambrian rocks are present either directly at outcrop or at depth, beneath a cover of younger rocks. Cratons account for more than half of the world's emerging land, and at the same time, they are areas that have been very little drilled since most of the world's drilling has been done in Phanerozoic sedimentary basins. This opens up a vast field of exploration in terms of the potential for natural H$_2$.

Discovering H$_2$ accumulations requires exploration based on the concept of a hydrogen system, as is done in a classic way of analyzing petroleum systems in the search for HC. However, there are important differences between HC and H$_2$ systems, linked to the specificity of H$_2$. H$_2$ is a highly diffusive and reactive gas. Also, whereas HC gases are practically inert in the subsurface, in the absence of O$_2$ or sulfate, H$_2$ is highly reactive, and capable of reacting with its mineral environment, in particular with oxidized rocks. But, H$_2$ is also widely naturally used by microorganisms, since it provides a source of energy, necessary for the life of the entire deep biomass hidden in subsurface in the geological environment.

Over the last century, but especially in recent decades, much progress has been made in our understanding of hydrogen systems in the geological environment. Much of this work has been based on surface observations and analyses (see references above). Gradually, however, studies carried out at the subsurface level, in mines or boreholes, have provided us with valuable insights into the genesis, migration and, above all, accumulation and preservation of H$_2$ at depth. On the basis of this experience, we propose in this chapter to discuss some lessons we had from these experiences and also the most favorable conditions for the presence of potentially exploitable H$_2$ in the subsurface.

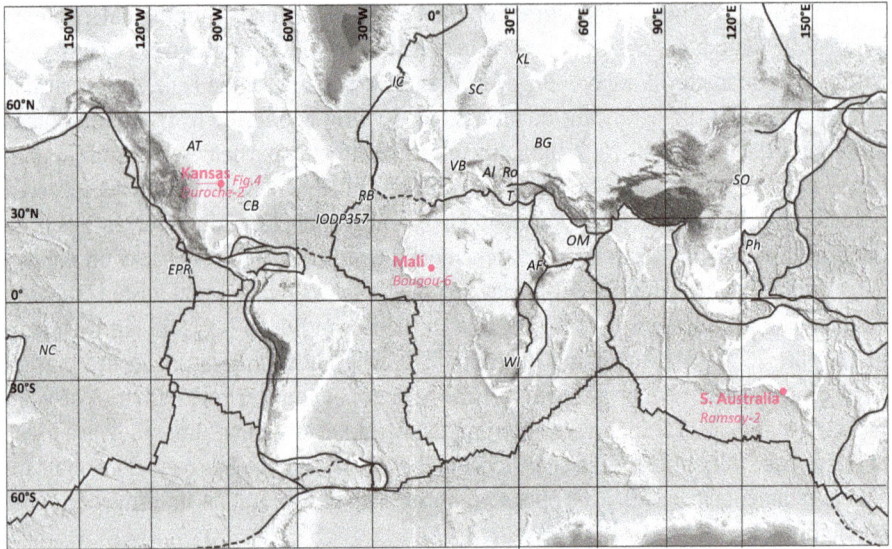

Figure 2.1: Location of the sites mentioned in the text. KL, Kola peninsula; IC, Iceland; SC, Scandinavia; VB, Vaul-en-Bugey; Al, Albania; Ro, Romania; AT, Athabasca; RB, Rainbow site on the Mid-Atlantic Ridge; IODP357, drilling site along the Mid-Atlantic Ridge; SO, Songliao Basin; OM, Oman; Ph, Philippines; AF, Afar; WI, Witwatersrand Basin. Black lines correspond to tectonic plate boundaries.

2.2 Sources of H_2 in subsurface

2.2.1 Oxidation of transition metal-rich minerals

The generation of H_2 in the subsurface takes a number of different routes but the processes most often invoked for the generation of natural H_2 are linked to the alteration of rocks by oxidation of minerals rich in transition metals, particularly Fe^{II}. Such an interpretation is commonly proposed in the case of ophiolitic massifs, notably with the presence of partly serpentinized peridotites (see [7] and references therein). Based on drilling results, this has been also more recently proposed in the case of Precambrian basement rocks, considering the type of rocks discovered in wells below H_2-rich reservoirs in Kansas [30–32] and in Mali [25]. These proven H_2-generating processes are particularly active during serpentinization of basic and ultrabasic rocks, especially mantle rocks and peridotites. These processes take place in oceanic domains, where they have been observed in boreholes [33]. They are also active in ophiolitic massifs, where they have been observed in the subsurface, either in mines [7, 20, 21] or by drilling [34, 35]. Fe^{II}-rich minerals like olivine, $(Mg,Fe)_2SiO_4$, commonly present in basic rocks and forming the dominant mineral in ultrabasic mantle rocks are prone to be weathered by subsurface water generating H$_2$ as a by-product of oxidation processes. Indeed, the Fe^{II}-rich endmember (fayalite) of a solid solution of olivine reacts with water to form serpentine, magnetite, and H$_2$, according to the following classical reaction:

$$6\,Fe_2SiO_4 + 7\,H_2O \rightarrow 3\,Fe_3Si_2O_5(OH)_4 + Fe_3O_4 + H_2$$

Such a weathering process (serpentinization) occurs when a water phase is present at lower temperature than supercritical conditions. Optimum conditions, in terms of kinetics of serpentinization, are in the range 300–320 °C, but the process is also active at lower temperature even lower than 50 °C.

Other minerals containing Fe^{II} are also prone to generate H$_2$ during weathering or thermal destabilization. For instance, biotite weathering produces H$_2$, with the precipitation of hematite, magnetite, or chamosite [36]. Fe^{II}-rich garnet (almandine) weathering also produces H$_2$. Similarly, siderite (carbonate) thermal destabilization at temperature >300 °C, in the presence of water-forming magnetite and carbon oxides, also produces H$_2$ [37]:

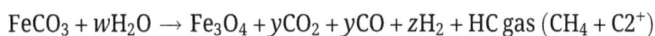

$$FeCO_3 + wH_2O \rightarrow Fe_3O_4 + yCO_2 + yCO + zH_2 + HC\,gas\,(CH_4 + C2^+)$$

2.2.2 H$_2$S oxidation

Many geological contexts, favorable to the presence of natural H$_2$, are devoid or very poor in hydrogen sulfide, like ophiolitic or cratonic areas. However, it is not the case for volcanic systems, or organic-rich sediments interfering with sulfate. Many wells drilled in volcanic provinces are characterized by fluxes of H$_2$S associated with H$_2$, notably in Iceland [16]. This can be explained by the fact that when magmatic gas, which very generally contains H$_2$S, rises and cools down (<400 °C) in volcanic conduits, it interferes with the hydrosphere and host rocks. When rocks are altered by hydrothermal fluids, iron can dissolve to form aqueous Fe^{2+}, and hydrogen sulfide can be oxidized with iron to form pyrrhotite, pyrite, and H$_2$ [38, 39]:

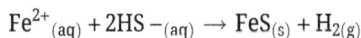

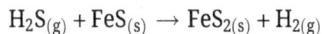

$$Fe^{2+}_{(aq)} + 2HS^-_{(aq)} \rightarrow FeS_{(s)} + H_{2(g)}$$

$$H_2S_{(g)} + FeS_{(s)} \rightarrow FeS_{2(s)} + H_{2(g)}$$

H$_2$S can also react notably with hematite and magnetite in the host rocks to form pyrite:

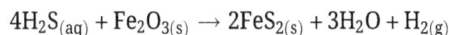

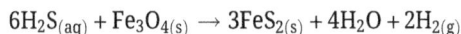

$$4H_2S_{(aq)} + Fe_2O_{3(s)} \rightarrow 2FeS_{2(s)} + 3H_2O + H_{2(g)}$$

$$6H_2S_{(aq)} + Fe_3O_{4(s)} \rightarrow 3FeS_{2(s)} + 4H_2O + 2H_{2(g)}$$

Moreover, in sediments affected by hypoxic/anaerobic conditions, sulfate-reducing bacteria can oxidize organic compounds by sulfate dissolved in water, with the release of H_2S. The thermal reduction of sulfate is also prone to generate H_2S at depth. In both cases, the H_2S that is produced is susceptible to react with iron, providing H_2. Indeed, whatever is the context, once hydrogen sulfide is present in the subsurface, it is prone to be oxidized in the presence of Fe^{2+} ions or Fe^{II}-rich minerals to form H_2. Thus, in all these cases, H_2S oxidation, which is a process very frequently observed in boreholes drilled in volcanic provinces or in organic-rich sedimentary basins where sulfate is available, is prone to the generation of H_2.

2.2.3 The question of water radiolysis

Radiolysis of water is often cited as a possible source of H_2 in continental domains [22, 40–44] and many others. Notably, it has been shown in Athabasca (Canada) uranium mines that, indeed, U-rich radioactive shale levels are associated with the presence of H_2 [45]. Radiolysis is responsible for the dissociation of water into different molecular products and free radicals (e_{aq}^-, H, OH, H_3O^+, H_2, and H_2O_2 [46–55]), forming a very unstable mix of globally balanced oxidizing and reducing agents. However, in bulk water within porous mediums, notably in fracture systems when moving away from radioactive sources, these compounds are very likely to recombine to reform water. This back process has been very well-known in nuclear research and industry for now eight decades, since the Manhattan project. The main process is the reactivity in bulk water of dissolved H_2 and H_2O_2 by the well-known Allen's reaction chain, giving back water [56–59]. In fact, in the case of the Canadian uranium mines, no important contents of H_2 were found outside, notably above, radioactive levels, in other places of this area of Athabasca. It is therefore not possible to demonstrate that radiolysis can actually form an efficient source of H_2, capable of migrating and accumulating in the geological environment away from the radioactive sources. More generally, as we have just seen, poorly radioactive rocks such as ophiolites and notably peridotites, through oxidation of Fe^{II}-rich minerals, are effectively a source of H_2, capable of migrating in the geological environment. It therefore seems difficult to invoke iron oxidation as the main source of H_2 in ophiolites and not consider it in cratonic environments, where Fe^{II} oxidation phenomena are also highly active (see above). Moreover, in very general terms, we can note that the places where surface flows or deep H_2 accumulations have been discovered in cratonic areas, do not correspond to the places where radioactivity is most significant on the continents. Also, heat flow is a direct proxy of the global radioactivity in the subsurface at a surface location and in fact, cratons correspond to the least radioactive continental domains; this is well-illustrated by the fact that they are characterized by lower radiogenic geothermal fluxes than in areas of more recent intraplate continental crust [60–62]. Also, in many places, H_2 presence is correlated with magnetic and gravimetric positive anomalies, which are not characteristic of very radiogenic rocks. Because of these several arguments, available data suggest that radiolysis cannot be favored as an efficient H_2

source at depth [28]. Finally, to the best of our knowledge, there is no direct indication, notably by drilling data, that free H_2-rich gas accumulation can be sourced by radiolysis processes.

2.2.4 Biological production of H_2 and late thermal cracking of organic-rich sediments

Some bacteria, notably acetogenic bacteria, produce H_2 in subsurface by dark fermentation of organic matter, without any photosynthetic process. At shallow depths, in soils in particular, it is very common to detect the presence of H_2 generated by biological activity (Figure 2.2). It is generally accepted that these organisms live in symbiosis with communities of methanogens that limit the concentration of H_2 in the environment, which is necessary for their development. In this way, H_2 is transferred between species, limiting it to transient low levels. Biological activity is therefore unlikely to generate significant quantities of H_2.

During the thermal cracking of organic matter, the thermogenic gas generated partly comprises H_2 during the final stages of maturation, at temperatures above ~200 °C. Indeed, the progressive thermal breaking of molecular bonds generates increasingly smaller molecules, including H_2 (Figure 2.2). This affects both organic matter through primary cracking, and thermogenic HC (oil and gas) through secondary cracking. Thus, the ultimate fate of thermal cracking of the most mature HC dry gas, methane, is to generate graphite and H_2 (CH_4 pyrolysis). A good example of this process is the in the deep Songliao Basin in China. In this basin, it has been shown that the presence of HC gases is mainly expressed above 6,000 m, while there is the presence of hydrogen below 6,000 m [19]. This fact is perfectly compatible with an H_2 origin linked to the ultimate primary or secondary thermal cracking of organic matter and HC in the deepest part of this sedimentary basin.

2.2.5 Other processes

Finally, it should be noted that other natural H_2 genesis processes have been proposed, but even if they do exist, it was not possible to demonstrate from drilling results that they are effective and they hardly appear to be capable of generating significant H_2 accumulations of economic interest. However, it is worth noting that magmatic gases all contain significant levels of H_2: the hotter the magma, the higher the H_2 content [63]. This is the result of H_2/H_2O chemical equilibrium processes at high temperatures and could explain H_2 presence in geothermal wells related to outgassing toward the surface in the vicinity of active magma chambers [64]. Such processes are limited in scope on a global scale but could be of some importance when considering the valorization of hydrothermal fluids in the case of geothermal opera-

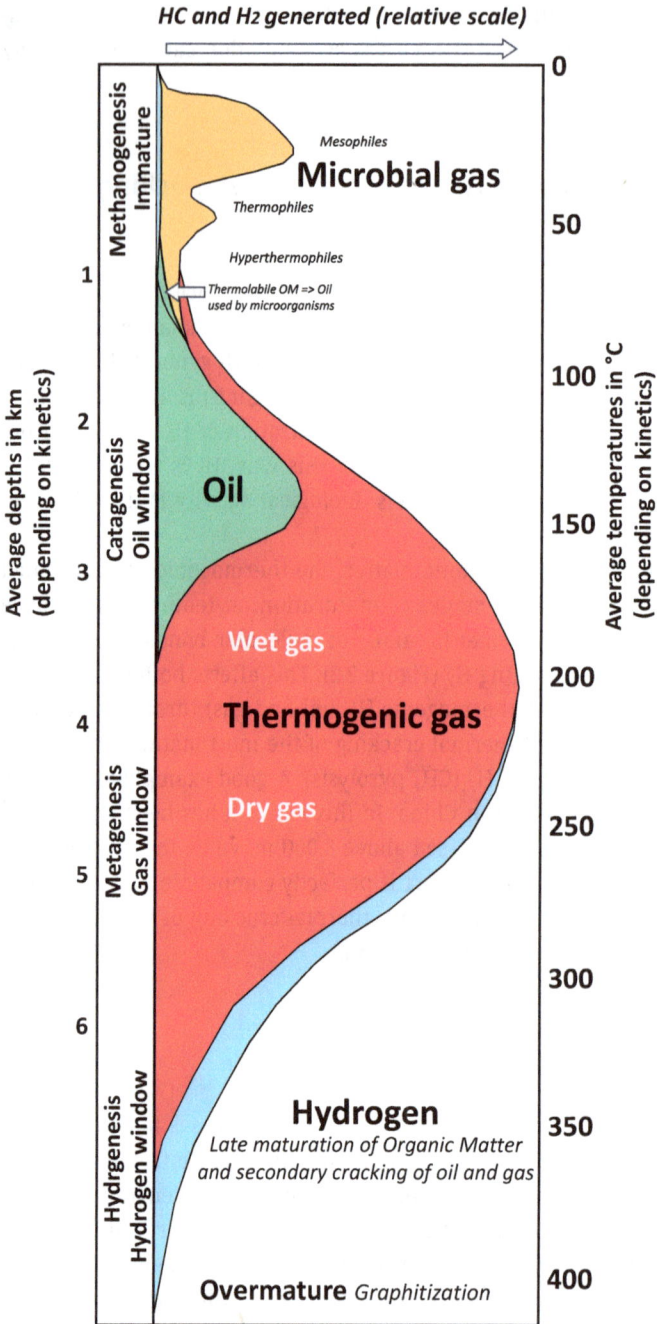

Figure 2.2: Conceptual log of organic matter degradation versus depth by microorganisms' activity and thermal cracking.

tions. The recent discovery of "dark" oxygen in the deep waters of the eastern Pacific has been interpreted as the consequence of seawater electrolysis processes, meaning that H$_2$ is also generated by electrolysis [65]. In the present state of knowledge, such processes appear to be fairly limited on a global scale. Mechanical crushing (catacla-sis) of silica-rich rocks by tectonic processes, in particular, is sometimes mentioned as a source of H$_2$ along active faults. Here again, it is unlikely that this could generate very large quantities of H$_2$. The origin of H$_2$ by outgassing of the primordial hydrogen present at the Earth's core has sometimes been invoked [66]. However, this interpre-tation is hampered by the fact that H$_2$ would have to be associated with ^{3}He. Given the very low concentrations of ^{3}He found on the Earth, it seems difficult to suppose that large quantities of primordial H$_2$ could still be outgassing in any significant way. An origin of H$_2$ by thermal stabilization of ammonium, which is present in clays in particular, has been proposed [67]. However, as this process can only occur at high temperatures, it is unlikely that such an origin is compatible with the H$_2$ present in the main accumulations discovered to date.

2.3 Migration of H$_2$

2.3.1 Gas diffusion versus free gas advection

In static aquifers, dissolved H$_2$ diffusion in porous media obeys Fick's classical first law of diffusion, which states that a linear relationship exists between the flow of matter and its concentration gradient from places with high concentration to places of low concentration as

$$J_{H_2} = -D_{H_2} \nabla C_{H_2} \tag{2.1}$$

J_{H2} is the diffusion flux of H$_2$ in water, D_{H2} is the diffusion coefficient or diffusivity of H$_2$ in water, and C_{H2} is the concentration of H$_2$ in water. The diffusion coefficient is dependent on the salinity and temperature, and it is nearly independent of pressure [68, 69]. If we compare the diffusion of H$_2$ with that of other common gases in subsur-face in the range 25 to 50 °C, H$_2$ is more diffusive in water than methane and nitrogen in water within the pore network of subsurface rocks, but it is less diffusive than he-lium (Table 2.1; Figure 2.3). In pure water, at 25 °C, the diffusion coefficient of H$_2$ is 5.13×10^{-9} m^2/s and for CH$_4$, it is 1.85×10^{-9} m^2/s. The diffusion coefficient increases with temperature. At 50 °C, the diffusion coefficient of H$_2$ is 7.29×10^{-9} m^2/s and the one of CH$_4$ is $\sim 3 \times 10^{-9}$ m^2/s. The diffusion coefficients of both H$_2$ and CH$_4$ decrease with salinity. For instance, the H$_2$ diffusion coefficient decreases up to 38% as the sa-linity increases from 1 to 5 molal. As a consequence, for similar environmental condi-tions, if we compare the diffusion of H$_2$ with that of methane, we can roughly assume that, in the 20–50 °C range, H$_2$ is about three times more diffusive in water than methane, for the same concentration (Table 2.1). However, even if H_2 is more diffusive than HC gas, its diffusion rate in water is still much lower than the migration rate of H_2 as a gas phase. Advection processes follow Darcy's law (1856) in a fine-grained poorly porous medium, where the laminar gas flow Q is expressed as follows:

$$Q = -\frac{K}{\eta} \left[\vec{\nabla}P - \rho_w \vec{g} \right]$$
(2.2)

with K being the permeability, η the gas viscosity, $\vec{\nabla}P$ the pressure gradient, and $\rho_w \vec{g}$ the hydrostatic component of the pressure gradient. The pressure due to the H_2 gas inside a porous medium invaded with water is expressed as follows:

$$P_{H_2} = (\rho_w - \rho_{H_2})\, g\, h$$
(2.3)

where ρ_w is the density of water, ρ_{H2} is the density of H_2 at the depth considered, g is the gravitational acceleration, and h is the height of gas accumulation.

Considering the basic laws of gas migration in permeable medium, as well as the average physical values, as presented in Table 2.1, the existence of a gas phase is therefore much more effective for significant H_2 migration flows.

Moreover, in the most superficial domains, the solubilization of H_2 decreases more and more, so that the less H_2 can solubilize; the lower its concentration, the lower is its diffusive flux. The corollary is that in the most superficial domains, if H_2 cannot solubilize and if a H_2 recharge is maintained, the more H_2 will tend to form a gas phase and therefore migrate more easily. We will see later that this has very strong implications for H_2 retention processes in the subsurface.

Table 2.1: Diffusion coefficient, density and viscosity of hydrogen, methane, nitrogen, and He at normal temperature (20 °C) and pressure (1.013 bar) (data from [69]).

NTP (1.013 bar, 20 °C)	H_2	CH_4	N_2	He
Diffusion coefficient in pure water (cm^2/s^{-1})	4.58×10^{-9}	1.60×10^{-9}	1.88×10^{-9}	6.71×10^{-9}
Density (kg/m^3)	0.083	0.67	1.17	0.178
Viscosity (Pa s)	0.88×10^{-5}	1.10×10^{-5}	1.76×10^{-5}	1.96×10^{-5}

2.3.2 The role of hydrodynamics

While H_2 migration is more efficient in the gas phase than in the dissolved phase, we must not overlook the fact that many aquifers are not static and that if H_2 is present in the dissolved phase in an aquifer, then it is likely to be transported in the water, simply by hydrodynamics over long distances, and as the water migrates into more superficial domains, it may then pass into the gas phase. A mechanism of this type is likely occurring in the Kansas example, which is characterized by very significant

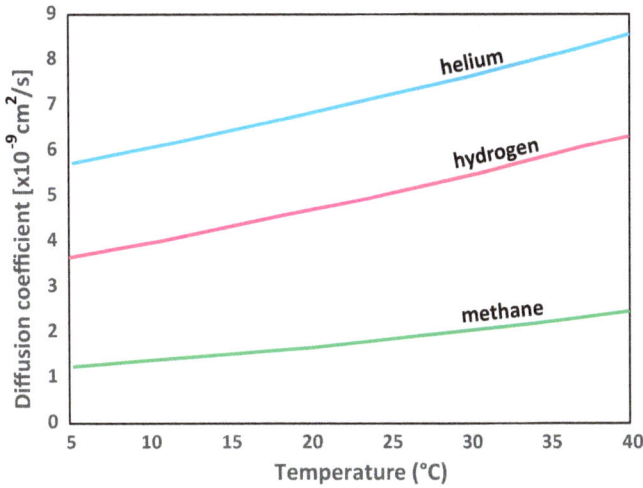

Figure 2.3: Diffusion coefficient in water of CH_4, H_2, and He versus temperature (data from [67, 69]).

large-scale hydrodynamics due to hydraulic recharge in mountain systems and the migration of water present in aquifers eastwards over very long distances (>300 km) with the existence of artesian systems [28] (Figure 2.4).

2.3.3 Migration of dissolved Fe^{2+} versus migration of H_2

Among the results obtained from the study of drilling fluids collected from H_2-rich wells in Kansas, it was found that artesian wells expel water very rich in dissolved iron Fe^{2+} [28]. As mentioned above, in many cases, Fe^{II} oxidation is considered as the source of H_2 at depth. But iron in an oxidation state 2+ is not necessarily present only in solid mineral matter. Iron can solubilize as Fe^{2+} ions over a wide range of pH and Eh conditions. This is well illustrated in the classical Pourbaix diagram for iron (Figure 2.5). Once iron has solubilized, it is highly susceptible to migration in the water in which it is dissolved by hydrodynamic processes. When pH and Eh conditions vary, particularly when pH rises, Fe^{2+} in solution is likely to oxidize in a location far from its initial source. The oxidizing iron can then reduce the water, forming H_2, and precipitate in the mineral form Fe^{III} as iron oxides or hydroxides. And so, when it comes to migration, the question arises as to whether it is the source element of the H_2 that has been most affected by the migration, or whether it is the H_2 once formed that has migrated, either in dissolved or gaseous form.

In other environments, notably in cratonic contexts, the pH of water associated with the presence of H_2 can be also close to neutral. This is notably the case in the H_2 field of Bourakébougou in Mali [25].

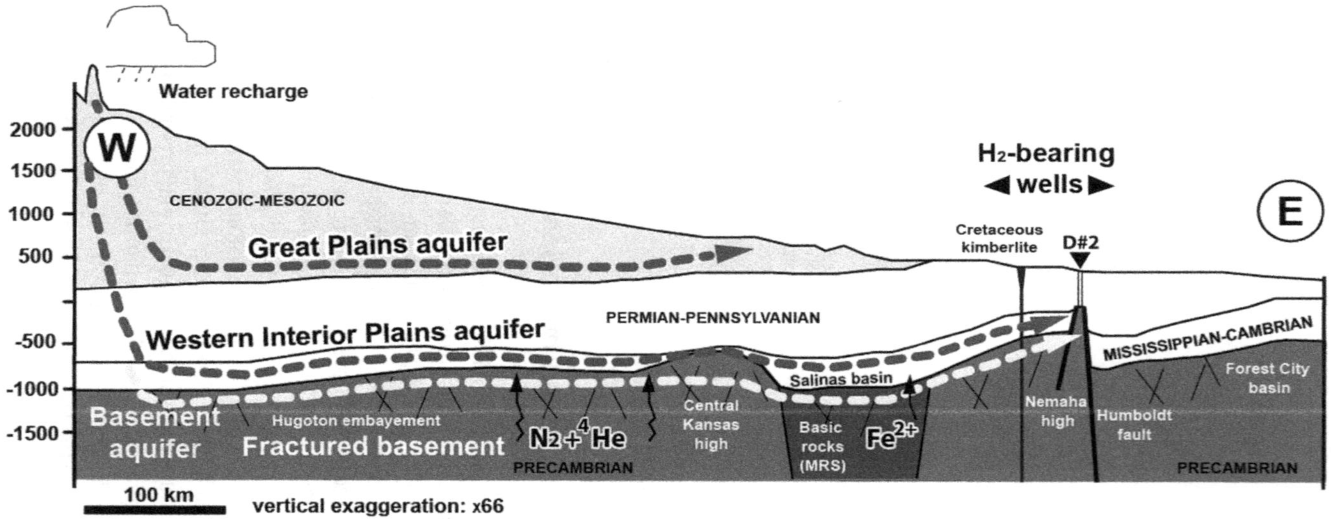

Figure 2.4: Simplified geological cross section of Kansas, illustrating long-distance hydrodynamic processes and their relationship with H_2 generation and migration.

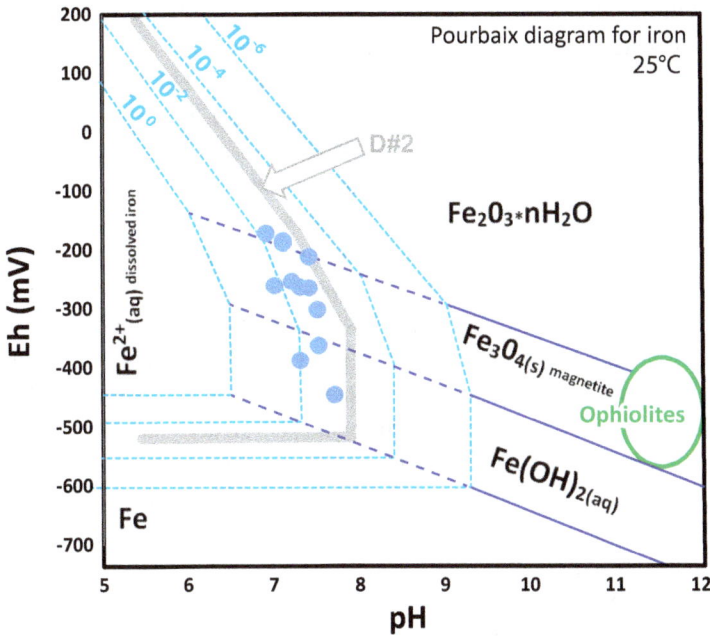

Figure 2.5: Eh-pH Pourbaix diagram for iron at atmospheric pressure and 25 °C. Blue lines represent domain limits for different concentrations of Fe^{2+}. Gray line represents Fe^{2+} domain limit for the concentration measured at well Sue Duroche#2 (modified from [28, 30]).

In ophiolitic rocks, water flowing through fractures generally has a high pH (between 10 and 12). In this case, iron oxidation is accompanied by the direct precipitation of iron oxides (see above). In volcanic contexts, pH levels are always very low. So, for neutral or acidic pH, the question of iron solubilization arises directly. In this case, it is possible that dissolved Fe^{2+} has migrated significant distances through the hydrodynamics of deep aquifers.

2.4 Natural H$_2$ reservoir

If we want to explore H$_2$ accumulations in the subsurface, the question immediately arises as to which rock types can form an efficient H$_2$ reservoir in the geological environment. Indeed, many questions remain as to how H$_2$ can accumulate in the subsurface. Can H$_2$ accumulate only in conventional reservoirs, as understood in hydrogeology or the oil industry, that is rocks with relatively high porosity, either matrix porosity or fractures in the rock environment? However, its very specific properties also raise the question of whether H$_2$ can simply accumulate in the matrix of rocks that do not constitute conventional porous reservoirs. Another notable point of uncertainty lies in knowing whether H_2 can accumulate only in the dissolved form in water, or whether it can accumulate in reservoirs as a gas phase.

A recent study carried out on the Bourakebougou natural H_2 field in Mali, where a lot of information is available, showed that H_2 is accumulated in the gas phase in very conventional reservoirs, mainly in a relatively shallow reservoir, at depths of 90 and 110 m, consisting of karstified carbonates, mainly dolomites with marble levels [25] (Figures 2.6–2.8). Deeper down, the H_2-bearing reservoirs are mainly sandstone, with porosities varying between 4.5% and 6.4%. In these porous sandstone levels, according to available data, H_2 is present in the gas phase down to a depth of around 800 m, whereas it is mainly dissolved below this depth [25] (Figures 2.6 and 2.7).

In Kansas, various studies have shown that H_2 is present in Permian-Carboniferous carbonate sediment reservoirs, but from the available data, it would appear that the main reservoir richest in H_2 is located in the porosity of the fractured and altered Precambrian basement. Above, in sedimentary reservoirs, H_2 is present mainly in dissolved form, and in particular, the production of a Carboniferous artesian karst reservoir is accompanied by H_2 and N_2 releases. In the basement, free gas was released during the drilling of the Duroche-2 well [28].

So, based on current knowledge, it seems that the main natural accumulations of H_2 in the natural environment are formed in classic porous reservoirs, comparable to hydrogeological or petroleum reservoirs.

In ophiolitic rocks, water flowing through fractures generally has a high pH (between 10 and 12). In this case, iron oxidation is accompanied by the direct precipitation of iron oxides (see above). On the other hand, in other environments, notably in cratonic contexts, the pH of water, associated with the presence of H_2, can be close to neutral [25, 28]. In volcanic contexts, pH levels are always low. So, for neutral or acidic pH, the question of iron solubilization arises directly. In Kansas, for example, we know that some artesian boreholes producing water and H_2 have very high levels of dissolved iron. In this case, it is possible that dissolved Fe^{2+} has migrated significant distances through the hydrodynamics of deep aquifers.

2.5 Retention of H_2 in subsurface

A key point about H_2 retention at depth is that H_2 is more soluble in water than methane at deeper depths than 1,000 m (i.e., pressure >100 bar), but the lower the pressure toward the surface, the less soluble H_2 becomes in water. At atmospheric pressure, the possibility of solubilizing H_2 in water becomes very small (maximum 0.0003 mol H_2/kg H_2O). The consequence is that toward the surface, H_2 concentration gradients necessarily become weaker and weaker, becoming minimal at very shallow depths (Figure 2.9). If H_2 can almost no longer dissolve, the more difficult it is for it to diffuse in water. As per Fick's law, if ∇CH_2 decreases, the flux JH_2 also decreases. In fact, the

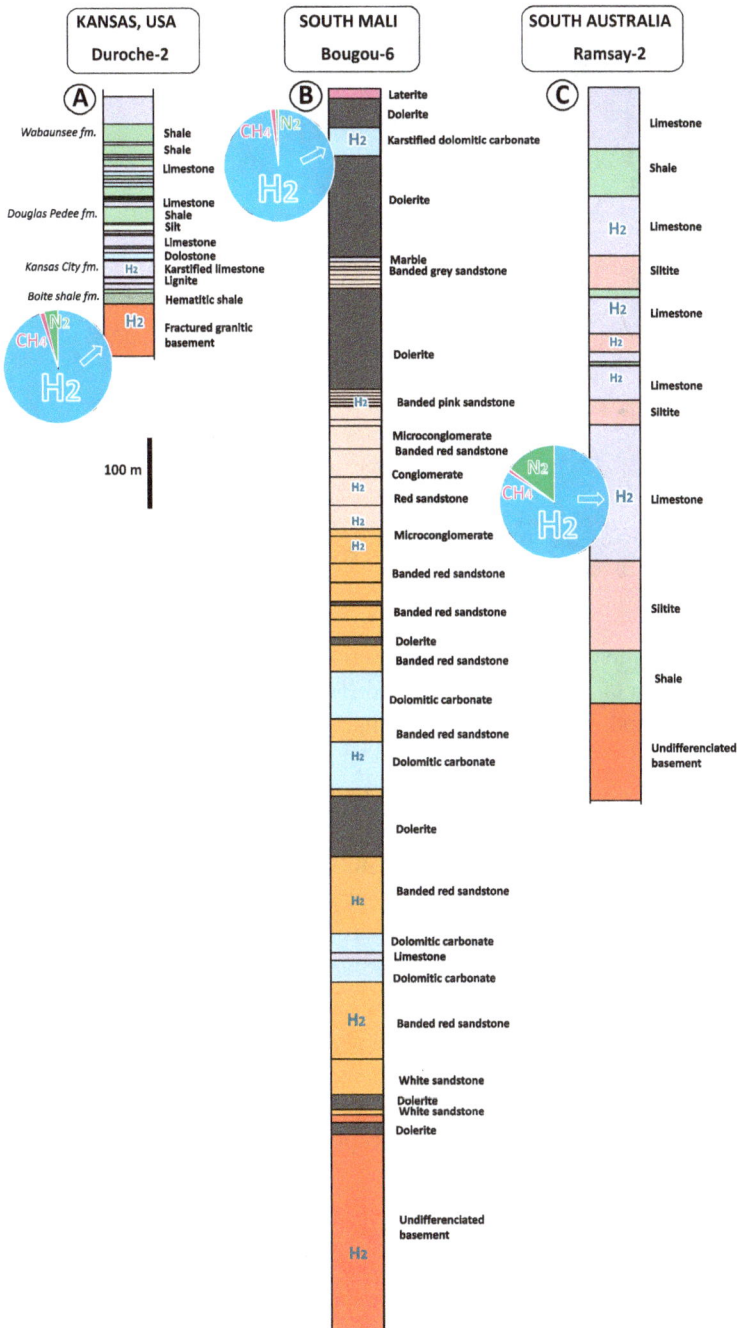

Figure 2.6: Examples of three different synthetic logs of wells where free H₂-rich gas has been discovered at depth (location in Figure 2.1; (A) from [28, 30]; (B) from [25, 27]; (C) from [1, 27, 29]).

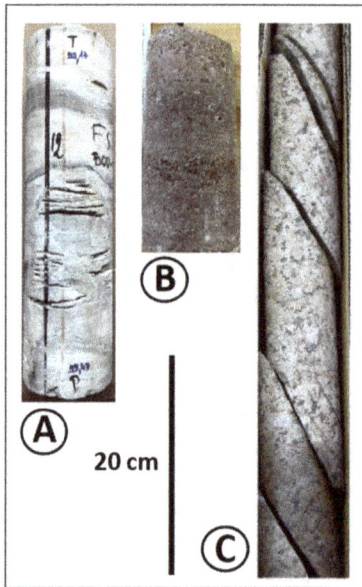

Figure 2.7: Different cores illustrating the main different types of reservoirs in which H_2-rich gas (>90 mol%) has been discovered. (A) Carbonate karstic reservoir in Bourakébougou in Mali; (B) coarse sandstone reservoir in Bourakébougou in Mali; fractured Precambrian basement rocks in Kansas (A and B modified from [25, 27]).

upper part of the hydrosphere acts as a kind of seal on H_2 diffusion. The consequence is that this allows a form of retention of dissolved H_2 at depth, and if there is a constant flux issued from depth, then at shallow depth, H_2 tends to form a free gas phase. It is therefore easy to understand why, to date, gas-phase H_2 reservoirs have been discovered at shallow depths (much shallower than the majority of large methane accumulations).

When H_2 forms a gas phase, it becomes very difficult to retain in the rock environment, as H_2 is a small molecule capable of easily penetrating even the smallest rock pore network.

Due to the low density of H_2, the pressure gradient generated by the buoyancy of H_2 as a gas phase eq. (2.3) within a geological environment invaded with water is high (at atmospheric pressure, 20 °C, H_2 density $\rho_{H_2} = 0.083 \, kg/m^3$; water density $\rho_w = 998 \, kg/m^3$; at 10 bar, 20 °C, H_2 density $\rho_{H_2} = 0.408 \, kg/m^3$; and water density $\rho_w = 999 \, kg/m^3$).

The gas is trapped when the gravitational force is less than the capillary force. Leakage occurs when P_{H2}>Pe, with Pe being the capillary pressure [72]. The Laplace law states that

$$Pe = 2\gamma \, cos\Theta/R \tag{2.4}$$

where γ is the interfacial tension, Θ is the contact angle between the fluid and the solid matrix, and R is the radius of the gas bubble. At 20 °C and 55 bar, γ is equal to ~0.051 N/m for H_2 and ~0.065 N/m for CH_4 [73–76].

As the gravitational forces for a column of H_2 are very strong (8.43 times stronger than for a column of CH_4, for example), capillary pressure must be very high to limit

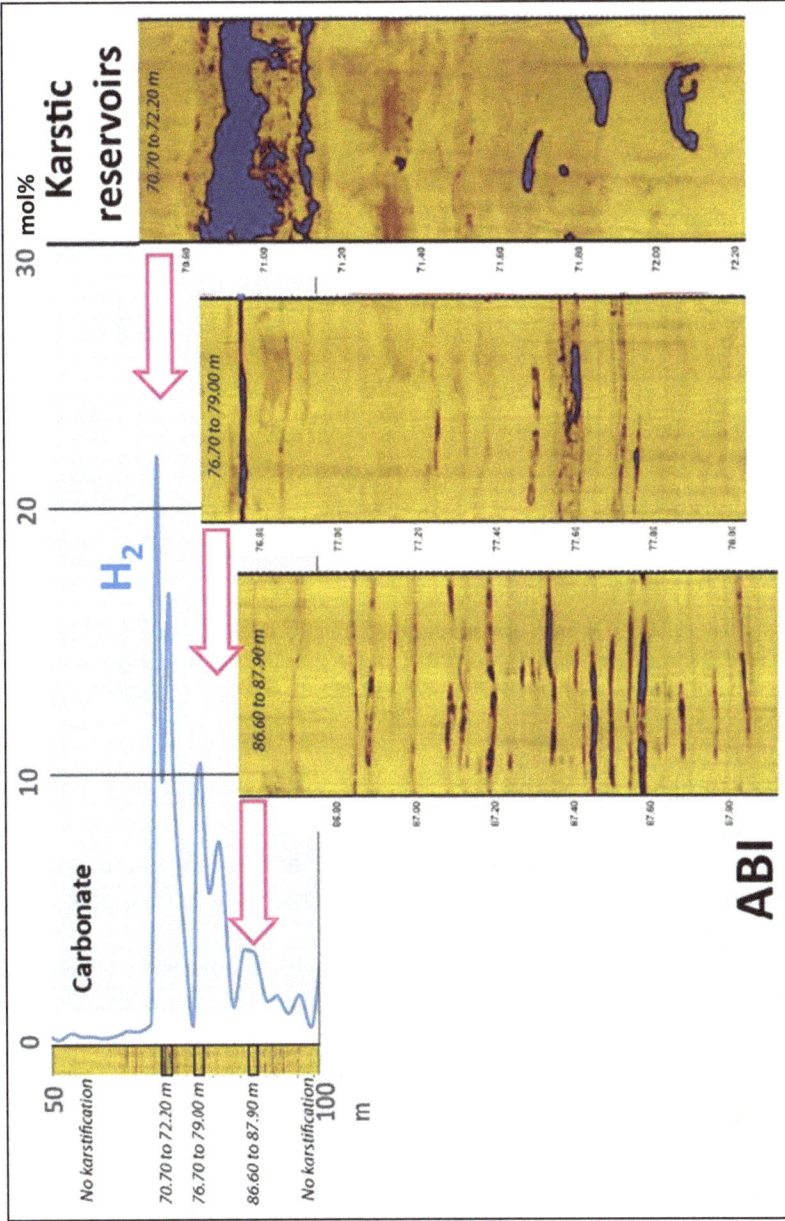

Figure 2.8: Borehole imagery *versus* gas logging (mol%) showing H₂ gas presence within the karstic system in the Bougou 1 well in Mali (modified from [25, 27]).

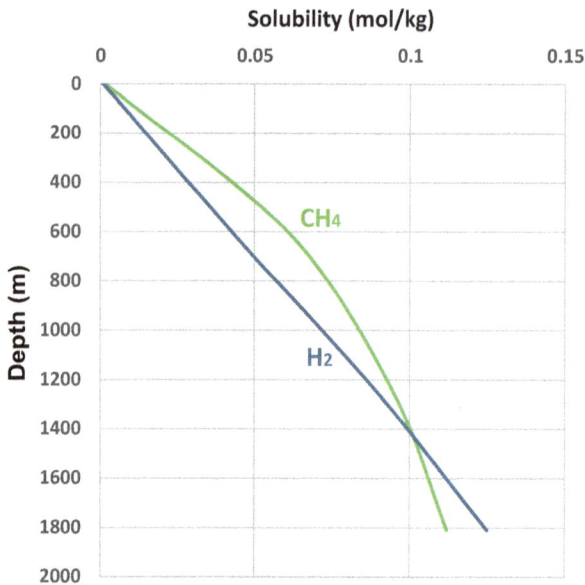

Figure 2.9: Compared solubility of H_2 and CH_4 *versus* depth, according to the pressure and temperature conditions recorded in Bougou-6 well in Mali (from Maiga, 2023 [25]). Solubility data are from [70, 71]. This graph shows that H_2 is less soluble than CH_4 above a depth of 1,400 m, whereas it becomes more soluble than methane below 1,400 m.

H_2 leakage, so seals must be very effective with rocks with extremely low porosity. As capillary pressure is inversely proportional to pore size, rocks must be much less porous and permeable to retain a gas-phase H_2 column than an HC gas column. The buoyancy of H_2 in the gas phase is very high, which contributes to favor the migration of H_2 toward the surface by Darcy flux.

In addition to the pressure gradient, which contributes to leakage, H_2 viscosity is very low ($\sim 0.88 \times 10^{-5}$ Pa.s NTP), as a consequence, according to Darcy's law, the permeability needs to be extremely low to prevent any H_2 leakage trough seal rocks. This is the crucial point to retain H_2, which is a very small and very diffusive molecule. Indeed, one of the major lessons we learned from drilling results is that free H_2 gas accumulations only exist under excellent seals. This is indeed what is observed in the case of the main H_2 accumulations known to date. For example, the emblematic Bourakébougou H_2 field (98%) in Mali is preserved under a seal of exceptional quality, formed by a massive volcanic sill (dolerites) that is almost unfractured, and the only minor fractures that can be detected are not open fractures [25]. In Kansas, in the Junction City region, the seal is a very impermeable shale formation [28]. Similarly, in southern Australia, the seal of H_2-rich gas accumulations corresponds to thick shale layers [1]. In Kangaroo Island, South Australia, there is 48 m of clay, and in the Ramsay-1 well, there are 90 m of complex layers, mainly shale, phyllite, and slate, with intercalation of sandstone and limestone. In Australia's Amadeus Basin, a gas mixture containing H_2 and He is trapped beneath a salt seal [77]. In all cases, the levels are thick and very low in permeability (Figure 2.6). The need for highly efficient seals capable of retaining H_2 also helps to explain the frequent association between H_2 and He, both of which have a small volume and low mass, making them highly diffusive compounds with high buoyancy in the gas phase (high gravity forces). This association would therefore be mainly due to physical characteristics and not to a common origin.

In the case of hydrothermal volcanic systems, which by their very nature are open systems, there is generally no effective seal to trap the gas. In this case, we have no choice but to try to valorize the hydrothermal fluids by recovering the hydrogen present in the drilling fluids. Similarly, in the case of ophiolitic massifs, there is generally no seal to trap the gas effectively, and the gas leaks extensively into the fracture systems [8].

2.6 H₂ reactivity

The preservation of natural H_2 in geological reservoirs is not only dependent on the physical processes of leakage from the reservoirs but also on the chemical and biological reactivity of H_2. This is crucial to understanding the integrity of H_2 in geological reservoirs. Various processes are responsible for H_2 reactivity in the subsurface. Some processes are purely chemical, generally at high temperatures, notably in volcanic systems (magmatic WGS, hydrothermal Sabatier). Others are linked to microbial activity, in particular the activity of methanogens.

2.6.1 Purely chemicals processes

In volcanic systems, at high temperatures (>400 °C), as magma gas rises, H_2 content is controlled by temperature, sulfur content, pressure, and redox conditions. SO_2 tends to be converted to H_2S as magma gas cools to maintain fluid-rock equilibrium, according to the following reaction [78, 79]:

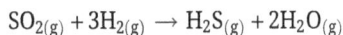

$$SO_{2(g)} + 3H_{2(g)} \rightarrow H_2S_{(g)} + 2H_2O_{(g)}$$

The process affects the H_2/H_2O ratio, which gradually decreases with temperature as H_2 is consumed. For example, in the Afar triangle, this interpretation is consistent with the fact that SO_2 was present at higher levels than H_2S (SO_2/H_2S ratio between 20 and 70) in the magmatic gas from the Ardoukoba eruption (between Lake Assal and Ghoubbet Bay in Djibouti, in November 1978), and in magmatic gas from the Erta'Ale volcano in Ethiopia, while SO_2 has not been detected at high levels in wells, nor at the surface in active fumaroles and hot springs in the Afar region (see discussion in [77, 79]).

When H_2 is present in sufficient quantities, it can react with CO_2 to form CH_4, generally below 400 °C, according to the Sabatier reaction defined in 1902. The overall mechanism of this reaction is unclear, either CO_2 methanation reacts all at once, or it dissociates to form CO intermediates which in turn react to produce CO methanation by reverse water gas shift. Indeed, the Sabatier reaction is either an inverse direct steam-reforming reaction, which is hydrogenation of carbon dioxide:

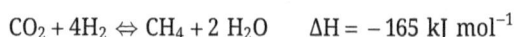

$$CO_2 + 4H_2 \Leftrightarrow CH_4 + 2\,H_2O \qquad \Delta H = -165 \text{ kJ mol}^{-1}$$

or the linear combination of two reactions, the reverse water-gas shift reaction,

$$CO_2 + H_2 \Leftrightarrow CO + H_2O \qquad \Delta H = 41 \text{ kJ mol}^{-1}$$

and the reverse methane steam reforming reaction, in which syngas (CO and H_2) produces CH_4 and water by hydrogenation of carbon monoxide,

$$CO + 3H2 \Leftrightarrow CH_4 + H_2O \qquad \Delta H = -206 \text{ kJ mol}^{-1}$$

The sum of the reactions gives the Sabatier reaction and its enthalpy of reaction. The Sabatier reaction can also occur via formate, acetate, and methanol [80]. The reaction of CO-rich magmatic gases with H_2, according to the reverse methane steam-reforming reaction, may in fact be a more widespread CH_4-generating process than the Sabatier reaction. As such, it could be the most efficient way of generating CH_4 and consuming H_2 in volcanic gases, since CO is barely detectable in hydrothermal systems, whereas it is present in magmatic gases along with H_2 [77]. This CO-consuming reaction can be considered a Fisher-Tropsch (FTT) polymerization reaction (with $n = 1$ in the Fisher-Tropsch process: $(2n + 1)\,H_2 + nCO \rightarrow C_nH_{2n+2} + nH_2O$). FTT reactions operate mainly between 300 and 150 °C. Since the reverse SR reaction of methane and the Sabatier reaction are both largely exothermic, very high temperatures (>400 °C) are unfavorable to the reaction of carbon oxides to form CH_4, while low temperatures limit the kinetics of these reactions below 150 °C. Due to the stoichiometry of these reactions (lower number of moles in the products than in the reactants), these reactions favor CH_4 yields at high pressure. These reactions are optimally active in the 400–150 °C range and at high pressure, which are common conditions in hydrothermal systems.

2.6.2 Biological processes

If H_2 tend to form accumulations at relatively shallow depths in the presence of water, they are all the more prone to being exposed to reaction processes linked to microbiology. Indeed, many methanogens exist via metabolic processes linked to H_2 consumption (hydrogenotrophs [81–86] and many others). The resulting global chemical reaction chain is the reduction by H_2 of hydrogenocarbonate dissolved in water (formed by solubilization of CO_2 or carbonate rocks), resulting in the formation of CH_4 and water:

$$HCO_3{}^- + H^+ + 4H_2 \rightarrow CH_4 + 3H_2O$$

This reaction occurring in an aqueous medium is equivalent to the Sabatier gaseous reaction (methanation of CO_2 or reverse CH_4 wet reforming) but in that case, the process is occurring at a moderate temperature (viable for microorganisms), thanks to biological catalysts.

For example, such processes were well demonstrated during the storage of a gas mixture containing H_2, CO_2, and CO in a geological aquifer in the Czech Republic (Lobodice site [87]). Within 6 months, a significant proportion of the H_2 (~20%) was consumed by methanogens, producing CH_4 with fairly high $\delta^{13}C$ values (between −44 and −40 per mil). The process was associated with a significant reduction in stored gas volume and the production of organic acids (formic, acetic, and propionic) and aldehydes. Mass balance, with consideration of possible migration or leakage of H_2 (or other components) from the reservoir, was done using the chemical composition of the stored gas. The balance was based on the content of the inert component (N_2). This has shown that simple leakage of H_2 from the reservoir did occur, but could not explain the changes in composition, which was readily explicable by consumption of H_2 and reduction of $HCO_3{}^-$ and by reverse CH_4 steam-reforming catalytic CO reduction.

Microorganisms are therefore highly efficient at consuming H_2. If we wish to define locations where this likelihood of H_2-consuming microbial reactions is minimal, one possibility might be the case of purely gaseous reservoirs, thus minimizing water-free gas interactions. Microbial activity is then reduced and H_2 consumption minimized. This may be the case in Mali's upper Bourakébougou reservoir. Indeed, in this case, the karst cavities are generally well isolated from gas-water contacts (Figure 2.8).

Another possibility would be the case of areas where there are natural poisons in the geological environment preventing the activity of microorganisms. For example, with regard to drilling in Kansas, it has been shown that the water of the Precambrian basement contains significant levels of natural poison like arsenic [28, 30]. This type of poison could contribute to minimizing biological activity. Another case, which could be more frequent, is the presence of salt in the geological environment, which causes aquifers to become saturated with NaCl, a well-known process responsible for slowing down biological activity. This might be the case, for instance, of the deep H_2-bearing reservoirs of the Amadeus basin in Australia [75, 77].

Finally, another possibility for restricting the biological activity of hydrogenotrophs is simply the carbon oxides restriction in the environment, which in cratonic or ophiolitic contexts is a common feature. This is indeed the case in the waters of certain ophiolite massifs, such as the ophiolites of Oman, where the only carbon present is in the form of dissolved CH_4, but where concentrations of HCO_3^- and CO_3^{2-} are equal to zero [6].

2.7 Production

In the example of the Bourakébougou H_2 field in Mali, it has been shown that the H_2 system is dynamic, and that pressure does not drop during production; even an increase in gas pressure was observed after a production cycle [25]. This shows that there is a H_2 recharge in the reservoir during and after production cycles. H_2 recharge probably comes from H_2 dissolved at depth in aquifers. Indeed, it has been shown that underneath free gas accumulations, there are aquifers with significant quantities of dissolved H_2 capable of degassing, in response to gas withdrawal to balance reservoir pressure [25]. This aspect could be important in terms of production perspectives, since production could be concentrated on small free gas reservoirs that could be progressively recharged from below significantly during production.

2.8 Conclusion

Considering natural hydrogen systems, the available subsurface data to date about the origin of H_2 in accumulations favor mostly an origin through oxidation processes of iron-rich minerals (olivine, magnetite, garnets, micas, etc.). In the case of volcanic systems, an H_2S oxidation origin is also highly probable.

Drilling results have shown that H_2 systems are in all cases dynamic systems. In the case of volcanic systems and most of ophiolitic systems, these are open systems where the gas is, as far as we know, not stored in reservoirs but constantly seeping through fracture systems. But in any case, even when free gas is stored in a geological reservoir, the H_2 systems are also dynamic, with leakage compensated by progressive recharge. There is also a need for certain hydrodynamics of water if H_2 is indeed issued from H_2O (source) to allow the progressive recharge of the reservoirs.

Subsurface H_2 migration is mostly efficient as a gas phase, but in the case of aquifers containing dissolved H_2, it is very likely that hydrodynamics plays a significant role in H_2 migration, the dissolved gas being transported within the water. In some cases, where pH levels are neutral or low, the question also arises as to whether it is not dissolved Fe^{2+} that is affected by significant migration and, depending on variations in pH and Eh, may be responsible for oxidizing away from the initial iron-rich source rock, generating H_2.

The reservoirs where the main H_2 accumulations known to date have been discovered are classic reservoirs: karstified and/or fractured carbonates, porous sandstones, and fractured basement. However, it should be emphasized that a significant proportion of H_2 can be stored in dissolved form in deep aquifers. This dissolved gas enables free gas reservoirs to be recharged during production. Possibly, most of the produceable H_2 might be present in a dissolved form.

In all cases, the seals of H_2 accumulations are of excellent quality (massive volcanic sills, salt, thick shale, etc.). Smaller H_2 accumulations are more easily retained than larger ones due to lower gravitational forces.

Aspects of H_2 preservation, notably by minimizing consumption by methanogens, are also important to consider like reservoirs, where free gas-water interaction is minimized. Environments unfavorable to the development of life (hypersaline and/or with natural poisons) are also a way of limiting microbial activity. Finally, carbon-restricted environments are also conducive to the preservation of H_2 in reservoirs.

Regarding production, free gas production is the most favorable in terms of economic viability and technical feasibility. Indeed, to date, we don't really know how to produce dissolved gas. However, even if only free gas is produced, it is highly likely, as in the Mali example, that reservoirs will be affected by a progressive recharge from dissolved gas in the aquifers underlying the free gas accumulations, and that this recharge possibly occur on comparable timescale as reservoir production.

References

[1]   Ward L. K. Inflammable gases occluded in the Pre-Palaeozoic rocks of South Australia. Transactions of the Royal Society of South Australia. 1933, 57: 42–47. https://www.biodiversitylibrary.org/bibliography/16197

[2]   Milkov A. V. Molecular hydrogen in surface and subsurface natural gases: Abundance, origins and ideas for deliberate exploration. Earth-Science Reviews. 2022, 230: 104063. https://doi.org/10.1016/j.earscirev.2022.104063

[3]   Welhan J. A., Craig H. Methane and hydrogen in East Pacific Rise hydrothermal fluids. Geophysical Research Letters. 1979, 6: 829–831. https://doi.org/10.1029/GL006i011p00829

[4]   Charlou J. L., Donval J. P., Fouquet Y., Jean-Baptiste P., Holm N. G. Geochemistry of high H_2 and CH_4 vent fluids issuing from ultramafic rocks at the Rainbow hydrothermal field (36°14'N, MAR). Chemical Geology. 2002, 191: 345–359. https://doi.org/10.1016/S0009-2541(02)00134-1

[5]   Sleep N. H., Meibom A., Fridriksson T., Coleman R. G., Bird D. K. H_2-rich fluids from serpentinization: Geochemical and biotic implications. Proceedings of the National Academy of Sciences. 2004, 101: 12818–12823. https://doi.org/10.1073/pnas.0405289101

[6]   Neal C., Stanger G. Hydrogen generation from mantle source rocks in Oman. Earth and Planetary Science Letters. 1983, 66: 315–320. https://doi.org/10.1016/0012-821X(83)90144-9

[7]   Vacquand C., Deville É., Beaumont V., Guyot F., Pillot D., Sissmann O., Arcilla C., Prinzhofer A. Reduced gas seepages in ophiolitic complexes: Evidence for multiple origins of the H_2-CH_4-N_2 gas mixtures. Geochimica et Cosmochimica Acta. 2018, 223: 437–461. https://doi.org/10.1016/j.gca.2017.12.018

[8]   Zgonnik V., Beaumont V., Larin N., Pillot D., Deville E. Diffused flow of molecular hydrogen through the Western Hajar mountains, Northern Oman. Arabian Journal of Geosciences. 2019, 12: 71. https://doi.org/10.1007/s12517-019-4242-2

[9]   Abrajano T. A., Sturchio N. C., Bohlke J. K., Lyon G. L., Poreda R., Stevens C. Methane-hydrogen gas seeps, Zambales Ophiolite, Philippines: Deep or shallow origin? Chemical Geology. 1988, 71(1–3): 211–222. https://doi:10.1016/0009-2541(88)90116-7

[10] Deville E., Prinzhofer A. The origin of $N_2/H_2/CH_4$ -rich natural gas seepages in ophiolitic context: A major and noble gases study of fluid seepages in New Caledonia. Chemical Geology. 2016, 440: 139–147. https://doi.org/10.1016/j.chemgeo.2016.06.011

[11] Ikorsky S., Gigashvili G. M., Lanyov V., Narkotiev V. D., Petersilye I. A. The investigation of gases during the Kola Superdeep borehole drilling. Geologische Jahrbuch. 1999, D107: 145–152.

[12] Larin N., Zgonnik V., Rodina S., Deville É., Prinzhofer A., Larin V. N. Natural molecular hydrogen seepages associated with surficial, rounded depression on the European craton in Russia. Natural Resources Research. 2015, 24(3): 363–383. https://DOI:10.1007/s11053-014-9257-5

[13] Zgonnik V., Beaumont V., Deville E., Larin N., Pillot D., Farrell K. Evidences for natural hydrogen seepages associated with rounded subsident structures: The Carolina bays (Northern Carolina, USA). Progress in Earth and Planetary Science. 2015, 2(31): 1–15. https://DOI:10.1186/s40645-015-0062-5

[14] Moretti I., Brouilly E., Loiseau K., Prinzhofer A., Deville E. Hydrogen emanations in intracratonic areas: New guidelines for early exploration basin screening. Geosciences. 2021, 11(3): 145. https://doi.org/10.3390/geosciences11030145

[15] D'Amore F., Giusti D., Abdallah A. Geochemistry of the high-salinity geothermal field of Asal, Republic of Djibouti, Africa. Geothermics. 1998, 27: 197–210. https://doi.org/10.1016/S0375-6505(97)10009-8

[16] Arnason B., Sigfusson T. Application of geothermal energy to hydrogen production and storage. 2007. https://www.researchgate.net/publication/237613011

[17] Deville E., Prinzhofer A. L'hydrogène naturel: Une source potentielle d'énergie propre et renouvelable? Géologues. 2015, 185: 105–110.

[18] Deronzier F., Giouse H. Vaux-en-Bugey (Ain, France): The first gas field produced in France, providing learning lessons for natural hydrogen in the sub-surface? BSGF – Earth Sciences Bulletin. 2020, 191: 7. https://doi.org/10.1051/bsgf/2020005

[19] Wang L., et al. The occurrence pattern of natural hydrogen in the Songliao Basin, P.R. China: Insights on natural hydrogen exploration. International Journal of Hydrogen Energy. 2024, 50(Part B): 261–275. https://doi.org/10.1016/j.ijhydene.2023.08.237

[20] Baciu C., Etiope G. A direct observation of a hydrogen-rich pressurized reservoir within an ophiolite (Tisovita, Romania). International Journal of Hydrogen Energy. 2024, 73: 402–406. https://doi.org/10.1016/j.ijhydene.2024.06.065

[21] Truche L., et al. A deep reservoir for hydrogen drives intense degassing in the Bulqizë ophiolite. Science. 2024, 383(6683): 618–621. https://DOI:10.1126/science.adk9099

[22] Sherwood-Lollar B., Onstott T. C., Lacrampe-Couloume G., Ballentine C. J. The contribution of the Precambrian continental lithosphere to global H_2 production. Nature. 2014, 516: 379–382. https://doi:10.1038/nature14017

[23] Cook A. P. The occurrence, emission and ignition of combustible strata gases in Witwatersrand gold mines and Bushveld platinum mines and means of ameliorating related ignition and explosion hazards. Safety in Mines Research Advisory Committee. Itasca Africa (Pty) Ltd. Project Number: GAP. 1996, 504: 89.

[24] Coveney R. M., Goebel E. D., Zeller E. J., Dreschhoff G. A. M., Angino E. E. Serpentinization and the origin of hydrogen gas in Kansas. American Association of Petroleum Geologists Bulletin. 1987, 71: 39–48. https://doi.org/10.1306/94886D3F-1704-11D7-8645000102C1865D

[25] Newell K. D., Doveton J. H., Merriam D. F., Sherwood Lollar B., Waggoner W. M., Magnuson L. M. H$_2$-rich and hydrocarbon gas recovered in a deep Precambrian well in northeastern Kansas. Natural Resources Research. 2007, 16(3): 277–292.

[26] Prinzhofer A., Cisse C. S. T., Diallo A. B. Discovery of a large accumulation of natural hydrogen in Bourakebougou (Mali). International Journal of Hydrogen Energy. 2018, 43: 19315–19326. https://doi.org/10.1016/j.ijhydene.2018.08.193

[27] Maiga O., Deville E., Laval J., Prinzhofer A., Diallo A. B. Characterization of the Bourakebougou natural hydrogen reservoirs in Mali. Nature, Scientific Reports. 2023, 13: 11876. https://doi.org/10.1038/s41598-023-38977-y

[28] Maiga O., Deville E., Laval J., Prinzhofer A., Diallo A. B. Trapping processes of large volumes of natural hydrogen in the subsurface: The emblematic case of the Bourakebougou H$_2$ field in Mali. International Journal of Hydrogen Energy. 2024, 50: 640–647. https://doi.org/10.1016/j.ijhydene.2023.10.131

[29] Gold Hydrogen. Ramsay 2 update: very high hydrogen concentrations up to 86% purity found along with the very high helium concentrations. 2023. https://www.goldhydrogen.com.au/

[30] Guélard J., Beaumont V., Guyot F., Pillot D., Jezequel D., Ader M., Newell K. D., Deville E. Natural H$_2$ in Kansas: Deep or shallow origin? G^3 Geochemistry, Geophysics and Geosystems. 2017, 18: 1841–1865. https://DOI:10.1002/2016GC006544

[31] Combaudon V., Sissman O., Bernard S., Viennet J. C., Megevand V., Le Guillo C., Guélard J., Martinez I., Guyot F., Derluyn H., Deville E. Are the Fe-rich clay veins in the igneous rock of the Kansas (USA) Precambrian crust of magmatic origin? Lithos. 2024, 474–475: 107583. https://doi.org/10.1016/j.lithos.2024.107583

[32] Kularatne K., Senechal P., Combaudon V., Darouich O., Subirana M. A., Proietti A., Delhaye C., Schaumlöffel D., Sissmann O., Deville E., Derluyn H. X-ray micro-computed tomography-based approach to quantify natural H$_2$ generation by Fe^{II} oxidation in the intracratonic lithologies. International Journal of Hydrogen Energy. 2024, 78: 861–870. https://doi.org/10.1016/j.ijhydene2024.06.256

[33] Fruh-Green, et al. Magmatism, serpentinization and life: Insights through drilling the Atlantis Massif (IODP Expedition 357). Lithos. 2018, 323: 137–155. https://doi.org/10.1016/j.lithos.2018.09.012

[34] Ellison E. T., Templeton A. S., Zeigler S. D., Mayhew L. E., Keleman P. B., Matter J. M., and the Oman Drilling Project Science Party. Low-temperature hydrogen formation during aqueous alteration of serpentinized peridotite in the Samail Ophiolite. Journal of Geophysical Research. 2021, 126. https://doi.org/10.1029/2021JB021981

[35] Templeton, et al. Accessing the subsurface biosphere within rocks undergoing active low-temperature serpentinization in the Samail Ophiolite (Oman Drilling Project). JGR Biogeosciences. 2021, e2021JG006315. https://doi.org/10.1029/2021JG006315

[36] Murray J., Clément A., Fritz B., Schmittbuhl J., Bordmann V., Fleury J. M. Abiotic hydrogen generation from biotite-rich granite: A case study of the Soultz-sous-Forêts geothermal site, France. Applied Geochemistry. 2020, 119: 104631. https://doi.org/10.1016/j.apgeochem.2020.104631

[37] McCollom T. M. Formation of meteorite hydrocarbons from thermal decomposition of siderite (FeCO$_3$). Geochimica et Cosmochimica Acta. 2003, 67(2): 311–317.

[38] Syverson D. D., Ono S., Shanks W. C., Seyfried W. E. Multiple sulfur isotope fractionation and mass transfer processes during pyrite precipitation and recrystallization: An experimental study at 300 and 350 °C. Geochimica et Cosmochimica Acta. 2015, 165: 418–434.

[39] Stefansson A., et al. Gas chemistry of Icelandic thermal fluids. Journal of Volcanology and Geothermal Research. 2017, 346: 81–94.

[40] Lin L.-H., Slater G. F., Lollar B. S., Lacrampe-Couloume G., Onstott T. C. The yield and isotopic composition of radiolytic H$_2$, a potential energy source for the deep subsurface biosphere. Geochimica et Cosmochimica Acta. 2005, 69: 893–903. https://doi:10.1016/j.gca.2004.07.032

[41] Blair C. C., D'Hondt S., Spivack A. J., Kingsley R. H. Radiolytic hydrogen and microbial respiration in subsurface sediments. Astrobiology. 2007, 7: 951–970.

[42] Warr O., Giunta T., Ballentine C. J., Sherwood-Lollar C. J. Mechanisms and rates of ^{4}He, ^{40}Ar, and H_2 production and accumulation in fracture fluids in Precambrian Shield environments. Chemical Geology. 2019, 530: 119322. https://doi.org/10.1016/j.chemgeo.2019.119322

[43] Sauvage J. F., Flinders A., Spivack A. J., Pockalny R., Dunlea A. G., Anderson C. H., Smith D. C., Murray R. W., D'Hondt S. The contribution of water radiolysis to marine sedimentary life. Nature Communications. 2021, 12. https://doi.org/10.1038/s41467-021-21218-z

[44] Boreham C. J., Edwards D. S., Czado K., Rollet N., Wang L., Van der Wielen S., Champion D., Blewett R., Feitz A., Henson P. A. Hydrogen in Australian natural gas: Occurrences, sources and resources. APPEA Journal. 2021, 61(1): 163–191. https://doi.org/10.1071/AJ20044

[45] Truche L., Joubert G., Dargent M., Martz P., Cathelineau M., Rigaudier T., Quirt D. Clay minerals trap hydrogen in the Earth's crust: Evidence from the Cigar Lake uranium deposit, Athabasca. Earth and Planetary Science Letters. 2018, 493: 186–197. https://doi.org/10.1016/j.epsl.2018.04.038

[46] Debierne A. Recherches sur les gaz produits par les substances radioactives. Décomposition de l'eau. Annales de Physique (Paris). 1914, 2: 97–127.

[47] Burton M. Radiation chemistry. The Journal of Physical and Colloid Chemistry. 1947, 51(2): 611–625. https://doi:10.1021/j150452a029.ISSN0092-7023

[48] Allen A. O. The radiation chemistry of water and aqueous solutions. Van Nostrand, New York, 1961.

[49] Appleby A., Schwarz H. A. Radical and molecular yields in water irradiated by gamma rays and heavy ions. Journal of Physical Chemistry. 1937, 1969, 73: 6.

[50] LaVerne J. A., Schuler R. H. Decomposition of water by very high linear energy transfer radiations. Journal of Physical Chemistry. 1983, 87: 4564–4565.

[51] Spinks J. W. T., Woods R. J. An introduction to radiation chemistry. New York, USA: John Wiley, 1990, 574.

[52] Rotureau P., Renault J. P., Lebeau B., Patarin J., Mialocq J. C. Radiolysis of confined water: Molecular hydrogen formation. ChemPhysChem. 2005, 6: 1316–1323.

[53] Ershov B. G., Gordeev A. V. Model for radiolysis of water and aqueous solutions of H_2, H_2O_2 and O_2. Radiation Physics and Chemistry. 2008, 77(8): 928–935.

[54] Le Caër S. Water radiolysis: Influence of oxide surfaces on H_2 production under ionizing radiation. Water. 2011, 3: 235–253. https://doi:10.3390/w3010235

[55] Crumière F., et al. LET effects on the hydrogen production induced by the radiolysis of pure water. Radiation Physics and Chemistry. 2013, 82: 74–79.

[56] Pastina B., LaVerne J. A. Effect of molecular hydrogen on hydrogen peroxide in water radiolysis. Journal of Physical Chemistry A. 2001, 105: 9316–9322.

[57] Lertnaisat P., Katsumura Y., Mukai S., Umehara R., Shimizu Y., Suzuki M. Simulation of the inhibition of water α-radiolysis via H_2 addition. Journal of Nuclear Science and Technologies. 2014, 51: 9. https://doi.org/10.1080/00223131.2014.907548

[58] Horne G. P., Pimblott S. M., LaVerne J. A. Inhibition of radiolytic molecular hydrogen formation by quenching of excited state water. Journal of Physical Chemistry B. 2017, 121: 5385–5390. https://DOI:10.1021/acs.jpcb.7b02775

[59] Fang Z., Cao X., Tong L., Muroya Y., Whitaker G., Momeni M. An improved method for modelling coolant radiolysis in ITER. Fusion Engineering and Design. 2018, 127: 91–98.

[60] Pollack H. N., Hurter S. J., Johnston J. R. Heat loss from the Earth's interior: Analysis of the global data set. Review of Geophysics and Space Physics. 1993, 31: 267–280. https://doi.org/10.1029/93RG01249

[61] Stein C. Heat flow of the world. 1995. https://doi:10.1029/RF001p0144

[62] Lucazeau F. Analysis and mapping of an updated terrestrial heat flow data set. Geochemistry, Geophysics, Geosystems. 2019, 20: 4001–4024. https://doi.org/10.1029/2019GC008389

[63] Fischer T., Chiodini G. Volcanic, magmatic and hydrothermal gases. In: The encyclopedia of volcanoes. Elsevier Inc, 2015. http://dx.doi.org/10.1016/B978-0-12-385938-9.00045-6

[64] abc, 12345, null ref

[65] Sweetman A. K., et al. Evidence of dark oxygen production at the abyssal seafloor. Nature Geoscience. 2024, 17: 737–739. https://doi.org/10.1038/s41561-024-01480-8

[66] Toulhoat H., Zgonnik V. Chemical differentiation of planets: A core issue. The Astrophysical Journal. 2022, 924(2): 83. https://doi.ff10.3847/1538-4357/ac300bff

[67] Jacquemet N., Prinzhofer A. The association of natural hydrogen and nitrogen: The ammonium clue? International Journal of Hydrogen Energy. 2022, 50(Part B): 161–174. https://DOI:10.1016/j.ijhydene.2023.07.265

[68] Cussler E. L. Diffusion: Mass transfer in fluid systems, 2nd éd. New York: Cambridge University Press, 1997, 600.

[69] Wang S., Zhou T., Pan Z., Martin Trusler J. P. Diffusion coefficients of N$_2$O and H$_2$ in water at temperatures between 298.15 and 423.15 K with pressures up to 30 MPa. Journal of Chemical & Engineering Data. 2023, 68: 1313–1319. https://doi.org/10.1021/acs.jced.3c00085

[70] Duan Z., Moller N., Greenberg J., Weare J. H. The prediction of methane solubility in natural waters to high ionic strength from 0 to 250 °C and 0 to 1600 bar. Geochemica Cosmochimica Acta. 1992, 56: 1451–1460.

[71] Zhu Z., Cao Y., Zheng Z., Chen D. An accurate model for estimating H$_2$ solubility in pure water and aqueous NaCl solutions. Energies. 2022, 15: 5021. https://doi.org/10.3390/en15145021

[72] Vavra C. L., Kaldi J. G., Sneider R. M. Geological applications of capillary pressure: A review. American Association of Petroleum Geologists Bulletin. 1992, 76: 840–850. https://doi.org/10.1306/BDFF88F8-1718-11D7-8645000102C1865D

[73] Chow Y. T., Maitland G. C., Trusler J. P. M. Interfacial tensions of (H$_2$O + H$_2$) and (H$_2$O + CO$_2$ + H$_2$) systems at temperatures of (298 to 448) K and pressures up to 45 MPa. Fluid Phase Equilibria. 2018, 475: 37–44. https://doi.org/10.1016/j.fluid.2018.07.022

[74] Hashemi L., Glerum W., Farajzadeh R., Hajibeygi H. Contact angle measurement for hydrogen/brine/sandstone system using captive-bubble method relevant for underground hydrogen storage. Advances in Water Resources. 2021, 154: 103964. https://doi.org/10.1016/j.advwatres.2021.103964

[75] van Rooijen W. A., Habibi P., Xu Dey P., Vlugt T. J. H., Hajibeygi H., Moultos O. A. Interfacial tensions, solubilities, and transport properties of the H$_2$/H$_2$O/NaCl system: A molecular simulation study. Journal of Chemical & Engineering Data. 2023. https://doi.org/10.1021/acs.jced.2c00707

[76] Tian H., Fan J., Yu Z., Liu Q., Lu X. The high-pressure methane/brine/quartz contact angle and its influence on gas reservoir capillarities. Minerals. 2023, 13: 164. https://doi.org/10.3390/min13020164

[77] Leila M., Loiseau K., Moretti I. Controls on generation and accumulation of blended gases (CH$_4$/H$_2$/He) in the Neoproterozoic Amadeus Basin, Australia. Marine and Petroleum Geology. 2022, 140: 105643. https://doi.org/10.1016/j.marpetgeo.2022.105643

[78] Giggenbach W. F. Redox processes governing the chemistry of fumarolic gas discharges from White Island, New Zealand. Applied Geochemistry. 1987, 2: 143–161. https://doi:10.1016/0883-2927(87)90030-8

[79] Deville E., Mohamed Hassan K., Moussa Ahmed K., Prinzhofer A., Pelissier N., Guélard J., Noirez S., Mohamed Magareh H., Omar Said I. H$_2$ generation versus H$_2$ consumption in volcanic gas systems: A case study in the Afar hot spot in Djibouti. Applied Geochemistry. 2023, 156: 105761. https://doi.org/10.1016/j.apgeochem.2023.105761

[80] McCollom T. M., Seawald J. S. Abiotic synthesis of organic compounds in deep-sea hydrothermal environments. Chemical Reviews. 2007, 107(2): 382–401. https://doi:10.1021/cr0503660

[81] Jones W. J., Nagle D. P. Jr, Whitman W. B. Methanogens and the diversity of archaebacteria. Microbiological Reviews. 1987, 51: 135–177.

[82] Ward J. A., Slater G. F., Moser D. P., Lin L.-H., Lacrampe-Couloume G., Bonin A. S., Davidson M., Hall J. A., Mislowack B., Bellamy R. E. S., Onstott T. C., Sherwood Lollar B. Microbial hydrocarbon gases in the Witwatersrand Basin, South Africa: Implications for the deep biosphere. Geochimica Cosmochimica Acta. 2004, 68: 3239–3250. https://doi.org/10.1016/j.gca.2004.02.020

[83] Sinha N., Nepal S., Kral T., Kumar P. Survivability and growth kinetics of methanogenic archaea at various pHs and pressures: Implications for deep subsurface life on Mars. Planetary and Space Science. 2017, 136: 15–24. https://doi.org/10.1016/j.pss.2016.11.012

[84] Hug L. A., Baker B. J., Anantharaman K., Brown C. T., Probst A. J., Castelle C. J., Butterfield C. N., Hernsdorf A. W., Amano Y., Ise K., Suzuki Y., Dudek N., Relman D. A., Finstad K. M., Amundson R., Thomas B. C., Banfield J. F. A new view of the tree of life. Nature Microbiology. 2016, 1: 16048. https://doi.org/10.1038/nmicrobiol.2016.48

[85] Colman D. R., Poudel S., Stamps B. W., Boyd E. S., Spear J. R. The deep, hot biosphere: Twenty-five years of retrospection. Proceedings of the National Academy of Science. 2017, 114: 6895–6903. https://doi.org/10.1073/pnas.1701266114

[86] van Dam F., Kietavainen R., Drake H. Dissolved microbial methane in the deep crystalline crust fluids-current knowledge and future prospects. Geofluids. 2022. https://doi.org/10.1155/2022/3945073

[87] Buzek F., Onderka V., Vančura P., Wolf I. Carbon isotope study of methane production in a town gas storage reservoir. Fuel. 1994, 73: 747–752. https://doi:10.1016/0016-2361(94)90019-1

Gabor C. Tari
Chapter 3
Natural hydrogen exploration: some similarities and differences with oil and gas exploration

Abstract: The fundamental question in natural hydrogen exploration is whether it is emerging as a process comparable to established practices in the hydrocarbon industry. There are papers published just in the last few years suggesting that many items in our collective industry and academic toolbox could be readily applied to natural hydrogen exploration. The consensus appears to be that most of the petroleum systems elements the industry tends to focus on in exploration projects largely overlap with those of the natural hydrogen system.

From an exploration point of view, several play types for natural hydrogen indeed appear to be very similar to what the oil and gas industry is used to. These include cases where there is a functioning trap, due to effective top seals. Numerous examples can be found in pre-salt traps worldwide where hydrogen has been documented for a long time as part of existing natural gas accumulations. Another, but unusual trapping style has been documented in the first hydrogen field discovery in Mali where the top seal is a set of dolerite dykes. In these cases, one expects finite hydrogen resources to be in place and the exploration approach has indeed some resemblance to that of hydrocarbon prospecting.

Another group of natural hydrogen targets revolve around large mega-seeps (fairy circles) and geometrically smaller but pronounced fault-controlled seepages to the surface. These hydrogen occurrences seemingly have no traps or seals and, therefore, do not find a proper analogue in oil and gas exploration workflows. Strictly speaking, these are not yet hydrogen plays as there are no commercial discoveries associated with them. The hydrogen fluxing along fault planes requires a fresh look at the exploitation of various fault architectures if shallow drilling would target conductive (or "leaky") faults at shallow depth. The promise of this set of plays is that if these seeps really correspond to ongoing charge in a dynamic, truly renewable system in a steady-state process, tapping successfully into them would provide sustainable resources via a low-flux hydrogen "farming" process.

It is quite likely that natural hydrogen exploration, if it becomes economically successful at one point, will look much more different than similar to hydrocarbon exploration.

Gabor C. Tari, OMV Energy, Trabrennstrasse 6–8, 1020 Vienna, Austria, e-mail: gabor.tari@omv.com

https://doi.org/10.1515/9783111437040-003

Keywords: hydrocarbon versus hydrogen systems, natural hydrogen exploration, fault control on fluxing hydrogen fluids, soil gas measurements, gas species

3.1 Introduction

There is a growing interest in natural hydrogen as a potential new source of energy with a negligible carbon footprint, especially compared to all the other human-made hydrogen species. Especially the white (or natural, geologic or geogenic) and orange hydrogen became the focus of intense international research, having the potential to become viable and probably significantly cheaper alternatives to the currently used black, gray, blue, pink, and green hydrogen, which all have a significant carbon footprint.

The primary goal of this chapter is to address the fundamental question about natural hydrogen exploration, that is, how different is it going to look compared to what we are used to in the petroleum industry? After many decades of very little consideration given to natural hydrogen as an exploration target with some notable exceptions (e.g., [1–3]) there are many papers and presentations published lately suggesting that data collecting methods, interpretation techniques, and workflows developed mostly by the hydrocarbon industry for more than a century could be readily applied to find natural hydrogen in the subsurface (e.g., [4–13]). The consensus appears to be that three out of four of the petroleum systems elements (sensu [14]) the hydrocarbon industry tends to focus on in exploration projects are still going to play pivotal roles (i.e., migration, trapping, and sealing) and it is only the generation/charge part, which follows very different rules for hydrogen systems. Other but similar schemes (e.g., [15, 16]) listed source, migration pathway, reservoir, trap/seal, preservation, and timing as critical and essential components of petroleum and hydrogen systems, respectively [13]. Some even suggested that the same hydrocarbon exploration style risking process can be adopted for hydrogen leads/prospects (e.g., [17]).

This chapter first briefly highlights the typical legacy data constraints energy companies face in the process of hydrogen exploration followed by a short summary devoted to some of the differences between hydrocarbon versus hydrogen systems. Building on this, the overlapping dimensions of petroleum and hydrogen exploration are discussed looking at aspects of generation, migration, trapping, and preservation in a few examples. The case of the only known natural hydrogen field discovered in Mali [18–20] is mentioned briefly to highlight its nature with some hints of being an outlier for a hydrogen find.

The subsequent examples describing hydrogen mega-seeps, conductive fault zones, soil gas measurements, and gas mixtures, however, capture the unique facets of hydrogen systems. These features are meant to illustrate the technical challenges of a simple copy-and-paste application of the well-established exploration tools and workflows of the petroleum industry to the new field of searching for natural hydrogen.

It is also very important to highlight the current nontechnical challenges for progressing with natural hydrogen exploration. While oil and gas as fossil fuels are one of a kind, with no competing alternatives as to energy content by volume or weight, hydrogen could be produced not only from natural sources but rather artificially. In particular, green hydrogen represents an alternative to and therefore competing product for natural hydrogen even if it comes with a larger carbon footprint and higher production costs. The risk and current lack of know-how for finding natural hydrogen make an exploration business case very difficult as it stands. Especially the worldwide lack of economically producing natural hydrogen fields, which could be used as analogues, is a major hindrance. In contrast, in the case of green hydrogen there is no risk as the production technology and the metrics of the value creation are already established; therefore investments tend to focus more on this hydrogen species at present.

Finally, the conclusions offer a personal and forward-looking perspective as to how much we can rely on our collective hydrocarbon exploration experience going forward, when it comes to finding and producing natural hydrogen in the subsurface.

3.2 Legacy data on natural hydrogen available in the E&P sector of energy companies

Building on previous partial databases [21], provided the first truly global data compilation on natural hydrogen occurrences. Using mostly global hydrocarbon industry data sets, our own data compilation at OMV confirmed the robustness and validity of his reporting. Systematic data mining of legacy data in various basins focusing on hydrogen tends to provide numerous finds, as was shown, for example, by Lefeuvre et al. [22] in the case of the Paris Basin. However, legacy well databases with hydrogen mentions come with significant challenges as was shown in Brazil [23].

As an independent comparison, we have analyzed OMV's well database specifically in the Vienna Basin with more than 100 legacy recordings of hydrogen in various sampling contexts (up to 20% in gas blends). However, based on a closer inspection of these data points, many of them turned out to be misleading, as cases for artificial hydrogen. These include gas measurements obtained shortly after side-wall core tests (designated as "detonation gas" in vintage well reports in Austria), drilling through shales with high total organic content (TOC) or routine gas composition analysis from producing wells after a few years of operation corresponding to corroding steel casings in the borehole, etc. Another major source of misleading hydrogen signals is drill-bit metamorphism (DBM), which is a phenomenon that has little impact on hydrocarbon exploration projects, but it is clearly very crucial for natural hydrogen projects (e.g., [24]). Artificial generation of hydrogen gas was observed during the drilling of hard basement rocks as the process overheats the oil-based mud (OBM) used.

In the field and experimental drilling data, artificially generated hydrogen gas has been observed in several volume % of the drilling OBM gas [24]. In our experience, DBM-related hydrogen can even reach tens of volume % based on monitoring drilling fluids using headspace gas analysis. This may lead to a false positive for hydrogen signal or can mask a subtle existing natural hydrogen flux in the borehole.

The hydrocarbon industry, in general, does not have much in its databases as to the "source rocks" for natural hydrogen, such as ophiolites and basement rocks rich in radioactive materials, as these rock sequences were/are rarely targeted during drilling for oil and gas. Processes of serpentinization, hydration, and radiolysis, for example, have been almost exclusively studied by the academia (e.g., [25, 26]) and some aspects are considered by the mining industry (e.g., [27]). Consequently, there is a large knowledge gap in the petroleum industry when it comes to the estimation of the hydrogen-generating potential in areas where there may not be any relevant subsurface data to initiate even a screening evaluation.

3.3 Hydrocarbon versus hydrogen systems

Prinzhofer and Cacas-Stentz [28] compiled the overlapping and contrasting elements of hydrocarbon versus hydrogen systems (see also the overview by Jackson et al. [12]). The obviously similar elements include generation in a "kitchen" followed by migration of the gas (and/or oil) through sufficiently permeable rocks to reach a subsurface trap where porous reservoir rocks can retain an accumulation sealed by non-permeable units on top. This scheme has all the petroleum system elements in it [14], that is, generation, migration, trapping, and sealing.

The range of source rocks for natural hydrogen generation is surprisingly wide ranging from radioactive basement rocks to ophiolites going through serpentinization. The particulars of these source rocks are covered by other chapters in this book and will not be described here. The only overlapping source rock between hydrocarbons and hydrogen is organic matter in sediments, which can generate both but at different temperatures. Hydrogen can be generated from overmature organic matter such as shales and coals and its yield per unit rock volume is comparable to that of economic shale gas [29]. Importantly, hydrogen is generated after hydrocarbon generation is over, at greater depth and temperature, extending to about 450 °C in deep sedimentary basins.

Migration of hydrogen gas from a source rock volume cannot be modeled the same way as that of hydrocarbons due to its different physical and chemical parameters complicated by its solubility changes with depth and tendency to migrate in a gas mixture rather than a pure gas phase (e.g., [30]). In general, given its characteristics, hydrogen cannot be modeled simply as "just another gas" as shown by Arrouvel and

Prinzhofer [31]. Basin modeling software packages developed with hydrocarbons in mind need to be substantially modified (e.g., [32]).

Specifically, upward migration along fault zones appears to be more pronounced and selective given the small kinematic diameter of molecular hydrogen [6]. Trapping assumes the presence of effective lateral and top seals in a certain subsurface geometry, which could be achieved by a broader range of lithologies and faults for hydrocarbons than for hydrogen. The most effective sealing lithologies known for hydrocarbons such as evaporites do seem to seal and therefore trap hydrogen as well, as will be discussed later.

Hydrocarbon traps may maintain their integrity for up to hundreds of millions of years while natural hydrogen accumulations are expected to have orders of magnitudes shorter life span due to the extreme agility of this gas species. However, there is no data set confirming this statement, again, due to the lack of known hydrogen fields globally.

In comparison, perhaps the most crucial difference between hydrocarbon and hydrogen systems revolves around the static versus dynamic nature of these systems, including the residence time, the fluxing rate, and the ubiquitous temporal change in a subsurface gas blend with hydrogen in it. The one aspect that will be highlighted in the following is the ubiquitous nature of strong seepage associated with hydrogen systems, at least 4–5 orders of magnitude larger than in the case of hydrocarbon systems [28].

Given the substantial differences between these systems could we expect to apply the same thinking, tools, and workflows, well established in hydrocarbon exploration, to hydrogen exploration?

For this brief overview, natural hydrogen plays were subdivided into two major categories in the context of their similarity to hydrocarbon exploration (Figure 3.1). Plays which are identical to or largely overlapping with classic petroleum plays are either (a) associated with evaporites or (b) have some unusual seals effective on a geologically very short timeframe. Natural hydrogen manifestations, which appear very unusual and distinct from a hydrocarbon exploration perspective and are not proven resource plays yet, are categorized into (c) large mega-seepage (fairy circles or subcircular depressions) and d) fault-controlled gas fluxing to the surface (Figure 3.1).

3.4 Overlapping aspects of hydrocarbons versus natural hydrogen exploration

These aspects are subdivided into two broad categories here, both with functioning traps due to an effective seal. In the first category, the seal is an evaporite one, which is frequently the "super-seal" in hydrocarbon exploration. In the second category, the

SIMILAR TO HYDROCARBON EXPLORATION

*FUNCTIONING TRAPS
DUE TO EFFECTIVE TOP SEALS*

SPECIFIC TO HYDROGEN EXPLORATION

*NO TRAPS, NO SEALS
ONLY (MEGA-)SEEPS*

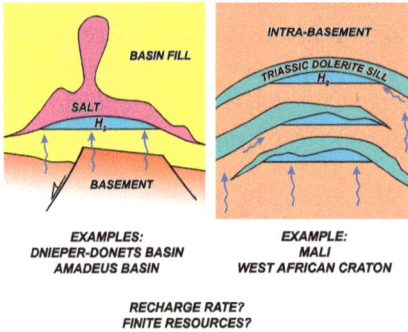

EXAMPLES:
DNIEPER-DONETS BASIN
AMADEUS BASIN

EXAMPLE:
MALI
WEST AFRICAN CRATON

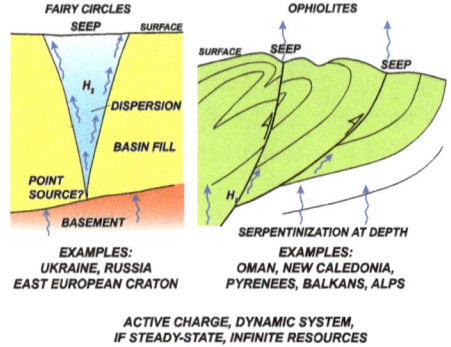

EXAMPLES:
UKRAINE, RUSSIA
EAST EUROPEAN CRATON

EXAMPLES:
OMAN, NEW CALEDONIA,
PYRENEES, BALKANS, ALPS

*RECHARGE RATE?
FINITE RESOURCES?*

*ACTIVE CHARGE, DYNAMIC SYSTEM,
IF STEADY-STATE, INFINITE RESOURCES*

Figure 3.1: Cartoonish overview of documented hydrogen occurrences, which appear to be similar (left) versus different (right) from play types known from petroleum exploration [33].

single, but important case is that of the Bourakébougou hydrogen field in Mali where the somewhat unusual seal is provided by dolerite sills.

One of the first French natural gas fields, Vaux-en-Bugey, discovered in 1906, had an effective Triassic evaporite seal trapping 5% hydrogen in its initial gas mix [34]. Similarly, in dozens of hydrocarbon fields in the Dnieper-Donets rift basin of Ukraine considerable hydrogen contents (a few percent on average, but up to 20%) were recorded as early as the 1960s [35]. Many of the fields in this petroleum superbasin are sealed by Devonian or Permian salt [36].

Another example of the salt super-seal can be found in the Amadeus Basin, in the interior of Australia [37]. In a deep pre-salt well drilled here He- and H_2-rich gases were reported (He ~9%, H_2 ~11.4%) beneath a Proterozoic evaporite system with traces of hydrocarbons. The large relative proportion of He (even more agile gas than hydrogen) in the gas blend shows that the evaporite seal was/is effective for this accumulation on a scale of millions of years [37].

Another illustration of evaporites acting in the same sealing way for hydrogen (and helium) as they do for hydrocarbons is the Lokachy gas field in western Ukraine (Figure 3.2). There are relatively thin (10–20 m thick) anhydrite layers in this gas field forming multiple seal levels [38] but they are efficient to define separate gas reservoir units with independent gas-water contacts. The source of the helium and hydrogen is the Proterozoic basement underneath, due to radiolysis. The overall upward decreasing amount of helium and hydrogen shows the relative efficiency of the multiple seal horizons. The present-day situation in this gas field (Figure 3.2) is probably the result of gradual changes in the $CH_4/H_2/He$ ratios in the various gas pools through geologic time when the relative proportion of hydrogen tends to decrease in the trap over time [28]. The relatively small proportions of hydrogen in the gas blend may point to a

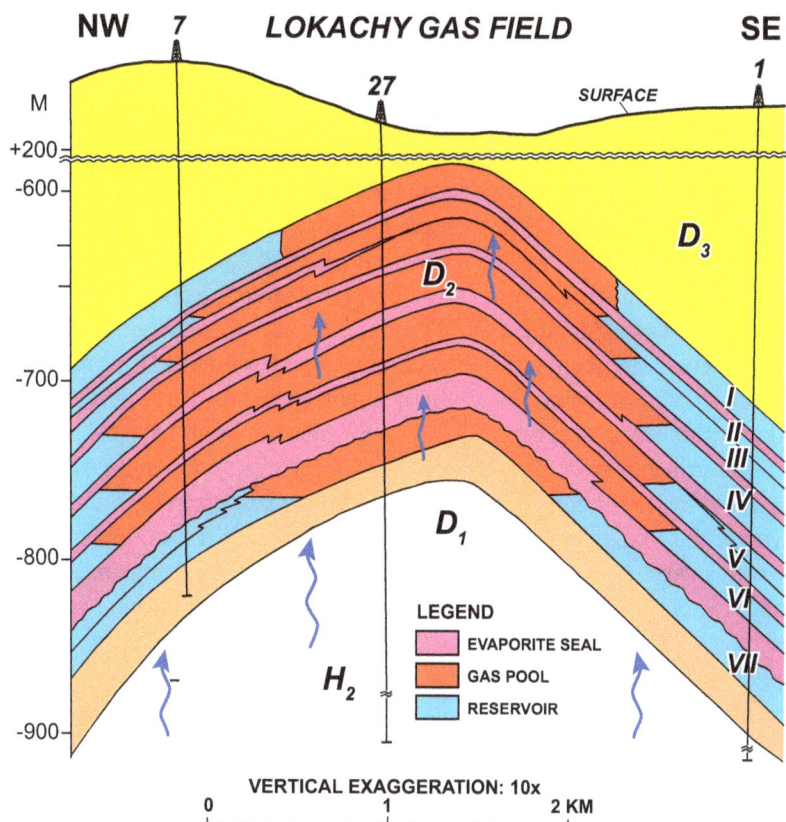

Figure 3.2: The mature Lokachy gas field, western Ukraine, illustrating the role of relatively thin Devonian anhydrite layers serving as seals for both hydrogen and helium in intercalated carbonate-clastic reservoirs (redrawn from [38]).

Reservoir	CH_4	C_2H_6	C_3H_8	$i-C_4H_{10}$	$n-C_4H_{10}$	C_5H_{12}	CO_2	He	Ar	N_2	H_2
I	92,96	2,000	0,540	0,115	0,087	-	0,145	0,281	-	3,97	-
II	93,35	0,513	0,089	0,027	-	-	0,579	0,226	-	5,22	-
III	95,35	0,237	0,019	-	0,022	-	0,222	0,303	-	3,65	-
IV	92,35	1,178	0,423	0,094	0,122	0,069	0,608	0,210	0,392	4,49	0,071
V	93,26	0,823	0,207	0,086	0,066	0,018	0,387	-	-	5,15	-
VI	93,71	1,317	0,364	0,051	0,049	0,071	0,459	0,170	0,097	3,45	0,265
VII	95,26	0,837	0,098	0,039	0,016	0,016	0,365	0,132	0,118	2,98	0,143

long-term depletion of this gas component reflecting the limitations of the anhydrite layers as seals.

Based on these case studies, it is not surprising that several, so far undrilled natural hydrogen targets rely on evaporite super-seals in the same way as in petroleum exploration projects. One of these undrilled hydrogen/helium targets is the Monzon prospect in the Spanish Pyrenees foreland, trapped under Triassic (Keuper) salt [39].

Another prominent undrilled hydrogen project is located on the French side of the Pyrenees [6] where supra-salt targets are considered reliable analogues to the nearby giant Lacq gas field, which has a non-evaporitic shale seal. However, as [40] argued, the ultimate test of this hydrogen system would be the drilling beneath the Triassic salt to reach a mantle body (i.e., the source due to serpentinization) located in a relatively shallow depth. This play type will require a challenging and expensive drilling program down to about 7 km, which would be considered a high-risk, high-reward exploration effort in the petroleum industry.

So, what could be the undrilled upside for hydrogen exploration in mature hydrocarbon basins where a lot of drilling has been achieved already? Due to the unavoidable interaction between oil and hydrogen migrating into the same trap, the question only applies to basins dominated by natural gas. As an example, returning to the Dnieper-Donets Basin, the fundamental difference between natural gas and hydrogen systems should be considered focusing on the source rock element. Pre-salt traps in the "marginal steps" along the margins of this Paleozoic rift basin have been considered risky for natural gas, given the lack of underlying source rock sequences (Figure 3.3). However, the hydrogen documented in this basin [35] is sourced from the Proterozoic basement of the Ukrainian Shield and could charge pre-salt traps in the marginal segments of this very large basin, which saw very limited drilling compared to the more central parts. The generic play cartoon (Figure 3.3) may apply to other salt basins in a rift setting, and not only onshore, but to offshore ones as well, like the Red Sea, for example.

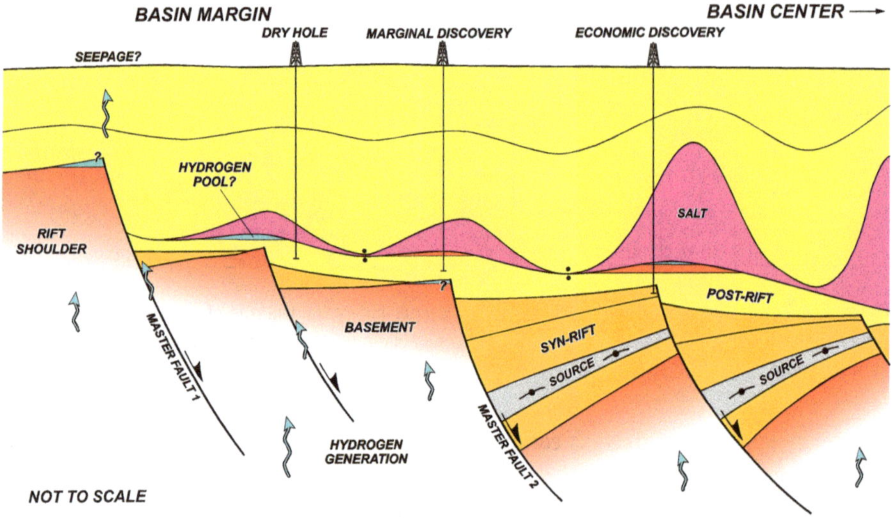

Figure 3.3: Play type cartoon of pre-salt traps in marginal "terraces" along the flanks of the Dnieper-Donets Basin of Ukraine. While these traps for hydrocarbons are not considered viable due to the lack of underlying source rocks, it may not be the case for hydrogen accumulations.

As for expanding the hydrogen exploration efforts offshore, there is a consensus that the geoenergy industry needs to be successful onshore first in finding economic discoveries. As a sidenote, the world's first "overwater" oil well was completed on Caddo Lake, Louisiana, in 1911, many decades after the historic oil find by Colonel Drake in Pennsylvania in 1859.

So, is there drillable prospectivity for natural hydrogen offshore, which could be considered as a follow-up on successful prospecting onshore?

A potential example of an offshore hydrogen play fairway may exist south of Cyprus [41]. Ongoing serpentinization and hydrogen fluxing in the outcropping Troodos ophiolites is proven by head-space gas samples collected from several hyper-alkaline springs (Figure 3.4, our own unpublished data). Therefore, an active hydrogen "kitchen" is inferred offshore where the gas derived from the serpentinization of the Mesozoic ophiolites at the depth of about 3–5 km may be trapped beneath the super-seal Messinian evaporites. As the evaporites may directly overlie the ophiolite sequence, there is a reservoir risk associated with the pre-salt sequence. Fractured serpentinites, however, can have reasonable porosity and permeability characteristics and there are examples where there is oil production from them (e.g., some Cuban oil fields [42]).

For hydrocarbon exploration, this area is considered not prospective given the lack of a sedimentary basin with source and reservoir rocks beneath the proven Messinian seal. Indeed, farther to the south, in the deeper water where there is a well-developed sedimentary basin beneath the salt, natural gas finds were predicted (e.g., [43]) and found during the last decade. So, in this case, there is a clear spatial separation of prospective areas between hydrocarbons and hydrogen. This spatial separation is somewhat similar to the situation shown in Figure 3.3. At any rate, natural hydrogen exploration at present needs to see a breakthrough onshore before it could possibly be considered offshore as an economically viable effort. However, potentially, an extension of successful play fairways from onshore to offshore may eventually work the same way as in hydrocarbon exploration.

3.5 The first natural hydrogen field, Bourakébougou (Mali)

The accidental natural hydrogen discovery at Bourakébougou [18–20] provided a case study, which also highlights the similarity between hydrocarbon and hydrogen exploration. Whereas the initial discovery well was drilled for water in 1987 without much pre-drill subsurface interpretation work, the subsequent drilling campaign of about 30 wells in 2017–2019 made it clear that there are at least five gas reservoirs situated at very shallow (between 30 and 135 m) to relatively greater depth (1,125–1,500 m). The unusual trapping element is that the seal for the gas pay zones is a set of uppermost Triassic dolerite dykes emplaced in a Proterozoic sequence of carbonate and

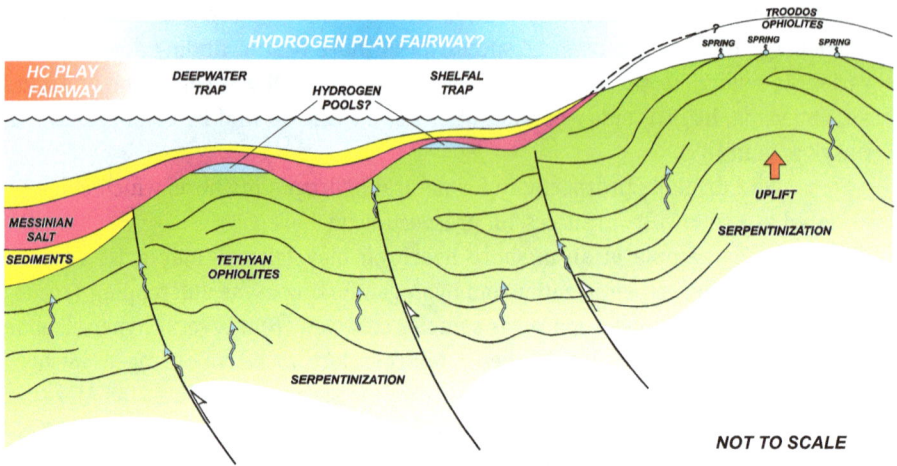

Figure 3.4: Play concept for a potential offshore hydrogen exploration project in Cyprus. The overall structural geometry along the southern margin of Cyprus in the region is simplified after [44]. The hydrocarbon play fairway in the deep water does not overlap with the area of a potential natural hydrogen play in the shallow water and onshore.

clastic reservoirs. The sealing properties of the various dolerites are determined by the extent of fractures in them [20].

The gas blend has up to 98% hydrogen, suggesting that the Bourakébougou field may be just a "one-off" given its exceptional purity. Currently, there is just one well producing in this possibly 50–100 bcf H_2 accumulation with a rate of about 45,000 scf/ d. The pressure is quite low, that is, around 4 bar pressure and after about 7 years of stable production there are no signs of depletion.

In cases like Bourakébougou, one expects finite hydrogen resources to be in place in multiple traps and the exploration approach has indeed lots of resemblance to that of hydrocarbon prospecting. The presence of gas chimneys and subtle surface depressions (fairy circles, [18]) in the broader basin indicate that the generation, and/or migration of the hydrogen gas are still active. Modeling of the gas blend suggests that the time it required to fill the traps at Bourakébougou, was extremely short, on the scale of just a few hundred years [28] emphasizing the highly dynamic nature of the hydrogen system. Therefore, the seemingly transient, short-lived (possibly as young as only 500 years), accumulation of Bourakébougou, may be just a lucky "snapshot" of a hydrogen trap, which may not be representative of other natural hydrogen accumulations globally at all.

3.6 Geological and geophysical aspects unique to hydrogen exploration

These aspects are subdivided into two categories here (Figure 3.1), both distinctly different from what we are used to in the hydrocarbon exploration industry and related to the very leaky nature of hydrogen systems. First, large mega-seeps, fairy circles (or subcircular depressions), are discussed followed by the description of the fluxing of hydrogen toward the surface specifically along faults.

Fairy circles as mega-seeps are related to hydrogen fluxing to the surface in a particular manner. First documented by Larin et al. [45], these spectacular features quickly became the "poster child" for natural hydrogen systems due to their ubiquitous presence in many continents across the globe, for example, in Carolinas (USA) [46], Brazil [18, 47], Namibia [48, 49], Australia [50, 51], and Colombia [52].

Many dismiss these features as exploration targets even though a fairy circle was already targeted in Nebraska [53]; however, the results of this drilling effort remain unpublished.

The subsurface geometry and formation of the fairy circles are not understood at this point in time. It is proposed here that an analogy with leaking gas pipelines (Figure 3.5) may be useful to have a 3D spatial model to be tested. As a case study, vegetation damage associated with a leaking pipeline was documented by Smith [54]. Patches of decreased growth of barley were observed above a Victorian (!) 12-inch cast iron main gas pipe located at a depth of approximately 0.7 m below the surface as part of the North Lincolnshire Grid [54]. The pipeline used to supply coal (town) gas, but now it transports natural gas.

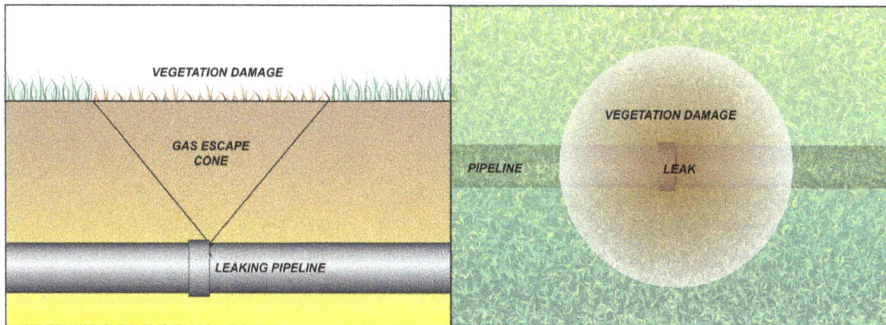

Figure 3.5: A useful analogue for the genesis of fairy circles is provided by leaking gas pipelines. The subcircular vegetation damage on the surface is caused by the oxygen-displacement from the soil surrounding the root system of the plants.

The patches were approximately 2 m in diameter and were spaced at about 10-m intervals that coincided with the joints in the pipeline, which were inferred to leak the gas (Figure 3.5). The main damaging effect on plant growth of natural gas leaking from pipelines into the soil is the displacement of oxygen essential for the plants [54]. This process indicates an inverse cone-shaped upward dispersion of the gas from a point source below (Figure 3.5a), which manifests itself on the surface as a subcircular zone of diminished plant growth (Figure 3.5b).

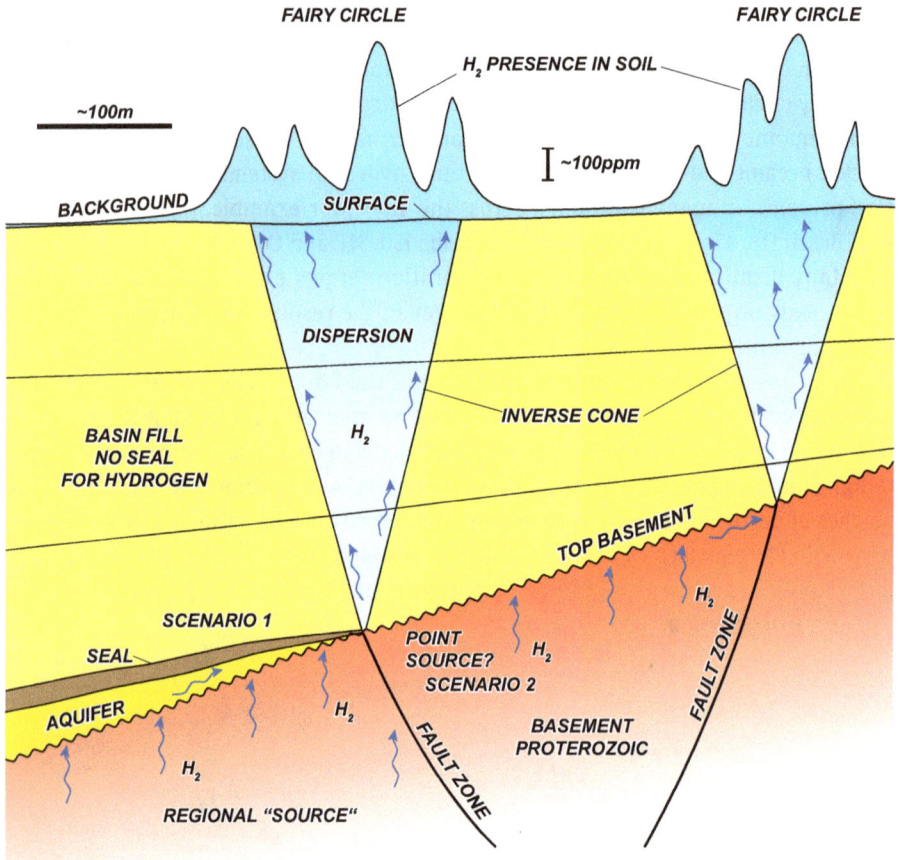

Figure 3.6: Conceptual model for the subsurface context of fairy circles. This implies individual and likely transient point sources either (a) along a lateral pinch-out of a seal sequence or (b) along a fault zone at the base of the sedimentary fill of a basin (assuming that the hydrogen is generated in the underlying basement).

Taking this analogue and upscaling it to the typically observed dimensions of fairy circles on the scale of hundreds of meters to a few kilometers [45], a preliminary subsurface model is proposed here (Figure 3.6). Whether the point sources for the fairy circles are related to the pinch-out of an aquifer on top of the hydrogen-generating basement (scenario #1) or it is more fault-controlled (scenario #2) remains to be seen.

On a smaller scale, that is, a few to tens of meters, mofettas (volcanic discharge consisting chiefly of carbon dioxide) also appear to be underlain by an inverse cone-shaped gas migration based on ambient seismic noise techniques, which can image the vertically migrating fluid channels directly (e.g., [55, 56]).

Another characteristic of the fairy circles is a subtle surface depression (e.g., [57]), which was shown to be different from other similar surface features, such as dolinas [7]. Indeed, preliminary modeling using a two-phase flow can also produce a slight surface depression corresponding to gas migrating up from a point source below Figure 3.7.

Remote sensing techniques are the easiest and fastest to identify fairy circles in a region looking for their map-view expressions, but they could be misleading if they are the only source of data for the interpretation. Field checks are essential to establish the anomalous presence of hydrogen in the soil, as some other geomorphic features such as dolinas could also have similar map-view geometries [48]. In our experience, evaporitic karsts, loess-related "steppe saucers," abandoned peat mines, even historic, but abandoned and recultivated rock quarries, etc. could be misinterpreted as fairy circles using satellite data only.

Another important aspect of the seepage identified in the fairy circles is that they are likely to be very short-lived, transient features, for example., the Elektrostal in Russia [45], perhaps with an active life span of fluxing to the surface on the order of just a few hundred years. Afterwards, they become inactive (diminished or no fluxing) leaving a slowly decaying morphological footprint on the surface. Therefore, the temporal element of this extremely fast (geologically speaking) evolution is best studied by time-lapse data. Remote sensing techniques like InSAR (interferometric synthetic aperture radar) can detect the ground motion for individual fairy circles with their early relative rise (i.e., "soufflé" effect) followed by subsidence during the fluxing phase [58]. These ongoing studies are vital to the proper understanding of fairy circles as a seemingly unique feature of natural hydrogen systems. Also, from the point of view of monitoring engineered geological hydrogen storage, natural hydrogen seeps, including fairy circles provide useful analogues [59].

The gas seepage associated with fairy circles does not find a close analogue in the hydrocarbon exploration experience. Whereas seafloor seepage associated with oil or gas leakage is relatively common above hydrocarbon fields (i.e., pockmarks), their spatial and temporal, emanation rate characteristics (e.g., [60]) are different from those of the fairy circles described so far only onshore. Furthermore, in oil and gas exploration workflows the presence or absence of pockmarks is not considered a critical aspect; however, in hydrogen exploration, it seems to be much more important as evidence for an actively generating "kitchen" in the subsurface.

displacement

gas flux

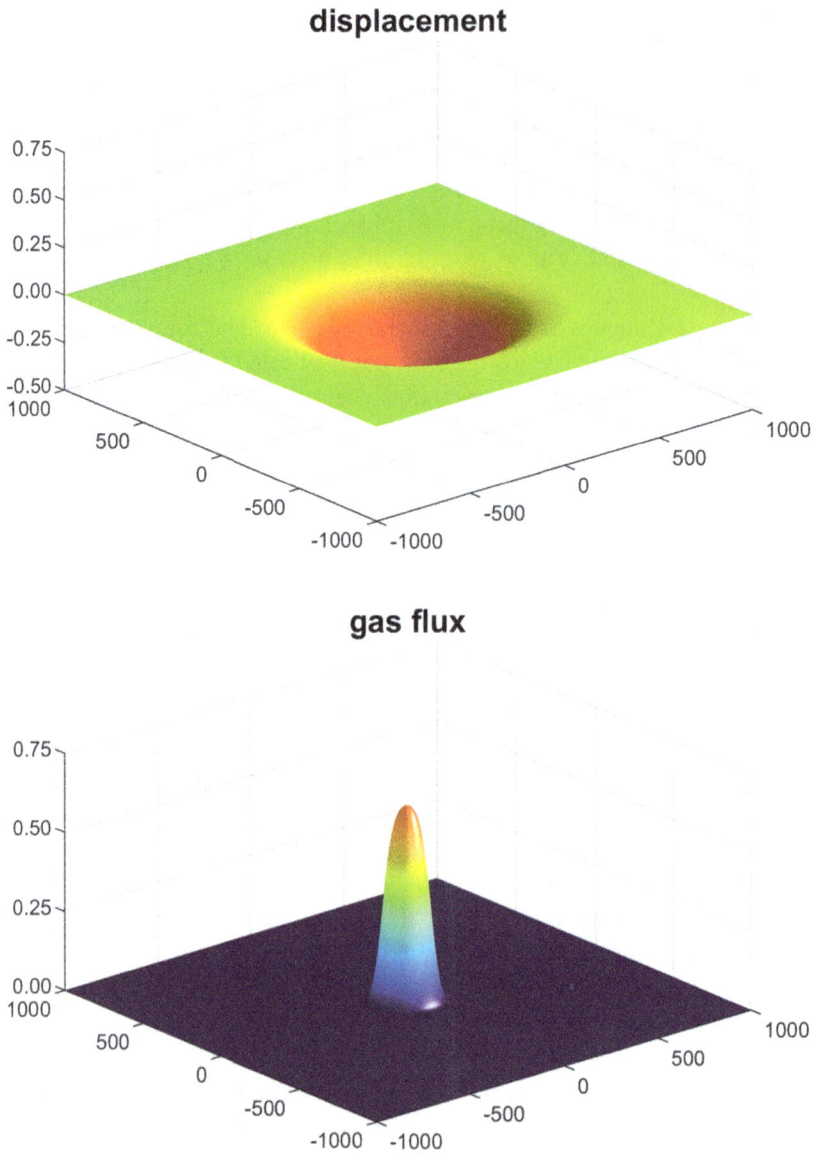

Figure 3.7: Numerical modeling of a fairy circle using a point source of gas at depth. Given certain assumptions about the gas rate and the lithological composition of the sedimentary succession between the point source and the surface, a slight topographic depression can be modeled on par with the observed geometry of actual field examples (Martin Schöpfer, personal communication, 2024). Horizontal scale in meters, vertical scale in relative units.

3.7 Fault control on fluxing hydrogen fluids

The large faults segmenting an ophiolite succession (Figure 3.1) are meant to empha-size the crucial role of faults becoming the main migration routes (or "highways," Alain Prinzhofer personal communication, 2023) for the exceptionally agile hydrogen (and helium) molecules to reach the surface. The oil and gas industry is very familiar with faults in the context of fault-seal analysis (e.g., [61]). But hydrogen fluxing along fault planes requires a fresh look at the exploitation of various fault architectures (e.g., fault core versus damage zone) not from the sealing but from rather the leaking point of view. In hydrogen exploration drilling would selectively target conductive (or leaky) faults for potential hydrogen "farming" at shallow depth.

The preferential migration of gases along a fault zone and the fact that gases present in fracture zones of active faults could be characterized by a high concen-tration of H_2 and/or CO_2 have been known for decades (e.g., [62, 63]). Hydrogen gas concentration along active faults associated with historical earthquakes usually amounts to as high as several percent at maximum, whereas the concentration of hydrogen from inactive faults may be orders of magnitudes smaller, in the 100-ppm range.

Upward migration of gases in a fault damage zone requires some porosity and permeability and active or neotectonic faults tend to have these as diagenetic pro-cesses did not have time yet to plug the pore space. Existing brittle fault evolution models assume a cyclic permeability evolution of fault zones through geological time (e.g., [64]). Typically, the core zone acts as a fluid conduit; however, cementation by precipitation of minerals and grain growth decreases permeability within the core zone and forces the gas flow into the adjacent damage zones flanking the plugged-up fault. Only renewed fault could initiate a new fluid flow evolution cycle by opening up the core fault zone by brittle deformation [65].

The possibility of directly tapping into fault zones appears to be a unique play type in hydrogen exploration as opposed to hydrocarbon exploration (Figure 3.1). There are more and more studies emphasizing the exploration importance of a gas mixture containing hydrogen migrating along brittle fault zones (e.g., [6, 33, 66, 67]). Therefore, proper imaging and targeting of fault planes at relatively shallow depths, (i.e., 200–1,000 m, e.g., [68]) is much more important than in hydrocarbon projects (Figure 3.8a). A better understanding of the fault architecture is needed not from the sealing perspective but rather from the point of view of conductive processes associ-ated with them.

The anticipated flow rates from a maximum few tens of meters thick complex fault zone at low pressure (given the shallow depth) are not going to provide the gas production rates in a single vertical well we are accustomed to in the petroleum busi-ness. However, the horizontal stacking of numerous shallow (i.e., hundreds of meters deep) and low-cost wells, on the scale of tens of them, systematically following a fault

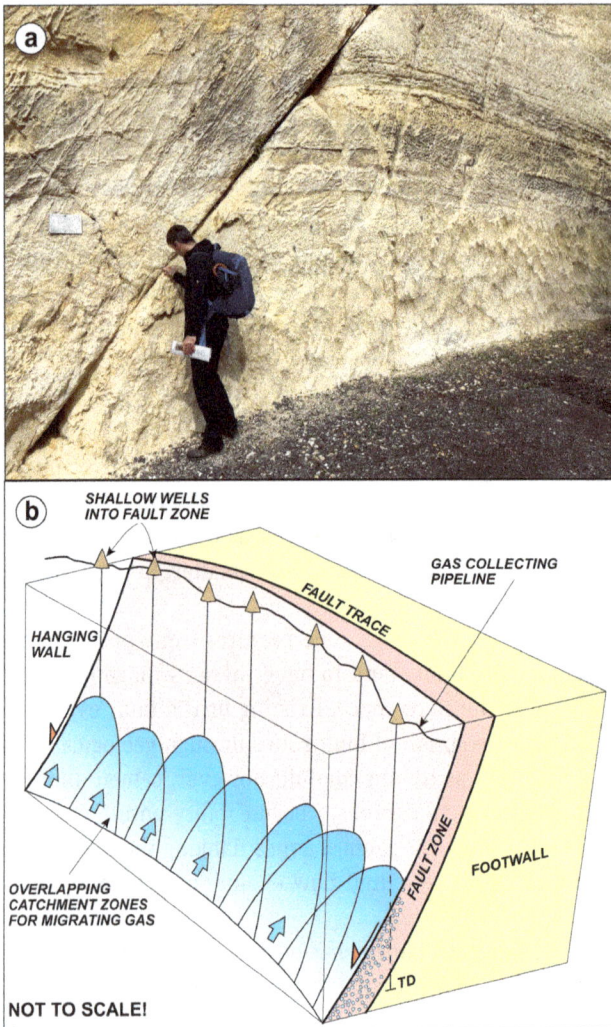

Figure 3.8: (a) An outcrop example of a brittle normal fault zone across Miocene calcarenites, Rust Ridge, NE Austria. In this case, the main displacement zone is very well defined, but many auxiliary structures developed within a few meters from the main fault, which has an offset on the order of tens of meters. Fault zone architecture (fault core versus damage zone) plays a key role in a migrating gas blend of various components (e.g., hydrogen, helium, nitrogen, methane, and carbon dioxide) with very different kinetic diameters. (b) Conceptual cartoon of natural hydrogen farming by targeting a major fault zone with numerous shallow (few hundred meters deep) wells along strike. See the text for an explanation.

plane (Figure 3.8b) could provide a combined commercially viable flow rate. The low-cost and simple surface facilities for the envisioned hydrogen farming scheme (Figure 3.8b) will require technological solutions, which do not exist in the hydrocarbon industry but may be more reminiscent of shallow geothermal production schemes for district heating. The most attractive part of hydrogen farming would be its sustainable character something impossible to achieve in hydrocarbon production given the recharge rates.

An important risk for a shallow accumulation or fluxing fault zone is the consumption of hydrogen by microorganisms in the near-surface zone. This challenge is somewhat similar to the biodegradation issue in hydrocarbon fields. However, given the dynamic and fast nature of hydrogen fluxing along a fault, depletion due the bioconsumption (e.g., [69]) and/or oxidation (e.g., [4]) may not be so overwhelming as in the case of hydrogen advection and diffusion toward the surface without faults.

The importance of fluxing along various fault planes may translate to different play type summaries than customary in the hydrocarbon industry. A hydrogen exploration play type diagram for a thrust fold belt and its foreland may look quite different, with much more focus on faults and the surface expression of migrating gases (Figure 3.9). 2D basin modeling by Palmowski et al. [30] indicates a very different world for hydrogen generation and migration in a setting like this where the interaction between the different kinds of fluids, stratigraphic units, and faults is not nearly as intuitive as in hydrocarbon exploration.

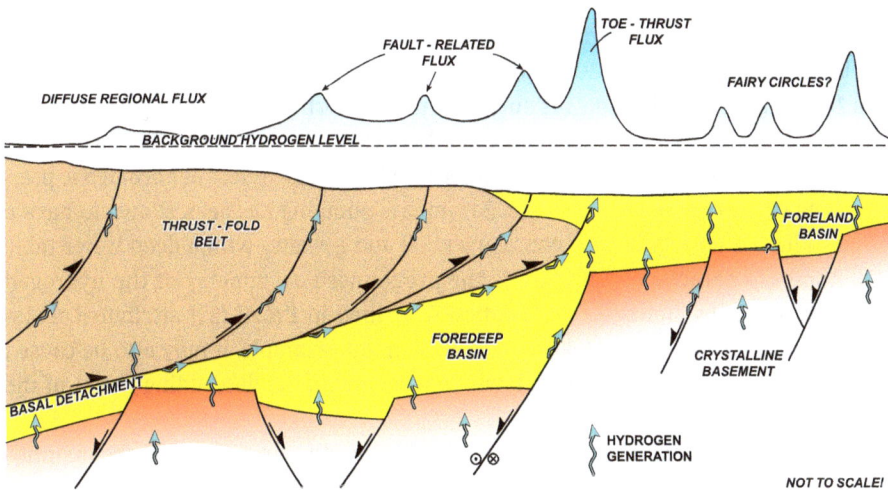

Figure 3.9: A fictitious hydrogen exploration play type cartoon for a thrust-fold belt. The emphasis in this case is on the migration of the hydrogen along fault systems and its surface manifestation rather than its accumulation in traps like in petroleum exploration.

3.8 Soil gas measurements as a key part of hydrogen exploration

The at least 4–5 orders of magnitude larger seepage rates of hydrogen systems, compared to hydrocarbon systems [28], underline the focus on surface exploration tools, which are designed to detect and quantify the hydrogen signal (e.g., [5, 7, 47, 57, 70, 71]).

The presence or absence of surface seepages is not a critical part of hydrocarbon exploration workflows; however, it is of paramount importance for hydrogen exploration [5, 6]. In fact, this was highlighted by the early work by Larin et al. [45] where soil gas sampling at the depth of 1 or 2 m became the critical evidence for hydrogen fluxing to the surface in the studied fairy circles.

Monitoring of the temporal evolution of hydrogen presence in the soil is key [47]. However, the results show a very complex "breathing" system affected by many factors including air versus soil temperature, barometric pressure, capillary pore pressure in the soil, soil wetness and composition, and wind, all of them occurring in the context of daily and/or seasonal variations. Hydrogen breathing patterns are now being found not only associated with fairy circles (e.g., [57]), but in other settings as well.

As an example, a soil hydrogen gas monitoring effort in the Eastern Alps of Austria is shown here (Figure 3.10). The soil hydrogen signal at 1-m depth was recorded by a 1-min high-resolution soil gas autonomous monitoring device without any pumping for a couple of weeks. In the three-day long example reproduced here the influence of the air temperature in a daily pattern is obvious, the effect of barometric pressure is much more subtle in this case. Barometric pumping has been shown to have a major influence on fugitive gas emissions at oil and gas sites with a deep water table [72]. Another characteristic feature is the pronounced asymmetry of the hydrogen peaks, which looks identical to those first described in Brazil [57] attributed tentatively to biogenic effects by Myagkiy et al. [73]. In our example, it may also be caused by the barometric pressure change lows coinciding with the descending flanks of the hydrogen peaks.

Given the complexities mentioned above, it is hardly surprising that the quantification of hydrogen fluxing in an area, where soil gas measurements indicate the presence of a "kitchen" at depth, remains a challenge. Intuitively, the hydrogen signal observed at or near the surface (i.e., 1–2 m depth) is seen as providing a minimum value due to the biogenic consumption of the ascending hydrogen gas by bacteria (e.g., [73]).

There is an ongoing debate as to what constitutes a reliable natural (i.e., geologic) hydrogen signal and how this could be distinguished from the one simply caused by fermentation (biologic hydrogen) in the soil (e.g., [74]). Similarly, the relative role of diffusion versus advection for a gas blend containing hydrogen in near-surface conditions remains a topic for further studies.

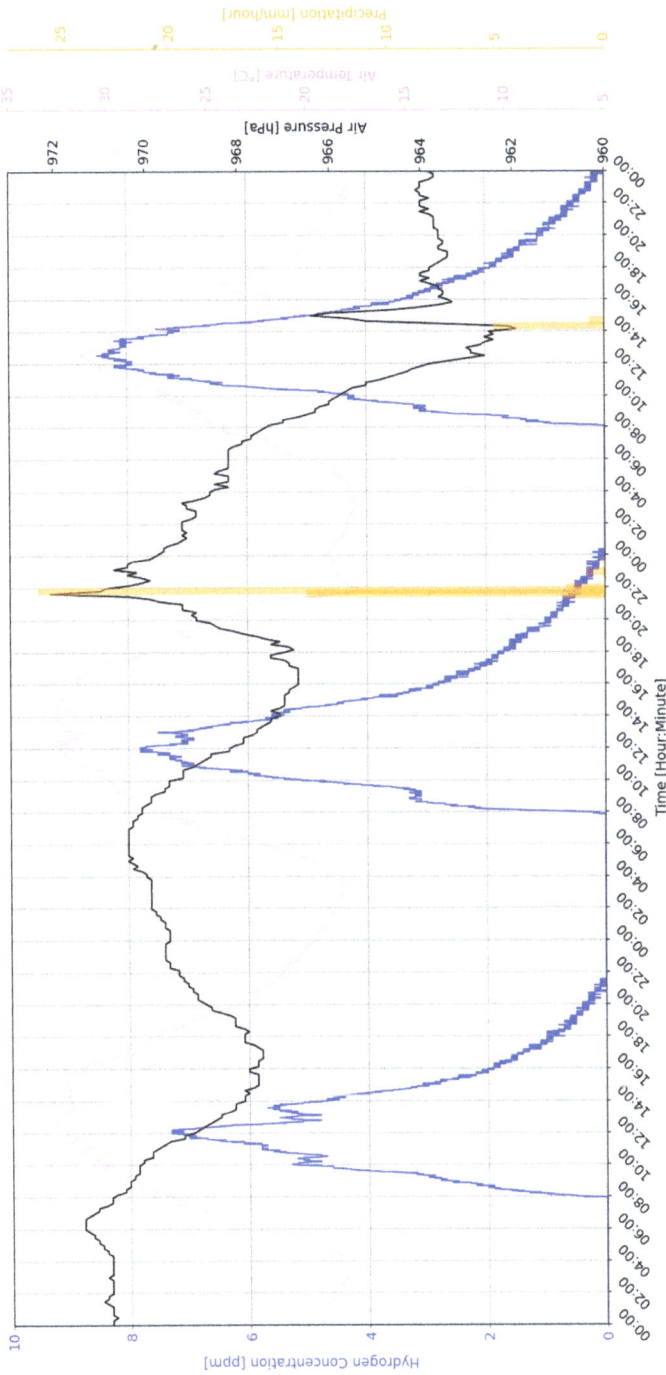

Figure 3.10: Raw data was acquired by a high-resolution soil-gas monitoring device (with no pump) deployed in Austria measuring the presence of hydrogen in ppm at 1-m depth in every minute during a 3-day period. Note the very close correlation of the minor, ppm-scale hydrogen signal with the air temperature and the slightly shifted negative correlation with air pressure. Whether the observed "breathing" hydrogen signal corresponds to a deep geologic signal or rather it is the expression of biologic hydrogen generated in the soil (or both) remains to be seen. As to the origin of the hydrogen, an interpretation cannot be decisive using just this data set in isolation.

A lot of progress is expected in the next few years in the field of surface detection of natural hydrogen, specifically in the soil. This aspect, again, is shaping up very differently from that of petroleum exploration where, even after a century and a half of collective industry experience, surface geochemistry (e.g., [75]) is seen as a tool of marginal importance.

3.9 Emphasis on numerous gas species, not only on hydrogen

Another major difference between hydrocarbon and hydrogen exploration is the deliberate focus on a wide range of gas species not just on natural gas or hydrogen, respectively.

The pre-drill prediction of non-HC gases in a natural gas prospect is important in the context of finding undesirable inert gases or components such as CO_2 or, in the worst case, H_2S. This challenge typically occurs at a relatively late stage of the exploration cycle.

The context appears to be different for natural hydrogen exploration. The presence and proportions of other gases in a gas blend in legacy well tests, seeps, or in soil gas are important clues during the early stage of the exploration cycle. The presence or absence of helium could clarify the nature of the hydrogen kitchen, that is, rocks generating hydrogen at depth via radiolysis or serpentinization, respectively. Abiotic methane is a useful proxy for hydrogen generation (e.g., [76]) and the importance of nitrogen in the context of natural hydrogen was just recently underlined [77]. In lieu of existing hydrogen and helium data sets in a given exploration area other gas data should be used as a proxy for the presence of gases migrating from the deep subsurface.

For example, work done specifically on CO_2 as a fugitive gas [78] utilized various soil gas cross-plots, such as CO_2 versus O_2 to determine the presence of exogeneous gas in a given area. The relationship between O_2 and CO_2 can be used to distinguish the origin of CO_2 from in situ background processes in the vadose zone or from exogeneous deep gas sources. This technique was used as a proxy for deep gas migrating to the surface in the French Pyrenees [5]. In our experience working in the Moravian region of the Czech Republic, such information on gases other than hydrogen itself could be indeed effectively used to screen for potential seepage sites from a deep source in an area (Figure 3.11).

Radon is also a gas species, which is frequently measured at the near surface, mostly in the context of health hazards associated with buildings and water supplies. Therefore, there are extensive public data sets available, which could be used as a proxy for radioactive basement rocks, which could also generate hydrogen (and helium) via radiolysis. Radon in soil gas has been also studied with earthquake prediction in mind and mapping of the radon signal along fault traces in various tectonic

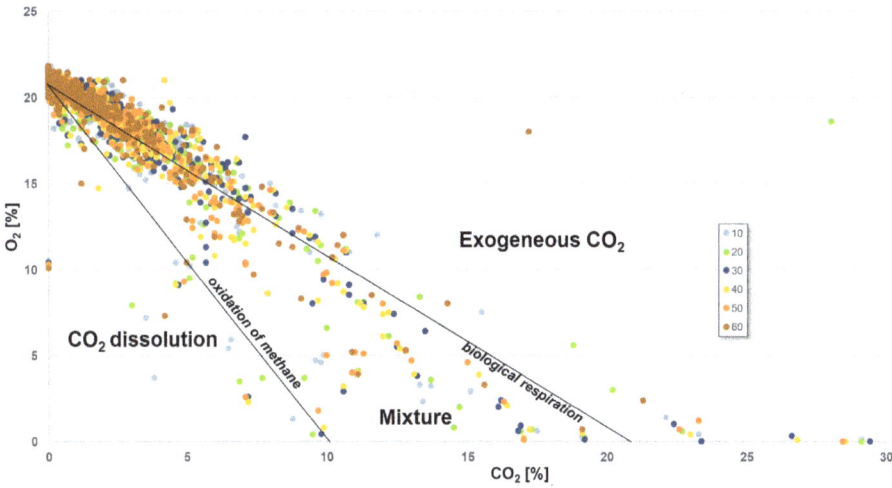

Figure 3.11: Soil gas measurement results in Moravia (Czech Republic). This data set with more than 500 data points was used to identify potential gas seeps on the surface associated with a deep source. This indirect technique to screen for potential hydrogen seepage was used by Lefeuvre et al. [5]. The various colors stand for relative sampling depths (in cm) in the soil.

settings (i.e., extensional, compressional, transtensional, or transpressional) provided different soil gas anomaly patterns at the surface [79]. Even seismically inactive faults were found to have radon anomalies associated with them (e.g., [6, 80]). Therefore, radon measurements can also provide a proxy for the hydrogen-conducting potential of both active and inactive faults.

Overall, there is seemingly much more emphasis on the integration of various surface and subsurface gas data sets in ongoing hydrogen exploration programs compared to the everyday hydrocarbon exploration practice.

3.10 Nontechnical challenges for natural hydrogen exploration and production

The global demand for hydrogen is steadily increasing, mostly driven by its use as a feedstock for producing ammonia, methanol, and various materials. Hydrogen is not used at scale yet as an energy source although it is seen as the green replacement of fossil fuels, eventually. The promise of natural hydrogen is to have a very low production cost (less than USD 1/kg) and negligible carbon footprint (total CO_2 intensity below 0.2/kg) compared to other hydrogen types. Despite these major advantages, in practice, it is very challenging to make a business case for a natural hydrogen project at present, for several reasons. Some of the above-ground challenges are discussed below.

First, various compilations on the proportion of hydrogen in natural gas blends, both in subsurface gas fields and in surface seeps, show that it is quite often only in the 2–20% range (e.g., [21, 38, 59]). The 98% pure hydrogen found in the Bourakébougou is seen as an exception, which could be quite misleading when predicting average hydrogen content in yet-to-find accumulations.

Since hydrogen needs to be separated from the other gases at extra costs, smaller percentages of hydrogen in a gas blend quickly erode the profitability of a project [81]. The financials of hydrogen production were modeled and even though commerciality can be achieved even if the hydrogen component of a gas discovery is very small (i.e., less than 5%), the incremental separation cost could make natural hydrogen more expensive than competing products, notably green or blue hydrogen (Figure 3.12).

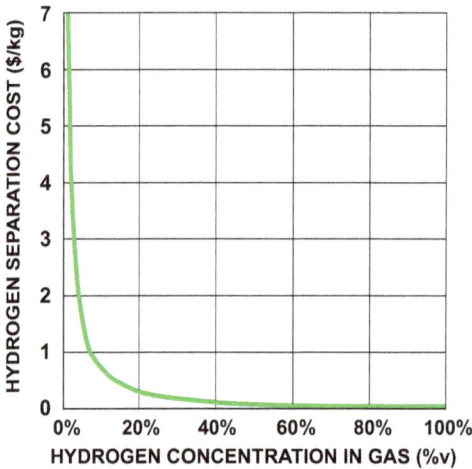

Figure 3.12: Low concentrations (1–5% by volume) in a gas mix make natural hydrogen similarly priced to green hydrogen (i.e., USD 3–6/kg) due to the incremental costs of separation leaving only the lower carbon footprint as a competitive edge (courtesy of Rob West of Thunder Said Energy).

Therefore, there is a real threat to the establishment of commercial natural hydrogen exploration and production from green or blue hydrogen. In the case of green hydrogen, the technology of splitting water molecules using electrolysis is well known. Importantly, the process consumes significant energy with a potentially high carbon-footprint depending on how the electricity is generated. However, the average USD 3–8/kg production costs of green hydrogen are not burdened by poorly understood subsurface risks like in the case of natural hydrogen at this point in time. In one possible scenario, green hydrogen will become the dominant product in the hydrogen market in the next decade not leaving room for natural hydrogen to get established in a timely manner to compete.

Second, location does matter. Transporting hydrogen over long distances is expensive and emission-intensive, therefore finding and producing natural hydrogen, close to demand markets is critical, preferably within less than tens of kilometers. The Bourakébougou hydrogen field in Mali is located very far from potential demand hindering its full-scale development. Therefore, it cannot provide a functioning and economically attractive analogue in a business case for other natural hydrogen projects elsewhere. Another dimension of the location challenge is the possibility of storing hydrogen nearby at scale, an option which is not always available depending on the geology of the area (e.g., [82]).

Third, regulations for natural hydrogen exploration and production are typically not in place with some notable exceptions (e.g., France and Australia). Therefore, hydrogen may be lumped together with hydrocarbons in a legislation context or treated as a critical mineral under existing mining codes. Either way in most countries it remains a problem to start an exploration project specifically on natural hydrogen.

Fourth, lack of systematic and sufficient investments. Due to the lack of producing natural hydrogen fields with demonstrated monetary value creation there is no real competition yet. Most energy companies are only monitoring this space waiting for a breakthrough by others. Business cases may be dismissed simply by quoting the worldwide absence of successful hydrogen exploration projects as credible analogues at the time of writing this chapter (January 2025).

All these challenges are fundamentally different from the very well-established and profitable petroleum business.

3.11 Conclusions and outlook

Many claim that existing hydrocarbon exploration workflows can be easily adjusted to fit the search for natural hydrogen. However, based on the collective efforts by the academia and the energy industry during the last few years, it does not seem to be the case. While thinking of petroleum systems as an analogue is indeed a useful starting point, even the early trends show that exploration for natural hydrogen if proven to be economically successful eventually, will have its own markedly different workflows. These will only partially overlap those designed specifically for the search for petroleum.

Coming from the angle of petroleum exploration, many envision finding hydrogen fields in a sedimentary basin fill (e.g., [5, 39]). It is more likely, however, that intra-basement targets may become more common in hydrogen exploration. It is to be noted that even the Bourakébougou hydrogen field in Mali is practically located within a basement unit even if the Neoproterozoic (!) reservoir rocks are sedimentary in nature [18–20]. There are cases of intra-basement petroleum finds in fractured crystalline reservoirs, but these remain the exception rather than the rule.

Since the most common sources of natural hydrogen are typically located in the basement, looking for intra-basement traps may become a unique aspect of hydrogen exploration. In fact, a new play considered in geothermal exploration focusing on pro-lific permeable zones in deep crystalline settings globally [83], may be a useful ana-logue for intra-basement hydrogen plays too.

In fact, the natural hydrogen exploration approach may end up being closer to the practices of geothermal energy development or the mining industry, both of which are accustomed to resource exploration in different rocks as pointed out by Yedinak [84]. While a solid understanding of geology and geophysics will remain a must, new surface and subsurface technologies specifically developed for natural hy-drogen exploration are becoming commonplace.

Quite differently from fossil fuels, natural hydrogen could be a sustainable energy source in the case of the hydrogen farming scenario described above. If a cost-efficient technology could be to exploit hydrogen-conducting fault zones with very low pressure and low yields at shallow depth, their production could be sustainable. This would lead, instead of centralized schemes seen in the hydrocarbon industry, to a broadly distributed energy structure [85]. The distributed nature of smallish and local energy generation is already present in the pattern of some of the geothermal energy solutions (e.g., shallow heat pumps).

The fact that natural hydrogen tends to co-occur with other gases will be a deci-sive factor in the next decade or so. In particular, exploration for helium (e.g., [86]) appears to be a major driving force as the frequent association of helium with hydro-gen (e.g., [67]) may make a strong business case by co-producing them. The case of natural hydrogen as a single target in a gas blend with other gases, such as nitrogen and carbon-dioxide, translates to a very challenging business case at present.

Natural hydrogen exploration by and large is already different from hydrocarbon exploration and the differences are expected to become even more pronounced than the similarities in the near future. The energy industry and academia need to work together to reach a breakthrough before artificial hydrogen species start to dominate the market advancing toward a hydrogen society. To paraphrase and slightly alter the message of [87], natural hydrogen molecules may indeed change the world.

Acknowledgments: This overview paper offers a personal perspective by a geologist with more than three decades of experience in hydrocarbon exploration who has been following natural hydrogen exploration efforts only for 3 years. It is too early to have strong opinions about natural hydrogen exploration as it is clearly in its infancy.

Conversations about natural hydrogen with Alain Prinzhofer, Eric Gaucher, Nico-las Lefeuvre, Daniel Palmowski, and Giuseppe Etiope were very helpful in this rapidly emerging field. Working together on several potential natural hydrogen projects with lot of colleagues in Central and Eastern Europe also shaped my thinking. For confi-dentiality reasons I list here only my Austrian colleagues: Reinhard Sachsenhofer, Clemens Zach, Erich Österreicher, Martin and Katrina Schöpfer, Alex Pengg, Johannes

Weitz, Johannes Wegscheider, Alex Kovacs, Moritz Kottulinsky, Nils Bezwoda, Bernhard Grasemann, and David Misch.

This chapter benefited from two thorough reviews by Reza Rezaee, which is gratefully acknowledged. Thanks also due to both editors, Reza Rezaee and Brian J. Evans, for their efforts putting this book together.

References

[1] Smith N. J. P. It's time for explorationists to take hydrogen more seriously. First Break. 2002, 20: 246–253.

[2] Smith N. J. P., Shepherd T. J., Styles M. T., Williams G. M. Hydrogen exploration: A review of global hydrogen accumulations and implications for prospective areas in NW Europe. In: Dore A. G., Vining B. A., eds., Petroleum geology: North-West Europe and global perspectives – proceedings of the 6th petroleum geology conference. London: Petroleum Geology Conferences Ltd. Published by the Geological Society, 2005, 349–358.

[3] Prinzhofer A., Deville É. Hydrogène naturel. La prochaine revolution énergétique? Une energie inepuisable et non polluante. Belin, Paris, 2015, 199.

[4] Gaucher E. C. New perspectives in the industrial exploration for native hydrogen. Elements. 2020, 16: 8–9.

[5] Lefeuvre N., Truche L., Donzé F. V., Ducoux M., Barré G., Fakoury R. A., Calassou S., Gaucher E. C. Native H2 exploration in the western Pyrenean foothills. Geochemistry, Geophysics, Geosystems. 2021, 22(8): e2021GC009917.

[6] Lefeuvre N., Truche L., Donzé F. V., Gal F., Tremosa J., Fakoury R. A., Calassou S., Gaucher E. C. Natural hydrogen migration along thrust faults in foothill basins: The North Pyrenean Frontal Thrust case study. Applied Geochemistry. 2022, 145: 105396.

[7] Moretti I., Brouilly E., Loiseau K., Prinzhofer A., Deville E. Hydrogen emanations in intracratonic areas: New guidelines for early exploration basin screening. Geosciences. 2021, 11: 145. doi: 10.3390/geosciences11030145

[8] Rigollet C., Prinzhofer A. Natural hydrogen: A new source of carbon-free and renewable energy that can compete with hydrocarbons. First Break. 2022, 40: 78–84.

[9] Rigollet C., Lefeuvre N. Methodological guidelines for natural H2 exploration. In: AAPG Europe conference, Budapest, 3–4 May 2022, abstract.

[10] Lévy D., Roche V., Pasquet G., Combaudon V., Geymond U., Loiseau K., Moretti I. Natural H 2 exploration: Tools and workflows to characterize a play. Science and Technology for Energy Transition. 2023, 78: 27.

[11] Zhao H., Jones E. A., Singh R. S., Ismail H. H. B., WahTan S. The hydrogen system in the subsurface: Implications for natural hydrogen exploration. SPE 216710, 2023, ADIPEC, Oct 2023.

[12] Jackson O., Lawrence S. R., Hutchinson I. P., Stocks A., Barnicoat A. C., Powney M. Natural hydrogen: Sources, systems and exploration plays. Geoenergy. 2024, 2: geoenergy2024–002.

[13] Saucier H. Rethinking the resource potential of geologic hydrogen, vol. 46. AAPG Explorer, 2025, 28–33.

[14] Magoon L. B., Dow W. G. The petroleum system, vol. 60. AAPG Memoir, 1994, 3–24.

[15] Ellis G. S., Gelman S. E. Model predictions of global geologic hydrogen resources. Science Advances. 2024, 10: eado0955.

[16] Gelman S. E., Hearon J. S., Ellis G. S. Prospectivity mapping for geologic hydrogen (No. 1900). US Geological Survey, 2025, 43. https://doi.org/10.3133/pp1900

[17]   Zamora G., Loma R., Monge A., Masini M., Vayssaire A., Olaiz A. Can native hydrogen be part of the energy transition? In: 83rd EAGE annual conference & exhibition. 2022, abstract.

[18]   Prinzhofer A., Cissé C. S. T., Diallo A. B. Discovery of a large accumulation of natural hydrogen in Bourakébougou (Mali). International Journal of Hydrogen Energy. 2018, 43(42): 19315–19326.

[19]   Maiga O., Deville E., Laval J., Prinzhofer A., Diallo A. B. Characterization of the spontaneously recharging natural hydrogen reservoirs of Bourakébougou in Mali. Scientific Reports. 2023, 13: 11876.

[20]   Maiga O., Deville E., Laval J., Prinzhofer A., Diallo A. B. Trapping processes of large volumes of natural hydrogen in the subsurface: The emblematic case of the Bourakébougou H2 field in Mali. International Journal of Hydrogen Energy. 2024, 50: 640–647.

[21]   Zgonnik V. The occurrence and geoscience of natural hydrogen: A comprehensive review. Earth-Science Reviews. 2020, 203: 103140.

[22]   Lefeuvre N., Thomas E., Truche L., Donzé F. V., Cros T., Dupuy J., Pinzon-Rincon L., Rigollet C. Characterizing natural hydrogen occurrences in the Paris basin from historical drilling records. Geochemistry, Geophysics, Geosystems. 2024, 25(5): e2024GC011501.

[23]   De Freitas V. A., Prinzhofer A., Françolin J. B., Ferreira F. J. F., Moretti I. Natural hydrogen system evaluation in the São Francisco Basin (Brazil). Science and Technology for Energy Transition. 2024, 79: 95.

[24]   Strapoc D., Ammar M., Abolins N., Gligorijevic A. Key role of regearing mud gas logging for natural H2 exploration. In: SPWLA annual logging symposium. Stavanger, Norway, 2022, D031S001R003. June 10–15, 2022.

[25]   Klein F., Bach W., McCollom T. M. Compositional controls on hydrogen generation during serpentinization of ultramafic rocks. Lithos. 2013, 178: 55–69.

[26]   Sherwood Lollar B. S., Onstott T. C., Lacrampe-Couloume G., Ballentine C. J. The contribution of the Precambrian continental lithosphere to global H2 production. Nature. 2014, 516: 379–382.

[27]   Goskolli E. New challenges to the deep development of the Bulqiza chrome mines. Mining Revue. 2022, 282: 29–34.

[28]   Prinzhofer A., Cacas-Stentz M. C. Natural hydrogen and blend gas: A dynamic model of accumulation. International Journal of Hydrogen Energy. 2023, 48: 21610–21623.

[29]   Horsfield B., Mahlstedt N., Weniger P., Misch D., Vranjes-Wessely S., Han S., et al. Molecular hydrogen from organic sources in the deep Songliao Basin, PR China. International Journal of Hydrogen Energy. 2022, 47: 16750–16774. doi: 10.1016/j.ijhydene.2022.02.208

[30]   Palmowski D., Fernandez N., Lefeuvre N., Kleine A., Tari G. Modelling natural hydrogen systems. Extending the basin modeler's comfort zone. In: 85th EAGE annual conference & exhibition (including the workshop programme). European Association of Geoscientists & Engineers, 2024, 1–5.

[31]   Arrouvel C., Prinzhofer A. Genesis of natural hydrogen: New insights from thermodynamic simulations. International Journal of Hydrogen Energy. 2021, 46(36): 18780–18794.

[32]   Ellis G. S., Palmowski D., Lefeuvre N. Application of reactive transport modelling to natural hydrogen exploration: An example from the French Pyrenean foreland. In: 3rd annual H-Nat natural hydrogen summit. Perth, Western Australia, 2023, November 27–28, 2023. https://www.hnatworld summit.com/en/conference-program

[33]   Tari G. Natural (gold) hydrogen exploration: Some similarities and differences with oil and gas exploration. AAPG Madrid Conference. 2023, November 7–8, Abstract Volume: 283–284.

[34]   Deronzier J. F., Giouse H. Vaux-en-Bugey (Ain, France): The first gas field produced in France, providing learning lessons for natural hydrogen in the sub-surface? Bulletin de la Societe Geologique de France. 2020, 191(1).

[35]   Bagriy I. D., Riepkin O., Paiuk S., Kryl I., Gafych I. Scientific justification of spatial distribution and mapping of anomalous manifestations of white hydrogen – Energy raw material of the XX century. Ukrainian Raw Materials. 2023, 1: 4–10.

[36] Ulmishek G. F. Petroleum geology and resources of the Dnieper-Donets basin. US Geological Survey Bulletin, 2001. http://pubs.usgs.gov/bul/2201/E/2201eE

[37] Leila M., Loisseau K., Moretti I. Controls on generation and accumulation of blended gases (CH4/ H2/He) in the Neoproterozoic Amadeus Basin, Australia. Marine and Petroleum Geology. 2022, 140: 105643. doi: 10.1016/j.marpetgeo.2022.105643

[38] Kusznir S., Kost M., Pankiv R., Seniv O., Kozak R. Geochemical and hydrogeological peculiarities of the Lokachy gas field (Lviv Paleozoic Basin). Geology and Geochemistry of Fossil Fuels. 2013, 3–4: 108–124. in Ukrainian with English abstract.

[39] Atkinson C., Matchette-Downes C., García-Curiel S. Natural hydrogen in the Monzon-1 well, Ebro basin, northern Spain. Geologues. 2022, 213: 96–102.

[40] Gaucher E., Moretti I., Pélissier N., Burridge G., Gonthier N. The place of natural hydrogen in the energy transition: A position paper. European Geologist 2023, 5–9.

[41] Tari G. Natural hydrogen potential in the East Med: A quick look. Larnaca, Cyprus: AAPG Europe Conference, 2022, 23–24 May, 2023, abstract.

[42] Echevarria-Rodriguez G., Hernandez-Perez G., Lopez-Quintero J. O., Lopez-Rivera J. G., Rodriguez-Hernandez R., Sanchez-Arango J. R., Socorro-Trujillo R., Tenreyro-Perez R., Yparraguirre-Pena J. L. Oil and gas exploration in Cuba. Journal of Petroleum Geology. 1991, 14: 259–274.

[43] Montadert L., Lie Ø., Semb P. H., Kassinis S. New seismic may put offshore Cyprus hydrocarbon prospects in the spotlight. First Break. 2010, 28(4).

[44] Symeou V., Homberg C., Nader F. H., Darnault R., Lecomte J. C., Papadimitriou N. Longitudinal and temporal evolution of the tectonic style along the Cyprus Arc system, assessed through 2-D reflection seismic interpretation. Tectonics. 2018, 37: 30–47.

[45] Larin N., Zgonnik V., Rodina S., Deville E., Prinzhofer A., Larin V. N. Natural molecular hydrogen seepage associated with surficial, rounded depressions on the European craton in Russia. Natural Resources Research. 2015, 24: 369–383.

[46] Zgonnik V., Beaumont V., Deville E., Larin N., Pillot D., Farrell K. M. Evidence for natural molecular hydrogen seepage associated with Carolina bays (surficial, ovoid depressions on the Atlantic Coastal Plain, Province of the USA). Progress in Earth and Planetary Science. 2015, 2: 1–15.

[47] Moretti I., Prinzhofer A., Françolin J., Pacheco C., Rosanne M., Rupin F., Mertens J. Long-term monitoring of natural hydrogen superficial emissions in a Brazilian cratonic environment. Sporadic large pulses versus daily periodic emissions. International Journal of Hydrogen Energy. 2021, 46: 3615–3628. doi: 10.1016/j.ijhydene.2020.11.026

[48] Moretti I., Geymond U., Pasquet G., Aimar L., Rabaute A. Natural hydrogen emanations in Namibia: Field acquisition and vegetation indexes from multispectral satellite image analysis. International Journal of Hydrogen Energy. 2022, 47(84): 35588–35607. doi: 10.1016/j.ijhydene.2022.08.135

[49] Roche V., Geymond U., Boka-Mene M., Delcourt N., Portier E., Revillon S., Moretti I. A new continental hydrogen play in Damara Belt (Namibia). Scientific Reports. 2024, 14: 11655.

[50] Rezaee R. Assessment of natural hydrogen systems in Western Australia. International Journal of Hydrogen Energy. 2021, 46(66): 33068–33077.

[51] Frery E., Langhi L., Maison M., Moretti I. Natural hydrogen seeps identified in the north Perth basin, western Australia. International Journal of Hydrogen Energy. 2021, 46(61): 31158–31173.

[52] Patino C., Piedrahita D., Colorado E., Aristizabal K., Moretti I. Natural H2 transfer in soil: Insights from soil gas measurements at varying depths. Geosciences. 2024, 14(11): 296.

[53] Petrowiki. Exploration for natural hydrogen. Exploration for Natural Hydrogen – PetroWiki. Last accessed on January 18. 2025.

[54] Smith K. L. Remote sensing of leaf responses to leaking underground natural gas unpublished. PhD thesis 2002, University of Nottingham, 241.

[55] Estrella H. F., Umlauft J., Schmidt A., Korn M. Locating mofettes using seismic noise records from small dense arrays and matched field processing analysis in the NW Bohemia/Vogtland Region, Czech Republic. Near Surface Geophysics. 2016, 14: 327–335.

[56] Umlauft J., Estrella H. F., Korn M. Imaging fluid channels within the NW Bohemia/Vogtland region using ambient seismic noise and MFP analysis. In: EGU general assembly conference abstracts. 2016, EPSC2016-2134.

[57] Prinzhofer A., Moretti I., Françolin J., Pacheco C., d'Agostino A., Werly J., Rupin F. Natural hydrogen continuous emission from sedimentary basins: The example of a Brazilian H2-emitting structure. International Journal of Hydrogen Energy. 2019, 44: 5676–5685.

[58] Békési, E., Szárnya, C., Prinzhofer, A., Twaróg, A., Porkoláb, K., and Tari, G.: Exploration of "fairy circles" associated with natural hydrogen seepages with synthetic aperture radar interferometry and backscatter analysis, EGU General Assembly 2025, Vienna, Austria, 27 Apr–2 May 2025, EGU25-8374, https://doi.org/10.5194/egusphere-egu25-8374, 2025.

[59] McMahon C. J., Roberts J. J., Johnson G., Edlmann K., Flude S., Shipton Z. K. Natural hydrogen seeps as analogues to inform monitoring of engineered geological hydrogen storage. Geological society, vol. 528. London: Special Publications, 2023, 461–489.

[60] Coughlan M., Roy S., O'Sullivan C., Clements A., O'Toole R., Plets R. Geological settings and controls of fluid migration and associated seafloor seepage features in the north Irish Sea. Marine and Petroleum Geology. 2021, 123: 104762.

[61] Knipe R. J., Fisher Q. J., Jones G., Clennell M. R., Farmer A. B., Harrison A., Kidd B., McAllister E. R. P. J., Porter J. R., White E. A. Fault seal analysis: Successful methodologies, application and future directions. Norwegian Petroleum Society Special Publications. 1997, 7: 15–38.

[62] Sugisaki R., Ido M., Takeda H., Isobe Y., Hayashi Y., Nakamura N., Satake H., Mizutani Y. Origin of hydrogen and carbon dioxide in fault gases and its relation to fault activity. The Journal of Geology. 1983, 91: 239–258. https://doi.org/10.1086/628769

[63] Ware R. H., Roecken C., Wyss M. The detection and interpretation of hydrogen in fault gases. Pure and Applied Geophysics. 1984, 392–402. https://doi.org/10.1007/BF00874607

[64] Indrevær K., Stunitz H., Bergh S. G. On Palaeozoic–Mesozoic brittle normal faults along the SW Barents Sea margin: Fault processes and implications for basement permeability and margin evolution. Journal of the Geological Society. 2014, 171: 831–846. https://doi.org/10.1144/jgs2014-018

[65] Pei Y., Paton D. A., Knipe R. J., Wu K. A review of fault sealing behaviour and its evaluation in siliciclastic rocks. Earth-Science Reviews. 2015, 150: 121–138. https://doi.org/10.1016/j.earscirev.2015.07.011

[66] Truche L., Donzé F. V., Goskolli E., Muceku B., Loisy C., Monnin C., Dutoit H., Cerepi A. A deep reservoir for hydrogen drives intense degassing in the Bulqizë ophiolite. Science. 2024, 383(6683): 618–621.

[67] Prinzhofer A., Rigollet C., Lefeuvre N., Françolin J., de Miranda P. E. V. Maricá (Brazil), the new natural hydrogen play which changes the paradigm of hydrogen exploration. International Journal of Hydrogen Energy. 2024, 62: 91–98.

[68] Peignard L., Hauville B. Multiphysics for industrial gas exploration – The Fonts-Bouillants case study. In: E3S web of conferences, vol. 504. EDP Sciences, 2024, 01003.

[69] Menez B. Geologic hydrogen and methane as fuel for life. Elements. 2020, 16: 39–46.

[70] Boreham C. J., Edwards D. S., Czado K., Rollet N., Wang L., van der Wielen S., Champion D., Blewett R., Feitz A., Henson P. A. Hydrogen in Australian natural gas: Occurrences, sources and resources. The APPEA Journal. 2021, 61: 163–191.

[71] Mainson M., Heath C., Pejcic B., Frery E. Sensing hydrogen seeps in the subsurface for natural hydrogen exploration. Applied Science. 2022, 12: 6383. https://doi.org/10.3390/app12136383

[72] Forde O. N., Cahill A. G., Beckie R. D., Mayer K. U. Barometric-pumping controls fugitive gas emissions from a vadose zone natural gas release. Scientific Reports. 2019, 9(1): 14080.

[73]   Myagkiy A., Brunet F., Popov C., Krüger R., Guimarães H., Sousa R. S., Charlet L., Moretti I. H2 dynamics in the soil of a H2-emitting zone (São Francisco Basin, Brazil): Microbial uptake quantification and reactive transport modelling. Applied Geochemistry. 2020, 112: 104474.

[74]   Etiope G., Ciotoli G., Benà E., Mazzoli C., Röckmann T., Sivan M., Squartini A., Laemmel T., Szidat S., Haghipour N., Sassi R. Surprising concentrations of hydrogen and non-geological methane and carbon dioxide in the soil. Science of the Total Environment. 2024, 948: 174890.

[75]   Abrams M. A., Schumacher D. Surface geochemistry methods for petroleum exploration. Encyclopedia of petroleum geoscience. Encyclopedia of Earth Sciences Series. 2020, 1–11.

[76]   Etiope G., Samardžić N., Grassa F., Hrvatović H., Miošić N., Skopljak F. Methane and hydrogen in hyperalkaline groundwaters of the serpentinized Dinaride ophiolite belt, Bosnia and Herzegovina. Applied Geochemistry. 2017, 84: 286–296.

[77]   Jacquemet N., Prinzhofer A. The association of natural hydrogen and nitrogen: The ammonium clue? International Journal of Hydrogen Energy. 2024, 50: 161–174.

[78]   Romanak K. D., Bennett P. C., Yang C., Hovorka S. D. Process-based approach to CO2 leakage detection by vadose zone gas monitoring at geologic CO2 storage sites. Geophysical Research Letters. 2012, 39(15).

[79]   Sun X., Yang P., Xiang Y., Si X., Liu D. Across-fault distributions of radon concentrations in soil gas for different tectonic environments. Geosciences Journal. 2018, 22: 227–239.

[80]   Benà E., Ciotoli G., Ruggiero L., Coletti C., Bossew P., Massironi M., Mazzoli C., Mair V., Morelli C., Galgaro A., Morozzi P. Evaluation of tectonically enhanced radon in fault zones by quantification of the radon activity index. Scientific Reports. 2022, 12(1): 21586.

[81]   Thunder Said Energy. Energy transition research & technologies – Thunder Said Energy, last accessed on January 22, 2025.

[82]   Duffy O., Hudec M., Peel F., Apps G., Bump A., Moscardelli L., Dooley T., Fernandez N., Bhattacharya S., Wisian K., Shuster M. The role of salt tectonics in the energy transition: An overview and future challenges. Tektonika. 2023, 1: 18–48.

[83]   Bischoff A., Heap M. J., Mikkola P., Kuva J., Reuschlé T., Jolis E. M., Engström J., Reijonen H., Leskelä T. Hydrothermally altered shear zones: A new reservoir play for the expansion of deep geothermal exploration in crystalline settings. Geothermics. 2024, 118: 102895.

[84]   Yedinak E. M. The curious case of geologic hydrogen: Assessing its potential as a near-term clean energy source. Joule. 2022, 6(3): 503–508.

[85]   Lapi T., Chatzimpiros P., Raineau L., Prinzhofer A. System approach to natural versus manufactured hydrogen: An interdisciplinary perspective on a new primary energy source. International Journal of Hydrogen Energy. 2022, 47: 21701–21712.

[86]   Gluyas J. G., Fowler N. The future of geoenergy – A perspective. Geoenergy. 2024, 2: geoenergy2023-058.

[87]   Truche L., McCollom T. M., Martinez I. Hydrogen and abiotic hydrocarbons: Molecules that change the world. Elements: An International Magazine of Mineralogy, Geochemistry, and Petrology. 2020, 16: 13–18.

Kanchana Kularatne and Isabelle Moretti

Chapter 4
Geological and geochemical pathways of onshore natural hydrogen generation

Abstract: Molecular hydrogen (H_2) of abiotic origin is prevalent across various geological settings. The generation of H_2 in these environments is gaining increasing attention, particularly due to the rising demand for cheap and clean hydrogen as an energy source. For decades, it was believed that H_2 generation primarily occurred via high-temperature serpentinization, a process involving the hydrothermal alteration of ultramafic rocks in mid-oceanic ridges. While hydrothermal serpentinization produces large fluxes, it remains beyond the scope of current natural H_2 exploration efforts since they are offshore and too far for the customers.

In contrast, natural H_2 found within ophiolites and in the cratonic areas, "onshore," presents significant potential for exploration. These occurrences are not necessarily associated with active tectonics but are found within the stable interiors of continents. The discovery of natural H_2 seepages in Archean and Neoproterozoic areas is relatively recent, and the mechanisms underlying its generation in these settings are not yet well understood.

This chapter provides an exhaustive review of the current knowledge and relevant case studies, offering a comprehensive understanding of the geological and geochemical pathways responsible for the generation of abiotic H_2 onshore. This insight is crucial for the prospecting of natural H_2 reservoirs and for advancing the utilization of natural H_2 as a sustainable energy resource.

Keywords: natural hydrogen, geological and geochemical pathways, ophiolites, cratonic areas

4.1 Introduction

The energy market has undergone a transformative evolution, moving from solid fuels like coal to liquid hydrocarbons such as oil, followed by a significant shift to gaseous energy sources like natural gas. Each transition has aimed at improving efficiency, reducing emissions, and adapting to technological advancements. Today, the

Kanchana Kularatne, Laboratoire des Fluides Complexes et leurs Réservoirs, Université de Pau et des pays de l'Adour, France, e-mail: kanchana.kularatne@univ-pau.fr
Isabelle Moretti, Laboratoire des Fluides Complexes et leurs Réservoirs, Université de Pau et des pays de l'Adour, France

https://doi.org/10.1515/9783111437040-004

focus is expanding toward cleaner alternatives, with hydrogen (H_2) emerging as a promising frontier. H_2 is a raw material commercially produced, mainly through steam methane reforming and coal gasification. Water electrolysis is slowly emerging as a lower carbon alternative but it consumes energy and water. However, when produced sustainably, H_2 becomes a zero-emission energy carrier, complementing renewables like wind and solar in an integrated energy mix. This diversification underscores the global drive toward decarbonization, fostering resilience and adaptability in energy systems, while meeting the growing demand for sustainable, low-carbon energy. In this context, natural H_2 is gaining increasing attention since it is a low-carbon energy source that could potentially offer a viable alternative to meet energy needs with a lower carbon footprint.

Natural H_2 is prevalent in the Earth. In the past decades, it was believed that the hydrogen in Earth is primarily generated via serpentinization, a process involving the hydrothermal alteration of ultramafic rocks into an assemblage of serpentine, brucite, and magnetite, releasing molecular hydrogen [1–6]. Hydrothermal serpentinization typically occurs along one of the Earth's most tectonically active settings, Mid Oceanic Ridge (MOR), where the ultramafic oceanic lithosphere encounters hydrothermal water circulating through the fractured oceanic lithosphere at spreading centers, at temperatures between 150 and 350 °C. This process has been widely documented, particularly in the hydrothermal fields such as Lost city [5, 7], Rainbow [8, 9], Trans-Atlantic Geotraverse [10], East Pacific Rise [11], Lucky Strike [12], and Juan de Fuca [13]. Serpentinization generates large fluxes of hydrogen in Earth, which serves as the prime source of energy to sustain deep life where sunlight does not penetrate [14, 15]. Due to the high reactivity, hydrogen generated within the hydrothermal fields [5–13] often reacts with carbon dioxide, nitrogen, and sulfur, forming a variety of molecules that are often found in the fluids sampled from hydrothermal vents such as methane, ammonia, reduced sulfur species (H_2, CO_2, N_2, CH_4, NH_3, H_2S, and H_2SO_4), and other reduced hydrocarbon molecules. H_2 creates a sufficiently reduced environment for the synthesis of such abiotic molecules, including the molecules that are found in primitive life forms, providing evidence for the hypothesis of abiotic origin of life forms. Hydrogen formed in the deep MOR, however, does not reach the atmosphere [16] due to (i) its high reactivity and therefore conversion into other molecules and (ii) microbial consumption. Consequently, although the largest fluxes of natural H_2 is generated via serpentinization in the MOR, they remain beyond the scope of current natural hydrogen exploration efforts.

Natural H_2 also occurs in the middle of stable continental crust. Unlike MOR, these areas are not necessarily associated with active tectonic or volcanic activity. In this chapter, we use the term *onshore hydrogen*. Compared to natural hydrogen generated in the MOR, onshore hydrogen presents significant potential for exploration. Onshore, natural hydrogen has been documented decades ago. For example, hydrogen gas was found in the early 1980s in the wells drilled in the mid-continental rift system in the USA, Oman ophiolite, and in New Caledonia and Mali at Bourakebougou in 1987 [17], but the idea of exploration was not yet established [18–22]. The scarcity of hydrogen in millions of wells drilled for oil and natural gas in sedimentary basins initially led to the belief that natural hydrogen accumulations were rare [23]. However, the discovery of an almost pure hydrogen reservoir (over 98% H_2) in a well in the village of Bourakebougou, in 2007 challenged this belief and sparked interest in exploring natural hydrogen resources [17, 24]. Since then, hydrogen seepages have been discovered in every continent, notably in countries such as Turkey, Oman, Japan, the Philippines, New Zealand, Australia, New Caledonia, Greece, Namibia, Spain, Italy, France, Bosnia, California, and Canada. In the fluids sampled from these seepages, hydrogen is often the sole or the dominant component of gas mixtures (H_2, CO_2 CH_4, and NH_3), though in some locations, hydrogen is found to be less abundant than methane.

Onshore natural hydrogen seepages primarily occur in three forms: (i) gaseous hydrogen: gaseous hydrogen has been detected venting to the surface, such as in Bolivia [25], or burning upon contact with oxygen, forming flames, as observed at the Chimaera site in Turkey [26]. It is also associated with subcircular depressions (SCDs), also referred to as fairy circles, found in regions of Africa, Australia, and Brazil [27–29]; Gaseous hydrogen has also been identified as trapped within minerals and fluid inclusions [30, 31] (ii) hydrogen dissolved in water: Hydrogen has been reported dissolved in deep groundwater, aquifer water, or surface water [32]. For a review of hydrogen occurrences and quantities across these forms, readers may consult Zgonnik [33], but many other occurrences have been discovered since this publication.

Onshore natural hydrogen seepages have been identified across a wide range of geological settings, lithologies, and depths, highlighting the diversity of processes involved in hydrogen generation within the continental crust. Unlike the uniform hydrogen production mechanism at MOR, which is directly linked to serpentinization, onshore hydrogen generation appears to result from a broader array of processes. In this chapter, we explore the various origins of onshore natural hydrogen, focusing on abiotic pathways, and discuss the different generation mechanisms associated with specific source rocks, as reported in the literature to date. The authors aim to present all possible abiotic onshore hydrogen-generating mechanisms, and then discuss whether each of them is considered for exploration or not. This comprehensive approach underscores the complexity of hydrogen generation in continental settings and its significant potential for future exploration.

4.2 Geological and geochemical pathways of onshore natural hydrogen generation

As mentioned earlier, onshore natural hydrogen has been reported from diverse geological settings, lithologies, depths, and physicochemical conditions. These variations undoubtedly point to multiple origins of hydrogen. We believe that within hydrogen reservoirs, hydrogen originating from various pathways may coexist. Various authors have proposed different classifications. In this chapter, we present the main known pathways that could generate natural H_2 onshore, except the processes that generate hydrogen through microbial activities [34]. We have classified them into six distinct categories (Figure 4.1). Among them, (i) radiolysis of water, (ii) organic maturation, and (iii) redox reaction between Fe^{2+} and water are the main pathways that generate onshore natural hydrogen with strong exploration potential. In contrast, (iv) primordial hydrogen, (v) volcanic degassing, and (vi) mechanoradicles represent other potential hydrogen-generating pathways that are not the direct target of current exploration efforts. The radiolysis of water, organic maturation, and hydrogen generation via mechanoradicles are straightforward pathways and all published papers up to date agree with the reactions associated with these. However, when considering hydrogen generation via the reduction of water by Fe-rich rocks, some authors use the word serpentinization instead of oxidation-reduction, while some categorize, depending on the origin of the Fe-rich source rock (oceanic lithosphere, sedimentary BIF, intrusion). In this chapter, we focused on the reactions of water reduction by iron oxidation as a single category, and different reactions that can occur in different rock types will be further described. Finally, some authors consider the degassing of primordial hydro-

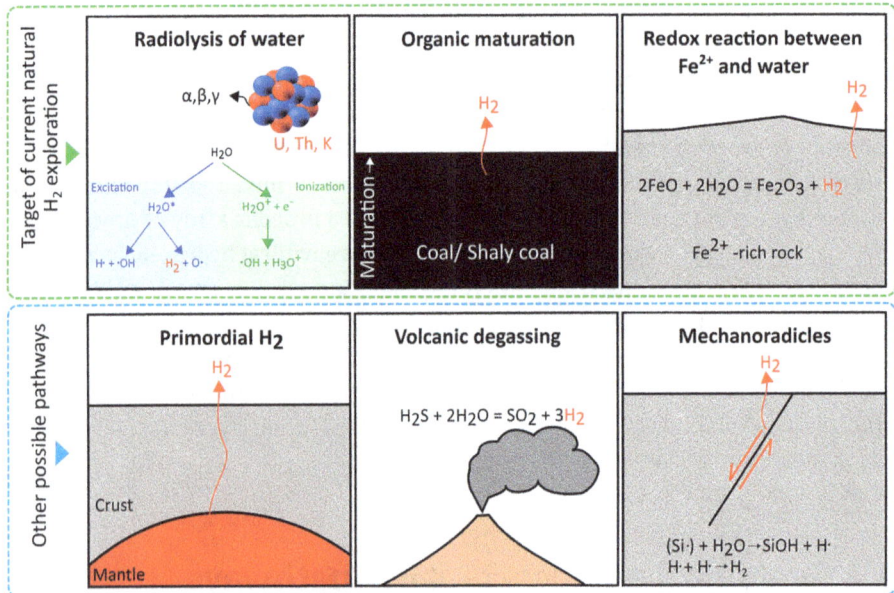

Figure 4.1: Potential origins of onshore natural hydrogen, categorized into six pathways (adapted from Klein et al. [16]). Current exploration efforts primarily target radiolysis of water [35], organic maturation, and redox reactions between Fe^{2+} and water, while primordial sources, volcanic degassing, and mechanoradicals receive less focus. It is important to note that hydrogen generated through these latter pathways can often be found mixed with hydrogen originating from the former pathways.

gen under the same category as volcanic degassing. But, here we consider that the primordial molecular hydrogen is emanated without a volcanic eruption and the volcanic hydrogen is generated essentially during a volcanic eruption. Below is a detailed elaboration of these six pathways.

4.2.1 Radiolysis of water

The natural decay of radioactive nuclides of potassium (^{40}K), thorium (^{232}Th), and uranium (^{235}U and ^{238}U) is involved in natural hydrogen generation via radiolysis of water. The decay of these radionuclides generates α, β, and γ radiation, which are ionizing radiation. Let's first consider the decay of the radionuclides ^{40}K, ^{232}Th, and ^{238}U.

Decay of ^{40}K: Potassium ^{40}K decays into stable calcium and argon as follows:

$$^{40}K \rightarrow {}^{40}Ca + \beta - + ve$$

$$^{40}K + e^- \rightarrow {}^{40}Ar + ve$$

Decay of ^{232}Th: Like uranium, thorium's decay continues through a series of alpha and beta decays, involving fourteen intermediate isotopes, until it stabilizes as ^{208}Pb. The main decay step include:

$$^{232}Th \rightarrow {}^{238}Ra + \alpha$$

Decay of ^{238}U: Uranium (^{238}U) undergoes a complex decay chain, eventually forming stable lead ^{206}Pb. The main decay step includes:

$$^{238}U \rightarrow {}^{234}Th + \alpha$$

The interaction of this ionizing radiation with water can split the water molecule into free radicles of hydrogen (H•) and hydroxyl (OH•), instead of H^+ and OH^-. The combination of two hydrogen-free radicles forms molecular hydrogen [35]:

$$H_2O \xrightarrow{a,\beta,\gamma} (H\bullet) + (OH\bullet)$$

$$(H\bullet) + (H\bullet) \rightarrow H_{2(g)}$$

In addition, the interaction can also generate oxidizing species such as H_2O_2, O_2, and O^-. Estimates show that the radiolysis generates about 4.7×10^{10} mol/a of hydrogen [36]. The radiolysis process has no dependence on pressure and temperature and is known to occur at all temperatures at which water is stable, even in the forms of ice, vapor, or hydrated salts [16]. In the cratonic setting, radiolysis of water is exclusive to lithologies with natural radioactive nuclides or the minerals containing them, such as potassium-containing minerals: orthoclase, $KAlSi_3O_8$; microcline, $KAlSi_3O_8$; muscovite, $KAl_2(AlSi_3O_{10})$

$(OH)_2$; biotite, $K(Mg, Fe)_3(AlSi_3O_{10})(OH)_2$. Thorium-containing minerals: thorite, $ThSiO_4$; monazite, $(Ce, La, Nd, Th)PO_4$; allanite, $(Ca, Ce, La, Y, Th)(Al, Fe)_3(SiO_4)_3(OH)$; thorianite, $(Th, U)O_2$. Uranium-containing minerals: uraninite, UO_2; coffinite, $U(SiO_4)_{1-x}(OH)_{4x}$; brannerite, $(U, Ca, Y, Ce)(Ti, Fe)_2O_6$; carnotite, $K_2(UO_2)_2(VO_4)_2 \cdot 3H_2O$; autunite, $Ca(UO_2)_2(PO_4)_2 \cdot 10\text{--}12H_2O$; torbernite, $Cu(UO_2)_2(PO_4)_2 \cdot 8\text{--}12H_2O$. These minerals can be associated with various lithologies such as granites, including Precambrian continental lithosphere, pegmatites, basalt, and sedimentary basins, and metamorphic rocks such as quartzite. Moreover, as mentioned earlier, hydrolysis refers to the splitting of any form of water, including water molecules bound to hydrated salts. Experiments show that salts such as $CaCl_2 \cdot 2H_2O$, $CaCl_2 \cdot 6H_2O$, $MgCl_2 \cdot 2H_2O$, and $MgCl_2 \cdot 6H_2O$ are capable of generating hydrogen via interaction with gamma radiation and that the yield of H_2 is independent of the cation or the degree of hydration [37]. The quantities of hydrogen vary <1 molecule/100 eV. However, field measurements on natural hydrogen seepages originated via the hydrolysis of hydrated salts are not known.

The generation of hydrogen via radiolysis depends on several factors:

(i) Concentration of radionuclides: The rate of hydrogen production is directly proportional to radionuclide concentration, as it determines the number of decay events and the resulting α, β, and γ radiation emitted within the formation [38].

(ii) Water availability in pores and fractures: The presence of water molecules in pore spaces and fractures is essential for radiolysis. In dry rock with no water in these spaces, H_2 generation via radiolysis is theoretically impossible. In sedimentary or crystalline rock, water is typically confined to pores and fractures, making rock permeability and porosity critical parameters for estimating H_2 production in regions with known radionuclide concentrations [38].

(iii) Salinity: Radiolysis is more efficient in salt solutions than in distilled water, as demonstrated by Wang et al. [39]. When salts dissolve, covalent bonds form between cations and oxygen, and between anions and hydrogen. For example, in NaCl, sodium bonds with oxygen and chlorine with hydrogen, while in $MgCl_2$, magnesium bonds with oxygen and chlorine with hydrogen. These interactions weaken the H–O bond energy in water, reducing the energy required to break the bond and enhancing H_2 production under irradiation [16].

This interplay of factors highlights the complex relationship between geological and chemical properties in determining H_2 radiolysis potential. Radiolysis can also generate hydrocarbons, however, this is out of the scope of this chapter [39].

Occurrences of hydrogen that has been formed via radiolysis are closely associated with two settings:

(i) Precambrian lithosphere: The Precambrian lithosphere comprises ancient foundational rock units of the Earth's crust, formed during the Precambrian Eon, which spans from the Earth's formation, approximately 4.6 billion years ago to 541 million years ago. These rocks are predominantly igneous and metamorphic in origin and are often deeply buried beneath younger sedimentary layers. They host ancient saline fracture waters in the subsurface, with groundwater residence times ranging from millions to billions of years [36]. The hydrogen dissolved in these fracture waters is primarily attributed to radiolysis. Estimates based on the relationship between H_2/He isotopes, concentrations of K, U, and Th, as well as rock density and porosity, suggest that the Precambrian continental lithosphere generates between 0.36 and 2.27×10^{11} moles of hydrogen per year. This production rate is comparable to the hydrogen generated at MOR hydrothermal fields, indicating that the Precambrian basement may represent one of the largest potential hydrogen reservoirs in the continental crust [36, 38]. However, the cratons also contain BIF and the H_2 content in that context may also come form oxidoreduction of the Fe^{2+} as we will discuss.

(ii) Formations that host radioactive mineral deposits: Radioactive mineral deposits are hosted in diverse geological formations, reflecting the varied processes that concentrate uranium, thorium, and other radioactive elements [40]. Granitic rocks and pegmatites often host primary uranium and thorium minerals like uraninite and monazite, while sandstone formations are significant for uranium deposits formed through leaching and precipitation under reducing conditions. High-grade uranium is also found in unconformity-related deposits at the interface of older basement rocks and sedimentary layers, such as in the Athabasca Basin, Canada. Quartz pebble conglomerates, like those in South Africa's Witwatersrand Basin, host ancient placer uranium deposits. Carbonatites and alkaline complexes, such as Mountain Pass in the USA, are important for thorium and rare-earth elements, whereas volcanic rocks and black shales host uranium associated with acidic magmatic processes and organic-rich sediments, respectively. Additional settings include phosphorites, marine deposits enriched with uranium and thorium, and glacial or fluvial deposits, where weathering redistributes radioactive minerals. Collectively, these formations highlight the complex geologic and geochemical conditions necessary for concentrating radioactive elements and hence generation of hydrogen via radiolysis. In these formations, hydrogen is often found trapped in fluid inclusions. For example, fluid inclusions form quartz and dolomite veins from five unconformity-related uranium deposits (Athabasca basin, Canada) [41]. The most common gases found in these inclusions are H_2, O_2, CO_2, CH_4, C_2H_6, and N_2. Some brine inclusions composed of "NaCl-rich" and "$CaCl_2$-rich" brine have similar gas contents [41, 42]. Another study reports fluid inclusions in quartz from three Precambrian uranium deposits: Oklo (Gabon), Rabbit Lake, and Cluff Lake D (Saskatchewan, Canada), where co-existing H_2 and O_2 have been found [43].

In the review by Levy et al. [24], the rocks containing radionuclides or the minerals rich in radionuclides are categorized into H2_GR3. The distribution of uranium-bearing rocks is given in Figure 4.2. These rocks are targets for current hydrogen exploration.

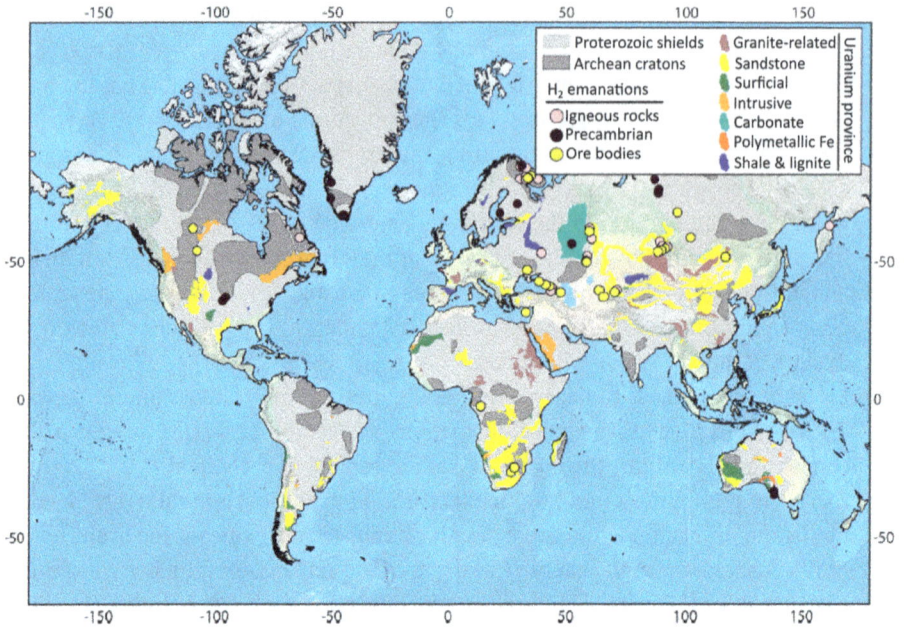

Figure 4.2: Spatial distribution of uranium-bearing rocks with hydrogen generation potential through radiolysis. While the map highlights regions rich in uranium, it does not include other radionuclide-bearing rocks or minerals that may also contribute to hydrogen generation [24].

4.2.2 Organic maturation

Hydrogen (H_2) generation during organic maturation is a rather well-known process closely tied to the transformation of organic matter under thermal stress during burial [44, 45]. These transformations are driven by chemical reactions, including cracking, cyclization, and aromatization, which alter the composition and structure of organic matter while releasing hydrogen at various stages. Various processes are involved in hydrogen generation:

(i) Aromatization: As organic matter matures, the transformation of kerogen involves the growth of aromatic ring systems. This reaction is critical in reducing the hydrogen-to-carbon (H/C) ratio of the kerogen, as hydrogen is partitioned between hydrogen-rich volatile compounds and an increasingly aromatic residue. Simultaneously, the breaking of chemical bonds (C–O and C–C) releases alkyl chains from kerogen by a process called cracking. These chains undergo reduction, consuming some of the hydrogen liberated during aromatization.

(ii) Cyclization and ring-opening: Cyclization, or the formation of ring structures, is another pathway that liberates hydrogen. At the same time, ring-opening reactions may occur, which require hydrogen input. These reactions collectively balance hydrogen production and consumption as the molecular structure of kerogen evolves.

(iii) Late-stage reactions: At very high levels of thermal maturity, kerogen undergoes further transformations. Methyl groups, which represent the last remnants of reactive carbon in highly mature organic matter, are stripped away. This "late gas" phase can release free hydrogen. Annulation, the process of forming larger aromatic stacks, also contributes to hydrogen generation in these advanced stages.

(iv) Dehydrogenation: Dehydrogenation simply means the removal of hydrogen from an organic molecule. But more specifically, this means the conversion of alkanes to olefins, and then aldehydes, alcohols, and also aromatics.

(v) Decarboxylation and decarbonylation: These are two processes that indirectly generate hydrogen.

Decarboxylation is a chemical reaction in carboxylic acids, by which the molecule is separated into an alkyl group and acid group, later forming an alkane and CO_2:

$$ROOH \rightarrow R-H + CO_2$$

Decarbonylation is a chemical reaction in aldehydes and formic acid, leading to the formation of alkanes, CO, and water:

$$In\ aldehydes:\ RCHO \rightarrow RH + CO$$

$$In\ formic\ acid:\ HCOOH \rightarrow CO + H_2O$$

These two processes release CO_2 or CO, which is later processed in water-gas shift reactions to produce additional hydrogen, as given below:

$$CO + H_2O \rightleftharpoons CO_2 + H_{2(g)}$$

Hydrogen is then generated all over the organic maturation process but it will be likely used to form lighter hydrocarbon in the first kilometers and will only remain as a free gas when all liquid HC have been already expelled. The kinetic of this maturation has been quantified by Moretti et al. [46], and they estimated that the temperature for peak hydrogen generation is around 250 °C. Horzfield et al. [47–49] proposed a lower temperature (around 200 °C), but globally, all authors agree on the yield of the rich TOC rock, which is ¼ of the TOC [46, 50]. As a result, deeply buried coal veins are an excellent hydrogen-generating rock.

Despite hydrogen generation, the overall hydrogen content of organic matter decreases as maturity progresses [44, 45]. This loss is due to increasing aromatization and condensation, which consolidate carbon structures into more stable, less reactive forms. For instance, in coal, hydrogen content and volatile matter yields decline as maturity advances, correlating with higher concentrations of protonated aromatic carbon at intermediate stages, followed by hydrogen loss during annulation. Similar patterns are observed in shale and crude oil, where the aromaticity and condensation of nitrogen, sulfur, and oxygen (NSO)-containing compounds increases with maturation.

The rocks that generate hydrogen via organic maturation have been categorized into H2_GR4 by Levy et al. [24]. The distribution of rocks that are capable of generating hydrogen by organic maturation is given in Figure 4.3 [24].

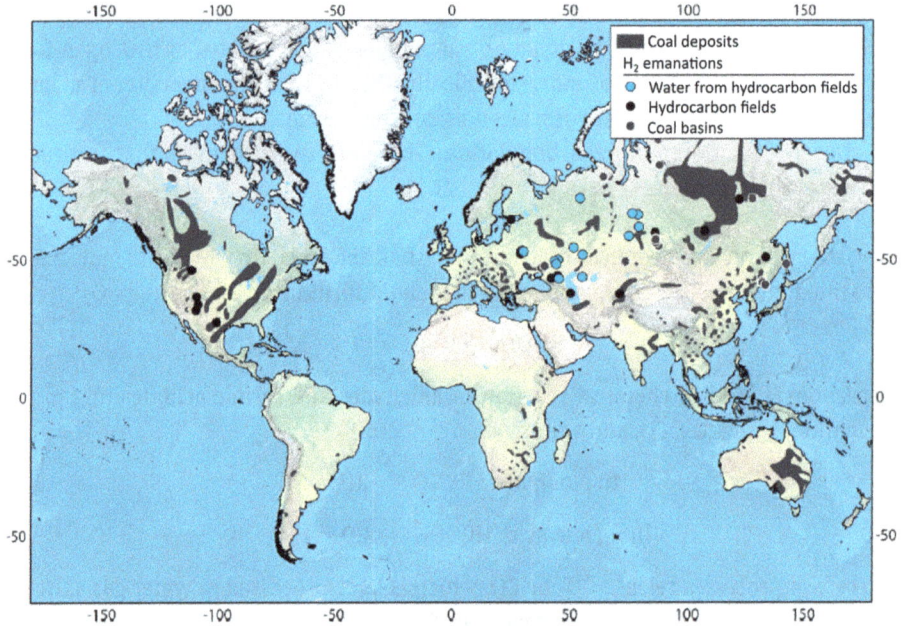

Figure 4.3: Spatial distribution of hydrogen generating rocks via organic maturation. Highlighted on the map are coal deposits, hydrocarbon fields, and locations where water samples with hydrogen concentrations exceeding 10% have been identified. These features collectively indicate regions with potential for hydrogen production linked to organic processes [24, 33].

4.2.3 Redox reaction between Fe^{2+} and water

The generation of natural hydrogen through the redox reaction between ferrous iron (Fe^{2+}) and water is exclusive to the Fe-rich lithologies. This reaction involves the oxidation of ferrous irons (Fe^{2+}) in the Fe-rich primary minerals into ferric irons (Fe^{3+}) while reducing water into molecular hydrogen (H_2). The generalized reaction can be expressed as [51]:

$$2Fe^{2+} + 2H_2O \rightarrow 2Fe^{3+} + 2OH^- + H_{2(g)}$$

In this generalized reaction, Fe^{2+} represents the divalent iron present in any of the Fe-rich primary minerals such as olivine and pyroxene. Fe^{3+} represents the trivalent iron produced by the reaction, which is also hosted in a Fe-rich secondary mineral, such as the most common magnetite and Fe-rich clays. The redox reaction above is, therefore, typically a dissolution–precipitation type fluid-rock interaction. We have identified four Fe-rich lithologies in which natural hydrogen has been reported [51]. They are: (i) ophiolitic massifs and exposed mantle rocks in rift basins, (ii) granites, (iii) banded iron formations (BIFs), and (iv) diorites. We also included a fifth category to include rocks that are not necessarily identified: (v) unspecified Fe-rich cratonic rocks. In the recent classification of hydrogen-generating source rocks by Lévy et al. [24], the ophiolite massifs have been classified under H2_GR1, whereas the granites, BIFs, and igneous intrusions such as diorites are classified under H2_GR2. Depending on the Fe^{2+}-hosting primary mineral/s and the Fe^{3+}-hosting secondary mineral/s involved, the abovementioned generalized reaction can take many forms. For example, serpentinization is one of the most common forms of the above equation, where olivine is the Fe^{2+}-hosting primary minerals, whereas, serpentine (can host both Fe^{2+} and Fe^{3+}), and magnetite (can host both Fe^{2+} and Fe^{3+}) are the Fe^{3+}-hosting secondary phases. The generalized equation for the redox reaction between Fe^{2+} and water has a stoichiometric ratio of Fe^{2+}:H_2 of 2:1; therefore, the amount of hydrogen generated by Fe^{2+} and water redox reaction can be estimated by the quantity of Fe^{2+} present in the rock [51]. However, this ratio doesn't hold to all alternative equations, because the Fe^{2+}:H_2 ratio differs, depending on the mineralogy of the primary and the secondary mineral. Below, we will discuss the various forms of this redox reaction for above mentioned Fe-rich lithologies:

(i) Ophiolitic massifs and exposed mantle rocks
An ophiolitic massif is a section of the Earth's oceanic crust and the underlying upper mantle that has been uplifted and exposed above sea level, often within a continental crust setting. Some authors refer to ophiolites as "ancient seafloors." The ophiolites are often fully or partially serpentinized ultramafic rocks. Natural hydrogen seepages have been reported in association with ophiolite massifs all over the world (Figure 4.4). Some well-documented ones are the Samail ophiolite in the Sultanate of Oman [52–54], New Caledonian ophiolite [55], Zambales ophiolite in the Philippines [56], Turkey [57, 58], Dinarides ophiolite in Balkans [24, 59, 60], and the Lonzo ophiolite in the Western Italian Alps [31]. The composition of these ophiolites can be lherzolite, harzburgite, dunite, or troctolite [61, 62], and the Fe-rich minerals that commonly present in these lithologies are olivines, orthopyroxenes and clinopyroxene. Similar rocks can be found also in the exposed mantle rocks in paleo-rift basins involved in later compression such as in the Pyrenees, France [63–65]. Both in the ophiolite massifs and in the near-surface mantle rocks, one of the most possible hydrogen generation reactions is the serpentinization of olivine and pyroxene [1, 66–70], as given by the equation below:

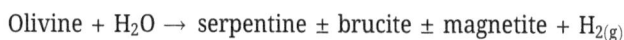

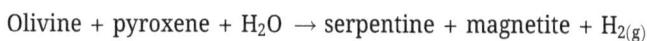

$$\text{Olivine} + H_2O \rightarrow \text{serpentine} \pm \text{brucite} \pm \text{magnetite} + H_{2(g)}$$

$$\text{Olivine} + \text{pyroxene} + H_2O \rightarrow \text{serpentine} + \text{magnetite} + H_{2(g)}$$

Although ophiolites cover relatively small areas compared to cratons, they hold significant potential for serpentinization and, consequently, hydrogen generation.

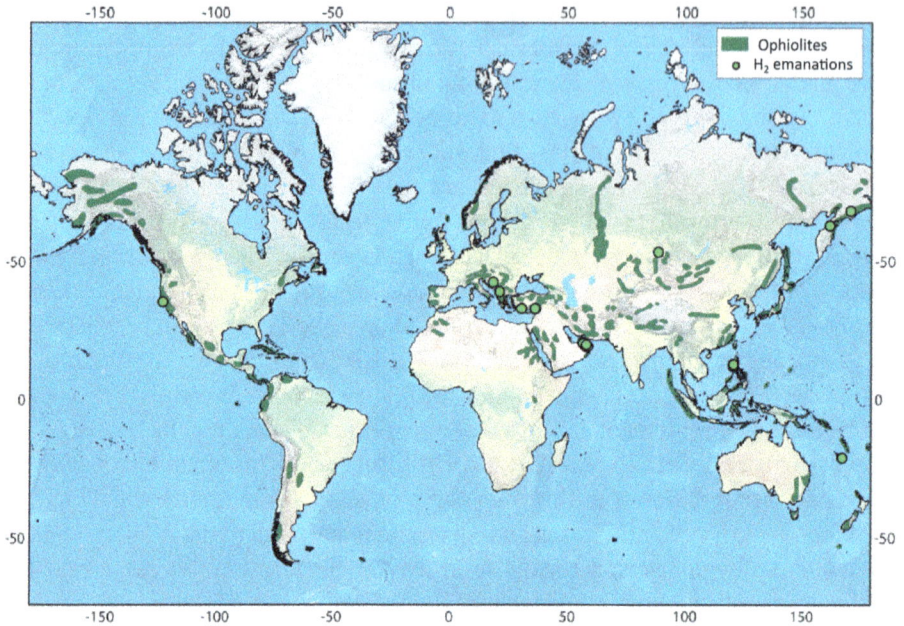

Figure 4.4: Spatial distribution of ophiolite massifs [24].

(ii) Granites
Peralkaline granites have been documented as a source of natural H_2 [71] in Strange Lake in Canada [72, 73], Lovozero and Khibiny in Russia [74, 75], and Ilımaussaq in Greenland [76]. The Fe-rich phases reported in peralkaline granites are arfvedsonite (Na-amphibole) and biotite mica, and their alterations reported to date are given below [77–80]:

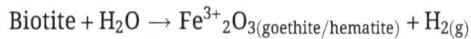

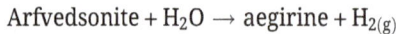

$$\text{Biotite} + H_2O \rightarrow Fe^{3+}{}_2O_{3(\text{goethite/hematite})} + H_{2(g)}$$

$$\text{Biotite} + H_2O \rightarrow Fe_3O_{4(\text{magnetite})} + H_{2(g)}$$

$$\text{Arfvedsonite} + H_2O \rightarrow \text{aegirine} + H_{2(g)}$$

(iii) Fe-rich sedimentary formations and BIFs
Natural hydrogen emissions in the sedimentary formation have been documented in Bourakebougou in Mali [81], Namibia [82], in the Sao Fransisco Basin in Brazil [28, 83] and in the Solimões basin, also in Brazil [84]. The origin of natural hydrogen in the sedimentary sequence in Mali is attributed to many different processes, but oxidation of Fe-rich minerals is considered to be the main contribution [85]. The Fe-rich phases reported in the sedimentary formations are siderite and hematite.

Siderite alteration [86, 87]:

$$3FeCO_{3(siderite)} + H_2O \rightarrow Fe_3O_{4(magnetite)} + 3CO_{2(g)} + H_{2(g)}$$

Magnetite alteration [82, 88]:

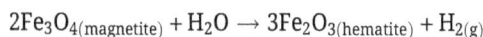

$$2Fe_3O_{4(magnetite)} + H_2O \rightarrow 3Fe_2O_{3(hematite)} + H_{2(g)}$$

Natural hydrogen emissions have been reported in association with BIF in Australia [27, 89, 90], Namibia [29], and Brazil [81, 91]. The spatial distribution of BIFs is given in Figure 4.5. These hydrogen seepages are often spotted as subcircular depressions or "fairy circles," identified from the aerial photographs [92]. The Fe-rich phases reported in BIF are riebeckite (Na-amphibole), minnesotite (talc group mineral), and magnetite [90].

Alterations in BIFs [90, 92]:

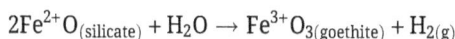

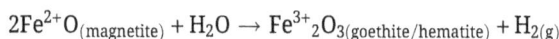

$$2Fe^{2+}O_{(silicate)} + H_2O \rightarrow Fe^{3+}O_{3(goethite)} + H_{2(g)}$$

$$2Fe^{2+}O_{(magnetite)} + H_2O \rightarrow Fe^{3+}{}_2O_{3(goethite/hematite)} + H_{2(g)}$$

(iv) Diorite
The studies by Combaudon et al. [93] and Kularatne et al. [51] are the first to report hydrogen emissions associated with a diorite in the Precambrian basement of Kansas, USA. The Fe-rich phases present in the studied gabbro are fayalite, pyroxene, and magnetite [51, 93]:

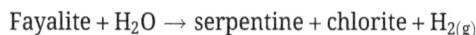

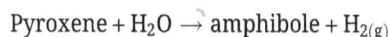

$$Fayalite + H_2O \rightarrow serpentine + chlorite + H_{2(g)}$$

$$Pyroxene + H_2O \rightarrow amphibole + H_{2(g)}$$

(v) Unspecified Fe-rich crustal rocks:
Hydrogen (H_2) generation and iron reduction are associated with the formation of pyrite in Fe-rich cratonic rocks within the upper 20 km of the crust, which is called "pyritization." Thermodynamic models show that as cratonic areas cool over geological time (e.g., a decrease of about 0.1 °C per million years), sulfide species react with iron oxides (such as hematite and magnetite) to form pyrite. This involves the reduction of ferric iron (Fe^{3+}) to ferrous iron (Fe^{2+}) and the oxidation of sulfur, producing H_2 as a byproduct, via several pathways [88]:

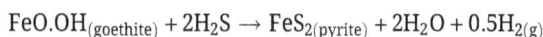

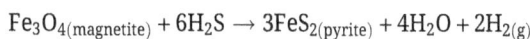

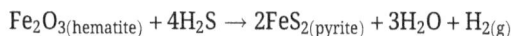

$$FeO.OH_{(goethite)} + 2H_2S \rightarrow FeS_{2(pyrite)} + 2H_2O + 0.5H_{2(g)}$$

$$Fe_3O_{4(magnetite)} + 6H_2S \rightarrow 3FeS_{2(pyrite)} + 4H_2O + 2H_{2(g)}$$

$$Fe_2O_{3(hematite)} + 4H_2S \rightarrow 2FeS_{2(pyrite)} + 3H_2O + H_{2(g)}$$

H_2 generated by this process is reportedly smaller (approximately 0.0002 m^3 of $H_2/m^2/$ year) in typical cratonic rocks compared to the fluxes from other studied hydrogen systems (~0.02–0.04 m^3 of $H_2/m^2/$year) [88]. However, in massive iron oxide formations like BIFs, H_2 fluxes could reach similar magnitudes. Thus, while craton cooling and pyrite formation contribute to H_2 production, they are not the primary mechanisms of geological hydrogen generation. Nonetheless, these processes are crucial for iron reduction and the geochemical cycling of iron in both the continental crust and the oceanic upper mantle.

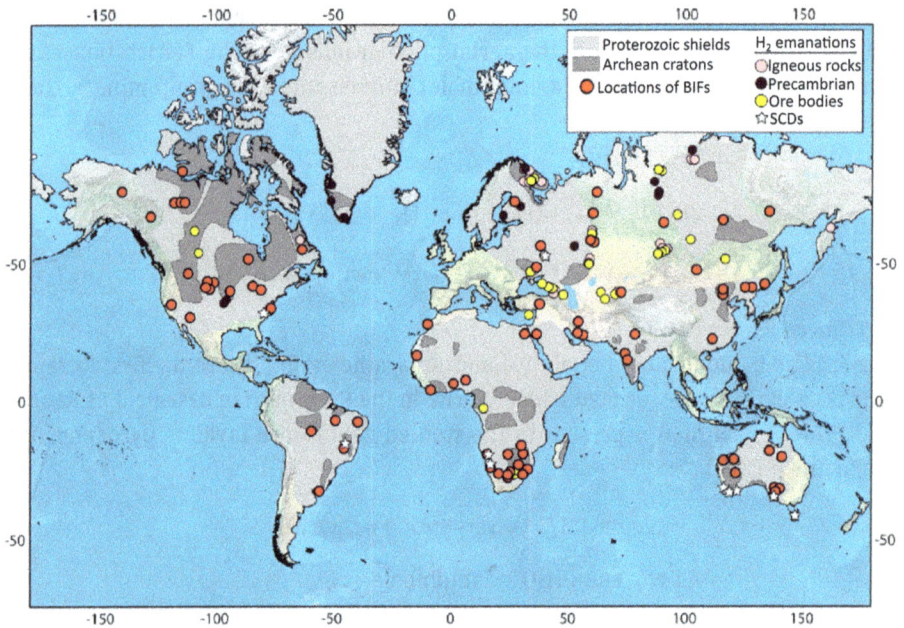

Figure 4.5: Spatial distribution of BIFs and hydrogen emanations associated with igneous rocks, Archean cratons, Precambrian shields, ore bodies and fairy circles [24].

4.2.4 Primordial hydrogen

Primordial hydrogen refers to molecular hydrogen that has been trapped within the Earth since its accretion and is sometimes referred to as "fossil hydrogen," "mantle degassing," or "deep-seated hydrogen" by various authors. But where exactly is this hydrogen stored within the Earth? Indirect evidence, such as anomalies in electrical conductivity, suggests the presence of hydrogen in the Earth's lower mantle, near the core-mantle boundary [94]. Additionally, density models of the Earth's core propose that it may contain approximately 1% hydrogen [95, 96]. This highlights the deep origins of primordial hydrogen and its potential role in Earth's geodynamic and geochemical processes. We propose that primordial hydrogen may be regarded as a form of "deep-seated hydrogen"; however, we emphasize that "deep-seated hydrogen" can also originate from other processes or sources such as subduction-related volcanism [97] (please see Section 4.2.5 for further information).

Primordial hydrogen, formed during the Earth's accretion, should theoretically have the same δD isotopic composition as that of the solar system. It has been reported in the literature that the primordial hydrogen is depleted with deuterium and the typical compositions are (δD) −60% to −70% [98]. Over billions of years, tectonic processes could have moved portions of this hydrogen closer to the crust, allowing for gradual release in certain geological settings. However, this does not explain the presence of primordial hydrogen in the cratonic areas. One can argue that the potential migration of primordial hydrogen in cratonic regions, which are the stable portions of the continental crust, is unlikely. Because cratons are largely undisturbed by recent tectonic activity, any hydrogen released in these areas could indicate the presence of ancient reservoirs that have remained isolated since the early stages of Earth's development. Vidavskiy and Rezae [99] describe that the hydrogen migration through the lithosphere to the surface is indeed influenced by various geological complexities, including sedimentary overburden, tectonic dislocations, folding, intrusions, and metamorphism, which create significant obstacles. Despite these challenges, hydrogen's volatility and its ability to penetrate matter as "proton gas," often accompanied by reactive silanes, enable its unique degassing system. Factors such as the depth of the solid basement, petrophysical properties of overburden, tectonic activity, and thermal-chemical interactions govern the dispersion or concentration of hydrogen. In favorable conditions, hydrogen flows remain concentrated near the surface, while in more complex scenarios, the flow dissipates due to structural heterogeneity and deeper basement depths. Some studies have suggested a primordial origin for hydrogen found in deep wells, whereas some believe that the SCD's found on the Earth's surface are considered clear evidence of hydrogen gas emanating from deeper geospheres [99, 100].

The deepest human-made borehole on the Earth is the Kola superdeep well in the Kola Peninsula, near the Russian border with Norway. The true vertical depth of this well is 12,262 m (12 km), and the literature reports that the drilling mud came out was "boiling" with an unexpected level of hydrogen [101, 102]. The well was drilled into metamorphosed granite and Archean rocks. However, the origin of hydrogen has not been confirmed to be primordial or any other source.

Another study suggesting a deep-seated origin is by McCarthy et al. [103], which reported hydrogen detected in the soil near a petroleum exploration well in Kansas, USA. The well, drilled to depths of 640–670 m, contained a gas mixture comprising 50% H_2, 50% N_2, and less than 1% hydrocarbons. However, the study did not conclusively establish the origin of the hydrogen detected in the soil, which was used as a proxy for the subsurface hydrogen in the wells. Similarly, the study by Coveney [19] suggests a multiple origin of hydrogen in Kansas wells, including a deep-seated origin, but not specifically to be primordial, or of mantle origin, and serpentinization.

Deep-seated origin has been also suggested from multiple places in Russia [33]. Primordial hydrogen is, however, not a target of current exploration efforts.

4.2.5 Volcanic degassing

Volcanic degassing is the emission of hydrogen during or immediately after a volcanic eruption [16]. Some authors refer to this as "magmatic hydrogen" [16]. Volcanic degassing emits hydrogen of multiple origins:

(i) Molecular hydrogen from magma underneath: Emission of molecular hydrogen has been reported from many volcanoes [104–107]. For example, a study conducted at Stromboli (Italy) during 2009–2010 analyzed volcanic gases from craters using both discontinuous dry gas collection and continuous monitoring, and reported emanation of H_2 and CO_2 [107]. They reported that the emitted hydrogen has a similar composition to the hydrogen observed within magma underneath and the quantity of hydrogen measured was closely correlated with the atmospheric pressure.

(ii) Hydrogen generated via hydration of mantle wedge at the oceanic–continental subduction zones: As the oceanic plate is subducted, it undergoes dehydration reactions. These reactions release water that has been stored within the oceanic crust and sediments in the form of hydrous minerals. The released water migrates into the overlying mantle wedge, where it reacts with the peridotite rock to hydrate the mantle, forming hydrous minerals such as serpentine, chlorite, and talc [97]. This hydration process is accompanied by redox reactions that produce molecular hydrogen under high-temperature and high-pressure conditions. A recent study conducted in the Pampean slab in South America showed that the flat subduction zones are significantly more productive in generating hydrogen than steep subduction zones due to their wider mantle hydration zones [97]. The larger spatial extent of water–rock interactions in flat subduction provides sustained conditions for hydrogen production, whereas the confined hydration zone in steep subduction zones restricts the volume of hydrogen generated. The authors also claim that the deep hydrogen production could be particularly sensitive to the amount of subducted sediments, regardless of whether subduction is flat or steep.

(iii) Hydrogen generated via reactions occurring during the degassing: Several studies report that hydrogen can be generated via the reaction between H_2S and water, during the eruption of volcanos, such as Mount Etna, Hawaii in the USA, Costa Rica, and White Island volcano in New Zealand [16]. The most common feature observed in these volcanoes is the strong correlation between SO_2 and H_2 concentrations, measured during the degassing. Further, they report that the redox conditions and the H_2/H_2O ratios are governed by the SO_2-H_2S equilibrium prevailing during the degassing. In these conditions, hydrogen is generated via the reaction between H_2S and water as given below:

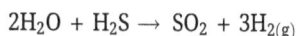

$$2H_2O + H_2S \rightarrow SO_2 + 3H_{2(g)}$$

The generation of hydrogen by this process is controlled by the pressure. Ascending magma has a high pressure and during the eruption into the atmosphere, the pressure suddenly drops to the atmospheric pressure, causing sulfur to partition more strongly into the gaseous phase than dissolved Magma, which shifts the equilibrium toward the right-hand side, facilitating the forward reaction, generating hydrogen.

Volcanic hydrogen or magmatic hydrogen is not considered to be a focus of the current exploration effort, however, the recent study by Gauthier et al. [97] emphasizes that the flat subduction zones related to volcanism could be promising for future hydrogen exploration.

4.2.6 Mechanoradicles

Mechanoradicals are free radicals formed when mechanical forces, such as shear stress, grinding, or fracturing, break chemical bonds in a material, leaving unpaired electrons or charged surface species. In geological contexts, mechanoradicals are commonly generated during the comminution of silicate rocks [108, 109]. The mechanical forces involved in fault activity can break Si–O covalent bonds in silicate minerals in two ways:

(i) Homolysis: which creates surface radicals such as $\equiv Si\bullet$ and $\equiv SiO\bullet$

$$Si-O \rightarrow \, \equiv Si\bullet + \, \equiv SiO\bullet$$

(ii) Heterolysis: which creates charged radicals such as $\equiv Si^+$, H^+, and $-O-Si\equiv$ [16, 109]

$$Si-O \rightarrow \, \equiv Si^+ + \, ^-O-Si \equiv$$

These surface radicals either recombine to form siloxane bonds (Si-O-Si) or react with water [16]. The latter leads to formation of molecular hydrogen, as shown below:

$$\equiv Si\bullet + H_2O \rightarrow SiOH + H\bullet$$

$$H\bullet + H\bullet \rightarrow H_{2(g)}$$

Mechanoradical hydrogen generation occurs wherever silicate-bearing rocks are mechanically fragmented into smaller particles. This process is particularly common in active fault zones associated with orogenic belts, subduction zones, continental rifts, and transform boundaries. Beyond tectonic activity, rock comminution beneath glaciers also generates H_2. Additional processes that contribute to H_2 generation through rock fragmentation include frost and salt wedging, thermal contraction or expansion, abrasion by rivers or wind, gravitational impacts such as landslides or rockfalls, meteorite impacts, and reaction-driven fracturing.

Laboratory experiments show that hydrogen generation increases with frictional work (i.e., earthquake magnitude) and that a hydrogen concentration of more than 1.1 mol/kg of fluid can be achieved in a fault zone after earthquakes of even small magnitudes [108].

Overall, in the scope of current hydrogen exploration, hydrogen generated by mechanoradicles is not a target. A publication however suggested this source for a structural trap on fault in the Jura mountain [110].

4.3 Concluding remarks

This chapter has explored the diverse pathways responsible for abiotic onshore hydrogen generation, highlighting the geological and geochemical processes that give rise to this promising resource. A recent estimation by the USGS estimates that the resources may cover thousands of years of the world consumption [111]. Before this synthesis, an estimation of about 23 MT/year has been proposed both by Zgonnik [33] and Worman et al. [112], but they only take into account the serpentinization and the radiolysis. Horsfield et al. [49] also arrived at huge numbers just for the Sangliao basin in China, coming only from the late maturation of the organic matter. And Hirose et al. [108] also proposed millions of tons/year from the faults. The H_2 that may come from the sedimentary and intrusive iron-rich rocks have not yet been estimated, but taking into account the large part of the continental crust that contains Archean and Neoproterozoic rocks [24], it is definitively larger than the ophiolites resources. These numbers will surely be firmed up in the coming years, but at the beginning of 2025, we can consider that there is no longer any question about natural H_2 resources. The figures show that the Earth, like all the other planets and stars, contains and emits an enormous amount of H_2. The question is which one of these reserves are economic reserves? Are we capable of recovering this flow of H_2 that comes out of the Earth? Today, except for Mali, the only numbers are from the one form of Gold Hydrogen in South Australia, based on the two wells (Ramsay 1 and 2) already drilled, where they announced a best estimate of 1.3 Mt of H_2 (high estimate 8.8 Mt) with a high content in Helium (41 billion standard cubic feet (bscf) as the best estimate). These high values confirm the economic interest of the structure and companies are currently doing the delineation.

Among the six pathways identified, radiolysis of water, organic maturation, and redox reactions between Fe^{2+} and water emerge as key mechanisms for prospecting onshore natural hydrogen reservoirs. The largest hydrogen reservoirs that have been reported to date have been hosted by formations with a high concentration of radioactive elements, high quantities of organic matter, and high quantities of iron (Fe^{2+}), which favor the three processes mentioned above. Therefore, future natural H_2 exploration must also be focused on these formations. In contrast, H_2 originated from primordial sources, volcanic degassing, and mechanoradicals do not seem to contribute to large natural H_2 reservoirs, therefore they are not directly relevant to current exploration efforts. However, considering the large heterogeneities in the natural systems, it is highly likely that the natural hydrogen discovered in the reservoirs originates from a combination of multiple sources [113].

Similar to petroleum systems, a hydrogen system encompasses the processes of generation, migration, and accumulation of hydrogen within geological settings. However, as demonstrated in this chapter, hydrogen generation systems exhibit far greater variability and complexity compared to their petroleum counterparts. This complexity arises from the diverse pathways of hydrogen production and the wide range of geological conditions under which they occur, underscoring the critical importance of detailed system characterization. Future research must address key knowledge gaps, particularly in delineating hydrogen reservoirs, improving exploration technologies, and estimating recoverable quantities. At this stage, we cannot rule out the possibility that it is the limits of our understanding that lead to this apparent complexity and variability.

Unlike petroleum generation, which may unfold over millions of years through gradual organic transformation, hydrogen generation can often be considered instantaneous, driven by dynamic and ongoing geological and geochemical processes. This immediacy introduces significant uncertainties regarding the rate, extent, and flux of hydrogen regeneration, posing challenges to both exploration and long-term resource management. Especially, Section 4.2.3 describes the various pathways of the redox reaction between Fe^{2+} and water, showing equations. Currently, the correct kinetics of these reactions are not well known. The reaction kinetics plays a key role in understanding the regeneration rate of hydrogen. Addressing these issues, together with understanding the geological settings, thermodynamic conditions, and geochemical markers associated with each pathway, is critical to unlocking this resource's full potential. Collaboration between geoscientists, energy companies, and policymakers will be essential to responsibly harness this energy source, while minimizing environmental impact.

References

[1]   McCollom T. M., Klein F., Ramba M. Hydrogen generation from serpentinization of iron-rich olivine on Mars, icy moons, and other planetary bodies. Icarus. 2022, 372.
[2]   Moody J. B. Serpentinization: A review. Lithos. 1976, 9: 125–138.
[3]   Preiner M., et al. Serpentinization: Connecting geochemistry, ancient metabolism and industrial hydrogenation. Life. 2018, 8.
[4]   Mével C. Serpentinisation des péridotites abysales aux dorsales océaniques. Comptes Rendus – Geoscience. 2003, 335: 825–852.

[5] Proskurowski G., et al. Abiogenic hydrocarbon production at lost city hydrothermal field. Science (80-.). 2008, 319: 604–607.

[6] Klein F., et al. Iron partitioning and hydrogen generation during serpentinization of abyssal peridotites from 15°N on the Mid-Atlantic Ridge. Geochimica et Cosmochimica Acta. 2009, 73: 6868–6893.

[7] Kelley D. S., et al. A serpentinite-hosted ecosystem: The Lost City hydrothermal field. Science (80-.). 2005, 307: 1428–1434.

[8] Charlou J. L., Donval J. P., Fouquet Y., Jean-Baptiste P., Holm N. Geochemistry of high H2 and CH4 vent fluids issuing from ultramafic rocks at the Rainbow hydrothermal field (36°14′N,MAR). Chemical Geology. 2002, 191: 345–359.

[9] Seyfried W. E., Pester N. J., Ding K., Rough M. Vent fluid chemistry of the Rainbow hydrothermal system (36°N, MAR): Phase equilibria and in situ pH controls on subseafloor alteration processes. Geochimica et Cosmochimica Acta. 2011, 75: 1574–1593.

[10] Früh-Green G. L., Orcutt B. N., Green S. Expedition 357 scientific prospectus: Atlantis massif serpentinization and life. International Ocean Discovery Program. 2015.

[11] Francheteau J., et al. 1 Ma East Pacific Rise oceanic crust and uppermost mantle exposed by rifting in Hess Deep (equatorial Pacific Ocean). Earth and Planetary Science Letters. 1990, 101: 281–295.

[12] Konn C., et al. Extending the dataset of fluid geochemistry of the Menez Gwen, Lucky Strike, Rainbow, TAG and Snake Pit hydrothermal vent fields: Investigation of temporal stability and organic contribution. Deep Sea Research Part I: Oceanographic Research Papers. 2022, 179: 103630.

[13] Buatier M. D., Früh-Green G. L., Karpoff A. M. Mechanisms of Mg-phyllosilicate formation in a hydrothermal system at a sedimented ridge (Middle Valley, Juan de Fuca). Contributions to Mineralogy and Petrology. 1995, 122: 134–151.

[14] Russell M. J., Hall A. J., Martin W. Serpentinization as a source of energy at the origin of life. Geobiology. 2010, 8: 355–371.

[15] Plümper O., et al. Subduction zone forearc serpentinites as incubators for deep microbial life. Proceedings of the National Academy of Sciences of the United States of America. 2017, 114: 4324–4329.

[16] Klein F., Tarnas J. D., Bach W. Abiotic sources of molecular hydrogen on earth. Elements. 2020, 16: 19–24.

[17] Diallo A., Cissé C. S. T., Lemay J., Brière D. J. La découverte de l'hydrogène naturel par Hydroma, un « Game Changer » pour la transition énergétique. Annales des Mines – Réalités Industrielles. 2022, Novembre 2: 154–160.

[18] Goebel E. D., Coveney R. M., Angino E. E., Zeller E. J., Dreschhoff G. A. Geology, composition, isotopes of naturally occurring H2/N2 rich gas from wells near Junction city, Kansas. Oil and Gas Journal. 1984, 82(19): 215–222.

[19] Coveney R. M., Goebel E. D., Zeller E. J., Dreschhoff G. A. M., Angino E. E. Serpentinization and the origin of hydrogen gas in Kansas (USA). American Association of Petroleum Geologists Bulletin. 1987, 71.

[20] Johnsgard S. K. The fracture pattern of north-central Kansas and its relation to hydrogen soil gas anomalies over the midcontinent rift system. University of Kansas, Geology, 1988.

[21] Angino E. E., Zeller E. J., Dreschhoff G. A. M., Goebel E. D., Coveney R. M. Jr Spatial distribution of hydrogen in soil gas in central Kansas, USA. Geochemistry of Gaseous Elements and Compounds. 1990, 485–493.

[22] Newell K. D., et al. H2-rich and hydrocarbon gas recovered in a deep Precambrian well in Northeastern Kansas. Natural Resources Research. 2007, 16. https://doi.org/10.1007/s11053-007-9052-7

[23] Gaucher E. C. New perspectives in the industrial exploration for native hydrogen. Elements. 2020, 16: 8–9.

[24] Lévy D., et al. Natural H2 exploration: Tools and workflows to characterize a play. Science and Technology for Energy Transition. 2023, 78.

[25] Moretti I., Baby P., Alvarez Zapata P., Mendoza R. V. Subduction and hydrogen release: The case of Bolivian Altiplano. Geoscience. 2023, 13.

[26] Etiope G., Schoell M., Hosgörmez H. Abiotic methane flux from the Chimaera seep and Tekirova ophiolites (Turkey): Understanding gas exhalation from low temperature serpentinization and implications for Mars. Earth and Planetary Science Letters. 2011, 310: 96–104.

[27] Frery E., Langhi L., Maison M., Moretti I. Natural hydrogen seeps identified in the North Perth Basin, Western Australia. International Journal of Hydrogen Energy. 2021, 46.

[28] Prinzhofer A., et al. Natural hydrogen continuous emission from sedimentary basins: The example of a Brazilian H 2 -emitting structure. International Journal of Hydrogen Energy. 2019, 44: 5676–5685.

[29] Moretti I., Geymond U., Pasquet G., Aimar L., Rabaute A. Natural hydrogen emanations in Namibia: Field acquisition and vegetation indexes from multispectral satellite image analysis. International Journal of Hydrogen Energy. 2022, 47: 35588–35607. doi: 10.1016/j.ijhydene.2022.08.135

[30] Hall D. L., Bodnar R. J. Methane in fluid inclusions from granulites: A product of hydrogen diffusion? Geochimica et Cosmochimica Acta. 1990, 54: 641–651.

[31] Vitale Brovarone A., et al. Massive production of abiotic methane during subduction evidenced in metamorphosed ophicarbonates from the Italian Alps. Nature Communications. 2017, 8: 1–13.

[32] Chavagnac V., Monnin C., Ceuleneer G., Boulart C., Hoareau G. Characterization of hyperalkaline fluids produced by low-temperature serpentinization of mantle peridotites in the Oman and Ligurian ophiolites. Geochemistry, Geophysics, Geosystems. 2013, 14: 2496–2522.

[33] Zgonnik V. The occurrence and geoscience of natural hydrogen: A comprehensive review. Earth-Science Reviews. 2020, 203: 103140.

[34] Gupta S., Fernandes A., Lopes A., Grasa L. Microbes and parameters influencing dark fermentation for hydrogen production. 2024, 1–26.

[35] Reeves K. G., Kanai Y. Electronic excitation dynamics in liquid water under proton irradiation. Scientific Reports. 2017, 7: 1–8.

[36] Lollar B. S., Tullis C. O., Lacrampe-couloume G., Ballentine C. The contribution of the Precambrian continental lithosphere to global H2 production. Nature. 2014, 516: 379–382.

[37] LaVerne J. A., Tandon L. H 2 and Cl 2 production in the radiolysis of calcium and magnesium chlorides and hydroxides. Journal of Physical Chemistry A. 2005, 109: 2861–2865.

[38] Lin L. H., Slater G. F., Sherwood Lollar B., Lacrampe-Couloume G., Onstott T. C. The yield and isotopic composition of radiolytic H2, a potential energy source for the deep subsurface biosphere. Geochimica et Cosmochimica Acta. 2005, 69: 893–903.

[39] Wang W., et al. Radioactive genesis of hydrogen gas under geological conditions: An experimental study. Acta Geologica Sinica (English Edition). 2019, 93: 1125–1134.

[40] Bruneton P., Cuney M. Geology of uranium deposits. In: Uranium for nuclear power. Woodhead Publishing, 2016, 11–52.

[41] Richard A. Radiolytic (H2, O2) and other trace gases (CO2, CH4, C2H6, N2) in fluid inclusions from unconformity-related U deposits. Procedia Earth and Planetary Science. 2017, 17: 273–276.

[42] Chi G., Haid T., Quirt D., Fayek M., Blamey N., Chu H. Petrography, fluid inclusion analysis, and geochronology of the End uranium deposit, Kiggavik, Nunavut, Canada. Mineralium Deposita. 2017, 52: 211–232.

[43] Dubessy J., et al. Radiolysis evidenced by H2-O2 and H2-bearing fluid inclusions in three uranium deposits. Geochimica et Cosmochimica Acta. 1988, 52: 1155–1167.

[44] Durand B. Kerogen: Insoluble organic matter from sedimentary rocks. Editions Technip, 1980.

[45] Tissot B. P., Welte D. H. Petroleum formation and occurrence. Springer Science & Business Media, 2013.

[46] Moretti I., Bouton N., Ammouial J., Carrillo Ramirez A. The H2 potential of the Colombian coals in natural conditions. International Journal of Hydrogen Energy. 2024, 77: 1443–1456.

[47] Horsfield B., Nelskamp S., Sachse V., Sośnicka M., Boreham C., Mahlstedt N., Hartwig A. Natural H2 from sedimentary organic matter: Is It All It is Cracked up to be? 85th EAGE Annual Conference & Exhibition. 2024, 2024(1): 1–5. European Association of Geoscientists & Engineers.

[48] Boreham C. J., et al. Modelling of hydrogen gas generation from overmature organic matter in the Cooper Basin, Australia. APPEA J. 2023, 63: S351–S356.

[49] Horsfield B., et al. Molecular hydrogen from organic sources in the deep Songliao Basin, P.R. China. International Journal of Hydrogen Energy. 2022, 47: 16750–16774.

[50] Li X., Krooss B. M., Weniger P., Littke R. Liberation of molecular hydrogen (H2) and methane (CH4) during non-isothermal pyrolysis of shales and coals: Systematics and quantification. International Journal of Coal Geology. 2015, 137: 152–164.

[51] Kularatne K., et al. X-ray micro-computed tomography-based approach to estimate the upper limit of natural H2 generation by Fe2+ oxidation in the intracratonic lithologies. International Journal of Hydrogen Energy. 2024, 78: 861–870.

[52] Neal C., Stanger G. Hydrogen generation from mantle source rocks in Oman. Earth and Planetary Science Letters. 1983, 66.

[53] Ellison E. T., et al. Low-temperature hydrogen formation during aqueous alteration of serpentinized peridotite in the Samail Ophiolite. Journal of Geophysical Research: Solid Earth. 2021, 126.

[54] Sano Y., Urabe A., Wakita H., Wushiki H. Origin of hydrogen-nitrogen gas seeps, Oman. Applied Geochemistry. 1993, 8.

[55] Deville E., Prinzhofer A. The origin of N2-H2-CH4-rich natural gas seepages in ophiolitic context: A major and noble gases study of fluid seepages in New Caledonia. Chemical Geology. 2016, 440.

[56] Abrajano T. A., et al. Methane-hydrogen gas seeps, Zambales Ophiolite, Philippines: Deep or shallow origin? Chemical Geology. 1988, 71.

[57] Hosgörmez H. Origin of the natural gas seep of Çirali (Chimera), Turkey: Site of the first Olympic fire. Journal of Asian Earth Sciences. 2007, 30.

[58] Hosgormez H., Etiope G., Yalçin M. N. New evidence for a mixed inorganic and organic origin of the Olympic Chimaera fire (Turkey): A large onshore seepage of abiogenic gas. Geofluids. 2008, 8.

[59] Etiope G., et al. Methane and hydrogen in hyperalkaline groundwaters of the serpentinized Dinaride ophiolite belt, Bosnia and Herzegovina. Applied Geochemistry. 2017, 84.

[60] Maravelis A. G., Koukounya A., Tserolas P., Pasadakis N., Zelilidis A. Geochemistry of Upper Miocene-Lower Pliocene source rocks in the Hellenic Fold and Thrust Belt, Zakynthos Island, Ionian Sea, western Greece. Marine and Petroleum Geology. 2015, 66.

[61] Arai S., Kadoshima K., Morishita T., Arai S., Kadoshima K., Morishita T. Widespread arc-related melting in the mantle section of the northern Oman ophiolite as inferred from detrital chromian spinels. Journal of the Geological Society London. 2006, 163: 869–879.

[62] Ulrich M., et al. Multiple melting stages and refertilization as indicators for ridge to subduction formation: The New Caledonia ophiolite. Lithos. 2010, 115.

[63] Lefeuvre N., et al. Native H2 exploration in the Western Pyrenean Foothills. Geochemistry, Geophysics, Geosystems. 2021, 22: 1–20.

[64] Loiseau K., Aubourg C., Petit V., Bordes S., Lefeuvre N., Thomas E., . . . Moretti I. Hydrogen generation and heterogeneity of the serpentinization process at all scales: Turon de Técouère lherzolite case study, Pyrenees (France). Geoenergy. 2024, 2(1).

[65] Lefeuvre N., et al. Natural hydrogen migration along thrust faults in foothill basins: The North Pyrenean Frontal Thrust case study. Applied Geochemistry. 2022, 145.

[66] McCollom T. M., Seewald J. S. Serpentinites, hydrogen, and life. Elements. 2013, 9: 129–134.

[67] McCollom T. M., Bach W. Thermodynamic constraints on hydrogen generation during serpentinization of ultramafic rocks. Geochimica et Cosmochimica Acta. 2009, 73: 856–875.

[68] McCollom T. M., Klein F., Moskowitz B., Solheid P. Experimental serpentinization of iron-rich olivine (hortonolite): Implications for hydrogen generation and secondary mineralization on Mars and icy moons. Geochimica et Cosmochimica Acta. 2022, 335.

[69] Klein F., et al. Fluids in the Crust. Experimental constraints on fluid-rock reactions during incipient serpentinization of harzburgite. American Mineralogist. 2015, 100: 991–1002.

[70] Dugamin E., Truche L., Donzé F. V. Natural hydrogen exploration guide. Geonum. 2019, 1(16).

[71] Marks M. A. W., Markl G. A global review on agpaitic rocks. Earth-Science Reviews. 2017, 173. https://doi.org/10.1016/j.earscirev.2017.06.002

[72] Salvi S., Williams-Jones A. E. Reduced orthomagmatic C-O-H-N-NaCl fluids in the Strange Lake rare-metal granitic complex, Quebec/Labrador, Canada. European Journal of Mineralogy. 1992, 4.

[73] Salvi S., Williams-Jones A. E. Alteration, HFSE mineralisation and hydrocarbon formation in peralkaline igneous systems: Insights from the Strange Lake Pluton, Canada. Lithos. 2006, 91.

[74] Potter J., Salvi S., Longstaffe F. J. Abiogenic hydrocarbon isotopic signatures in granitic rocks: Identifying pathways of formation. Lithos. 2013, 182–183.

[75] Nivin V. A. Occurrence forms, composition, distribution, origin and potential hazard of natural hydrogen–hydrocarbon gases in ore deposits of the Khibiny and Lovozero massifs: A review. Minerals. 2019, 9. https://doi.org/10.3390/min9090535

[76] Krumrei T. V., Pernicka E., Kaliwoda M., Markl G. Volatiles in a peralkaline system: Abiogenic hydrocarbons and F-Cl-Br systematics in the naujaite of the Ilímaussaq intrusion, South Greenland. Lithos. 2007, 95.

[77] Bernard C., Estrade G., Salvi S., Béziat D., Smith M. Alkali pyroxenes and amphiboles: A window on rare earth elements and other high field strength elements behavior through the magmatic-hydrothermal transition of peralkaline granitic systems. Contributions to Mineralogy and Petrology. 2020, 175.

[78] Bird S. Niches of the pre-photosynthetic biosphere and geologic preservation of Earth's earliest ecology. Geobiology. 2007, 5: 101–117.

[79] Murray J., et al. Abiotic hydrogen generation from biotite-rich granite: A case study of the Soultz-sous-Forêts geothermal site, France. Applied Geochemistry. 2020, 119.

[80] Truche L., et al. Hydrogen generation during hydrothermal alteration of peralkaline granite. Geochimica et Cosmochimica Acta. 2021, 308: 42–59.

[81] Prinzhofer A., Tahara Cissé C. S., Diallo A. B. Discovery of a large accumulation of natural hydrogen in Bourakebougou (Mali). International Journal of Hydrogen Energy. 2018, 43.

[82] Roche V., et al. A new continental hydrogen play in Damara Belt (Namibia). Scientific Reports. 2024, 14: 1–11.

[83] Azor de Freitas V., Prinzhofer A., Ba Francolin J., Fonseca Ferreira F. J., Moretti I. Natural hydrogen system evaluation in the São Francisco Basin (Brazil). Submitted To STET. 2024, 95.

[84] Milesi V., Prinzhofer A., Guyot F., Benedetti M., Rodrigues R. Contribution of siderite-water interaction for the unconventional generation of hydrocarbon gases in the Solimões basin, north-west Brazil. Marine and Petroleum Geology. 2016, 71.

[85] Maiga O., Deville E., Laval J., Prinzhofer A., Diallo A. B. Characterization of the spontaneously recharging natural hydrogen reservoirs of Bourakebougou in Mali. Scientific Reports. 2023, 1–13. doi: 10.1038/s41598-023-38977-y

[86] Milesi V., et al. Unconventional generation of hydrocarbons in petroleum basin: The role of siderite/water interface A new calibration of the carbonate clumped isotope thermometer based on synthetic calcites. Mineralogical Magazine. 2013, 77: 1661–1817.

[87] Milesi V., et al. Formation of CO_2, H_2 and condensed carbon from siderite dissolution in the 200–300 °C range and at 50MPa. Geochimica et Cosmochimica Acta. 2015, 154: 201–211.

[88] Arrouvel C., Prinzhofer A. Genesis of natural hydrogen: New insights from thermodynamic simulations. International Journal of Hydrogen Energy. 2021, 46.

[89] Boreham C. J., et al. Hydrogen in Australian natural gas: Occurrences, sources and resources. APPEA J. 2021, 61.

[90] Geymond U., Ramanaidou E., Lévy D., Ouaya A., Moretti I. Can weathering of banded iron formations generate natural hydrogen? Evidence from Australia, Brazil and South Africa. Minerals. 2022, 12.

[91] Moretti I., et al. Long-term monitoring of natural hydrogen superficial emissions in a Brazilian cratonic environment. Sporadic large pulses versus daily periodic emissions. International Journal of Hydrogen Energy. 2021, 46.

[92] Geymond U., et al. Reassessing the role of magnetite during natural hydrogen generation. Frontiers in Earth Science. 2023, 11: 1–11.

[93] Combaudon V., et al. Are the Fe-rich-clay veins in the igneous rock of the Kansas (USA) Precambrian crust of magmatic origin? Lithos. 2024, 474–475.

[94] Williams Q., Hemley R. J. Hydrogen in the deep Earth. Annual Review of Earth and Planetary Sciences. 2001, 29: 365–418.

[95] Walshe J. L., Hobbs B., Ord A., Regenauer-Lieb K., Barmicoat A. Mineral systems, hydridic fluids, the Earth's core, mass extinction events and related phenomena. Deposit Research: Meeting the Global Challenge: Proceedings of the Eighth Biennial SGA Meeting Beijing, China. 2005, 65–68. Springer Berlin Heidelberg.

[96] Poirier J. P. Light elements in the Earth's outer core: A critical review. Physics of the Earth and Planetary Interiors. 1994, 85: 319–337.

[97] Gauthier, A., Larvet, T., Le Pourhiet, L., & Moretti, I. (2024). Water budget in flat vs. steep subduction: implication for volcanism and potential for H2 production. BSGF-Earth Sciences Bulletin, 195(1), 26. https://doi.org/10.1051/bsgf/2024026.

[98] Loewen M. W., Graham D. W., Bindeman I. N., Lupton J. E., Garcia M. O. Hydrogen isotopes in high 3 He/ 4 He submarine basalts: Primordial vs. recycled water and the veil of mantle enrichment. Earth and Planetary Science Letters. 2019, 508: 62–73.

[99] Vidavskiy V., Rezaee R. To cite this article: Vitaly Vidavskiy, Reza Rezaee. Natural deep-seated hydrogen resources exploration and development: Structural features, governing factors, and controls. Journal of Energy and Natural Resources. 2022, 11: 60–81.

[100] Larin N., et al. Natural molecular hydrogen seepage associated with surficial, rounded depressions on the European craton in Russia. Natural Resources Research. 2015, 24.

[101] Ikorsky S. V. The investigation of gases during the Kola Superdeep borehole drilling (to 11.6 km depth). Geologisches Jahrbuch. 1999, D 107: 145–152.

[102] Bodén A., Eriksson K. G. Deep drilling in crystalline bedrock: The deep gas drilling in the Siljan impact structure. Springer-Verlag, 1988.

[103] McCarthy J. H. J., Cunningham K. I., Roberts A. A., Dietrich J. A. Soil gas studies around hydrogen-rich natural gas wells in northern Kansas. USGS Open-File Report. 1986, 86–461.

[104] Moussallam Y., et al. Characterisation of the magmatic signature in gas emissions from Turrialba Volcano, Costa Rica. Solid Earth. 2014, 5: 1341–1350.

[105] Aiuppa A., et al. Hydrogen in the gas plume of an open-vent volcano, Mount Etna, Italy. Journal of Geophysical Research: Solid Earth. 2011, 116: 1–8.

[106] Capaccioni B., et al. Hydrocarbons in gas-emissions geothermal from. Geochem. 1993, 27: 7–17.

[107] Di Martino R. M. R., Camarda M., Gurrieri S. Continuous monitoring of hydrogen and carbon dioxide at Stromboli volcano (Aeolian Islands, Italy). Italian Journal of Geosciences. 2021, 140: 1–16.

[108] Hirose T., Kawagucci S., Suzuki K. Mechanoradical H2 generation during simulated faulting: Implications for an earthquake-driven subsurface biosphere. Geophysical Research Letters. 2011, 38: 1–5.

[109] Liang Y., Cui W., Masuda Y., Hirose T., Tsuji T. Identifying General Reaction Conditions for Mechanoradical Natural Hydrogen Production. 2024.

[110] Deronzier J. F., Giouse H. Vaux-en-Bugey (Ain, France): The first gas field produced in France, providing learning lessons for natural hydrogen in the sub-surface? BSGF – Earth Sciences Bulletin. 2020, 191.

[111] Ellis G. S., Gelman S. E. Model predictions of global geologic hydrogen resources. Science Advances. 2024, 10: eado0955.

[112] Worman S. L., Pratson L. F., Karson J. A., Klein E. M. By serpentinization within oceanic lithosphere. 2016, 1–9. doi: 10.1002/2016GL069066.Received

[113] Leila M., Loiseau K., Moretti I. Controls on generation and accumulation of blended gases (CH4/H2/He) in the Neoproterozoic Amadeus Basin, Australia. Marine and Petroleum Geology. 2022, 140: 105643.

Part II: **Natural hydrogen generation mechanisms**

Javier García-Pintado* and Marta Pérez-Gussinyé

Chapter 5
Shallow peridotites at magma-poor rifted margins: occurrences, serpentinization, and H₂ generation

Abstract: Rifted margins are formed by the rifting of continental crust, a process that results in the formation of new oceanic crust. At so-called magma-poor rifted margins, the thinned continental crust is underlain by serpentinized mantle in the distal, deeper parts of the margin. This distal area transitions oceanward to the mantle that was also serpentinized and exhumed on the seafloor at the continent-ocean transition (COT) before steady-state oceanic crustal accretion began. Mantle exposure at the COT occurred over hundreds of kilometers and was accompanied by sedimentation. Serpentinization, which occurs during mantle hydration, is one of the major geological processes by which natural hydrogen is produced. This raises questions about how much hydrogen has been produced in the past, how much is currently accumulated under sedimentary successions, and finally, the current hydration state of the shallow mantle in order to assess its potential to produce additional hydrogen via stimulation. The serpentinization observed on these margins is specific to certain protolith rocks, namely lherzolitic peridotites. In addition, the occurrence of serpentinization depends on the presence of appropriate thermodynamic conditions in terms of pressure and temperature, as well as access to water. The latter requires the embrittlement of the rifted margin crust, which is also dependent on the rheological properties of the mantle and continental rocks, the thermal regime, and the tectonic stresses. Thus, the rates of extension in rifted margins exert a strong control on the onset of magmatism and the serpentinization of the shallow mantle underneath the thinned continental crust and the COT. The exploration of these offshore locations requires the input of several disciplines, including geophysics, geodynamics, and petrology. In this study, we focus on the fundamental thermodynamic processes and the use of geodynamic

Acknowledgments: This work would not have been possible without the financial support of Deutsche Forschungsgemeinschaft (DFG, German Research Foundation) under Germany's Excellence Strategy – EXC-2077 – 390741603, funding the Cluster of Excellence in MARUM: The Ocean Floor – Earth's Uncharted Interface.

*Corresponding author: Javier García-Pintado, MARUM Center for Marine Environmental Sciences, Universität Bremen, Bremen, Germany, e-mail: jgarciapintado@marum.de
Marta Pérez-Gussinyé, MARUM Center for Marine Environmental Sciences, Universität Bremen, Bremen, Germany

https://doi.org/10.1515/9783111437040-005

modeling to examine the sensitivity of thermodynamically constrained mantle hydration and H_2 generation to spreading rates in the distal and COT domains. We provide a brief overview of the associated seismic velocities, which serves as an indication of the multidisciplinary work required for comprehensive analyses of these new frontier margins. The ubiquitous presence of shallow peridotites in magma-poor margins suggests that should technology advance to the point of making them viable reserves of natural (possibly stimulated) H_2, they could make a significant contribution to meeting future H_2 demand in certain areas of the world.

Keywords: magma-poor margins, continent-ocean transition, serpentinization, natural hydrogen

5.1 Introduction

The exposure of mantle peridotite near or at the Earth's surface and its access to water leads to a number of mineral reactions, some of which involve the formation of abiotic molecular hydrogen, H_2, in association with serpentinization reactions. The mechanisms that regulate the generation of hydrogen during serpentinization are still subject to continuous study. The generation of H_2 is most commonly associated with magnetite formation, which contains the oxidized form of iron Fe(III). However, in peridotites with various degrees of serpentinization recovered during the Ocean Drilling Program (ODP) a remarkable variation in the abundance of magnetite has been found, ranging from nearly magnetite free (≤ 0.04 wt%) to as much as 6.15 wt% magnetite [1]. This magnetite is key to the formation of dissolved molecular hydrogen, $H_{2(aq)}$, via oxidation of iron: $2FeO + H_2O = Fe_2O_3 + H_{2(aq)}$. The largest occurrences of shallow mantle rocks on Earth close to coastal environments, with the potential to generate natural H_2, are magma-poor rifted margin with slow and ultraslow spreading [2]. These margins typically comprise hyperextended thinned continental crust in the distal sections, which are underlain by serpentinized mantle, and potentially a wide zone of exhumed serpentinized mantle, the continent-ocean transition (COT) [3–6]. A characteristic case for the presence of magma-poor margins is the southern North Atlantic Ocean. Figure 5.1 summarizes the location of magma-poor margins in the southern North Atlantic (see details in [2]) and the main structural elements in this type of margin (after [6]).

Much attention of the processes leading to H_2 production in submarine settings has been directed to slow-spreading mid-ocean ridges (MORs). Less well understood is serpentinization at rifted continental margins, where different protolith compositions and thermal regimes exist [7]. Within the COT, a number of non-volcanic rifted margins exhibit a band of partially serpentinized peridotites, which continue under the hyperextended, faulted continental crust, and are thought to represent ultramafic

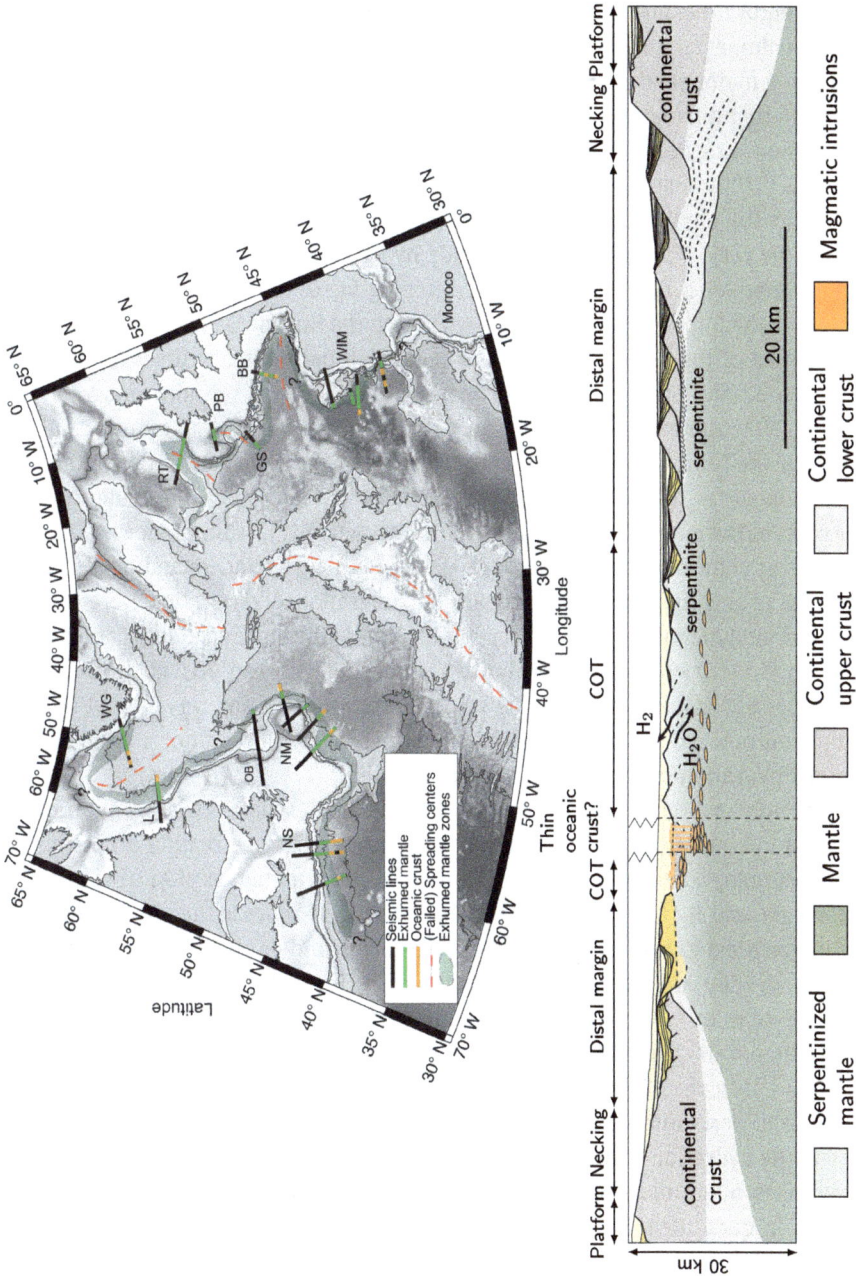

Figure 5.1: On top, magma-poor margins of the southern North Atlantic Ocean where exhumed and serpentinized mantle has been interpreted based on seismic velocity data; modified from [2]. At the bottom, general scheme of magma-poor margin architecture based on the West Iberia-Newfoundland margins; modified from [6].

subcontinental lithosphere, partially serpentinized by contact with water. Water in sufficient volumes in this region can only come from the surface, and it has been suggested that a major condition for the onset of serpentinization is the embrittlement of the entire crust during progressive extension and hence the development of active crustal penetrating faults acting as fluid conduits [8]. The continental crustal thinning during the COT in magma-poor margins, its embrittlement, and the thermal regime will provide the environmental conditions determining the hydration and oxidation of ultramafic rock. The early modeling study by Pérez-Gussinyé and Reston [8] found that the entire crust becomes brittle at stretching factors of ~3–5, depending on the strain rate, which for example compares well with the thickness of the crust observed just landward of the onset of partially serpentinized peridotites at the west of Iberia margin (WIM), and other magma-poor margins [9]. Later, 3D seismic tomography at the rifted WIM imaged the distribution of serpentinization in the mantle, and revealed that the local volume of serpentinite beneath the thinned, brittle continental crust is related to the amount of displacement along each fault, implying that sea water reaches the mantle only when the faults are active [10]. Thus, numerical models coupling geodynamic and geochemical processes can help in understanding the mechanisms and testing hypothesis of H_2 generation, migration, and trapping. Here we couple the thermodynamics of mantle serpentinization with a geodynamic model to evaluate the geometry of shallow peridotite emplacement in the COT of magma-poor margins, discuss the uncertainty in its serpentinization state and related past H_2 production, and point to the related seismic velocities, a key observation in the evaluation of the hydration state of mantle rocks.

Ultramafic mantle rocks also occur in orogenic belts, which represent sutures between tectonic plates. On the one hand, onshore settings are readily accessible to human technology, so they have received most of the attention as sites of potentially profitable natural H_2. On the other hand, shallow mantle rocks in the offshore domain cover larger areal expanses. The complex technology and costly investment involved in the exploration and potential exploitation of H_2 in offshore locations imply that, even more than at onshore sites, fundamental analyses are needed on the mechanisms leading to H_2 generation and accumulation in order to de-risk investments. In general, the presence of shallow peridotites can be seen from two viewpoints. Firstly, serpentinized peridotites are associated with past H_2 generation, which was concurrent with the hydration of the mantle at the time of rifting. A significant proportion of this hydrogen may still be trapped beneath the sedimentary cover, although this has not yet been quantified. Secondly, fresh peridotites in adequate locations (including under failed rifts) might be stimulated to generate molecular hydrogen.

5.2 Protolith bulk composition and thermodynamic databases

Let us consider the basic aspects of the hydration and oxidation of ultramafic and mafic rocks underlying the hyperextended continental crust and the COT in relation to the generation of abiotic natural hydrogen. Serpentinization is commonly associated with ultramafic rocks as protolith, although serpentinization of olivine in mafic rocks as troctolites and olivine gabbros also occurs. However, in mafic rocks, most of the olivine-replacing serpentines lack the mixing with brucite, which is commonly observed in peridotites [11], and the high amount of silica generally prevents the formation of Fe(III) and related H_2 formation. The relation between silica activity and serpentinization has been studied for more than two decades. Serpentines have the lowest silica activity of common crustal rocks, which also explains the occurrence of low-silica minerals, such as hydrogrossular, andradite, jadeite, and others in serpentinites or rocks adjacent to serpentinites [12]. Petrographic phase relations in hydrothermally altered peridotites in the Mid-Atlantic Ridge (MAR) with incomplete serpentinization show two major reaction pathways for seawater-peridotite interactions: first, the reaction of pyroxene to talc and tremolite, and second, the reaction of olivine to serpentine, magnetite, and brucite, with the latter pathway being favored at temperatures below 250 °C [13]. It is the oxidation to Fe(III), present in magnetite but also possible in serpentines, in the serpentinization process that assists the reduction of hydrogen to produce molecular H_2. At (or closer to) equilibrium, completely serpentinized peridotites may lack brucite and talc, as changes in pH fluid and silica activity during initial alteration may make these react to produce further serpentinite, potentially associated to H_2 generation depending on the pressure and temperature conditions in the secondary alteration. Under oxidizing conditions, it was reported by Bach et al. [13] that a substitution of brucite by abundant iowaite in MORs occurs. As a Fe(III) serpentine, the generation of iowaite is also associated with production of H_2. The presence of talc may also indicate the existence of previous serpentinization (and H_2 generation), as this may have been generated from serpentines under increased silica activity by secondary Si-metasomatism. Cronstedtite and greenalite, other minerals in the serpentine group, also contain ferric iron. Hisingerite, a low-temperature phyllosilicate-containing Fe(III), may also form solid solutions with serpentine at low temperature (~200 °C) [14].

A number of databases to support the calculation of geochemical thermodynamic equilibrium are publicly available, and they represent a compendium of current knowledge about the thermodynamic transformations that apply to rock metamorphic equilibria in general, and serpentinization and generation of natural hydrogen in particular. These databases are designed to support thermodynamic calculations either by law of mass action (LMA) or by minimization of the Gibbs free energy. Yet, the knowledge about serpentinization processes is still advancing. In addition to lack of experimental constraints for some minerals, experimentally derived thermody-namic data exhibit inherent systematic and random errors. A discussion on errors in thermodynamic analyses as applied to serpentinization is given by Klein et al. [15]. Ar-guably, the most characteristic example of uncertainty in relation to H_2 generation is cronstedtite. The thermodynamic properties of cronstedtite have not been directly mea-sured and, in the few (LMA) available databases that it is included, estimated standard Gibbs free energies of formation ($\Delta_f G^0$) differ from each other [16]. Also, due to this lack of experimental constraints, no database for Gibbs minimization has contained solid solutions for serpentines with ferric iron until the recent work by Eberhard et al. [17]. However, at low temperature, the increase in Fe(III)/Fe(II) in the rock during serpen-tinization substantially relies on the increase in cronstedtite in the rock (e.g., [1, 18]). Eberhard et al. [17] present solid solution models with Fe(III), and the Mg-serpentine endmembers lizardite, as low-temperature serpentine polymorph, and antigorite, the high-temperature (and high-pressure to ultra-high-pressure) polymorph.

As pointed by Albers et al. [7] in an experimental study of the magma-poor WIM, the protolith peridotite rocks at passive-continental margins able to produce H_2 are lherzolite, instead of the harzburgite/dunite at MORs, and the alteration temperatures tend to be lower. They analyzed samples from the WIM, and conducted thermody-namic calculations with a olivine composition fixed at X_{Mg} = Mg/(Mg + Fe) of 0.90, with the LMA software EQ3/6 v8.0 [19] and a customized database by Klein et al. [15] for equilibrium at 50 MPa assembled using SUPCRT92 [20]. The database considered solid solutions, and for serpentinites it included greenalite and cronstedtite as endmembers containing ferric iron. Their mass balance and thermodynamic calculations indicate that for the lherzolites at magma-poor rifted margins, serpentine hosts most Fe(III) with a distinct compositional trend towards a cronstedtite endmember, with a corre-sponding hydrogen yield from 120 to >300 mmol H_2 per kg rock. This contrasts with the higher production of 200–350 mmol H_2 for the silica-poorer harzburgites, where Fe(III) is allocated to magnetite formation at >200 °C at MORs.

When the analysis includes coupling thermodynamic equilibrium with geody-namics models, rather than customized at a specific pressure, the thermodynamic da-tabase needs to cover all possible ranges of pressure and temperature present during the simulated evolution of the case study. In addition, as we will expand below, there is a clear benefit in mapping the thermodynamic state of the geodynamic model into seismic velocities, as this allows for the joint evaluation of modeling results and seis-mic data. In this context, the most practical possibility is to apply publicly available general thermodynamic databases of metamorphic equilibrium. In this study, we have opted to conduct our thermodynamic equilibrium calculations with the Gibbs-minimizing software Theriak-Domino [21, 22] coupled to a geodynamic model. The thermodynamic databases included in Theriak-Domino are based on the database by Holland and Powell [23] with some recent updates, and the database JUN92d.bs, which also includes recent updates in the versions that allow for the calculation of elastic moduli and seismic velocities. Still, none of these databases have yet included the

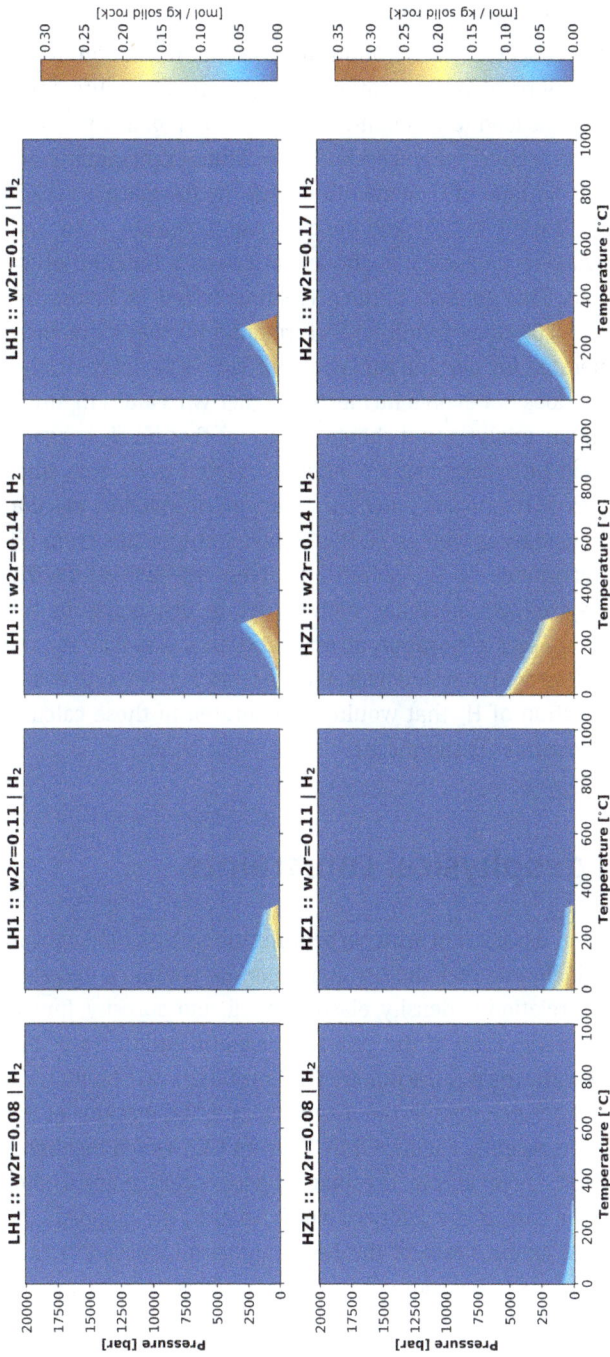

Figure 5.2: Comparison of production of H₂ generation as a function of temperature, pressure, and water-to-rock ratio (w/r), for a generic lherzolite (LH1), and generic harzburgite (HZ1) bulk compositions. Thermodynamic calculations conducted with Theriak-Domino and the database JUN92d.bs.

solid solutions with Fe(III) in serpentines published by Eberhard et al. [17]. Thus, calculations with these databases allocate H_2 generation to magnetite formation. Nonetheless, thermodynamic calculations of H_2 generation for a generic lherzolite and a generic harzburgite with JUN92d.bs lead to a maximum H_2 generation of 300 mmol H_2 and 350 mmol H_2 per kg rock, respectively (Figure 5.2). These values exactly match the results by Albers et al. [7]. Calculations of H_2 generation with the database TC-DS633 (a version of the [24] database as used in [25]), and the database JUN92d.bs yield similar results in terms of H_2 formation, including a similar pattern as a function of the water-to-rock (w/r) mass ratio. This general pattern is summarized in Figure 5.2, where it is shown that for w/r = 0.08 there is still no H_2 formation for lherzolite and a small amount starts to be generated for harzburgite, while for w/r = 0.17 full hydration is complete. The H_2 production does not evolve further for higher w/r values in any of these generic protoliths. A similar analysis (not shown) indicates that no H_2 is generated for a generic gabbro/basalt bulk composition, which is related to its high silica content. Summarizing, in terms of maximum potential hydrogen production, calculations with these databases, JUN92d.bs as well as TC-DS633, match the results from [7]. For low temperatures, the estimates of H_2 generation given the lack of Fe(III)-serpentines in these databases are likely partially compensated by displacing the calculated equilibrium towards a higher generation of magnetite in the search for the minimum Gibbs energy in the thermodynamic solver. Future work is needed to evaluate whether the additional fraction of H_2 that would be generated in these calculations by considering Fe(III)-serpentines is significant.

5.3 Observational geophysical constraints

Hydration of ultramafic rocks in association with serpentinization causes changes in the rock's physical properties, including rock density, magnetic susceptibility, and elastic moduli. Key observations, in relation to density, elastic moduli, and porosity, for the analysis of rifted margins and estimation of the general serpentinization status are those of seismic reflection and wide-angle refraction (RWAR) profiling and resulting velocity models [6]. A synthesis of seismic data in the magma-poor rifted margins of the southern North Atlantic has been recently conducted by Welford [26], which highlights the variable nature of the COT, even within the magma-poor rifted margin end-member case (see also [6]). In the case of the southern North Atlantic, the mapped COT zones show a symmetric lateral extent north of the Newfoundland–Azores Fracture Zone (NAFZ) and asymmetric extents to the south. The seismic profiles indicate a range of different basement types as exhumed mantle, anomalously thin oceanic crust, and thinned continental crust, sometimes including failed rift areas. Beyond the thinned hyperextended continental crust, exhumed serpentinized mantle is ubiquitous, with one exception being offshore Morocco, where no exhumed serpentinized mantle has been observed on seismic models published to date, indicating a fundamental asymmetry with its Nova Scotian conjugate margin ([26], and references therein).

Care is needed, though, in associating observations of physical properties with the generation of natural hydrogen. One illustrative example is the analysis of serpentinized ultramafic rocks by Cutts et al. [27], which shows that density has a strong (negative) correlation ($R^2 = 0.94$) with the serpentinization degree, while magnetic susceptibility is positively albeit weakly correlated ($R^2 = 0.31$) with serpentinization. Despite these correlations can be potentially exploited, for example, by combining gravity and magnetic data, the nuanced details of the serpentinization reactions need to be taken into account. Specifically, the most likely reason for the weak correlation between magnetic susceptibility and serpentinization degree found [27] lies in the evidence shown by previous studies (e.g., [1, 7, 18]) that low-temperature serpentinization (≤ 200 °C) often proceeds with low magnetite formation, with Fe(III) being partially in serpentine, as summarized in the previous section. That is, not all serpentinization processes result in strong magnetic signatures in hydrogen-rich systems at low temperatures. On the other hand, the formation of magnetite necessarily comes with simultaneous generation of natural hydrogen. Also the decrease in density and the elastic moduli with hydration, which affects results from gravimetry data and seismic velocities, is commonly associated with serpentinization, but the reduction in density and modification in mechanical properties are also associated with other hydrated minerals, as talc, brucite, or the amphibole and chlorite mineral groups. The migration of H_2 away from the Fe(III)-bearing minerals also needs to be taken into account.

In addition, hydration may lead to an overlap between the density and elastic moduli of the hydrated peridotite and those of hydrated gabbro/basalt at the shallow depths corresponding to the serpentinization window. Fresh, unhydrated ultramafic rocks (peridotites) have higher density and elastic moduli than fresh mafic rocks (gabbro/basalt). However, under shallow conditions at the COT (below ~350 °C and less than ~1,500 bar), the ultramafic rock can accept up to ~12–13 wt% of water content in the solid matrix when fully hydrated to serpentine and other minerals, whereas the mafic rock has less than ~5 wt% of water content when fully hydrated. This is summarized in the top row of Figure 5.3, which shows the amount of water in the rock when it is fully hydrated. The gradient of seismic velocities with depth is a common basis for distinguishing between ultramafic and mafic rocks in seismic data, with the general interpretation that exhumed and serpentinized mantle has a higher increase in seismic velocities from the surface to ~6–8 km depth than gabbro/basalt (e.g., see Box 1 in [6]). In practice, this distinction is open to interpretation, and in the shallow rocks within the serpentinization window, from ~4 km depth bsf to the seafloor, there is a large degree of overlap in seismic velocities for both bulk rock types due to hydration. The bottom row of Figure 5.3 shows a thermodynamic calculation of P-wave velocities for the solid matrix of fully hydrated equilibrium rocks with generic lherzolite, harzburgite, and gabbro compositions, for which the effective elastic moduli are obtained by the Voigt-Reuss-Hill estimation (see [28]). For the pressures and temper-

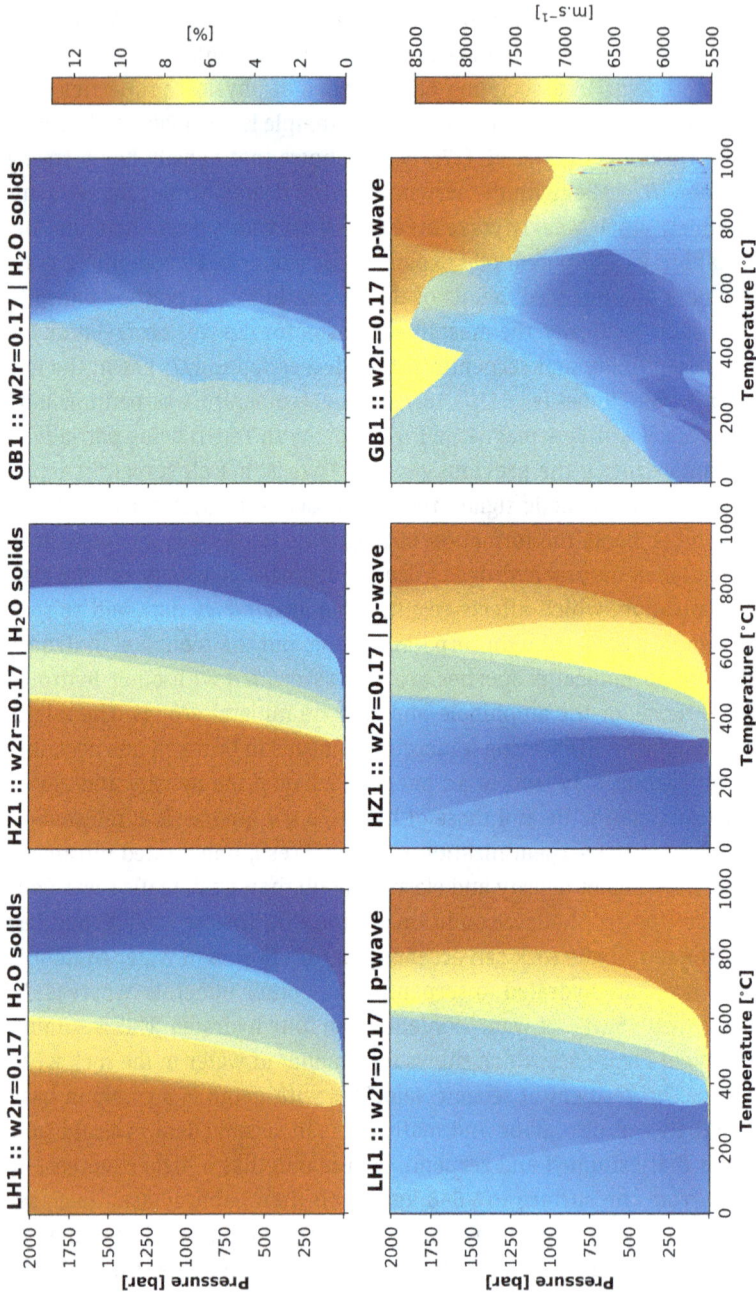

Figure 5.3: Comparison of water content in the solid rock (upper row) and P-wave velocities of the solid rock (lower row) as a function of temperature and pressure, at a water-to-rock ratio sufficient to fully hydrate the rock, for generic protolith compositions. From left to right: lherzolite (LH1), harzburgite (HZ1), and gabbro/basalt (GB1). Thermodynamic calculations as in Figure 5.2.

atures we can expect in extensional environments in distal margins and COTs in the serpentinization window (below ~350 °C and below ~1,500 bar), Figure 5.3 shows that P-wave velocities are in the 5,500–6,000 m/s range for both fully hydrated peridotites (LH1 and HZ1) and fully hydrated mafic rocks (GB1). Thus, under fully hydrated conditions, the overlap in P-waves between lherzolite and gabbro at the COT is significant. However, if the peridotites are not hydrated, they reach P-wave velocities in this pressure and temperature window of ~7,000–7,500 m/s (not shown). And especially in the shallowest domain, the effect of porosity (including fractures) adds further uncertainty to the inversion of seismic data. The combination of thermodynamic calculations with geodynamic models can help reduce the uncertainty in these interpretations. This points to the application of improved numerical modeling efforts (e.g., [9, 29]) jointly with observational constraints, preferably via data assimilation techniques [6, 30], as an avenue for future research on continental margins.

5.4 Numerical simulation of H_2 generation in magma-poor rifted margins

5.4.1 Geodynamic model

Here we apply the geodynamic model Rift2Ridge; a Lagrangian finite-element model with non-Newtonian visco-elasto-plastic rheologies, isostatic boundary conditions, and a free surface [6, 31–33]. The model includes sedimentation and erosion processes in subaerial and submarine environments, magma production, and a parameterization for magma emplacement in the crust. A recent use of the model applies to the analysis of thermal fields in rifted margins [34], which is relevant for the study of H_2 generation in these settings. In this study we use Theriak-Domino to account for metamorphic solid-rock thermodynamic equilibrium, as a function of pressure, temperature, and w/r ratios. Background static porosity, specific for oceanic and continental crust, decays with depth, for which we apply parametric curves based on the synthesis by Kuang and Jiao [35]. Then, porosity is further modulated by a poro-elastic approach after [36] and [37], being dynamic permeability sensitive to porosity variations via a Kozeny-Carman relation. To provide an estimate of the circulating fluid, Darcy flux is approximated by assuming state-state conditions for a fluid pressure solver, using the equations of state (EOS) for pure water by Wagner and Pruß [38] for the fluid density and dynamic viscosity. Here, we consider thermal conduction and advection of the thermal fields with the moving Lagrangian mesh, but not the effect of thermal advection by fluid flow. For magma generation, we follow the formulation by Phipps Morgan [39], where the solidus evolves as a function of mantle depletion. The implementation of magma production is strongly relevant to the estimates of serpentinization. Not only does the bulk composition of mantle peridotites evolve with depletion as magma is generated, but also the accretion of magmatic material to form new oceanic crust provides a limiting factor for the occurrence of shallow peridotites.

5.4.2 Thermodynamic considerations

Here we assume that the initial bulk mantle composition in the model is 100% lherzolite and that it gradually evolves, reaching a 100% harzburgite composition, when its degree of depletion by melt release reaches 25%. With this consideration, and the amount of mafic (gabbro/basalt) melt emplacement, the mineral composition at the mesh nodes is obtained by linear combination of the mineral assemblages obtained from thermodynamic equilibrium of three rock endmembers: lherzolite, harzburgite, and gabbro/basalt. For each of these endmembers we assume a generic bulk composition in term of oxides, and pre-calculate the corresponding mineral assemblage for thermodynamic equilibrium as a function of pressure, temperature, and w/r ratios. The w/r ratio axis varies from 0.0 to 0.17, beyond which there is no further metamorphic transformation. The thermodynamics of peridotites is calculated in the NCFMASH system (Na_2O-CaO-FeO-MgO-Al_2O_3-SiO_2-H_2O). Mafic rocks in addition include K_2O, MnO, P_2O_5, and TiO_2. We focus on serpentinization and H_2 generation and do not perform thermodynamic calculations for the continental crust. We use the indicated database JUN92d.bs. As the geodynamic model evolves, the pre-calculated thermodynamic equilibrium arrays are mapped into the model space by considering the pressure, temperature, and w/r ratio at the model nodes.

5.4.3 Uncertainty considerations

Model uncertainties include those of rock material properties and boundary conditions. In addition, the models presented in this study consider thermodynamics for metamorphic petrology but not an explicit reactive transport of hydrothermal fluids coupled to the tectonic evolution. Thus, they do not allow for the evaluation of the saturation state of dissolved minerals in the pore fluids and metasomatism. This brings further uncertainty to the hydration and redox conditions of mantle peridotites subject to hydrothermal flow. To embrace a range of hydration possibilities, we encapsulate model uncertainty in a number of tests by perturbing the time of the circulating water estimated by the Darcy flux approximation. The product of this circulation time and the fields of Darcy flux module result in the w/r ratio fields applied to evaluate the system thermodynamics. This is a simple proxy for both the uncertainty in the porosity and permeability fields, which in turn depend on rheological parameters and tectonic deformation as well as for the degree of saturation of dissolved minerals in the circulating water. The resulting water-to-rock ratios in the following tests are positively correlated with higher amounts of generated natural hydrogen. The timescales applied to the Darcy flux range from 10 days to 6 h. This ranges from scenarios with a high degree of hydration up to scenarios where the bulk of the mantle remains unaltered, that is, low-hydration scenarios. Our previous sensitivity analyses indicate that the given uncertainty bounds on the degree of hydration in the mantle and mafic rocks from magma emplacement are reasonable.

5.4.4 Transition to a magmatic oceanic crust

As the spreading velocity increases in rifting margins the upwelling of the mantle is promoted, with the subsequent faster decompression and higher magma generation rates. Thus, magma supply is strongly correlated with spreading rates, and high magma supply leads to magmatic oceanic crust formation, hindering the opportunity for shallow peridotite rock to occur. Also, variations in the inherited fertility of the mantle and mantle potential temperatures play a role in magma supply and in the configuration of the COT transition and general seafloor morphology [40].

Let us consider an ensemble of numerical model runs at full spreading velocity of 15, 20, 25, and 30 mm/year, with the mantle potential temperature set to 1,350 °C, as a generic representation of slow-spreading magma-poor rifted margins. The model domain is initially 200 km wide and 150 km deep and grows laterally with spreading. The mantle rock is initially fertile (lherzolite) and transitions into a more depleted harzburgite composition, as oceanic spreading evolves. For illustration, Figure 5.4 shows snapshots of rifting, and the opening of a new ocean for a 20 mm/year simulation, in a zoom on a 100-km wide area around the 0.0 in the horizontal axis. The embrittlement of the continental crust at thinning prior to continental breakup, can be observed by the location of the active plastic deformation (shaded red) reaching and even crossing the mantle at 3–4 Myr in the distal margin, as previously described by simpler numerical models (e.g., [8, 41]) and seismic tomography (e.g., [4, 10]). Oceanward, beyond the distal margin, after ~5 Myr starts the development of the COT. At the point of continental breakup, there is a limited amount of magma emplacement (pink area), such that there is an opportunity for the mantle to appear at shallow locations, either below the thinned hyperextended continental crust or even as exhumed mantle, before the establishment of a stable mafic oceanic crust. This simulation results in reasonably symmetric margins, with the left margin having some more areas of exhumed mantle than the right margin. This figure also illustrates that after initiation of a mafic oceanic accretion, long detachment faults at ultraslow-spreading settings may allow for further areas of exhumed mantle to emerge at the ocean bottom. This is exemplified by the continuous area of mafic crust at 6 Myr, which is then sectioned by a fault, leading to the leftmost block of mafic rock to be detached from the rest of the mafic domain at 7 Myr.

From model initiation and for the set of spreading velocities indicated above, the situation at which full stretching reaches ~200 km, including rifting and posterior oce-

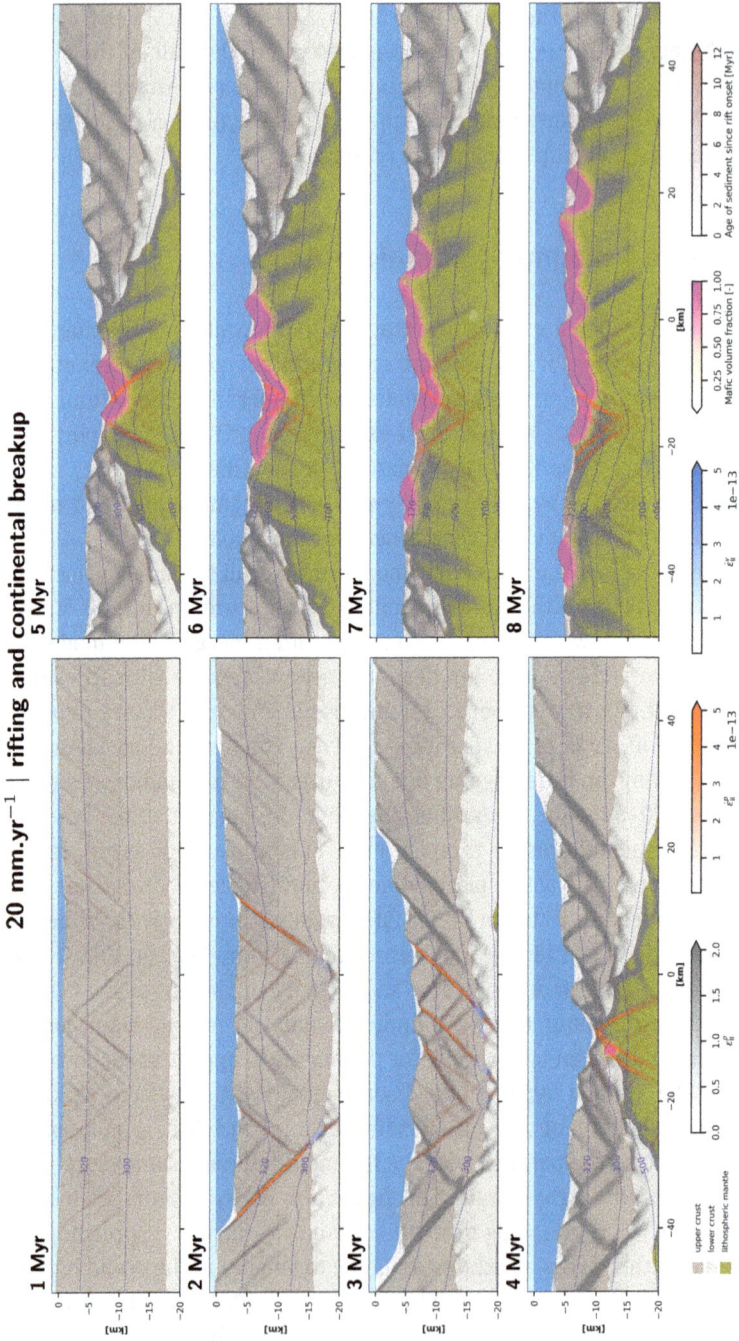

Figure 5.4: Sequence of continental rifting and initiation of oceanic spreading for a 20 mm/year simulation; zoom on a 20-km deep, 100-km wide window. The pink area represents mafic rocks. Shaded grey represents the plastic strain, shaded red the active plastic strain rate, and shaded blue the active viscous strain rate. The age of the sediments is also indicated. On top, the blue area is the ocean and cyan is the atmosphere.

anic spreading, gives a temporal snapshot where the basement morphology in the margins has generally stabilised. Some tectonic stresses are still present at the COT, and sedimentation plus thermal relaxation continues, but the active deformation in this situation, as well as the bulk generation of natural H_2, has migrated to the recently formed oceanic ridge. The simulation times corresponding to 200 km of stretching and the above spreading velocities (15, 20, 25, and 30 mm/year) are 13.3, 10.0, 8.0, and 6.6 Myr, respectively. The spatial integration of the rate of magma generation in the model domain, in m^3/s, divided by the spreading rate gives an estimate of the average magmatic oceanic crust thickness, whose evolution for a set of simulations is shown in Figure 5.5. With some variations, the general pattern as a function of spreading rate is systematic for other ensemble members with similar spreading rates. This figure shows that a sharp increase in magmatic production in these passive margins is concurrent with the final thinning of continental lithosphere (the hyperextended continental margin domain) and continental breakup. This provides the window of opportunity for the occurrence of subcontinental peridotites in the COT and mantle exhumation before the establishment of a stable oceanic magmatic crust. At the spatial scale, the model runs in Figure 5.5 correspond to 100 km of half-spreading distance, at which point all of the curves have already flattened to their corresponding stable magma production rates, now being emplaced at ~60 km from the continental breakup location, represented by the diamonds. At the timescale, the timing of the transition towards an oceanic magmatic crust is negatively correlated with the spreading rates, with a clear delay in the onset of magmatism for lower spreading rates. However, at the spatial scale, the distances nearly collapse, as indicated by the mapping of the average mafic crustal thickness into the half-spreading distances. Still, there are ~5 km of additional space allocated to rifting and crustal thinning in the

Average oceanic crustal thickness

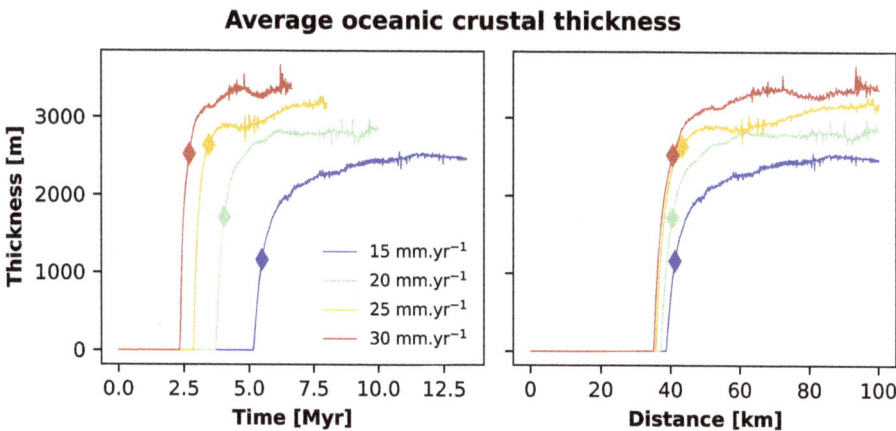

Figure 5.5: Average thickness of the generated magmatic oceanic crust as a function of time since model initiation, and mapping into half-spreading distances. Diamonds represent the time/location of continental breakup.

COT in the 15 mm/year scenario with respect to the 30 mm/year simulations. The onset of magmatism is then stronger for faster spreading rates, and by the time of continental breakup the amount of emplaced magma for 30 mm/year is about twice that for 15 mm/year. This experiment exemplifies the general trend of higher magma production for increasing spreading rates, in turn leading to a corresponding stronger limit to the spatial window of shallow peridotite material. In summary, the opportunity for shallow peridotite emplacement increases for the lowest spreading rates.

5.4.5 Sensitivity to spreading rate and hydration

At each spreading velocity, we evaluated a small ensemble (five members) to consider the uncertainty in hydration of the protolith. Figure 5.6 shows the relation between spreading rate, as a control factor, and the generation of natural H_2 for the member with the lowest amount of hydration allowed at each evaluated spreading rate. At this stage, H_2 refers to the total amount of molecular hydrogen, irrespective of its state as a gas, $H_{2(gas)}$, or dissolved in liquid water, $H_{2(aq)}$. The units consider that the model domain is 1-m thick perpendicular to the 2D section. The sequence shown in Figure 5.6 is consistent with the summary of Figure 5.5 in the general sense that the lowest spreading rates have a higher proportion of shallow peridotite in the COT. This includes both areas of unhydrated fresh peridotite and areas of serpentinized peridotite, in turn leading to higher amounts of generated natural hydrogen. In these simulations, all production of H_2 is associated with production of magnetite. Shallow peridotite fractions, and also H_2 production, are either underlying the hyperextended continental crust in distal sections of the margins, or in areas of exhumed mantle covered (at this stage) by thin packs of sediments. Figure 5.7 is as Figure 5.6, but instead shows the members in the ensembles with the highest degree of hydration. The general behavior, which is that the proportion of shallow peridotite in the COT is higher for slower spreading rates, is maintained. Another pattern shown by these simulations is that for the highest evaluated spreading rates of 30 mm/year the magmatic emplacement is enough to cover both conjugate COTs, marking the transition towards a more magmatic spreading style. Although there is some H_2 generation at 30 mm/year, it is not localized, but distributed across the profiles (Figures 5.6 and 5.7). For lower spreading rates (10, 15, and 20 mm/year), the magmatic emplacement shows a more variable behaviour. For these, in the members shown in Figure 5.6, magma is quite symmetrically emplaced, but in the members shown in Figure 5.7 the magma preferentially emplaces in one of the margins. This grouping of asymmetric behaviour is not related to the hydration of the mantle but rather a coincidence in these simulation sets, and other ensemble members with different degrees of hydration do not adhere to this grouping. In either case, even in the cases of strongly asymmetric magmatic emplacement, there is a spatial window with shallow peridotites at both conjugate margins in all simulations, with the COT having a higher amount of local

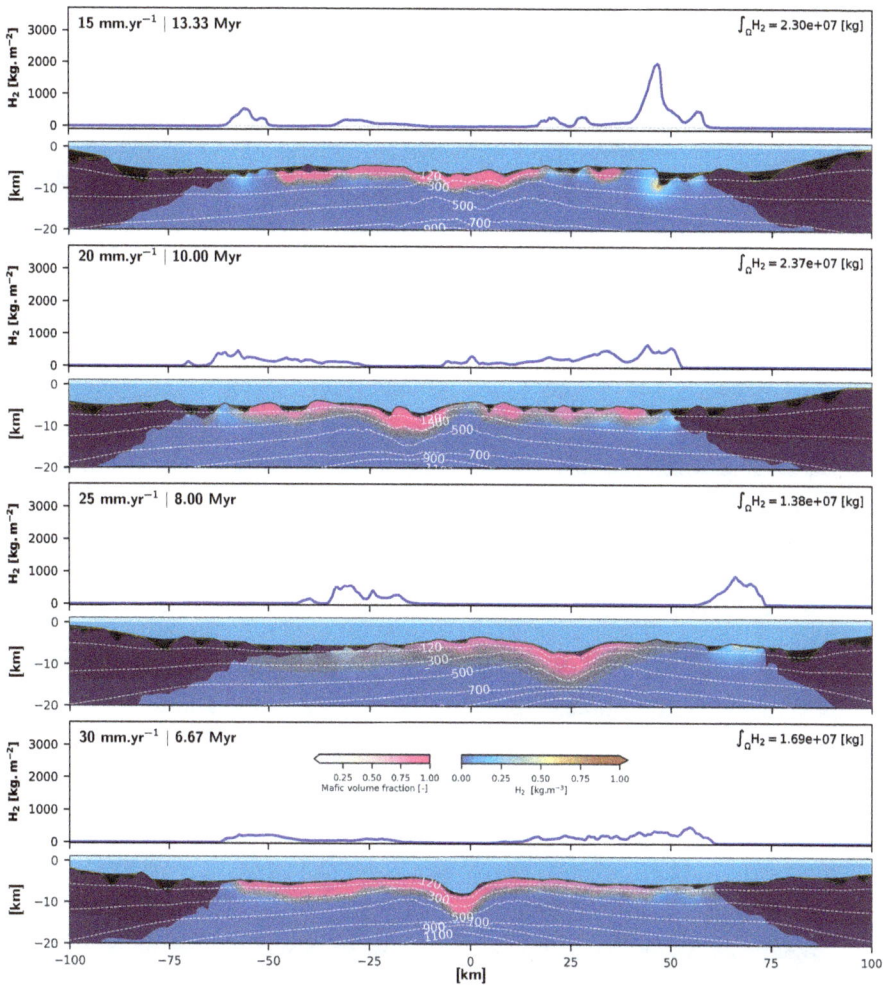

Figure 5.6: Examples of the relation between spreading rate as control factor, and generation of natural H_2 for the scenario where the lowest degree of hydration is allowed. At each spreading rate, the plot on top shows the integral along the vertical axis of generated H_2, in kg/m^2 of surface area, and a notation of the total integral of H_2 in the (1-m thick) domain. Then, the plot below shows the hydrogen field and the volume fraction of gabbro/basalt rock over the total rock volume, with the remaining rock fraction being peridotite. The continental crust is masked in dark violet, where the orange isochrones indicate the sedimentary cover.

concentration of generated H_2 for 15 and 20 mm/year. By comparison, 20 mm/year is the spreading velocity of the WIM.

In general, with increasing spreading rates, there is not only an increase in magma production, but the upwelling of mantle material is also faster and the host environment in the COT is warmer at the times of active deformation, associated with

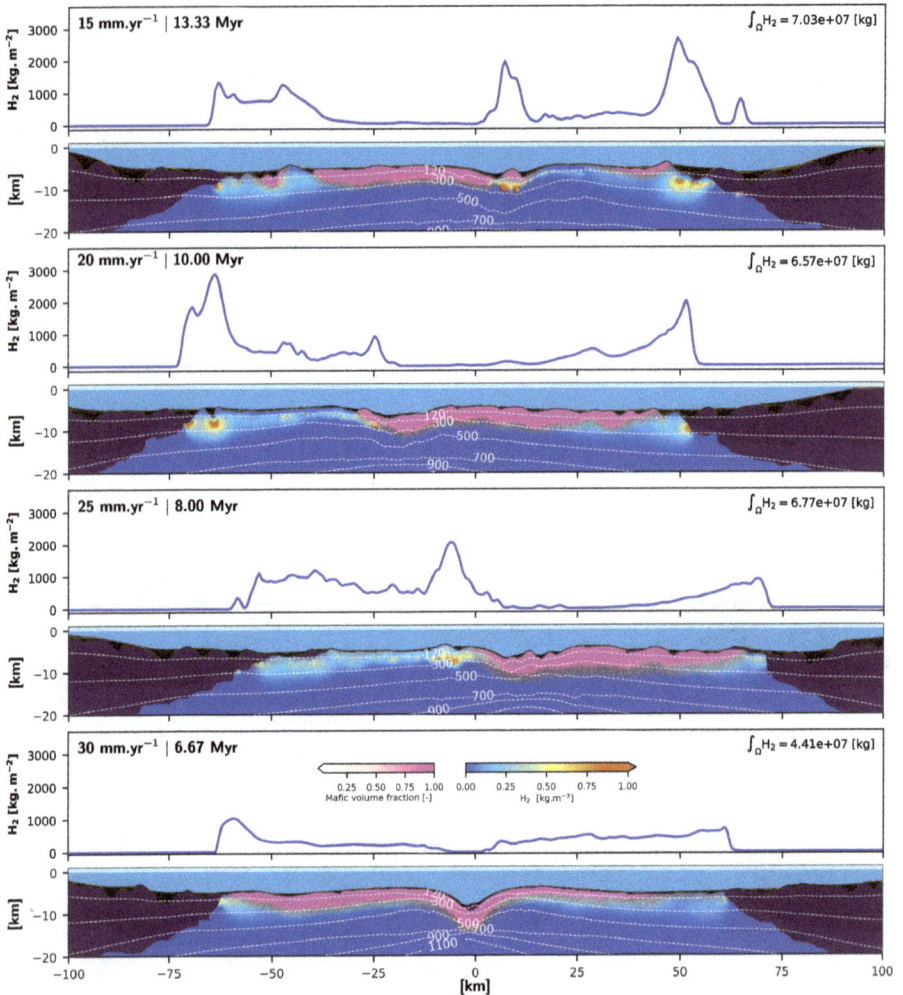

Figure 5.7: Examples of the relation between spreading rate as control factor, and generation of natural H_2 for the scenario where the highest degree of hydration is allowed (details as in Figure 5.6).

most of the serpentinization. A too-hot environment may well contain water and an adequate protolith but not be thermodynamically suited for H_2 formation. Beyond these simulations, alternative mantle potential temperatures may displace the optimal strain rate conditions regarding the generation of H_2. In addition, the figures shown are likely lower bounds, as additional H_2 may be expected from low-temperature serpentinization, via hosting of Fe(III) in serpentine minerals, mainly cronstedtite. Interestingly, from the morphological perspective, the simulations at 30 mm/year also appear to develop a simpler more abrupt abutting of the continental crust. While variation in mantle temperatures will surely influence the lag between breakup and initiation of oceanic crustal growth by magmatic accretion, it seems difficult that at this spreading velocity (and faster) the adequate conditions for past H_2 generation or for emplacement of shallow peridotite susceptible to stimulation, are given.

5.4.6 Relation between hydration and seismic velocities

As summarized above, the hydration of the mantle leads to a decrease in the elastic moduli and corresponding seismic velocities. The thermodynamic calculations make it possible to conduct a forward mapping of the model state into seismic velocities for comparison with seismic tomography, which is a main source of information for revealing major faults under the ocean (e.g., [42]), patterns of ultramafic rocks and mafic intrusions (e.g., [43–45]), and the serpentinization status (e.g., [46]). In seeking natural H_2 and shallow peridotites in magma-poor margins, preferential sites would be those at ultra-thinned hyperextended continental crust, including failed rifts, and the COT. In the southern North Atlantic, instances of failed rifts have been evidenced by seismic data in the NE Newfoundland margin (in the Orphan Basin), in the northern Irish Atlantic margin, or in the southeastern Newfoundland and northwestern Iberian margins. Systematically, underneath the thinned continental crust in failed rifts, low seismic velocities in the upper mantle occur, which indicate some degree of serpentinization (e.g., [47–51]).

As a generic simulation example, Figure 5.8 refers to an ensemble member for 20 mm/year full spreading velocity with an intermediate degree of hydration. Accordingly, the integral of the natural hydrogen produced in the domain at 10 Myr of simulation, 4.7e7 kg H_2, lies in between the low-hydration and high-hydration members at 20 mm/year shown in Figures 5.6 and 5.7, respectively. Let us choose this as a likely scenario. In this case, the distribution of the mafic oceanic crust is mostly symmetric between the conjugate margins, with spreading accommodated by the classic segmented magmatism imaged in ultraslow-spreading (e.g., [52, 53]). Equally symmetric is the distribution of the areas of shallow peridotite emplacement below the thinned continental crust and as exhumed mantle. The distribution of generated H_2 follows that of the magnetite, with higher areas of H_2 generation being at the transition between the low-temperature serpentines (chrysotile/lizardite), shown in Figure 5.8, and high-temperature serpentine (antigorite). The mapping into P-wave velocities for the oceanic crust shown in the bottom plot of Figure 5.8 is conducted by a simple hierarchical application of the Voigt-Reuss-Hill (VRH) averaging. First, the VRH approximation is applied to estimate the effective velocity for each rock endmember, considering fresh mantle (lherzolite), depleted mantle (harzburgite), and gabbro as the three endmembers, as done for Figure 5.3. Second, the degree of depletion of the mantle peridotite and the fraction of mafic intrusion are applied, again with VRH, to obtain a P-wave estimate of the bulk solid rock. Third, the porosity is taken into account to weight, again via VRH, the moduli of water and those of the solid rock. An improved

approach to map lithologies into seismic velocities would be to apply effective medium theory (EMT), as an approximation to estimate the effective elastic moduli of a
composite medium (solid matrix + inclusions), given the elastic moduli of the matrix,
concentration (porosity), and shape of the inclusions [54]. Still, the VRH approach reveals a clear sensitivity of the vertical gradients of P-wave velocities to the degree of
hydration. As seen, the P-wave velocity plot in Figure 5.8 reveals low P-wave anomalies in relation to the high H_2 areas at the COT in both margins (x-coordinates around
−70 km and 52 km). As indicated above, mantle velocities decrease with increasing
degrees of serpentinization. Figure 5.8 exemplifies how alternative mafic and ultramafic zones between the conjugate margins result in close P-wave values as a result
of mantle hydration and porosity effects. Joint evaluation of seismic data with numerical models, preferably with assimilation (history-matching) capabilities (e.g., [30])
will surely provide further insight into the analysis of magma-poor margins in a future. Currently, most of the marine seismic lines available worldwide and specifically
in magma-poor margins have been conducted with forward-model ray tracing (see
[26]). More advanced tomographic inversion and specifically full-waveform inversion
(FWI) require much more closely spaced ocean-bottom seismometers, and, for exam-

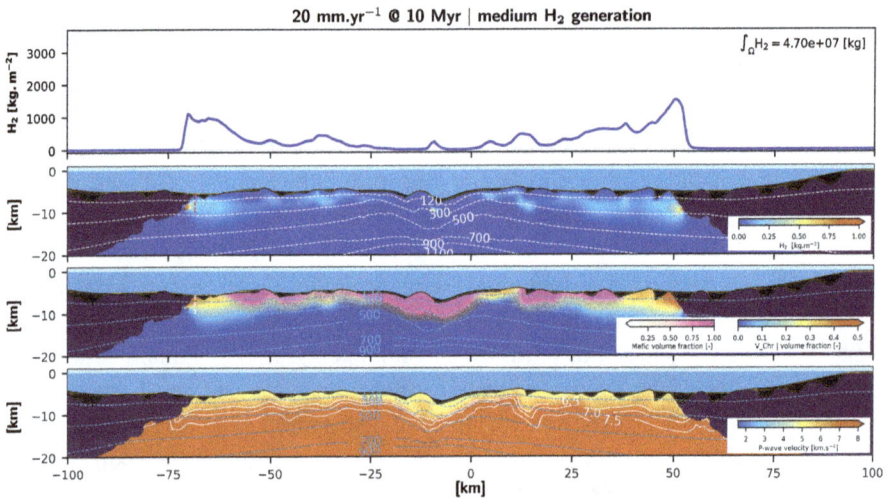

Figure 5.8: Simulation with 20 mm/year full spreading velocity at 10 Myr after the initiation of the
simulation, for a simulation member with intermediate hydration level at the COT. The plot on top shows
the integral along the vertical axis of generated H_2, in kg/m² of surface area, and a notation of the total
integral of H_2 in the domain. Then, from top to bottom: (a) field of produced natural H_2; (b) field of the
volumetric fraction of intruded mafic rock (gabbro/basalt) over the total rock volume, and field of the
volumetric fraction of chrysotile/lizardite over the total rock volume; and (c) P-wave velocity field, with
isolines indicated in white. In the field plots, the continental crust is masked in dark violet, where the
orange isochrones indicate the sedimentary cover. Isotherms are indicated in dashed white in the H_2 field,
and in blue in the other fields. The vertical lines at −69.4 km in the hydrogen and lizardite plots depict a
transect evaluated in the following section.

ple, only one FWI profile in the Nova Scotia margin by Jian et al. [55] was available for the synthesis by Welford [26]. FWI reduces the uncertainty in the interpretation of serpentinization, and its more common application will also prove highly valuable.

5.4.7 Thermophysical properties

For exploration and exploitation purposes it is also relevant to have background information about both the material properties of the rock and sedimentary cover, as well as the thermophysical properties of the fluid phases hosting the generated natural hydrogen. These will also support the evaluation of migration dynamics. Primary serpentinization and natural hydrogen generation is a relatively fast process when an adequate protolith is present and suitable temperature, pressure, and water availability are given. The environmental conditions leading to H_2 generated in the past will have likely evolved up to the present time, and a proper accounting for the processes over geological timescales would preferably couple models of H_2 generation with multiphase migration models, concurrently simulating both. There are still substantial uncertainties in these two and an adequate coupling is a matter of future work.

In a specific real-case scenario a location for exploration could be chosen based on joint analyses of seismic data and history-matched numerical models (e.g., [30]). For illustration purposes in this generic modeling case, let us evaluate some general environmental conditions in the simulations at 20 mm/year full spreading and at 10 Myr. From this ensemble, we select a member where an intermediate amount of H_2 is generated in order to approach a likely situation, the one depicted in Figure 5.8 and focus on a transect that in principle we could deem as suitable for exploration. Let us choose the vertical transect indicated in the left margin in Figure 5.8. This is located at the landward side of the COT, and was early covered by sediments, with a convexity at their base, which could potentially serve as a seal and location for accumulation of H_2 migrated from deeper levels. Figures 5.9 and 5.10 summarize some material properties, including the total amount of generated H_2 and thermophysical variables along this transect at 10 Myr. At this time in the simulation, the amount of generated H_2 has stabilised both in the shallow subcontinental domain affected by serpentinization and in the near-coast exhumed mantle. The situation can then be considered as approximating the initial conditions leading to the migration of H_2 in the multiphase system and potential trapping below a sealing sedimentary pack. We can expect that at this time the generated H_2 will have likely suffered some degree of upward migration, so that the maximum accumulations of H_2 would be likely shallower than those corresponding to Figures 5.9 and 5.10. Still, these plots serve to illustrate general considerations. At this transect the protolith is mainly lherzolite, as hardly any mantle depletion by melt generation has occurred at this location. Figure 5.9 shows that the maximum generation of H_2 closely follows the profile of the magnetite volumetric fraction, resulting from the previously described thermodynamic database conditions. The sea

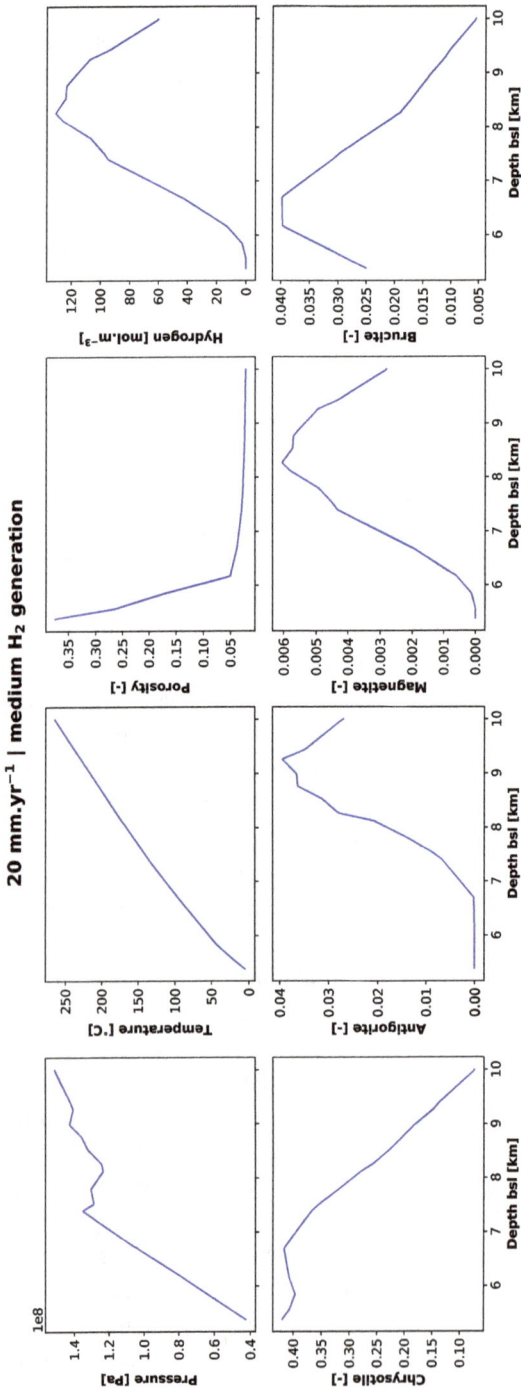

Figure 5.9: Selected variables for the vertical transect shown in Figure 5.8, at 10 Myr simulation time. Mol concentrations of molecular hydrogen H_2 refer to cubic meter of bulk rock plus voids. The lower row indicates volumetric fractions in the solid rock of selected mineral species resulting from serpentinization.

20 mm.yr⁻¹ | medium H₂ generation

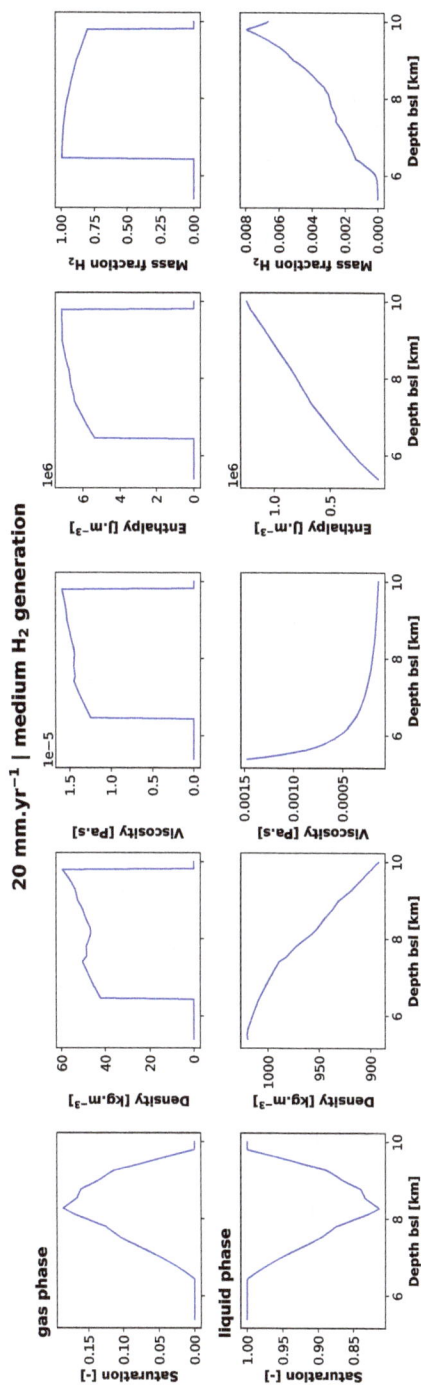

Figure 5.10: Selected thermophysical variables for the same profile as in Figure 5.9, with gas phase in the top row and liquid phase in the bottom row. Given the high difference in values for the gas and liquid phase, plots for the same variable have independent scales for each phase.

floor at this location is 5.39 km below sea level (bsl). The generation of H_2 associated with magnetite formation peaks at ~8.3 km bsl (~2.9 km below sea floor), which lies in between the domain of the maximum generation of the low-temperature Mg-endmember serpentine (chrysotile/lizardite) and the high-temperature serpentine polymorph (antigorite). As seen, a substantial part of the shallowest chrysotile/lizardite is not related to H_2 production. If the thermodynamic database had included Fe(III)-holding serpentine, the estimated amount of H_2 in this shallow subdomain could have been larger. Brucite, which is involved in the reaction leading to magnetite and H_2 generation, follows the general pattern of the low-temperature serpentine polymorphs rather than that of antigorite. Last, the porosity considered at this time, which strongly decays over the shallowest kilometer below the seafloor as it deepens through the still thin sedimentary cover, will soon later decay by aging through mineral precipitation and thickening of the sedimentary pack. Note that at this time, some degree of hydrothermal circulation will occur in this area, likely supporting mineral precipitation.

The maximum content of H_2 generation in this transect is 130 mol/m^3. At this location, the bulk density of the rock is 2,878 kg/m^3, corresponding to a substantial but not complete hydration. Thus, the corresponding yield of H_2 per kilogram of (hydrated) solid rock at this location is 45 mmol/kg. This is far from the maximum value that could be produced for lherzolite according to the thermodynamic database and previous publications (300 mmol/kg solid rock; see Figure 5.2). Note that at a similar depth, 2.5 km to the left of the selected transect there is an area of much higher H_2 production (see Figure 5.8), where H_2 yields reach 174 mmol/kg. If the sedimentary cover provides adequate trapping, it is possible that the generated H_2 from these neighboring source areas converges along its migration towards the surface.

Considering that water and H_2 are miscible in both gas and liquid phase, and with the simplifying assumption that the pore pressure equals the dynamic pressure of the solid rock, the total amount of generated molecular H_2 shown in the upper-right plot in Figure 5.9 leads in the simulation conditions to a two-phase fluid in part of the transect, with a gas phase containing a mixture of $H_{2(gas)}$ and water vapor, and a liquid phase also containing dissolved $H_{2(aq)}$, as summarized in Figure 5.10. The calculation of the curves in Figure 5.10 considers the dissolution of hydrogen into water using Henry's law [56] and CoolProp [57] to provide the equations of state (EOS) of hydrogen by Leachman et al. [58] with viscosity calculated with the formulation given in [59] and the EOS of water by Wagner and Pruß [38] with viscosity given by the formulation in [60]. For the gas phase, the density, dynamic viscosity, and mass-specific enthalpy for $H_{2(gas)}$ are calculated using the partial pressure of H_2 in the gas phase following Dalton's law, and for the water vapor the partial pressure is evaluated with Raoult's law, with the evaluation of composite bulk gas properties following [61, 62]. The phase saturation plots indicate that only in the central part of the transect, where H_2 concentrations are higher, a two-phase system exists, with the gas phase reaching 19% of the pore volume for the indicated maximum of 130 mol/m^3 total H_2, at ~2.9 km below the ocean floor. Regarding the mass fraction, most of the gas phase is hydrogen, with some amount of water vapor for the deeper section of the transect. Note that despite the high temperature, the density of the gas phase (mostly hydrogen) at these depths (7–10 km bsl) in the shown transect is ~50 kg/m^3. For comparison, the density of H_2 at ambient conditions of 20 °C and 0.1 MPa is 0.084 kg/m^3, that is, 595 times lower. At the pressure and temperature conditions in the multiphase domain (~1,400 bar) the dynamic viscosity of $H_{2(gas)}$ is around 1.5e-5 Pa.s. Note that we do not consider here the gases associated with magma intrusions. It is also possible that some amount of gases (H_2, CH_4, He, CO_2, etc.) from magma mixes with the H_2 from serpentinization in the COT.

5.5 Conclusions

Magma-poor rifted margins are a unique class of tectonic boundaries with enormous potential for the generation of natural and/or stimulated hydrogen. However, their accessibility is limited with respect to potential onshore locations of natural hydrogen accumulations, and it remains unclear whether and under what conditions a significant fraction of the H_2 generated at the distal margins and COT on geological scales is currently trapped. Here we present the context and general aspects related to the potential exploration of natural hydrogen in these tectonic boundaries. Ultimately, the aim of this work is to highlight the potential of these margins as producers of molecular hydrogen and to illustrate various aspects that need to be considered to de-risk exploration in these environments, including the awareness of protolith conditions, tectonic considerations, and the multiphase state of the hydrogen-water mixture. Significant seismic data gaps exist and a clearer understanding of optimal geophysical and petrological data acquisition, as well as H_2 generation, migration, and trapping mechanisms, guided by numerical models, is needed to improve the characterization of these margins to support exploration.

References

[1]   Klein F., Bach W., Humphris S. E., Kahl W. A., Jöns N., Moskowitz B., et al. Magnetite in seafloor serpentinite – Some like it hot. Geology. 02 2014, 42(2): 135–138.
[2]   Liu Z., Pérez-Gussinyé M., García-Pintado J., Mezri L., Bach W. Mantle serpentinization and associated hydrogen flux at North Atlantic magma-poor rifted margins. Geology. 01 2023, 51: 284–289.
[3]   Brun J., Beslier M. Mantle exhumation at passive margins. Earth and Planetary Science Letters. 1996, 142(1): 161–173.
[4]   Minshull T. A. Geophysical characterisation of the ocean–continent transition at magma-poor rifted margins. Comptes Rendus Geoscience. 2009, 341(5): 382–393.

[5]   Reston T. The structure, evolution and symmetry of the magma-poor rifted margins of the north and central Atlantic: A synthesis. Tectonophysics. 2009, 468(1): 6–27. Role of magmatism in continental lithosphere extension continental lithosphere extension.

[6]   Pérez-Gussinyé M., Collier J. S., Armitage J. J., Hopper J. R., Sun Z., Ranero C. R. Towards a process-based understanding of rifted continental margins. Nature Reviews Earth & Environment. 2023, 4(3): 166–184.

[7]   Albers E., Bach W., Pérez-Gussinyé M., McCammon C., Frederichs T. Serpentinization-driven H_2 production from continental break-up to mid-ocean ridge spreading: Unexpected high rates at the West Iberia Margin. Frontiers in Earth Science. 2021, 9.

[8]   Pérez-Gussinyé M., Reston T. J. Rheological evolution during extension at nonvolcanic rifted margins: Onset of serpentinization and development of detachments leading to continental breakup. Journal of Geophysical Research: Solid Earth. 2001, 106(B3): 3961–3975.

[9]   Pérez-Gussinyé M., Reston T. J., Phipps Morgan J. Serpentinization and magmatism during extension at non-volcanic margins: The effect of initial lithospheric structure. In: Wilson R., Whitmarsh R., Taylor B., F N. (eds.), Non-volcanic rifting of continental margins: A comparison of evidence from land and sea, London: Geological Society, Jan 2001, 187: 551–576.

[10] Bayrakci G., Minshull T., Sawyer D., Reston T., Klaeschen D., Papenberg C., et al. Fault-controlled hydration of the upper mantle during continental rifting. Nature Geoscience. 2016, 9(5): 384–388.

[11] Nozaka T., Wintsch R. P., Meyer R. Serpentinization of olivine in troctolites and olivine gabbros from the hess deep rift. Lithos. 2017, 282–283: 201–214.

[12] Frost B. R., Beard J. S. On silica activity and serpentinization. Journal of Petrology. 05 2007, 48(7): 1351–1368.

[13] Bach W., Garrido C. J., Paulick H., Harvey J., Rosner M. Seawater-peridotite interactions: First insights from ODP Leg 209, MAR 15°N. Geochemistry, Geophysics, Geosystems. 2004, 5(9).

[14] Tutolo B. M., Evans B. W., Kuehner S. M. Serpentine–hisingerite solid solution in altered ferroan peridotite and olivine gabbro. Minerals. 2019, 9(1): 47–60.

[15] Klein F., Bach W., McCollom T. M. Compositional controls on hydrogen generation during serpentinization of ultramafic rocks. Lithos. 2013, 178: 55–69. Serpentinites from mid-oceanic ridges to subduction.

[16] Zolotov M. Y. Formation of brucite and cronstedtite-bearing mineral assemblages on ceres. Icarus. 2014, 228: 13–26.

[17] Eberhard L., Frost D. J., McCammon C. A., Dolejš D., Connolly J. A. D. Experimental constraints on the ferric fe content and oxygen fugacity in subducted serpentinites. Journal of Petrology. 09 2023, 64(10): egad069.

[18] Bach W., Paulick H., Garrido C. J., Ildefonse B., Meurer W. P., Humphris S. E. Unraveling the sequence of serpentinization reactions: Petrography, mineral chemistry, and petrophysics of serpentinites from MAR 15°N (ODP Leg 209, Site 1274). Geophysical Research Letters. 2006, 33(13): L13306.

[19] Wolery T. W., Jarek R. L. Software user's manual. EQ3/6, version 8.0. Sandia National Laboratories – U.S. Dept. Of Energy Report. 2003.

[20] Johnson J. W., Oelkers E. H., Helgeson H. C. Supcrt92: A software package for calculating the standard molal thermodynamic properties of minerals, gases, aqueous species, and reactions from 1 to 5,000 bar and 0 to 1,000 ° C. Computers & Geosciences. 1992, 18(7): 899–947.

[21] De Capitani C., Petrakakis K. The computation of equilibrium assemblage diagrams with Theriak/Domino software. American Mineralogist. 2010, 95(7): 1006–1016.

[22] deCapitani C. T.-D. https://titan.minpet.unibas.ch/minpet/theriak/prog20220528/, 2022.

[23] Holland T. J. B., Powell R. An internally consistent thermodynamic data set for phases of petrological interest. Journal of Metamorphic Geology. 1998, 16(3): 309–343.

[24] Holland T. J. B., Powell R. An improved and extended internally consistent thermodynamic dataset for phases of petrological interest, involving a new equation of state for solids. Journal of Metamorphic Geology. 2011, 29(3): 333–383.

[25] Holland T. J. B., Green E. C. R., Powell R. Melting of peridotites through to granites: A simple thermodynamic model in the system KNCFMASHTOCr. Journal of Petrology. 05 2018, 59(5): 881–900.

[26] Welford J. K. Magma-poor continent–ocean transition zones of the southern North Atlantic: A wide-angle seismic synthesis of a new frontier. Solid Earth. 2024, 15(6): 683–710.

[27] Cutts J. A., Steinthorsdottir K., Turvey C., Dipple G. M., Enkin R. J., Peacock S. M. Deducing mineralogy of serpentinized and carbonated ultramafic rocks using physical properties with implications for carbon sequestration and subduction zone dynamics. Geochemistry, Geophysics, Geosystems. 2021, 22(9): e2021GC009989.

[28] Berryman J. G. Mixture theories for rock properties. In: Ahrens T. (ed.), Rock physics & phase relations: A handbook of physical constants, Livermore, California: American Geophysical Union (AGU), 1995. 205–228.

[29] Brune S., Kolawole F., Olive J. A., Stamps D. S., Buck W. R., Buiter S. J. H., et al. Geodynamics of continental rift initiation and evolution. Nature Reviews Earth & Environment. 2023, 4(4): 235–253.

[30] Liu Z., Pérez-Gussinyé M., Rüpke L., Muldashev I. A., Minshull T. A., Bayrakci G. Lateral coexistence of ductile and brittle deformation shapes magma-poor distal margins: An example from the west iberia-newfoundland margins. Earth and Planetary Science Letters. 2022, 578: 117288.

[31] Andrés-Martínez M., Pérez-Gussinyé M., Armitage J., Morgan J. P. Thermomechanical implications of sediment transport for the architecture and evolution of continental rifts and margins. Tectonics. 2019, 38(2): 641–665.

[32] Pérez-Gussinyé M., Andrés-Martínez M., Araújo M., Xin Y., Armitage J., Morgan J. P. Lithospheric strength and rift migration controls on synrift stratigraphy and breakup unconformities at rifted margins: Examples from numerical models, the Atlantic and South China Sea margins. Tectonics. 2020, 39(12): e2020TC006255.

[33] Mezri L., García-Pintado J., Pérez-Gussinyé M., Liu Z., Bach W., Cannat M. Tectonic controls on melt production and crustal architecture during magma-poor seafloor spreading. Earth and Planetary Science Letters. 2024, 628: 118569.

[34] Pérez-Gussinyé M., Xin Y., Cunha T., Ram R., Andrés-Martínez M., Dong D., et al. Synrift and postrift thermal evolution of rifted margins: A re-evaluation of classic models of extension, London: Geological Society, Special Publications 2024, 547.

[35] Kuang X., Jiao J. J. An integrated permeability-depth model for earth's crust. Geophysical Research Letters. 2014, 41(21): 7539–7545.

[36] Detournay E., Cheng A. Fundamentals of poroelasticity. In: Fairhurst C. (ed.), Comprehensive rock engineering: Principles, practice and projects, vol ii, analysis and design method, New York: Pergamon Press, 1993. 113–171.

[37] Chen Z., Zhang Y. Well flow models for various numerical methods. International Journal of Numerical Analysis and Modelling. 2009, 6(3): 375–388.

[38] Wagner W., Pruß A. The IAPWS formulation 1995 for the thermodynamic properties of ordinary water substance for general and scientific use. Journal of Physical and Chemical Reference Data. 2002, 31(2): 387–535.

[39] Phipps Morgan J. Thermodynamics of pressure release melting of a veined plum pudding mantle. Geochemistry, Geophysics, Geosystems. 2001, 2(4).

[40] Tucholke B. E., Parnell-Turner R., Smith D. K. The global spectrum of seafloor morphology on mid-ocean ridge flanks related to magma supply. Journal of Geophysical Research: Solid Earth. 2023, 128(12): e2023JB027367.

[41] Ros E., Pérez-Gussinyé M., Araújo M., Thoaldo Romeiro M., Andrés-Martínez M., Morgan J. P. Lower crustal strength controls on melting and serpentinization at magma-poor margins: Potential implications for the South Atlantic. Geochemistry, Geophysics, Geosystems. 2017, 18(12): 4538–4557.

[42] Zhao M., Canales J. P., Sohn R. A. Three-dimensional seismic structure of a Mid-Atlantic Ridge segment characterized by active detachment faulting (Trans-Atlantic Geotraverse, 25°55′N–26°20′N). Geochemistry, Geophysics, Geosystems. 2012, 13(11).

[43] Canales J. P., Collins J. A., Escartín J., Detrick R. S. Seismic structure across the rift valley of the mid-atlantic ridge at 23°20′ (mark area): Implications for crustal accretion processes at slow spreading ridges. Journal of Geophysical Research: Solid Earth. 2000, 105(B12): 28411–28425.

[44] Canales J. P., Dunn R. A., Arai R., Sohn R. A. Seismic imaging of magma sills beneath an ultramafic-hosted hydrothermal system. Geology. 05 2017, 45(5): 451–454.

[45] Xu M., Zhao X., Canales J. P. Structural variability within the Kane oceanic core complex from full waveform inversion and reverse time migration of streamer data. Geophysical Research Letters. 2020, 47(7): e2020GL087405.

[46] Horning G., Sohn R. A., Canales J. P., Dunn R. A. Local seismicity of the rainbow massif on the mid-atlantic ridge. Journal of Geophysical Research: Solid Earth. 2018, 123(2): 1615–1630.

[47] Hauser F., O'Reilly B. M., Jacob A. W. B., Shannon P. M., Makris J., Vogt U. The crustal structure of the rockall trough: Differential stretching without underplating. Journal of Geophysical Research: Solid Earth. 1995, 100(B3): 4097–4116.

[48] Reston T., Pennell J., Stubenrauch A., Walker I., Perez-Gussinye M. Detachment faulting, mantle serpentinization, and serpentinite- mud volcanism beneath the porcupine basin, southwest of Ireland. Geology. 07 2001, 29(7): 587–590.

[49] Reston T., Gaw V., Pennell J., Klaeschen D., Stubenrauch A., Walker I. Extreme crustal thinning in the south porcupine basin and the nature of the porcupine median high: Implications for the formation of non-volcanic rifted margins. Journal of the Geological Society. 09 2004, 161(5): 783–798.

[50] Pérez-Gussinyé M., Ranero C. R., Reston T. J., Sawyer D. Mechanisms of extension at nonvolcanic margins: Evidence from the Galicia interior basin, west of Iberia. Journal of Geophysical Research: Solid Earth. 2003, 108(B5): 2245.

[51] O'Reilly B., Hauser F., Ravaut C., Shannon P., Readman P. Crustal thinning, mantle exhumation and serpentinization in the porcupine basin, offshore Ireland: Evidence from wide-angle seismic data. Journal of the Geological Society. 09 2006, 163(5): 775–787.

[52] Corbalán A., Nedimović M. R., Louden K. E., Cannat M., Grevemeyer I., Watremez L., et al. Seismic velocity structure along and across the ultraslow-spreading Southwest Indian Ridge at 64°30′E showcases flipping detachment faults. Journal of Geophysical Research: Solid Earth. 2021, 126(10): e2021JB022177.

[53] Robinson A. H., Watremez L., Leroy S., Minshull T. A., Cannat M., Corbalán A. A 3-D seismic tomographic study of spreading structures and smooth seafloor generated by detachment faulting – The ultra-slow spreading Southwest Indian Ridge at 64°30′E. Journal of Geophysical Research: Solid Earth. 2024, 129(9): e2024JB029253.

[54] Kuster G. T., Toksöz M. N. Velocity and attenuation of seismic waves in two-phase media: Part i. theoretical formulations. Geophysics. 1974, 39(5): 587–606.

[55] Jian H., Nedimović M. R., Canales J. P., Lau K. W. H. New insights into the rift to drift transition across the northeastern Nova Scotian margin from wide-angle seismic waveform inversion and reflection imaging. Journal of Geophysical Research: Solid Earth. 2021, 126(12): e2021JB022201.

[56] IAPWS. Guidelines on the Henry's constant and vapour liquid distribution constant for gases in H_2O and D_2O at high temperatures. Tech. Rep., IAPWS, 2004.

[57] Bell I. H., Wronski J., Quoilin S., Lemort V. Pure and pseudo-pure fluid thermophysical property evaluation and the open-source thermophysical property library CoolProp. Industrial & Engineering Chemistry Research. 2014, 53(6): 2498–2508.

[58] Leachman J. W., Jacobsen R. T., Penoncello S., Lemmon E. W. Fundamental equations of state for parahydrogen, normal hydrogen, and orthohydrogen. Journal of Physical and Chemical Reference Data. 2009, 38(3): 721–748.

[59] Muzny C. D., Huber M. L., Kazakov A. F. Correlation for the viscosity of normal hydrogen obtained from symbolic regression. Journal of Chemical & Engineering Data. 2013, 58(4): 969–979.

[60] Huber M. L., Perkins R. A., Laesecke A., Friend D. G., Sengers J. V., Assael M. J., et al. New international formulation for the viscosity of H_2O. Journal of Physical and Chemical Reference Data. 04 2009, 38(2): 101–125.

[61] Pruess K., Oldenburg C., Moridis G. Tough2 user's guide, version 2.0. Tech. Rep. LBNL-43134, Lawrence Berkeley National Laboratory, Berkeley CA, USA, 1999.

[62] Wilkins A., Green C. P., Ennis-King J. PorousFlow: A multiphysics simulation code for coupled problems in porous media. Journal of Open Source Software. 2020, 5(55): 2176.

Humberto Reis and Olivier Lhote

Chapter 6
Exploring natural hydrogen in the oldest nuclei of continents: why do cratons matter?

Abstract: An increasing number of natural hydrogen emissions and discoveries have been reported and studied worldwide. Their detailed examination reveals a dominant distribution in the interior of continents, within geological domains known as cratons. Comprising the oldest nuclei of continents, cratons host some of the largest hydrographic basins in the world and correspond to abnormally thick and cold lithospheric domains containing rocks as old as c. 4.4 Ga. These lithospheric domains became mechanically strong and buoyant after a long history of accretion and differentiation and, for this reason, are able to resist multiple crustal recycling events since the earliest evolutionary stages of Earth. Their strength also allowed the preservation of unique rock assemblages and tectonic architectures in the Archean to Paleoproterozoic basement, as well as the multiple basin-cycle records that form the Proterozoic to Phanerozoic intracratonic basins. Recent advances have shown that their complex geological characteristics and tectonically stable conditions make cratons promising settings for natural hydrogen exploration. In this chapter, we analyze the relationship between the known natural hydrogen shows, emissions, and discoveries and the composition and architecture of the oldest cratonic nuclei of continents. Key examples from West Africa, South Australia, and Brazil (São Francisco craton) reveal a close relationship between the main occurrences reported so far and a wide range of potential sources in the basement, as well as reservoirs, seals, and traps in the overlying intracratonic basins. Besides gathering favorable conditions for natural hydrogen generation through hydration of Fe^{2+}-rich basement assemblages, radiolysis, and overmaturation of organic-rich sedimentary rocks, these old lithospheric domains ex-

Acknowledgments: The authors thank Storengy/ENGIE France and HR Consulting Energy and Geosciences for the support and permission for publishing part of the ideas presented here. The authors are also grateful to Tiphaine Fargetton, Chaterine Formento, Tiphayne Tual, Pierre Levin, Stephane Galibert, and Fernando Alkmim for the insightful and exciting discussions over the last years. The revisions and suggestions from Reza Rezaee helped to improve the original version. They thank him and Brian J. Evans for the invitation.

Humberto Reis, HR Consulting Energy and Geosciences Ltda., Rua Nunes Vieira, 35, 401, Belo Horizonte, Minas Gerais 30350-120, Brazil; Departamento de Ciência da Computação/Instituto de Geosciencias, Universidade Federal de Minas Gerais (UFMG), Campus Pampulha, Belo Horizonte, Minas Gerais, Brazil, e-mail: hr@hrenergyandgeosciences.com, reis.humbertols@gmail.com
Olivier Lhote, ENGIE, Paris La Défense, France, e-mail: Olivier.LHOTE@storengy.com

https://doi.org/10.1515/9783111437040-006

hibit continental-scale structures that may comprise efficient migration pathways connecting deep sources and different types of reservoirs and structural traps sealed by igneous intrusions or low-permeability sedimentary rocks. The surface and subsurface water availability through geological time has probably enhanced natural hydrogen generation and, in some cases, its migration and storage. A similar relationship between natural hydrogen emissions and discoveries can also be observed in the interior of the North American craton and other cratons around the world. In spite of the serendipity of many natural hydrogen discoveries, our examples demonstrate that their occurrence within the oldest nuclei of continents is most likely controlled by typical craton features rather than coincidence. Thus, understanding the geological architecture and evolution of these ancient lithospheric domains is essential for defining natural hydrogen plays and constructing successful exploration strategies. Advancing the definition of the still poorly understood processes and elements of the natural hydrogen system is also needed. Regardless of the many open questions, current knowledge points toward the first-order importance of cratons in the nascent hydrogen research and exploration.

Keywords: natural hydrogen, craton, Archean, Proterozoic, intracratonic basin, ultramafic rocks, radiolysis, exploration play

6.1 Introduction

A few billion years ago, only a dense cloud of dust and hydrogen gas existed in the region of the Universe that later became our solar system (e.g., [1]). In just a few million years, this primitive hydrogen-rich cloud collapsed and agglutinated to form planetesimals, as well as gaseous, and rocky planets. As one of these planets, the 4.56-Ga-old Earth evolved from an uninhabitable primitive form to its current dynamic and differentiated state. This process involved the evolution and amelioration of its early, toxic, and hot atmosphere, from which large volumes of hydrogen gases were lost to space or incorporated into other compounds that comprise different earthly spheres. The secular changes in molecular hydrogen concentrations in the atmosphere were apparently induced by their higher reactivity, lower size and density, and the balance between local and external supply [2]. Most of the hydrogen atoms on modern Earth are stored in water molecules of the hydrosphere, as well as in organic and inorganic compounds that are part of the biosphere and lithosphere, such as proteins, carbohydrates, hydrocarbons, coal, and hydrated minerals (e.g., [3, 4]). Experimental and theoretical evidence suggests that large amounts of primordial H_2 were also trapped in the Earth's core [5, 6].

The understanding of Earth's evolution and dynamics, as depicted above, has been accompanied over the last decades by the misleading conception that no hydrogen accumulations could exist in the subsurface. Combined with the higher risk and lack of commercial interest in this compound, this misconception has also prevented its massive exploration up to now. Nevertheless, the emergence of hydrogen as a key-element in the transition to a global low-carbon economy [7], the recent discovery of subsurface natural hydrogen accumulations in Mali [8], the numerous reports and studies revealing anomalous concentrations in subsurface reservoirs, and the higher rates of natural production at the surface in different geological settings have changed this scenario [9–12]. Currently, billions of dollars are being invested by government agencies, start-ups, and major oil and gas companies in natural hydrogen research and exploration.

When carefully examined, the majority of known natural hydrogen occurrences and studies conducted thus far are concentrated in the interior of continents, within their oldest crustal domains and sedimentary covers (Figure 6.1). Referred to as cratons, these domains exhibit unique geological characteristics that allowed them to survive multiple crustal recycling events through geological time and to preserve a rock record as old as 4.5 Ga [1]. This rock record stands out as the major direct evidence of the evolution of the planet Earth, from its earliest extreme conditions to its modern habitable state [13]. The curious and close relationship between worldwide occurrences and cratons seems to reveal promising tectonic settings to find economic accumulations or production fields. Nevertheless, this connection is still unclear.

In this chapter, we explore the connection between cratons and naturally occurring hydrogen shows and accumulations. After discussing the definition, architecture, and evolution of these ancient continental domains through the geological time, we analyze how the typical elements inherited from their long evolutionary history make cratons promising settings for exploring natural hydrogen. We address examples from South and North America, Africa, Europe, and Australia, which represent the best-known cases to date. By discussing these connections, we also contribute to advances in the understanding of continent-related natural hydrogen systems and plays.

6.2 Cratons

Cratons comprise the oldest segments of the continental crust and make up most of the Precambrian shields and platforms. Underlying continent-wide lowland areas, these crustal domains are well-drained and host some of the largest aquifers and hydrographic basins in the world, from which their names are usually derived (Figures 6.1 and 6.2).

Two major architectural concepts have been adopted in the last centuries to define cratons and their limits; however, recent scientific advances have led to modern definitions purely based on geodynamic aspects and their whole lithospheric structure [17]. In this view, cratons can be defined as old, thick, cold, and stable continental lithospheres that may exhibit thicknesses exceeding 300 km and show melt-depleted, mechanically resistant, and buoyant mantles (e.g., [17, 18]). These characteristics result

a)

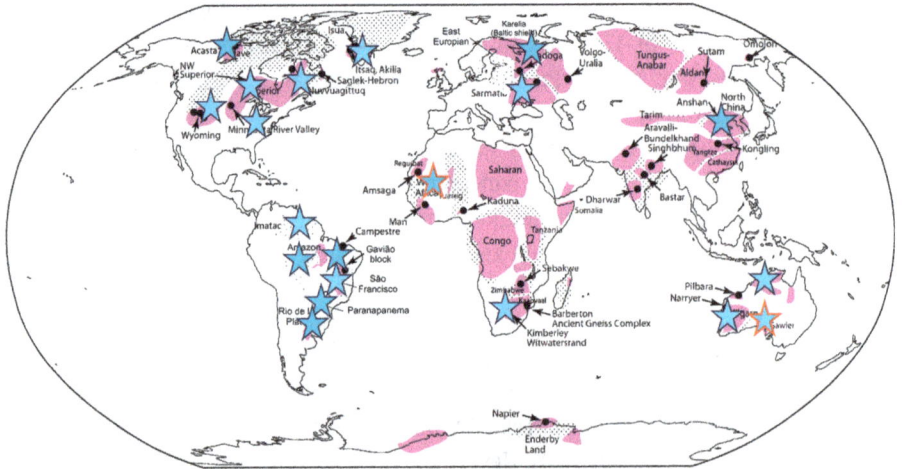

b)

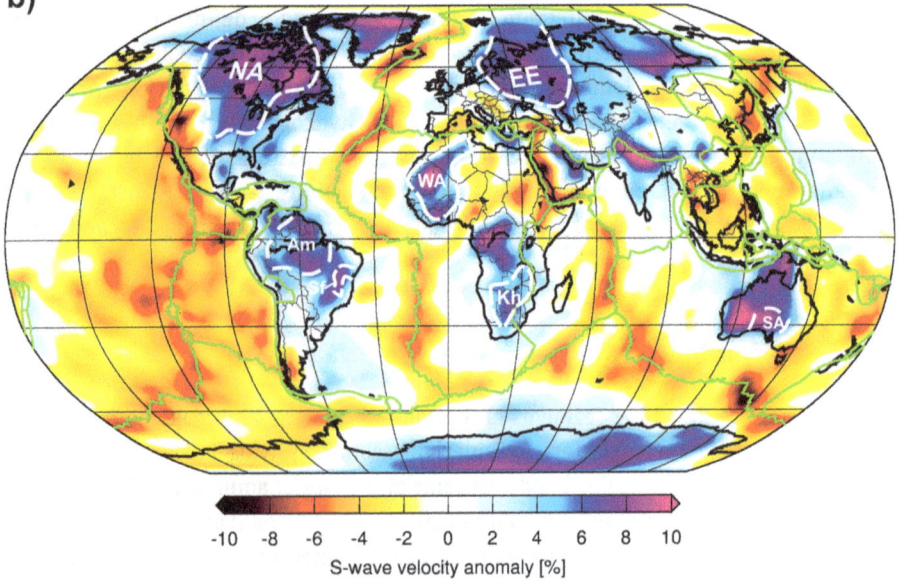

Figure 6.1: (a) Cratons of the world and natural hydrogen occurrences (blue stars) documented by the authors and available in the literature (e.g., [11, 14]) as surface emissions, shown in old hydrocarbon and new H$_2$ wells, as well as fluid inclusions in Precambrian basement rocks. The red-contoured blue stars indicate the Mali and South Australia discoveries. The first is the only field producing H$_2$ known so far. The world map of cratons was modified from [15] and extracted from [13]. (b) Global tomographic seismic

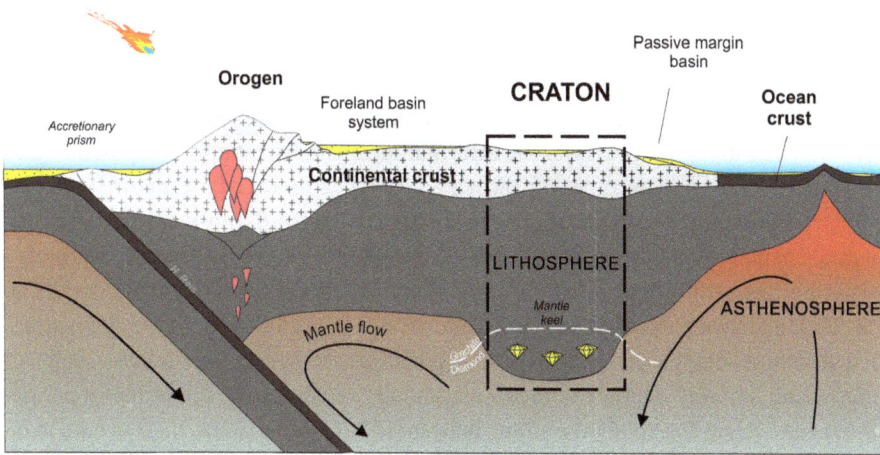

Figure 6.2: Sketch illustrating the cratonic lithospheric structure compared with other sedimentary settings. See text for further details. Based on Alkmim (2004).

from a few billion years of evolution and have allowed their survival through multiple crustal recycling events. Consequently, cratons preserve a Hadean to Cenozoic geological record, show heat fluxes and geothermal gradients two to four times lower than surrounding settings, and exhibit limits invariably defined by deep-seated structures of Precambrian to modern orogenic and rift-to-passive margin systems [19–23].

Available information reveals that worldwide cratons acquired their tectonic stability during two main global crustal recycling episodes, which occurred in the Neoarchean and Paleoproterozoic eras [24, 25]. Together with contrasting concepts adopted in earlier studies, these factors have led to different definitions in the literature that often distinguish Archean and (Paleo)Proterozoic cratons. In this chapter, we adopted the geodynamic- and lithospheric-based definition, in which cratons are continental-scale domains that encompass adjacent lithospheric domains stabilized during both the Archean and Proterozoic eons. For the sake of clarification, other names are used herein only to distinguish these domains. For instance, while the North American craton includes the Archean Superior, Slave, and Wyoming nuclei, the Kalahari craton includes the Archean Kaapvaal and Zimbabwe nuclei. Both cratons are also composed of Archean lithospheric domains partially reworked during the early to mid-Proterozoic orogenies and contemporaneous juvenile rocks (e.g., [24, 26]) (Figure 6.1).

Figure 6.1 (continued)

model showing the S-wave velocity anomaly at 100 Km-depth. Modified from [16]. Bluish areas indicate the thickest lithospheric domains underlain by anomalously thick cratonic mantle keels. The North American (NA), East European (EE), Kalahari (Kh), and South Australian (SA) cratons include multiple large-scale older nuclei stabilized during the Archean (also known as Archean cratons), as shown in (a). WE, West African craton; Am, Amazon craton; Sf, São Francisco craton.

6.2.1 Anatomy of cratons and their evolution through time

Cratons are defined by two major lithospheric components: (i) a differentiated mantle and (ii) a heterogeneous crust containing a Precambrian basement overlain by Phanerozoic rock assemblages. Overall, information on their mantle composition and structure is derived from xenoliths recovered from deep-sourced magmatic rocks, petrological modeling and deep geophysical surveys. These data reveal that cratonic mantles usually form thick roots beneath continents and are dominantly composed of peridotites, with minor amounts of pyroxenites and eclogites (Figures 6.1 and 6.3) [18, 27, 28]. Showing densities ranging from circa 3.36 to 3.31 kg/m^3, they exhibit stratified structures defined by a highly iron-depleted upper portion and a lower segment affected by variable degrees of melt metasomatic replacement. While the upper depleted zone seems to explain the strong buoyant behavior of cratons, their overall mantle composition and structure make them mechanically strong lithospheric domains [29] (Figure 6.3).

Cratonic crusts exhibit thicknesses ranging between approximately 30 and 50 km and usually comprise an Archean to Paleoproterozoic basement overlain by intracratonic sedimentary basins, Mesozoic fossil rifts, and associated assemblages [17, 21]. The basement is composed of assemblages that include Hadean to Archean metamorphic rocks derived from granitoids of the trondhjemite-tonalite-granodiorite series[1] (also referred to as TTG complexes) and mafic- to ultramafic-rich igneous rocks associated with felsic and chemical to clastic sedimentary units of the greenstone belt successions. The Hadean to Archean cratonic assemblages form unique tectonic architectures referred to as dome-and-keels, in which large TTG-dominated domes are surrounded by troughs filled with greenstone belt successions [30–33]. These assemblages are unconformably overlain by Paleoproterozoic continental to passive margin strata and intruded by more evolved granitoids. While greenstone belts are known for containing world-class orogenic gold and other metal deposits, the Paleoproterozoic metasedimentary units host the largest Lake Superior-type iron ore deposits known so far, such as those from the Brazilian Carajás and Quadrilátero Ferrífero mineral provinces [34–36].

The basement of cratons has been portrayed as a record of the extreme conditions prevailing in the early Precambrian Earth. In this scenario, the higher mantle potential temperatures during the Hadean and Archean eons would have: (i) prevented the emergence of long-lived and steep subduction systems; (ii) induced crustal recycling processes marked by gravitational overturns and plume activity; and (iii) hampered the formation of intermediate andesitic magmas [13 and references

---

1 The TTG suites are typical Archean assemblages that correspond to silica-rich (commonly 70.0 wt% or greater) rocks, with high Al_2O_3 (15.0–16.0 wt%) and Na_2O (3.0–7.0 wt%) contents. They also exhibit low K_2O/Na_2O (<0.5) ratios and ferromagnesian oxides (≤5.0 wt%), fractionated rare earth element (REE) patterns, and low heavy REE contents [25].

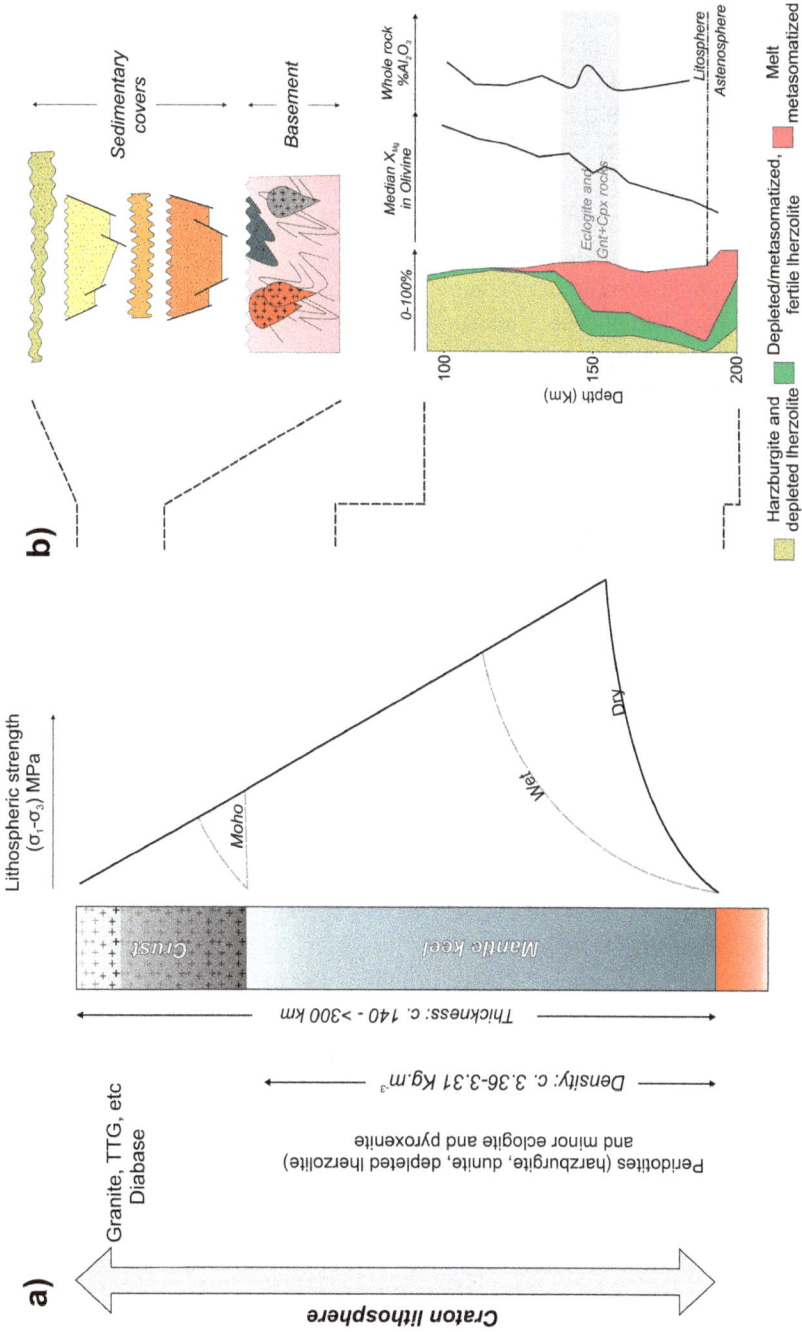

Figure 6.3: (a) Simplified column showing the main lithospheric elements of cratons, their average composition, density, and thickness. The illustration also depicts the lithospheric strength of cratons (black line) compared with wetter, non-cratonic lithospheres. (b) Sketch of the typical craton crusts

therein]. These processes seem to have favored the formation of silica- and sodium-rich TTG suites in shallow, short-lived subduction zones (which later gave rise to stable, modern-like slabs) and the formation of ultramafic extrusive rocks (i.e., komatiites) locally associated with tholeiitic metabasalts of the greenstone belt successions [37–41]. The Archean to early Proterozoic thermal conditions may have also favored the formation of dome-and-keel architectures, either through gravitational inversion or accretion followed by orogenic collapse [30, 31, 33, 42–44].

The formation of Siderian Lake Superior-type banded iron formations is closely related to profound changes occurring in the Earth system between 2.0 and 2.5 billion years ago. Following the uptake of iron from the Fe-saturated oceans due to the emergence of oxygenic photosynthesis in the biosphere, the oxygen-poor atmosphere-ocean system shifted to O_2-richer conditions during the 2.3–2.2 Ga Great Oxidation Event (GOE). These conditions marked coeval changes in sedimentary systems and apparently followed the oldest-known Snowball Earth glaciation, referred to as the 2.45–2.30 Ga Huronian Glaciation (Figure 6.4) [13, 35, 45–48].

Intracratonic basins are common elements of craton crust and correspond to polyhistoric depocenters that host multiple and superposed basin-cycle records [49, 50]. They are mostly circular- to oval-shaped in map-view and may exhibit saucer and steer's head geometries in depth [21, 50, 51]. Showing a long subsidence history that often started with a Proterozoic or Paleozoic rifting episode, these depocenters are typically filled with shallow marine- to continental-dominated strata composed of siliciclastics, carbonate rocks, and evaporites (Figure 6.5) [21, 50, 52, 53]. The basin-fill units may reach up to several kilometers-thick and range from Proterozoic to Phanerozoic in age. Well-known examples include the Amazonas, Solimões, Parnaíba, São Francisco, and Parecis basins of Brazil [54], the Taoudeni and Congo basins of Africa [52, 55], the Ordos Basin of China [22], the Mid-Continent Rift System, and the Illinois, Michigan, and Williston basins of North America [50, 56] (Figure 6.5).

Different mechanisms have been proposed as controlling the initiation of intracratoic basins: (i) rifting, (ii) plume activity, (iii) mantle underplating, (iv) reactivation of preexisting structures under flexural subsidence, among others [50 and references therein]. Nevertheless, the available tectono-stratigraphic record indicates that these basins were formed during major Precambrian to Phanerozoic plate reorganization episodes, either involving mantle plumes or not [21, 50, 51, 58]. Their evolution since then has been punctuated by multiple subsidence and uplift episodes, invariably involving the reactivation of Archean to Proterozoic crustal structures of the basement [59]. The craton-related stability and their long evolutionary history have allowed intracratonic basins to preserve important tectonic and climatic events

---

Figure 6.3 (continued)
composed of TTG domes (pink), greenstone belt keels (green), associated intrusions (red and gray), and Paleoproterozoic iron-rich strata, often overlain by polyhistoric intracratonic basins. Below, chemotomography of the Slave craton lithospheric mantle [18]. Modified from [18, 27].

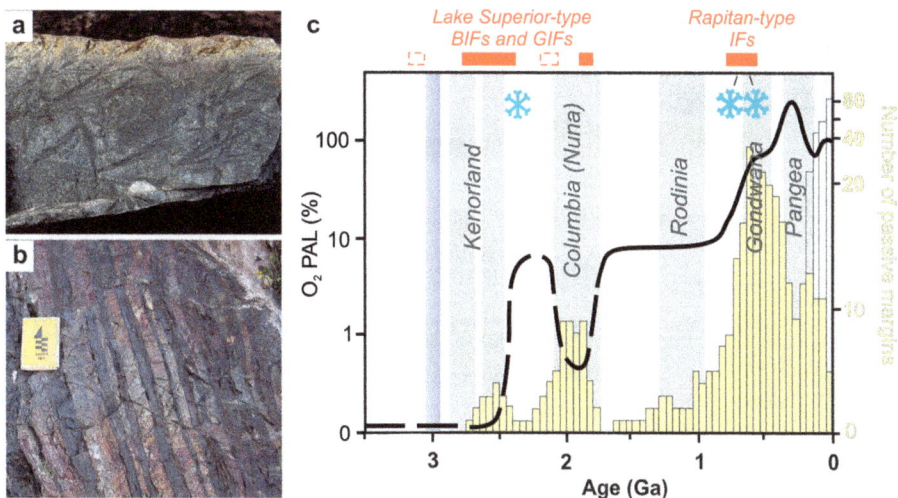

Figure 6.4: (a) Neoarchean komatiites of the Rio das Velhas greenstone belt showing typical spinifex structures, and (b) deformed and metamorphosed Siderian Lake Superior-type banded iron formation (BIF) of the Quadrilátero Ferrífero Mineral province (Brazil). (c) Evolution of passive margins, episodes of enhanced deposition of iron formations (red bars), global glacial events (Paleoproterozoic Huronian and middle to late Neoproterozoic Snowball Earth), oxygen atmospheric levels, and supercontinents (gray shaded bars). Modified from [57] and references therein. Abbreviations: BIF, banded iron formation; GIF, granular iron formation; IF, iron formation; PAL, Present-day level. The blue ice symbols represent the global glaciations known to date.

affecting the Earth system since the Proterozoic, including the Neoproterozoic Snowball Earth glaciations, the Gondwana and Pangea assembly and break-up, and Cretaceous climate and anoxia events [60, 61]. Brazilian and African intracratonic basins, such as the Parnaíba, Paraná, Tacutu, Taoudeni, and Cape-Karoo, also contain widespread igneous intrusions that were emplaced during the Central and South Atlantic events (Figure 6.5) [51, 54, 58].

Mesozoic fossil rifts that were formed during the Pangea supercontinent break-up are often found associated with cratonic lithospheres. These rifts predominantly host fine- to coarse-grained siliciclastic deposits and, in some cases, thick evaporite strata [54 and references therein]. Fossil rifts such as the Tacutu and Reconcavo-Tucano-Jatobá of South America and the Central to West Africa rift systems connect the cratonic lithospheres with the ocean and may contain widespread igneous intrusions formed during the Mesozoic Central and South Atlantic events [54]. In contrast to their counterparts developed in extra-cratonic lithospheres, these basins are characterized by a tectonic architecture and sedimentary dispersal directly controlled by the higher cratonic mechanical strength and its Precambrian to early Paleozoic inherited structures [17, 62]. The Mesozoic extensional far-field stresses also induced the formation of fossil rift basins in the continent's interior. In the São Francisco Basin of Brazil, the Cretaceous opening of the South Atlantic Ocean culminated in the forma-

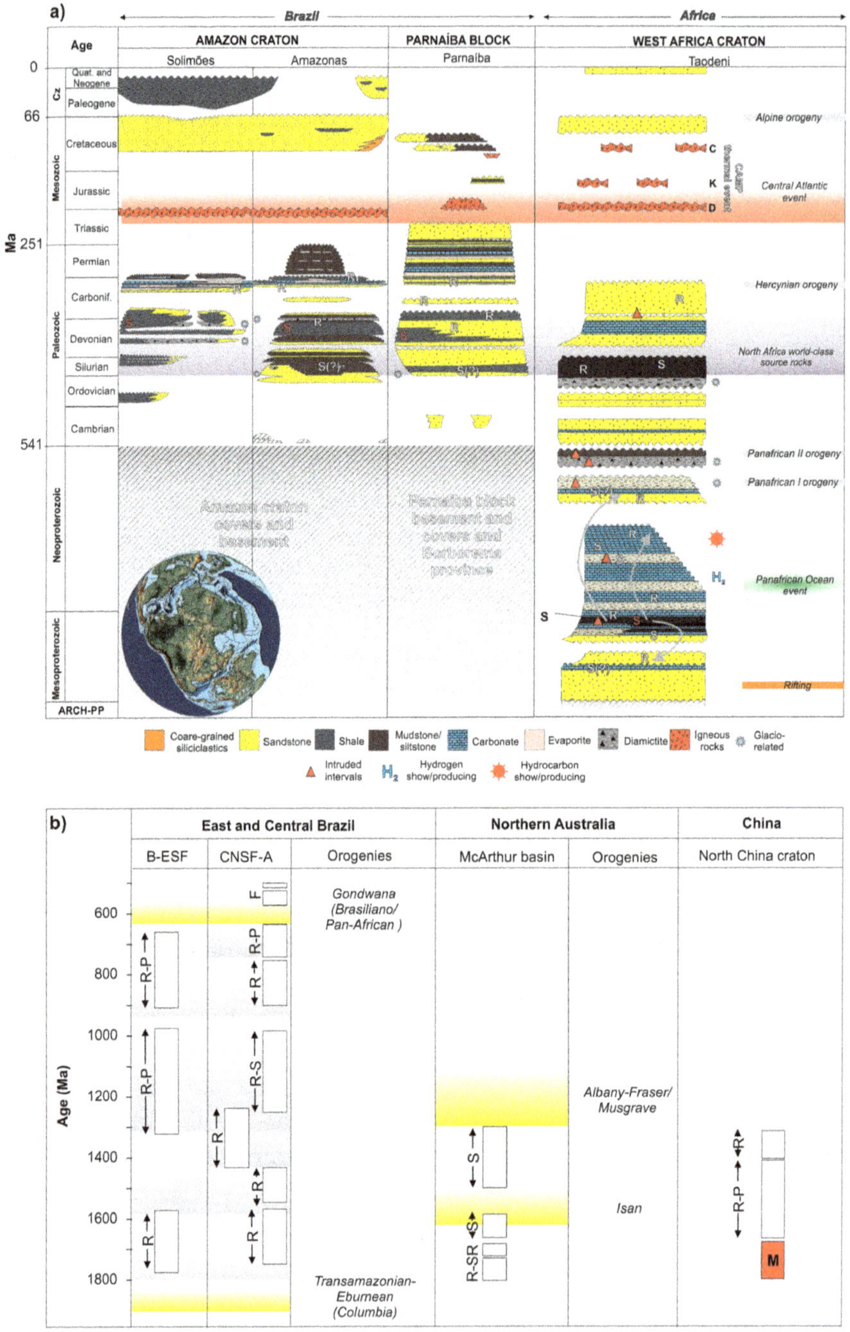

Figure 6.5: (a) Stratigraphic chart correlating Brazilian and West African intracratonic basins, indicating petroleum system elements, natural gas shows, and natural hydrogen discoveries. Modified from [54 and references therein] and [69]. In the lower right, a tectonic reconstruction depicts the connection between

tion of a relatively small fossil rift whose structures and sedimentary dispersal were almost entirely controlled by the reactivation of preexisting Neoproterozoic to early Paleozoic structures [21, 63]. However, no clear and perennial connections with contemporaneous continental margin basins have been identified to date.

Intracratonic basins and Mesozoic fossil rifts have been known for decades for their remarkable hydrocarbon reserves [64]. In northern African, South American, and North American basins, petroleum and natural gas accumulations are usually found in continental to marine siliciclastic and carbonate reservoirs of different ages. These accumulations are associated with structural and stratigraphic traps, sealed by shales and/or evaporites, and have been mostly sourced by Ordovician to Cretaceous organic-rich strata [60]. Precambrian petroleum systems and unconventional reservoirs are also common components of Asian, Australian, American, and African intracratonic basins such as the East Siberia, McArthur, Michigan, Taoudeni, and São Francisco [65–67]. The latter hosts the only unconventional petroleum system known so far that is exclusively composed of Precambrian to early Paleozoic elements and processes [68].

6.2.2 On the tectonic evolution of cratons and the supercontinent cycle

Although it is still a matter of debate, the rock assemblages and tectonic architectures preserved in cratons seem to depict different Earth dynamics during the Hadean to early Archean [25]. Different models based on these records, numerical modeling and the observation of other solar system planets have been proposed to explain the formation of the earliest crust and its interaction with the mantle and core. Most of these models converge on pre-plate tectonic modes controlled by the higher temperatures and lower viscosity of the mantle, and dominated by the nucleation and premature breakoff of transient subduction systems, higher mantle plume activity, and higher rates of gravitational overturns [71, 72]. Different lines of evidence, however, reveal that the planet's ameliorating conditions led to the growth and preservation of increasingly higher volumes of crust through the Archean. It is likely that at least 60–70% of the present-day continental crust had already formed by the Mesoarchean (Figure 6.7) (e.g., [73, 74]). From this time onward, stable and steep subduction zones seem to have become increasingly common, allowing mantle and crust differentiation and, ultimately, the stabilization of the oldest cratonic lithospheres in the Neoarchean and early Proterozoic. Once stabilized, these lithospheric pieces grew as the nuclei of

---

Figure 6.5 (continued)
the two continents and the correlated basins 160 million years ago [70]. (b) Correlation chart between the intracratonic Proterozoic strata of the São Francisco (Brazil), McArthur (Australia), and the North China craton (China). Based on [21, 53].

the first continents and began to interact with other lithospheres to form the earliest large landmasses [75, 76] (Figure 6.6).

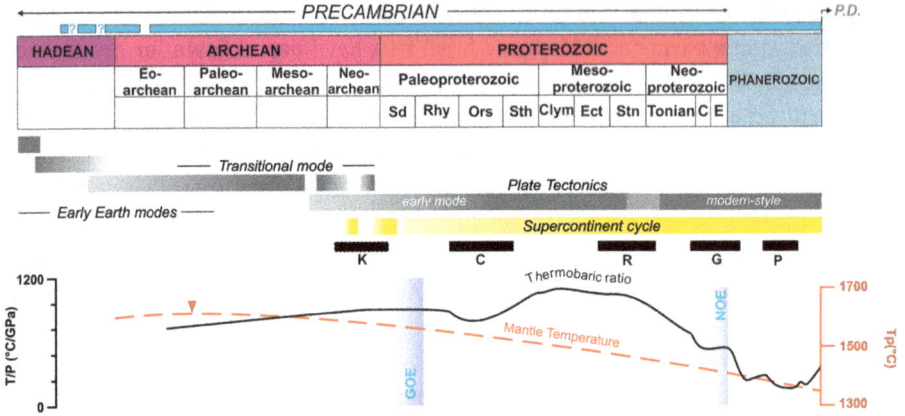

Figure 6.6: Earth's tectonic modes through geological time, as indicated by the oldest mineral and rock records preserved in cratons, numerical modeling and the observation of solar system planets and satellites (after [75]). Below, the thermobaric (black) and potential mantle temperature (red) from [76, 77], respectively. Supercontinents: K, Kenorland; C, Columbia; R, Rodinia; G, Gondwana; P, Pangea. Abbreviations: P.D., present day; Sd, Siderian; Rhy, Rhyacian; Ors, Orosirian; Sth, Statherian; Ect, Ectasian; Stn, Stenian; C, Cryogenian; E, Ediacaran; GOE, Great Oxidation Event; NOE, Neoproterozoic Oxidation Event. The blue bar above indicates the probable time interval in Earth evolution during which surface oceans existed. Figure modified from [13].

The available geological record indicates that complete cycles of formation and destruction of oceans (i.e., Wilson Cycles) started just after the stabilization of the first landmasses in the Archean. At least, five supercontinents have been formed since then: (i) the Archean Kenorland, (ii) the Paleoproterozoic Columbia, (iii) the Meso to Neoproterozoic Rodinia, (iv) the late Neoproterozoic to early Paleozoic Gondwana, and (v) the Paleozoic to Mesozoic Pangea (Figure 6.6) [75]. Composing the nuclei of these continents, worldwide cratons grew and were partially destroyed since the late Archean, preserving similar and correlative rock assemblages that can be found in their basement and sedimentary covers (Figure 6.5). Consequently, their geological characteristics, formation processes, and natural economic resources can also be tracked through paleotectonic reconstructions and direct rock correlations.

6.3 Natural hydrogen in the oldest nuclei of continents

Many of the worldwide discoveries and natural hydrogen emissions reported thus far are intrinsically associated with typical tectonic elements of cratons. Although still poorly understood, these occurrences seem to be associated with: (i) the composition of basement assemblages and their sedimentary covers, (ii) inherited tectonic structures, (iii) the record of major magmatic episodes, among others. Table 6.1 lists selected shows and discoveries, their geological settings, possible natural hydrogen system elements and processes and related aquifers/hydrographic basins. The following sections explore three of these cases, which are discussed in terms of exploratory plays and natural hydrogen system.

6.3.1 West African craton of Africa: the first natural H_2 discovery

To date, the West African craton hosts the only known natural hydrogen subsurface accumulation that has produced for a long period [8, 10]. After the accidental discovery of the gases during the drilling of a shallow water well in 1987, an exploratory campaign exclusively focused on natural hydrogen was initiated in 2012 by a small company named Hydroma Inc. (formerly Petroma). The old and new wells produced gases containing up to 98% H_2 and minor amounts of N_2, CH_4, and He [10], which were used to supply a CO_2-free, pilot power plant constructed at the Bourakebougou village. The plant supplied the small village with electricity for the first time and has been operating for more than 7 years. The wildcat hydrogen well produced gas to the surface from subsurface reservoirs for decades with an approximate flow rate of 1,500 m^3/day and a pressure of 4 bars [8]. No pressure decline has been reported and, surprisingly, current measurements indicate an increase to 5 bars [8].

The West African craton is one of the largest African cratons, with a maximum length and width of 2,800 km and 1,600 km, respectively. The craton exhibits a thick lithosphere that may exceed 250 km, as well as relatively colder mantle temperatures and crustal thickness varying between c. 20 km and 40 km [78, 79]. To the west, east, and south, respectively, it is bounded by the Pan-African Trans-Saharan orogenic system, the Hercynian Mauritanides fold-thrust belt, and the passive margin systems developed during the opening of the central-southern Atlantic Ocean. To the north, its limit coincides with the intracontinental Atlas fold-thrust belt and the Mesozoic basins of the Moroccan southern provinces and Western Meseta [80–83]. The Malian natural hydrogen discovery is located in the southern portion of the craton, within the border of the intracratonic Taoudeni Basin. The basin contains Proterozoic to Cenozoic sedimentary successions that unconformably overlie the basement assemblages exposed in the Man-Leo shield [8, 84]. The shield is formed by Archean TTG gneisses and gran-

Table 6.1: List of selected hydrogen occurrences and discoveries according to their geological settings, reservoirs, possible sources and seals, and hydrographic basins. Three of them are described in detail in the text, where sources are provided. The data from North America are based on [86–88].

	West Africa	South Australia	São Francisco	North America
Hydrogen occurrences and discoveries	The Mali discovery showing 97% purity and neighboring. H_2-emitting fairy circles.	High H_2-rich gases from old wells, whose occurrence and concentrations were confirmed by new wells show up to 86% and 17% of H_2 and He, respectively. Fairy circles are present.	Wells show with high H_2 content (up to approximately 40%) and positive emissions on several fairy circles, some of them measured over the long term.	Wells showing up to 91.7% of free H_2 content in close to the Mid-Continent Rift system, and commonly associated with higher amounts of N_2 and He (up to 3%) and variable to absent concentrations of CH_4 and CO_2. Fairy circles with natural hydrogen emissions measured locally and old soil measurements indicating anomalous concentrations of H_2, He, and CH_4.
Geological setting	Intracratonic basin	Intracratonic basin	Intracratonic basin	Intracratonic basin (Mid-Continent Rift System)

Possible sources, seals, reservoirs, and migration pathways	Sources: Hydration reactions on ferrous minerals in mafic intrusions, basement ultramafics, and Fe₂-rich units in the sedimentary covers as sources. Reservoirs: Carbonates and sandstones of the Neoproterozoic Souroukoto and the Bakoye (Koniakari) groups Seals: intrusions, shale or salt Migration pathways: Deep-seated faults(?)	Source: Water radiolysis (?) induced by Proterozoic granites Reservoirs: Mostly dolomites, limestones and siliciclastics of the Cambrian Kulpara and Parara formations Seals: (?) Migration pathways: (?)	Sources: Hydration reactions ferrous minerals in basement ultramafics, mafic intrusions and banded iron formations (?), and water radiolysis. Reservoirs: Fractured shales of the Ediacaran-Cambrian Bambuí Group Seals: Proterozoic shales, diagenetic and mafic intrusions. Migration pathways: Deep-seated faults	Sources: Hydration of ferrous minerals in ultramafic and mafic rocks and banded iron formations (?) of the basement and MCRS, as well as water radiolysis, serpentinization of smaller Cretaceous alkaline intrusions and artificial tubing-related reactions Reservoirs: Precambrian arkoses, Paleozoic sandstones and limestones Migration pathways: Precambrian to Phanerozoic, steeply-dipping faults
Hydrographic basins and aquifers	Hydrographic basin: Senegal watershed Aquifer: Deep aquifers linked to subsurface intrusions and H₂ reservoirs in Mali	Hydrographic basin: South Australian Gulf drainage basin Aquifers: Deeper aquifers hosted by the Cambrian Kulpara and Parara formations	Hydrographic basin: São Francisco river watershed Aquifer: Deep aquifers in the Bambuí and the underlying Proterozoic strata	Hydrographic basin: Mississippi River Basin Aquifer: Western Interior plain and Fractured basement

ites and greenstone belt successions locally intruded by younger granitoids and which were partially reworked during the 2.27–1.96 billion years old Transamazonian or Birimian orogeny [84, 85].

The West African craton basement hosts multiple fossil rifts apparently formed during the Mesoproterozoic and Neoproterozoic eras, which are filled with siliciclastic-dominated deposits passing upward into carbonate-rich passive margin strata [69]. These successions are overlain by different syn-orogenic, unconformity-bounded strata formed during the Neoproterozoic Pan-African orogenesis, as well as Ordovician to Carboniferous sedimentary successions recording continental to marine settings and containing the correlatives of world-class petroleum source rocks found in northern Africa (Figure 6.5). Together with older basin-fill strata, these deposits were partially deformed during the late Paleozoic assembly of the Pangea supercontinent, which led to the formation of the Mauritanides fold-thrust belt to the west [80]. Triassic to Jurassic widespread basic igneous intrusions cut the Taoudeni basin's sedimentary successions and record the Mesozoic break-up of the supercontinent and the opening of the Central Atlantic Ocean [89]. These intrusions are apparently correlative to the formation of the Gorouma Rift, which is preserved in the eastern West African craton and is part of the West Africa Rift System [52].

The natural hydrogen elements and processes of the Malian discovery were discussed in detail by [8] and [10]. According to surface and subsurface data, the system is mostly associated with the Tonian-Cryogenian Souroukoto and the Ediacaran Bakoye (Koniakari) groups, which partially fill the ENE-trending Tambaoura rift [52]. In the area, these units unconformably overlie granites, granodiorites, diorites, syenites, and aplites and are intruded by thick dolerite sills likely formed during the Central Atlantic event. The natural hydrogen is hosted by different levels of tight sandstones and dolomites showing porosities ranging from less than 1% to 14% and that were apparently sourced by hydration reactions in iron-rich sandstones interbedded in the strata, serpentinization reactions induced by intrusion-related fluid migration, among others. Geological maps and cross-sections shown by [8] strongly suggest an accumulation associated with a wide anticline in which the reservoirs are sealed by thick dolerite intrusions. Due to the tight nature of the reservoirs, a diagenetic seal component can also be considered. The data presented by [8] reveal a dynamic system apparently sourced by aquifer degassing in relatively short timescales. Geochemical model considering the typical mixture of gases found in association with H_2 (He, CH_4, and N) and their range of concentrations suggest that the Mali natural hydrogen would be 500 years old [90].

Natural hydrogen emissions associated with anomalous helium concentrations were measured in the northern faulted margin of the West African craton, within the southern coastal basins of Morocco [91]. Good indications of hydrogen have also been detected in the Western Meseta domain. These areas were formed during the Mesozoic Central Atlantic event and host thick sedimentary successions containing evaporite strata and widespread igneous basic intrusions (e.g., [80, 83]). These geological elements are thought to represent good ingredients for an active natural hydrogen system.

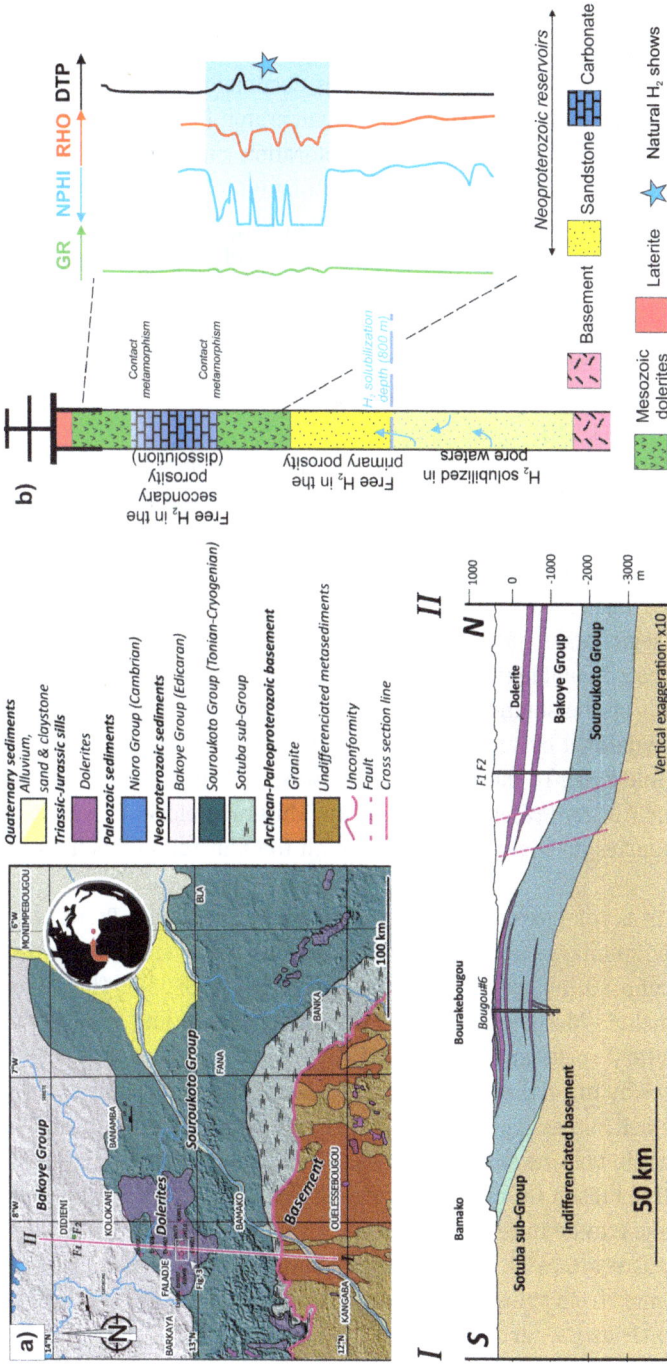

Figure 6.7: (a) Geological map of the southern border of the Taoudeni intracratonic basin, West African craton. The cross-section below shows the subsurface distribution of the Neoproterozoic intracratonic covers intruded by Mesozoic dolerites. (b) Illustrative well-log showing the relationship between the natural hydrogen shows, the sandstone and carbonate (limestones and dolomites) reservoirs and subsurface aquifer under natural pressure gradients. To the right, a detailed view of H₂-rich zone in a carbonate reservoir and its characteristic signature in gamma-ray (GR), neutron porosity (NPHI), density (RHO), and sonic (DT) logs. Modified from [8].

6.3.2 The South Australian craton

One of the most relevant natural hydrogen discoveries documented in the literature was made on the York Peninsula, in the South Australian craton [92]. In the area, the Ramsay 1 well was drilled during a hydrocarbon exploration campaign in 1931 and discovered multiple discharges of natural gases containing up to approximately 80% molar concentrations of H_2, along with minor amounts of CH_4, N_2, and CO_2 [93]. The ongoing re-evaluation of the discovery, conducted by the Gold Hydrogen company, confirmed large amounts of natural hydrogen (approximately 90%) and crustal helium (>10%) at depths of up to 1,000 m. These discoveries are part of a large number of natural hydrogen occurrences and emissions that have been thoroughly documented and analyzed on the surface and subsurface throughout the entire of Australia (Figure 6.8) (e.g., [12, 93, 94]).

The South Australian craton is one of the several cratonic domains of Australia and, although its limits are still poorly known due to the widespread Phanerozoic cover, it occupies an area of more than 500,000 km^2 in the southern portion of the country. Its basement is composed of multiple domains, which include the Archean to Mesoproterozoic Gawler (also referred to as the Gawler craton), the Paleoproterozoic Curnamona Province, segments of Paleo to Mesoproterozoic orogens and the Neoproterozoic to early Paleozoic Adelaide superbasin (Figure 6.8) (e.g., [95–97]). The craton is bounded to the south by the southern Australia passive margin, while its eastern and northern limits are defined by structures of the Ediacaran to Cambrian Delamerian Orogen (Adelaide fold belt). To the west and northwest, it is apparently bounded by the Mesoproterozoic Albany-Fraser Orogen and the Musgrave block. However, in these areas, the boundaries are deeply buried beneath the Centralian Basin strata (Figure 6.8).

The Gawler domain is composed of an Archean core associated with Proterozoic to Mesoproterozoic sedimentary successions and intrusions that were partially deformed and metamorphosed during the early Mesoproterozoic [95, 96]. The oldest known assemblages include Mesoarchean granites exposed locally and widespread late Archean sedimentary rocks and volcanics, which were metamorphosed into gneisses and migmatites by the end of the Neoarchean, as well as correlative komatiite-bearing greenstone belt successions [95, 96] The Archean rocks are unconformably overlain by passive margin volcano-sedimentary rocks and banded iron formations of the Paleoproterozoic Hutchinson Group and its correlatives. Widespread Paleoproterozoic to Mesoproterozoic granite intrusions and felsic to mafic volcanics are also common elements in the eastern Gawler domain [95]. On the Yorke Peninsula, these assemblages include granites from suites dated to approximately 1.8 to 1.6 Ga, as well as major exposures of siliciclastics, carbonate rocks, graphite-rich rocks, and volcanics of the 1.7 Ga Wallaroo Group [95]. Proterozoic rocks of the eastern Gawler domain are well known for their world-class IOCG deposits, including the Olympic Dam.

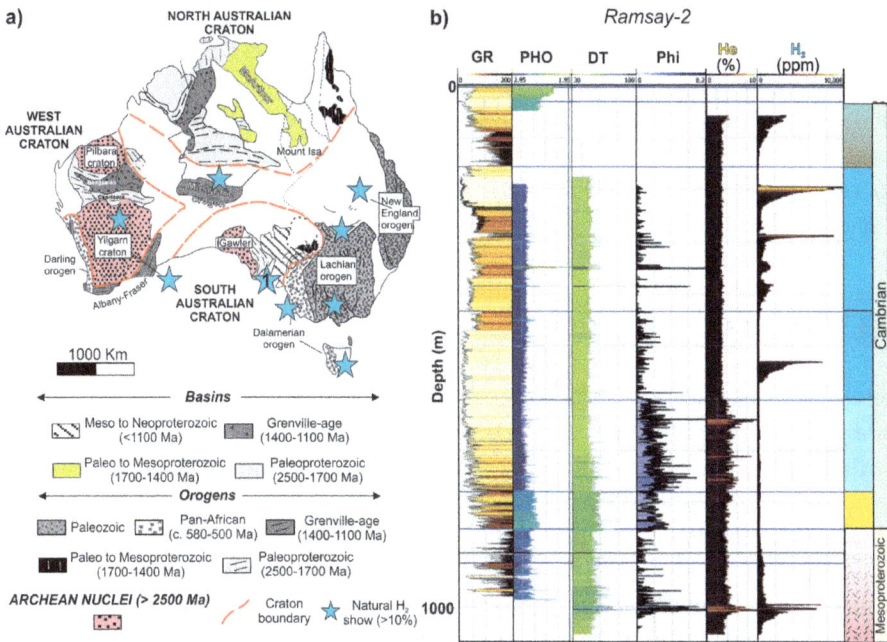

Figure 6.8: (a) Cratons of Australia [modified from 98]. The Archean nuclei are indicated for the West and South Australian cratons. The natural hydrogen shows are historical data compiled and analyzed by [93]. (b) Well log of a well drilled by Gold Hydrogen in the Yorke Peninsula, showing multiple hydrogen-rich reservoirs associated with remarkable amounts of crustal helium. The well followed the drilling of the 1930s Ramsay well (1) and perforated the Cambrian strata of the Stansbury Basin and Mesoproterozoic granites partially altered in the basement. Well logs: GR, gamma ray; PHO, density; DT, sonic; Phi, neutron porosity; He, Helium; H_2, natural hydrogen. Modified from [92].

The Adelaide superbasin [97] strata unconformably overlie basement rocks of the South Australian craton and record more than 300-Myr-long evolution, which started with a rifting episode during the Neoproterozoic breakup of the Rodinia supercontinent. On the Yorke Peninsula, the early to mid-Cambrian siliciclastics and carbonate-dominated successions of the Stansbury Basin mark the younger, passive margin episode recorded by the Adelaide superbasin. This episode was followed by the late stages of the Delamerian Orogeny, which partially deformed Neoproterozoic to early Paleozoic deposits in the easternmost South Australian craton (Figure 6.8) [97, 98].

The natural hydrogen discoveries in the Yorke Peninsula are hosted mostly by dolomites and limestones of the Cambrian Kulpara and Parara formations of the Stansbury Basin (Figure 6.8) [92, 93]. While in the Ramsay 1 well, it was found in a relatively shallow fractured zone affecting the Parara limestone, the nearby Ramsay 2 reported H_2 in both formations through mud gas samples. In the 2024 release, Gold Hydrogen also reported remarkable amounts of helium in the lower Kulpara Formation, the underlying Winulta Formation, and basement rocks (Figure 6.8). Recently, both wells were tested and sampled. The natural hydrogen and associated helium gases are thought to have been formed through water radiolysis induced by radioactive granites of the basement and migrated upward through fracture systems and faults [93]. Due to the composition of basement rocks in the Yorke Peninsula, hydration reactions associated with Proterozoic magmatic and iron-rich rocks are not discarded. In this context, copper deposits such as those of the Olympic Dam may also have played a role.

6.3.3 São Francisco craton

Located in eastern Brazil, the São Francisco craton encompasses some of the best-studied cases and datasets of natural hydrogen occurrences and surface emissions [99–104]. Subsurface exploration campaigns have been conducted in its sedimentary covers since the early twentieth century, within a geological domain referred to as the São Francisco intracratonic basin. Although these campaigns primarily focused on petroleum and natural gas accumulations, old hydrocarbon wells drilled in the 1980s by Petrobras S.A. reported natural gases with anomalous concentrations of helium. Large-scale hydrocarbon exploration campaigns conducted by the Brazilian Petra and Petrobras, other majors such as Shell and different smaller companies three decades later culminated in the serendipitous discovery of gases containing up to c. 40% of natural hydrogen, helium concentrations exceeding 1%, and variable amounts of inorganic alkanes and nitrogen [68, 101, 102]. Subsequent surface studies afterward reported multiple and continuous natural hydrogen emissions in subcircular surface depressions [99, 100] while gas geochemistry surveys identified possible sources for the H_2 and He [103].

The São Francisco craton exhibits a horse's head-shaped geometry and extends for more than 500,000 km^2 in the eastern Brazil. The craton represents the South American counterpart of the African Congo craton, from which it separated during the Mesozoic opening of the South Atlantic Ocean [17, 24]. The craton is bounded to the southeast, south, west, and north by Brasiliano/Pan-African orogenic belts formed between the late Neoproterozoic and early Paleozoic, and to the east by the Brazilian South Atlantic margin (Figure 6.9) [105] The craton comprises an Archean to Paleoproterozoic basement that is covered by sedimentary successions currently preserved in three major morphotectonic domains: (i) the intracratonic São Francisco Basin, (ii) the Paramirim aulacogen, and (iii) the Reconcavo-Tucano-Jatobá rift system [17, 54].

The basement is well exposed along the southernmost portion of the craton, where it is composed of Archean TTG gneisses and migmatites, greenstone belt successions containing variable volumes of ultramafic volcanics, and younger granite suites (Figure 6.4) [24]. These rock assemblages are unconformably overlain by Paleoproterozoic passive margin to collisional strata containing world-class banded iron formation deposits, which are exposed in the Quadrilátero Ferrífero mineral province (Figure 6.4) [36]. The

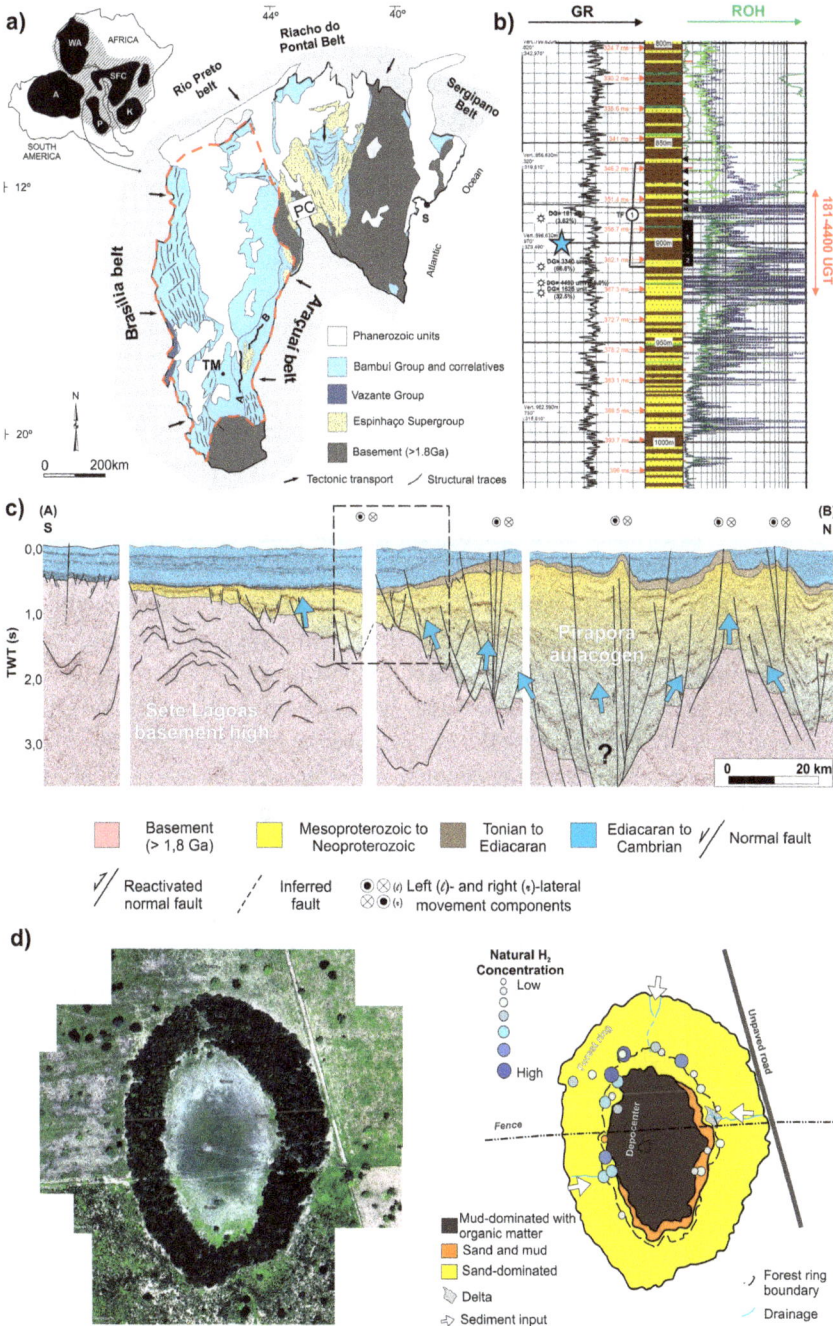

Figure 6.9: (a) Simplified geological map of the São Francisco craton and the São Francisco Basin (red dashed line), eastern Brazil. In the upper left, the paleotectonic reconstruction shows Brazilian and African cratons in West Gondwana, before the Mesozoic opening of the South Atlantic Ocean. Modified from [21].

São Francisco craton lithosphere stabilized after the Paleoproterozoic Transamazonian orogenic cycle, during which older and contemporaneous assemblages were deformed and metamorphosed under variable conditions [30, 36].

From around 1.8 Ga, the São Francisco craton lithosphere underwent multiple basin-forming episodes, which led to the deposition of [21]:

i) the Meso to Neoproterozoic, siliciclastic-dominated rift to passive margin strata of the Paranoá Group and the Espinhaço Supergroup;

ii) the Neoproterozoic, diamictite-bearing rift successions of the Jequitaí Formation; and

iii) the carbonate and siliciclastic strata of the Bambuí Group, marking an Ediacaran to Cambrian foreland basin system.

Besides overlaying an older Paleoproterozoic(?) succession observed only in seismic sections, the Meso to Neoproterozoic rift strata are cut by different generations of mafic dykes and sills, and their deposition culminated in the formation of a NW-trending graben referred to as the Pirapora aulacogen (Figure 6.9). The graben is currently preserved in the central portion of the basin and was successively reactivated after its nucleation [21, 24, 59]. The foreland basin cycle took place contemporaneously with the formation of the Brasiliano/Pan-African belts that bound the São Francisco craton, whose NS-trending, external portions captured the intracratonic covers (Figure 6.9). Thinner Mesozoic sedimentary and volcanic-related successions exposed in the central to northern portions of the basin mark the intracontinental effects of the South Atlantic opening [63]. These strata are partially contemporaneous with the intrusion of continental-scale dyke swarms and alkaline suites that are exposed in the southwestern São Francisco [21].

H_2-rich natural gases were discovered in the eastern and western portions of the São Francisco Basin, usually associated with up to 1% crustal He and variable concentrations of organic to inorganic alkanes and nitrogen. Higher concentrations were identified in fractured fine-grained siliciclastics of the Ediacaran to Cambrian Bambuí Group, especially, within the partially inverted segment of the Pirapora aulacogen (Figure 6.9). In the eastern São Francisco Basin, the Ediacaran-Cambrian Araçuaí foreland fold-thrust belt comprises large-scale, double-plunging drape folds that were

Figure 6.9 (continued)

(b) Well log of an old hydrocarbon well drilled in basins, which produced hydrogen-rich gases to the surface during drill stem tests (blue star). The approximate location of the well is shown in the composite 2D seismic section (c) within the black dashed rectangle. High-amplitude reflectors cutting the rift strata within the Pirapora aulacogen commonly correspond to different generations of mafic dykes and sills. Blue arrows represent possible migration pathways. The trace of the section is shown in (a). Modified from [59]. (d) Example of an H_2-emitting fairy circle in the São Francisco Basin exhibiting typical attributes such as a forest ring, sub-rounded map-view geometry, and very low relief. The sedimentary deposits shown in the right map record an intermittent lacustrine to deltaic system, and the H_2 concentrations measured at the surface may exceed 1,000 ppm.

formed through the partial inversion of preexisting normal faults. These folds are locally cut by NW- to NE-trending strike-slip zones that can extend for hundreds of kilometers and developed through the reactivation of major Pirapora aulacogen faults (Figure 6.9) [59]. Recent analyses combining the geochemistry of major and noble gases associated with natural hydrogen indicate it is derived from both hydration reactions of Fe_2-rich basement rocks and water radiolysis [103]. Based on a large and basin-wide geochemical data of alkanes and the tectono-stratigraphic architecture of the H_2-rich natural gas discoveries, Reis and Fonseca [102] identified a strong component derived from overmature hydrocarbon source rocks. The available information suggests a natural hydrogen system dominantly composed of low-permeability (and fractured) reservoirs fed by different basement and intracratonic cover sources. Potential accumulation zones are likely associated with drape folds and partially inverted normal faults showing steep dips and strike-slip components, which are able to connect the Archean to Paleoproterozoic basement with shallower zones [104, 106]. In the central and eastern portions of the basin, H_2-emitting fairy circles exhibiting concentrations that may exceed 1,000 ppm were monitored and appear to represent leakage zones of subsurface reservoirs [99, 100].

6.4 Exploring natural hydrogen systems in cratonic domains

Although both the natural hydrogen system and plays are still poorly known, experimental data and case studies in hydrothermal systems indicate that hydrogen can be naturally generated through hydration reactions involving Fe^{2+}-rich rocks, water radiolysis, overmaturation of organic-rich rocks, and shearing along active fault zones [11, 14, 107–114]. Assuming an advective-driven system, it is possible to assume that hydrogen gases generated at depth could migrate to the surface or accumulate in subsurface porous and fractured rocks through open tectonic structures, carrier beds, and pressurized aquifers [115]. Although critical questions regarding the seal rock efficiency in the H_2 source-accumulation-preservation model remain unresolved, limited H_2 exploration data, well-known cases from the hydrocarbon industry, and field- and laboratory-based studies indicate different types of rock media that may act as effective seals or permeability barriers [116–118]. Furthermore, physicochemical modelings point toward a subsurface water behavior either as a reservoir or seal. This assumption is based on the solubility of hydrogen, which is higher under pressurized conditions and lower under shallower burial [8].

Other models argue in favor of natural hydrogen systems that are remarkably different, sustained by the deep mantle and core degassing of primordial gases over geological time and their diffusive migration through planetary- to lithospheric-scale pathways (e.g., [6, 11, 115]). Part of these models is based on Larin's concept of the Primordially Hydridic Earth [6] and relies on the existence of large amounts of primordial hydrogen stored in the deep Earth during its accretion 4.56 billion years ago.

By examining the natural hydrogen occurrences described in the previous sections and from cratons worldwide (Table 6.1), it is possible to track several common elements, such as: (i) low-porosity sedimentary or igneous reservoirs; (ii) the association between H_2, N_2, inorganic and organic CH_4 and crustal-derived He; (iii) the close relationship with widespread mafic igneous intrusions ranging from Proterozoic to Mesozoic in age; (iv) the direct or indirect association with large-scale rifts, basement-rooted, steeply-dipping faults, and drape folds; (vii) the typical craton basement composition containing both (naturally radioactive) granitic rocks and Fe_2-bearing units; and (viii) widespread H_2-emitting fairy circles on the surface (Figures 6.7–6.9 and Table 6.1). These elements may occur associated with different types of aquifers and strongly suggest a natural hydrogen system, in some aspects, similar to the well-known petroleum systems. Since many of these characteristics are typical of cratonic crusts, the above-mentioned elements can be found in virtually all cratons of the world (Figures 6.1, 6.3–6.5, and 6.10).

The potential for Precambrian crusts to generate natural hydrogen and associated reaction pathways has been investigated by various authors. It has been demonstrated that serpentinization reactions in olivine-rich ultramafic rocks may generate hydrogen under a wide range of temperatures, varying from near-ambient to conditions exceeding 200 °C. These reactions are typically favored by the amount and oxidation state of Fe^{2+} in serpentine minerals and the presence of metal catalysts (e.g., [111, 119–121]). Laboratory experiments and detailed analyses conducted by Shibuya et al. [122] and Tamblyn and Hermann [123] revealed the potential to produce significant amounts of natural hydrogen from synthetic komatiites and Archean samples from Australian, African, Chinese, and North American cratons [113]. Additionally studies showed that H_2 generation from magnetite-rich banded iron formations may occur under temperature conditions as low as 80 °C (and possibly near-ambient), with gas concentrations directly related to the magnetite concentration in the rocks. A series of open-system pyrolysis experiments conducted on organic-rich shales and coals of different types and maturation states revealed a systematic production of molecular hydrogen under overmature conditions, with temperatures ranging mainly from 300 to 700 °C [124, 125]. Other experiments conducted on siderite minerals (found in carbonate-rich rocks) demonstrated the production of significant amounts of hydrogen gas under hydrothermal conditions between 200 and 300 °C [126].

Although it is still difficult to apply the simulated parameters and production rates to natural conditions, cratons host almost all rock assemblages that can naturally generate H_2. These rocks can be found in the Archean to Paleoproterozoic komatiite-bearing greenstone belt successions, iron-rich and carbonate-rich banded iron formations, and intracratonic coals and overmature petroleum source rocks. The latter have been found in different intracratonic basins of North America and Brazil, composing unconventional hydrocarbon systems [68]. For instance, most of the Meso-

a)

Hydrogen occurrences
Mesozoic igneous rocks
Phanerozoic orogens
Paleozoic orogens
Archean to early Paleozoic
Cratons and stable continental crust

b)

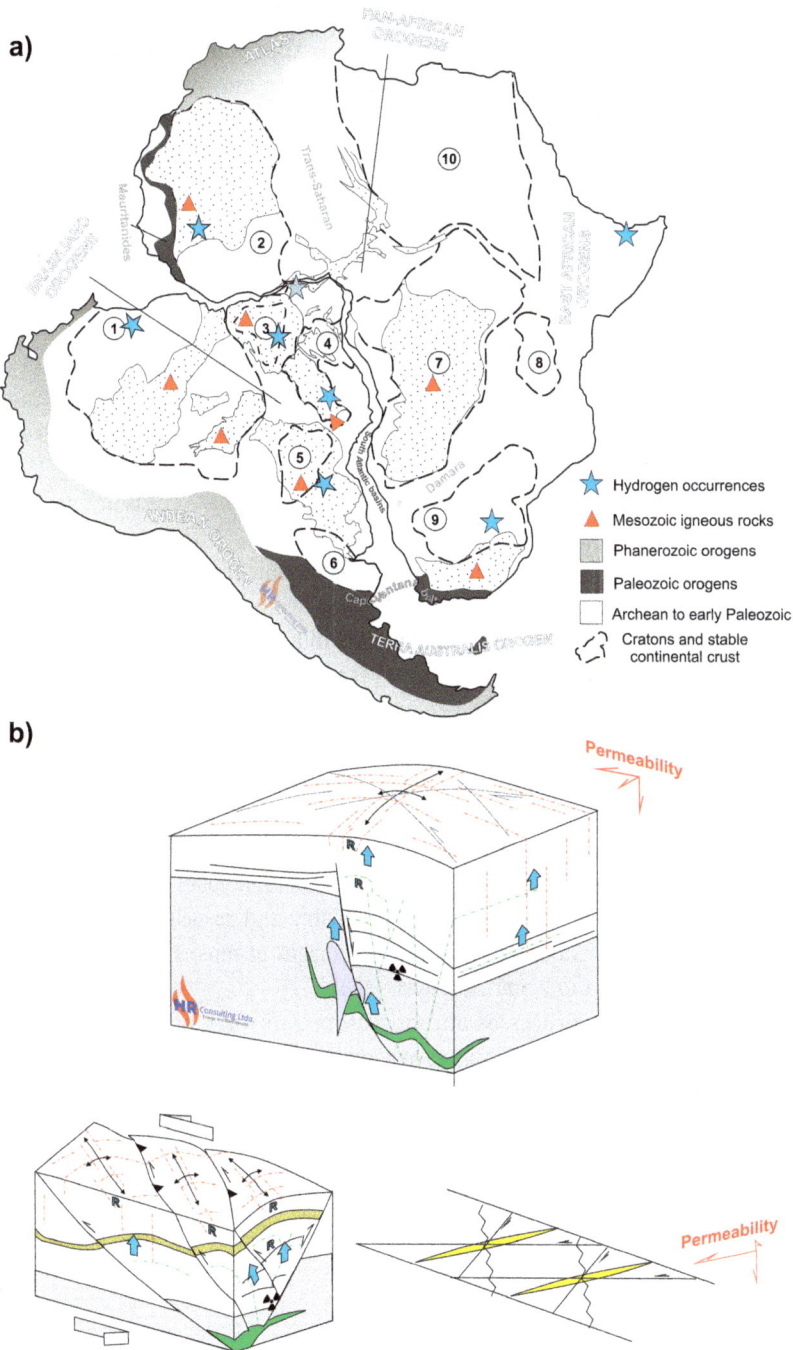

Figure 6.10: (a) West Gondwana reconstruction showing the South American and African cratons, intracratonic basins (dotted areas) and known occurrences of natural hydrogen. In almost all cases,

proterozoic to Ediacaran source rocks of the São Francisco Basin reported in the literature underwent extreme thermal conditions [68]. This apparently culminated in the formation of typical gases and isotopic signatures found in unconventional overmature systems and, as a consequence, those conditions expected for natural hydrogen generation [68, 102, 125]. Cratons are also known to preserve important iron-rich siliciclastic to carbonate deposits formed during extreme climate conditions affecting the Earth system in the dawn of the Phanerozoic [13, 127], thus, siderite might also be a common Fe(II)-rich mineral within cratonic covers. As lowland and relatively stable tectonic areas, cratons currently host continental-scale river basins bounded by higher topographic areas that are commonly underlain by the roots of Precambrian to Paleozoic mountain chains (Table 6.1). Assuming these conditions may have prevailed throughout most of their evolutionary history, it is possible to presume hydraulic gradients and a large surface and subsurface water availability to react with iron-rich source rocks over geological time.

Precambrian granitic rocks and clasts from different cratons were found to host fluid inclusions with high amounts of naturally generated H_2 [128]. Their natural radioactivity, on the other hand, may induce the coupled formation of hydrogen and crustal helium through water radiolysis. Based on an exhaustive compilation of natural H_2 occurrences worldwide and assuming the coupled contribution of radiolysis and hydration reactions [14], it was estimated that the Precambrian continental lithosphere would generate $0.36–2.27 \times 10^{11}$ mol of H_2 per year. Further modeling based on empirical data from rocks of the North American craton showed comparable global rates $(0.4–5.4 \times 10^{10}$ mol/year), in which hydration reactions and radiolysis would contribute $0.1–4.8 \times 10^{10}$ and $0.7–1.2 \times 10^{10}$ mol/year, respectively [129]. In the Australian Ramsay well, radiolytic reactions induced by the local Mesoproterozoic granites are strong candidates for forming the reported natural hydrogen and helium gases (Figure 6.8) [93]. In the São Francisco Basin, the natural gas geochemistry and associated noble gases support the role of these reactions and the coupled origin of most H_2 from both the hydration of basement rocks and water radiolysis [103].

The hydrogen migration processes and migration pathways in continental settings are still debated, and only a few studies are available. Nevertheless, the common occurrence of these gases and associated compounds within positive basement structures and close to intracratonic fossil rifts, such as the Taoudeni and São Francisco cratons, are strong indications of fault-related migration processes (Figures 6.7 and 6.9). Similar conditions are also found in the North American craton, where the main H_2-rich natural gases found in Kansas occur over the Nemaha uplift and a few

---

Figure 6.10 (continued)
widespread Proterozoic to Mesozoic intrusions are widespread. Cratons: 1, Amazon; 2, West Africa; 3, Parnaíba block; 4, São Francisco; 5, Paranapanema; 6, Rio de La Plata; 7, Congo; 8, Tanzania; 9, Kalahari; 10, Sahara meta-craton. (b) Possible natural hydrogen plays that can be explored in craton settings, according to the current knowledge of natural hydrogen system elements and processes.

kilometers east of the Mid-Continent Rift System axis [88, 130]. Recent studies have demonstrated the close spatial relationship between major He-rich gas fields in the Four Corners area of the USA, basement faults, and intrusions, indicating that the upward advective flow of noble gases formed in the basement is facilitated by tectonic structures. Since the same processes forming crustal-derived He can also form natural H_2, we consider that similar migration mechanisms and pathways are feasible for natural hydrogen systems. As with the He-rich gases of the Four Corners area, many of the reported natural hydrogen occurrences in cratons also contain alkanes and (crustal-derived?) N_2, which are often considered to have migrated via fault- and fracture-related advective flow [8, 10, 68, 88]. The analysis of tectonic structures in the São Francisco Basin, indicates that, when these faults reach the surface or near-surface areas, the H_2 formed at depth may leak into surface subcircular depressions (Figure 6.9) [99, 100]. Regardless of the tectonic setting, diffusion can also contribute to the vertical migration of hydrogen due to its very small size and buoyancy.

Once the appropriate conditions are reached, H_2-rich gases can fill porous rocks, become trapped, and accumulate along their migration pathway. This process requires the presence of a reservoir rock sealed by an effective impermeable layer, which is capable of slowing down its upward flow (permeability barrier) or preventing the hydrogen from rising vertically to the surface. If present, helium could accumulate in the same trapped reservoir, along with N_2 and abiotic CH_4. Depending on the seal's efficiency, the migration mechanisms, and H_2 production rates, dynamic and continuously replenished hydrogen accumulations could also be possible.

By studying the natural hydrogen subsurface accumulations of the Taoudeni Basin (West African craton) [116], it was demonstrated that the thicker and less fractured the dolerite sills are, the more efficient they will be at trapping H_2 in the subsurface. These trapping mechanisms seem to be enhanced by shallow aquifers in which hydrogen solubility is very low. On the other hand, the natural hydrogen subsurface occurrences of the São Francisco Basin (São Francisco craton) are intrinsically associated with its petroleum system elements. In this system, some of the most important sealing rocks are multiple intervals of Ediacaran to Cambrian non-fractured shales that may be tracked over almost the entire basin [68]. Studies conducted by [117] in the Cigar Lake uranium deposit of the North American craton (Canada) revealed that shales containing chlorite (sudoite) and illite clay minerals can retain remarkable amounts of hydrogen through adsorption. Although less common in fine-grained sedimentary rocks [118], studies demonstrate that sepiolite and palygorskite clay minerals may adsorb even higher amounts of hydrogen. Clay-rich sealing rocks commonly comprise widespread intervals in the intracratonic covers worldwide, whereas widespread mafic sills and dykes can be found in virtually all cratons of the world. For instance, almost all cratons of South America and Africa exhibit natural hydrogen subsurface shows and/or seeps and intracratonic sedimentary covers cut by both Proterozoic and/or Mesozoic mafic intrusions. The latter are widespread and mark the Mesozoic South Atlantic opening (Figures 6.5, 6.7, 6.9, and 6.10).

According to the available data, it is possible to assume that tight sedimentary and non-sedimentary reservoirs are common elements of natural hydrogen systems in cratons, as well as carbonate reservoirs showing secondary porosity (e.g., [8, 68, 92, 93]). Even though the importance of conventional reservoirs cannot be ruled out, it seems that the characteristics of these media offer additional means for storing and retaining natural hydrogen-rich gases over time. Tight carbonate and sandstone reservoirs can be found in intracratonic basins of North America, China, South America, among others, and are known for their low to very low permo-porosities, smaller pore throats, variable tortuosities, complex diagenetic evolution, and heterogeneity [131, 132]. Based on the detailed stratigraphic and petrographic analysis of Mesoproterozoic tight sandstone reservoirs of the São Francisco Basin [132], researchers have demonstrated permo-porosity patterns resulting from a complex post-depositional and hydrothermal history coupled with the intrusion of Proterozoic (?) mafic sills and the evolution of the hosting depocenter. If combined, these characteristics may allow the accumulation of natural hydrogen and offer an additional diagenetic-related sealing component.

6.4.1 On the natural hydrogen plays

The exploratory plays and conditions under which natural hydrogen can be produced economically are still contentious. Besides surface surveys based on the detection and monitoring of natural hydrogen and helium seeps (e.g., [100]), the few active exploration programs developed after the H_2 discovery in Mali (West African craton) have mostly focused on reassessing shows and discoveries previously reported in old hydrocarbon and water wells (e.g., [23, 88, 92]). Nevertheless, current knowledge allows for the assumption of two main exploratory play concepts based on (i) the generation-migration-accumulation-trapping and (ii) the dynamic and continuous production. The first concept depends on critical and still unclear factors, such as the capacity of H_2 to migrate through advective flows and the efficiency of natural seals, while the latter depends mostly on high and short-term rates of generation (regardless of a crustal or deep mantle/core origin).

In cratons, the relationship between known gas occurrences and potential elements and processes of the natural hydrogen system have induced exploratory campaings strongly based on a generation-migration-accumulation-trapping concept (Table 6.1 and Figures 6.7–6.9). Considering the architecture and evolutionary history of cratons (Figures 6.1–6.5), exploratory plays in which components enhance the connection between deep basement sources and shallower reservoirs and seals should be preferred. Once hosting reservoirs and successions capable of retaining natural hydrogen in the subsurface, the intracratonic sedimentary cover is the best option to allow the accumulation of significant gas volumes and increase the exploration chance of success. Thus, the following natural hydrogen plays seem promising (Figure 6.10):

- fault-related anticlines formed through the positive reactivation of steeply-dipping normal faults
- positive flower structures
- four-way-dipping and non-fractured mafic sills
- folded mafic sills associated at depth with reactivated normal faults
- salt- or sill-sealed pinch-outs

In all above-mentioned plays, deep-rooted and steeply dipping fault structures are potential migration pathways, and either conventional or unconventional reservoirs sealed by non-fractured mafic sills, evaporites, and shales are potential targets. In the case of the four-way-closing sill plays, it is possible to envisage conditions similar to those explored in the intracratonic Parnaíba Basin of northeast Brazil, where widespread Mesozoic intrusions play a dual role in the hydrocarbon system. Besides offering the heat necessary to generate natural gas, their systematic stratigraphic-level changes allow the formation of traps to retain the gas in subsurface Paleozoic siliciclastic reservoirs [133] (Figure 6.10). These atypical plays are informally referred to as "sill jumps" and can be found in almost all cratons that once formed the West Gondwana/Pangea supercontinent (Figure 6.10). Due to its genetic link and common occurrence with natural hydrogen in cratons [14, 129], helium can be considered an upside in most types of plays.

Although it is still controversial, a few lines of evidence seem to point toward relatively high rates of natural hydrogen generation and/or replenishment in nature. For instance [8], reported that, even after decades of production, the reservoirs in Mali preserved stable pressures for a long period and ultimately showed a slight increase. This implies a constant replenishment of the produced hydrogen at equal or slightly higher rates, which could be related to the local aquifer degassing [8]. If supplying fault- and igneous-related plays such as those discussed above, the short-term natural replenishment modeled by [90] would imply a hybrid and even more promising natural hydrogen play.

6.5 The emerging natural hydrogen exploration: why do cratons matter?

The detailed examination of natural hydrogen occurrences in cratons worldwide seems to reveal an intrinsic relationship between these gases and the unique elements forming the oldest continental nuclei. These elements can be tracked in virtually all cratons of the world and have been formed and preserved through their 4 billion years long and complex evolution (Figures 6.1 and 6.10). They include possible sources comprising Archean to Paleoproterozoic ultramafic and mafic successions and other Fe(II)-rich assemblages of the basement, as well as a large volume of granitic and nuclide-rich rocks (Figures 6.1–6.6). The same strong mechanical characteristics that allowed cratonic lithospheres to survive multiple crustal recycling events over time also favored the preservation of these basement assemblages since the Archean and the subsequent formation of a plentiful set of tectonic structures, fluid migration pathways, and intracratonic sedimentary successions able to act as reservoirs, organic-rich sources, and sealing rocks for natural hydrogen (Figure 6.5). Their geological characteristics and trajectory through time also culminated in the preservation of widespread igneous intrusions that seem to play a major role in continental natural hydrogen systems (Figure 6.7). The conditions would have been fueled through time by favorable topographic gradients, water infiltration, and basin-wide surface and subsurface water systems necessary to promote hydration and radiolytic reactions.

Although it is difficult to precisely predict the future of hydrogen exploration, it is clear that emerging efforts are becoming increasingly focused on cratons. Besides hosting the majority of natural hydrogen occurrences known to date, the only known area that has produced this resource so far is located within the Taoudeni Basin of the West African craton [8, 10]. On the other hand, one of the main recent achievements in exploration comes from the South Australian craton of Australia [92], with other promising cases being explored in the North American craton. Combined with their typical characteristics and other cases presented here, cratons represent some of the most interesting geological settings for exploring natural hydrogen and, perhaps, producing economic volumes in the near future. Due to their long evolutionary history and multiple interactions over time, many of their characteristics – and, thus, natural hydrogen system elements – can be tracked across virtually all continents (Figure 6.10). Answering critical open questions, however, is mandatory to advance. These questions include: (i) what are the rates and optimum conditions to generate hydrogen in nature?; (ii) are these rates sufficient to sustain continuous production to the surface over time?; (iii) what mechanisms control the natural hydrogen migration through the lithosphere? (iv) do rocks like mafic sills, evaporites, and shales containing chlorite, illite, and other clay minerals work as efficient seals for hydrogen?; (v) how long and under what conditions can accumulations of natural hydrogen survive over geological time?; (vi) what is the role of local microbial consumption in hydrogen-saturated subsurface reservoirs?; among others. Only after addressing these gaps we will be able to advance on the natural hydrogen exploration in cratons and elsewhere. Meanwhile, cratons remain significant.

References

[1]   Halla J., Noffke N., Reis H., Awramik S., Bekker A., Brasier A., et al. Ratification of the base of the ICS geological time scale: The Global Standard Stratigraphic Age (GSSA) for the hadean lower boundary. EPISODES: Journal of International Geoscience. 2024, 47(2): 381–389. doi: 10.18814/epiiugs/2024/024002

[2]   Catling D. C., Zahnle K. J. The archean atmosphere. Science Advances. 2020, 6: eaax1420. doi: 10.1126/sciadv.aax1420

[3]   Lide D. Handbook of chemistry and physics, 84th ed. Boca Raton, Florida, USA: CRC Press LLC, 2003.

[4]   Etiope G., Sherwood Lollar B. Abiotic methane on earth. Reviews of Geophysics. 2013, 51(2): 276–299.

[5]   Tagawa S., Sakamoto N., Hirose K., et al. Experimental evidence for hydrogen incorporation into Earth's core. National Communications. 2021, 12: 2588. doi: https://doi.org/10.1038/s41467-021-22035-0

[6]   Vidavskiy V., Rezaee R. Natural deep-seated hydrogen resources exploration and development: Structural features, governing factors and controls. Journal of Energy and Natural Resources. 2021, 11(3): 60–81.

[7]   IEA (International Energy Agency). Global hydrogen review. Paris, France, IEA, 2024. Accessed January 22, 2025 at https://www.iea.org/reports/global-hydrogen-review-2024

[8]   Maiga O., Deville E., Laval J., Prinzhofer A., Diallo A. B. Characterization of the spontaneously recharging natural hydrogen reservoirs of Bourakebougou in Mali. Scientific Reports. 2023, 13: 11876. doi: https://doi.org/10.1038/s41598-023-38977-y

[9]   Smith N. J. P., Shepherd T. J., Styles M. T., Williams G. M. Hydrogen exploration: A review of global hydrogen accumulations and implications for prospective areas in NW Europe. In: Doré A. G., Vining B. A., eds., Petroleum geology: North-West Europe and global perspectives – proceedings of the 6th petroleum geology conference. London, UK: Geological Society, 2005, 349–358.

[10] Prinzhofer A., Cissé C. S. T., Diallo A. B. Discovery of a large accumulation of natural hydrogen in Bourakebougou (Mali). Intenational Journal of Hydrogen Energy. 2018, 43(42): 19315–19326. doi: https://doi.org/10.1016/j.ijhydene.2018.08.193

[11] Zgonnik V. The occurrence and geoscience of natural hydrogen: A comprehensive review. Earth-Science Reviews. 2020, 203: 103140. doi: https://doi.org/10.1016/j.earscirev.2020.103140

[12] Moretti I., Brouilly E., Loiseau K., Prinzhofer A., Deville E. Hydrogen emanations in intracratonic areas: New guide lines for early exploration basin screening. Geosciences. 2021, 11(3): 145. doi: https://doi.org/10.3390/geosciences11030145

[13] Reis H. L. S., Sanchez E. A. M. Precambrian. In: Alderton D., Elias S. A., eds., Encyclopedia of geology, Vol. 202, 2nd ed. Academic Press, 2021, 23–54.

[14] Sherwood Lollar B., Onstott T., Lacrampe-Couloume G., Ballentine C. J. The contribution of the Precambrian continental lithosphere to global H2 production. Nature. 2014, 516: 379–382. doi: https://doi.org/10.1038/nature14017

[15] Condie K. C. Earth's oldest rocks and minerals. In: Van Kranendonk M. J., Bennet V. C., eds., Earth's oldest rocks, 2nd edn. Amsterdam, Netherlands: Elsevier, 2019, 239–253.

[16] Schaeffer A. J., Lebedev S. Global shear-speed structure of the upper mantle and transition zone. Geophysical Journal International. 2013, 194(1): 417–449. doi: doi:10.1093/gji/ggt095

[17] Alkmim F. F. O que faz de um craton um cráton? O cráton do São Francisco e as revelações almeidianas ao delimitá-lo. In: Mantesso-Net V., Bartorelli A., Carneiro C. D. R., Brito-Neves BBde, org. Geologia do Continente Sul Americano: Evolução da obra de Fernando Flávio Marques de Almeida. São Paulo, Beca, 2004, 17–35.

[18] Perchuk A. L., Gerya T. V., Zakharov V. S., Griffin W. L. Building cratonic keels in Precambrian plate tectonics. Nature. 2020, 586(7829): 395–401. doi: 10.1038/s41586-020-2806-7

[19] Jaupart C., Mareschal J. C., Bouquerel H., Phaneuf C. The building and stabilization of an Archean Craton in the Superior Province, Canada, from a heat flow perspective. Journal of Geophysical Research: Solid Earth. 2014, 119(12): 9130–9155. doi: 10.1002/2014jb011018

[20] Podugu N., Ray L., Singh S. P., Roy S. Heat flow, heat production, and crustal temperatures in the Archaean Bundelkhand craton, north-central India: Implications for thermal regime beneath the

Indian shield. Journal of Geophysical Research: Solid Earth. 2017, 122(7): 5766–5788. doi: 10.1002/2017jb014041

[21] Reis H. L. S., Alkmim F. F., Fonseca R. C. S., Nascimento T. C., Suss J. F., Prevatti L. D. The São Francisco Basin. In: Heilbron M., Cordani U. G., Alkmim F. F., eds., São Francisco Craton. Eastern Brazil: Tectonic Genealogy of a Miniature Continent. Springer International Publishing, 2017, 117–143.

[22] Gao P., Qiu Q., Jiang G., et al. Present-day geothermal characteristics of the Ordos Basin, western North China Craton: New findings from deep borehole steady-state temperature measurements. Geophysical Journal International. 2018. doi: 10.1093/gji/ggy127

[23] Sun G., Liu S., Santosh M., Gao L., Hu Y., Guo R. Thickness and geothermal gradient of Neoarchean continental crust: Inference from the southeastern North China Craton. Gondwana Research. 2019, 73: 16–31. doi: 10.1016/j.gr.2019.02.001

[24] Heilbron M., Cordani U. G., Alkmime F. São Francisco Craton, eastern Brazil: Tectonic genealogy of a miniature continent. Germany, Regional Geology Reviews. 2017. doi: 10.1007/978-3-319-01715-0

[25] Brown M., Johnson T., Gardiner N. J. Plate Tectonics and the Archean Earth. Annual Review of Earth and Planetary Sciences. 2020, 48(1): 291–320. doi: 10.1146/annurev-earth-081619-052705

[26] Jacobs J., Pisarevsky S., Thomas R. J., Becker T. The Kalahari Craton during the assembly and dispersal of Rodinia. Precambrian Research. 2008, 160(1–2): 142–158. doi: 10.1016/j.precamres.2007.04.022

[27] Hyndman R. D., Currie C. A., Mazzotti S. P. Subduction zone backarcs, mobile belts, and orogenic heat. GSA Today. 2005, 15(2). doi: 10.1130/1052-5173

[28] Sudholz Z. J., Yaxley G. M., Jaques A. L., et al. Multi-stage evolution of the South Australian Craton: Petrological constraints on the architecture, lithology, and geochemistry of the lithospheric mantle. Geochemistry, Geophysics, Geosystems. 2022, 23(11). doi: 10.1029/2022gc010558

[29] Paul J., Ghosh A. Evolution of cratons through the ages: A time-dependent study. Earth and Planetary Science Letters. 2020, 531. doi: 10.1016/j.epsl.2019.115962

[30] Alkmim F. F., Marshak S. Transamazonian orogeny in the Southern São Francisco craton region, Minas Gerais, Brazil: Evidence for Paleoproterozoic collision and collapse in the Quadrilátero Ferrífero. Precambrian Research. 1998, 90(1): 29–58. doi: https://doi.org/10.1016/S0301-9268(98)00032-1

[31] Marshak S. Deformation style way back when: Thoughts on the contrasts between Archean/Paleoproterozoic and contemporary orogens. Journal of Structural Geology. 1999, 8–9,1175–1182. doi: https://doi.org/10.1016/S0191-8141(99)00057-7

[32] Van Kranendonk M. J., Smithies R. H., Griffin W. L., et al. Making it thick: A volcanic plateau origin of Palaeoarchean continental lithosphere of the Pilbara and Kaapvaal cratons. In: Geological society, Vol. 389(1). London: Special Publications, 2014, 83–111. doi: 10.1144/sp389.12

[33] Wiemer D., Schrank C. E., Murphy D. T., Wenham L., Allen C. M. Earth's oldest stable crust in the Pilbara Craton formed by cyclic gravitational overturns. Nature Geoscience. 2018, 11(5): 357–361. doi: 10.1038/s41561-018-0105-9

[34] Goldfarb R. J., Groves D. I., Gardoll S. Orogenic gold and geologic time: A global synthesis. Ore Geology Reviews. 2001, 18(1): 1–75. doi: https://doi.org/10.1016/S0169-1368(01)00016-6

[35] Hagemann S. G., Angerer T., Duuring P., et al. BIF-hosted iron mineral system: A review. Ore Geology Reviews. 2016, 76: 317–359. doi: 10.1016/j.oregeorev.2015.11.004

[36] Alkmim F. F., Teixeira W. The Paleoproterozoic Mineiro belt and the Quadrilátero Ferrífero. In: Heilbron M., Cordani U., Alkmim F., eds., São Francisco Craton, Eastern Brazil. Regional geology reviews. Cham: Springer. doi: https://doi.org/10.1007/978-3-319-01715-0_5

[37] Martin H., Moyen J.-F. Secular changes in tonalite-trondhjemite-granodiorite composition as markers of the progressive cooling of Earth. Geology. 2002, 30(4): 319–322.

[38]   Martin H., Smithies R. H., Rapp R., Moyen J. F., Champion D. An overview of adakite, tonalite–trondhjemite–granodiorite (TTG), and sanukitoid: Relationships and some implications for crustal evolution. Lithos. 2005, 79(1–2): 1–24. doi: 10.1016/j.lithos.2004.04.048

[39]   Halla J. Highlights on geochemical changes in archaean granitoids and their implications for early Earth geodynamics. Geosciences. 2018, 8(9). doi: 10.3390/geosciences8090353

[40]   Barnes S. J., Arndt N. T. Distribution and geochemistry of komatiites and basalts through the Archean. In: Van Kranendonk M. J., Bennett V. C., Hoffmann J. E., eds., Earth's oldest rocks, 2nd ed. Elsevier, 2019, 103–116. doi: https://doi.org/10.1016/B978-0-444-63901-1.00006-X

[41]   Pirajno F., Huston D. L. Paleoarchean (3.6–3.2 Ga) Mineral systems in the context of continental crust building and the role of mantle plumes. In: Van Kranendonk M. J., Bennett V. C., Hoffmann J. E., eds., Earth's oldest rocks, 2nd ed. Elsevier, 2019, 187–208. doi: doi: https://doi.org/10.1016/B978-0-444-63901-1.00009-5

[42]   Van Kranendonk M. J., Hugh Smithies R., Hickman A. H., Champion D. C. Review: Secular tectonic evolution of Archean continental crust: Interplay between horizontal and vertical processes in the formation of the Pilbara Craton, Australia. Terra Nova. 2007, 19(1): 1–38. doi: 10.1111/j.1365-3121.2006.00723.x

[43]   de Wit M. J., Furnes H., Robins B. Geology and tectonostratigraphy of the Onverwacht Suite, Barberton Greenstone Belt, South Africa. Precambrian Research. 2011, 186(1): 1–27. doi: https://doi.org/10.1016/j.precamres.2010.12.007

[44]   Lana C., Tohver E., Cawood P. Quantifying rates of dome-and-keel formation in the Barberton granitoid-greenstone belt, South Africa. Precambrian Research. 2010, 177(1–2): 199–211.

[45]   Eriksson P. G., Altermann W., Nelson D. R., Mueller W. U., Catuneanu O. The precambrian earth: Tempos and events. Amsterdam: Elsevier, 2004, 941.

[46]   Bekker A., Slack J. F., Planavsky N., et al. Iron formation: The sedimentary product of a complex interplay among mantle, tectonic, oceanic, and biospheric processes. Economic Geology. 2010, 105(3): 467–508. doi: 10.2113/gsecongeo.105.3.467

[47]   Young G. M. Precambrian: Overview. Reference Module in Earth Systems and Environmental Sciences. Elsevier, 2017, 1–7. doi: https://doi.org/10.1016/B978-0-12-409548-9.10498-2017.

[48]   Crockford P. W., Kunzmann M., Bekker A., Hayles J., Bao H., Halverson G. P. Claypool continued: Extending the isotopic record of sedimentary sulfate. Chemical Geology. 2019, 513: 200–225. doi: 10.1016/j.chemgeo.2019.02.030

[49]   Martins-Neto M. A. Sequence stratigraphic framework of Proterozoic successions in eastern Brazil. Marine and Petroleum Geology. 2009, 26(2): 163–176.

[50]   Allen P. A., Armitage J. J. Cratonic basins. In: Busby, C, Azor,A. (eds.) Tectonics of Sedimentary Basins: Recent Advances, 1st ed, Wiley, 2011, 602–620. doi: https://doi.org/10.1002/9781444347166.ch30

[51]   Daly M. C., Fuck R. A., Julià J., Macdonald D. I. M., Watts A. B. Cratonic basin formation: A case study of the Parnaíba Basin of Brazil. Geological Society, London, Special Publications. 2018, 472(1): 1–15. doi: 10.1144/sp472.20

[52]   Villeneuve M. Paleozoic basins in West Africa and the Mauritanide thrust belt. Journal of African Earth Sciences. 2005, 43(1–3): 166–195. doi: 10.1016/j.jafrearsci.2005.07.012

[53]   Allen P. A., Eriksson P. G., Alkmim F. F., et al. Chapter 2 Classification of basins, with special reference to Proterozoic examples. In: Geological society, Vol. 43(1). London, Memoirs, 2015, 5–28. doi: 10.1144/m43.2

[54]   Alkmim F. F., Reis H. L. S. Brazil and the Guianas. In: Alderton D., Elias S. A., eds., Encyclopedia of geology, 2nd ed. Academic Press, 2021, 27–46.

[55]   Delvaux D., Maddaloni F., Tesauro M., Braitenberg C. The congo basin: Stratigraphy and subsurface structure defined by regional seismic reflection, refraction and well data. Global and Planetary Change. 2021, 198. doi: 10.1016/j.gloplacha.2020.103407

[56] Stein C. A., Stein S., Merino M., Randy Keller G., Flesch L. M., Jurdy D. M. Was the Midcontinent Rift part of a successful seafloor-spreading episode? Geophysical Research Letters. 2014, 41(5): 1465–1470. doi: 10.1002/2013gl059176

[57] Cawood P. A., Hawkesworth C. J. Earth's middle age. Geology. 2014, 42(6): 503–506.

[58] Linol B., de Wit M. J., Kasanzu C. H., Da Silva Schmitt R., Corrêa-Martins F. J., Assis A. Correlation and paleogeographic reconstruction of the Cape-Karoo Basin sequences and their equivalents across central West Gondwana. Origin and Evolution of the Cape Mountains and Karoo Basin. Regional Geology Reviews. 2016, 183–192.

[59] Reis H. L. S. Neoproterozoic Evolution of the São Francisco basin, SE Brazil: Effects of tectonic inheritance on foreland sedimentation and deformation. Ouro Preto, Brazil: Universidade Federal de Ouro Preto, PhD thesis, 2016.

[60] Macgregor D. S. The hydrocarbon systems of North Africa. Marine and Petroleum Geology. 1996, 13(3): 329–340. doi: https://doi.org/10.1016/0264-8172(95)00068-2

[61] Kendall C., Chiarenzelli J., Hassan H. S. World source rock potential through geological time: A function of basin restriction, nutrient level, sedimentation rate, and sea-level rise. Search and Discover Article #40472. 2009. Accessed January 2025 at www.searchanddiscovery.com/documents/2009/40472kendall/ndx_kendall.pdf)

[62] Gordon A., Destro N., Heilbron M. The Recôncavo-Tucano-Jatobá rift and associated Atlantic continental margin basins. In: Heilbron M., Cordani U., Alkmim F., eds., São Francisco Craton, Eastern Brazil. Regional geology reviews. Cham: Springer, 2017, 171–185. doi: https://doi.org/10.1007/978-3-319-01715-0_9

[63] Rodrigues R. T., Alkmim F. F. D., Reis H. L. S., Piatti B. G. The role of tectonic inheritance in the development of a fold-thrust belt and superimposed rift: An example from the São Francisco basin, eastern Brazil. Tectonophysics. 2021, 815: 228979. doi: https://doi.org/10.1016/j.tecto.2021.228979

[64] Selley R. C., Sonnenberg A. S. Elements of petroleum geology, 3rd ed. . Boston, USA: Academic Press, 2015.

[65] Craig J., Biffi U., Galimberti R. F., et al. The palaeobiology and geochemistry of Precambrian hydrocarbon source rocks. Marine and Petroleum Geology. 2013, 40,1–47. doi: 10.1016/j.marpetgeo.2012.09.011

[66] Frolov S. V., Akhmanov G. G., Bakay E. A., et al. Meso-Neoproterozoic petroleum systems of the Eastern Siberian sedimentary basins. Precambrian Research. 2015, 259: 95–113. doi: 10.1016/j.precamres.2014.11.018

[67] Cox G. M., Sansjofre P., Blades M. L., Farkas J., Collins A. S. Dynamic interaction between basin redox and the biogeochemical nitrogen cycle in an unconventional Proterozoic petroleum system. Scientific Reports. 2019, 9(1): 5200. doi: 10.1038/s41598-019-40783-4

[68] Reis H. L. S. Gás natural. In: Pedrosa-Soares A. C., Voll E, Cunha EC, coord. Recursos minerais de minas gerais. Belo Horizonte, Brazil: Companhia de Desenvolvimento de Minas Gerais, 2018, 1–35.

[69] Martín-Monge A., Baudino R., Gairifo-Ferreira L. M., et al. An unusual Proterozoic petroleum play in Western Africa: The Atar Group carbonates (Taoudeni Basin, Mauritania). In: Geological society, vol. 438(1). London: Special Publications, 2016, 119–157. doi: 10.1144/sp438.5

[70] Scotese C. R., Wright N. PALEOMAP Paleodigital Elevation Models (PaleoDEMS) for the Phanerozoic PALEOMAP Project, 2018. Accessed 2021 at https://www.earthbyte.org/paleodem-resourcescotese-and-wright-2018

[71] Bédard J. H. Stagnant lids and mantle overturns: Implications for Archaean tectonics, magmagenesis, crustal growth, mantle evolution, and the start of plate tectonics. Geoscience Frontiers. 2018, 9(1): 19–49. doi: 10.1016/j.gsf.2017.01.005

[72] Stern R. J., Gerya T., Tackley P. J. Stagnant lid tectonics: Perspectives from silicate planets, dwarf planets, large moons, and large asteroids. Geoscience Frontiers. 2018, 9(1): 103–119. doi: 10.1016/j.gsf.2017.06.004

[73] Dhuime B., Hawkesworth C. J., Cawood P. A., Storey C. D. A change in the geodynamics of continental growth 3 billion years ago. Science. 2012, 335(6074): 1334–1336. doi: 10.1126/science.1216066

[74] Hawkesworth C., Cawood P. A., Dhuime B. Rates of generation and growth of the continental crust. Geoscience Frontiers. 2019, 10(1): 165–173. doi: 10.1016/j.gsf.2018.02.004

[75] Cawood P. A. Earth Matters: A tempo to our planet's evolution. Geology. 2020, 48(5): 525–526. doi: 10.1130/focus052020.1

[76] Brown M., Kirkland C. L., Johnson T. E. Evolution of geodynamics since the Archean: Significant change at the dawn of the Phanerozoic. Geology. 2020, 48(5): 488–492. doi: 10.1130/G47417.1

[77] Herzberg C., Condie K., Korenaga J. Thermal history of the Earth and its petrological expression. Earth and Planetary Science Letters. 2010, 292(1–2): 79–88. doi: 10.1016/j.epsl.2010.01.022

[78] Celli N. L., Lebedev S., Schaeffer A. J., Gaina C. African cratonic lithosphere carved by mantle plumes. National Communications. 2020, 11(1): 92. doi: 10.1038/s41467-019-13871-2

[79] Finger N. P., Kaban M. K., Tesauro M., Mooney W. D., Thomas M. A Thermo-Compositional Model of the African Cratonic Lithosphere. Geochemistry, Geophysics, Geosystems. 2022, 23(3). doi: 10.1029/2021gc01029

[80] Michard A., Frizon de Lamotte D., Saddiqi O., Chalouan A. An outline of the geology of morocco. In: Michard A., Saddiqi O., Chalouan A., Lamotte D., eds., Continental evolution: The geology of morocco: Structure, stratigraphy, and tectonics of the Africa-Atlantic-Mediterranean triple junction. Berlin, Heidelberg: Springer, 2008, 1–31.

[81] Jessell M. W., Liégeois J.-P. 100 years of research on the West African Craton. Journal of African Earth Sciences. 2015, 112: 377–381. doi: 10.1016/j.jafrearsci.2015.10.008

[82] Albert-Villanueva E., Permanyer A., Tritlla J., Levresse G., Salas R. Solid hydrocarbons in proterozoic dolomites, taoudeni basin, Mauritania. Journal of Petroleum Geology. 2015, 39(1): 5–27. doi: 10.1111/jpg.12625

[83] Skikra H., Amrouch K., Soulaimani A., Leprêtre R., Ouabid M., Bodinier J.-L. The intracontinental High Atlas belt: Geological overview and pending questions. Arabian Journal of Geosciences. 2021, 14(12). doi: 10.1007/s12517-021-07346-2

[84] Grenholm M., Jessell M., Thébaud N. A geodynamic model for the Paleoproterozoic (ca. 2.27–1.96 Ga) Birimian Orogen of the southern West African Craton – Insights into an evolving accretionary-collisional orogenic system. Earth-Science Reviews. 2019, 192: 138–193. doi: 10.1016/j.earscirev.2019.02.006

[85] Rollinson H. Archaean crustal evolution in West Africa: A new synthesis of the Archaean geology in Sierra Leone, Liberia, Guinea and Ivory Coast. Precambrian Research. 2016, 281: 1–12. doi: 10.1016/j.precamres.2016.05.005

[86] Coveney R. M. Jr, Goebel E. D., Zeller E. J., Dreschhoff G. A. M., Angino E. E. Serpentinization and the origin of hydrogen gas in Kansas1. AAPG Bulletin. 1987, 71(1): 39–48.

[87] Newell K. D., Doveton J. H., Merriam D. F., Sherwood Lollar B., Waggoner W. M., Magnuson L. M. H2-rich and hydrocarbon gas recovered in a deep precambrian well in Northeastern Kansas. Natural Resources Research. 2007, 16(3): 277–292. doi: 10.1007/s11053-007-9052-7

[88] Guélard J., Beaumont V., Rouchon V., et al. Natural H2 in Kansas: Deep or shallow origin? Geochemistry, Geophysics, Geosystems. 2017, 18(5): 1841–1865. doi: 10.1002/2016gc006544

[89] Gouiza M., Bertotti G., Hafid M., Cloetingh S. Kinematic and thermal evolution of the Moroccan rifted continental margin: Doukkala-High Atlas transect. Tectonics. 2010, 29(5). doi: 10.1029/2009tc002464

[90] Prinzhofer A., Cacas-Stentz M.-C. Natural hydrogen and blend gas: A dynamic model of accumulation. International Journal of Hydrogen Energy. 2023, 48(57): 21610–21623. doi: https://doi.org/10.1016/j.ijhydene.2023.03.060

[91] Sadki O., Prinzhofer A., Berkat N. E., Aver S. Natural hydrogen exploration in the South of Morocco. Online Conference, H-Nat 2022 Conference. 2022.

[92] Hydrogen G. Annual report. Brisbane, Australia: Gold Hydrogen Limited, 2024, 2024.

[93] Boreham C. J., Edwards D. S., Czado K., et al. Hydrogen in Australian natural gas: Occurrences, sources and resources. The APPEA Journal. 2021, 61(1). doi: 10.1071/aj20044

[94] Rezaee R. Assessment of natural hydrogen systems in Western Australia. International Journal of Hydrogen Energy. 2021, 46(66): 33068–33077. doi: 10.1016/j.ijhydene.2021.07.149

[95] Fraser G., McAvaney S., Neumann N., Szpunar M., Reid A. Discovery of early Mesoarchean crust in the eastern Gawler Craton, South Australia. Precambrian Research. 2010, 179(1–4): 1–21.

[96] Betts P. G., Armit R. J., Stewart J., et al. Australia and nuna. In: Geological society, Vol. 424(1). London: Special Publications, 2015, 47–81. doi: 10.1144/sp424.2

[97] Lloyd J. C., Blades M. L., Counts J. W., et al. Neoproterozoic geochronology and provenance of the Adelaide Superbasin. Precambrian Research. 2020, 350. doi: 10.1016/j.precamres.2020.105849

[98] Zhu R., Zhao G., Xiao W., Chen L., Origin T. Y. Accretion, and Reworking of Continents. Reviews of Geophysics. 2021, 59(3). doi: 10.1029/2019rg000689

[99] Prinzhofer A., Moretti I., Françolin J., et al. Natural hydrogen continuous emission from sedimentary basins: The example of a Brazilian H2-emitting structure. International Journal of Hydrogen Energy. 2019, 44(12): 5676–5685. doi: 10.1016/j.ijhydene.2019.01.119

[100] Moretti I., Prinzhofer A., Françolin J., Pacheco C., Rosanne M., Rupin F., Mertens J. Long-term monitoring of natural hydrogen superficial emissions in a Brazilian cratonic environment: Sporadic large pulses versus daily periodic emissions. International Journal of Hydrogen Energy. 2021, 46(5): 3615–3628.

[101] Reis H. L. S. Minas gerais. In: Delgado F., Santos E., orgs. O desenvolvimento da exploração de recursos não-convencionais no Brasil: Novas óticas de desenvolvimento regional. Rio de Janeiro, Brazil, FGV, 2021, 87–95.

[102] Reis H., Fonseca R. Does the unusual geochemical composition of the São Francisco basin natural gas (E Brazil) reveal typical characteristics of ancient and overmature petroleum systems? Goldschmidt2021 Abstracts – Goldschmidt2021. European Association of Geochemistry Virtual. 2021. doi: 10.7185/gold2021.6383

[103] Flude S., Magalhães N., Warr O., Bordmann V., Fleury J.-M., Reis H. L. S., Trindade R. I. F., Sherwood Lollar B., Ballentine C. J. Origin of H2 and CH4 gases in the Eastern São Francisco basin, Brazil. Goldschmidt2021 Abstracts – Goldschmidt2021. European Association of Geochemistry. 2021. Virtual. doi: https://doi.org/10.7185/gold2021.6052

[104] Freitas V. A., Prinzhofer A., Françolin J. B., Ferreira F. J. F., Moretti I. Natural hydrogen system evaluation in the São Francisco Basin (Brazil). Science and Technology for Energy Transition. 2024, 79. doi: 10.2516/stet/2024091

[105] Alkmim F. F., Marshak S., Pedrosa-Soares A. C., Peres G. G., Cruz S. C. P., Whittington A. Kinematic evolution of the Araçuaí-West Congo orogen in Brazil and Africa: Nutcracker tectonics during the Neoproterozoic assembly of Gondwana. Precambrian Research. 2006, 149(1): 43–64. https://doi.org/10.1016/j.precamres.2006.06.007

[106] Reis H. L. S. Craton, tectonics and natural hydrogen. Perth, Australia, H-Nat, 2023.

[107] Sugisaki R., Ido M., Takeda H., et al. Origin of hydrogen and carbon dioxide in fault gases and its relation to fault activity. The Journal of Geology. 1983, 91(3): 239–258.

[108] Sleep N. H., Meibom A., Fridriksson T., Coleman R. G., Bird D. K. H2-rich fluids from serpentinization: Geochemical and biotic implications. Proceedings of the National Academy of Sciences. 2004, 101(35): 12818–12823. doi: 10.1073/pnas.0405289101

[109] Suzuki N., Saito H., Hoshino T. Hydrogen gas of organic origin in shales and metapelites. International Journal of Coal Geology. 2017, 173: 227–236. doi: 10.1016/j.coal.2017.02.014

[110] Truche L., Bourdelle F., Salvi S., Lefeuvre N., Zug A., Lloret E. Hydrogen generation during hydrothermal alteration of peralkaline granite. Geochimica et Cosmochimica Acta. 2021, 308: 42–59. doi: https://doi.org/10.1016/j.gca.2021.05.048

[111] Tutolo B. M., Seyfried W. E. Jr, Tosca N. J. A seawater throttle on H(2) production in Precambrian serpentinizing systems. Proceedings of the National Academy of Sciences of the United States of America. 2020, 117(26): 14756–14763. doi: 10.1073/pnas.1921042117

[112] Gaucher E. C., Moretti I., Pélissier N., Burridge G., Gonthier N. The place of natural hydrogen in the energy transition: A position paper. European Geologist. 2023, 55. doi: https://doi.org/10.5281/zenodo.8108239

[113] Geymond U., Briolet T., Combaudon V., et al. Reassessing the role of magnetite during natural hydrogen generation. Frontiers in Earth Science. 2023, 11. doi: 10.3389/feart.2023.1169356

[114] Moretti I., Bouton N., Ammouial J., Carrillo Ramirez A. The H2 potential of the Colombian coals in natural conditions. International Journal of Hydrogen Energy. 2024, 77: 1443–1456. doi: 10.1016/j.ijhydene.2024.06.225

[115] Lodhia B. H., Peeters L., Frery E. A review of the migration of hydrogen from the planetary to basin scale. Journal of Geophysical Research: Solid Earth. 2024, 129(6): e2024JB028715. doi: https://doi.org/10.1029/2024JB028715

[116] Maiga O., Deville E., Laval J., Prinzhofer A., Diallo A. B. Trapping processes of large volumes of natural hydrogen in the subsurface: The emblematic case of the Bourakebougou H2 field in Mali. International Journal of Hydrogen Energy. 2024, 50: 640–647. doi: 10.1016/j.ijhydene.2023.10.131

[117] Truche L., Joubert G., Dargent M., Martz P., Cathelineau M., Rigaudier T., Quirt D. Clay minerals trap hydrogen in the Earth's crust: Evidence from the Cigar Lake uranium deposit, Athabasca. Earth and Planetary Science Letters. 2018, 493: 186–197. doi: https://doi.org/10.1016/j.epsl.2018.04.038

[118] Wang Y., Cao Z., Peng L., et al. Secular craton evolution due to cyclic deformation of underlying dense mantle lithosphere. Nature Geoscience. 2023, 16(7): 637–645. doi: 10.1038/s41561-023-01203-5

[119] McCollom T. M., Bach W. Thermodynamic constraints on hydrogen generation during serpentinization of ultramafic rocks. Geochimica Et Cosmochimica Acta. 2009, 73(3): 856–875. doi: https://doi.org/10.1016/j.gca.2008.10.032

[120] Klein F., Bach W., Jöns N., McCollom T., Moskowitz B., Berquó T. Iron partitioning and hydrogen generation during serpentinization of abyssal peridotites from 15°N on the Mid-Atlantic Ridge. Geochimica Et Cosmochimica Acta. 2009, 73(22): 6868–6893. doi: 10.1016/j.gca.2009.08.021

[121] Neubeck A., Duc N. T., Bastviken D., Crill P., Holm N. G. Formation of H2 and CH4 by weathering of olivine at temperatures between 30 and 70 °C. Geochemical Transactions. 2011, 12(1): 6. doi: 10.1186/1467-4866-12-6

[122] Shibuya T., Yoshizaki M., Sato M., et al. Hydrogen-rich hydrothermal environments in the Hadeanocean inferred from serpentinization of komatiites at 300 °C and 500 bar. Progress in Earth and Planetary Science. 2015, 2(1). doi: 10.1186/s40645-015-0076-z

[123] Tamblyn R., Hermann J. Geological evidence for high H2 production from komatiites in the Archaean. Nature Geoscience. 2023, 16(12): 1194–1199. doi: 10.1038/s41561-023-01316-x

[124] Li X., Krooss B. M., Weniger P., Littke R. Liberation of molecular hydrogen (H2) and methane (CH4) during non-isothermal pyrolysis of shales and coals: Systematics and quantification. International Journal of Coal Geology. 2015, 137: 152–164. doi: https://doi.org/10.1016/j.coal.2014.11.011

[125] Horsfield B., Mahlstedt N., Weniger P., et al. Molecular hydrogen from organic sources in the deep Songliao Basin, P.R. China. International Journal of Hydrogen Energy. 2022, 47(38): 16750–16774. doi: https://doi.org/10.1016/j.ijhydene.2022.02.208

[126] Milesi V., Guyot F., Brunet F., et al. Formation of CO2, H2 and condensed carbon from siderite dissolution in the 200–300 °C range and at 50MPa. Geochimica Et Cosmochimica Acta. 2015, 154: 201–211. doi: 10.1016/j.gca.2015.01.015

[127] Caxito F., Uhlein G. J., Uhlein A., Pedrosa-Soares A. C., Kuchenbecker M., Reis H. L. S., et al. Chapter Three – Isotope stratigraphy of Precambrian sedimentary rocks from Brazil: Keys to unlock Earth's hydrosphere, biosphere, tectonic, and climate evolution. In: Montenari M., ed., Stratigraphy & timescales. Academic Press, 2019, 73–132.

[128] Parnell J., Blamey N. Global hydrogen reservoirs in basement and basins. Geochemical Transactions. 2017, 18(1): 2. doi: 10.1186/s12932-017-0041-4

[129] Warr O., Giunta T., Ballentine C., Sherwood Lollar B. Mechanisms and rates of 4He, 40Ar and H2 production and accumulation in fracture fluids in Precambrian shield environments. Chemical Geology. 2019, 530: 119322. doi: https://doi.org/10.1016/j.chemgeo.2019.119322

[130] Halford D. T., Karolytė R., Andreason M. W., Cathey B., Cathey M., Dellenbach J. T. Probabilistic determination of the role of faults and intrusions in helium-rich gas fields formation. Geochemistry, Geophysics, Geosystems. 2024, 25(6). doi: 10.1029/2024gc011522

[131] Zou C. Unconventional petroleum geology. Elsevier, 2013.

[132] Paiva J. M. S., Drummond Chicarino Varajão A. F., Reis H. L. S., Gomes N. S., Suss J. F. The post-depositional evolution of Mesoproterozoic to early Neoproterozoic tight sandstone reservoirs, São Francisco basin, Brazil. Journal of South American Earth Sciences. 2025, 151: 105278.

[133] Miranda F. S., Vettorazzi A. L., Cunha P., et al. Atypical igneous-sedimentary petroleum systems of the Parnaíba Basin, Brazil: Seismic, well logs and cores. In: Geological society, vol. 472(1). London: Special Publications, 2018, 341–360. doi: 10.1144/sp472.15

John Hanson
Chapter 7
A possible origin of organic natural hydrogen

Abstract: A new theoretical interpretation of subsurface hydrocarbon generation based on the decomposition and rearrangement of organic matter by thermocatalytic matrix reactions focuses on the diverse and prolific role that carbon catalysts play. These catalysts have high magnetic susceptibilities and can abstract carbon, hydrogen, and heteroatoms, particularly oxygen, aiding the decomposition of organics. Scholl reactions cause cyclodehydrogenation and are responsible for ring closure and intermolecular coupling reactions, promoting growth. However, when stacking of these moieties occurs, dislocation of the aromatic electron cloud contributes to their exotic behavior and superior catalytic abilities. Carbon catalysts are active at all points in the hydrocarbon system, evolving from biomarkers and polycyclic aromatic hydrocarbons through to amorphous carbon. Desorption of CO and CO_2 as water shift reactions rejuvenate carbon catalysts as well as creating hydrogen. The significance of the progressively increasing reactivity and strange behavior of carbon catalysts are only now being understood within the context of graphene research. They can better explain hydrocarbon evolution and methane decomposition along with natural hydrogen creation and its preservation at the low temperatures of the subsurface. Further work on carbon catalysts may provide future insight into new low-temperature subsurface catalytic methods for synthesis of petrochemicals and carbon-free natural hydrogen required for the energy transition.

Keywords: organic natural hydrogen, carbon catalysts, methane decomposition

7.1 Introduction

A gargantuan effort is underway to explore natural hydrogen, typically looking at generation mechanisms associated with reactions of iron and water, with migration and trapping ideas borrowed from the hydrocarbon system [1–7]. However, logic dictates the possibility of an organic-rich sediment source for hydrogen generation too; potentially opening up more basinal areas to explore [8]. After all, most anthropogenic hydrogen is currently created by mechanisms and processes closely associated with our knowledge of grey hydrogen generation from methane [9]. This paper expands on my previous work on the subject attempting to identify overlooked carbon catalysts involved in natural organic hydrogen generation [10], leading to novel insights into hydrocarbon generation [11].

John Hanson, Independent Correspondence, e-mail: jcph100@gmail.com

https://doi.org/10.1515/9783111437040-007

Despite extensive research for an alternate model, the classical thesis of hydro-carbon generation in the subsurface still prevails, having been successfully tested on numerous occasions. It relies heavily on a series of time- and temperature-dependent kinetic reactions extrapolated from very high temperatures of 600–900 °C to subsurface basinal temperatures of 50–250 °C. This temperature disparity and lack of natural hydrogen have been weaknesses in an otherwise strong model. The counterpoint to this is often framed as matrix reactions of clays, alkali metals, or alkali earth metals that achieve the same results but within the observed subsurface temperature and pressure regime, but this model remains incomplete and unadopted [12–20]. Matrix reactions act in a linked and overlapping series of co-catalytic reactions, not only producing hydrocarbons but also, occasionally, natural hydrogen [21]. However, source rocks and oil deposits are also rich in a diverse array of carbon catalysts, from diagenesis to metagenesis, which are rarely mentioned as catalysts. They are either modified biomarker products, surviving organic matter decomposition, or are formed in situ from the rearrangement of precursor reactants that may be an overlooked alternative. It is hoped that an introduction to carbon catalysts may shift opinion and unify the subsurface realm with anthropogenic production. For example, the asphaltenes that have been extensively studied are also known as polycyclic aromatic hydrocarbons (PAHs) and are world-class nanographene catalysts. Classic models do not use these or the other abundant carbon catalysts observed in the classic model to drive the reactions and processes, leaving observations of natural hydrogen unresolved. Research into carbon catalytic reactions with organic matter and kerogen may resolve a few issues with the classical theory of hydrocarbon generation as well as foster new ideas for the generation of natural hydrogen [22–24].

7.2 Catalysts in the subsurface

Metals (alkali and alkali earth) and clays acting as co-catalysts have previously been reported as matrix reactants, but are seen as contributing only in a minor way to the subsurface hydrocarbon and natural hydrogen generation process. The reactions of the alkali and alkali earth metal cations are important, particularly Na^+ [25]. They create hydroxyls from clays, vitally important for the breakup of large geo molecules such as organic matter, proto-kerogen, and kerogen, by hydrolysis. They are important for the isomorphous substation reactions of clays that mediate hydroxyl anion release, thereby increasing the number of active sites available for inter-layer reactions. They are involved in the complexion of acetates, which are widely dispersed in hydrocarbon-forming basins globally [26]. They are key reactants within the saponification reactions within kerogen, which inhibit microbe consumption of natural hydrogen. It has also been noted that these cations may reduce the temperature at which other organic reactions occur, for example, addition reactions that contribute to aromatization [25, 27, 28]. Clay catalysts are very important within the petrochemical industry and have been extensively researched from hydrocarbon and hydrogen evolution perspectives. The catalytic clay conversion series [29], from smectite to illite to chlorite and then onto low-grade Barrovian metamorphic aluminosilicates, have been previously alluded to. These catalysts have been researched and highlighted as a pathway to regioselective production of hydrogen during clay catalysis, with the familiar smectite to illite and/or chlorite conversion [30]. By contrast, carbon catalysts have not been considered as important, perhaps, as at the time of the development of the classical thesis of hydrocarbon generation, the seminal works on graphene and follow-up work of carbon catalysis had not yet occurred.

The matrix reactions of carbon catalysts are probably the most important and are responsible for hydrocarbon and hydrogen generation at low temperatures and pressures. Structurally, they are composed of an aromatic ring as a core, but develop into polycyclic aromatic moieties as they evolve and may have one or more side chains or alkyl groups attached. On the basis of their structure, they are said to be either island type (continental if larger) or archipelago type. In the classical model, they are referred to as aromatics, if they do not possess a side chain, or asphaltenes, if they have a side chain. Structural differences mean that island types are less polar, as they have fewer NSO atoms or groups, whilst archipelago types are more likely to be polar, as they contain more NSOs. Topological differences mean both types having varying high magnetic susceptibilities, making them good catalysts, but archipelago types, more so due to additional polar effects [31].

There are two broad evolutionary types found within the hydrocarbon generation system [31]. The first is the familiar fossil molecules, often grouped together as biomarkers. During the decomposition of organic matter, proto-kerogen, and kerogen, a vast array of proto-asphaltenes and other polycyclic moieties, often with NSO functional groups, are released. They are classed as fossil molecules and thought to be biomolecular fragments that have been preserved with minor modifications that can be matched to original molecules observed in organisms. Initial modification may consist of dechelation and opening of cyclic amine ring structures by adsorption and chemisorption of oxygen. The second type is authigenic in situ moieties, formed by reactions of smaller precursor molecules and atoms remaining from organic matter decomposition and rearrangement [32]. These evolve into primitive cyclic species and then grow into more complex aromatic structures. These forms mature and may look very similar or even duplicates of some biomarker species [33]. Evolutions of both types coalesce in structure to form one indistinguishable group known as PAHs, and are found in similar locations throughout generations. They have highly reactive surfaces, acting as molecular sponges by adsorbing polar organic molecules and heteroatoms, which adds to their already highly reactive nature [34, 35]. This class of catalysts has been marginalized due to its annoying habit of aggregating and inhibiting large-scale petrochemical reactions by the production of coke [36–38]. Due to the extensive research into graphene, a new focus within physics and chemistry is intensively studying carbon catalysts. Hence, a new look at the hydrocarbon and hydrogen generation process is required as the generation process is full of interesting carbon catalytic forms, from beginning to end.

To understand the possible evolution of carbon catalysts, it is instructive to take inspiration from research on the evolution of partially combusted carbon pollutants, seen in most cities, for a guide to what might be developing within the subsurface [39, 40]. The development of carbon pollutants begins with precursor molecules reacting by radical reaction, acetylene additions, or vinylacetylene additions under standard conditions or by polycyclic molecules with low vapor pressure as the starting point. Nucleation of these leads to the creation of "parent PAHs," which react further, producing larger molecules with greater numbers of aromatic rings, making them ever more carcinogenic. Increasingly larger compounds are formed, ultimately ending in particulate soot formation. It is thought that both biomarker and in situ authigenic carbon catalysts evolve into parent PAHs as their evolution coalesces and then evolve in a similar manner in the subsurface [41]. Both biomarker and in situ forms then develop along the same path, eventually stacking together, creating large 3D allotropes of carbon [42], such as amorphous carbon. The approximate series from small to large size catalysts is polar alicyclic alkenes, polar cyclic/aromatics, polycyclic aromatic hydrocarbons (PAHs) [43] (known as aromatics and asphaltenes by geochemists some forms of which are biomarkers), stacked HMW PAHs, randomly stacked giant carbon macromolecules, pyrobitumen, activated carbon, carbon black, and graphite. The termination of the process being carbon blacks transformation into graphite, which neatly links into the various forms of carbon observed during low-grade metamorphism [44, 45], may be helpful to better understand the process prior to graphitization in anthracitic coals between 500 and 600 °C [46].

Carbon catalysts owe their remarkable reactivity to diverse bonding hybridization forms (sp, sp2, and sp3) [47], π-π stacking ("pancake bonding") in a parallel orientation, and π-σ interaction in a T-shape orientation plus the additional complexity of Stone-Wales defects within 3D crystalline structures [48–50]. Numerous topologies can be achieved during evolution by the combination of other arenes, alkyl groups, or other side chains, plus the abstraction of NSO atoms, to build up the surface structure increasing reactivity [51, 52]. Chelation, complexion, or adsorption of metal atoms (alkali and alkali earth metals are of interest here), into the aromatic structures may also play an important role in the reactive ability of these catalysts [53]. However, it is the effect that stacking has on carbon catalysts which is the most interesting phenomenon exhibited [50].

Recent work on graphene has shown that stacked stepping arrangement and flexure of the molecules in the order of 1–50 can enhance electron distortions, further creating odd effects such as Cooper pairs, Phonons, and moiré patterns [54]. There is some evidence that this may be linked to molecular self-assembly [55], superconductivity [56], and super electronegativity [57]. This exotic behavior may well be the insight required to better understand the process of decomposition and rearrangement of organic matter and kerogen, and allow the extra energy required for bond break-ing and formation of difficult reactions. The stacked staircase arrangement mentioned would fit into the early observations by Tissot and Welte, displayed in their electro-micrographs of kerogen [58]. Here, stacked planar carbon-rich molecules are at first randomly orientated in immature kerogen and later clustered with similar orienta-tions. This insight is novel and needs further research and development.

Catalytic activity of immature carbon catalysts is initially low due to the low num-ber of aromatic rings and low magnetic susceptibility, imbuing an inability to reacti-vate at original active sites. As carbon catalysts mature by increasing the number of aromatic rings, the catalytic ability and the ability of the catalyst to sorb increase as the size, structural heterogeneity, and ability to reactivate annealed reaction sites in-crease. Reactivity follows the same general form, in which carbon catalysts evolve ex-cept that activated carbon is considered the peak of catalytic reactivity and carbon black and graphite have a reduced catalytic activity in comparison, due to annealing or solid-state rearrangement, eliminating active sites, as their molecular structures become more ordered. Furthermore, the complication is that not all structural forms that exist are active catalysts due to topology, annealing, or steric arrangement. Also, despite differing carbon catalysts coexisting at the same time, they may not be the dominant catalyst active at the time, some being more reactive or polar than others. Also, it is thought that the timing of the development of each overlap acting as a co-catalyst. Despite structural molecular similarities, some PAHs may have lesser activity compared to the dominant catalyst-aiding reactions at that time. For example, smaller polar benzocarbazoles that are PAHs with a cyclic amine in the structure may be pref-erentially fractionated, as it has less reactive cores in its structure [59]. They may look structurally very similar compared to some of the other less active slightly polar PAHs, but may reactivate at some future episode again by the abstraction of new mol-ecules into the surface or edge structure. The less reactive fraction of these forms end up on the periphery of the main colloidal suspension, along with petroleum species, and are more prone to expulsion. Highly polar forms, on the other hand, are more likely to abstract NSO atoms or groups through adsorption and chemisorption, which generally means that they grow larger and stay bound to organic matter within the source rock [60].

This series of catalysts are considered as being highly reactive due to their struc-tural arrangement and accompanying high magnetic susceptibility [32, 61]. The ad-sorptive nature of these catalysts means that in the subsurface, they exist in a range of colloidal suspensions or sols to solid forms [32]. Colloidal carbon catalysts act as centers of adsorption and reaction, causing fractionation of less reactive species out of the sol into the free phase, which are more likely to be expulsed [62, 63]. Replicating the colloidal nature under bulk experimental conditions to research these catalysts is difficult and until recently, characterization of individual molecules or groups was hard, as techniques had not been developed to discern them. Comprehension of gen-eration mechanisms was unclear due to their complex nature and lack of techniques for study [64]. Some work on residues of carbon such as pyrobitumen [35] or activated carbon [65] have been studied and shown to have extensive catalytic capabilities. The issue with carbon catalysts has always been the tendency to be extremely reactive initially and then exhibit a sharp drop off in reactivity, and a tendency to anneal active sites [66]. Reactivation of this catalytic group is often seen as time-consuming and expensive [67]. This has contributed to the probability of carbocatalytic reactive mechanisms being responsible for hydrocarbon, and hydrogen generation in the subsurface being envisaged as low. Due to their extensive nanostructure, they are the principal preservers of natural hydrogen in the subsurface [68] and are also powerful catalysts that are able to crack ethane and methane into hydrogen [65].

The reaction mechanism of carbon catalysts is dominated by the abstraction of atoms from less reactive molecules, which can contribute to the evolutionary growth of the carbon catalyst, decomposing kerogen and cracking of saturated molecules [32, 60, 69]. Also by Scholl reactions, they act intramolecularly to close open aromatic cycles, releasing natural hydrogen, or intermolecularly, as a Friedel-Crafts coupling reaction, to joining two or multiple aromatic carbon-based molecules together [70, 71]. The key to the way in which carbon catalysts work in the decomposition of organic matter and kerogen is demonstrated by the oxidation and decomposition of asphaltenes in bitumen road surfaces, leading to pothole development [72–74]. They have the ability to adsorb, chemisorb, and abstract oxygen from kerogen, water, CO_2, and other oxides by asphaltenes (and larger carbon catalysts) into the side chain and also the primary ring structures [75]. Chemisorption of oxygen may leave the aromatic ring intact or destabilize the aromatic structure, fragmenting it into an array of products for rearrangement. This process is insightful, giving an understanding of the decomposition of both kerogen and asphaltenes within the subsurface system [75–78].

Oxidation of road asphalt takes place at standard temperature and pressure, may be mediated by ultraviolet light, and may also be assisted by gritting with salt such as NaCl contributing Na^+ [79–81]. Archipelago motifs of the asphaltenes within road asphalt binder show a reduction in the alkyl structures or side chains, adopting a more island (or continental, if a larger number of rings) core aromatic structure. The more reactive archipelago molecules abstract oxygen into the side chain at the benzylic carbon, creating a carbonyl. π-π stacking of other asphaltenes in the form of a dimer or turbostratic arrangement may cause the spatial extent of the π electron cloud to decentralize or shift away from the oxidized benzylic carbon, leading to bond weakness. The carbonyl reacts further, producing carboxyl or alcohol groups; this may cause scission, releasing CO_2, H_2O, and volatiles, including alkanes, decomposing the asphaltene into a less mature form. The process of oxidation in asphaltene binder is seen to be reversible by the addition of a bio-oil rejuvenator (pig slurry), which is high in nitrates. Rather than stacking almost identical asphaltene molecules, nitrogen-rich bio-oil species may interact in a similar way preferentially, abstracting at the side chain and releasing nitrogen oxide. Asphaltenes are replenished from the resin and aromatic fraction in the bitumen binder by a similar adsorption and abstraction process.

A similar oxidation process by adsorption of CO_2 within enhanced recovery reservoirs rich in asphaltenes [82] may be responsible for the observations seen in their aggregation-hindering production [83]. Incorporation of oxygen from CO_2 into the structure of asphaltene releases CO into the free phase, which in turn reacts with water to release natural hydrogen.

Within the subsurface, a similar set of reactions are envisaged, mediated with moderate heat rather than ultraviolet light. Asphaltenes created at the end of diagenesis form colloids around organic matter, which transforms it into kerogen by abstracting oxygen and other heteroatoms from it, causing rearrangement. Early π-π stacking of immature asphaltenes, perhaps in conjunction with biomarker forms with higher nitrogen content, gives high reactivities to these structures, allowing ease of abstraction of oxygen from other molecules [50]. Decomposition of the organics leads to the formation of carboxylic acids, alkanes, from the destruction of asphaltene side chains and also by aromatic ring fragmentation. This releases resins, aromatics, and more immature asphaltenes as products for in situ PAH formation, and as products for the maturation of other asphaltenes. Oxidation of the aromatic core structures occurs, resulting in ring cleavage products, amongst which are useful intermediary diketones such as 4-oxo-2-hexenedial [84].

Asphaltene cores grow larger and in doing so, the center of the electron cloud shifts further from the margins, leading to bond scission occurring more frequently at benzylic carbons [72]. More mature asphaltenes, therefore, have the appearance of a shortened side chain structure, compared to immature forms. Post carboxylic acid production oxygen is depleted within the subsurface system as CO_2 and acetates migrate out during secondary migration. The amount of oxygen within the system stays the same as it is abstracted and recycled by the various carbon catalysts as decomposition of the LMW PAHs, resins, and aromatics, creating more asphaltenes. The oxygen becomes more concentrated in the more mature carbon catalysts, which stack even more, reducing distances between stacks. The carbon catalysts mature and evolve into stronger catalysts with higher magnetic susceptibilities with larger perimeters to their aromatic cores, which means a greater number of different points of attachment for side chains. This means electron cloud density in the aromatic core can hold onto more numerous shortened side chains by virtue of steric hindrance. Chemisorption of oxygen is too great for the release of acids, oxygen being swapped between the stronger and weaker of the catalysts. The resin and aromatics get utilized in the process until asphaltenes stack and aggregate in turbostratic clusters of carbon in the macromolecular form. These 3D species are coated in oxygen and other NSO on the surface. Contemporaneous to these, the numerous alkyl sidechains and fragments from ring fragmentation that have undergone scission are left as alkanes, some of which have oxygen attached to others in branched form and are in solution with biomarker alkanes that have been released by the decomposition of kerogen. Alkanes are cracked by the highly reactive 3D forms from pyrobitumen, but by abstraction of carbon from the n-alkane. The abstraction of carbon into the catalyst means that no hydrogen is utilized in this process. This means that natural hydrogen produced during subsurface reaction is more likely to be preserved or migrate. The classical model uses free radicals to perform cracking of alkanes, which consume additional hydrogen in the process, leading to the assumption that organic natural hydrogen preservation in the subsurface is impossible.

Abstraction of NSO by these catalysts provides a mechanism of catalytic decomposition of organic matter, but also of growth by abstraction and accumulation of new carbon molecules into the molecular structure. However, there is a further mechanism of ring closure and coupling provided by Scholl reactions, evolving hydrogen by cyclodehydrogenation. This occurs contemporaneously as the abstraction reactions are happening. There is debate as to whether these are free radicals mediated by Lewis and Brønsted acid sites or via protonation mechanisms from cations such as H^+ and Na^+ or larger. Two bonded hydrogens being released as covalently bonded carbon-to-carbon bonds are formed during cyclodehydrogenation [85, 86]. Intramolecular ring closure occurs as opposing hydrogens in close proximity are removed and carbon-to-carbon bonds are created [87]. Polycondensation coupling reactions occur by a similar means but are intermolecular, joining two distinctly separate molecules together by forming a carbon-to-carbon bond at the expense of hydrogen [88]. Scholl reactions are one of the principal cyclodehydrogenation reactions creating PAHs for industrial purposes, in combination with Brønsted and Lewis acids, and therefore seem a likely route to the creation of macromolecular structures from existing smaller carbon catalysts in the subsurface [87, 89–94]. Cyclodehydrogenation using Na^+ may also be instrumental as a cation intermediary in the Scholl reaction, which is responsible for the generation of the carbon catalysts [70]. Lewis and Brønsted acids and metal cation exchange are familiar concepts in the other matrix reactions, particularly clays, which makes the idea of Scholl reactions an appealing reaction route [39, 40, 95, 96].

Carbon catalysts therefore can decompose kerogen by abstraction of oxygen, increasing their reactivity, and crack hydrocarbons by abstraction of carbon, and are instrumental in the formation, efficient saving, and preservation of natural hydrogen. Post cyclodehydrogenation and condensation of new PAHs, the C:H ratio increases with natural hydrogen, exiting the system either by preservation in nanoporosity or migration.

7.3 Foundational model of organic hydrogen production

7.3.1 Preamble

For ease of explanation, it is assumed that the examples that follow is an organic shale source rock; the different reactions and processes have been split into separate chronological steps to allow simplification, but not all of the steps are shown, allowing the explanation of the overall process to flow and the story to unfold. Within each step, an attempt to characterize the stage of kerogen and bitumen rearrangement, along with the subsurface generation and mechanisms associated with some of the hydrocarbons and natural hydrogen generated at that point is suggested, which is summarized later in Figure 7.2 and aided by Figure 7.1.

Organic-rich sediments, ranging from terrigenous environments to black shale and coal, are created during anoxic or dysoxic conditions, which preferentially preserve decaying organic matter [97, 98]. Organic matter preserved within them is a complex mixture of biopolymers, mainly composed of proteins, carbohydrates, lipids, and lignin. Partial fragments of these delicate macromolecules are preserved within kerogen as biomarkers, acting as pointers to their origin [99]. Proteins and carbohydrates are the most easily decomposed [100], but remnants of these have been preserved by steric effects, leading to isolated fragments encapsulated in lipid and lignin-like macromolecules reformed around them. Perishable organic matter gradually decomposes, releasing biomarkers, and is at the same time reformed through physiochemical processes during decay, creating kerogen and acids. It is thought that the decomposition of kerogen by various reactions follows a broad evolutionary series, dominated by kerogen, PAHs, bitumen, oxygen-rich carbon molecules, and branched alkanes. Rearrangement of kerogen/bitumen ceases, as they are fully decomposed, and subsequent products are converted into n-alkanes. Reactions continue by the cracking of alkane molecules, producing HMW n-alkanes, then ever smaller LMW n-alkanes, methane, and finally natural hydrogen.

7.3.2 Step 1

During diagenesis, decomposition of organics is achieved by catalysis from clays, releasing hydroxyls during clay rearrangement and also by microbes and bacteria feeding on organic matter. As it decays, it releases proto-asphaltenes in the form of biomarkers, which are rich in nitrogen compounds and metal complexes [101]. Stacking and aggregation of heterocyclic amines in some of these compounds cause instability in the molecule [82] and interaction, resulting in alteration of the spatial extent of the electron cloud and abstraction of oxygen from organic matter into cyclic amine, releasing nitrogen oxides and sulfur oxides [102, 103]. This induces ring cleavage and eventually converts biomarkers to the various acyclic and cyclic compounds found in humin, humic, and fulvic acids to asphaltene biomarkers. For example, chlorophyll-a is converted into pheophytin, then pheophorbide-a, and the cyclic structure is opened up by stacking and oxygenolytically cleaving the ring structure, ultimately releasing phytol [84, 104, 105]. Clays exchange alkali and alkali earth metals with proto asphaltene biomarkers, for example Mg^{2+} in chlorophyll dechelates, reacting with clay interlayers, releasing more hydroxyls and causing the decomposition of organics. Eventually, through cyclization and aromatization, sufficient asphaltenes are formed and create a protective colloidal coating around the organic matter. The colloidal layer prevents further microbial and bacterial attacks, acting as a partial barrier from other reactions happening in the source rock [106]. The asphaltenes (PAHs) are immature with perhaps 1–4 aromatic cores. Colloid formation allows direct interaction between asphaltenes and organic matter, which results in direct abstraction of oxygen by asphaltenes. Rearrangement transforms left over rearrangement products into in situ authigenic PAHs and convert the organics into kerogen inside the colloidal suspension [107]. Organic matter, which does not generate sufficient asphaltenes, does not create a protective colloidal film and is open to complete decomposition, leaving behind carbon-rich organic "fossil molecules" [99, 100, 106]. These are classified as diagenetic solid bitumen by petrographers [108] and may, at a future point, form the basis of primary carbon-based catalytic reaction sites due to their high aromatic content. In this way, large volumes of asphaltenes are lost to the system and expulsed as immature oils. Although, natural hydrogen is released during diagenesis, it is consumed by microbes and bacteria, and utilized in reactions.

7.3.3 Step 2

Kerogen is difficult to characterize [109, 110] due to its tendency to aggregate and its habit of existing in a colloid form. It can be thought of as a series of clumped linear organic aliphatic strands interspersed with cyclic or aromatic groups, with cross-linkages formed from functional groups containing NSO heteroatoms binding the structure together [111]. This hints at the rearranged structure made up of biomolecules rich in biomarkers early stacking and preferential release of sulfur-enriched molecules [112]. The giant geo-organic molecules are highly unstable and prone to decomposition and reformation, caused by the abstraction of NSO groups (particularly oxygen) by surrounding asphaltenes. Constant decomposition of the kerogen structure means that asphaltenes continuously fragment too and their products reform or mature within their colloids. This leads to fractionation at an early stage and throughout rearrangement, separating high from lower magnetic susceptibilities, plus highly polar and less polar groups of molecules during expulsion events [113–115]. Steric shielding means there is hysteresis of biomarker release, with decomposition and reformation reactions predominantly initiated by reactions that occur at the margins of these giant molecules. Release of hydroxyl and sulfate radicals from biomarker rearrangement within the colloid penetrates deeper into the organic molecular structure than the peripheral asphaltenes and fragments biomarkers (as well as asphaltenes), prior to reformation and rearrangement of the new kerogen. Other reformation reactions of kerogen include nucleophilic addition, dehydration, and oxidative dehydrogenation. Both decomposition and reformation reaction suites are influenced by electron charge and bonding imbalances at active NSO heteroatom sites within the giant geo-structure. Decomposition in the main occurs initially at sulfur functional groups as sulfur-to-carbon bonds can be preferentially broken at lower temperatures [116]. Once sulfur-rich groups are expulsed from the system as immature oils, the polar carbonyl functional group locations that are physically accessible become the focus of decomposition and rearrangement. Linking esters that join larger macromolecules of kerogen together are particularly susceptible to reaction. This creates separate molecular rafts or molecular fragments, which may be final products such as immature asphaltenes or else aggregate or are broken down and reformed further [117].

PAHs (asphaltenes) either create cookie cutter fashion as biomarkers or instigate world-class catalysts, particularly when stacked, and their high NSO content may even make parts of their structure into N-doped super catalysts [118]. Immature PAHs react at this early stage, abstracting oxygen, making carbonyl groups, which go on to form alcohols and carboxylic acids. This may anneal the carbon-active sites, temporarily transforming them to lower reactive annealed forms, meaning they are more likely to be expulsed from the colloid into the source rock, and migrate. Immature PAHs continue to be released post-diagenesis due to hysteresis and their relatively smaller size despite pore membrane effects on fractionation release [112, 119]. The remaining kerogen is depleted in sulfur, but oxygen and nitrogen functional groups are held, to a greater extent, within the kerogen, with only minor changes [113].

Oxygen abstraction by PAHs is extensive and associated with nonhydrocarbon fluids formation dominated by a series of high to low-molecular-weight (LMW) alicyclic to cyclic carboxylic acids. This is a similar route to chlorophyll decomposition, with phytol perhaps being turned into a carboxylic acid [120], which is released during decomposition, acting as a stored source for future hydrocarbon generation. The excessive generation of acid and CO_2 leads to frequent expulsion events and early release of these immature hydrocarbons. The constant breakdown and renewal of PAHs, utilizing resin, and as part of acid creation during decomposition and rearrangement, leads to leads to instability and, in some instances, rupture of the colloid [63].

There are two possibilities for natural hydrogen evolution by carbocatalytic groups. The first is by the adsorption and the chemisorption of water at active sites, abstracting oxygen, resulting in natural hydrogen and retaining oxygen within the carbon structure. The second is from the abundant CO_2 in the system, which absorbs onto reactive site surfaces, and then chemisorbs into the catalyst, splitting into two lots of CO, part of which is released and reacts with water to form hydrogen and CO_2 in a water-gas shift reaction. The other part of the CO remains sorbed at the reactive site of the catalyst as a point of weakness or growth. Microbial action may consume some of the natural hydrogen, which connects to the recent understanding of smectite evolution having a microbial signature, to its early development. Hydrogen may also be used in the hydrogenolysis reactions, which decompose kerogen [121].

7.3.4 Step 3

The dominant process can be loosely thought of as solid-solid buffering occurring between the colloidal kerogen and clays, leading to fundamental chemical rearrangement in each. This occurs when the colloid film has been breached due to the rearrangement weakness of the colloid, allowing clay-kerogen complexes to be formed [14]. Organic molecules can adsorb onto the clay surface and react predominantly with the Lewis and Brønsted active sites [13, 29]. The Lewis and Brønsted active sites mediate Scholl reactions, which ramp up significantly the coupling reactions of PAHs and kerogen. On one hand, kerogen is decomposed by hydrolysis reactions as OH^- is released from clays and the decomposition of kerogen by PAHs creates protons by decarboxylation of acids, modifying the interlayer space of the clays. Further acid attack, in turn, causes more clay rearrangement, releasing more OH^- by the cationic exchange or total destruction of the clay interlayer. Equilibrium is constantly changing in the pore fluids of the clay and primary pore space of the other mineral particles, likewise in the micro- and nanospace of the interlayer of clays following Le Chatelier's principle.

Clay rearrangement works by a process of isomorphic substitution in the interlayer of the clays, as H^+ is a more reactive cation than most of the alkali and alkali earth metals in the original clay structure. It causes preferential release and dissolution of OH^- anions into interlayer fluids [122–124]. Protons react directly in cationic exchange to minimize the charge imbalances at the active sites of smectites, but can also act to distribute its charge as a more passive bridging species between both sides of the crystal layers by complexion with water [125]. To achieve bridging, H^+ is in the form of hydronium and may exist initially as larger more exotic species, such as eigen and zundel forms, to passively bridge the large spaces of both sides of the clay interlayer. The large interlayer space means LMW hydrocarbon species can also enter the interlayer and react directly at clay-catalytic active sites. The labile acids and alkanes in the interlayer space of the clay, along with cation complexes, particularly hydronium, undergo reactions, creating large quantities of volatile C_1–C_3 hydrocarbons, natural hydrogen, and CO_2, whilst promoting more reactive sites on the interlayer surfaces [126, 127]. The wider interlayers of smectite allow the active sites internally to participate in secondary cracking directly in a similar fashion to zeolites [128]. These reactions are regioselective and an unknown proportion will result in volumes of diatomic hydrogen. The convention is for hydrogen to be reacted away in radical reactions, however, the nanostructure of stabilized illite clays may act to shield some small volumes, preserving it.

As both sides of the clay crystal become increasingly charged and rebalanced by the incorporation of hydronium and other large cations, there is a corresponding decrease in interlayer space as more cations interact with them, pulling the two sides of the space ever closer together. The more familiar H_3O^+ hydronium species may then become dominant in the smaller interlayer space. Both Brønsted and Lewis acid sites of clay catalyze carbanion and carbocation reactions, breaking down kerogen further and also producing intermediate acid forms, which react to produce more natural hydrogen and CO_2. The interlayer space of the clays undergoes substantial modification through cationic exchange, releasing hydroxyl groups and creating secondary porosity. The observation by Goldstein [129] that the addition of water-to-clay-catalyzed reactions reduces the rate of reaction but also enhances the overall yield, being a result of hydronium release, is insightful here. Carbocation and carbanion reactions, initiated by hydronium, are suggested to be one of the principal cracking mechanisms at this point; this may be similar to hydronium creation in zeolites [130, 131]. Pressure from the produced volatiles and gases exceeds rock strength, and secondary migration occurs, causing rock fabric to collapse and primary porosity to reduce. Pore fluids continually act as a transport mechanism, exchanging hydronium and OH^- between kerogen and clay to maintain equilibrium.

Due to size constraints and high molecular weight (HMW), carboxylic acids, proto-hydrocarbons, and alkanes are unable to enter the interlayer space and participate in the reaction inside the colloid. However, larger molecules have the potential to be cracked by Whitmore reactions to produce LMW acids, and alkene and alkane fragments. H^+, in the form of hydronium molecules, act as mobile Brønsted Acid active sites within the primary pore space, causing Whitmore protonation reactions throughout the fluid in the pore network. Whitmore reactions are instrumental in splitting the long-chain aliphatic-like molecules at this juncture [97, 98, 132]. Scission of HMW alkanes during Whitmore reactions produces an alkane and olefin doublet [133, 134]. The excess generation of olefins by this process and others such as disproportionation leads to cyclic [135] and aromatic molecule formation [136–140], creating in situ PAHs precursors outside of colloids, which are turned into a series of expanded PAHs by cyclodehydrogenation [60]. This is part of the Scholl reaction that is aided by Lewis and Brønsted acid sites of clays. Whitmore reactions decrease acidity in the primary pores and, overall, the pH of the system has a tendency to increase, as acid-forming reactions utilize protons. Acidity is also weakened by the slowing down of the proton-hopping Grotthuss mechanism [141] of hydronium within pores, as protons are utilized in Whitmore exchange reactions far outside the clay lattice [142, 143].

It is conjectured that HMW carboxylic acids undergo Whitmore reactions occurring at molecular sites, distal to the alpha carbon and polar carboxylic acid group, in a similar fashion to the protonation of alkanes. Fractionation of HMW and LMW carboxylic acids occurs as they dissolve in different solvents [144]. Oils and proto-hydrocarbons surrounding the kerogen cores dissolve HMW acids, whilst LMW acids dissolve more easily in water and are dispersed more widely. The LMW acids are labile and readily dissolved in aqueous solution. They become widely distributed basinally via secondary migration [145]. Fractionation, based on aqueous dissolution, accounts for the high concentrations of acetic and other LMW acids, observed with more stable HMW molecules confined to their local kerogen environment [146]. Large amounts of LMW hydrocarbons that are small enough to escape along with hydrogen are also expelled, as they have such small polarities that they cannot be absorbed by kerogen.

Rearrangement inside the colloid increases the number and size of early developed PAHs, and stacks may be 4–6 units of thickness, enhancing the stepped staircase structure and reactivity. Although these are 3D in nature, there is little to no nanopore development due to the tightly packed structure. The proximity to each stack decreases and the random orientation reduces aligning in sympathy with each other due to prevailing stresses.

The bulk of natural hydrogen generation is caused by cyclodehydrogenation due to the closure of ring structures or coupling of Scholl reactions. Other contributions come from the regioselectivity of olefins and protons from acids reacting with carbanions from early carbon catalyst development at Lewis acid sites. It is assumed that any hydrogen produced migrates or is preserved in clays or else is reacted away; the split between each is unknown.

7.3.5 Step 4

Kerogen is in an advancing stage of maturity, demonstrated by lower content of active oxygen groups with increased aromatic and cyclic development; concomitantly, PAHs have matured also. Stacking of PAHs continues with the previous grouping and orientation trends. Petrographically, this takes the form of solid carbon and liquids, which is classed as initial oil–solid bitumen. This perhaps reflects the breakup of the colloidal structure due to fragmentation of the surrounding PAHs. As a result of this, lower reactive PAHs, in conjunction with olefins and alkanes, exist in the free phase, away from colloidal centers that still function awaiting expulsion. Kerogen has become almost completely exhausted of oxygen, whilst the PAHs have become rather oxygen-rich. The amount of oxygen within the system is fairly constant but is recycled between PAHs by abstraction, causing their fragmentation and rearrangement. Direct competition between the various PAHs for oxygen leads to a differentiation in outcomes. Some PAHs abstract oxygen and proceed to fragment at the side chain, the ring structure, or at both, whilst other PAHs are merely stripped of the side chain and grow at the core, eventually adding a much smaller side chain. Constantly changing high electronegativity of organic catalysts act as reaction-initiation points to polymerize ring structures [147, 148]. Scholl reactions create and add to them and grow [60, 70], with Na^+ ions enhancing ring formation [28]. The cyclodehydrogenation process eliminates hydrogen to create carbon-to-carbon bonds and build a carbon nanostructure. Oxidation of the aromatic core structures, resulting in ring cleavage products, are intermediary molecules that aid in situ authigenic PAH generation. Cores grow larger by Scholl reaction and abstract oxygen into the aromatic ring, shifting the electron cloud in cores fragmentation, and eventually creating HMW alkanes [149].

It is significant though that towards the end of this stage, stacked planar carbon forms and groupings of carbon clusters develop into 3D volumes by π-σ interaction, resulting in a T-shape orientation. Scholl reactions cause cross linkages between molecules to increase enhancing the overall catalytic ability. The new 3D arrangement of the larger carbon catalysts appears as an aggregated haphazard arrangement (turbostratically stacked), forming significant 3D volumes rather than approximately planar forms of stacked PAHs. This gives rise to defects in the carbon structure that are not only exceptionally reactive but can also act as areas of nanoporosity, acting as preservation for diatomic hydrogen. The beginning of preservation inside the nanostructure of the carbon catalysts, rather than just in clays, is a significant development, allowing hydrogen to be shielded from external reactions that make hydrocarbons. Researchers suggest that carbon catalysts in the subsurface rock system during conversion perform secondary cracking, post the clay catalytic peak, but there are only a few references [35]. Meanwhile, the catalytic ability of the clays reduces as smectite converts, and there are larger volumes of illite and chlorite that are less reactive.

Cyclodehydrogenation of PAHs, creating carbon catalysts, evolves natural hydrogen, which is not consumed in the reaction. Natural hydrogen creation by chemisorption of water-abstracting oxygen, incorporating it into the catalytic structure and releasing diatomic hydrogen, is the main generation method. Abstraction of oxygen onto carbon catalysts evolves CO, which reacts with water-producing natural hydrogen.

7.3.6 Step 5

Large quantities of turbostratically arranged clusters of large carbon macromolecules form pyrobitumen, signifying high maturity and existing in a glassy or super-glassy state [108, 150–152]. Long-chained crosslinked oxygen-rich methyl alkanes, from fragmentation along with biomarker alkane moieties, are interspaced between crystallites and source rock. Resins have almost entirely been converted into PAHs and alkanes [151, 153]. Aromatics have either been converted into PAHs or else have been utilized in polycondensation, to create pyrobitumen. The oxygen-rich methyl alkanes are actively being transformed into *n*-alkanes within the oil window, as hydrocarbons and natural hydrogen generation proceed by abstraction of NSO. Scholl reactions catalyze oxidized ring structures of the carbon crystallites to form new PAHs. Cyclodehydrogenation continues as carbon catalysts mature and are rearranged into larger un-structured turbostratic forms, chemisorbing oxygen to the periphery of the molecule and creating natural hydrogen. The larger-sized crystallites are pyrobitumens and have higher electronegativities. They are able to absorb smaller and smaller chain length alkanes for cracking, and release shorter ones to fractionate the overall hydro-carbon system. Hydrogen is not consumed in this process, unlike radical reactions.

The smectite clay catalyst eventually becomes exhausted and turns to illite-trapping hydronium within the interlayer as part of its rearranged lattice during transformation [154, 155]. The collapse of smectite to form illite sees a sharp drop in cation exchange capacity and a reduction in porosity that leads to expulsion of some alkanes, but potentially preserving hydrogen [156, 157]. The process of smectite to illite conversion has recently focused on hydronium incorporation into the illite lattice by some authors [154]. This may be the remnant of the hydronium creation process pos-ited above and preserved in the illite crystal lattice. This would tie in with the obser-vation made earlier of smectite shrinking and dewatering as it transforms to illite and reduces the interlayer size. The weak clay catalyst is no longer of sufficient catalytic activity to continue cracking short-chained alkanes, which have ever higher bond strength and need more intensive catalysts for them to be cracked. The onset of chlo-rite is often noted as occurring prior to and in parallel to the total illitization of smec-tite. With pH increase, this trend in the development of chlorite increases. Chlorite development reduces the volume of water in the pore space as it is incorporated into the chlorite lattice. Carboxylic acid generation is very low, consequently, the pH of the pore fluids gradually increases as less hydronium is produced from maturing ker-ogen as H^+ is reacted away [131, 158]. This also means CO_2 production from decarbox-ylation decreases, which means gas shift reactions by chemisorption of CO_2 due to this process have reduced also. The peak of illitization of smectite occurs, plus or minus chlorite development, and is usually associated with T_{max} from Rock Eval ex-periments.

The high electronegativity of carbon catalysts means that they attract any remain-ing molecules with polar functional groups, sorb them into their structure, and react with them locally. This includes the very marginally polar shorter-chain alkanes that would not have been sorbed by carbon catalysts such as PAHs, enabling an ever-finer fractionation to occur. This tendency means that they act as a type of molecular sieve, separating more polar hydrocarbons on their surface to be cracked and allowing pri-marily alkanes to escape by expulsion from the rock matrix. Figure 7.1 shows that during carbon-catalyzed reactions to crack alkanes into smaller aliphatic molecules, a carbon atom is abstracted from the larger alkane into the carbon catalyst and releases two smaller alkanes. Surface reactions at the pore water/carbon macromolecule inter-face cause water-gas shift reactions, which create CO [69]. CO, desorbed and released from the macromolecules, is dissolved in pore fluids and reacts with water to create CO_2, releasing diatomic hydrogen. Because of the release of CO from macromolecules, new reaction sites are created within the carbon macromolecule. This is pivotal in carbon catalyst development. as before this point, only annealing of pre-existing reaction sites occurs and no mechanism is allowed for reactivation unless by the addition of NSO or carbon into the structure. This process causes the carbon catalyst to maintain its size or grow at a higher rate, than noted previously, whilst remaining active. This is seen as a series of multilayered turbostratic carbon structures in a more amorphous form. Additionally, the 3D structure creates more extensive nanoporosity, which allows the preservation of natural hydrogen within the carbon catalysts.

This switch to nucleated microcrystallite growth of in situ carbon catalyst development, away from large PAHs and large macromolecular clumps, is instrumental not only in natural cracking and development of hydrocarbon generation but also in the creation of natural hydrogen by regioselectivity and more cyclodehydrogenation. Figure 7.1 attempts to show an idealized reaction pathway of carbon catalysts becoming more effective by creating new reaction sites, by combining and desorption of new atoms into their structure, allowing reactivation. The resulting annealing and reactivation of the carbon catalyst boosts their catalytic abilities to a higher degree. Ethane is used as the molecule that is being decomposed, and shows how the abstraction of CO keeps the carbon catalyst from annealing. It could be a much larger molecule that undergoes cracking via several possible regioselective pathways. From Figure 7.1, it is shown that when water is fully utilized within the shale, the process ceases. There may be other pathways that cease the process, including reduction of temperature, as the mechanism is considered to be thermocatalytic in nature.

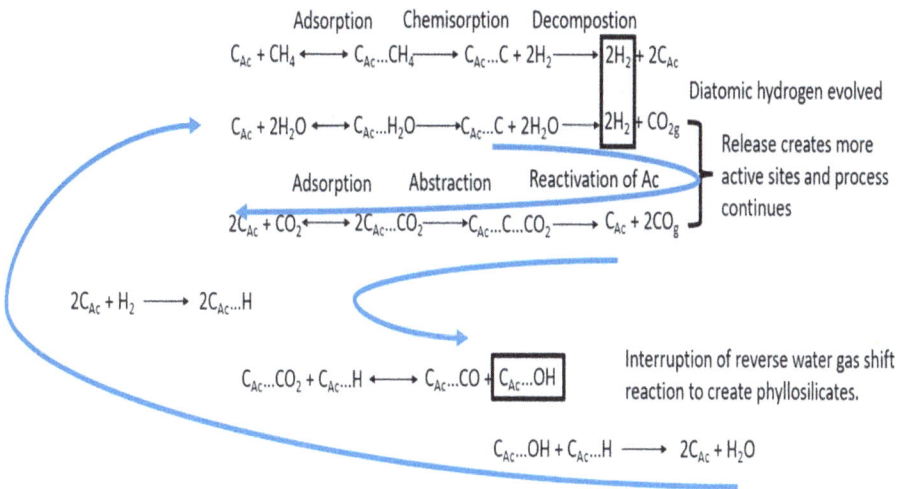

Figure 7.1: Insertion of carbon from methane decomposition annealing the catalyst and simplified equations showing reactivation of amorphous carbon catalyst.

7.3.7 Step 6

Pyrobitumen was previously formed in situ from the polycondensation of microcrystalline carbon macromolecules, which are powerful enough to crack the C_7 to C_3. Still, some smaller ring structures and heteroatoms progress in maturity, due to hysteresis, and Scholl reactions couple the larger crystallites form into amorphous carbon crystallites, which grow. LMW alkanes dominate until only ethane and methane predominate. As the reactions are considered local in nature and confined to the surface of the carbon crystallite, the tendency to form methane and hydrogen is higher than when considering Scholl reactions and the possibilities of products. Ethane and methane are finally cracked or decompose to release natural hydrogen [159, 160]. Absorption and then adsorption within the nano- and microporosity of the carbon catalyst are key to this process. Ethane and methane are finally able to be sorbed and held by amorphous carbon, surrounded by clays in a similar habitat to gas shales that exist in the Permian shale gas basin in Texas. This may indicate that hysteresis occurs, perhaps spanning millions of years, as reactants adsorb, chemisorb, and eventually desorb over the intervening period, and appear to be dormant.

As amorphous carbon grows within the shale, imperfections in the carbon lattice create numerous chemically active sites [45, 68]. Additionally, adsorption and chemisorption of other polar molecules with diverse functional groups (particularly the heteroatoms) are expressed at the surface and create a graphene oxide-type structure, amplifying catalytic ability [161]. The carbon catalysts act as a molecular sponge in many respects and allow expulsed hydrocarbons to be relatively free of heteroatoms. Coincidentally, this process of activating the surface of carbon catalysts with other functional groups or heteroatoms is like the methods used to reactivate or enhance the ability of amorphous carbon catalysts in laboratory experiments. This highly reactive surface structure has the ability to adsorb various reactants, especially the mobile light hydrocarbons that are normally relatively stable and unreactive in the subsurface. Amorphous carbon is in the temperature window of 2–250 °C for the conversion of dry gases that would be difficult to crack, to form methane, and finally produce vast amounts of hydrogen, as per Figure 7.1.

The C_1 and C_2 hydrocarbon fraction of the dry gas phase is the last to be cracked, which occurs once carbon catalysts have exclusively taken the form of highly reactive amorphous carbon. Amorphous carbon crystallites form from the differing allotropes of carbon and develop highly heterogeneous structures. This has excellent catalytic abilities with sp, sp^2, and sp^3 hybridization reaction sites, high cation exchange capacity (CEC), high pore volumes, and high specific surface area [47]. Due to proximity and the high electronegativity, envisaged in the compressed shale light, hydrocarbons are sorbed many molecules thick on the carbon surface, creating an island structure as the Langmuir-Hinshelwood mechanism posits [162]. This allows the remaining short-chained alkanes to be cracked by means of abstraction of carbon from the alkane into the amorphous carbon lattice [163, 164]. The absorption of C_1 and C_2 gas onto amorphous carbon is perhaps suggestive of the work by Mango [165], positing an equilibrium between carbon and methane in source rocks.

Preservation is highly dependent upon the quantities of organic matter within the source rock, the structural form of the carbon catalyst and the degree of micro- and nanopore space developed within the organics. As the carbon catalysts are instrumental in the Scholl reaction process, at this point, hydrogen and remaining light ends are not all immediately reacted away. Preservation of gases and smaller volatiles are adsorbed onto the surface and into nano- and micropores, becoming sorbed and trapped in the amorphous carbon catalysts that are developing. At other locations, the natural hydrogen is under high pressure due to solid state rearrangement, and catastrophically escapes the crystallite. This volatile escape makes pressure waves, causing localized atomic rearrangement, making local sp_3 diamondoid-like streaks within the amorphous carbon [166]. Due to the very large surface areas and extensive pore volumes of some of these, the preservation may be considerable [167]. Residual carbon catalysts in coal formations can be large, and preserved gases are a known mining hazard when they are released during mining operations as fire damp. Organic-rich shales with poor residual carbon development that are widely dispersed throughout the rock will result in poorer preservation of early-stage intermediate gases and volatiles species, most of which are conveniently adsorbed onto the authigenic amorphous carbon [68]. It is suggested that this occurs around the 2–250 °C range at the wet to dry gas phase of cracking [168, 169].

Lab experiments using fluidized bed reactors attempting to decompose methane over amorphous carbon suffer from continuous annealing of the carbon catalyst, inhibiting continuous reaction. The decomposition reaction of alkanes is dependent upon the rejuvenation of the amorphous carbon catalyst in situ or else the reaction would deactivate due to the annealing of active reaction sites [66, 170]. As shown in Figure 7.1, it is believed the reactivation process begins with the adsorption of water onto the surface of amorphous carbon and, via chemisorption, produces diatomic hydrogen and CO_2 [67, 171] In turn, CO_2 and CO are released from the carbon surface. As they desorb, they are re-adsorbed at an alternate location. This causes new unstable bonding locations on the amorphous carbon lattice, creating new active sites and keeping the process going. Reverse water-gas shift occurs as a side reaction to recycle water, so that the overall reactions may continue. The reaction does eventually stop, as it is governed by the creation of chlorite from illite. Illite mops up any remaining water in the source rock to create chlorite and as more chlorite is created, the reverse water-gas shift reaction is interrupted, deactivating the amorphous carbon and leading to the creation of carbon black [30]. The change from amorphous carbon to carbon black anneals numerous nanopore spaces, leading to the expulsion of remaining methane and diatomic hydrogen.

7.3.8 Step 7

The source rock is now in the metagenic zone of hydrocarbon development and is under more anhydrous conditions and subject to higher temperatures. The nature of prior reactions with the carbon catalyst has morphed the structure of amorphous carbon, either by deposition of carbon atoms onto pre-existing carbon structures or by solid state rearrangement. It is anticipated that petrologically, this can be observed as zoned crystallites of carbon, the inner cores of which will be structurally different to the outer zone. This sterically shields the more reactive allotrope of carbon in the core but also preserves any remaining volatiles.

As amorphous carbon changes to carbon black, the overall conversion system to hydrogen within the source rock becomes slightly less active or may stop. There is the possibility that this occurs because of the prior process of hydrogen generation, which means no volatile hydrocarbon residues remain to be converted. Complete conversion of alkanes may lead to the expulsion of hydrogen and secondary migration or may lead to preservation within the nanostructure of the carbon or surrounding shales. Equally, there is the possibility that water removal is complete during these anhydrous-dominated conditions, and this stops the process in Figure 7.1 from proceeding further. Large quantities of carbon black remain, along with associated preserved volatiles and C_1 and C_2 gases. This may be suggestive of the situation seen in some of the more mature anthracitic coal deposits. Also, there is the possibility that the volatiles are sterically shielded, post absorption onto the carbon structure, and hence preserved within the nanopore structure of the carbon, and are not being desorbed to allow primary migration and other reactions to occur at the conclusion of hydrogen generation.

During the metagenesis phase, chlorite undergoes solid state rearrangement as temperatures increase, until around 370 °C, which is when water reaches its supercritical state. At this juncture, chlorite rearranges its structure, transforming to create garnet, and dehydrates, releasing supercritical free water into the anhydrous shale. Highly mobile supercritical water acts as a non-polar solvent and reacts with pyrobitumen, other hydrocarbon residues, and carbon black [172–175]. The hydrocarbon residues decompose and are dissolved by the supercritical water, forming a secondary gas cap, and the carbon black becomes reactivated and once again, morphs into an amorphous carbon form [176]. A second thermo-catalytic series of reactions, similar to stages 3–5, produce C_7–C_1 alkanes that are cracked and then finally create natural hydrogen. Contemporaneously, garnet continues to develop, as hydrogen is expelled from the rock system. Eventually, the reverse water-gas shift reaction is interrupted by authigenic phyllosilicate development; this time, it is biotite that governs the reaction. As the garnet transforms to biotite, water is incorporated into the biotite lattice. This time, the carbon catalysts change from amorphous carbon to carbon black anneals completely, as there are no rogue atoms to create imperfections. It undergoes solid state rearrangement, transforming into a more ordered graphite form. The solid-state rearrangement to graphite produces a more ordered lattice, in which there are fewer flaws in the crystalline structure. This results in the expulsion of all hydrogen from the nanopores, the process being very similar to graphitization of anthracitic coals [177].

7.4 Discussion

The existence of natural hydrogen in the subsurface is still relatively new to Western science, perhaps having been studied the most in the former Soviet Union countries, including Russia, for a lot longer, with ideas gradually crossing over. As the novel nature of it is incorporated into the existing cannon of western knowledge, revision of what we believe to be true and confirmation of it is still ongoing. For this investigation, the starting point has been the review by Zgonnik [24], cataloguing the methods and locations of known seeps and hydrogen accumulations worldwide. He notes the locations where natural hydrogen and organics are associated with each other cover coal, oil, and methane deposits. This may be suggestive that hydrogen that has evolved from organics may not migrate far or migrate at the same time as hydrocarbons, rather than the coincidental reservoiring of hydrogen in the same trap. The idea that hydrogen in coal measures migrated and is reservoired in the coal, post-coalification, seems odd. Additionally, this work demonstrates that if hydrogen does evolve from organics, then it survives the process of hydrocarbon generation existing in its diatomic form, and there are mechanisms for migration that do not preclude destruction en route. Either way, it is highly suggestive of the idea of organic natural hydrogen generation as a process.

Field work of Suzuki et al. on the southeast coast of Japan may have demonstrated the finger print of the above process, showing tiny volumes of paleo gases remaining in shales and metapelites [178]. The field work carried out demonstrates the transition of gases in samples collected from transects through shales into metapelites and CO_2 to alkanes to hydrogen, demonstrating the increase in hydrogen concentration with increased paleo temperature, moving from shales into metamorphic zones. Data for residual methane and hydrogen certainly seems to follow in a similar manner, outlined in the model above and summarized in Figure 7.2, which may give further insights into the results of Suzuki's work. A graphical summary of the overall process highlights two separate phases of diatomic hydrogen generation. One is the end point for traditional hydrocarbon generation, giving rise to the notional methane preservation limit. The second, post garnet formation, is associated with hotter low-grade metapelites. The duality peaks of methane and hydrogen generation are outlined in my work, which is perhaps to some degree corroborated by the Suzuki team paper. Mechanistically, the big difference is my suggestion of carbocatalysts, as opposed to simple heating for hydrogen generation.

The work carried out by Boreham et al. [179] shows good data from the Cooper Basin in Australia with varying concentrations of natural hydrogen. Thermogenic models of organic matter from land plant, derived organic matter in shales and coals, demonstrate that kinetic models can be constructed that yield natural hydrogen as the source of the natural hydrogen. The authors posit that high volumes of natural hydrogen may occur, similar to those found in gas shale deposits, at deeper levels in the Nappamerri Trough. It is interesting to see natural hydrogen in well bores being connected to a possible source derived from organics. This demonstrates that there are viable migration pathways for natural hydrogen in a sedimentary basin setting. A different model for natural hydrogen is utilized in modelling this data, but perhaps a similar exercise could be carried out using information enclosed in this work that may yield similar conclusions.

Some investigations have been carried out, such as [35], which seems to suggest that pyrobitumen had the positive effect of accelerating the cracking of wet gases; C_6 cracked, producing C_5–C_1, which seems to confirm the higher temperature and stronger catalyst portion of the subsurface thermocatalytic process outlined above. The idea of partitioning the longer-chained reactants by absorption on pyrobitumen and the subsequent release of shorter-chained products, post-cracking, was also established. This current thesis attempts to build and extrapolate these ideas up and down the carbon catalytic series of the differing absorption and cracking abilities at various stages. The thesis is also trying to stimulate discussion around mechanisms of generation and catalytic properties. Do free radical reactions truly happen throughout the whole of the fluid mix in hydrocarbon generation or is it instead "local Scholl reaction" that takes place, governed by adsorption onto ring systems, with the abstraction of heteroatoms and carbon from the cracked reactant into the structure of the catalyst? This would alter the number of routes for the generation of products, and also the kinetics. Likewise, the preservation and release of natural hydrogen requires more investigation, as some authors believe that early hydrogen generation is not possible. Yet, the counter to that is from the observation of natural hydrogen in early coal generation, which is on the other side of peak oil generation.

A cogent argument can be made for a carbocatalytic model of subsurface hydrocarbon and organic natural hydrogen generation. By suggesting an alternative, several issues of the classic model can be resolved. It may help to select a few of these as examples of the positive outcomes of adopting a new perspective. The most important are the low temperatures and pressures at which these reactions occur, particularly for the cracking of lighter alkanes. This has not been able to be replicated in the lab by radical reaction [18, 22]. To paraphrase Burnham [147], "are the kinetics for each differing step the same? Are the E_a distributions required to describe the differing steps necessary, or can they be lumped together?". Research into the decomposition of methane by amorphous carbon shows initiation temperatures around 2–250 °C [180–184], which makes the cracking of lighter alkanes a distinct possibility. Annealing of carbon catalysts is an issue, but it has been shown that these can be resolved by the addition of CO_2 [67], which seems to fit in neatly with water-gas shift reactions for hydrogen generation.

It has always previously been assumed that natural hydrogen will be completely consumed during the hydrocarbon generation process due to the dominance of radical reactions. Consequently, computer-programmed reaction models may have hardwired preselects to react hydrogen away by its combination with char to produce methane [147]. In preselecting for natural hydrogen's absence as a final product, to create methane opportunities, alternative narratives become restricted and may lead to less inspirational science. However, the organic natural hydrogen system is driven by the abstraction and Scholl reactions rather than being dominated by free radical reactions. Abstraction of carbon to crack lighter alkanes by carbon catalysts uses less hydrogen than the free radical mechanisms, currently envisaged. This aids the likelihood of the generation and survival of natural hydrogen during the generation process.

At higher vitrinite reflectance maturity levels, hydrocarbon generation in the classic model stops at the methane preservation limit [185–188]; the new model extends this to greater depths. The complex nanoporous 3D structure that some carbon catalysts exhibit is ideal for the generation and preservation of LMW alkanes plus natural hydrogen. The work by Zgonnik [24], has shown that hydrogen is preserved alongside oil, gas, as well as coal deposits, globally. As the carbon crystallites evolve 3D heterogeneity, their ability to sterically shield hydrogen in nanopores increases. This is highly dependent upon the volume and maturity of initial organic material in the rock system. Coal has high total organic carbon (TOC) content and low clay content, whilst organic shales have the reverse–TOC content of around 2–7%, encapsulated by significant volumes of clay. This is probably why hydrogen has been known as fire damp in coal mines for a greater period than the potential discovery of natural hydrogen in shales. The disparity in volumes of carbon in the source rocks gives natural differences in early-stage volumes of hydrogen preservation. The interspersed nature of the 3D carbon crystallites in shales means a greater distance for hydrogen in the free phase to travel in the rock fabric during primary migration for hydrogen, before safely being preserved. However, a high proportion of hydrogen is created on the surface of carbon crystallite itself and hence, there is no distance to be travelled, prior to preservation. Also, the maturity of carbon crystallites within high organic carbon rocks, such as coal, often comes with a low hydrogen index (HI) source. This means that the transformation to carbon might be expected to be quicker than for higher HI kerogens, as less cyclodehydrogenation has to occur before transformation to carbon crystallites. The greater speed to maturity of coals might mean earlier preservation and greater volumes of preservation of hydrogen, as highly porous carbon crystallites are more ubiquitous. The downside is it is anticipated that fewer liquid and gaseous hydrocarbons will be developed due to the low HI and thus resulting in a low yield of hydrogen.

The stacked nature of carbon catalysts is the real feature of this new mode by providing catalytic abilities that have not been widely recognized previously [50]. Understanding the minutiae of these reactions is key to unlocking their potential in other areas such as stimulated hydrogen from immature oils. Does it take a long time for these randomly disordered molecules to rearrange and stack in the perfect way to allow circumvention of normal reaction hurdles? Does an inability to stack by some PAHs result in their early expulsion as immature oil? Does stacking of chiral dimers featured in PAHs alter catalytic ability, along with compaction of sediment, to provide the necessary flexures in the structure that promote exotic features such as Cooper pairs, phonons, and moiré patterns? The process may consist of a series of punctuated sub-systems that occur rapidly, once the correct carbon catalyst has been created, with hysteresis occurring as new carbon catalysts slowly develop and stack before the next phase ensues. Kinetic modelling of systems, as currently derived, may already include catalytic reactions in a kind of blended average of purely thermal and catalytic mechanisms, so perhaps these need to be reevaluated. However, if the breakthroughs are at the quantum level, then maybe not, as new science will be required.

Work by Chinese researchers on supercritical water and oil shales with high residual hydrocarbon content is artificially creating the secondary gas cap, simulating the late-stage hydrogen event, proposed in Figure 7.2, and releasing hydrogen, methane, and other short-chain alkanes and olefins [176, 189–194]. This does not fit well with the classic model of hydrocarbon generation but some dovetail quite nicely with the ideas outlined above. Their work shows that abstraction of oxygen, in this case from super critical water, fragments the PAHs and releases useful hydrocarbon, and produces natural hydrogen. This acts as a demonstration of the large potential volumes that may be yielded from the oxidation of immature PAHS as well as the late-stage natural hydrogen process. Also, it points to a method for the stimulation of PAHs to reactivate active sites and stimulate in situ hydrogen development, on demand, at scale. This also points to the possibility that there may be economic quantities of natural hydrogen preserved in gas shales, particularly ones with higher TOC content and enhanced nanopore development that aids preservation. The method for attaining a substantial release of hydrogen, maybe a frack using supercritical water. This is why the work of Suzuki is also so interesting [178]. We do not have observations of hydrogen being produced along with methane from the gas shales. This may be observational bias, but equally because there may be no primary hydrogen migration occurring, leaving it trapped in nanopores of carbon crystallites. Alternatively, it may be limited with hydrogen remaining sorbed to the highly electronegative carbon catalyst. This may be a function of our limited knowledge of the optimal production of hydrogen from carbon-rich rocks. Pure water acts as a baffle and sometimes even as a seal to hydrogen migration within water-filled zones in the Barakoubugou Field in Mali [195, 196]. A water coating over the nanopore throat conduits in gas shales may be what is preventing expulsion and production from the carbon crystallites. The current production technique of fracking these reservoirs to increase permeability with water might also be actively hindering hydrogen flow by water baffling. This is why during coal bed methane extraction, the hydrogen preserved there can flow, as the wells are typically water unloaded, allowing hydrogen into the production string. Is water blocking the reason for production volumes from the current crop of hydrogen exploration wells have results that are sub-optimal?

Although lots of investigation of carbon catalytic reactions have been recently undertaken due to the interest in graphene generation, more needs to be incorporated into the geochemical sphere to answer numerous questions. The study of niche or specialized carbon catalysts is of interest, as they can be potentially formed during hydrocarbon and hydrogen generation, generating enhanced carbon catalysts in situ. The incorporation of NSO into the lattice of amorphous carbon increases reactivity, being seen as an enhancement in laboratory experiments, when compared to the pure carbon catalyst, by creating more reaction sites at imperfections in the lattice. Likewise, the observation of the incorporation of Na^+ into kerogen during Scholl reactions or by rearrangement early in the saponification phase of kerogen breakdown is interesting, as this may lead to carbon catalysts doped with sodium. Sodium-laden carbon structures have been shown to have special properties, acting as low-temperature superconductors. What part, if any, superconductivity plays during the subsurface generation of natural hydrogen is currently unknown. That Scholl reactions are more extensive and can ensure many multiple couplings happen en masse could point to superconductivity playing a part by propagation over several different molecules. The differentiation or fractionation that occurs during hydrocarbon and natural hydrogen generation has been previously recognized and discussed, however this seems to be somewhat restricted to the latter stages of hydrocarbon and natural hydrogen generation, even though some carbon catalysts are super-electrophiles [57, 197–199]. During the entire hydrocarbon generation system, highly electronegative carbon species are able to abstract NSO, and later carbon into their structures. Perhaps this needs to be factored into bond breaking and formation as part of calculations, since they have not being previously considered.

Figure 7.2 is an attempt to synthesize the hydrocarbon generation process along with the generation opportunities for organic natural hydrogen, carbon catalyst development, and clay conversion into phyllosilicates. Early developed carbon catalysts, PAHs, macromolecules, and pyrobitumen are lumped together as carbon residues. Amorphous carbon, carbon black, and graphite are split out individually to make inspection easier to comprehend. Natural hydrogen is shown to evolve from early in the process, but due to preservation being reliant on the development of 3D carbon crystallites, only small quantities are shown. Two main peaks of hydrogen generation are shown. One, at the end of primary dry gas development, somewhat equivalent to the old methane preservation limit. And, a second deeper hydrogen peak, which is dependent upon the temperature at which water is in a super critical state. The bulk of hydrogen generation is anticipated to occur post-methane development at the end of hydrocarbon generation. After the decomposition of ethane and methane to create

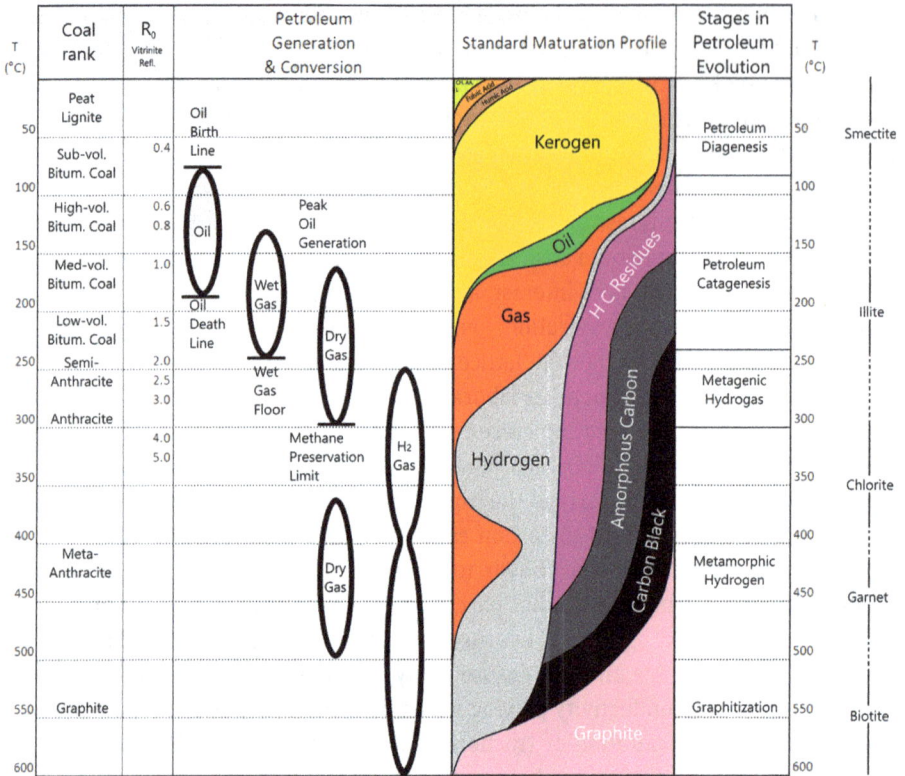

Figure 7.2: Proposed inclusion of hydrogen into hydrocarbon generation model.

hydrogen, there are few other chemical reactions that occur, as there are so few reactants. Steric shielding is not as important to allow the preservation of hydrogen, as there is nothing for hydrogen to react with. Hydrogen will migrate and fill nano- and micropores of both shale and carbon crystallites but will also be expelled and migrate further afield. Hydrogen that is sorbed into carbon crystallites will be difficult to desorb due to the high electronegativity remaining in the carbon catalyst.

Current natural hydrogen exploration is focused on the evolution mechanisms by oxidation of iron-rich minerals as a source rock and drilling targets that can be potentially quite deep. The benefit of pursuing a model of organic natural hydrogen is that the source, reservoir, and seals are comparable to the hydrocarbon industry norms. Drillable targets should be shallow by comparison with iron-rich sources. The exploration effort is focused on areas that are already known to geoscience and have world-class datasets covering them. Active exploration efforts currently focused on the Pyrenees, Spain, may use this model as a natural hydrogen source rather than iron oxidation sources, which may be restricted to the French side of the Pyrenees. Due to the well log data, providing erroneous responses, and deeper gas chimneys being seen as processing artifacts on seismic, there are still viable targets among the intensively drilled basinal regions of the world. These same basinal areas have existing infrastructure that could be modified for export purposes and are close to already established markets and production centers. However, it is noted that locations where we anticipate natural hydrogen to be located from this new model are lacking. It is a mystery why the large gas fields of the Southern North Sea, UK, for example, are devoid of natural hydrogen, containing no known trace, whilst the Polish Southern Permian Basin has natural hydrogen [200]. This seems odd as one would expect the evolution of both methane and hydrogen from the maturation of coals in the carboniferous source rocks and a good seal to be provided by the salt. Could the inversion in this basin be the key to the non-preservation of natural hydrogen or is there a data bias?

With there being no statutory obligation to record or report hydrogen occurrences previously, the scientific and economic value was overlooked, leading to data bias. It is proposed that large volumes of early-stage hydrogen exist adsorbed onto or absorbed into nanopores in a similar fashion to those already produced for methane, in places such as the Permian Basin in Texas, USA, and highlighted by Boreham et al. [179].

7.5 Conclusion

The idea of natural hydrogen as part of the hydrocarbon generation process within basinal areas could be the focus of a new "dash for gas" in high-maturity shales. Equally and more likely is that carbon catalysts will be recognized as pivotal in the stimulation of natural hydrogen from immature oils. New hydrogen exploration concepts developing in basinal areas, in addition to those in hard rock cratonic realms, increase the probability in exploration scenarios. This change in focus would mean that the vast data sets and experience garnered from earlier oil and gas exploration can be utilized and re-evaluated in its discovery. Additionally, the legacy infrastructure, logistics, and market can be re-utilized to reduce costs to the ultimate consumer when economical reserves are brought to market.

Anticipated volumes of hydrogen developed via this mechanism are currently ill-defined and require further investigation, but could range from moderate to extremely large, like current methane exploration from shale gas. The notional doublet effect of the secondary cracking of alkanes creates two routes to hydrogen via thermo-catalysis. One from the early-stage initial gas cap, post-oil development, and the other late-stage from the deeper secondary gas cap. It is conjectured that these subsurface reactions can be replicated in the lab on a like for like basis of temperature and pressure, foregoing the requirement for high-temperature experiments currently employed. Other research chemists focused on organic catalysts for energy transition appear to hold great store in metal-doped carbon catalysts for the hydrogen evolution reaction, amongst other things, so perhaps geochemists need to take heed. A revision in thinking may be beneficial and lead to experimental work that can exactly duplicate subsurface processes and mechanisms when all matrix reactants are accounted for. It may ultimately lead to lower energy requirements to carry out current petrochemical processes and may allow tighter selectivity options for any new processes that are developed. Perhaps this would allow the higher standards of living to which we all aspire in a more environmentally friendly way.

References

[1]   Gaucher E. C. New perspectives in the industrial exploration for native hydrogen. Elements. 2020, 16(1): 8–9.

[2]   Boreham C., et al. Hydrogen in Australian natural gas: Occurrences, sources and resources. The APPEA Journal. 2021, 61: 163.

[3]   Ball P. J. *Natural hydrogen: The new frontier*. In: K C. (Editor.), *Geoscientist*, London: Geological Society of London, 2022.

[4]   Mainson M., et al. Sensing hydrogen seeps in the subsurface for natural hydrogen exploration. Applied Sciences. 2022, 12(13): 6383.

[5]   Milkov A. V. Molecular hydrogen in surface and subsurface natural gases: Abundance, origins and ideas for deliberate exploration. Earth-science Reviews. 2022, 230: 104063.

[6]   Asahi I., et al. Remote sensing of hydrogen gas concentration distribution by Raman lidar. In: Proceedings of SPIE-The International Society for optical engineering, 2012.

[7]   Magoon L. B., Dow W. G. The petroleum system – From source to trap. In: Memoir, ed. A.A.o.P. Geologists, Vol. 60. 1994, AAPG.

[8]   Smith N. J. P., et al. Hydrogen exploration: A review of global hydrogen accumulations and implications for prospective areas in NW Europe. In: Petroleum geology conference series. 2005. London: Geological Society.

[9]   Fowles M., Carlsson M. Steam reforming of hydrocarbons for synthesis gas production. Topics in Catalysis. 2021, 64: 856–875.

[10]   Hanson J., Hanson H. Hydrogen's organic genesis, Unconventional Resources, 2024. 4, 100057. Elsevier B.V.

[11]   Quigley T. M., Mackenzie A. S. The temperatures of oil and gas formation in the sub-surface. Nature. 1988, 333(6173): 549–552.

[12]   Johns W. D. Clay mineral catalysis and petroleum generation. Annual Review of Earth and Planetary Sciences. 1979, 7: 183–198.

[13]   Bu H., et al. Effects of complexation between organic matter (OM) and clay mineral on OM pyrolysis. Geochimica Et Cosmochimica Acta. 2017, 212: 1–15.

[14]   Cai J., et al. Control of clay mineral properties on hydrocarbon generation of organo-clay complexes: Evidence from high-temperature pyrolysis experiments. Applied Clay Science. 2022, 216: 106368.

[15]   Du J., et al. Diversified roles of mineral transformation in controlling hydrocarbon generation process, mechanism, and pattern. Di Xue Qian Yuan. 2021, 12(2): 725–736.

[16]   Du J., et al. The effect of diagenetic environment on hydrocarbon generation based on diagenetic mineral assemblage in mudstone. Petroleum Science and Technology. 2018, 36(24): 2132–2142.

[17] Espitalié J., Senga Makadi K., Trichet J. Role of the mineral matrix during kerogen pyrolysis. Organic Geochemistry. 1984, 6(C): 365–382.

[18] Goldstein T. P. Geocatalytic reactions in formation and maturation of petroleum1. AAPG Bulletin 1983, 67(1): 152–159.

[19] Greensfelder B. S., Voge H. H., Good G. M. Catalytic and thermal cracking of pure hydrocarbons: Mechanisms of reaction. Industrial & Engineering Chemistry. 1949, 41(11): 2573–2584.

[20] Song D., et al. A comprehensive study on the impacts of rock fabric on hydrocarbon generation and pore structure evolution of shale under semi-confined condition. Marine and Petroleum Geology. 2021, 124: 104830.

[21] Vander Wal R., Nkiawete M. Carbons as catalysts in thermo-catalytic hydrocarbon decomposition: A review. Journal of Carbon Research. 2020, 6(2): 23.

[22] Mango F. D. The light hydrocarbons in petroleum: A critical review, Oxford: Elsevier Ltd, 417–440, 1997.

[23] Yu S., et al. Incorporation of wet gases to kerogen in petroleum formation and evolution. Organic Geochemistry. 2023, 180: 104605.

[24] Zgonnik V. The occurrence and geoscience of natural hydrogen: A comprehensive review. Earth-Science Reviews. 2020, 203: 103140.

[25] Yan L.-J., et al. Effects of alkali and alkaline earth metals on the formation of light aromatic hydrocarbons during coal pyrolysis. Journal of Analytical and Applied Pyrolysis. 2016, 122: 169–174.

[26] Drummond S. E., Palmer D. A. Thermal decarboxylation of acetate. Part II. Boundary conditions for the role of acetate in the primary migration of natural gas and the transportation of metals in hydrothermal systems. Geochimica Et Cosmochimica Acta. 1986, 50(5): 825–833.

[27] Wang W., et al. Review on the catalytic effects of alkali and alkaline earth metals (AAEMs) including sodium, potassium, calcium and magnesium on the pyrolysis of lignocellulosic biomass and on the co-pyrolysis of coal with biomass. Journal of Analytical and Applied Pyrolysis. 2022, 163: 105479.

[28] Wu D., et al. Reaction molecular dynamics study on the mechanism of alkali metal sodium at the initial stage of Naphthalene Pyrolysis evolution. Energies (Basel). 2023, 16(17): 6186.

[29] Sposito G., et al. Surface geochemistry of the clay minerals. Proceedings of the National Academy of Sciences – PNAS, 1999. **96**(7): p. 3358–3364.

[30] Meng J., et al. Conversion reactions from dioctahedral smectite to trioctahedral chlorite and their structural simulations. Applied Clay Science. 2018, 158: 252–263.

[31] Acevedo S., Castillo J. Asphaltenes: Aggregates in terms of A1 and A2 or Island and Archipelago structures. ACS Omega. 2023, 8(5): 4453–4471.

[32] Strausz O. P., Mojelsky T. W., Lown E. M. The molecular structure of asphaltene: An unfolding story. Fuel. 1992, 71(12): 1355–1363.

[33] Morimoto M., et al. Synthetic asphaltene for green carbon material. Fuel. 2024, 358: 130293.

[34] Wu L., et al. Formation of pyrobitumen from different types of crude oils and its significance: Insight from elemental composition analysis. Marine and Petroleum Geology. 2023, 152: 106227.

[35] Pan C., et al. The effects of pyrobitumen on oil cracking in confined pyrolysis experiments. Organic Geochemistry. 2012, 45: 29–47.

[36] Wang F., et al. Coke formation of heavy oil during thermal cracking: New insights into the effect of olefinic-bond-containing aromatics. Fuel (Guildford). 2023, 336: 127138.

[37] Guisnet M., Magnoux P. Organic chemistry of coke formation. Applied Catalysis. A, General. 2001, 212(1): 83–96.

[38] Mahapatra N., et al. Pyrolysis of asphaltenes in an atmospheric entrained flow reactor: A study on char characterization. Fuel. 2015, 152: 29–37.

[39] Reizer E., Viskolcz B., Fiser B. Formation and growth mechanisms of polycyclic aromatic hydrocarbons: A mini-review. Chemosphere (Oxford). 2022, 291(Pt 1): 132793–132793.

[40] Richter H., Howard J. B. Formation of polycyclic aromatic hydrocarbons and their growth to soot – A review of chemical reaction pathways. Progress in Energy and Combustion Science. 2000, 26(4): 565–608.

[41] Borisova L. S., Timoshina I. D. Regular trends in variation of the Asphaltene composition and structure in Dia- and Catagenesis. Petroleum Chemistry. 2022, 62(2): 229–239.

[42] Mastalerz M., et al. Origin, properties, and implications of solid bitumen in source-rock reservoirs: A review. International Journal of Coal Geology. 2018, 195(C): 14–36.

[43] Ben Amor N., Konate S., Simon A. Electronic excited states of planar vs bowl-shaped polycyclic aromatic hydrocarbons in interaction with water clusters: A TD-DFT study. Theoretical Chemistry Accounts. 2023, 142(8): 74.

[44] Izawa E. Carbonaceous matter in some metamorphic rocks in Japan. Chishitsugaku Zasshi. 1968, 74(8): 427–432.

[45] Obeelin A., Boulmier J. L., Durnand B. Electron microscope investigation of the structure of naturally and artificially metamorphosed kerogen. Geochimica Et Cosmochimica Acta. 1974, 38(4): 647–650.

[46] Rodrigues S., et al. Catalytic role of mineral matter in structural transformation of anthracites during high temperature treatment. International Journal of Coal Geology. 2012, 93: 49–55.

[47] Gou H. Microstructural landscape of amorphous carbon. National Science Review. 2024, 11(5): p. nwae 125.

[48] Tiwari S. K., et al. Stone-wales defect in Graphene. Small. 2023, 19(44): e2303340.

[49] Brayfindley E., et al. Stone-Wales rearrangements in polycyclic aromatic hydrocarbons: A computational study. The Journal of Organic Chemistry. 2015, 80(8): 3825–3831.

[50] Igarashi M., et al. Parallel-stacked aromatic molecules in hydrogen-bonded inorganic frameworks. Nature Communications. 2021, 12(1): 7025.

[51] Figueiredo J. L., Pereira M. F. R. The role of surface chemistry in catalysis with carbons. Catalysis Today. 2010, 150(1): 2–7.

[52] Kiani D. A.-O., Wachs I. A.-O. X. Practical considerations for understanding surface reaction mechanisms involved in heterogeneous catalysis. 2024,14(22): 2155–5435. Print.

[53] Majoe N., et al. Catalytic influence of alkali and alkali earth metals in black liquor on the gasification process: A review. Biomass Conversion and Biorefinery. 2024 14(18).

[54] Wood C., Exotic new superconductors delight and confound. In: Quanta Magazine. 2024, Simons Foundation, 1. https://www.quantamagazine.org/exotic-new-superconductors-delight-and-con found-20241206/.

[55] Cui D., et al. Probing the thermodynamics of Moiré patterns in molecular self-assembly at the liquid–solid interface, Chemistry of Materials, 2022, 34(5): 2449–2457.

[56] Kubozono Y., et al. Superconductivity in aromatic hydrocarbons. Physica. C, Superconductivity. 2015, 514: 199–205.

[57] Colquhoun H. M., et al. Superelectrophiles in aromatic polymer chemistry. Macromolecules. 2001, 34(4): 1122–1124.

[58] Tissot & Welte - Tissot B. P. a. W. D. H., Petroleum formation and occurence, 2nd ed. 1984, Springer.

[59] Bennett B., et al. Fractionation of benzocarbazoles between source rocks and petroleums. Organic Geochemistry. 2002, 33: 545–559.

[60] Ripani L., et al. Electron transfer in polyaromatic hydrocarbons and molecular carbon nanostructures. Current Opinion in Electrochemistry. 2022, 35: 101065.

[61] Gafurov M., et al. High-Field (3.4 T) electron paramagnetic resonance, 1H Electron-Nuclear double resonance, ESEEM, HYSCORE, and relaxation studies of Asphaltene solubility fractions of bitumen for structural characterization of intrinsic carbon-centered radicals. Nanomaterials. 2022, 12(23): 4218.

[62] Alafnan S., Sultan A. S., Aljaberi J. Molecular fractionation in the organic materials of source rocks. ACS Omega. 2020, 5(30): 18968–18974. vol. 2020, 2470–1343, Electronic.

[63] Mousavi M., et al. The influence of asphaltene-resin molecular interactions on the colloidal stability of crude oil. Fuel. 2016, 183: 262–271.

[64] Burdelnaya N. S., et al. Geochemical significance of the molecular and supramolecular structures of Asphaltenes (A review). Petroleum Chemistry. 2023, 63(1): 31–51.

[65] Muradov N. Catalysis of methane decomposition over elemental carbon. Catalysis Communications. 2001, 2(3): 89–94.

[66] Al-Hassani A. A., Abbas H. F., Wan Daud W. M. A. Production of COx-free hydrogen by the thermal decomposition of methane over activated carbon: Catalyst deactivation. International Journal of Hydrogen Energy. 2014, 39(27): 14783–14791.

[67] Pinilla J. L., et al. Hydrogen production by thermo-catalytic decomposition of methane: Regeneration of active carbons using CO2. Journal of Power Sources. 2007, 169(1): 103–109.

[68] Ansón A., et al. Porosity, surface area, surface energy, and hydrogen adsorption in nanostructured carbons. The Journal of Physical Chemistry B. 2004, 108(40): 15820–15826.

[69] Shi P., et al. Unveiling the interaction regimes between atomic oxygen and amorphous carbon surface depending on incident energy. Carbon. 2024, 226: 119229.

[70] Jassas R. S., et al., Scholl reaction as a powerful tool for the synthesis of nanographenes: A systematic review. RSC Advances. 2021, 11(51): 32158–32202.

[71] Agranat I., et al. The linkage between reversible Friedel–Crafts acyl rearrangements and the Scholl reaction. Structural Chemistry. 2019, 30(5): 1579–1610.

[72] Pahlavan F., et al. Characterization of oxidized asphaltenes and the restorative effect of a bio-modifier. Fuel. 2018, 212: 593–604.

[73] Zadshir M., Hosseinnezhad S., Fini E. H. Deagglomeration of oxidized asphaltenes as a measure of true rejuvenation for severely aged asphalt binder. Construction and Building Materials. 2019, 209 (0): 416–424.

[74] Zadshir M., et al. Investigating bio-rejuvenation mechanisms in asphalt binder via laboratory experiments and molecular dynamics simulation. Construction and Building Materials. 2018, 190: 392–402.

[75] Medina O. E., et al. Chemical and structural changes of asphaltenes during oxygen chemisorption at low and high-pressure. Fuel. 2025, 379: 133000.

[76] Jung H., Bielawski C. W. Asphaltene oxide promotes a broad range of synthetic transformations. Communications Chemistry. 2019, 2(1): 113.

[77] Premović P. I., Bojić J., Tonsa I. R. Kerogenization of asphaltenes by air oxygen: The heimar (bold petroleum seepage) sandstone from the dead sea basin (Israel). Journal of Scientific & Industrial Research. 1999, 58: 443–449.

[78] Valadi F. M., et al. Competitive adsorption of CO2, N2, and CH4 in coal-derived asphaltenes, a computational study. Scientific Reports. 2024, 14(1): 7664.

[79] Nassar N. N., et al. Kinetics of the catalytic thermo-oxidation of asphaltenes at isothermal conditions on different metal oxide nanoparticle surfaces. Catalysis Today. 2013, 207: 127–132.

[80] Siddiqui M. N., Ali M. F. Studies on the aging behavior of the Arabian asphalts. Fuel. 1999, 78(9): 1005–1015.

[81] Glaser R. R., et al. Low-temperature oxidation kinetics of asphalt binders. Transportation Research Record. 2013, 2370(1): 63–68.

[82] Afra S., et al. Alterations of asphaltenes chemical structure due to carbon dioxide injection. Fuel. 2020, 272: 117708.

[83] Liu B., et al. Mechanism of asphaltene aggregation induced by supercritical CO2: Insights from molecular dynamics simulation. RSC advances. 2017, 7(80): 50786–50793.

[84] Van Buren J., et al. Ring-cleavage products produced during the initial phase of oxidative treatment of Alkyl-Substituted aromatic compounds. Environmental Science and Technology. 2020, 54: 8352–8361. 1520–5851, Electronic.

[85] Zhai L., et al. Probing the arenium-ion (proton transfer) versus the cation-radical (electron transfer) mechanism of Scholl reaction using DDQ as oxidant. The Journal of Organic Chemistry. 2010. 75: 1520–6904. Electronic. 4748–4760.

[86] Rempala P., Kroulík B. T., Fau – King J., King B. T. A slippery slope: Mechanistic analysis of the intramolecular Scholl reaction of hexaphenylbenzene, Journal of American Chemistry Society. 2004, 126: 15002–15003.

[87] Rempala P., Kroulík J., King B. T. Investigation of the mechanism of the intramolecular Scholl reaction of contiguous phenylbenzenes. The Journal of Organic Chemistry. 2006, 71(14): 5067–5081.

[88] Little M. S., S.g.y. A. A. A., Heard K. W. J., Raftery J., Edwards A. C., Parry A. V. S., Quayle P. Insights into the Scholl coupling reaction: a key transformation of relevance to the synthesis of graphenes and related systems. European Journal of Organic Chemistry. 2017, (13): 1694–1703.

[89] Biesaga J., Szafert S., Pigulski B. 1,2,3-Triarylazulenes as precursors of azulene-embedded polycyclic aromatic hydrocarbons. Organic Chemistry Frontiers. 2024, 11(21): 6026–6035.

[90] Grzybowski M., et al. Comparison of oxidative aromatic coupling and the Scholl reaction, Angewandte Chemie International Edition. 2013. 38: 1521–3773. Electronic: 9900-9930.

[91] King B. T., et al. Controlling the Scholl reaction, Journal of Organic Chemistry. 2007, 72: 2279–2288. 0022–3263, Print.

[92] Cheung K. M., et al. Negatively curved molecular nanocarbons containing multiple heptagons are enabled by the Scholl reactions of macrocyclic precursors. Chemistry. 2023, 9(10): 2855–2868.

[93] Miao Q. Rearrangements come to Scholl. Nature Reviews Chemistry. 2021, 5(9): 602–603.

[94] Ormsby J. L., et al. Rearrangements in the Scholl oxidation: Implications for molecular architectures. Tetrahedron. 2008, 64(50): 11370–11378.

[95] Chen J., et al. Bound hydrocarbons and structure of pyrobitumen rapidly formed by asphaltene cracking: Implications for oil–source correlation. Organic Geochemistry. 2020, 146: 104053.

[96] Hou Y., et al. Structural evolution of organic matter and implications for graphitization in over-mature marine shales, south China. Marine and Petroleum Geology. 2019, 109: 304–316.

[97] Sáez R., et al. Black shales and massive sulfide deposits: Causal or casual relationships? insights from Rammelsberg, Tharsis, and Draa Sfar. Mineralium Deposita. 2011, 46(5): 585–614.

[98] Stow D. A. V., Huc A. Y., Bertrand P. Depositional processes of black shales in deep water. Marine and Petroleum Geology. 2001, 18(4): 491–498.

[99] Eglinton G., et al. Molecular preservation [and discussion]. Philosophical Transactions of the Royal Society of London Series B Biological Sciences. 1991, 333(1268): 315–328.

[100] Gupta N. S., et al. Rapid incorporation of lipids into macromolecules during experimental decay of invertebrates: Initiation of geopolymer formation. Organic Geochemistry. 2009, 40(5): 589–594.

[101] Borisova L. S. The origin of Asphaltenes and main trends in evolution of their composition during lithogenesis. Petroleum Chemistry. 2019, 59(10): 1118–1123.

[102] Umadevi D., Sastry G. N. Saturated vs. unsaturated hydrocarbon interactions with carbon nanostructures, 2296–2646. Print.

[103] Borisova L. S., Timoshina I. D. Geochemistry of Asphaltenes in organic matter of low thermal maturity. Geochemistry International. 2021, 59(3): 290–300.

[104] Durrett T. P., Welti R. The tail of chlorophyll: Fates for phytol. The Journal of Biological Chemistry. 2021, 296(0): 100802.

[105] Hörtensteiner S., et al. The key step in chlorophyll breakdown in higher plants. Cleavage of pheophorbide a macrocycle by a monooxygenase, Journal of Biological Chemistry. 1998, 273: 0021–9258. Print. 15335–15339.

[106] Navas-Cáceres O. D., Parada M., Zafra G. Development of a highly tolerant bacterial consortium for asphaltene biodegradation in soils. Environmental Science and Pollution Research. 2023, 30(59): 123439–123451.

[107] Czochanska Z., et al. Geochemical application of sterane and triterpane biomarkers to a description of oils from the Taranaki Basin in New Zealand. Organic Geochemistry. 1988, 12(2): 123–135.

[108] Sanei H. Genesis of solid bitumen. Scientific Reports. 2020, 10(1): 15595–15595.

[109] Derenne S., et al. Chemical evidence of kerogen formation in source rocks and oil shales via selective preservation of thin resistant outer walls of microalgae: Origin of ultralaminae. Geochimica Et Cosmochimica Acta. 1991, 55(4): 1041–1050.

[110] Bousige C., et al. Realistic molecular model of kerogen's nanostructure. Nature Materials. 2016, 15(5): 576–582.

[111] Rullkötter J., Michaelis W. The structure of kerogen and related materials. A review of recent progress and future trends. Organic Geochemistry. 1990, 16(4): 829–852.

[112] Kuangzong Q. Thermal depolymerization of kerogen and formation of immature oil. Organic Geochemistry. 1988, 13(4): 1045–1050.

[113] Han Y., et al. Fractionation of hydrocarbons and NSO-compounds during primary oil migration revealed by high resolution mass spectrometry: Insights from oil trapped in fluid inclusions. International Journal of Coal Geology. 2022, 254: 103974.

[114] Liao Y., Geng A. Stable carbon isotopic fractionation of individual n-alkanes accompanying primary migration: Evidence from hydrocarbon generation–expulsion simulations of selected terrestrial source rocks. Applied Geochemistry. 2009, 24(11): 2123–2132.

[115] Pan Y., et al. Characterization of free and bound bitumen fractions in a thermal maturation shale sequence. Part 1: Acidic and neutral compounds by negative-ion ESI FT-ICR MS, Organic geochemistry, 2019. 134, 1–15.

[116] Baskin D. K., Peters K. E. Early generation characteristics of a Sulfur-rich monterey kerogen. AAPG Bulletin. 1992. 76, 1–13.

[117] Baruah B., et al. TGA-FTIR analysis of Upper Assam oil shale, optimization of lab-scale pyrolysis process parameters using RSM. Journal of Analytical and Applied Pyrolysis. 2018, 135: 397–405.

[118] Chauhan A., et al. Metal-free N-doped carbon catalyst derived from chitosan for aqueous formic acid-mediated selective reductive formylation of quinoline and nitroarenes. ChemSusChem. 2022, 15(23): e202201560.

[119] Pelet R., Behar F., Monin J. C. Resins and asphaltenes in the generation and migration of petroleum. Organic Geochemistry. 1986, 10(1): 481–498.

[120] Summons R. E., Welander P. V., Gold D. A. Lipid biomarkers: Molecular tools for illuminating the history of microbial life. Nature Reviews Microbiology. 2022, 20(3): 174–185.

[121] Forsman J. P., Hunt J. M. Insoluble organic matter (kerogen) in sedimentary rocks. Geochimica Et Cosmochimica Acta. 1958, 15(3): 170–182.

[122] Nadeau P. H., et al. The conversion of smectite to illite during diagenesis: Evidence from some illitic clays from bentonites and sandstones. Mineralogical Magazine. 1985, 49(352): 393–400.

[123] Linares J., Huertas F., Barahona E. Conversion time from smectite to illite. A preliminary study. Applied Clay Science. 1992, 7(1): 125–130.

[124] Awwiller D. N. Illite/smectite formation and potassium mass transfer during burial diagenesis of mudrocks: A study from the Texas Gulf Coast Paleocene- Eocene. Journal of Sedimentary Petrology. 1993, 63(3): 501–512.

[125] Ferrage E., et al. Hydration properties and interlayer organization of water and ions in synthetic Na-smectite with tetrahedral layer charge. part 2. toward a precise coupling between molecular simulations and diffraction data. Journal of Physical Chemistry C. 2011, 115(5): 1867–1881.

[126] Kawamura K., et al. Volatile organic acids generated from kerogen during laboratory heating. Geochemical Journal. 1986, 20(1): 51–59.

[127] Ma W., et al. Interactions between mineral evolution and organic acids dissolved in bitumen in hybrid shale system. International Journal of Coal Geology. 2022, 260: 104071.

[128] Ma W., et al. Understanding brønsted-acid catalyzed monomolecular reactions of alkanes in Zeolite pores by combining insights from experiment and theory. ChemPhysChem. 2018, 19(4): 341–358.

[129] Goldstein T. P. Geocatalytic reactions in formation and maturation of petroleum. AAPG Bulletin. 1983, 67(1): 152–159.

[130] Wang M., et al. Genesis and stability of hydronium ions in Zeolite channels. Journal of the American Chemical Society. 2019, 141(8): 3444–3455.

[131] Shi H., et al. Tailoring nanoscopic confines to maximize catalytic activity of hydronium ions. Nature Communications. 2017, 8(1): 15442–15442.

[132] Fukushima K. Vacuum pyrolysis of recent sedimentary humic acids and kerogens. Geochemical Journal. 1982, 16(1): 43–49.

[133] Whitmore F. C. Mechanism of the Polymerization of Olefins by acid catalysts. Industrial & Engineering Chemistry. 1934, 26(1): 94–95.

[134] Yuan P., et al. Role of the interlayer space of montmorillonite in hydrocarbon generation: An experimental study based on high temperature–pressure pyrolysis. Applied Clay Science. 2013, 75–76: 82–91.

[135] Shi T.-H., Tong S., Wang M.-X. Construction of Hydrocarbon Nanobelts. Angewandte Chemie International Edition. 2020, 59(20): 7700–7705.

[136] Hoog H., Verheus J., Zuiderweg F. J. Investigations into the cyclisation (aromatisation) of aliphatic hydrocarbons. Transactions of the Faraday Society. 1939, 35: 993–1006.

[137] Abbott G. D., Maxwell J. R. Kinetics of the aromatisation of rearranged ring-C monoaromatic steroid hydrocarbons. Organic Geochemistry. 1988, 13(4): 881–885.

[138] Feng X., Pisula W., Müllen K. Large polycyclic aromatic hydrocarbons: Synthesis and discotic organization. Pure and Applied Chemistry. 2009, 81(12): 2203–2224.

[139] Khatymov R. V., Muftakhov M. V., Shchukin P. V. Negative ions, molecular electron affinity and orbital structure of cata-condensed polycyclic aromatic hydrocarbons. Rapid Communications in Mass Spectrometry. 2017, 31(20): 1729–1741.

[140] Wu D., et al. Two-dimensional nanostructures from positively charged polycyclic aromatic hydrocarbons. Angewandte Chemie International Edition. 2011, 50(12): 2791–2794.

[141] de Grotthuss C. J. T. Memoir on the decomposition of water and of the bodies that it holds in solution by means of galvanic electricity. Biochimica Et Biophysica Acta (BBA) – Bioenergetics. 2006, 1757(8): 871–875.

[142] Prakash M., Subramanian V., Gadre S. R. Stepwise hydration of protonated carbonic acid: A theoretical study. J Phys Chem A. 2009, 113(44): 12260–12275.

[143] Schmitt U. W., Voth G. A. The computer simulation of proton transport in water. The Journal of Chemical Physics. 1999, 111(20): 9361–9381.

[144] Barth T., Bjørlykke K. Organic acids from source rock maturation: Generation potentials, transport mechanisms and relevance for mineral diagenesis. Applied Geochemistry. 1993, 8(4): 325–337.

[145] Seewald J. S. Model for the origin of carboxylic acids in basinal brines. Geochimica Et Cosmochimica Acta. 2001, 65(21): 3779–3789.

[146] Tegelaar E. W., et al. A reappraisal of kerogen formation. Geochimica Et Cosmochimica Acta. 1989, 53(11): 3103–3106.

[147] Burnham A. K., Braun R. L. Development of a detailed model of petroleum formation, destruction, and expulsion from lacustrine and marine source rocks. Organic Geochemistry. 1990, 16(1): 27–39.

[148] Yang K., et al. Bay/ortho-Octa-substituted Perylene: A versatile building block toward novel polycyclic (hetero)aromatic hydrocarbons. Accounts of Chemical Research. 2024, 57(5): 763–775.

[149] Zhang P., Xu X., Luo X. Degradation pathways and product formation mechanisms of asphaltene in supercritical water. Journal of Hazardous Materials. 2024, 478: 135488.

[150] Van Speybroeck V., et al. The kinetics of cyclization reactions on polyaromatics from first principles. ChemPhysChem. 2002, 3(10): 863–870.

[151] Bernard S., et al. Formation of nanoporous pyrobitumen residues during maturation of the Barnett Shale (Fort Worth Basin). International Journal of Coal Geology. 2012, 103: 3–11.

[152] Wang X. Y., Yao X., Müllen K. *Polycyclic aromatic hydrocarbons in the graphene era.* Science China. Chemistry. 2019, 62(9): 1099–1144.

[153] Mastalerz M., Drobniak A., Stankiewicz A. B. Origin, properties, and implications of solid bitumen in source-rock reservoirs: A review. International Journal of Coal Geology. 2018, 195: 14–36.

[154] Escamilla-Roa E., Nieto F., Sainz-Dí Az C. I. Stability of the Hydronium cation in the structure of illite. Clays and Clay Minerals. 2016, 64(4): 413–424.

[155] Du J., et al. Variations and geological significance of solid acidity during smectite illitization. Applied Clay Science. 2021, 204: 106035.

[156] Zheng Y., et al. Exploring the coupling relationship between hydrocarbon generation of continental shale and nanopore structure evolution – A case study of Shahejie formation in Bohai Bay Basin. Journal of Petroleum Exploration and Production Technology. 2021, 11(12): 4215–4225.

[157] Ziemiański P. P., Derkowski A. Structural and textural control of high-pressure hydrogen adsorption on expandable and non-expandable clay minerals in geologic conditions. International Journal of Hydrogen Energy. 2022, 47(67): 28794–28805.

[158] Wang X., et al. Anomalous hydrogen evolution behavior in high-pH environment induced by locally generated hydronium ions. Nature Communications. 2019, 10(1): 4876–4878.

[159] Serrano D. P., et al. Hydrogen production by methane decomposition: Origin of the catalytic activity of carbon materials. Fuel (Guildford). 2010, 89(6): 1241–1248.

[160] Harun K., Adhikari S., Jahromi H. Hydrogen production: Via thermocatalytic decomposition of methane using carbon-based catalysts. RSC Advances 2020, 10(67): 40882–40893.

[161] Zakertabrizi M., et al. Insight from perfectly selective and ultrafast proton transport through anhydrous asymmetrical graphene oxide membranes under Grotthuss mechanism. Journal of Membrane Science. 2021, 618: 118735.

[162] Baxter R. J., Hu P. Insight into why the Langmuir–Hinshelwood mechanism is generally preferred. The Journal of Chemical Physics. 2002, 116(11): 4379–4381.

[163] Wang J., et al. Mechanism of methane decomposition with hydrogen addition over activated carbon via in-situ pyrolysis-electron impact ionization time-of-flight mass spectrometry. Fuel. 2020, 263: 116734.

[164] Xuan G., et al. Mechanism of improving the stability of activated carbon catalyst by trace H2S impurities in natural gas for hydrogen production from methane decomposition. Fuel. 2021, 299: 120884.

[165] Mango F. Methane and carbon at equilibrium in source rocks. Geochemical Transactions. 2013, 14: 5.

[166] Caro M. A., et al. Growth mechanism and origin of high Sp3 content in Tetrahedral Amorphous Carbon. Physical Review Letters. 2018, 120(16): 166101.

[167] Wei S., et al. Characteristics and evolution of pyrobitumen-hosted pores of the overmature lower Cambrian Shuijingtuo Shale in the south of Huangling anticline, Yichang area, China: Evidence from FE-SEM petrography. Marine and Petroleum Geology. 2020, 116: 104303.

[168] Serrano D. P., Botas J. A., Guil-Lopez R. H2 production from methane pyrolysis over commercial carbon catalysts: Kinetic and deactivation study. International Journal of Hydrogen Energy. 2009, 34(10): 4488–4494.

[169] Dunker A. M., Kumar S., Mulawa P. A. Production of hydrogen by thermal decomposition of methane in a fluidized-bed reactor – Effects of catalyst, temperature, and residence time. International Journal of Hydrogen Energy. 2006, 31(4): 473–484.

[170] Moliner R., et al. Thermocatalytic decomposition of methane over activated carbons: Influence of textural properties and surface chemistry. International Journal of Hydrogen Energy. 2005, 30(3): 293–300.

[171] Muradov N., Smith F., T-Raissi A. Catalytic activity of carbons for methane decomposition reaction. Catalysis Today. 2005, 102–103: 225–233.

[172] Brunner G. Near critical and supercritical water. Part I. Hydrolytic and hydrothermal processes. The Journal of Supercritical Fluids. 2009, 47(3): 373–381.

[173] Lu Y., et al. Comparative study on the pyrolysis behavior and pyrolysate characteristics of Fushun oil shale during anhydrous pyrolysis and sub/supercritical water pyrolysis. RSC Advances. 2022, 12(26): 16329–16341.

[174] Siskin M., Katritzky A. R. Reactivity of organic compounds in superheated water: General background. Chemical Reviews. 2001, 101(4): 825–836.

[175] Xie T., et al., Experimental investigation on the hydrocarbon generation of low maturity organic-rich shale in supercritical water., Tallinn, Estonia: Oil shale, 2022 vol. 39, 3 169–188, 1984.

[176] Yong T. L.-K., Matsumura Y. Kinetics analysis of phenol and benzene decomposition in supercritical water. The Journal of Supercritical Fluids. 2014, 87: 73–82.

[177] Bonijoly M., Oberlin M., Oberlin A. A possible mechanism for natural graphite formation. International Journal of Coal Geology. 1982, 1(4): 283–312.

[178] Suzuki N., Saito H., Hoshino T. Hydrogen gas of organic origin in shales and metapelites. International Journal of Coal Geology. 2017, 173: 227–236.

[179] Boreham C. J., et al. *Modelling of hydrogen gas generation from overmature organic matter in the Cooper Basin, Australia.* The APPEA Journal. 2023.

[180] Raza J., et al. Methane decomposition for hydrogen production: A comprehensive review on catalyst selection and reactor systems. Renewable & Sustainable Energy Reviews. 2022, 168: 112774.

[181] Msheik M., Rodat S., Abanades S. Methane cracking for hydrogen production: A review of catalytic and molten media pyrolysis. Energies (Basel). 2021, 14(11): 3107.

[182] Fan Z., et al. Catalytic decomposition of methane to produce hydrogen: A review. Journal of Energy Chemistry. 2021, 58: 415–430.

[183] Zhang J., et al. Hydrogen production by catalytic methane decomposition: Carbon materials as catalysts or catalyst supports. International Journal of Hydrogen Energy. 2017, 42(31): 19755–19775.

[184] Ahmed S., et al. Decomposition of hydrocarbons to hydrogen and carbon. Applied Catalysis. A, General. 2009, 359(1): 1–24.

[185] Krevelen V. Graphical-statistical method for the study of structure and reaction processes of coal. Fuel. 1950, 29: 269–284.

[186] Greenwell A. Analyses of British coals and cokes, The Chichester Press, 1907.

[187] Hou Q., et al. Structure and coalbed methane occurrence in tectonically deformed coals. Science China. Earth Sciences. 2012, 55(11): 1755–1763.

[188] Vu T. T. A., et al. The structural evolution of organic matter during maturation of coals and its impact on petroleum potential and feedstock for the deep biosphere, Organic geochemistry, 2013. vol. 62, 17–27.

[189] Ge Z., et al. Hydrogen production by non-catalytic partial oxidation of coal in supercritical water: Explore the way to complete gasification of lignite and bituminous coal. International Journal of Hydrogen Energy. 2013, 38(29): 12786–12794.

[190] Guo L., Jin H., Lu Y. Supercritical water gasification research and development in China. The Journal of Supercritical Fluids. 2015, 96: 144–150.

[191] Li L., et al. Potential and challenges for the new method supercritical CO_2/H_2O mixed fluid huff-n-puff in shale oil EOR. Frontiers in Energy Research. 2022, 10(0): 1.

[192] Tan X., et al. The supercritical multithermal fluid flooding investigation: Experiments and numerical simulation for deep offshore heavy oil reservoirs. Geofluids. 2021, 2021: 5589543.

[193] Xu H., et al. Molecular evidence reveals the presence of hydrothermal effect on ultra-deep-preserved organic compounds. Chemical Geology. 2022, 608: 121045.

[194] Zheng L., et al. Molecular dynamics simulation of sub- and supercritical water extraction shale oil in slit nanopores. The Journal of Supercritical Fluids. 2023, 195: 105862.

[195] Prinzhofer A., Tahara Cissé C. S., Diallo A. B. Discovery of a large accumulation of natural hydrogen in Bourakebougou (Mali). International Journal of Hydrogen Energy. 2018, 43(42): 19315–19326.

[196] Maiga O., et al. Characterization of the spontaneously recharging natural hydrogen reservoirs of Bourakebougou in Mali. Scientific Reports. 2023, 13(1): 11876.

[197] Klumpp D. A., Anokhin M. V. Superelectrophiles: Recent advances. Molecules (Basel, Switzerland). 2020, 25(14): 3281.

[198] Goumont R., et al. A criterion to demarcate the dual Diels–Alder and σ-complex behaviour of aromatic and heteroaromatic superelectrophiles. Tetrahedron Letters. 2005, 46(48): 8363–8367.

[199] Zhong G. H., Chen X. J., Lin H. Q. Superconductivity and its enhancement in polycyclic aromatic hydrocarbons. Frontiers in physics. 2019, 7(0): 1.

[200] Kotarba M. J., Bilkiewicz E., Hałas S. Mechanisms of generation of hydrogen sulphide, carbon dioxide and hydrocarbon gases from selected petroleum fields of the Zechstein main dolomite carbonates of the western part of polish Southern Permian Basin: Isotopic and geological approach. Journal of Petroleum Science and Engineering. 2017, 157: 380–391.

Vitaly Vidavskiy* and Nikolay Larin

Chapter 8
Natural hydrogen and the primordially hydridic earth concept

Abstract: The primordially hydridic Earth (PHE) concept offers a transformative view of Earth's formation, proposing a hydrogen-rich planetary composition with metal hydrides as a key component of the core. This chapter examines the chemical, physical, and geophysical implications of the PHE model, integrating laboratory experiments, field observations, and theoretical calculations. The model suggests that hydrogen, initially abundant in Earth's interior, played a crucial role in the planet's evolution. It explains phenomena such as the core's density deficit, mantle dynamics, and the formation of geospheres through hydrogen degassing from metal hydrides, which influenced lithosphere development and tectonic activity.

The chapter explores experimental data, including hydrogen partitioning between silicate and metallic phases, the behavior of hydride-enriched cores under high pressures, and the conductivity of such systems. It also introduces the concept of "hydrogen chimneys," vertical degassing structures that facilitate hydrogen migration to the surface. Using high-resolution satellite imagery, researchers have identified potential degassing channels, particularly on the East European Platform, and linked them to geological and tectonic structures.

Drawing on Vladimir Larin's pioneering work, the chapter underscores the presence of substantial natural hydrogen reserves, suggesting they could meet humanity's energy needs sustainably. It reviews the search for hydrogen flows, the structure of degassing channels, and methods for locating accessible, high-concentration hydrogen sources near potential consumers. The PHE model also contextualizes hydrogen's role in the broader framework of Solar System formation, emphasizing its compatibility with alternative mechanisms, such as element distribution by first ionization potential.

Keywords: natural hydrogen, hydridic Earth, solar system formation, Earth composition

*Corresponding author: Vitaly Vidavskiy, AVALIO, West Perth, 6005 WA, Australia; Western Australian School of Mines, Curtin University, Kensington, 6151 WA, Australia, e-mail: Vv@avalio.net
Nikolay Larin, Natural Hydrogen Energy (NH2E) LLC., Colorado, USA; The Schmidt Institute of Physics of the Earth, 10 B. Gruzinskaya Street, Moscow, 123242, Russian Federation; AVALIO Pty Ltd., West Perth, 6005, WA, Australia

https://doi.org/10.1515/9783111437040-008

8.1 Introduction

8.1.1 Solar system formation

A supernova explosion that occurred approximately 4.5 billion years ago triggered the instability of interstellar diffuse matter, causing it to start collapsing toward its center of gravity. Possessing some initial rotation, the collapsing matter began to spin faster and faster, eventually forming something akin to a biconvex lens with an equatorial radius of around 50 million km [2]. (NB: Mercury's orbit is about 55 million km.) Further contraction, driven by the conservation of angular momentum, led to an increase in rotational speed, and the "nebula" entered a state of rotational instability, where centrifugal forces balanced gravitational forces at the equator. Continuous contraction of the nebula resulted in an even faster rotational velocity, leading to the ejection of material in the equatorial plane as centrifugal forces exceeded gravitational attraction. Astrophysical calculations indicate that the rate of mass accumulation within the rotating nebula was highly uneven [3]. If we consider the time from the onset of contraction to the onset of rotational instability as 1 million years, then roughly 1% of the mass was accumulated during the first three-quarters of this time period, while the last 50% of the mass accumulated over approximately 1,000 years. It is worth noting that the supernova explosion preceding these events not only facilitated the ionization of diffuse matter but also acted as a powerful nucleosynthesis[1] event, enriching the material undergoing gravitational collapse with many short-lived isotopes that could sustain the matter in a plasma state for a sufficiently long period of time. (Here, the authors note that Larin's views on nucleosynthesis were inspired by those of F. Hoyle [5]).

Thus, the contraction and spinning of the nebula led to the dispersal (or shedding) of a disk (Figure 8.1), while the heating, combined with the plasma state of matter, caused the emergence of a dipole magnetic field in its central core (the Proto-Sun), induced by the rapidly rotating plasma. The magnetic field of the nebula played a crucial role in the further evolution of our rotating system. The nebula became reinforced by magnetic field lines, which allowed it to maintain angular momentum throughout its components. This solution addresses the long-standing and challenging issue in cosmology, which Fred Hoyle highlighted, particularly pointing to the role of the magnetic field in resolving this matter: "It is well known that the angular momentum per unit mass about the center of gravity of the solar system is on the average some 50,000 times greater for the planetary material than that for the solar material" [4].

After the disk was ejected, the central core – the "Proto-Sun" – slowed down, and the magnetic field shut off. This means that the magnetic field in the newborn solar system was activated to equalize the angular velocities within the disk, and then it was deactivated. As a result, because particles with the same charge moving in the

---

1 The authors note that Larin's views on the nucleosynthesis were inspired by those of F. Hoyle.

same direction attract each other, the disk began to separate into numerous distinct rings. With the reduction of the magnetic field intensity, due to the phenomenon of self-induction, a ring-shaped electric current was likely generated in the individual rings of the protoplanetary disk, triggering the pinch effect. This effect led to the rings' fragmentation into separated pieces, which then turned into spherical globules, densely packing the newly formed protoplanetary disk. The rapid condensation of plasma into solid particles was hindered by Coulomb repulsion forces, although gravitational forces still promoted some consolidation within the extended spherical globules, which were about a million kilometers in size.

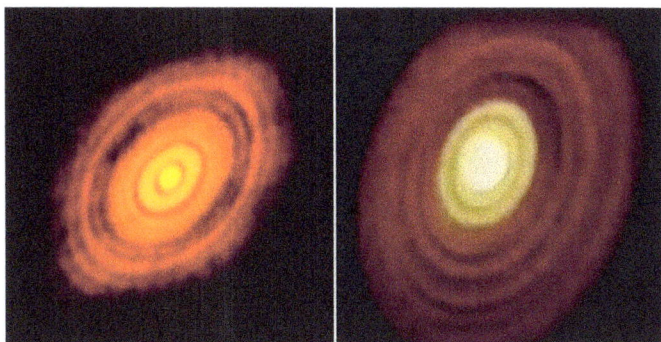

Figure 8.1: Protoplanetary disks. On the left: the protoplanetary disk surrounding the young star HL Tauri. ALMA observations [6] reveal substructures within the disk that have never been seen before and even show the possible positions of planets forming in the dark patches within the system (credit: ALMA (ESO/NAOJ/NRAO)). On the right is a shot from the popular science movie "Hypothesis" of 1984 [7] based on Dr. V. Larin's concept. The resemblance is striking.

Thus, we now have to determine how a disk composed of tens or hundreds of thousands of sparse spheres, rotating in the gravitational field of the Proto-Sun, should evolve. Any attempt at numerical modeling of the aggregation of solid particles often results in countless bodies of asteroid-like size, where further growth typically does not lead to the formation of a planet but rather to fragmentation. Calculations of collisions are further complicated by the fact that the encounter of two particles (whether grains or pebbles) is unlikely to result in their merging. More often, they may bounce away from each other, perturb each other's orbits due to gravitational attraction, fragment, and the list of possibilities could go on [1].

Almost half a century ago, two mathematicians, Eneev and Kozlov, simplified the computational process by setting the rotation of "droplets" in circular orbits and simulating their interactions based on the law of a perfectly inelastic collision (approach, gravitational bonding, tidal bulge formation, and merging). As a result of computer simulations, these researchers obtained the parameters characterizing the solar system: the characteristic number of planets and the proportions of their orbits, in accordance with the Titius-Bode law [8]. It should be noted that this mechanism assumes the complete "scooping up" of material by a forming planet. As the short-lived isotopes inherited from the supernova decayed, the gradual condensation of matter began, leading to a "gentle snowfall" of particles toward the center of gravity. Thus, over the course of 1.5–2o million years, the planets gradually formed. By this time, the Sun would have begun to shine, as the Proto-Sun's central core had reached the level required for thermonuclear reactions to be initiated.

More recently, the group of researchers working at the Atacama large millimeter/submillimeter array (ALMA) observatory noticed the "gaps" between the protoplanetary rings in the newly formed proto-stellar systems (Figure 8.2). They arrived at the conclusion that the process of planet formation, known as "accretion," is an integral part of the proto-stars' early evolution [9–13]. Larin formalized this statement [2].

The ALMA researchers stated that the "actual formation of the planetary system progresses rapidly in the 100,000 years to 1,000,000 years after star formation begins" [14].

According to PHE concept, following the proto-Sun's supernova explosion, it took 0.75 million years (Ma) for 1% of the Solar System's mass to consolidate, 1 Ma for the gravitational pull to accumulate the full mass, 1,000 years for the second half to consolidate, 100 years for the proto-planetary disk to spread into the equatorial plane, and approximately 1.5–2 Ma for the proto-planetary spheres to condense into solid planets. Remarkably, Larin's forecasts are accurate to the order of magnitude, especially considering that these predictions were made without the sophisticated instruments, such as orbital IR-range telescopes, available today.

8.1.2 Determining the Earth's initial composition

Recalling the hypothesis by Fred Hoyle regarding the critical role of the nebula's magnetic field during the separation phase of the protoplanetary disk, the material in the dispersing disk must have moved across the magnetic field lines, and if the charged particles were moving at "thermal" rather than relativistic speeds, they should have been captured by the magnetic field and brought to a halt within it (Figure 8.3).

Larin proposed that if Hoyle's ideas and framework are correct, a magnetic separation of elements based on their ionization tendencies should be observable: easily ionized elements would remain near the Sun, while atoms with high ionization potentials, remaining neutral, could bypass the magnetic separator and move to distant regions of the nebula. To test this hypothesis, we rely on the limited available data, including the composition of the Sun's photosphere (representing approximately 70% of the star's volume), the material of Earth's external geospheres, samples collected from the Moon, and recovered meteorites. By comparing pairs such as Earth and the Sun, Earth and the Asteroid Belt, and Earth and the Moon, and plotting the first ionization

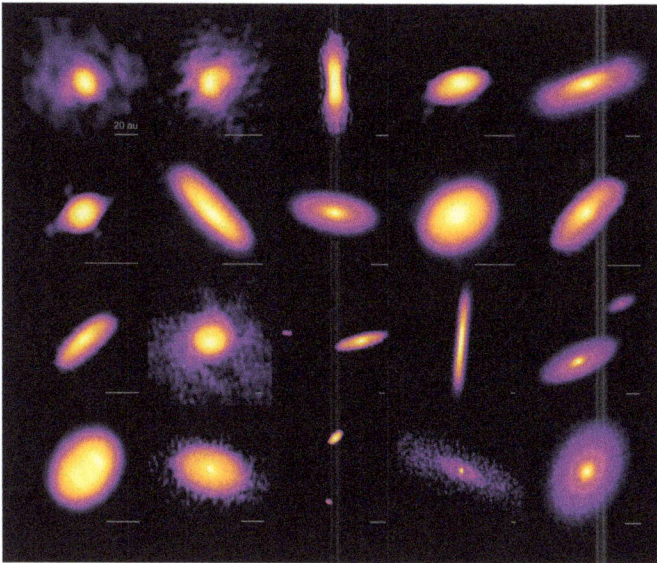

Figure 8.2: Images of disks around 19 protostars, including four binary systems observed with ALMA. For one binary system, disks around the primary and secondary are presented independently (second line, rightmost, and third line, leftmost). Disks are presented in the order of their evolutionary sequence (the one in the upper-left corner is the youngest, while the one in the lower-right corner is the oldest). The two oldest disks show faint ring-gap structures. A scale bar of 20 au (roughly the distance between the Sun and Uranus) is shown for each disk image. Credit: ALMA (ESO/NAOJ/NRAO) [14] and Ohashi et al. [9].

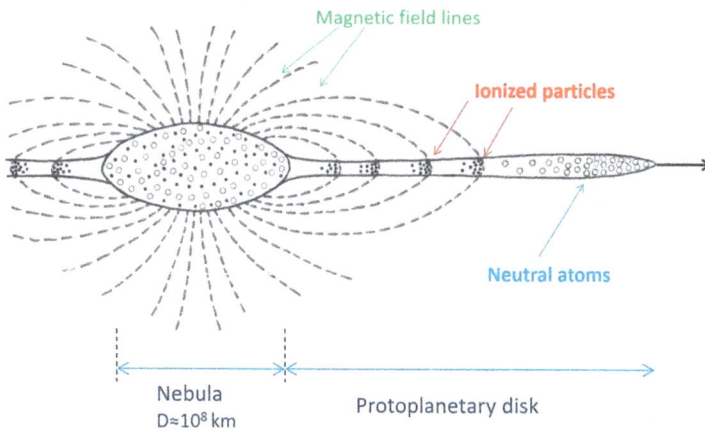

Figure 8.3: Magnetic differentiation of particles in the protoplanetary disk. The cross-section of the rotating nebula and the protoplanetary disk is shown (adapted from [2]).

potentials of elements on the *x*-axis against their relative abundances on the y-axis, we can analyze the results presented in Figures 8.4–8.6.

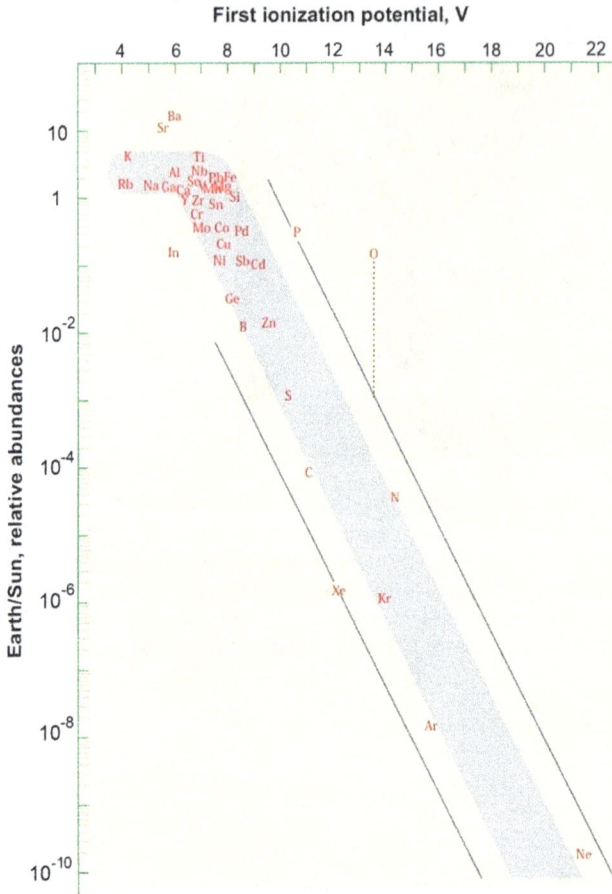

Figure 8.4: The Earth/Sun correlation (from [2]). Sun composition: Aller [15], Earth crust: Vinogradov [16], inert gases: Moroz [17].

Analysis of the graph in Figure 8.4 reveals that the carbon content on Earth is thousands of times higher than that on the Sun, while an apparent oxygen anomaly significantly deviates from the general trend. This anomaly raises critical questions: Why is oxygen the only element that falls outside the expected pattern? During the period when all matter was ionized, oxygen could not have been chemically bound in compounds, ruling out this as the cause. Instead, the anomaly is attributed to the purifying effect of hydrogen on the Earth's metallic mantle. Hydrogen, being highly effective at removing impurities from metals, played a key role in depleting the oxygen content from the deeper metallic layers.

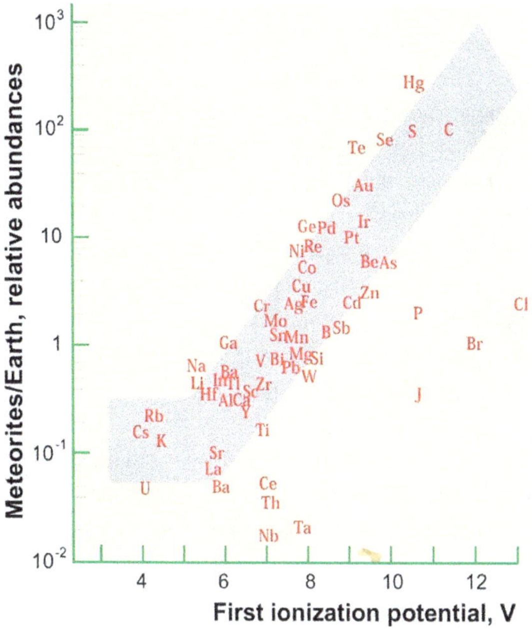

Figure 8.5: The meteorites/Earth correlation (from [2]). Meteoritic composition: Mason [18]'s crust: Vinogradov [16].

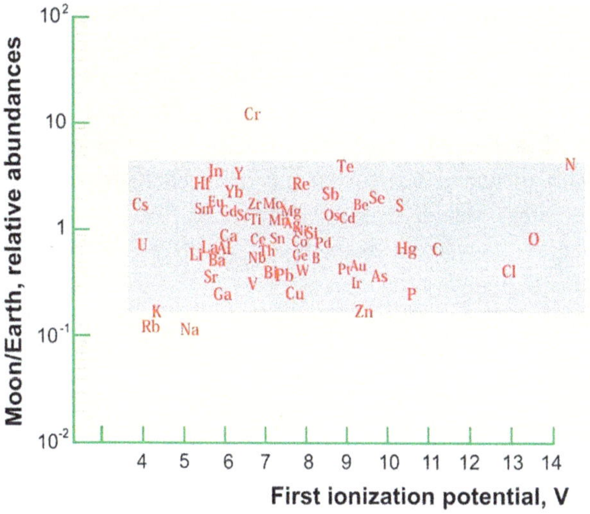

Figure 8.6: The Moon/Earth correlation (from [2]). The Moon: Samples from the Apollo 12 and Luna 16 missions.

As a result, the Earth's lithosphere – composed primarily of silicates and oxides – forms a thin "film" over the metallic mantle, with a thickness of up to 100 km (and as little as 3–5 km in the axial parts of the oceans). This distribution reflects the amount of oxygen allocated by the process of magnetic separation (refer Table 8.1). It is necessary to pay attention to the comparison of the compositions of meteorites with the Earth's outer crust (Figure 8.5). The asteroid belt, where the vast majority of stones falling onto the Earth originate, is three times farther from the Sun than our planet. Viewed through the lens of the magnetic separator prism, it becomes clear why, for instance, the mercury and carbon content in meteorites is two orders of magnitude higher than on Earth. Carbon is one of the lightest elements with an atomic mass of 12, while mercury is among the heaviest, with the atomic mass of 200. The idea of the solar wind, which postulates the distribution of chemical elements based on their mass, becomes irrelevant in this context.

When examining the Earth-Moon graph (Figure 8.6), it becomes evident that the idea of the Moon being "captured" is implausible. The magnetic separator worked equally for both, and it is most likely that we and our lunar neighbor are a pair of planets that split during the formation stage.

The graphs lead to a key conclusion: the distribution of chemical elements in the solar system depends on their first electron's ionization potential, known as 1IP. Any cosmogonic model must, at the very least, explain these observed regularities.

By analyzing the trend in Figure 8.4, we can infer Earth's initial composition as illustrated in Table 8.1. The primary petrogenic elements align roughly within an order of magnitude, but two significant deviations are observed: (1) an anomalously low oxygen content compared to commonly accepted models and (2) a hydrogen content that exceeds traditional estimates by two orders of magnitude. While the oxygen anomaly has been discussed, the hydrogen enrichment requires further explanation.

During the condensation of matter, the most refractory metallic "snowflakes" formed first and absorbed hydrogen due to its unique ability to dissolve into metals at volumetric ratios hundreds of times greater than the metal itself. These hydrogen-saturated metallic particles migrated toward the planet's center, where their compaction led to the formation of metal hydrides, forming the core of the nascent Earth. This concept, which underpins the PHE hypothesis, was first explored in foundational works [1, 19, 20].

8.1.3 A model of the modern Earth

The development of the PHE is a separate topic, detailed in the works of Larin. This section will attempt to provide a brief description of the evolution of the newborn planet:

When our planet is formed, it consisted of metal hydrides and was not hot. The concept of the "magma ocean," which suggests that planets were assembled from dust, rock, meteorite, and asteroid material, is rejected. The subsequent heating of the

Table 8.1: The primordial Earth: A comparison of calculated and "traditional,"
that is, generally accepted, chemical compositions (from [1]).

Element	Calculated (V. Larin)		Traditional	
	Atomic %	Weight %	Atomic %	Weight %
Si	19.5	45	14	15
Mg	15.5	31	16	14
Fe	2.5	12	15	32
Ca	0.9	3	1	1.5
Al	1.0	2	1.4	1.4
Na	0.7	1.5	0.1	0.1
Oxygen	0.6	1	49.8	30
C	0.03–0.3	0.03–0.3	0.08	0.04
S	0.01–0.1	0.03–0.3	0.0007	0.0004
Hydrogen	59	4.5	0.08	0.003

The "traditional" composition is from [24].

newborn planet occurred due to the decay of radioactive elements, which, at that time, were evenly distributed throughout the planet's volume.

The decomposition of hydrides occurred at relatively shallow depths, releasing hydrogen into space. As hydrides transformed into metals, they expanded in volume. This increase in volume led to cooling and a halt in the decomposition reactions of the hydrides. Radioactive decay would once again raise the temperature, initiating the cycle described previously and progressively increasing the volume of the metallosphere.

Each subsequent decomposition of hydrides caused hydrogen dissipation from the core through the metallic mantle. Here, it is essential to highlight a crucial point: hydrogen passing through metal can purify it of impurities, primarily oxygen, due to the exceptionally high binding energy of H–O [21–23].

This process led to the formation of the lithosphere, most of which had formed by the end of the Archean eon. Returning to Figure 8.4, the oxygen anomaly finds its explanation: all the oxygen, in the form of silicates and oxides, accumulated in the lithosphere as a result of hydrogen cleansing.

A comparison of two models – the initially hydridic Earth and the bulk silicate Earth – is presented in Figure 8.7.

New geochemical model

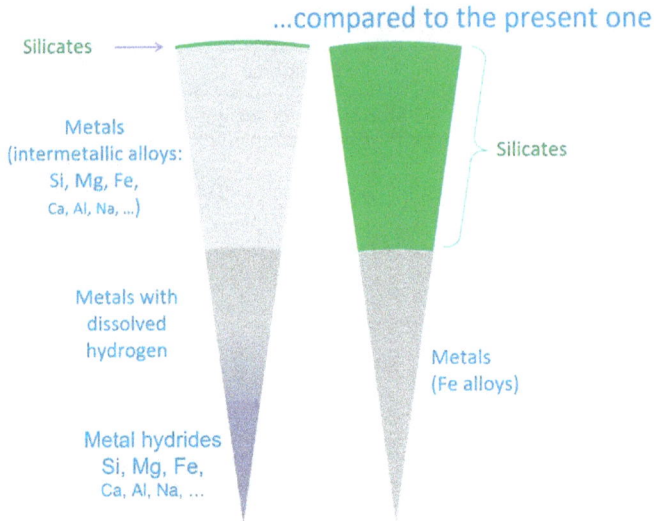

...compared to the present one

Silicates ──→

Metals
(intermetallic alloys:
Si, Mg, Fe,
Ca, Al, Na, ...)

Silicates

Metals with
dissolved
hydrogen

Metals
(Fe alloys)

Metal hydrides
Si, Mg, Fe,
Ca, Al, Na, ...

Figure 8.7: Comparison of geochemical models of Larin's Earth (left) and the conventional model (right).

8.2 Discussion

8.2.1 Overview of the current mainstream Earth formation and composition model challenges

The proofs of Larin's concept of the PHE continue to be published by multiple research groups [25–54]. The most illustrative examples are listed below in the text of the chapter.

For the purpose of this concept, we analyze a number of independent sources confirming the PHE concept as well as the deep-seated natural hydrogen generation model.

The first doubts about the classic all-iron core model emerged in the 1970s. The discrepancy between the calculated and observed density as well as the momentum of revolution, along with geophysical and electromagnetic (EM) data, called for a review of the model.

The PHE model challenges the existing mainstream model of Earth's core by proposing cold accretion, water in the form of primordial condensate rather than from chondrite meteorite bombardments, and a hydride core with conductivity potentially surpassing that of pure iron.

8.2.2 The Earth core composition

Stevenson [25] suggested that an alloy with a composition akin to FeH could account for the core's observed density. This proposed iron hydride would likely conduct electricity exceptionally well [26, 27]. Additionally, this alloy would have a lower melting point than iron, along with other characteristics that align with the current understanding of the Earth's core properties.

This statement correlates with the Earth's composition geochemical model suggested by Larin [1] through the PHE concept (Table 8.2).

Table 8.2: Geospheres' thicknesses and composition (from [1]).

Sphere	Depth interval	Composition
Lithosphere	0–150	Silicates and oxides
Metal-sphere	150–2,900	Alloys of silicon, magnesium, and iron compositions
Outer core	2,900–5,000	Metals with dissolved hydrogen and metal hydrides
Inner core	5,000–6,371	Metal hydrides

Walshe [28] concluded that the Earth's core likely serves as the primary reservoir for hydrogen. A heightened flow of hydrogen-bearing fluids is believed to activate the mantle and drive tectonic activity and metallogenesis over hundreds of millions of years.

According to the PHE concept, the inner core (IC) is composed of metal hydrides, with silicon and magnesium dominating (Table 8.2). The unusual compressibility of ionic hydrides suggests that at extremely high pressures (in the megabar range), hydride ions could become so compact that the lattice structure would consist of closely packed metal cations, with the significantly compressed hydride ions fitting into octahedral and tetrahedral voids between them. This structure provides a theoretical limit to how much ionic hydrides can be compressed. For instance, when magnesium transitions to Mg^{2+}, its atomic radius decreases from 1.6 to 0.66 Å (Figure 8.8), and for silicon transitioning from SiH_2 to Si^{2+}, the silicon atom diameter contracts from 1.34 to 0.55 Å. Under these conditions, the density of magnesium and silicon as ionic hydrides could theoretically increase up to 14 times when subjected to ultrahigh pressures.

Walshe [35] also suggested that chemically reducing anhydrous fluids, mainly consisting of H_2, CH_4, and H_2S are significant in forming Earth's large mineral deposits and provinces. He proposed that alkali-rich hydrogenic fluids, which contain notable amounts of alkali halides, could encourage alkali metasomatism by interacting with silicate melts or other aqueous or carbonic fluids in the crust and upper mantle. This process would generate reduced, alkali-enriched aqueous fluids, leading to reactions that are fast, non-reversible, and exothermic. Additionally, the stability of hydrides decreases with higher atomic numbers within a periodic group, favoring $Na \pm Li$ metasomatism

a) Normal pressure conditions

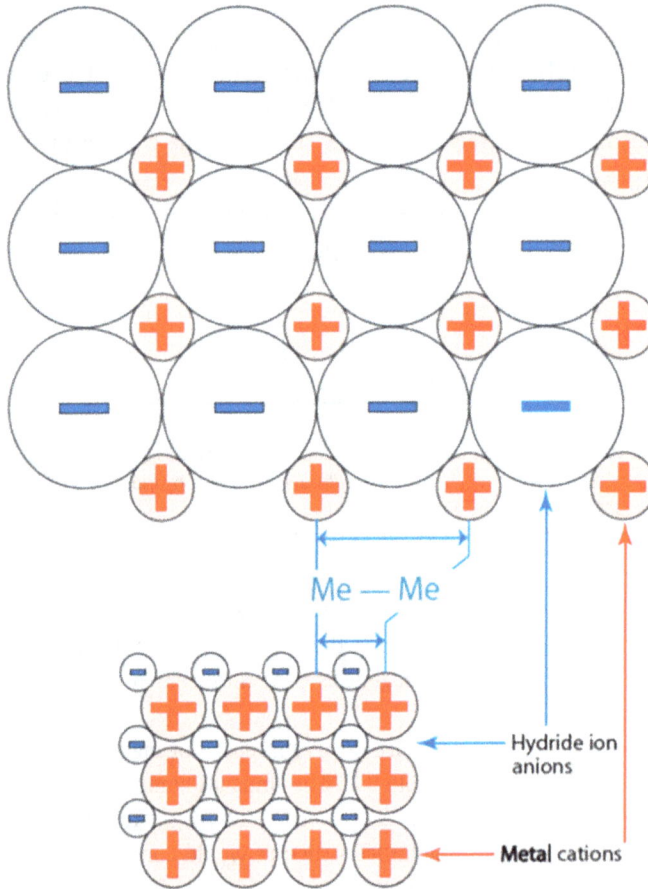

Me — Me

Hydride ion anions

Metal cations

b) Ultrahigh-pressure conditions

Figure 8.8: The transformation of the crystalline lattice of an ionic hydride under very high pressures: hydride-ion (−), metal (−), metal cation (+) (adapted from [2]).

over K metasomatism. At low pressures, hydrogen-rich fluids would likely transition into low-density HSCO fluids dominated by H_2, with traces of H_2S, CH_4, HCl, and CO.

This, in turn, correlates with Larin's PHE model for planet Earth [1, 2] (Figure 8.9), where advective hydrogen-based streams of fluids flow upward from the planet's interior toward the surface.

The Earth's core composition has been studied by several prominent groups of researchers, particularly those based in Japan. Tagawa [30] reached this conclusion after conducting a series of laboratory tests investigating hydrogen solubility in metals under extremely high pressures, performed using diamond anvils.

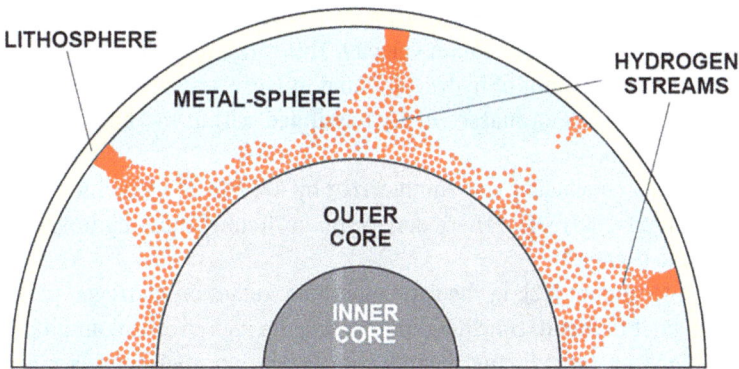

Figure 8.9: Hydrogen degassing from the core, composed of metal hydrides. Adapted from Larin [2].

The study experimentally examined the partitioning of hydrogen between molten iron and silicate melt at 30–60 GPa and 3,100–4,600 K, which are suggested to be the conditions present during the formation of the Earth's core. In particular, the researchers concluded that hydrogen is highly siderophile, or iron-loving, under these conditions. Core formation models suggest that around 0.3–0.6% by weight of hydrogen was integrated into the core, leaving only a minor amount of H_2O in the silicate mantle. This level of hydrogen content explains roughly 30–60% of the density deficit and the increased sound velocity observed in the outer core compared to pure iron.

The hydrogen content transferred from silicates to metals was estimated using three core formation models:

- Single-stage core formation model: This model proposes that core and mantle elements achieved chemical equilibrium in a single step at 50 GPa, 3,500 K, and under a specified oxygen fugacity condition. Though simplistic, it suggests that the core would contain between 0.32 and 0.61% (wt) H, with molten silicate containing approximately 687 ppm H_2O.
- Continuous core formation model: Here, core formation occurred through roughly 1,000 accretion events, with each impactor's core mixing with the magma ocean. This model implies that 0.3–0.6 wt% H in the core is necessary to maintain about 690 ppm H_2O in the bulk silicate Earth, provided water wasn't added solely during the final stage of accretion.
- Multi-stage core formation model: This model involves multiple stages, during which hydrogen transfer occurred in approximately 1,000 steps, with each impactor's metal core only partially equilibrating with the silicate melt at the base of the magma ocean. It suggests that the core would retain around 0.27–0.56 wt% H.

In conclusion, experimental data on hydrogen's behavior in metal-silicate partitioning suggest that the core likely contains 0.3–0.6 wt% H, leaving roughly 700 ppm H_2O in the early magma ocean unless most water arrived in the final stages of Earth's formation. This hydrogen content aligns with seismic data, accounting for 30–60% of the outer core's density deficit and higher sound velocity. This range (0.3–0.6 wt% H) corresponds to 37–73 times the amount of hydrogen found in Earth's oceans, with an additional approximately two ocean masses of H_2O assumed within the bulk silicate Earth, including the actual oceans.

The rather simplified conclusion was summarized by Kei Hirose in his interview with ScienceDaily (May 14, 2021) [31]: "There may be up to 70 times more hydrogen in Earth's core than in the oceans."

Yuan and Steinle-Neumann [32], in the course of their studies on hydrogen partitioning from silicate melt to metal, concluded that under the high-pressure and high-temperature conditions present during Earth's core formation, hydrogen is more readily absorbed by metal than by silicate. This makes it likely that the core contains a significant amount of hydrogen. Their research, which employed advanced quantum mechanical simulations on silicate and metallic melts, demonstrated that hydrogen becomes increasingly incorporated into metal rather than silicate under the high-pressure and high-temperature conditions in which Earth's core formed.

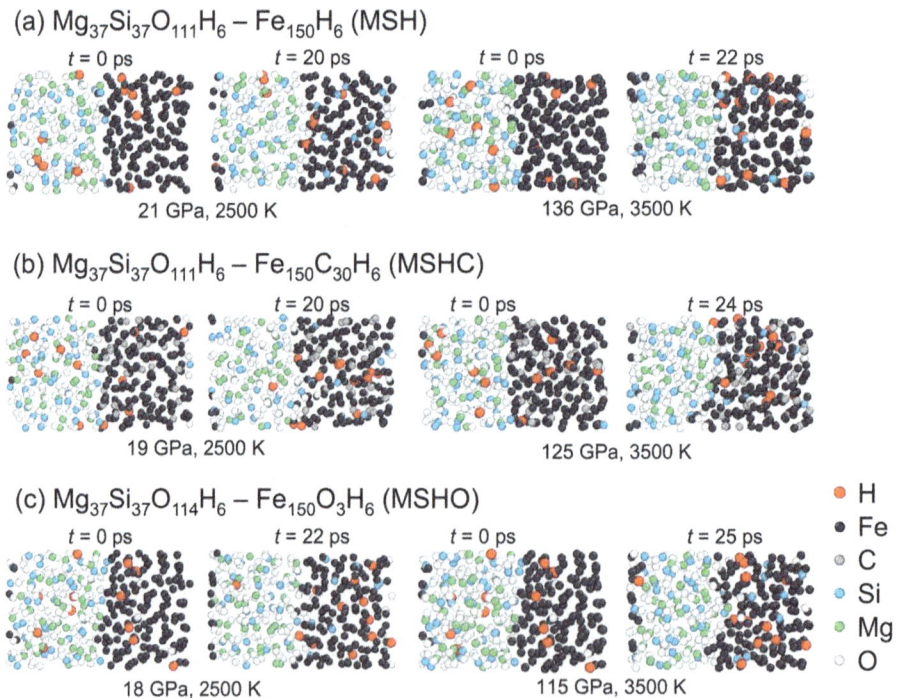

(a) $Mg_{37}Si_{37}O_{111}H_6 - Fe_{150}H_6$ (MSH)

(b) $Mg_{37}Si_{37}O_{111}H_6 - Fe_{150}C_{30}H_6$ (MSHC)

(c) $Mg_{37}Si_{37}O_{114}H_6 - Fe_{150}O_3H_6$ (MSHO)

Figure 8.10: Initial and final configurations for the two-phase simulations of silicate and metal liquids in contact (from Yuan and Steinle-Neumann [32]).

Two-phase simulations revealed a notable buildup of hydrogen in the metallic part of the simulation cell, quantified using an equilibrium constant, KD, determined through thermodynamic integration. Although there is some uncertainty, the results strongly support hydrogen's moderately siderophile behavior at low pressures (20 GPa and 2,500 K), aligning quantitatively with the experiments of Okuchi [33] and qualitatively with those of Clesi et al. [34] at similar pressures. Additionally, a pronounced rise in KD with increasing pressure suggests substantial hydrogen enrichment in the core-forming metal during magma ocean segregation.

The simulations also show that silicon is transferred from silicate to metallic liquid to a significant extent, while magnesium remains entirely within the silicate phase (Figure 8.10). This evidence supports a scenario in which silicon is incorporated into the Earth's core, potentially explaining the elevated Mg/Si ratio in the bulk silicate Earth compared to chondritic levels.

Under standard conditions, silicon and magnesium have densities of 2.33 and 1.74 g/cm^3, respectively. If these densities are scaled up by a factor of 14, they reach 32.62 g/cm^3 and 24.36 g/cm^3, surpassing the theoretical density of the Earth's IC at the planet's center (12.46 g/cm^3). This suggests that within a model of Earth with a hydridic IC, the extremely high core density isn't problematic. Instead, the issue is that the IC may lack sufficient density.

The core density subject was also addressed by the PHE in the following way: the above observation leads to a reconsideration of the density of the IC. Specifically, calculations indicate that the IC density might need to be increased to approximately 25 g/cm^3 (Figure 8.11). Additionally, to accurately match Earth's total moment of inertia, the asthenosphere density would need to be raised by about 0.2–0.25 g/cm^3, alongside adjustments to the outer core density distribution, while maintaining its total mass. Consequently, if there is a dense metallic layer beneath the lithosphere (with relatively low compaction gradients below 1,050 km), it would necessitate a significant increase in the planet's IC density.

In 2022, Wakamatsu et al. [35] performed X-ray diffraction (XRD) and picosecond acoustic measurements on iron hydrides synthesized in laser-heated diamond anvil cells at pressures up to 100 GPa and 300 K (Figure 8.12). Subsequently, the compressional wave velocities were determined, leading to the suggestion that a comparison between the preliminary reference Earth model and an extrapolation of Birch's law, which includes the temperature effects from the experimental V_p –ρ results, supports the hypothesis that hydrogen may be present in the Earth's IC.

According to Wakamatsu et al. [35], very recent studies claim that hydrogen in the *hcp* Fe lattice can transform into a superionic state under the conditions of the IC and exhibit V_S that is harmonic within the IC.

This suggestion was further studied by He et al. [36]. The conclusion concerned the superionic nature of the matter inside the IC. In particular, the density of the Earth's IC is lower than that of pure iron, suggesting that it contains lighter elements. These light elements, with liquid-like behavior, significantly reduce seismic velocities,

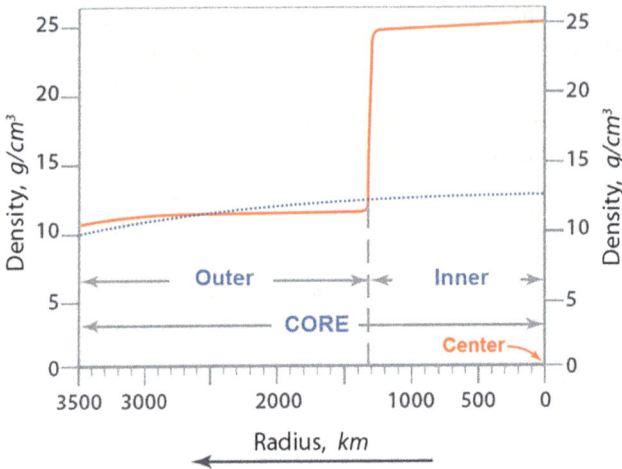

Figure 8.11: Density distribution in the planet's core: blue-dotted line – based on the traditional concept suggesting a mantle silicate composition; red line – according to the PHE model (from Larin [2]).

aligning with the seismological observations of the IC. The notable drop in shear-wave velocity helps explain the IC's soft nature. Furthermore, convection of light elements could impact both the seismological structure of the IC and the Earth's magnetic field (Figure 8.13).

This agrees with Larin's conclusion that within the IC, the mostly covalent bonds in hydrides between hydrogen and metals, such as iron, magnesium, and silicon, become more ionic due to extreme pressures. As pressure increases, the chemical bonding in hydrides is expected to become increasingly ionic, allowing the hydride ion's potential for compaction to be fully realized.

8.2.3 The role and dynamics of hydrogen within the D'-layer

For the geospheres situated further up and outward from the core, Lord et al. [37] resolved that the intermediate density of FeSi, which falls between that of the core and mantle, has led to its proposal as a component of the D'-layer (Figure 8.14), helping to explain the negative anomalies in compressional (V_p) and shear (V_s) velocities observed in this region.

The D'-layer was described in detail by Larin in his works [1, 2]. In particular, the substance of intermetallic diapirs is taken from the D'-layer immediately adjacent to the core (Figure 8.15) and, therefore, has quite recently been inside the planet's core. For this reason, it cannot immediately lose its initial oxygen content, as it has not been exposed to continuous purging with hydrogen like the older metal-sphere zones. However, it must also contain certain concentrations of hydrogen (since hydrogen always

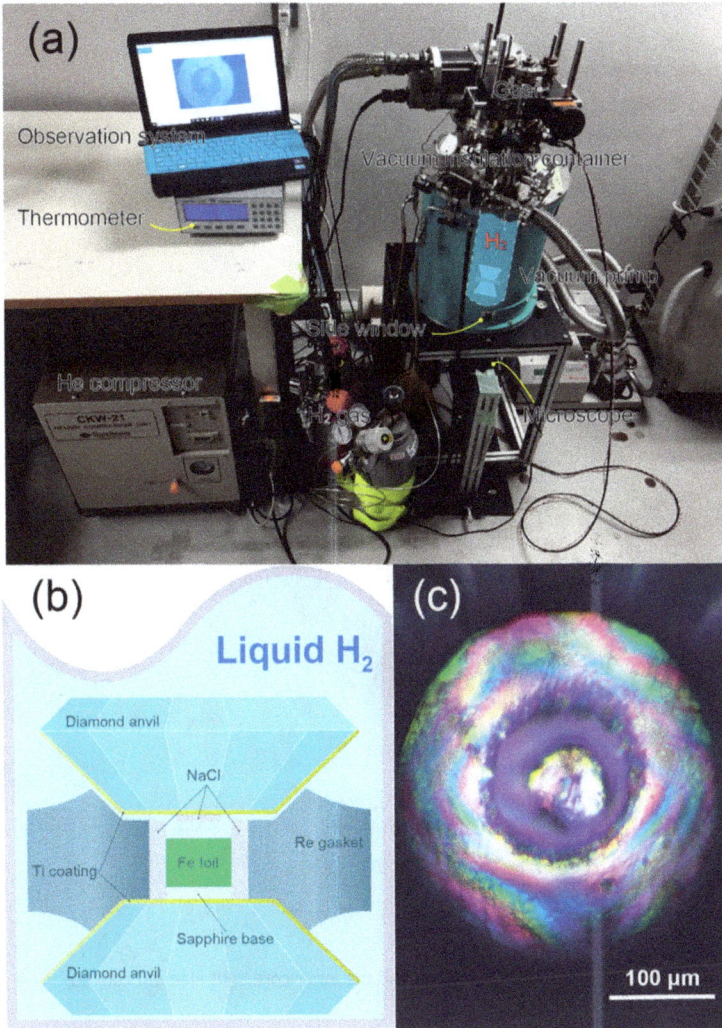

Figure 8.12: (a) A photograph of the liquid hydrogen loading system used in the study; (b) schematic illustration of the sample configuration inside the diamond anvil cell (DAC) during hydrogen loading at low temperatures; (c) microphotograph of the sample chamber just after recovery to room temperature from the liquid hydrogen loading system (from Wakamatsu et al. [35]).

exists in the D'-layer) effusing upward, inevitably causing oxygen to redistribute. As a result, in the diapir head part, intermetallic silicides have gradually been transformed into silicates due to "hydrogen purging" and oxygen removal from the deeper zones.

According to the PHE concept, the material of intermetallic diapirs protruding into the weakened areas of the crust toward the surface is sourced from the D'-layer, which is located directly on top of the core, meaning it has only recently been part of

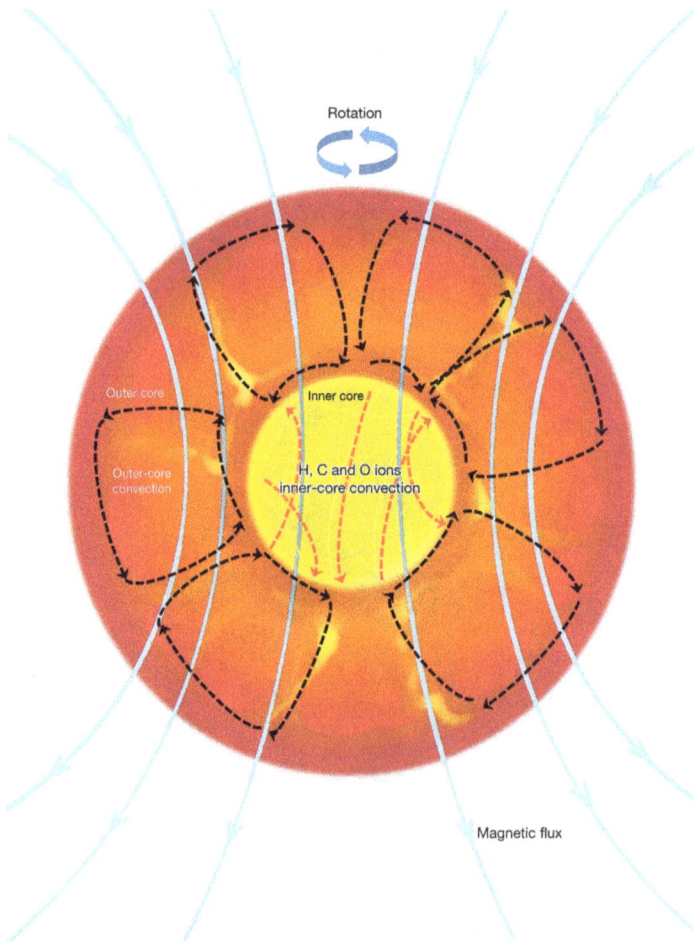

Figure 8.13: Schematic of the outer-core fluid convection and the inner-core light element convection (from He et al. [36]).

the planet's core. The metal-sphere known as the mantle is composed of intermetallic compounds and alloys formed by Si, Mg, and Fe, with the addition of Ca, Al, Na, and other metals. According to data available to Larin in the early 2000s, these pressure boundaries are found within the metal-sphere, with 90% of its volume composed of Mg_2Si FeSi, and metallic silicon. The relative abundance of these minerals can be expressed as a ratio: Mg_2Si:Si:FeSi = 6:3:1. Interestingly, the density of the Mg_2Si + Si + FeSi alloy in a 6:3:1 proportion is 2.64 g/cm^3, which matches the density of granite-gneisses in the upper levels of the continental crust. Therefore, if silicide diapirs protrude from below into the continental crust close to the surface, they should not be associated with high gravity anomalies.

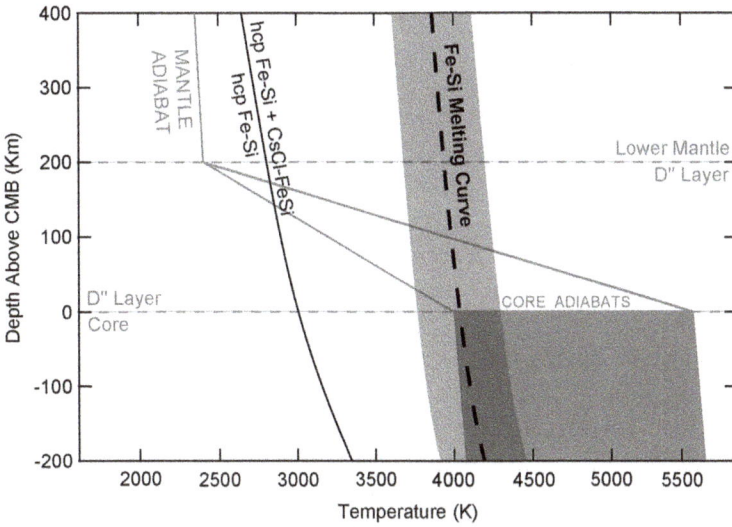

Figure 8.14: FeSi in the D'-layer. The solid gray line denotes the mantle adiabat, and the dark gray field represents the range of possible core adiabats; the two are connected by a simplified linear interpolation (dashed gray lines). The solid black line denotes the breakdown reaction hcp Fe-Si alloy → hcp Fe-Si alloy + CsCl-FeSi, and the heavy dashed black line represents the melting curve for FeSi from this study. The uncertainty in the melting curve is indicated by the light gray field (from Lord et al. [37]).

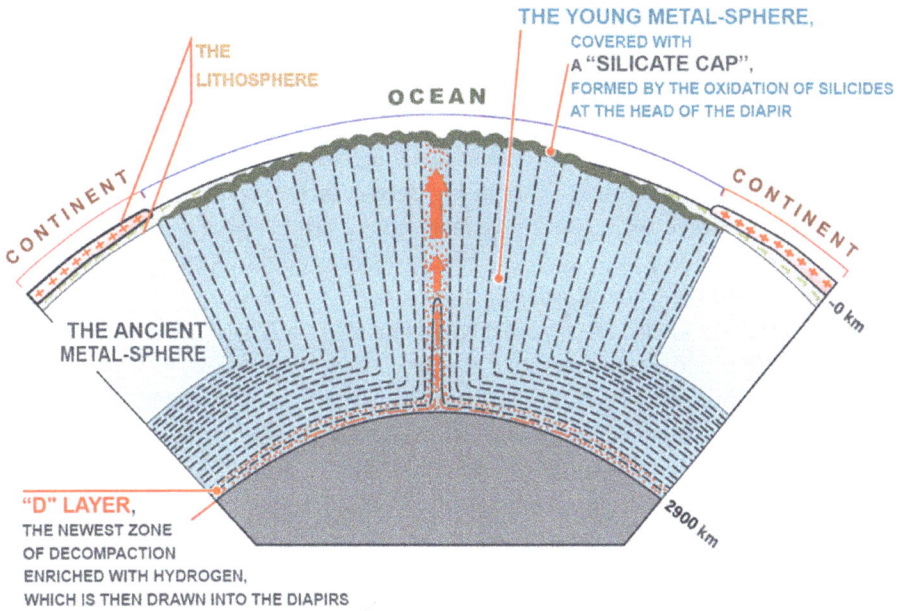

Figure 8.15: The deep structure of a modern ocean (adapted from Larin [2]).

8.2.4 Hydrogen in the asthenosphere

With regard to the asthenosphere, the diffusion rate of hydrogen in metals exceeds that in silicates and oxides by 6–7 orders of magnitude, making the lithosphere a barrier to hydrogen migration toward the Earth's surface. This barrier leads to hydrogen accumulation in the upper layers of the metasphere, just beneath the lithosphere. Given current insights into hydrogen-induced plasticity in metals, it stands to reason that the asthenosphere's flow would exhibit "viscous-plastic" behavior, promoting isostatic balance. This phenomenon is also responsible for reducing seismic wave velocities within the asthenosphere.

Contrary to common assumptions, the asthenosphere may remain "cold," with temperatures well below its melting threshold, as its capacity to flow doesn't necessitate high temperatures. Furthermore, the asthenosphere is unlikely to undergo decompaction; instead, it naturally gravitates toward compaction after its initial formation. It remains to be seen whether this compaction trend can be detected by geophysical methods. In the PHE model, the boundary where the silicate/oxide lithosphere ends and the asthenosphere begins signifies the top of the metal-sphere.

8.2.5 The Earth's mantle: the metal-sphere

The term "mantle" is well-established in Earth sciences and traditionally encompasses everything between the crust and the core. However, within the PHE concept framework, this term becomes somewhat limiting, as it consolidates geospheres of distinctly different compositions – namely, the silicate lithosphere's subcrustal layer and the metal-sphere beneath it.

The mantle composition was studied by Rohrbach et al. [38] and [39]. In 2007, Rohrbach [38] stipulated that, through high-pressure experiments, it was demonstrated that extensive regions of the asthenosphere are likely to be saturated with metals. Pyroxene and garnet synthesized at pressures above 7 GPa, in equilibrium with metallic iron (Fe), can incorporate enough ferric iron to suggest that the mantle at depths greater than 250 km is sufficiently reduced for a (Fe, Ni)-metal phase to remain stable. The findings indicate that the previously assumed oxidized state of the upper mantle does not accurately represent the entire upper mantle; instead, oxidation appears limited to a shallow layer only about 250 km thick. The overall oxygen fugacity (fO_2) trend in the mantle with depth appears to favor reduction, suggesting that not only the lower mantle and transition zone but also the deeper half of the upper mantle may be metal-saturated. A model composition enriched in FeO and depleted in olivine accurately captures reactions among natural mantle phases.

Using the mantle and core's highly siderophile element (HSE) abundances, a mass balance estimate suggests a (Fe, Ni)-metal phase concentration at depths of approximately 1,400 ppm. The upper mantle displays significant Ni and Co over-abundances, which, in absolute terms, are even more substantial than the HSE over-abundances.

If CH_4–H_2 fluids decompress, they can react with Fe_2O_3 released from majorite breakdown, producing CO_2 and H_2O, which lower the melting point and initiate redox melting. Recently discovered kimberlites exhibit reductions of up to 5 log units below the nickel–nickel oxide (Ni–NiO) buffer, approaching the saturation level for (Fe, Ni)-metal [38].

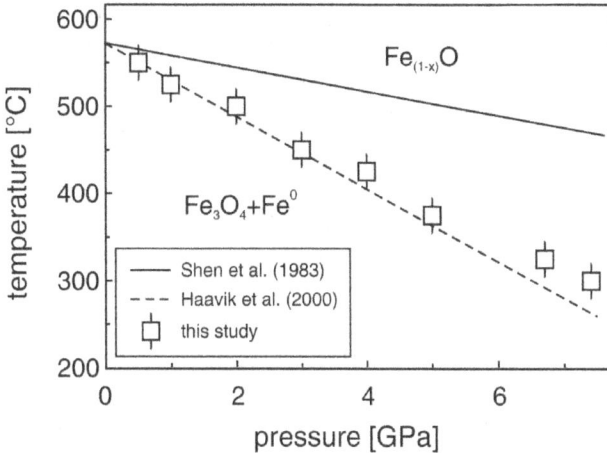

Figure 8.16: Magnetite + FeO and wustite phase boundaries in pressure-temperature space. The continuous line represents quench experiments, the dashed line corresponds to thermodynamic calculations, and the symbols illustrate the position of the phase boundary as determined by in situ X-ray diffraction. The univariant equilibrium (i.e., $4FeO = FeO + Fe_3O_4$) has a distinctly negative slope in pressure-temperature space, suggesting that Fe metal + wustite is the high-temperature assemblage [39].

In 2011, Rohrbach et al. [39] arrived at the more specific conclusion that experimental results suggest that, globally, an oxidized upper mantle in equilibrium with fayalite, ferrosilite, and magnetite (Figure 8.16) represents an exception rather than the norm. It is likely that over 75% of Earth's current mantle volume is saturated with metallic iron.

8.2.6 The PHE concept in nuclear physics: neutrino

A team of physicists studying terrestrial neutrinos was searching for a model of the Earth that could be utilized as a scientific basis to explain their observations. At one of the seminars, the head of the group, Bezrukov, explained that they had been sorting through one model after another, with no satisfactory result. Larin's model of the Earth's composition turned up next, and the group expected it to be rejected as well.

But instead, the PHE model was found to be the only applicable one and was eventually adopted as a basis.[2]

Bezrukov et al. [48] concluded that the hydride Earth model's projections regarding geoneutrino flux and the intrinsic heat flux of the Earth were examined in detail. The model's predicted geoneutrino flux demonstrates a capacity for alignment with empirical observations. However, the intrinsic Earth heat flux estimated by the model substantially exceeds the experimentally derived values, which are contingent on the assumption that thermal conductivity serves as the primary mechanism for heat transfer. To address this discrepancy, the research team proposed an alternative heat transfer mechanism within the Earth's crust: the conveyance of energy via hot gases generated at considerable depths. This hypothesis is bolstered by experimental evidence, including temperature profiles recorded in the Kola Superdeep Borehole, which were analyzed to support the proposed mechanism.

The PHE model serves as a robust framework for analyzing terrestrial phenomena, offering a unified approach to interpreting complex events. Within this model, we have demonstrated that diverse and intricate observations – such as geoneutrino fluxes, temperature profiles from super-deep boreholes, and the experimentally documented rise in ocean temperatures – can be coherently explained through a single, consistent theoretical foundation. This integrative capability underscores the model's explanatory power and versatility.

The PHE model posits substantial concentrations of ^{238}U, ^{232}Th, and ^{40}K within the Earth, suggesting their primordial presence in the Earth's core, adjusted for radioactive decay over time. Notably, potassium exhibits a particularly significant abundance compared to predictions from the bulk silicate Earth model, a discrepancy attributed to its relatively low ionization potential. This finding highlights the model's ability to reconcile geochemical and geophysical data with theoretical expectations.

A critical assumption of the PHE model is that current estimates of Earth's thermal flux, derived from temperature gradient measurements, fail to account for a substantial portion of heat transported by hot gases. The model predicts the generation of such gases at considerable depths, proposing an additional heat transfer mechanism beyond conventional thermal conductivity. This hypothesis provides a plausible explanation for the anomalously high temperatures observed at great depths in both continental and oceanic boreholes, offering new insights into Earth's thermodynamics.

---

2 From the personal comms between Dr. L. Bezrukov and Dr. V. Larin.

8.2.7 The attempt to evaluate the volumes of hydrogen degassed from the Earth's depths

The subject of the Moon slowly creeping away from Earth is quite popular in the media [49, 50]. This departure is estimated to be around 38 mm per annum.

It is generally accepted that the Earth's mass increases due to meteorite material falling on its surface and, at the same time, decreases due to the loss of hydrogen and helium. Moreover, until recently, it was believed that hydrogen leaving the atmosphere is formed due to the decay of water molecules under the influence of solar radiation. Quantitative estimates of the Earth's mass loss vary in the range of 50–100 thousand tons per year [51]. On the other hand, the quantity of cosmic dust falling to Earth is estimated to be between approximately 800 and 2,300 tons per year [52]. If we arm ourselves with ideas about the degassing of hydrogen from the bowels of the planet, we can attempt to re-evaluate the scale of this phenomenon.

To simplify the calculation, we will assume that the mass of the Earth decreases while the mass of the Moon, which has long since lost its atmosphere, remains unchanged.

Centrifugal force of the Moon: $F_1 = mv^2/R$

Force of attraction between the moon and earth: $F_2 = GMm/R^2$

Condition for the Moon to be in Earth's orbit: $F_1 = F_2$

$$mv^2/R = GMm/R^2$$

Now, let's examine how the decrease in the mass of the Earth is connected to the increase in the radius of the orbit of its satellite – the Moon. To achieve this, we transform the above equation into the following form:

$$\Delta M = v2\Delta R/G$$

and substitute reference data into it. The increase in the radius of the Moon's orbit, $\Delta R = 0.038$ m/year, the orbital velocity of the Moon, $v = 1{,}023$ m/s, and the gravitational constant

$$G = 6.674 \times 10^{-11} \, m^3/s^2 kg$$

The resulting value is impressive:

$$\Delta M = 5.96 \times 10^{14} \, kg/year$$

It turns out that, to explain the Moon's removal, the Earth must "lose weight" by 596 billion tons annually. (It should be noted that the calculations presented above are an "upper estimate" and do not take into account factors such as mass losses due to helium degassing processes, the tidal influence of the Moon, and other considerations.)

8.2.8 The expanding Earth model as an integral part of the PHE concept

Larin's model of Earth suggests that the planet has undergone a significant increase in volume throughout its geological history, attributed to the anomalously high compressibility of hydrides compared to metals. The decomposition of hydrides at the boundary between the hydride core and the metallic mantle is proposed to result in volume expansion. Over time, this process has led to an estimated 1.7-fold increase in Earth's radius, corresponding to a fivefold increase in volume. While this conclusion contrasts sharply with the traditional model of an iron core and silicate mantle, additional studies provide data that may support these findings. A group of researchers, employing space geodetic observations and gravimetric data, concluded that Earth's expansion rate is approximately 0.2 mm/year over recent decades [47]. Furthermore, Vanyo and Awramik [53] reported studies analyzing ancient stromatolites to interpret Earth's past rotational dynamics. Their findings indicate that during the late Proterozoic, there were a minimum of 410 days per year, with extrapolated values ranging between 410 and 485 days.

A recent publication [54] consolidates data from multiple sources to demonstrate an increase in Earth's day length over the past 2.5 billion years (Figure 8.17). This evi-

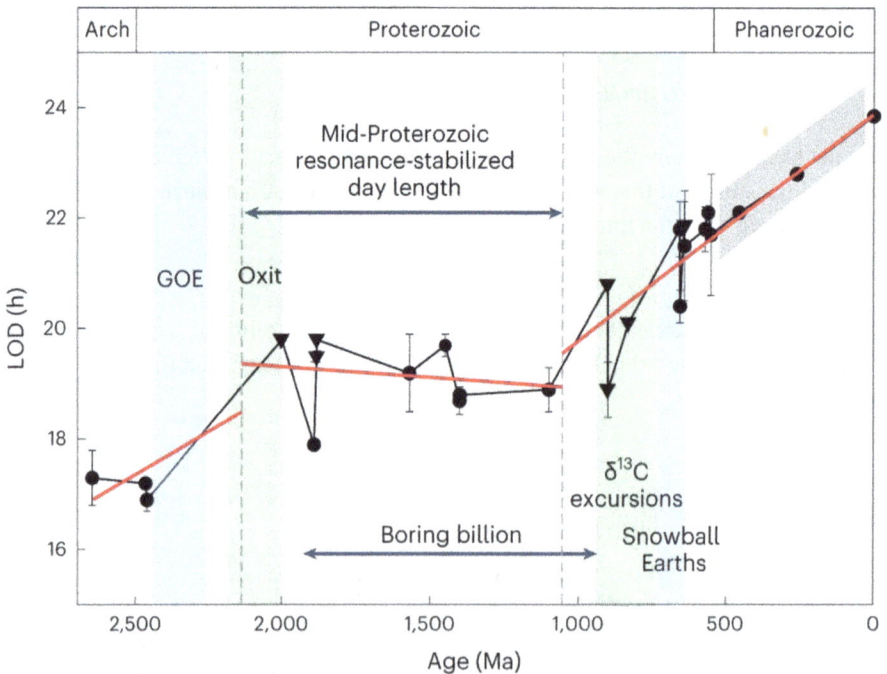

Figure 8.17: LOD constraints and trends [54].

dence strongly supports the conclusion that planetary expansion alters Earth's moment of inertia, leading to a gradual slowing of its rotation. Consequently, this results in fewer days per year and a progressive increase in the length of each day.

8.3 Conclusions

It is stipulated that the existing mainstream cosmogonic model may and shall be challenged. A different, alternative mechanism of Solar System formation is proposed. This mechanism, named by Larin as the PHE concept, suggests purely scientific – that is, chemical and physical – fundamental processes at the foundation of the chemical elements' distribution across the Solar System. For this purpose, a very well-known parameter, 1IP, or the first electron standard potential, is introduced and explained.

As a result, a new chemical composition of the planet Earth is proposed, with hydrogen playing a much more significant role in it: 59% (atom count) or 4.5 wt%. Consequently, it is concluded that hydrogen has been preserved in the Earth's core in the form of metal hydrides. This, among other things, resolves the long-standing issue of the Earth's core density deficit.

The composition and physical properties of the other deep geospheres, such as the D'-layer, the mantle, and the asthenosphere, are explained in new terms and supported by calculations and evidence.

Other aspects of the PHE concept, such as the expanding Earth model, are also reviewed and assessed through the lens of this new model of the chemical elements' original distribution as well as that of the planet's formation.

All of the arguments are supported by independent research – both in the laboratory and in the field (geophysical) [25–54].

The PHE concept presents a transformative perspective on Earth's formation, composition, and evolution. By proposing a hydrogen-rich planetary model, the PHE concept challenges traditional cosmogonic theories and offers an alternative explanation for the distribution of chemical elements across the solar system. Key postulates include the preservation of hydrogen in Earth's core as metal hydrides, resolving discrepancies in core density and conductivity, and the central role of hydrogen in the planet's dynamic processes.

The model integrates experimental and observational evidence, including the influence of hydrides on planetary expansion, hydrogen migration through Earth's geospheres, and the effects on geophysical phenomena such as tectonics and isostatic balance. Furthermore, the PHE concept reinterprets crustal hydrogen-generation mechanisms.

Through this chapter, the proposed model has been substantiated with geophysical, chemical, and theoretical data. While it disrupts established paradigms, it invites further investigation and critical evaluation to validate its claims. By challenging mainstream assumptions, the PHE concept opens new pathways for understanding Earth's origin and evolution, fostering a broader scientific discourse.

References

[1] Larin V. N. Hydridic earth: The new geology of our primordially hydrogen-rich planet. Calgary, Alberta, Canada: Polar Publishing, 1993.

[2] Larin V. N. Our earth. Moscow, Russia, Agar, 2005, Translation 2020, unpublished.

[3] Zwicky E. On the masses of nebulae and of clusters of nebulae. The Astrophysical Journal. 1937, 86: 3.

[4] Hoyle F. On the origin of the solar nebula. Quarterly Journal of the Royal Astronomical Society. 1960, Vol. 1, pp. 28–55.

[5] Hoyle F. The intelligent universe. New York: Holt, Rinehart and Winston, 1984.

[6] ALMA image of the protoplanetary disc around HL Tauri. European Southern Observatory, 2014 (Accessed December 4, 2024, at https://www.eso.org/public/images/eso1436a/).

[7] "The Hypothesis": The educational movie about Dr. V. Larin and the PHE Concept, 1984 (Accessed December 4, 2024, at https://youtu.be/ebrxLNfotBA).

[8] Eneev T. M., Kozlov N. N. A model of the accretion process in the formation of planetary systems. Astronomical Bulletin. 1981, 15: 2. (Rus.).

[9] Ohashi N., Tobin J. J., Jørgensen J. K., et. al. Early planet formation in Embedded Disks (eDisk). I. Overview of the program and first results. The Astrophysical Journal. 2023, 951: 8.

[10] Lin Z.-Y. D., Li Z.-Y., Tobin J. J., et. al. Early planet formation in Embedded Disks (eDisk). II. Limited dust settling and prominent snow surfaces in the edge-on class I disk IRAS 04302+2247. The Astrophysical Journal. 2023, 951: 9.

[11] Van 't Hoff M. L. R., Tobin J. J., Li Z.-Y., et. al. Early planet formation in Embedded Disks (eDisk). III. A first high-resolution view of submillimeter continuum and molecular line emission toward the class 0 protostar L1527 IRS. The Astrophysical Journal. 2023, 951: 10.

[12] Yamato Y., Aikawa Y., Onashi N., et. al. Early planet formation in Embedded Disks (eDisk). IV. The ringed and warped structure of the disk around the class I protostar L1489 IRS. Astrophysical Journal. 2023, 951: 11.

[13] Kido M., Takakuwa S., Saigo K., et. al. Early planet formation in Embedded Disks (eDisk). VII. Keplerian disk, disk substructure, and accretion streamers in the class 0 protostar IRAS 16544-1604 in CB 68. 2023, The Astrophysical Journal, Volume 953, p. 190 (22pp), Number 2. doi: 10.3847/1538-4357/acdd7a.

[14] ALMA digs deeper into the mystery of planet formation. Phys.org. 2023. Accessed December 3, 2024, at https://phys.org/news/2023-06-alma-deeper-mystery-planet-formation.html)

[15] Aller L.H. The abundance of the elements. 1961, New York, Interscience Publishers. 306pp.

[16] Vinogradov A. P. Differentiation of lunar material. Cosmochemistry of the moon and planets. In: Proceedings of the soviet-american conference on cosmochemistry of the moon and planets. Moscow, Nauka, 1975, pp. 5–28. Rus.

[17] Moroz VI. Physics of planets. Moscow, Nauka, main editorial office for physics and mathematics literature. 1967. 412 pages Rus.

[18] Mason B. Handbook of elemental abundances in meteorites. 1971. Gordon and Breach, 555 pages, ISBN-10: 0677149506, ISBN-13: 978-0677149509.

[19] Larin V. N. Hypothesis of a primordially hydridic Earth. Moscow, Nedra, USSR, Academy of Science of the USSR, the Ministry for Geology of the USSR, IMGRE, 1975. 101pp. (Rus.).

[20] Larin V. N. Hypothesis of a primordially hydridic earth, 2nd ed., revised and supplemented. Moscow, Nedra, 1980, Rus.

[21] Wolfe T. A., Jewett T. J., Gaur R. P. S. 1.06 – powder synthesis. In: Sarin V. K., ed., Comprehensive hard materials. Elsevier, Oxford, 2014 pp. 185–212. doi: https://doi.org/10.1016/B978-0-08-096527-7. 00006-4.

[22] M. S. Moats and W. G. Davenport, "Nickel and Cobalt Production," Treatise on Process Metallurgy, vol. 3, pp. 625–669, Elsevier Ltd., Jan 2014.

[23] Wu Z., Yang H., Wu L., et. al. Research progress in preparation of metal powders by pressurized hydrogen reduction. International Journal of Hydrogen Energy. 2021, 46: 35102–35120.

[24] Morgan J. W., Anders E. Chemical composition of Earth, Venus, and Mercury. Proceedings of the National Academy of Sciences of the United States of America. 1980, 77: 12.

[25] Stevenson D. J. Hydrogen in the Earth's core. Nature, 1977, Vol. 268, pp. 130–131.

[26] Stevenson D. J., ASHCROFT N. W. Conduction in fully ionized liquid metals. Physical Review A. 1974. Vol. 9. No. 2. pp. 782–789.

[27] Kvashnin A. G., Kruglov I. A., Semenok D. V., Oganov A. R. Iron superhydrides FeH5 and FeH6: Stability, electronic properties and superconductivity. Journal of Physical Chemistry C. 2018, 122: 4731–4736.

[28] Walshe J. L., Hobbs B., Ord A., Regehauer-Lieb K., Barnicoat A. Mineral systems, hydridic fluids, the Earth's core, mass extinction events and related phenomena. Chapter 1–17 in book: Mineral Deposit Research: Meeting the Global Challenge, 2005, pp. 65–68.

[29] Walshe J. L., Hobbs B. E., Ord A., Regenauer-Lieb K., Barnicoat A. C., Hall G. C. Hydrogen flux from the Earth's core giant ore deposits and related phenomena through Earth history. 2004. Goldsmith Abstracts, THEME 6: THE EARLY EARTH AND PLANETS. 6.4 Early giant ore bodies, 6.4.P07.

[30] Tagawa S., Sakamoto N., Hirose K., et. al. Experimental evidence for hydrogen incorporation into Earth's core. Nature Communications. 2021, 12: 2588.

[31] Where on Earth is all the water? ScienceDaily, 2021 (Accessed November 27, 2024, at https://www. sciencedaily.com/releases/2021/05/210514134102.htm).

[32] Yuan L., Steinle-Neumann G. Strong sequestration of hydrogen into the Earth's core during planetary differentiation. Geophysical Research Letters. 2020, 47. https://agupubs.onlinelibrary. wiley.com/doi/10.1029/2020GL088303

[33] Okuchi T. Hydrogen partitioning into molten iron at high pressure: Implications for Earth's core. Science. 1997, Vol. 278, pp. 1781–1784.

[34] Clesi V., Mohamed B. A., Nathalie B., et. al. Low hydrogen contents in the cores of terrestrial planets. Science Advances. 2018, 4: e1701876.

[35] Wakamatsu T., Ohta K., Tagawa S., Yagi T., Hirose K., Ohishi Y. Compressional wave velocity for iron hydrides to 100 gigapascals via picosecond acoustics. Physics and Chemistry of Minerals. 2022, 49: 17. doi: 10.1007/s00269-022-01192-8

[36] He Y., Sun S., Kim D. Y., Jang B. G., Li H., Mao H. Superionic iron alloys and their seismic velocities in Earth's inner core. Nature. 2022, 602: 258–262. https://doi.org/10.1038/s41586-021-04361-x

[37] Lord O. T., Walter M. J., Dobson D. P., Armstrong L., Clark S. M., Kleppe A. The FeSi phase diagram to 150 GPa. Journal of Geophysical Research Atmospheres. 2010, 115: B6. https://doi.org/10.1029/ 2009JB006528

[38] Rohrbach A., Ballhaus C., Golla-Schindler U., Ulmer P., Kamenetsky V. S., Kuzmin D. V. Metal saturation in the upper mantle. Nature 2007, Vol. 449(7161), pp. 456–458. doi: 10.1038/nature06183. PMID: 17898766.

[39] Rohrbach A., Ballhaus C., Ulmer P., Golla-Schindler U., Schonbohm D. Experimental evidence for a reduced metal-saturated upper mantle. Journal of Petrology. 2011, 52: 4. https://doi.org/10.1093/pe trology/egq101

[40] Benson W. N. The Origin of serpentine, a historical and comparative study. **American Journal** of **Science** – Fourth Series. 1918, XLVI: 276.

[41] Hess H. H. The problem of serpentinization and the origin of certain chrysotile asbestos talc and soapstone deposits. Society of Economic Geologists, Inc. Economic Geology. 1933, Vol. 28 (I7), pp. 634–657. https://doi.org/10.2113/gsecongeo.28.7.634.

[42] Iishi K., Saito M. Synthesis of antigorite. American Mineralogist. 1973, 58. Issue 9 - 10, pp. 915–919. Mindat Ref. ID 526140

[43] Drits V. A., Slonimskaya M. V., Stepanov S. S., Yurkova R. M., Dayniyak B. A. On the role of recovered fluids in the processes of ultrabasite serpentinization. GIN AN USSR, Lithology and Mineral Resources. Volume 5, 1983, pp. 102–113. ISSN 0024-497X (Rus.).

[44] Yurkova R. M. Mineral transformations of the ophiolite and associated volcanic-sedimentary complexes in the northwestern pacific fringing. Academy of Sciences of the USSR, Order of the Red Banner of Labour Geological Institute. Transactions. 1991, Vol. 464. ISSN 0002-3272, UDK 552.163.321 (Rus.).

[45] Murray J., Clement A., Fritz B., Schmittbuhl J., Bordmann V., Fleury J. M. Abiotic hydrogen generation from biotite-rich granite: A case study of the Soultz-sous-Forets geothermal site, France. Applied Geochemistry. 2020, 119: 104631.

[46] Pandit D. Chloritization in Paleoproterozoic granite ore system at Malanjkhand, Central India: Mineralogical studies and mineral fluid equilibria modelling. Current Science. 2014, 106: 4.

[47] Shen W. B., Sun R., Chen W. The expanding earth at present: Evidence from temporal gravity field and space-geodetic data. Ann Geophysics. 2011, 54: 4.

[48] Bezrukov L. B., Kurlovich A. S., Et. Al L. B. K. How Geoneutrinos can help in understanding of the Earth heat flux. Journal of Physics: Conference Series. 2017, Ser. 934 012011. DOI: 10.1088/1742-6596/934/1/012011.

[49] How the Moon is making days longer on Earth. BBC. 2024, Accessed January 16, 2025, at. https://www.bbc.com/future/article/20230303-how-the-moon-is-making-days-longer-on-earth)

[50] Will Earth ever lose its Moon? Live Science. 2023, Accessed January 16, 2025, at. https://www.live science.com/space/the-moon/will-earth-ever-lose-its-moon)

[51] Earth losses 50000 metric tons of mass every year. SciTechDaily. 2012, Accessed January 16, 2025, at. https://scitechdaily.com/earth-loses-50000-tonnes-of-mass-every-year/)

[52] Suttle M. D., Folco L. The extraterrestrial dust flux: Size distribution and mass contribution estimates inferred from the Transantarctic Mountain (TAM) micrometeorite collection. Journal of Geophysical Research: Planets. 2020, 125: e2019JE006241.

[53] Vanyo J. P., Awramik S. M. Stromatolites and earth – Sun – Moon dynamics. Precambrian Research. 1985, 29: 1–3.

[54] Mitchell R. N., Kirscher U. Mid-Proterozoic day length stalled by tidal resonance. Nature Geoscience. 2023, Vol. 16(7), pp. 567–569. https://doi.org/10.1038/s41561-023-01202-6.

Part III: **Hydrogen exploration and detection techniques**

Yashee Mathur and Tapan Mukerji

Chapter 9
Rock physics for quantitative geophysical interpretation of natural hydrogen resources

Abstract: Natural hydrogen systems formed through subsurface serpentinization of ultramafic rocks are an emerging focus area for low-carbon hydrogen production. Serpentinization, involving the hydrolysis of ferromagnesian minerals, alters rock properties, including density, elastic moduli, seismic velocities, and magnetic susceptibility, making it pivotal for geophysical exploration and characterization of subsurface hydrogen reservoirs. While geochemical insights into serpentinization are extensive, the geophysical interpretation using rock physics remains underexplored.

This study compiles over 1,000 samples from the literature from diverse geological environments to analyze physical property changes during serpentinization such as density and velocity reduction, and moduli variations. Regression models and cross-property relationships are developed to quantify these changes and link them to the degree of serpentinization. The findings reveal a predictable density decline (~3.3 g/cc for peridotites to ~2.6 g/cc for serpentinites), a significant velocity reduction (~40% for P-wave and ~60% for S-wave velocities), and distinct trends in moduli. These relationships are validated using differential effective medium (DEM) models, highlighting their utility in geophysical monitoring and exploration. Furthermore, cross-property plots, such as density versus magnetic susceptibility and velocity versus porosity, elucidate the interplay of rock composition, fluid presence, and porosity changes.

This comprehensive dataset and analysis enable the application of gravity, seismic, and electromagnetic methods to image hydrogen source rocks and reservoirs, advancing exploration strategies. The study also underscores the importance of integrating multi-physics approaches to mitigate interpretation pitfalls, offering a robust framework for natural hydrogen prospecting and real-time monitoring of stimulated hydrogen systems.

Keywords: rock physics, geophysics, natural hydrogen, stimulated hydrogen, rock properties

9.1 Introduction

Natural hydrogen is generated through various geological and geochemical processes in the sub-surface across a variety of geological environments, including the deep Earth [1–6]. Serpentinization of iron-rich rocks is one of the notable reactions respon-

Yashee Mathur, Tapan Mukerji, Department of Energy Science and Engineering, Stanford University

https://doi.org/10.1515/9783111437040-009

sible for the occurrence of natural hydrogen in the subsurface [7, 8] among others [3, 9]. The term "serpentinization" generically describes the hydrolysis and transformation of primary ferromagnesian minerals such as olivine ($(Mg,Fe)_2SiO_4$) and pyroxenes ($(Mg,Fe)SiO_3$), which may produce serpentine group minerals that include lizardite, chrysotile, and antigorite; magnetite (Fe_3O_4); Ni-Fe alloys; talc; chlorite; tremolite/actinolite; and the magnesium mineral brucite [10]. Serpentinization occurs in hydrothermal vents, mid-oceanic ridges, ophiolitic complexes, passive margins, subduction zones, forearc basins [3], and practically anywhere water and iron-rich rocks interact [11, 12]. Serpentinization is an exothermic reaction [12], and more than 1 Mt of H_2 is produced from various water-rock reactions [13]. The extent of hydrogen production during serpentinization is governed by various factors, such as the composition of the protolith, thermodynamic conditions, temperature, pressure, salinity, pH, water-rock ratio, and the presence of specific trace metals [11, 14–20]. Alongside natural serpentinization reactions, iron-rich rocks can also be geo-engineered chemically, biologically, mechanically, or thermally to produce hydrogen, which is termed "stimulated hydrogen" [12, 21].

During serpentinization, the rheology and physical properties of rocks, such as density, magnetic susceptibility, seismic velocity, electrical conductivity, and volume, change due to the formation of new mineral assemblages [22–27]. Since serpentinization is one of the major reactions occurring in mantle rocks, the changes in physical properties have been measured and studied extensively to understand the composition of the mantle, Earth's deep water cycle, and plate tectonics [28–31]. Thus, although the interest in serpentinization and ultramafic rocks from a natural hydrogen standpoint is recent, the data related to these changes has been available and studied for decades. Moreover, along with natural hydrogen prospectivity in recent times, ultramafic rocks have also been studied for metal resource exploration and carbon sequestration [12, 32–35].

The use of geophysical and rock physics properties for resource exploration has been carried out for a long time and has helped tremendously in identifying oil and gas resources, providing a link between fluid and rock reservoir properties, elastic properties, and seismic data [36–39]. For natural hydrogen, since the protolith that reacts to produce hydrogen has very different physical properties than the products, the physical property changes can be measured using gravimetric, aeromagnetic, or seismic imaging to further exploration [40–44]. It is argued that understanding the serpentinization of source rocks is an indirect way to identify hydrogen reservoirs as well. Moreover, for stimulated hydrogen production, real-time monitoring of the change in physical properties in comparison to background measurements can be done as a proxy to determine the occurrence and extent of serpentinization, as well as hydrogen generation. Although there is extensive literature available on changes in geochemical composition during serpentinization and controls on hydrogen generation, there exists very limited literature on using rock physics [40, 41, 43, 45, 46] and geophysics to explore natural hydrogen resources or monitor stimulated hydrogen systems.

In the current study, we compile more than 1,000 samples from the literature from different geological environments [22–24, 29, 47–57] to understand the changes in the physical properties of rocks during serpentinization. We analyze the effects of serpentinization on density, magnetic susceptibility, velocities, elastic properties, and porosities. Furthermore, we provide regression relationships to quickly ascertain the degree of serpentinization given physical property measurements or vice versa. We compare these relationships with widely used existing regression relationships between different physical properties. Moreover, we provide cross-property relationships between density-magnetic susceptibility, density-porosity, and velocity-porosity to delve deeper into the causes of changes in these properties. Finally, since physical properties come with their own non-uniqueness, we discuss common interpretation pitfalls. This is the first such holistic compilation and analysis of physical properties during serpentinization pertaining to natural hydrogen exploration.

9.2 Material and methods

From the literature, we collected data from different locations throughout the world, including but not limited to the Canadian Cordillera, the Mid-Atlantic Ridge south of the Kane Transform (MARK) fault, and ophiolites in Zedang, Greece, Oman, and France (Table 9.1). The geological settings include slow-spreading mid-oceanic ridges and ophiolite complexes. These samples span various lithologies, such as variably serpentinized peridotite rocks (e.g., harzburgites and lherzolites) as well as serpentinized pyroxenites including some samples of pure peridotite or dunite. With more than 1,000 samples, we analyzed different physical properties such as density, magnetic susceptibility (K), compressional (V_P) and shear (V_S) wave velocity, Poisson's ratio, bulk modulus, shear modulus, Young's modulus, and Lamé's constant. The degree of serpentinization in different samples was measured by the authors of the various studies via different methods including modal analysis of samples through X-ray diffraction or through electron probe microanalysis and petrographic analysis. Some sources used the regression relationship between density and the degree of serpentinization established in [22, 23]. In our compilation, most of the samples are at a pressure ranging from 20 to 50 MPa or lower, although measurements at much higher pressure and temperature are also available in the literature. We chose lower-pressure range measurements, as these would be the pressures at depths suitable for natural hydrogen exploration as well as stimulated hydrogen monitoring. To ensure consistent lithology across all the samples combined from different sources, we merged some rock types based on the ultramafic rock classification template provided by Le Maitre (2002) [58]. For instance, as per the classification, in our combined lithology, we would name harzburgite as peridotite and websterite as pyroxenite based on their olivine, orthopyroxene, and clinopyroxene composition. We do have some mafic rocks in our compilation, such as gabbro, metagabbro, or diabase, but we removed them from our main analysis, as the focus of this study is on serpentinization and physical property changes of ultramafic rocks. However, we included them later in the discussion on avoiding common interpretation pitfalls. See the references in Table 9.1 for more details on sampling methodologies for various physical properties and geological settings of different samples.

Table 9.1: Location, geological setting, rock types, distribution of physical properties, and number of samples for all the compiled data.

Location	Geological setting	Dominant lithology	Physical properties	No. of samples	References
Western Canadian Cordillera	Ophiolitic massifs	Ultramafic rocks	ρ, k	418	[57]
Zedang Ophiolite, Tibet	Ophiolites	Lherzolite, harzburgite	ρ, k	21	[45]
Mineoka Belt, Japan, and the South Mariana and Tonga Trenches	Accretionary prism and deep-sea floor	Lizardite, chrysotile	ρ, ϕ, V_P, V_S	4	[29]
Atlantis Massif, Mid-Atlantic Ridge (MAR)	Ophiolite complexes	Serpentinized peridotite	ρ, ϕ	4	[53]
Papua New Guinea	Ophiolitic rocks	Harzburgite, pyroxenite	V_P, V_S, ρ	12	[55]
Hida outer belt, Central Japan	Mantle wedge	Partially serpentinized peridotite	V_P, V_S, ρ	16	[54]
Pindos, Oman, France	Ophiolite complex	Variably serpentinized peridotites	ρ, ϕ, k	44	[52]
MAR and Hess Deep	Oceanic crust	Serpentinized abyssal peridotites	ρ, k	245	[23]
MAR, south of the Kane Transform Zone (MARK)	Mid-oceanic ridge	Serpentinized harzburgite, pyroxene-rich serpentinized harzburgite, metagabbro, gabbro, olivine gabbro, troctolite	V_P, V_S, ρ	66	[22]
Xigaze ophiolite (Tibet)	Ophiolite complex	Serpentinized harzburgite	ρ, ϕ, V_P, V_S	6	[56]

Table 9.1 (continued)

Location	Geological setting	Dominant lithology	Physical properties	No. of samples	References
Ronda peridotites, Southern Spain	Ophiolitic massifs	Dunite, harzburgite, lherzolite	V_P, V_S, ρ	14	[47]
Josephine peridotite (Oregon)	Ophiolites	Serpentinized harzburgites	ρ, k	39	[24]
Western United States	Ophiolites	Serpentinites, metagabbro, diabase, gabbro, pyroxenite	V_P, V_S, ρ	35	[48]
Mid-Atlantic Ridge	Oceanic crust	Serpentinites	V_P, V_S, ρ	3	[59]
Different locations across California, Washington, and Oregon	Ophiolites	Peridotite, dunite, partially serpentinized peridotite and dunite, serpentinite	V_P, V_S, ρ	11	[51]

ρ = density, k = magnetic susceptibility, V_P = compressional wave velocity, V_S = shear wave velocity, ϕ = porosity.

9.3 Physical property changes during serpentinization

Ultramafic rocks that undergo serpentinization and the products formed have clearly defined ranges of different physical properties. An increase in volume and a decrease in density are inevitable in isochemical serpentinization [24]. Broadly, during serpentinization, the density, velocity, and different elastic moduli decrease, while magnetic susceptibility and Poisson's ratio increase. In some cases, substantial volume increases of up to 40% occur during serpentinization, which can be accommodated by fracturing, thus further exposing fresh rock surfaces [25]. In terms of implications for hydrogen generation, during the serpentinization of peridotite, the $Fe^{3-}/(Fe^{3-} + Fe^{2-})$ ratios of serpentine are markedly higher than during the serpentinization of pyroxenite, leading to higher $H_2(aq)$ concentrations predicted for the serpentinization of peridotite [11]. In this section, we will explore the relationships of different physical properties with serpentinization along with relevant cross-property relationships.

9.3.1 Density and magnetic susceptibility versus serpentinization

9.3.1.1 Density versus serpentinization

The most studied and pervasive change during serpentinization is the marked reduction in density, as shown in Figure 9.1. The average density of pristine peridotites, approximately 3.1–3.3 g/cm^3, is much higher than the density of pure serpentinite, which is approximately 2.55–2.6 g/cm^3 [60]. Thus, the change in density has also been used to calculate the extent of serpentinization when direct measurements of the mineral constituents are not available. Figure 9.1 shows the decrease in density during serpentinization, with a least-squares regression fitted to the data and an R^2 value of 0.888. Popular regression relationships given by [22, 57] are also plotted. The *Miller and Christensen'97* line overlaps our line at higher densities, whereas the *Cutts et al. 21* line overlaps with our fit at lower densities. A few samples show densities lower than pure serpentinites, which might be a porosity effect. For instance, the lowest density sample is from the Tonga Trench region [29] and has a porosity of 25.8%.

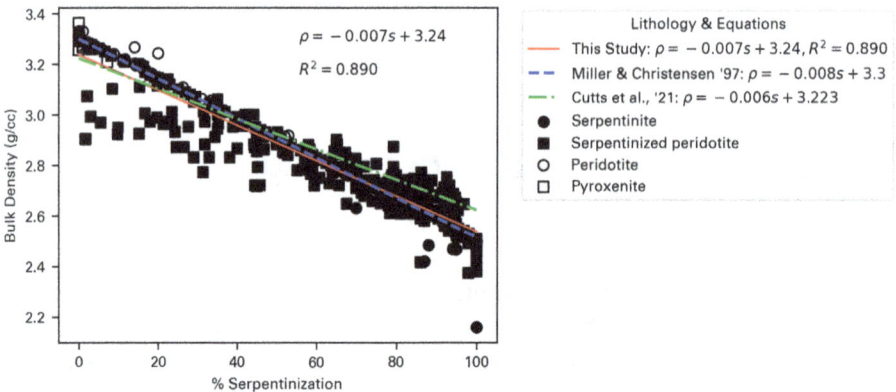

Figure 9.1: Bulk density versus serpentinization % (s), with fitted regression relationships from the current study as well as references [22, 57].

Density monitoring using well logs and gravity or muon tomography measurements [61] has been widely carried out for resource exploration. Thus, the change in density can be quickly applied to either image rocks undergoing natural serpentinization or monitor changes in density during stimulated hydrogen production to ascertain the extent of serpentinization.

9.3.1.2 Magnetic susceptibility versus serpentinization

Figure 9.2 shows the variation of magnetic susceptibility with serpentinization and the fitted relationship with an R^2 of 0.12. Broadly, magnetic susceptibility increases with serpentinization, and this increase is attributed to the formation of magnetite. In general, serpentinization of olivine typically leads to the formation of magnetite; however, the serpentinization of orthopyroxene and orthopyroxene-rich rocks often occurs without producing magnetite. Notably, olivine serpentinization does not invariably result in magnetite formation. For instance, hydrothermal experiments conducted on lherzolite at 200 °C yielded either trace amounts or no magnetite, while generating significant quantities of aqueous H_2 [62]. At lower temperatures, Fe is preferentially incorporated into brucite, suggesting a reduced thermodynamic drive for magnetite formation. Additionally, magnetite is not expected to form as part of the serpentinization assemblage when water-to-rock mass ratios (w/r) are very high [11]. The initial stages of serpentinization are marked by a relative scarcity of magnetite, which aligns with the enrichment of Fe in serpentine and brucite. This phase is subsequently followed by significant magnetite formation, accompanied by a characteristic decrease in fluid pH. The formation of magnetite also correlates with higher hydrogen generation during serpentinization. This can be measured, and hydrogen generation can be approximated using magnetic susceptibility [16]. However, the data cloud here also suggests that, based on the variable conditions of serpentinization, the formation and quantity of magnetite vary. Thus, using magnetic susceptibility alone as an indication of serpentinization might not be recommended.

9.3.2 Elastic and seismic properties

9.3.2.1 Velocity (V_P, V_S) and acoustic impedance versus serpentinization

Seismic properties of ultramafic rocks vary with their proportion of olivine to pyroxene and the abundances of accessory minerals formed during serpentinization such as brucite, magnetite, magnesite, tremolite, and talc [63]. Compressional wave velocities for monomineralic aggregates of olivine, pyroxene, and serpentine approximate 8.54, 7.93, and 5.10 km/s, respectively. Shear wave velocities for similar aggregates are 4.78, 4.65, and 2.35 km/s [51]. Along with density, both P and S-wave velocities, as well as acoustic impedance, decrease with serpentinization (Figures 9.3–9.5). There is an ~40% decrease in P-wave velocity, ~60% decrease in S-wave velocity, and ~60% decrease in acoustic impedance with 100% serpentinization. Velocity reductions are large and spatially sharp, as opposed to composition and temperature, whose effects are smaller or more gradual [31]. This velocity decrease has been used to estimate the degree of serpentinization in the oceanic mantle for quite some time [63]. One caution here is that at low pressures and temperatures, velocity not only depends on serpenti-

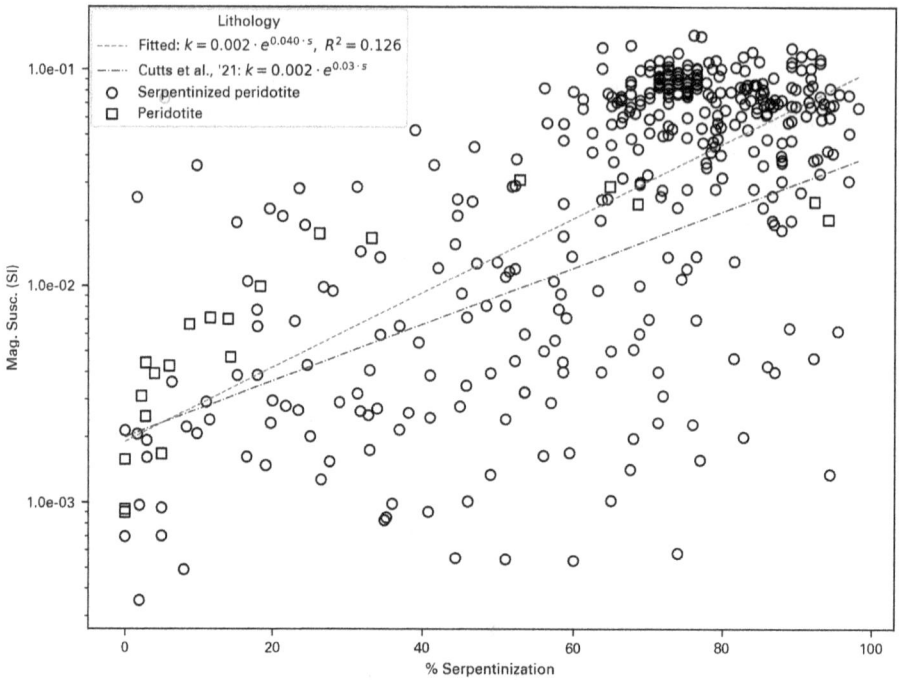

Figure 9.2: Magnetic susceptibility (K) versus % serpentinization(s) with regression relationship fitted from the current study and [57].

nization but also on fluid-filled porosity and preferred mineral orientation [29]. The compiled data is fitted using least square regression with an R^2 of 0.976, 0.937, and 0.973 for V_P, V_S, and acoustic impedance versus serpentinization, respectively. We rejected two peridotite samples that had very low velocity and were described as having very loose grains [51]. Similarly, two very high-velocity serpentinites were rejected as they contained antigorite, which is a high-temperature serpentinite and not relevant to our study. Thus, changes in velocities and acoustic impedance can also be used to seismically monitor serpentinization and explore source rocks for hydrogen production.

Changes in velocities and acoustic impedance are not just useful in identifying lithology but also in differentiating between different fluids within the same lithology. Rock physics amplitude versus offset (AVO) modeling has been used previously to differentiate between brine and hydrocarbon-saturated clastic lithologies [64, 65]. Similarly, it can be used to distinguish between hydrogen and other gases in the subsurface [43]. Moreover, other seismic parameters, such as attenuation – which increases with partial gas saturation – can also be studied and incorporated.

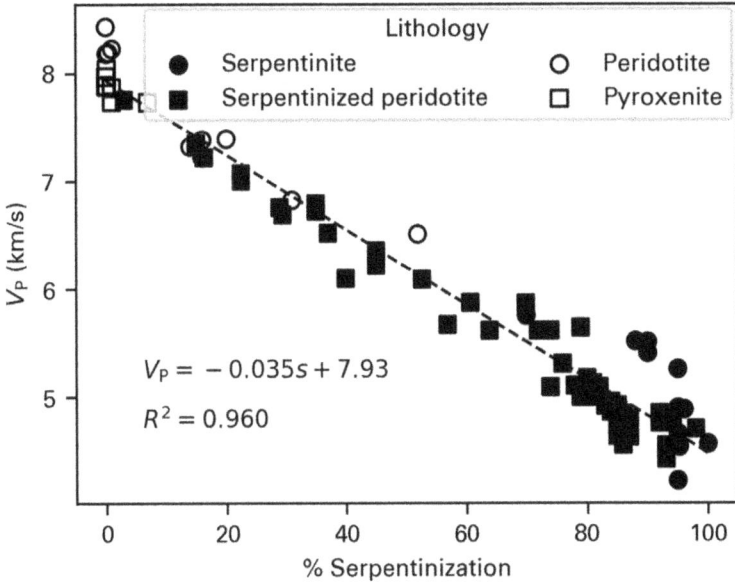

Figure 9.3: Compressional wave velocity (V_P) versus serpentinization % (s) and fitted regression relationship.

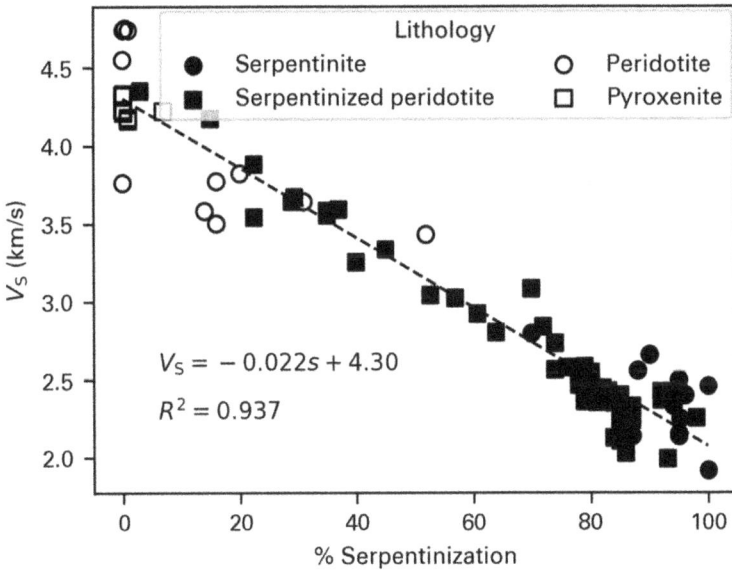

Figure 9.4: Shear wave velocity (V_S) versus serpentinization % (s) and fitted regression relationship.

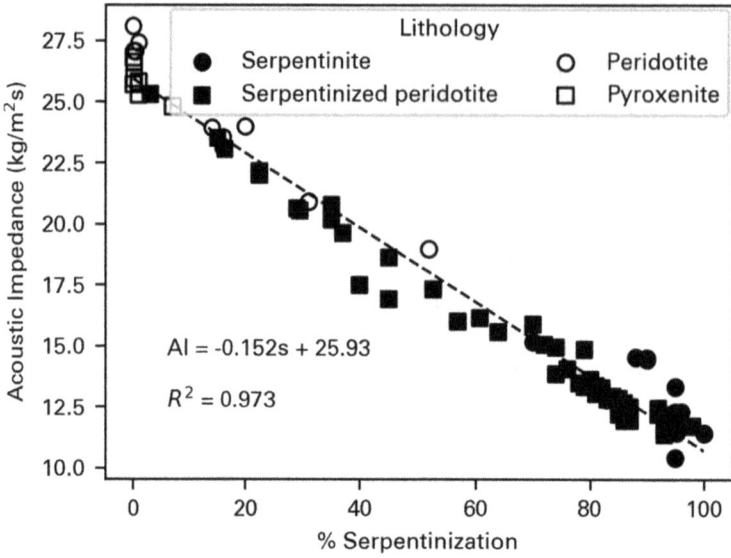

Figure 9.5: Acoustic impedance versus % serpentinization (s) and the fitted regression relationship.

9.3.2.2 Serpentinization versus Poisson's ratio

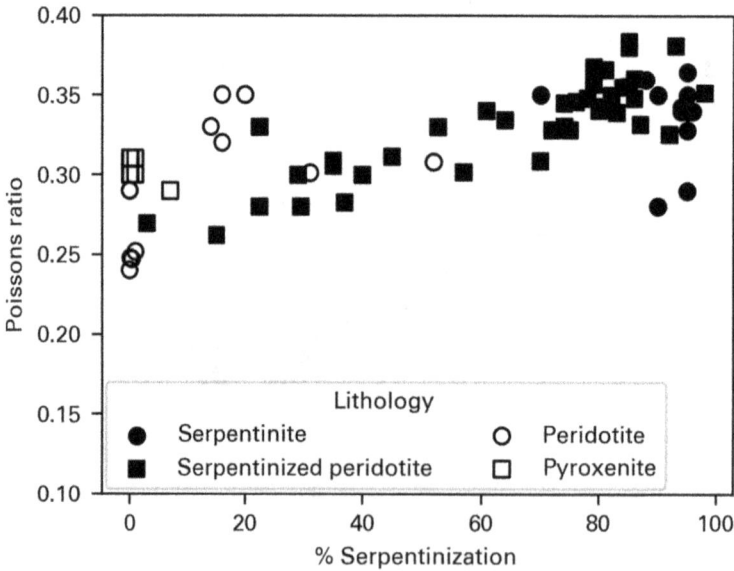

Figure 9.6: Poisson's ratio versus serpentinization %.

Poisson's ratio (ν), the ratio of transverse strain to axial strain under stress), is a key parameter in rock physics for characterizing the elastic behavior of materials. It can be represented in terms of velocity ratios, as follows:

$$\nu = \frac{1}{2} \cdot \frac{\left(\frac{V_P}{V_S}\right)^2 - 2}{\left(\frac{V_P}{V_S}\right)^2 - 1} \tag{9.1}$$

Serpentinized peridotites have low velocities and a high Poisson's ratio. Serpentinite has a distinctly high Poisson's ratio relative to most other silicate rocks, theoretically allowing it to be distinguished from other rocks in ophiolite complexes [22]. Poisson's ratio, in general, has been used to differentiate between differing lithologies, as it is not strongly affected by variations in temperature or pressure. Moreover, it is widely used to indicate fluids, infer stress distribution, and understand fracture distribution. Figure 9.6 shows the increase in Poisson's ratio with serpentinization from 0.25 to 0.3 for ultramafic rocks to 0.3–0.4 for highly serpentinized rocks. During serpentinization, the decrease in S-wave velocity is more pronounced than the decrease in P-wave velocity (Figures 9.3 and 9.4) and thus the increase in Poisson's ratio. A few serpentinite samples that show porosity >5% were rejected as high porosities reduce velocities, thus lowering the Poisson's ratio. However, this could also be a distinguishing factor for serpentinized peridotites with good porosities, as good porosities would be required for continuous access of fluid to fresh surfaces for serpentinization to occur. Apart from porosity, the presence of magnetite in highly serpentinized peridotites can also lower Poisson's ratio.

9.3.2.3 V_P/V_S ratio and acoustic impedance (AI)

The V_P/V_S ratio versus acoustic impedance plot is a standard rock physics template [66–68] and has been used to distinguish between different lithologies as well as different fluids. Moreover, V_P/V_S is not strongly affected by temperature or pressure. However, in the case of hydrogen [43], it has been shown that it is difficult to distinguish between hydrogen and methane-saturated reservoirs using the V_P/V_S versus AI template. However, it can be used to distinguish between hydrogen and brine pore fluids. As per Figure 9.7, we can see that the V_P/V_S ratio increases with serpentinization, except for some samples with low V_P/V_S and acoustic impedance, possibly due to very high porosity. Samples that have been less than ~50% serpentinized show a V_P/V_S of 2 and acoustic impedance >17.5 kg/m^{2}s whereas those samples that have undergone greater than 50% serpentinization are marked by high V_P/V_S of 2–2.3 and a lower acoustic impedance between 10 and 15 kg/m^{2}s. This provides a clear distinction between highly serpentinized samples that might have low potential to generate hydrogen versus samples that might yet produce hydrogen.

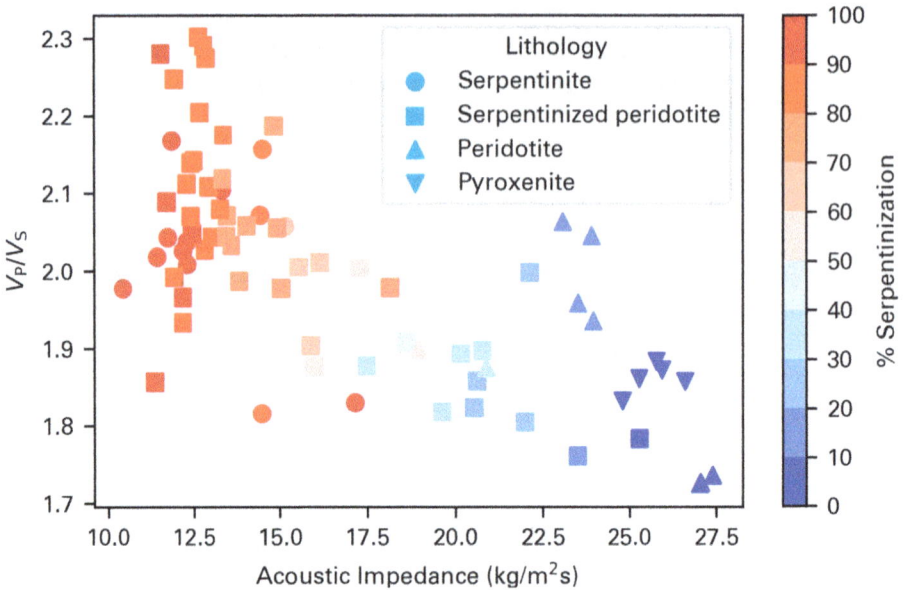

Figure 9.7: V_P/V_S versus acoustic impedance, colored according to the degree of serpentinization.

9.3.2.4 V_P versus V_S

In hydrocarbon exploration, relationships between V_P and V_S have been key to determining lithology and fluids, using, for example, AVO analysis. As a result, there is a wide variety of published V_P–V_S relationships for hydrocarbon reservoirs and cap rocks. The most reliable and often-used V_P–V_S relationships for hydrocarbon reservoirs are empirical fits to laboratory or well log data. Figure 9.8 shows the cross-plot of shear wave velocity versus compressional wave velocity for rocks relevant to natural hydrogen, fitted with a linear regression with an R^2 of 0.964. Three samples that showed high porosity (>5%) were not included in the fitting and plot outside the linear fit. Popular sandstone [70] and mudrock [69] lines are also shown in Figure 9.8, highlighting the difference between ultramafic rocks and sedimentary rocks, but at the same time indicating that the same templates used in hydrocarbon exploration, after adaptation, can be used for hydrogen exploration.

9.3.2.5 Rock elastic moduli changes

$$K = \rho \left(V_P{}^2 - \frac{4}{3} V_S{}^2 \right)$$

(9.2)

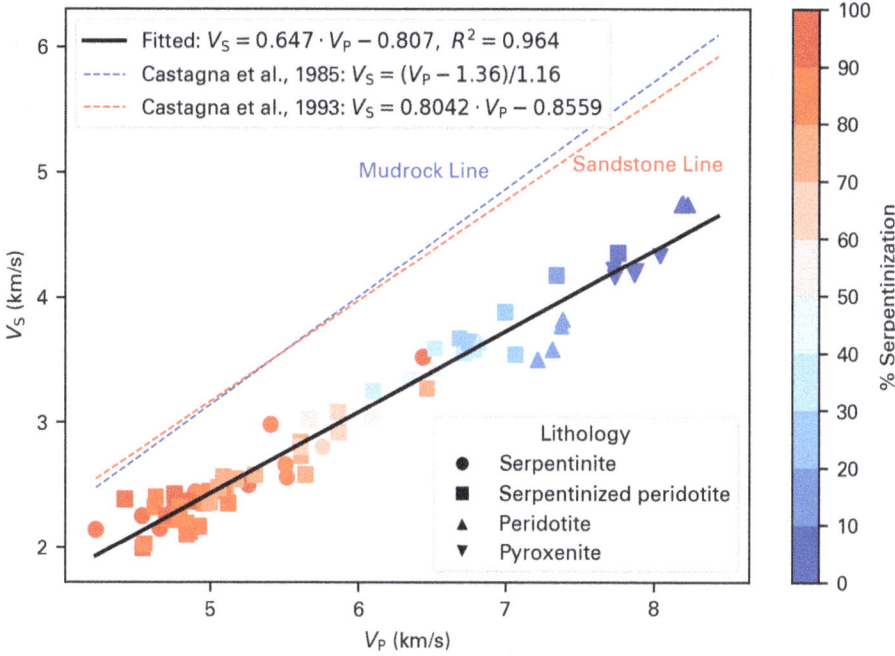

Figure 9.8: V_S versus V_P and their fitted regression, along with popular regression lines for sandstone and mud rock provided in [69, 70]. Points are colored according to the degree of serpentinization.

$$\mu = \rho V_S^2 \tag{9.3}$$

$$\lambda = \rho \left(V_P^2 - 2V_S^2\right) \tag{9.4}$$

$$E = \frac{9K\mu}{3K + \mu} \tag{9.5}$$

Bulk modulus (K), shear modulus (μ), Lame's constant (λ) and Young's modulus (E) are fundamental parameters used to quantify a rock's mechanical and elastic behavior under stress and have been extensively applied to infer geomechanical properties, lithology, and fluid content. These parameters can be expressed in terms of V_P and V_S and density ρ, as shown by equations (9.2)–(9.5). For instance, bulk modulus (K) is highly sensitive to the pore fluid type and is often used to differentiate between rocks saturated with different fluids (e.g., Gassmann fluid substitution [71]). Shear modulus, on the other hand, is unaffected by fluids (at low frequencies) and helps in determining rock rigidity and fracture propagation. Different moduli and Lame's constant show a decreasing trend with serpentinization, as expected (Figure 9.9).

Carlson [72] fitted Hashin-Shtrikman and Voigt-Reuss-Hill average models to bulk and shear modulus data for partially serpentinized peridotites from [48, 50, 51]. In this study, as shown in Figure 9.10, we fit a differential effective medium (DEM) inclu-

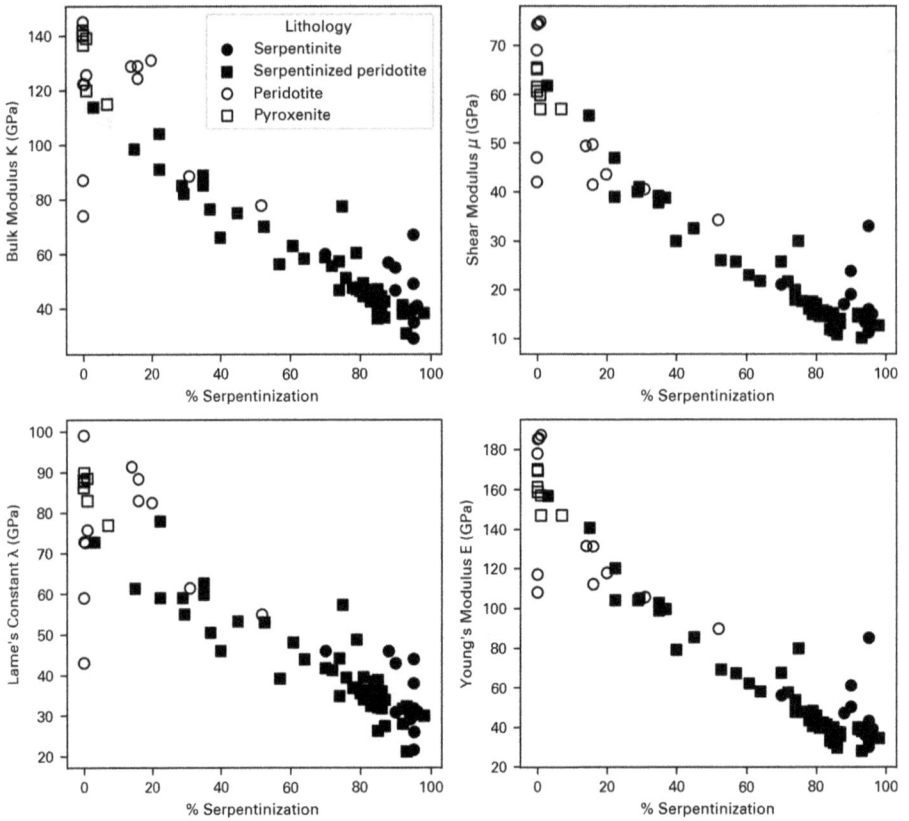

Figure 9.9: Changes in different rock moduli with serpentinization; bulk modulus (top left), shear modulus (top right), Lame's constant (bottom left), and Young's modulus (bottom right).

Table 9.2: Moduli for peridotite and serpentine used in DEM modeling.

Property	Olivine [reference]	Serpentine
Bulk modulus (GPa)	130 [75]	30
Shear modulus (GPa)	70	10
Aspect ratio		0.3

sion model [73, 74]. Inclusion models are helpful in determining the effective elastic properties of heterogeneous rocks, which is very pertinent to serpentinization, as we can then calculate the effective elastic properties at every stage as the reaction proceeds. The DEM model allows the addition of inclusions incrementally to the host material and tracks the effective properties of the combined medium progressively.

We start with the bulk and shear moduli of pure olivine, with the addition of serpentinite as inclusions with an aspect ratio of 0.3. The bulk and shear moduli of both olivine and serpentinite, taken for this study along with the aspect ratio, are given in Table 9.2. The DEM model fits well with the data that has been acquired at different locations and for variably serpentinized peridotites, which can be utilized to calculate effective elastic properties at various degrees of serpentinization.

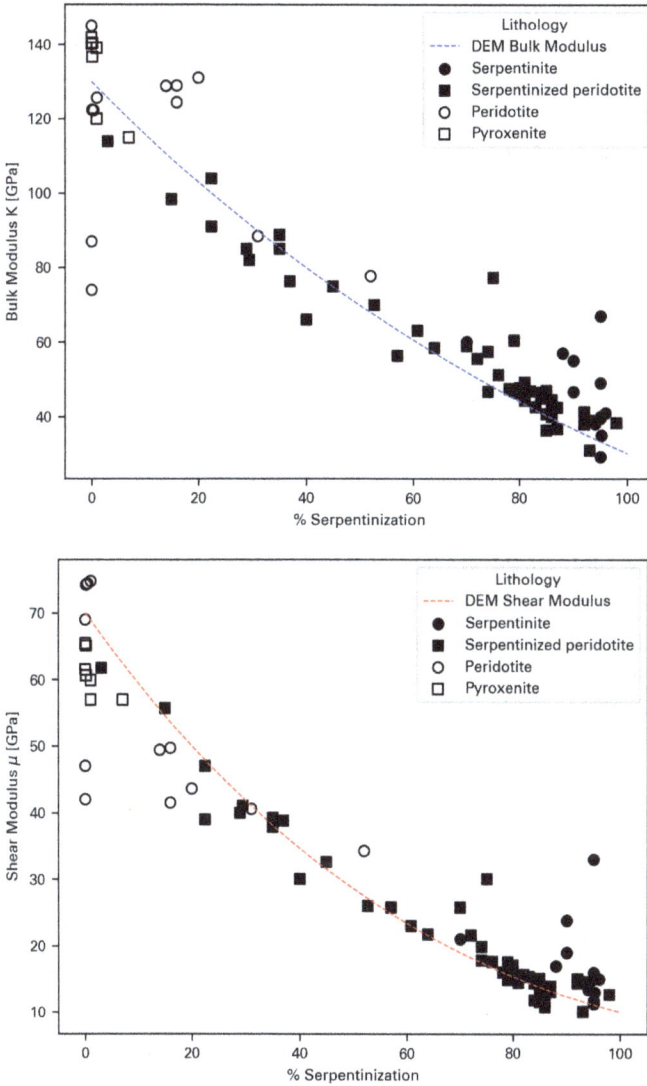

Figure 9.10: Bulk modulus (left) and shear modulus (right) versus serpentinization and effective rock moduli calculated using the DEM inclusion model [73, 74].

9.3.3 Cross-property relationships

9.3.3.1 Magnetic susceptibility versus density

Figure 9.11 shows that as density decreases, magnetic susceptibility increases during serpentinization. Changes in density during serpentinization are primarily linked to a rock's mineralogical changes, whereas changes in magnetic susceptibility are related to the concentration and distribution of ferro-magnetic minerals such as magnetite. As per Figure 9.11, <50% serpentinized samples do not show a magnetic susceptibility >20×10^{-3} (SI), inferring that magnetite might be formed in the later stages of serpentinization consistent with observations from [45] and [76]. However, in the samples from the Zedang ophiolites, magnetite formation was prevented above 50% serpentinization [45, 57]. It is argued that there are two magnetic susceptibility trends during serpentinization: one involves a 100-fold increase in magnetic susceptibility and is followed by most harzburgitic samples, whereas the second involves very little change in magnetic susceptibility and is followed by most dunitic samples and a minor proportion of harzburgitic samples [77]. Additionally, it is argued that above 60% serpentinization, the system transitions from a rock-dominated closed system to a fluid-dominated open system, resulting in an exponential increase in magnetite formation. This aligns well with the samples in Figure 9.11, where below 50–60% serpentinization, we observe a linear decrease in density and almost no increase in susceptibility. Above 60% serpentinization, the magnetic susceptibility increases exponentially. Broadly, as per Figure 9.11 and the discussion above, although the decrease in density is predictable, the formation of magnetite and the increase in magnetic susceptibility during serpentinization vary due to a wide range of parameters, namely protolith composition, temperature, and water-rock ratio. Therefore, while an increase in magnetic susceptibility indicates the formation of hydrogen and serpentinization, the converse might not be true. Thus, using magnetic susceptibility data as a proxy for serpentinization and hydrogen generation should be approached with caution.

9.3.3.2 Density versus velocity

The seismic P-wave velocity and density are positively correlated, and both decrease with serpentinization, as shown in Figure 9.12. The two distinct clusters in the density-velocity space are formed by the peridotites and serpentinites. The dataset is fitted using a power-law model as well as a linear model. The power-law model expresses density as a power-law function of velocity as given by [78]. The general formulations are provided by eqs. (9.6) and (9.7), and the constants a and b are empirical values determined from field data, depending on lithology and porosity. Both models show an $R^2 > 0.9$ and fit the data well:

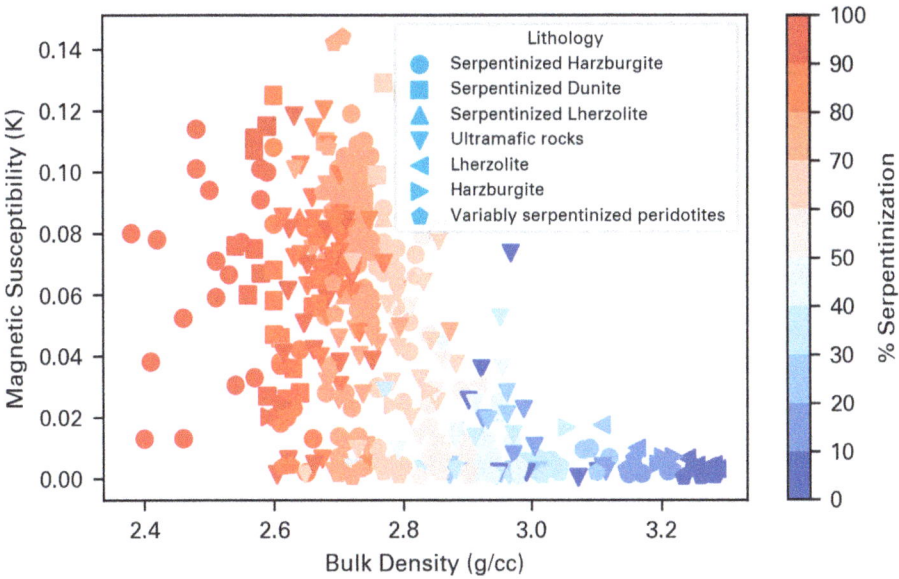

Figure 9.11: Magnetic susceptibility versus bulk density, colored according to the percentage of serpentinization.

$$\rho = a \cdot V_P{}^b \tag{9.6}$$

$$\rho = a + b \cdot V_P \tag{9.7}$$

Along with the fitted relationship for serpentinized peridotites using both models, we also plot popular relationships for sandstones with V_P in the range of 1.5–6 km/s, as given by Gardner et al. [78] for a power law model and by Godfrey et al. [79] for a linear model. These relationships can be very useful for a first-order approximation of density if seismic data is acquired for hydrogen exploration or to estimate velocity in the presence of gravity data. Density-porosity relationships can also be used to establish a link between well log scale data and regional scale data. Limitations include the assumption of uniform rock behavior.

9.3.3.3 Porosity versus density and velocity

As serpentinization progresses, both velocity and density decrease due to changes in mineralization, while porosity increases as a result of the creation of pore spaces during hydration (Figures 9.13 and 9.14). The further decrease in velocity and density during serpentinization is attributed to a porosity effect. Peridotites that have serpentinized <50% show porosities strictly less than 1%, whereas peridotites that have serpentinized >50% show porosities from 0.1 to 3% and occasionally >5%, reaching up to

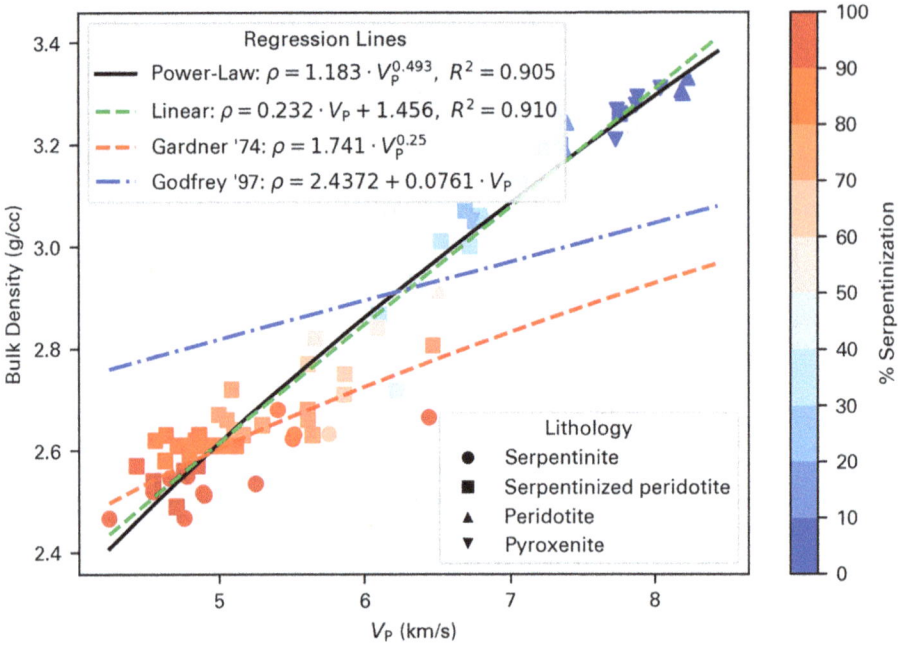

Figure 9.12: Bulk density versus P-velocity with the fitted regression relation plotted alongside popular relationships provided in [78, 79] for other lithologies. Data points are colored based on the percentage of serpentinization.

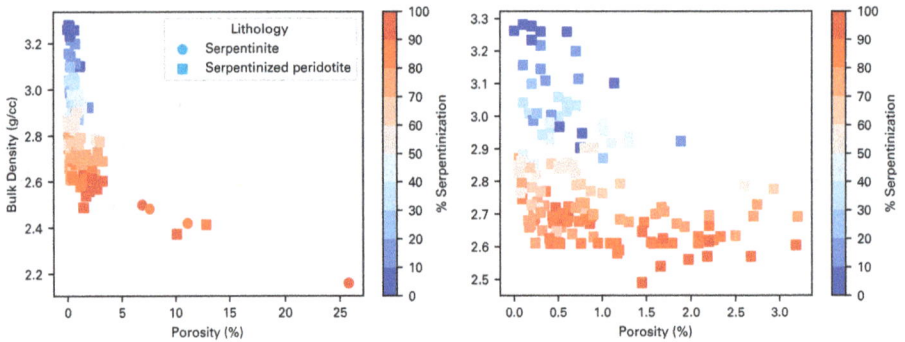

Figure 9.13: Bulk density versus porosity, colored according to % serpentinization; (left) the entire porosity range exhibited by the samples; (right) porosities <5%.

26%. This increase in porosity might be due to fluid-dominated serpentinization above 50%, which induces fracturing and further enhances porosity. Since serpentinites exhibit low permeabilities on the order of micro- to nano darcies, the high porosities might not be interconnected [80]. This could pose a bottleneck for hydrogen stimulation, as fresh surfaces would need to be constantly exposed to injected fluids.

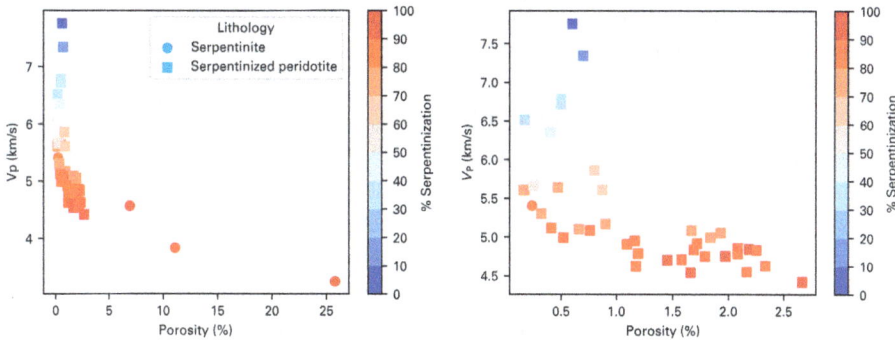

Figure 9.14: Velocity (V_p) versus porosity, colored according to % serpentinization; (left) the entire porosity range exhibited by the samples; (right) porosities <5%.

Moreover, Templeton et al. [21] argue that the H_2 generation potential of rocks with 1% porosity (i.e., a water-rock ratio of 0.01) will likely yield 20-fold less H_2 than rocks with water-rock ratios of 0.2. On the other hand, Tutolo et al. [81] suggest that even if serpentinites do not exhibit high porosity, the inherent porosity within the serpentine and brucite mineral structures may be sufficient to accommodate fluid flow and further serpentinization. Another property influenced by porosity is an increase in electrical conductivity. It has been demonstrated that serpentinized rocks can exhibit an increase in electrical conductivity by as much as 3–4 orders of magnitude compared to serpentine-free peridotite in the Indian Ocean ridge [26], although this increase might also be attributed to mineral conduction. Volume expansion during serpentinization can also induce fractures at various scales, creating secondary porosity.

9.4 Discussion

9.4.1 Implications for geophysical modeling and interpretation

This study and the established rock physics relationships represent the most comprehensive compilation of various geophysical and rock physics properties during serpentinization that can influence hydrogen generation and the exploration of hydrogen resources. There is a future possibility of developing or adapting logging tools capable of measuring the changes specific to hydrogen generation. Moreover, the established relationships presented here can support multiphysics interpretation workflows, including gravity, electromagnetic, electrical, and seismic data, to enhance the exploration of geologic hydrogen. The rock physics relationships can also provide constraints for field interpretations and help reduce uncertainty. In the early stages of natural hydrogen exploration, there will be a need to scan large areas in a cost-effective manner, and the preliminary understanding of geophysical properties might prove useful. Natural hydrogen exploration is still in its nascent stages, and exploration strategies are continuing to evolve, including the current use of various geophysical methods.

While electrical conductivity might be sensitive to the presence of hydrogen, we did not study it extensively here. Although a conductivity anomaly is non-unique, the presence of hydrogen has been argued to increase electrical conductivity in tectonic settings such as subduction zones [27]. Another factor extensively discussed in the literature is changes in seismic anisotropy during serpentinization [54, 63]. It is argued that the alignment of olivine crystals is weakened during serpentinization, leading to the formation of a relatively isotropic serpentine [56]. The current dataset did not show any such strong correlation, perhaps due to the presence of serpentine samples (e.g., antigorite) that also follow a preferred orientation and are formed at very high temperatures (>300 °C), which are not relevant to our study. A more detailed study on anisotropy can be conducted.

9.4.2 Interpretation pitfalls

Based on the composition of the protolith, a wide variety of serpentinization reaction pathways are possible [24]. Ophiolites, which are the most prospective geological setting for natural hydrogen exploration, not only contain ultramafic rocks but also co-exist with fresh or altered mafic rocks such as gabbro and metagabbro (Figure 9.15). All these changes in physical properties can be measured using gravity, aeromagnetic, electromagnetic, and seismic methods. On their own, each of these methods might provide a non-unique solution with different properties and can also lead to misinterpretation. However, integrating them together with the geology would minimize uncertainty and lead to a consistent prediction. As shown in Figure 9.15, serpentinites and serpentinized harzburgites show low values for both V_P and V_S, but partially serpentinized peridotites are interspersed with mafic rocks like diabase, spilite, olivine gabbro, and metagabbro. On the higher velocity end, peridotites are again separated from the rest of the rocks. This is consistent with the observations of the researchers [60, 82–84] who propose that P- and S-wave velocities and the density of partially serpentinized rocks are in the same range as fresh or altered mafic rocks. Thus, relying solely on one or two physical properties might not yield accurate predictions about serpentinization. However, during stimulated hydrogen generation, the lithology of the protolith would be well known, and then the rock physics properties and relationships can be used to infer the extent of serpentinization as well as hydrogen generation.

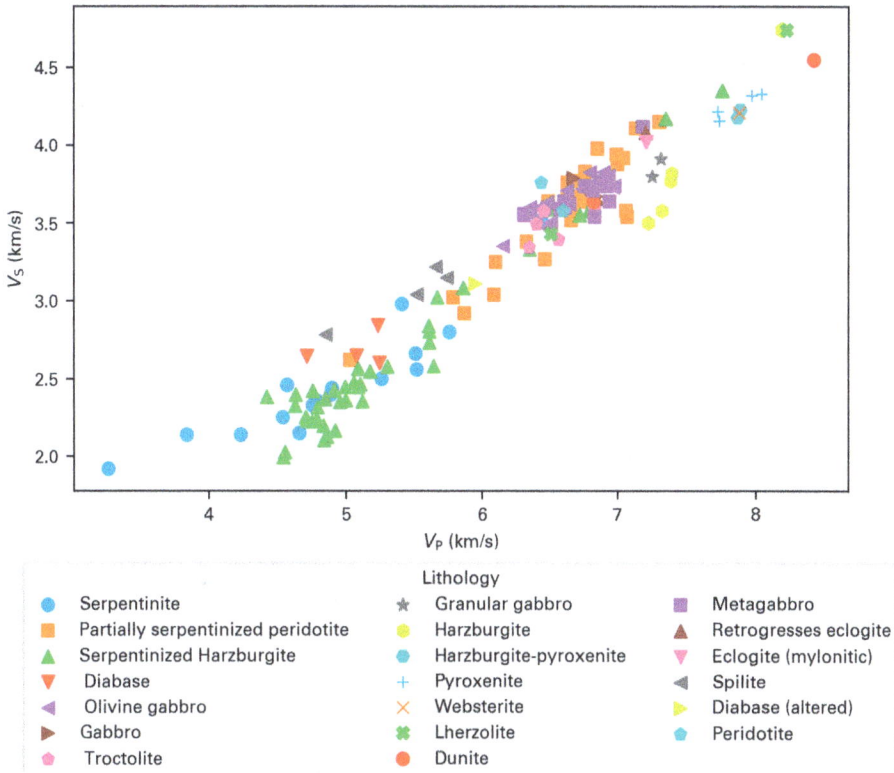

Figure 9.15: V_P versus V_S relationship for all the samples compiled, colored according to their lithology.

9.5 Summary and conclusion

This study investigates the role of rock physics in the geophysical interpretation of geologic hydrogen resources, with a focus on the serpentinization of ultramafic rocks as a natural mechanism for hydrogen generation. Serpentinization alters critical rock properties such as density, seismic velocities, porosity, and magnetic susceptibility, providing measurable indicators for identifying and characterizing hydrogen source rocks and reservoirs. By integrating and data mining over 1,000 rock samples from the literature across diverse geological settings, this work establishes robust rock property relationships and models the geophysical responses to serpentinization.

A comprehensive analysis of density trends reveals a predictable decline during serpentinization, transitioning from ~3.3 g/cc in unaltered peridotites to ~2.6 g/cc in fully serpentinized rocks. These changes are attributed to mineralogical transformations, including the replacement of olivine and pyroxene by serpentine, brucite, and magnetite. Similarly, seismic velocities (V_P and V_S) exhibit significant reductions of

~40% and ~60%, respectively, as serpentinization progresses. Elastic moduli, such as bulk modulus and shear modulus, also decrease systematically, reflecting the weakening of the rock framework due to mineral alterations and increased porosity. These trends are key indicators of serpentinization and can be directly linked to hydrogen production. All the rock physics relations established in this study are collated in Table 9.3.

Table 9.3: Regression relationships between different physical properties and their respective R-squared values. The equations are valid at pressures <50 MPa and at low temperatures.

Properties	Equation(s)	R^2
Density versus serpentinization	$\rho = -0.007s + 3.24$	0.888
P-wave velocity versus serpentinization	$V_P = -0.035s + 7.93$	0.96
S-wave velocity versus serpentinization	$V_S = -0.022s + 4.3$	0.937
Acoustic impedance versus serpentinization	$AI = -0.152s + 25.93$	0.973
Mag. susceptibility versus serpentinization	$K = 0.002 \cdot e^{0.04s}$	0.12
S-wave velocity versus P-wave velocity	$V_S = 0.647 \cdot V_P - 0.807$	0.964
Density versus P-wave velocity	$\rho = 1.183 \cdot V_P^{0.493}$	0.905
	$\rho = 0.232 \cdot V_P + 1.456$	0.91

Cross-property relationships play a pivotal role in the geophysical characterization of hydrogen reservoirs. For example, strong correlations between density and velocity, as well as magnetic susceptibility and density, allow for multi-parameter interpretations that enhance subsurface imaging. These findings are further validated using DEM modeling, which accurately captures the effects of serpentinization through the inclusion of geometry and mineralogical composition on bulk and shear moduli. The DEM framework, applied with appropriate aspect ratios, highlights the sensitivity of rock properties to microstructural changes, offering a powerful tool for interpreting heterogeneous rock systems.

The geophysical implications of these findings are significant for hydrogen exploration. Gravity surveys can be employed to detect density anomalies associated with serpentinized zones, while seismic methods are effective in imaging velocity and impedance contrasts. Additionally, electromagnetic techniques can map magnetic mineral content, providing complementary data for characterizing reservoirs. The integration of gravity, seismic, and electromagnetic methods reduces interpretation ambiguities and enhances the reliability of subsurface imaging, particularly in complex geological settings. These techniques also enable time-lapse monitoring of stimulated hydrogen systems, where fluid injections or fracturing promote hydrogen production.

Despite the advancements presented in this study, challenges remain. Variations in rock properties across different geological settings emphasize the need for localized calibration of empirical models. Furthermore, the effects of fluid composition and pressure variations on geophysical responses require more detailed investigations to improve predictive accuracy. The development of joint inversion techniques and advanced multiphysics modeling will be essential for fully exploiting the complementary strengths of geophysical methods in future studies. These challenges underscore the importance of continued research to refine methodologies and improve the interpretation of hydrogen systems.

In conclusion, this study provides a comprehensive framework for using rock physics to interpret geologic hydrogen reservoirs. The empirical relations, based on a large dataset, can help address key challenges in geophysical hydrogen exploration. The findings underscore the critical role of multi-parameter approaches in characterizing serpentinization and associated hydrogen production, advancing the field of hydrogen exploration. This work not only enhances the understanding of geophysical responses to serpentinization but also establishes a rock physics foundation for future research and technological innovation in natural hydrogen prospecting.

References

[1] Zgonnik V. The occurrence and geoscience of natural hydrogen: A comprehensive review. Elsevier B.V, Apr. 01 2020. doi: 10.1016/j.earscirev.2020.103140

[2] Williams Q., Hemley R. J. Hydrogen in the deep Earth. Annual Review of Earth and Planetary Sciences. 2024, 29: 365–418. doi: 10.1146/annurev.earth.29.1.365

[3] Klein F., Tarnas J. D., Bach W. Abiotic sources of molecular hydrogen on earth. Elements. 2020, 16(1): 19–24. doi: 10.2138/GSELEMENTS.16.1.19

[4] Mathur Y., Awosiji V., Mukerji T., Hosford A., Peters K. E. Soil geochemistry of hydrogen and other gases along the San Andreas fault. International Journal of Hydrogen Energy. 2024, 50: 411–419. doi: 10.1016/j.ijhydene.2023.09.032

[5] Tagawa S., Sakamoto N., Hirose K., et al. Experimental evidence for hydrogen incorporation into Earth's core. Nature Communications. Dec 2021, 12(1). doi: 10.1038/s41467-021-22035-0

[6] Isaev E. I., Skorodumova N. V., Ahuja R., Vekilov Y. K., Rje Johansson B. Dynamical stability of Fe-H in the Earth's mantle and core regions. Proceedings of the National Academy of Sciences. 2007, 104(22): 9168–9171.

[7] Truche L., McCollom T. M., Martinez I. Hydrogen and abiotic hydrocarbons: Molecules that change the world. Elements. 2020, 16(1): 13–18. doi: 10.2138/GSELEMENTS.16.1.13

[8] Coveney R. M. Jr., Goebel E. D., Zeller E. J., Dreschhoff G. A. M., Angino E. E. Serpentinization and the origin of hydrogen gas in Kansas1. American Association of Petroleum Geologists Bulletin. Jan 1987, 71(1): 39–48. doi: 10.1306/94886D3F-1704-11D7-8645000102C1865D

[9] Boreham C. J., Edwards D. S., Czado K., et al. Hydrogen in Australian natural gas: Occurrences, sources and resources. APPEA J. 2021, 61(1): 163. doi: 10.1071/aj20044

[10] Holm N. G., Oze C., Mousis O., Waite J. H., Guilbert-Lepoutre A. Serpentinization and the formation of H2 and CH4 on celestial bodies (Planets, Moons, Comets). Astrobiology. Jul 2015, 15(7): 587–600. doi: 10.1089/ast.2014.1188

[11] Klein F., Bach W., McCollom T. M. Compositional controls on hydrogen generation during serpentinization of ultramafic rocks. Lithos. 2013, 178: 55–69. doi: 10.1016/j.lithos.2013.03.008

[12] Osselin F., Soulaine C., Fauguerolles C., Gaucher E. C., Scaillet B., Pichavant M. Orange hydrogen is the new green. Nature Geoscience. 2022, 15(10): 765–769. doi: 10.1038/s41561-022-01043-9

[13] Lollar B. S., Onstott T. C., Lacrampe-Couloume G., Ballentine C. J. The contribution of the Precambrian continental lithosphere to global H2 production. Nature. Dec 2014, 516(7531): 379–382. doi: 10.1038/nature14017

[14] McCollom T. M., Klein F., Solheid P., Moskowitz B. The effect of pH on rates of reaction and hydrogen generation during serpentinization. Philosophical Transactions of the Royal Society A: Mathematical, Physical and Engineering Sciences. Feb 2020, 378(2165). doi: 10.1098/rsta.2018.0428

[15] McCollom T. M., Klein F., Robbins M., et al. Temperature trends for reaction rates, hydrogen generation, and partitioning of iron during experimental serpentinization of olivine. Geochimica et Cosmochimica Acta. 2016, 181: 175–200. doi: 10.1016/j.gca.2016.03.002

[16] Miller H. M., Mayhew L. E., Ellison E. T., Kelemen P., Kubo M., Templeton A. S. Low temperature hydrogen production during experimental hydration of partially-serpentinized dunite. Geochimica et Cosmochimica Acta. 2017, 209: 161–183. doi: 10.1016/j.gca.2017.04.022

[17] Lamadrid H. M., Rimstidt J. D., Schwarzenbach E. M., et al. Effect of water activity on rates of serpentinization of olivine. Nature Communications. Jul 2017, 8. doi: 10.1038/ncomms16107

[18] Huang R., Sun W., Song M., Ding X. Influence of ph on molecular hydrogen (H2) generation and reaction rates during serpentinization of peridotite and olivine. Minerals. 2019, 9(11). doi: 10.3390/min9110661

[19] Leong J. A., Nielsen M., McQueen N., et al. H2 and CH4 outgassing rates in the Samail ophiolite, Oman: Implications for low-temperature, continental serpentinization rates. Geochimica et Cosmochimica Acta. 2023, 347: 1–15. doi: 10.1016/j.gca.2023.02.008

[20] Andreani M., Daniel I., Pollet-Villard M. Aluminum speeds up the hydrothermal alteration of olivine. American Mineralogist. 2013, 98(10): 1738–1744. doi: 10.2138/am.2013.4469

[21] Templeton A. S., Ellison E. T., Kelemen P. B., et al. Low-temperature hydrogen production and consumption in partially-hydrated peridotites in Oman: Implications for stimulated geological hydrogen production. Frontiers in Geochemistry. 2024, 2(March): 1–20. doi: 10.3389/fgeoc.2024.1366268

[22] Miller D. J., Christensen N. I. Seismic velocities of lower crustal and upper mantle rocks from the slow-spreading Mid-Atlantic Ridge, south of the Kane Transform Zone (MARK). Proceedings of the Ocean Drilling Program Scientific Results Volume 153. 1997, 153. doi: 10.2973/odp.proc.sr.153.043.1997

[23] Oufi O. Magnetic properties of variably serpentinized abyssal peridotites. Journal of Geophysical Research. 2002, 107(B5). doi: 10.1029/2001jb000549

[24] Toft P. B., Arkani-Hamed J., Haggerty S. E. The effects of serpentinization on density and magnetic susceptibility: A petrophysical model. Physics of the Earth and Planetary Interiors. 1990, 65: 137–157.

[25] Klein F., Le Roux V. Quantifying the volume increase and chemical exchange during serpentinization. Geology. 2020, 48(6): 552–556. doi: 10.1130/G47289.1

[26] Stesky R. M., Brace W. F. Electrical conductivity of serpentinized rocks to 6 kilobars. Journal of Geophysical Research. Nov 1973, 78(32): 7614–7621. doi: 10.1029/jb078i032p07614

[27] Yoshino T., Manthilake G., Pommier A. Probing deep hydrogen using electrical conductivity. Elements. 2024, 20(4): 247–252. doi: 10.2138/gselements.20.4.247

[28] Birch F. Density and composition of the upper mantle: First approximation as an Olivine layer. Geophysical Monograph Series. 1969, 13: 18–36.

[29] Hatakeyama K., Katayama I. Pore fluid effects on elastic wave velocities of serpentinite and implications for estimates of serpentinization in oceanic lithosphere. Tectonophysics. 2019, 775 (November): 228309, 2020. doi: 10.1016/j.tecto.2019.228309

[30] Keppler H., Ohtani E., Yang X. The subduction of hydrogen: Deep water cycling, induced seismicity, and plate tectonics. Elements. 2024, 20(4): 229–234. doi: 10.2138/gselements.20.4.229

[31] Durand S., Juriček M. P., Fischer K. M. Hydrous melting and its seismic signature. Elements. 2024, 20(4): 241–246. doi: 10.2138/gselements.20.4.241

[32] Kelemen P. B., Matter J. In situ carbonation of peridotite for CO2 storage. Proceedings of the National Academy of Sciences of the United States of America. 2008, 105(45): 17295–17300. doi: 10.1073/pnas.0805794105

[33] Matter J. M., Kelemen P. B. Permanent storage of carbon dioxide in geological reservoirs by mineral carbonation. Nature Geoscience. 2009, 2(12): 837–841. doi: 10.1038/ngeo683

[34] Kularatne K., Sissmann O., Kohler E., Chardin M., Noirez S., Martinez I. Simultaneous ex-situ CO2 mineral sequestration and hydrogen production from olivine-bearing mine tailings. Applied Geochemistry. 2018, 95(May): 195–205. doi: 10.1016/j.apgeochem.2018.05.020

[35] Krevor S., Graves C., Van Gosen B., McCafferty A., G. S. (US). Mapping the mineral resource base for mineral carbon-dioxide sequestration in the conterminous United States. US Geological Survey. 2009.

[36] Avseth P., Mukerji T., Mavko G. Quantitative seismic interpretation: Applying rock physics tools to reduce interpretation risk. Cambridge university press, 2010.

[37] Mavko G., Mukerji T., Dvorkin J. The rock physics handbook. Cambridge university press, 2020.

[38] Dvorkin J., Derzhi N., Diaz E., Fang Q. Relevance of computational rock physics. Geophysics. 2011, 76(5): E141–E153.

[39] Bosch M., Mukerji T., Gonzalez E. F. Seismic inversion for reservoir properties combining statistical rock physics and geostatistics: A review. Geophysics. 2010, 75(5): 75A165–75A176.

[40] Frery E., Langhi L., Maison M., Moretti I. Natural hydrogen seeps identified in the North Perth Basin, Western Australia. International Journal of Hydrogen Energy. Sep 2021, 46(61): 31158–31173. doi: 10.1016/j.ijhydene.2021.07.023

[41] Lefeuvre N., Truche L., Donzé F. V., et al. Native H2 exploration in the Western Pyrenean Foothills. Geochemistry, Geophysics, Geosystems. Aug 2021, 22(8). doi: 10.1029/2021GC009917

[42] Zhang M., Li Y. The role of geophysics in geologic hydrogen resources. Journal of Geophysics and Engineering. 2024, 21(May): 1242–1253. doi: 10.1093/jge/gxae056

[43] Fuad M. I. A., Zhao H., Jaya M. S., Jones E. A. J. SPE-214789-MS rock physics modeling of hydrogen-bearing sandstone : Implications for natural hydrogen exploration and storage introduction. 2023. doi: 10.2118/214789-MS

[44] Meju M. A., Saleh A. S. Using large-size three-dimensional marine electromagnetic data for the efficient combined investigation of natural hydrogen and hydrocarbon gas reservoirs: A geologically consistent and process-oriented approach with implications for carbon footprint Redu. Minerals. May 2023, 13(6): 745. doi: 10.3390/min13060745

[45] Li Z., Moskowitz B. M., Zheng J., et al. Petromagnetic characteristics of serpentinization and magnetite formation at the Zedang Ophiolite in Southern Tibet. Journal of Geophysical Research: Solid Earth. Sep 2020, 125(9). doi: 10.1029/2020JB019696

[46] Lefeuvre N., Truche L., Donzé F. V., et al. Natural hydrogen migration along thrust faults in foothill basins: The North Pyrenean Frontal Thrust case study. Applied Geochemistry. Oct 2022, 145. doi: 10.1016/j.apgeochem.2022.105396

[47] Kern H., Tubia J. M. Pressure and temperature dependence of P- and S-wave velocities, seismic anisotropy and density of sheared rocks from the Sierra Alpujata massif (Ronda peridotites, Southern Spain). Earth and Planetary Science Letters. 1993, 119(1–2): 191–205. doi: 10.1016/0012-821X(93)90016-3

[48] Christensen N. I. Ophiolites, seismic velocities and oceanic crustal structure. Tectonophysics. 1978, 47(1–2): 131–157. doi: 10.1016/0040-1951(78)90155-5

[49] Miller D. J., Iturrino G. J., Christensen N. I. Geochemical and petrological constraints on velocity behavior of lower crustal and upper mantle rocks from the fast-spreading ridge at Hess Deep.

Proceedings of the Ocean Drilling Program Scientific Results Volume 147. 1996, 147. doi: 10.2973/odp.proc.sr.147.028.1996

[50] Christensen N. I., Salisbury M. H. Sea floor spreading, progressive alteration of layer 2 basalts, and associated changes in seismic velocities. Earth and Planetary Science Letters. 1972, 15(4): 367–375. doi: 10.1016/0012-821X(72)90037-4

[51] Christensen N. I. Elasticity of ultrabasic rocks. Journal of Geophysical Research. 1966, 71(24): 5921–5931.

[52] Bonnemains D., Carlut J., Escartin J., Mevél C., Andreani M., Debret B. Geochemistry, geophysics, geosystems. Geochemistry, Geophysics, Geosystems. 2016, 17: 1312–1338. doi: 10.1002/2015GC006205.Received

[53] Falcon-Suarez I., Bayrakci G., Minshull T. A., North L. J., Best A. I., Rouméjon S. Elastic and electrical properties and permeability of serpentinites from Atlantis Massif, Mid-Atlantic Ridge. Geophysical Journal International. 2017, 211(2): 686–699. doi: 10.1093/GJI/GGX341

[54] Watanabe T., Kasami H., Ohshima S. Compressional and shear wave velocities of serpentinized peridotites up to 200 MPa. Earth, Planets and Space. 2007, 59(4): 233–244. doi: 10.1186/BF03353100

[55] Kroenke L. W., Manghnani M. H., Rai C. S., Fryer P., Ramananantoandro R. Elastic properties of selected ophiolitic rocks from Papua New Guinea: Nature and composition of oceanic lower crust and upper mantle. 2013, 19(706): 407–421. doi: 10.1029/gm019p0407

[56] Horen H., Zamora M., Dubuisson G. Seismic waves velocities and anisotropy in serpentinized peridotites from Xigaze ophiolite: Abundance of serpentine in slow spreading ridge. Geophysical Research Letters. 1996, 23(1): 9–12. doi: 10.1029/95GL03594

[57] Cutts J. A., Steinthorsdottir K., Turvey C., Dipple G. M., Enkin R. J., Peacock S. M. Deducing mineralogy of serpentinized and carbonated ultramafic rocks using physical properties with implications for carbon sequestration and subduction zone dynamics. Geochemistry, Geophysics, Geosystems. 2021, 22(9): 1–23. doi: 10.1029/2021GC009989

[58] Le Maitre R. W., Streckeisen A., Zanettin B., Le Bas M. J., Bonin B., Bateman P. Igneous rocks: A classification and glossary of terms: Recommendations of the International Union of Geological Sciences Subcommission on the Systematics of Igneous Rocks. 2nd ed. Cambridge University Press, 2002.

[59] Christensen N. I. The abundance of serpentinites in the oceanic crust. 1972, 80(6): 709–719.

[60] Mével C. Serpentinisation des péridotites abysales aux dorsales océaniques. Comptes Rendus – Geoscience. 2003, 335(10–11): 825–852. doi: 10.1016/j.crte.2003.08.006

[61] Schouten D., Furseth D., van Nieuwkoop J. Muon tomography for underground resources. In: Muography: Exploring Earth's subsurface with elementary particles. Wiley, 2022, 221–235. doi: 10.1002/9781119722748.ch16

[62] Seyfried W. E., Foustoukos D. I., Fu Q. Redox evolution and mass transfer during serpentinization: An experimental and theoretical study at 200 °C, 500 bar with implications for ultramafic-hosted hydrothermal systems at Mid-Ocean Ridges. Geochimica et Cosmochimica Acta. Aug 2007, 71(15): 3872–3886. doi: 10.1016/j.gca.2007.05.015

[63] Christensen N. I. Serpentinites, peridotites, and seismology. International Geology Review. 2004, 46(9): 795–816. doi: 10.2747/0020-6814.46.9.795

[64] Ross C. P. Effective AVO crossplot modeling: A tutorial. Geophysics. 2000, 65(3): 700–711.

[65] Carcione J. M. AVO effects of a hydrocarbon source-rock layer. Geophysics. 2001, 66(2): 419–427.

[66] Chi X., Han D. Lithology and fluid differentiation using a rock physics template. Lead Edge. 2009, 28(1): 60–65.

[67] Odegaard E., Avseth P. Well log and seismic data analysis using rock physics templates. First Break. 2004, 22: 37–44.

[68] Avseth P., Mukerji T., Mavko G., Dvorkin J. Rock-physics diagnostics of depositional texture, diagenetic alterations, and reservoir heterogeneity in high-porosity siliciclastic sediments and

rocks – A review of selected models and suggested work flows. Geophysics. 2010, 75(5): 75A31–75A47.

[69]   Castagna J. P., Batzle M. L., Eastwood R. L. Relationships between compressional-wave and shear-wave velocities in clastic silicate rocks. Geophysics. 1985, 50(4): 571–581.

[70]   Castagna J. P., Batzle M. L., Kan T. K., Backus M. M. Rock physics – The link between rock properties and AVO response. Offset-Dependent Reflectivity: Theory and Practice of AVO Analysis SEG. 1993, 8: 135–171.

[71]   Gassmann F. Elastic waves through a packing of spheres. Geophysics. 1951, 16(4): 673–685.

[72]   Carlson R. L. The abundance of ultramafic rocks in Atlantic Ocean crust. Geophysical Journal International. 2001, 144(1): 37–48. doi: 10.1046/j.0956-540X.2000.01280.x

[73]   Zimmerman R. W. Compressibility of sandstones. 1990.

[74]   Norris A. N. A differential scheme for the effective moduli of composites. Mechanics of Materials. 1985, 4(1): 1–16.

[75]   Schön J. H. Physical properties of rocks: Fundamentals and principles of petrophysics. Elsevier, 2015.

[76]   Bach W., Paulick H., Garrido C. J., Ildefonse B., Meurer W. P., Humphris S. E. Unraveling the sequence of serpentinization reactions: Petrography, mineral chemistry, and petrophysics of serpentinites from MAR 15°N (ODP Leg 209, Site 1274). Geophysical Research Letters. 2006, 33(13). doi: 10.1029/2006GL025681

[77]   Maffione M., Morris A., Plumper O., van Hinsbergen D. J. J. Magnetic properties of variably serpentinized peridotites and their implication for the evolution of oceanic core complexes. Geochemistry, Geophysics, Geosystems. 2014, 15(4): 923–944. doi: 10.1002/2013GC004993

[78]   Gardner G. H. F., Gardner L. W., Gregory A. Formation velocity and density – The diagnostic basics for stratigraphic traps. Geophysics. 1974, 39(6): 770–780.

[79]   Godfrey N. J., Beaudoin B. C., Klemperer S. L. Ophiolitic basement to the Great Valley forearc basin, California, from seismic and gravity data: Implications for crustal growth at the North American continental margin. Geological Society of America Bulletin. 1997, 109(12): 1536–1562.

[80]   Hatakeyama K., Katayama I., Hirauchi K. I., Michibayashi K. Mantle hydration along outer-rise faults inferred from serpentinite permeability. Scientific Reports. 2017, 7(1): 1–8. doi: 10.1038/s41598-017-14309-9

[81]   Tutolo B. M., Mildner D. F. R., Gagnon C. V. L., Saar M. O., Seyfried W. E. Nanoscale constraints on porosity generation and fluid flow during serpentinization. Geology. 2016, 44(2): 103–106. doi: 10.1130/G37349.1

[82]   Carlson R. L., Miller D. J. ß Oceanic gabbro / diabase. Most. 1997, 24(4): 457–460.

[83]   Iturrino G. J., Christensen N. I., Kirby S., Salisbury M. H. 11. Seismic velocities and elastic properties of oceanic gabbroic rocks from Hole 735B. 1991, 118: 227–244.

[84]   Dewandel B., Boudier F., Kern H., Warsi W., Mainprice D. Seismic wave velocity and anisotropy of serpentinized peridotite in the Oman ophiolite. Tectonophysics. 2003, 370(1–4): 77–94. doi: 10.1016/S0040-1951(03)00178-1

Brian J. Evans

Chapter 10
The potential geophysical responses
of trapped natural hydrogen gas

Abstract: This chapter explores geophysical methodologies for natural hydrogen exploration. It begins with a discussion of the well-established gas 'bright-spot' seismic processing and interpretation technique for identifying gas-filled porous zones beneath dense caprocks, supported by case histories. Advanced three-component seismic methods for detecting subsurface fracturing and fracture orientation are also highlighted. For regions with magnetite-rich dolerite caprocks, air-borne and ground-based magnetic surveys are proposed as a complementary approach to soil sampling, while air-borne LiDAR techniques are briefly acknowledged.

The chapter emphasizes the value of reprocessing existing seismic and magnetic data using modern artificial intelligence techniques to enhance interpretation and streamline exploration. The role of deeper structural traps and faults in compressional Neoproterozoic basins containing evaporites is also examined, with seismic, magnetic, and gravity survey data used to detect direct hydrogen indicators and associated geophysical signatures, supported by two case histories. Finally, the challenges of data acquisition, interpretation, and integration are addressed, underscoring the need for a multidisciplinary geophysical approach to advance the understanding of natural hydrogen systems and enable efficient and sustainable exploration strategies.

Keywords: natural hydrogen, geophysics, seismic, DHI/DHHI, bright spot, magnetic and gravity surveys, crustal systems, reflectivity coefficient, AVO/AVA, three-component, crosswell tomography, artificial intelligence (AI)

10.1 Introduction

Exploration for natural hydrogen has become a topic of interest over the last few years, following the discovery and production of natural hydrogen in Bourakebougou, Mali, in 2011 [1]. The original well was drilled to produce water, passing through a shallow dolerite volcanic flow 112 m thick.

Acknowledgments: This chapter would not have been written without the WA State Department of Mines, Industry Regulation and Safety (DMIRS) open-file GeoVIEW.WA website data, for which the author is profoundly grateful. https://geoview.dmp.wa.gov.au/geoview/?Viewer=GeoView

Brian J. Evans, Curtin University, Perth, WA, Australia, e-mail: b.evans@curtin.edu.au

https://doi.org/10.1515/9783111437040-010

This is the situation where there is a major geophysical difference between a high-impedance rock (e.g., shale, carbonate, or igneous) and a low-impedance rock (e.g., sandstone), resulting in a large acoustic contrast between the rocks. If a seismic wave is transmitted toward this strong acoustic contrast, the seismic waves will either diffract (scatter) or have a major portion of their energy reflected. As a result, we can use seismic reflection methods in the exploration for gas since a porous sandstone containing gas, overlain by a competent caprock such as shale or crystalline rock such as dolerite, will result in a large amount of seismic energy reflecting from their interface, producing an amplitude anomaly we call a 'bright spot' in our seismic profile data. Such bright spots are one of the major reasons we can find more gas than oil because oil within a rock does not lower the acoustic impedance to such an extent as gas does.

So, we have existing tools to search for gas beneath a hard cap rock (the gas in this case being hydrogen, but it may also be mixed with helium since these two elements are often found together), and the next question still remains: can we detect fractures in rock which may contain hydrogen (and/or helium)? The answer is a limited 'yes.' The limitation is that we must use specialized techniques that require different non-standard seismic methods, and our seismic signal must be consistent, clear, and strong to ensure that our interpretation is correct (compared with 'bright spot analysis' which is a well-established practice and can be automated to a degree). The specialized techniques for fault and fracture detection include the use of three-component seismic geophones and, where possible, three-component seismic energy sources. The seismic source in this case must generate controlled compressional and shear-waves, which are needed to understand the presence and orientation of any faults or fractures.

We must also keep in mind that we are dealing with hard igneous rocks, not just soft sedimentary rocks. These igneous rocks often contain magnetite, which exhibits a magnetic response that, with sufficient magnetism, can hopefully be detected using either ground-based or air-borne magnetometer instruments. Used in conjunction, there is the potential to link the interpretation of a seismic 'bright spot' with a change in magnetism, allowing a 'joint seismomagnetic interpretation' to be performed to fit with the known geology of the area. However, we must first consider the geological setting before evaluating the use of any specific geophysical tools.

10.2 Geological settings for natural hydrogen

The locations around the world where natural hydrogen seeps occur seem to be in areas that are at the sedimentary edges of cratons [2] or beneath caprocks overlaying compressional zones such as those causing the Mali [3] deposit and the Pyrenees to develop [4]. Certainly, in Western Australia, there are surface depressions known as 'fairy circles,' which lie at the western edge of the Yilgarn craton, where the Darling Fault is considered to be the migration pathway [5].

Consider that the upper or lower crust of the Earth is under compressional stress as continental plates tectonically move toward each other, uplifting mountain chains or subducting. This results in the formation of volcanoes, surface basalt pavements, and subsurface dolerite sills within the soft regolith sedimentary layers. This is a possible model for Mali, whereas the Ramsay-2 well in Australia, which was recently drilled on the Yorke Peninsula, encountered 95% H_2 in fractured sediments at 530 m and 6.8% He at 900 m – considered to be just above the crustal fracture zone.

Of course, when a gas is capped at a shallow depth, its pressure is very low in the order of 4–5 bar (60–70 psi). If the sedimentary basin was deeper and such dyke/sill formation occurred, then the pressures might be around 200 bar (about 3,000 psi), which would be more suitable for commercial production as a high-pressure gas flow compared to the lower level at 5 bar.

10.2.1 A model for a deeper trap of hydrogen

In the search for a geological regime that will allow gas to be trapped at sufficient depth to make it highly pressurized, the seal must be considered as being completely impervious to the passage or escape of hydrogen. This is very difficult to achieve, especially since H_2 has the ability to pass through any standard oil and gas cap, such as shale, so a more competent rock than shale is needed. While a thick dolerite sequence is one solution (and there are many other rock sequences that can equally form competent and thick seals), another certain seal is that of evaporites (e.g., halite or salt). Salt readily moves, flows, and fills any gaps, faults, or fractures that it encounters, so it is of interest to review the types of deep sedimentary basins that have salt overlying sediments suitable as reservoirs such as sandstones.

The geology of Neoproterozoic-era basins provides an immediate answer to this search. During the Neoproterozoic 750–540 Ma, the supercontinent Rodinia, located south of the equator, contained several inland seas (where stromatolites began to emerge). However, it later experienced a series of breakup events that led to continental drift and a succession of glaciation events [6–9]. These inland seas began forming sedimentary basins within Rodinia, which rested upon a rigid continental crust 40–50 km thick. There was a series of north-dipping thrusts that extended down to the Moho, and large areas subsequently began to subside due to mantle instability. Some parts were uplifted, and large volumes of sand-sized clastic materials were deposited. Further uplift and deposition resulted in the formation of a large sand sheet at the base of the Neoproterozoic section.

Compressional tectonics then reactivated earlier thrust faults within the crust, causing uplift and the development of numerous smaller basins. With uplift and further deposition, large sheets of evaporites were deposited, followed by a period of glaciofluvial deposition [10]. Subsequently, the continual compressional stress-state of these basins resulted in the occasional development of further crustal fracture systems, encouraging lava flow to the surface and resulting in a mix of dolerite sills interleaved with salt, as is typically found in most of the Australian Neoproterozoic basins.

The geological model for this form of trapping mechanism is presented in Figure 10.1. In this model, in classic oil and gas parlance, we observe the presence of a source, seal, migration pathway, reservoir, and trap. The H_2 migrates upward, either through hydrostatic force or buoyancy, via iron-rich basement fractures, accumulating beneath a salt or dolomitic trap within a porous sandstone. The presence of fractures intersecting thinning salt or dolerite layering may provide a conduit for H_2 to pass through and continue migrating toward the surface. H_2, as a gas, is so mobile that it does not require a fracture to travel upward; it will migrate through any sedimentary section even if a fracture is unavailable, albeit at a much slower rate.

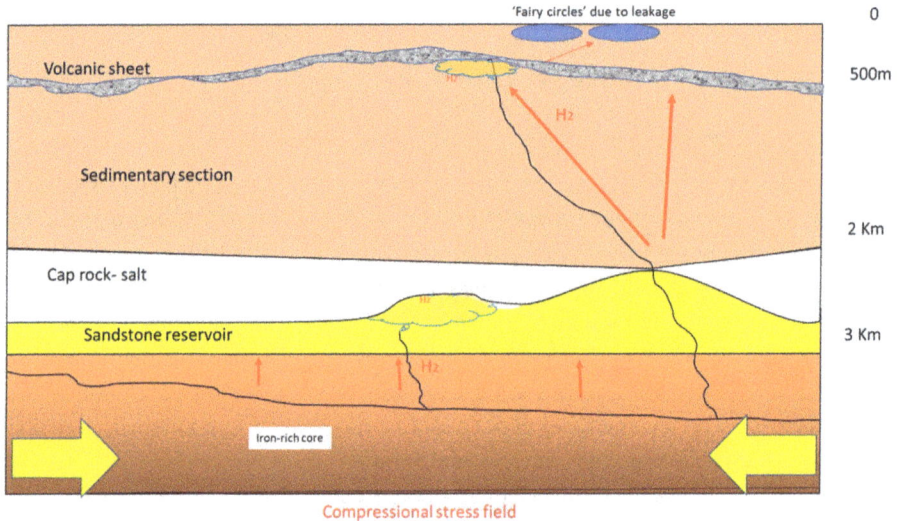

Figure 10.1: A model for the capture of natural H_2 by an evaporite or a volcanic sheet.

While there are many Neoproterozoic basins around the world that have experienced a compressional stress regime over their lifespan (Australia, Saudi Arabia, Oman, Europe, Brazil, China, etc.), one of the largest is the Officer Basin of Western and South Australia. The Officer Basin offers both shallow dolerite flows and deep salt with an underlying sandstone for potential H_2 trapping and storage. It is characterized by a complex stratigraphy, with a sandstone layer at the base of the basin overlaid by a salt sheet, followed by a sedimentary layer with erosional features, and finally crustal tectonics involving compression, rotation, and volcanism [11]. These geological processes have caused salt diapirism and the intrusion of igneous rocks, including dolerite dykes and sills (Figure 10.2).

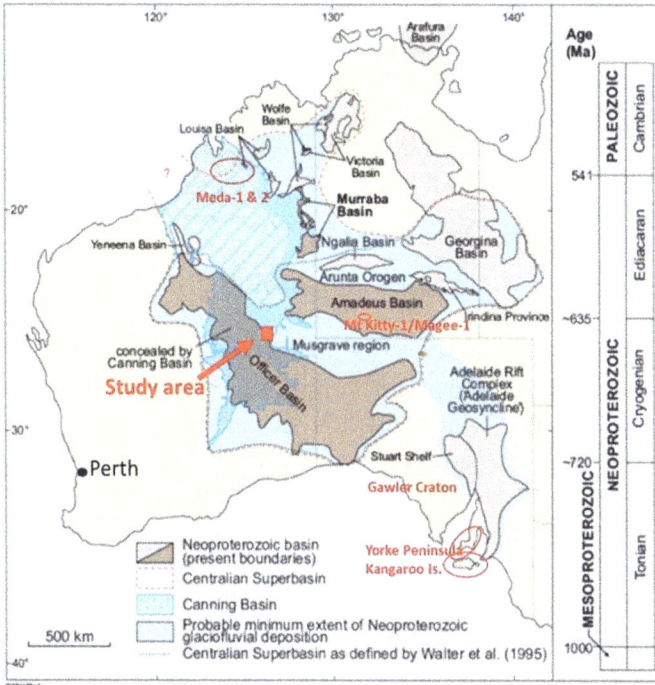

Figure 10.2: Example of the Neoproterozoic Officer Basin developed from the Centralian Superbasin, with drill hole areas exhibiting high levels of H$_2$ shown in red, as well as a study area designated as the geophysical study area (modified from Walter et al. [15]).

At the time of writing, there is ongoing drilling activity on Yorke Peninsula, proving-up both high H$_2$ and He content. Petroleum exploration wells Mt Kitty-1 and Magee-1 in the adjacent Amadeus Basin also recently produced high levels of H$_2$, while petroleum wells Meda-1 and Meda-2, located at the northern edge of the Canning Basin, recorded strong H$_2$ shows in 1958.

As mentioned, the Officer Basin contains all of the ingredients needed to be attractive for H$_2$ exploration and production. The problem with the basin is that there are no roads passing through it; instead, there are rough tracks. From 1979 to 1985, Shell conducted major seismic, gravity, and magnetic surveys through the area in the belief that it had similar geology to Oman, with major salt domes and oil. Gas was not a required commodity in those days.

The author of this chapter was a senior geophysicist with Shell at the time and speaks from personal experience. He was responsible for the operations of two seismic crews, and to have them work in the field, specialized seismic roads and tracks were constructed for hundreds of Kilometers across the country. Line spacing was typically 50 km. One issue that was very clear was that occasionally a line bulldozer would hit a surface basalt pavement, which extended for many hundreds of meters and so we were aware of the presence of basalt dykes and sills, typical of a Neoproterozoic basin. However, we were interested in oil targets at 2–3 km, and so such basalts were more a nuisance than a benefit. The area remains unexplored simply because of the lack of access, and it was so expensive to work that after five years, the budget was depleted with only two wells drilled – both only useful as stratigraphy data. However, this did refine the model as will be explained.

10.2.2 Refined model for a Neoproterozoic basin's structural geology

Figure 10.3 shows the refined basin model for the Officer Basin. Features that have become apparent with twenty-first-century seismic data processing include the crustal uplift zone, which exhibits fracturing along pathways through which H_2 may migrate upward to the basement surface. Above this, the Townsend Quartzite is variously distributed, and above it lie the intruded salt layers. Salt acts as an excellent seal for H_2 movement, filling any potential fractures. Consequently, H_2 may be trapped within the Townsend Quartzite where local fractures extend into the salt. Since the Officer Basin has undergone compression and rotation, any crustal impediments form triangulation zones, while diapirs of various shapes and sizes extend upward.

From a geophysical perspective, these gas traps should occur as seismic direct hydrocarbon indicators (DHIs) 'bright spots' beneath the salt—here, this term is replaced by 'direct hydrogen/helium indicators (DHHIs).' If data is processed properly, then conventional geophysical interpretations could be performed on existing seismic data, where the seismic provides structure, magnetics data provides a shallow basalt/dolerite guide, and gravity data provides an indicator of the presence of salt diapirs. We are looking for compressional basement fractures overlain by sandstone, which, in turn, is sealed by salt. Where the salt thins, we expect H_2 to migrate through, possibly to collect in sills.

Officer Basin has all of these tenets. It has a compressional basement with the Townsend Quartzite (which is tight, but at these pressures, that is acceptable since it may be fractured) covering it. This is overlain by a sheet of halite containing many salt diapirs of all shapes and sizes (although classic Gulf of Mexico salt domes do not seem to be present). Above the salt, from about 2 km and upwards, is the conventional sedimentary Proterozoic section. There are also the ever-present surface basalts, which caused such a problem for the seismic line bulldozers, so it is known that the 'Table Hill volcanics' not only breach the surface but also form dolerite sills in-transit to the surface. The Yowalga-2 well drilled through a very competent sheet of dolerite at 400 m depth, taking the drillers by surprise (they were using a standard roller-cone bit, another reason the drilling budget was blown).

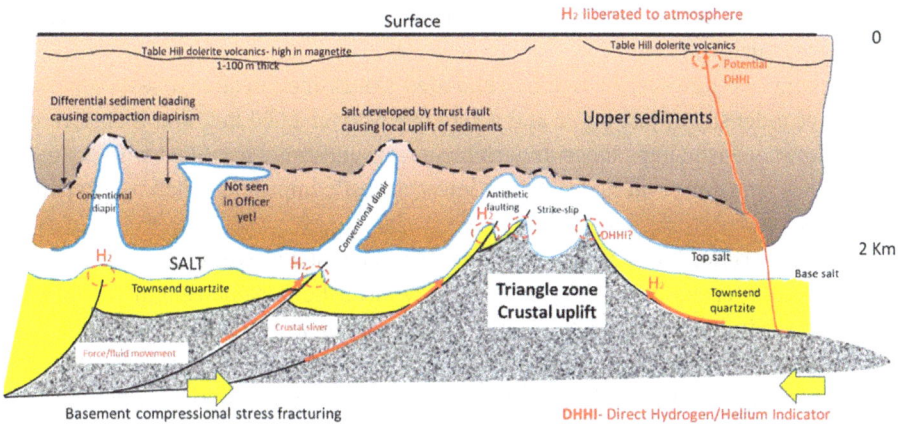

Figure 10.3: A refined model for Officer Basin geology and the development of geophysical indicators.

Consequently, all that is required now is to understand our geophysical responses and apply them to the existing public data to gain insight into its prediction of potential H_2 locations.

10.3 Geophysical indicators using seismic, gravity, and magnetic data

We use seismic reflection data to produce an image of a section through the Earth's rock layers, which is based on the reflectivity coefficient discussed earlier. We use gravity data, which responds to subtle changes in the Earth's gravity. Because salt diapirs are formed under extreme pressures in the basin, diapirs can be highly compacted and cause a local change in gravitational forces. We also use magnetics data because, if an intrusive rock (such as dolerite) contains magnetically intensive properties (such as magnetite), we can map its location as a function of the local change in the Earth's magnetic field. However, where gas (such as H_2) is concerned, the most useful tool is seismic data, and so we must first understand that.

10.3.1 Seismic amplitude response

In rock physics, a rock has the parameters of density and sonic velocity. If the rock has little permeability or porosity, it often has a high density (which is typical of dolerite, having a density of around 3.2 g/cc and a sonic velocity of over 5,000 m/s), as explained earlier. The multiplication of these two parameters – density (ρ) and velocity (*V*), represented as ρV –is known as the acoustic impedance, and in this case, it is considered 'high.'

By contrast, however, a weakly cemented sandstone, ideal as a reservoir rock, may have a density of just 2 g/cc and a sonic velocity of 2,000 m/s, and therefore will have a 'low' acoustic impedance. Invariably, most sandstones are filled with water, so a gas-filled sandstone will have a much lower acoustic impedance than a water-filled sandstone.

A seismic reflection is caused by the transmission of a pressure wave through the Earth to the interface between a rock with low acoustic impedance and a rock with high acoustic impedance. This is the same phenomenon as when a person shouts in a cavern—the noise travels through a low acoustic impedance medium (air) and is reflected back upon reaching the high acoustic impedance rock, causing 'echoes.' These are seismic reflections, and they occur when we pulse the ground (using explosives) or vibrate the ground (using vibrators). We then receive the reflected energy on geophones on land or hydrophones in water, which are very similar to the acoustic crystals that vibrate as sound passes through a mobile phone. In seismic profiling, we use numerous reflections to produce a 2D section through the Earth, much like viewing the image of a 2D CT scan—it operates on the same principle.

So, when a seismic reflection occurs, its location is at the junction of two rocks having different acoustic impedances. That is, the acoustic impedance of a rock, *Z*, is given by

$$Z = \rho.V, \tag{10.1}$$

where ρ represents density and *V* represents sonic velocity. Additionally, as the vibrating wave travels through one rock and is reflected at the intersection point with the next, the amount of energy that is reflected, known as the reflection coefficient *R*, is given by

$$R = \frac{(\rho_2 V_2) - (\rho_1 V_1)}{(\rho_2 V_2) + (\rho_1 V_1)}$$

Or simply

$$R = (Z_2 - Z_1)/(Z_2 + Z_1), \tag{10.2}$$

where ρ_1 is the density and Z_1 is the acoustic impedance of the first layer it passes through, while ρ_2 is the density and Z_2 is the acoustic impedance of the second layer [12].

When numbers are inserted, this shows how much energy is actually reflected at each interface and how much, therefore, is transmitted through the second rock. Because the further a wave travels, the more its energy level dissipates with each reflection, the reflected signal can become so weak that it needs to be enhanced. We do this by firing off many vibrations at different locations, referred to as different 'offset' recordings, and we can display them alongside each other. This data processing technique enhances the reflected signal.

10.3.2 Seismic wave characteristics

So, in hydrocarbon exploration, a good reflection amplitude occurs when a hard cap rock (e.g., shale) with a density of 3 g/cc and sonic velocity of 3,200 m/s (so $Z = 9,600$) caps a sandstone containing gas with a density of 2.6 gm/cc and velocity of 2,200 m/s (so $Z = 5,720$). This impedance contrast will produce a negative reflection in eq. (2) because the lower layer (Z_2) has a lower value than the upper cap rock (Z_1).

If a rock is a normal, competent shale, H_2 will migrate through the rock over time because its atoms are very small, whereas other gas accumulations tend to have larger atoms and, therefore, will take a very long time to pass through, if at all. However, if we introduce an evaporite rock such as halite (salt), it has zero pore space to allow the passage of atoms. Consequently, salt and anhydrite are excellent seals, as are some volcanic rocks such as dolerite, which may have no passage space.

In Mali, the sealing rock is a dolerite, which resulted from an intrusive dolerite dyke passing upward and trapping the rising hydrogen in a lateral sill, producing an impervious hydrogen trap at a depth of 112 m. By contrast, at the Monzón field in the Southern Pyrenees of Spain, the sealing rock is halite interbedded with shale, but this time at a depth of 3,500 m [4]. The acoustic impedance of dolomite varies, but generally, Z_d is around 14,000–20,000, which is at least twice as hard and compact as the average shale mentioned earlier. Halite, with a typical velocity of 4,600 m/s and a density of just over 2 g/cc, therefore has an acoustic impedance Z_h of around 9,200. This shows how dolerite can be twice as compact as halite (salt) or shale and would be an excellent seal, potentially better than any evaporite except anhydrite (in which $Z = 45,000$).

If a seismic wave travels downward through a lower acoustic impedance rock into a higher impedance rock (such as sandstone followed by more compacted claystone), eq. (2) will indicate that the energy of the reflection is positive because Z_2 (claystone) is overlain by Z_1 (sandstone), and this produces a typical seismic section where reflection traces kick in a positive direction. However, as explained with the gas trap, if Z_2 is less than Z_1, the result is a polarity reversal of the reflection.

10.3.2.1 Amplitude versus offset analysis

So far, we have considered a vertical wave arriving at the reflection interface between two rocks. If we now move the seismic source and receiver apart at the surface so that the arriving energy has a larger incidence angle than zero, the amplitude response will change.

We now have signal amplitude changing with different offsets, and variations in these amplitudes give an indication of the presence of water, say, compared with gas, as suggested in Figure 10.4. The longer the offset distance is, the greater the emphasis on whether the caprock is trapping gas or not.

Seismic reflection display

Brine   Oil   Gas

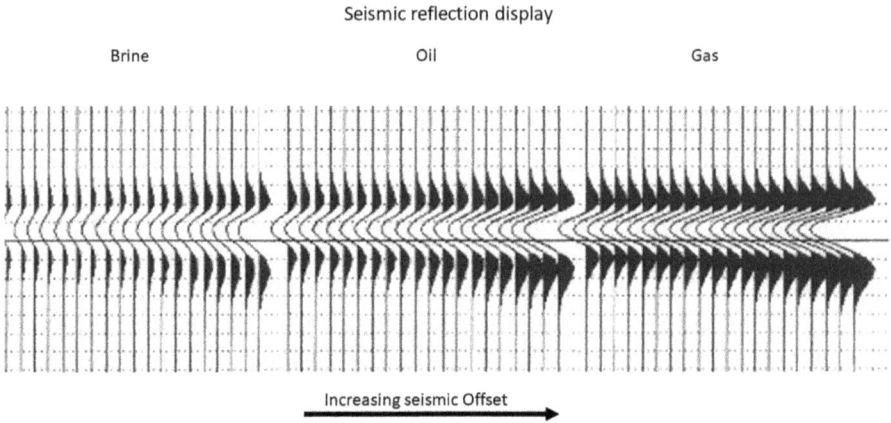

Increasing seismic Offset

Figure 10.4: If a gas is trapped beneath a high acoustic impedance rock, it appears with a higher amplitude than if it were water.

According to the well-established Zoeppritz equations [13], if there is gas in the lower layer causing Z_2 to reduce the reflection amplitude in value (and potentially reverse in phase), at larger angles this reflection energy will appear to increase in amplitude if Z_2 is caused by gas. Shuey [14] simplified the equations to the point where a plot could be drawn of the zero-reflection amplitude value (zero incidence angle) versus the longer offset reflection amplitude value. This became known as the amplitude versus offset/angle (AVO or AVA) plot and is described by the basic Shuey equation for incidence angles between 0° and 30° (the usual case in seismic surveying):

$$R_{(\theta)} = R_{(0)} + A \, Sin^2\theta \tag{10.3}$$

where $R_{(\theta)}$ is the reflectivity value at any angle, $R_{(0)}$ is the zero-offset value, and $A \, Sin^2 \theta$ is the gradient of the change as the angle increases (i.e., at the longer receiver station offsets). Applying this equation to seismic data can provide independent proof of the interpretation of a gas accumulation, which is more acceptable to management in the event that there is a proposal to drill [15].

This could work well for near-surface reflections, such as those at the shallow dolerite seal interface, but perhaps not so well for the deeper dolerites or salt diapirs, where a reflection's nonlinear move-out may cause this basic equation to be ineffective [16].

In hydrogen exploration, if a dolerite (Z_1) overlays a sandstone containing gas (Z_2), we can expect to see a strong amplitude reflection with a large amplitude anomaly beneath it. If the dolerite is very near or at the surface (as in Mali), then we can expect a strong reflection amplitude with a negative phase kick from the dolerite/gas interface. Being near the surface, if the seismic equipment is laid out with the nearest receiver station farther away than the gas trap's distance below the surface, then the amplitudes observed will only be those when the source travels over the top. Therefore, the best AVO analysis would occur if a shorter-spaced receiver survey recorded a specialized survey. The point is that the hydrogen reflection polarity from gas beneath a dolerite should be a negative phase, so at least this will help establish if a special survey is needed as proof.

Performing true amplitude recovery (TAR) processing of conventional seismic data and then computing the AVO response as a result has proven to be a great success in conventional gas exploration. This is the reason why so many more gas discoveries have been made since the mid-1980s (seismic sections can be delivered with red zones showing where the AVO is high). The large amplitude reflection has often been referred to as a 'bright spot,' while the trace reversal (indicating the presence of gas) has been called a 'dim spot' due to the 'phase change' as well as when this coincides with a low impedance difference, leaving a hole in the continuity of reflections [15]. To be complete, we also have 'flat-spots,' where hydrocarbons may be observed as flat reflections.

Typical published examples of locations where there are bright spots are shown in Figure 10.5 as DHIs, which are well established in the petroleum exploration industry. Where gas may be bubbling up at specific locations, it can disrupt the seismic signals that are passing through the gas, and since this disruption is very often vertical, these are known as 'gas-chimneys,' examples of which are shown in Figure 10.6. Further information on how these events appear on a seismic section will now be discussed.

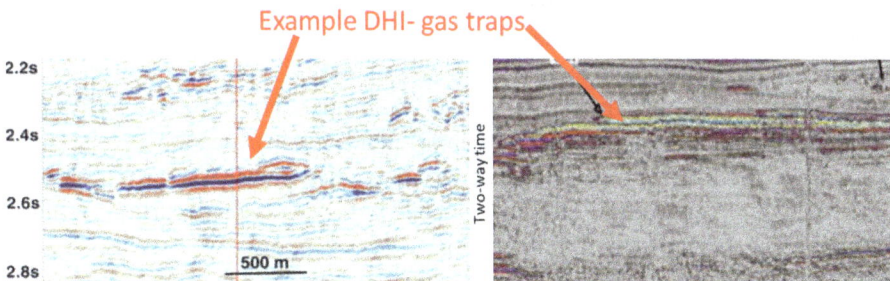

Figure 10.5: Typical examples of DHI seismic interpretations, where gas appears to accumulate. These bright spots would have been drilled to confirm their existence.

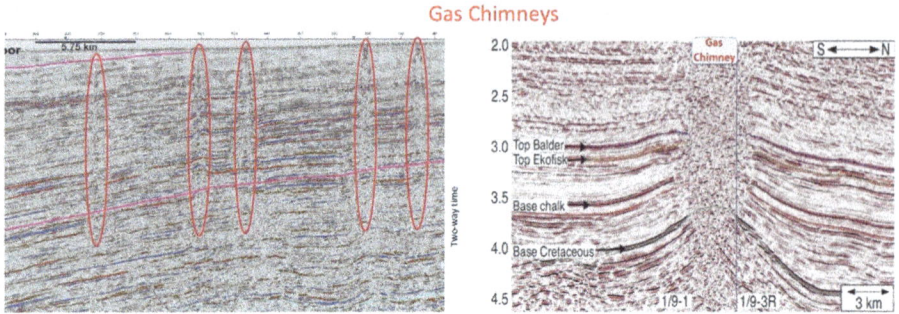

Figure 10.6: Another form of petroleum DHI is a gas chimney. Seismic reflections are disrupted by the constant leakage of gas to the surface. These could be indicators of the presence of an H_2 plume if observed leaking from the side of a bright spot gas accumulation (modified from Granli et al. [17]).

10.3.2.2 Seismic interpretation applied to Officer Basin study area

In Figure 10.2, a study area is indicated, which will now be used to emphasize some of the points being made about the interpretation of possible H_2 accumulations in this study area. It was stated that this area possesses all the requirements to demonstrate geophysical tools for the prediction of H_2.

The routine for interpreting conventional seismic sections, which may appear to show bright or dim spots, involves reprocessing the seismic data and performing an AVO analysis specifically on the section of the line that is of interest. TAR reprocessing may highlight other bright areas that were not apparent on other sections of the line. In the hydrocarbon exploration industry, these amplitude anomalies, along with vertical areas that do not receive reflections but are associated with gas-chimney effects, are generally referred to as DHIs. However, in H_2/He exploration, these are now referred to as *direct hydrogen/helium indicators* or DHHIs [18].

Because in H_2 gas traps we will always have a very high impedance (e.g., dolomite) overlying a very low impedance (H_2-saturated rock), and gas-chimneys may be prevalent along fractures or fault systems, we can then expect to first interpret the deep-seated faulting and then look through the deeper section for any bright/dim DHHI to be found in local high areas (bumps) immediately beneath an evaporite or salt seal. Next, search up any local faults for a migration pathway (or chimneys) and finally search near the surface for any possibility of a DHHI where surface fractures may serve as a further conduit for the escape of gas.

If a chimney is interpreted, an accumulation may be close by near its top, which is being fed by the chimney, or the chimney may be to one side and above the DHHI, which infers that either a deeper DHHI is leaking or a shallow accumulation has reached the 'fill and spill' point. The confirmation of a gas chimney may show a sag in the strata beneath it since the gas chimney reduces the seismic velocity in rocks beneath it. Therefore, conducting a velocity check is another approach. Either way, a high H_2 gas content in surface soil samples should provide the answer to the question of whether there is a near-surface leak.

If the dolerite is near the surface, it is possible to use magnetic mapping to determine where the fracture travels laterally and whether there is any gas signature wherever it travels. Meanwhile, gravity maps can provide a guide to whether there are near-surface salt accumulations or diapirs, which become very attractive if a chimney appears beneath or to the side of one.

Two lines from the northern Officer Basin were interpreted using existing data without further processing, utilizing open-file seismic, gravity, and magnetic data.

Figure 10.7 shows the 50.5 km of Line 83–05 running SW-NE, ending just short of the petroleum well Hussar-1. The tectonic concept is that a footwall under compressional stress exhibits forethrusting and backthrusting sections similar to the simple tectonic sand-box model developed by West Virginia University. The basement stress field is pushing upward but not quite over the footwall, while the backwall is working in the opposite direction. Strike-slip movement can occur due to stress rotation so the relative sedimentary levels of the thrusting on each side of the structure can differ from the model.

When interpreting this section, we observe the salt sheet dipping down toward the footwall, and the salt is at different levels on either side of two listric faults. Thus, we apparently have a strike-slip fault zone caused by the tectonic thrusting indicated by the arrows.

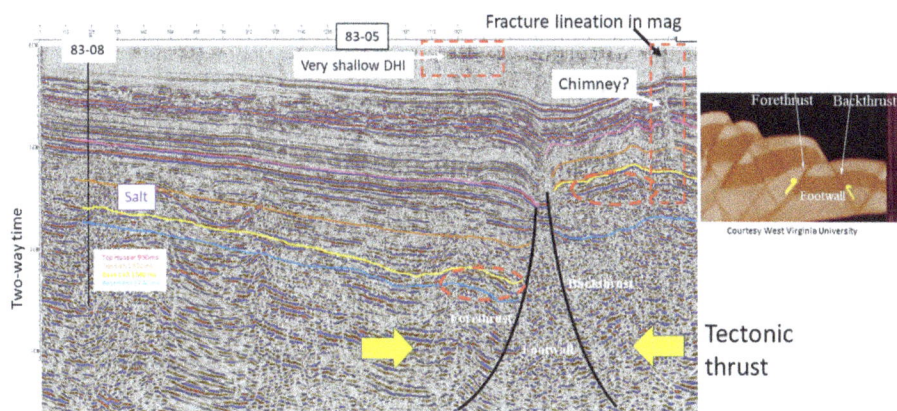

Figure 10.7: This open-file seismic line stopped short of the Hussar-1 well so the interpreted stratigraphy is widely accepted. A comparison of the line with that developed by tectonic modeling illustrates how a footwall is formed between forethrusting and backthrusting tectonic stresses.

In terms of the interpretation of an H_2 accumulation, we start by looking for local 'bumps' at the base of the salt (interpreted yellow horizon) and adjacent to any form of H_2 passage such as a fault or basement fracture. At the forethrusted fault, if we consider that H_2 may migrate up this listric fault, there is a local high that has been circled in red, so within this zone, it may be worthwhile to perform an AVO analysis on the reflection. On the other side of the footwall (backthrust), we can work our way up that fault and look for any local highs in the base of the salt near-by. Immediately, a good-quality strong reflection appears to the right of the backthrust fault, which could be an accumulation of H_2 that has migrated up the fault, met the base-salt barrier, and migrated up-dip, coming to rest in the area highlighted by the red dotted circle.

To support the notion that this area contains free H_2, the potential DHHI reflection looks solid and is located at just over 1,000 ms two-way time (about 1 km depth). More importantly, there appears to be a vertical disturbance in the seismic reflections to the right of this potential DHHI reflector, which exhibits the hallmarks of a gas chimney. This feature passes up through the strata via a distinct fault and appears to escape into the atmosphere. The conclusion drawn is that this is a gas reservoir at 1 km depth, which is at the fill-and-spill stage, actively leaking its H_2 into the atmosphere. This could potentially be the location of an H_2 'fairy circle,' so satellite photos would need to be checked for confirmation.

If this is a fault or fracture extending up to the surface, it may possibly be visible on the magnetics map. Another feature is also apparent at the top end of the seismic line. There appears to be a very shallow (within 200 m) bright spot to the left of the footwall faulting (labeled in Figure 10.7 as a very shallow DHI). This could be a dolerite sill trapping an H_2 accumulation, and again, this is a reasonable interpretation since magma may well have passed up either side of the fault system, up through fractures above the footwall, and spread out as a dolerite sheet below the surface (the Yowalga-2 well drilled into such a sheet at 400 m, and this line is not far from that well site).

Therefore, an AVO analysis could test this hypothesis, and better still, the areal extent of this seismic anomaly should initially be tested by a soil sampling survey. It now becomes necessary to review the magnetics map for the area since we want to determine if there could be a strong magnetic field in this area and, if so, use the magnetics map to see if there are faults or fractures associated with the interpreted gas chimney.

10.3.2.3 Magnetics and gravity interpretation applied to Officer Basin study area

In Figure 10.8, the air-borne magnetics map for this study area shows a shallow lineation (a probable fracture) passing from south to north across the northern tip of the seismic line and south of the well Hussar-1. Seismic line (Figure 10.7) indicates that where this fracture passes across the line is the location of the interpreted gas chimney, so it is likely that H_2 leakage could be found here.

Figure 10.8 shows line 83–05, colored in red, passing across an area of high magnetic intensity (shown as dark shading). Comparing this with seismic Figure 10.7, the very shallow DHHI is located on the west side of the footwall. This is interpreted as having a shallow Table Hill dolerite blanket, highlighted as a bright spot where it is trapping the gas. The seismic section in Figure 10.7 only shows the DHHI signature of trapped gas in one confined area rather than as a reflection across the full extent of the top of the section.

Turning to the air-borne gravity map for the study area in Figure 10.8 (at the same scale), high-density salt diapirs and walls appear as light blue, and the walls of salt and diapirs are seen pushing upward through the sediments as a result of tectonic compression from the northeast. Consequently, walls of salt are observed trending northwest-southeast. In this area, there is further tectonic stress influence from the granite Musgrave Block overthrust zone, which has its principal stress affecting the local salt movement. This block, shown in the gravity map in a strong green color (high gravity), causes the local salt wall to meander, as indicated by the white-dotted line. A local salt diapir is visible on the eastern side of the seismic line, which explains why the salt to the east of the backthrust fault is shallower than to the west, and therefore the footwall must run northwest-southeast.

Consequently, in deep salt-prone basins such as those of the Neoproterozoic, joint interpretation of seismic, gravity, and magnetics can provide an invaluable story, validating the interpretation of H_2 locations. To prove this is a correct interpretation, a short seismic grid is needed over the top of the shallow dolerite area, along with a soil sample survey across it. The seismic grid would confirm its areal extent, while the soil survey would confirm any H_2 leakage. Whichever way it is viewed, it is a very shallow drilling target, similar in depth to Mali.

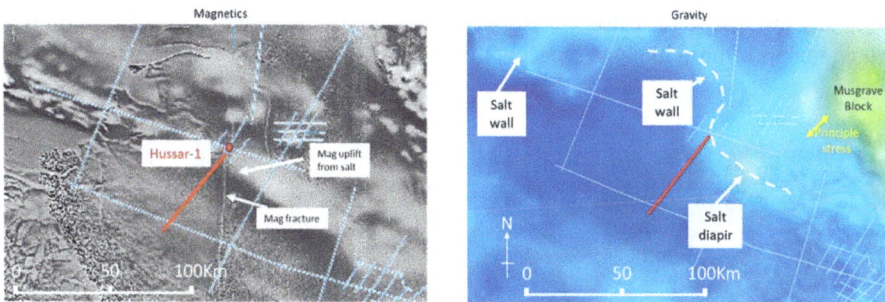

Figure 10.8: Interpretation of magnetics and gravity data. The magnetics map shows near-surface faults and fractures, while the gravity data indicate the locations of salt diapirs and walls.

10.3.2.4 A joint seismic, gravity, and magnetic interpretation

The next chosen seismic line for study is 83–06, which is 141 km long and passes directly through the well Hussar-1, as shown in Figure 10.9. This line travels from NW-SE over the edge of uplifted salt, then, according to gravity data, over a deeper sedimentary section, and finally over the flanks of the diapir to the east. It passes close to the end of the previous line, with the well providing the horizon tie-data interpreted in the figure, and the general principal stress orientation is at an angle to the line.

The seismic data suggests strike-slip faulting along the center of the line, with a high at the location of the Hussar-1 well. The well was a petroleum exploration well originally drilled in 1982 to a total depth of 2,040 m on a geological high, as can be seen on the seismic data. It was terminated when halite was intersected, with no oil or gas shows, and as was common practice in those days, neither oil nor gas would be expected beneath a salt high. There was no interest in gas at that time anyway, but it is arguable that there may be natural hydrogen beneath this salt blanket since there appears to be a bright spot beneath a salt cap. Four further bright spots are possible at the base of the salt and adjacent to interpreted basement fault tips (up which H_2 would have migrated, coming to rest in the Townsend Quartzite reservoir).

Of interest to the right of Hussar-1 are two potential halite diapirs, with the seismic possibly imaging just a slice through two vertical salt walls. Adjacent to the wall on the east side (right), broken reflections appear above 800 ms (about 800 m depth), suggesting that this may be another gas chimney. It is then interesting to note that, to the east of this, at around 100 ms (100 m), a strong reflection occurs, indicating another bright spot. It is assumed that this is the Table Hill dolerite, which is trapping H_2 and is being fed by the chimney beneath it. This could be an outstanding Mali look-alike, worthy of drilling. If not H_2, it may alternatively contain CH_4 but at low pressure. The point is that, if this is the case, it is being constantly fed by a gas chimney, which makes it more likely to be economic.

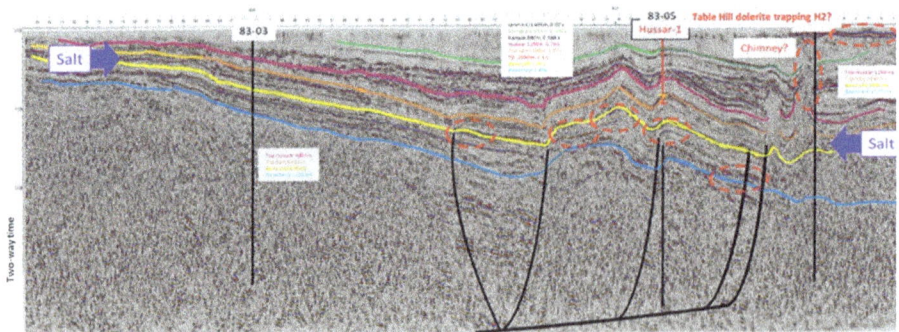

Figure 10.9: Seismic line showing potential deep bright spots, a chimney, and a shallow Table Hill trap.

The magnetic map shown in Figure 10.10 indicates that, to the west, the line travels over the top of a discordant magnetic ridge, which breaks-up half-way along (not imaged on the seismic because it is not trapping gas). It then appears again east of Hussar-1 as it passes over the edge and flanks of the salt wall. It is in this section that the Table Hill dolerite appears to have a bright spot.

The gravity map simply shows the line traveling from a relatively shallow (400 m) salt, which then deepens to the east until the Hussar-1 well is encountered, at which point it passes over the flanks of a salt wall and continues. It is worthwhile noting that the magnetics map shows the magnetic data weakening at the east end of the line and then strengthening again, so it is possible that gas may be leaking out at the end of this line.

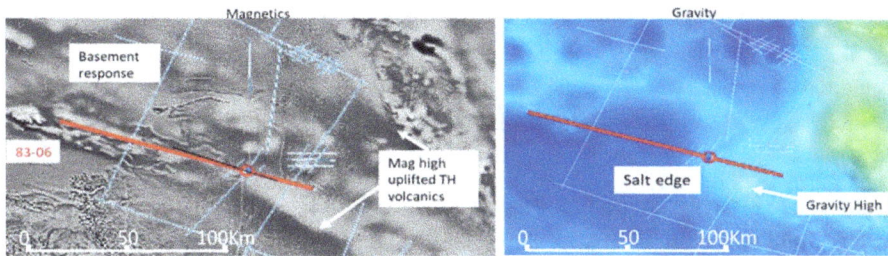

Figure 10.10: Interpretation of gravity and magnetic data.

This joint seismic, gravity, and magnetics interpretation of these two lines has demonstrated how useful geophysical mapping of a geological area can be, particularly in regions with salt tectonics, fracturing, and shallow dolerite flows trapping gas. When a bright spot appears to be encountered in the seismic data, an AVO study can confirm the presence of gas, whether it is shallow or deep-seated. However, the key point is that this is an independent study involving the reprocessing of existing data, rather than necessarily collecting new data.

The magnetic survey map in Figure 10.11 shows how useful magnetics can be in corroborating the presence of a shallow seismic bright spot, while the gravity map is useful in cases where salt may introduce a complexity that needs to be accounted for.

If there is no salt, a gravity survey would not be expected to provide much additional information. Therefore, in cases where we are drilling into crustal fractures to produce H_2, a different form of seismic survey may be required. This will be discussed later.

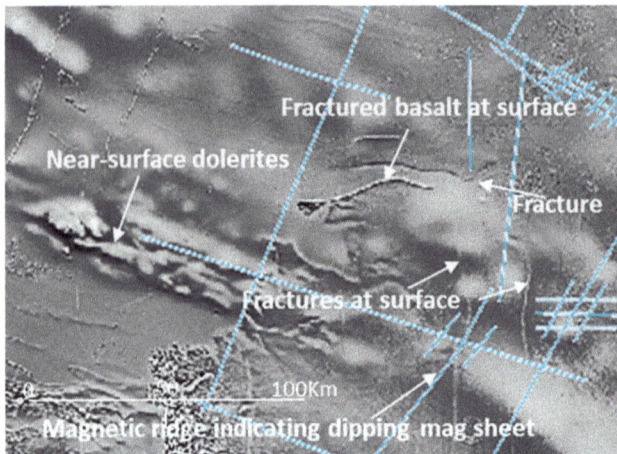

Figure 10.11: Magnetic intensity map shows near-surface fractures, dolomite dykes/sills, and basalt flows. Magnetic ridges may indicate dipping near-surface magnetic sheets or sills.

10.3.2.5 Use of artificial intelligence in searching for DHHIs in seismic and magnetic data

A seismic anomaly known as a *DHII* will, hopefully, be clearly defined on any seismic section as a reflection bright spot or local amplitude high. In the event that it is caused by H_2 being beneath a dolerite sill, it will likely appear horizontal. If it is deeper in the section and there is known to be a deeper trap similar to an evaporite, such as salt, then it may appear horizontal or angular.

In order to gather confidence that this is indeed a gas trap, we perform an amplitude-versus-offset or amplitude-versus-angle (as some call it) analysis, which takes the original reflection data processed for true relative amplitude, puts the traces together in a gather, and then looks at how the amplitudes change with offset/angle. This can be made into an automated machine learning process [19], whereby instructions are provided to a controller, which looks through the final stacked data, works out which amplitudes appear suddenly larger (within given thresholds), and then accesses the pre-stack data to perform the gather/amplitude analysis process. The output will give a probability of whether this may be a real gas anomaly or just some sort of noise anomaly (which happens when caustic points occur due to the geology focusing seismic energy on a specific point) as opposed to the 'gas effect.'

When this method provides such an alert, the next step may be to check the magnetic intensity of the area. In the case of a near-surface DHHI anomaly (within the top 500 m) and there are known dolerite dykes/sills or surface basalts, it would be of geological interest to examine the magnetics data covering the seismic anomaly. In a magnetic intensity black-and-white-shaded map like Figure 10.11, we can interpret a hair-thin straight or wandering line as a near-surface fracture. If the DHHI appears to be close to a fracture or fault according to the magnetics, this may indicate that a fracture is leaking H_2 to the surface, which can be verified with a soil sampler. If the soil sampler detects leakage, then the dolerite may be covering a reservoir that has 'filled and spilled.' Optimistically, this means it has potential as a producer of H_2, albeit at low pressure. If the surface appears mottled, it may indicate that the reworking of soils containing remnant basalt has occurred. Magnetic ridges can also suggest an old basalt flow dipping down into the sedimentary rock strata.

If, for example, a DHHI appears adjacent to a dipping dolerite sheet, it would be reasonable to assume that there may be a local high in the shallow dolerite sill or dyke and that a gas chimney may be close by. It would not be unusual to then look for confirmation of any form of gas chimney feeding H_2 into the DHHI. Again, a surface check with a soil sampler may add some additional information to the knowledge base.

The next step is to apply the seismic and magnetic intensity knowledge to pattern recognition (also known as artificial intelligence – AI) software, which could scan the top 500 ms of data in the seismic database. If it locates any possible DHHIs, it could then review the magnetic intensity map of the area within a 2 km radius to determine if there are any unexpected lineations or other anomalies that exist. The software would then provide an alert indicating the presence of a specific DHHI with associated magnetic anomalies in the vicinity.

If the seismic section shows a deeper DHHI (in areas, perhaps, where there is salt), then the same routine could be adopted; however, the software would not only be required to search for a deep-seated DHHI but also for vertical lines of noise where reflections in the section mis-tie or simply show discordant horizontal features (associated with a gas chimney).

So, this is a method of applying AI techniques to search existing data bases in order to minimize the effort required to locate such bright spots. A level of probability could be assigned to the results of the seismic and magnetic interpretations so that if there was a high probability, the software could then perform a rapid AVO/AVA check and return the answers. It would then be possible to provide a base map of automated interpretation, which a ground crew could subsequently go out and verify. Applying this to a marine survey may prove fruitful where the seismic data is cleaner, although one problem is that it is rare to find high-quality, high-resolution marine magnetics data, while another issue is that exploring for offshore hydrogen may not be financially desirable anyway. Whichever form of exploration is undertaken, AI interpretation can only truly be effective with high-quality seismic and magnetic data. Gravity data is really only useful as a guide to the presence of salt or a dome as a caprock, with many Neoproterozoic basins offering such trapping mechanisms.

10.3.3 Seismic sensing of fracture systems

For many decades, seismic surveys were used to image horizontal geological bedding on the basis that reflections were received from seismic impedance contrasts. But what happens when you have an interest in sub-vertical geology and fracturing? This became a topic of interest during the 1990s when the petroleum exploration industry became interested in producing gas from fractured shales and coal seams, as well as in understanding seismic anisotropy, which changes when fractures pass through flat-layer geological beds. It was also of interest to understand faulting between producing wells in case there were any missed productive zones trapped by fractures or faults.

10.3.3.1 Compressional and shear-wave characteristics

So far, this chapter has only discussed what happens when a seismic wave is transmitted through rocks, but that is only part of a larger, more complex story. When a seismic wave is generated, either by an impact or a vibratory source, it produces two types of seismic waves: a compressional wave (the simplest case, which has been discussed so far) and a shear wave, which travels at about half the speed of the compressional wave, with particle motion orthogonal (in two directions) to the compressional wave, as indicated in Figure 10.12.

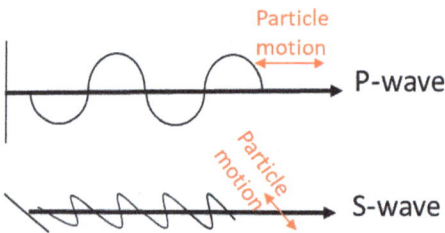

Figure 10.12: While a compressional or 'P-wave' is the normal seismic wave used in reflection seismology to image geology, there is also a shear or 'S-wave,' which travels at half the velocity, often with half the amplitude in the same direction as the P-wave but with orthogonal particle motion to the P-wave.

Imagine that we have a compressional P-wave traveling through a fracture, with the shear S-wave following behind it at half its velocity. The P-wave, with particle motion along the line of its travel direction, will travel straight through the fracture but may be slowed down slightly if the fracture is filled with water or gas, though not enough to make a significant difference. Meanwhile, the S-wave is following behind it at half the P-wave's velocity, but since the S-wave has particle motion orthogonal (at right angles) to the P-wave, it actually spends more time in the fracture gap than the faster P-wave.

This time spent in the gap does make a real difference to its speed because now it slows down perceptibly, but its particle motion also retains the direction in which it is vibrating (unless it passes through another fracture, reorienting it).

10.3.3.2 Application of shear waves to determine fracture presence and characterization

We can use this knowledge to gain a sense of the presence of fractures and their orientation. For example, in mining, consider that we have drilled an exploration hole looking for mineral ore and then tunneled underground to mine it. If we want to check that there are no fractures in the roof of the mine tunnel, we can make a strong seismic impact on the surface to cause a seismic wave to pass down through the fracture, and we place geophone seismic receivers in the roof of the tunnel to receive the arriving P- and S-waves. The surface impact will generate both a P-wave traveling at the sonic velocity of the rock and an S-wave traveling travelingat about half that speed, as indicated in Figure 10.13.

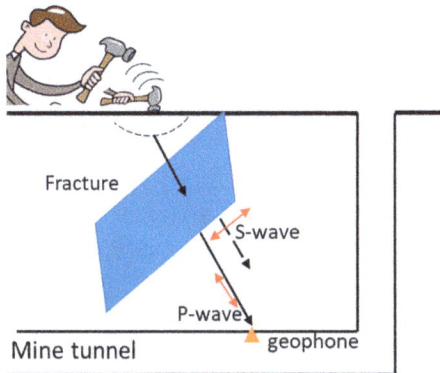

Figure 10.13: To determine if there is a fracture above the roof of a mine tunnel, a miner may test this by generating a seismic pulse at the surface. If a geophone is cemented onto the roof, the arriving S-wave will reach the roof much later than the arriving P-wave, with particle motion generally aligned with the fracture orientation and exhibiting lower amplitude.

Of particular importance in this method of fracture sensing is detecting the S-wave arrival versus the P-wave arrival and then checking the particle direction of the arriving S-wave. The S-wave would likely have particle motion along the direction of the fracture it has passed through because the fracture gap (whether filled with water or gas) influences the particle motion as it vibrates along the fracture direction. The P-wave particle motion would vibrate along the direction of travel and, as such, would not be affected much by the fracture's presence. So, let us apply that understanding in the search for a fracture containing H_2.

10.3.3.3 A method for determining the presence of a fault or fracture and its orientation

When drilling into fractures to produce H_2, the fracture generally has the following properties:
- It is listric or angular/inclined in general shape (i.e., not vertical).
- It is sinuous and has a ribbon-like form.
- It passes through a hard, dense, magnetic, heavy iron ore containing other minerals, with a high sonic P-wave velocity of around 5,000 m/s, an S-wave velocity of around 2,500 m/s, and a density of around 4 g/cc.
- If a P-wave passes through a fracture, its amplitude is slightly reduced, whereas the S-wave experiences a greater reduction in amplitude due to its particle motion spending more time within the fracture. The received P-wave should be clearly observed as the primary arriving wave, whereas a reduced-amplitude S-wave is seen as the secondary (arrival) wave, especially if the fracture is filled with H_2 since its amplitude would be significantly reduced compared with that of the P-wave.
- If an S-wave passes through the fracture, its particle direction will predominantly align with the direction of the fracture's orientation.
- The fracture soles-out at other complex fractures.
- If a fracture has associated sub-parallel fracture swarms, it is probable that the received P-wave amplitude will be far lower while the S-wave may completely disappear, its energy having been absorbed by the fracture swarm.
- The fracture's magnetic signature at or near the surface may be observed in airborne magnetometer data if it is a simple vertical fracture (at the surface).

A seismic reflection survey uses seismic reflections, which may not be apparent when attempting to image a sub-vertical fault or fracture. However, past experience in the oil and gas industry shows that if a company has a depleted oil and gas field and wants to locate any missed-pay zones, they must instead look for low-cost methods to identify them. Experiments have been conducted with a number of dry production wells, where imaging the missed-pay zones would require an expensive in-fill 3D seismic survey.

Instead, many petroleum exploration companies in the 1990s tested the direct transmission travel-time survey method between production wells by placing a seismic source down one hole and a marine seismic receiver cable or wall-locking geophones down an adjacent hole [20]. The time taken for a seismic wave to travel from one well to the other, along with their arriving amplitudes, would be affected by any oil or gas in-place and so by mathematical manipulation of the received waves (known as 'inversion'), an image could be produced, not dissimilar to an X-ray scan. Any oil or gas zones would appear as blank areas.

This method became known as *crosswell or crosshole tomography* [21], effectively scanning between two wells as indicated in Figure 10.14. In the past, the hardest problem has been the energy source, with sparkers and airguns dropped down the source well. For the generation of sufficient seismic energy received at the receiver well to enable seismic inversion, the exploding energy needs to be reasonably high if the seismic waves are to cross a sedimentary sand/shale geological column from one production well to another, with a typical distance being around 2 km. The issue was that it was not possible to raise the energy too high for fear of collapsing the source well. Consequently, the crosswell method was always beset by energy problems, resulting in poor data. Better data could be achieved by bringing the two wells closer together, thereby increasing the transmitted signal level, but in the oil and gas industry, having two wells closer than 2 km was rare.

On some occasions, it was possible to place the source down one or two water injection wells and the receivers down an oil or gas production well to try to obtain an image of missed pay-zones. However, it was rare to achieve a good image due to various source energy problems, though there were occasional successes. Today, 3D reflection surveys tend to be preferred because they are more consistent.

With this method, a number of receiver geophones or a marine seismic cable would be positioned down a receiver well, hard-wired to recording instruments at the surface. Source shots would then be fired at a number of positions down the source well. In Figure 10.14, as each source point is independently fired, each time the seismic compressional wave moves out rapidly, passing through the geological layering. Here, a faulted, dipping strata is used as an example, with a missed gas zone in the center.

Some parts of each shot's outgoing wave, indicated by seismic ray paths, pass from each shot to each receiver, with some passing through the gas zone. Those rays that pass through the gas zone will be reduced in amplitude since the gas absorbs the wave's particle motion, reducing its arriving amplitude and slowing it down. By comparison, a ray path that has not encountered the gas will pass at the normal sonic velocity of the rock. Therefore, those rays that do not pass through the gas will arrive without delay compared to those that have passed through it. These arrival times and amplitudes are entered into a database, after which the mathematical inversion process is performed to produce a general picture of the geology between the two wells.

By comparison, however, when working in a hard-rock geological environment (rather than the soft, permeable layers of an oil/gas field), such energy issues will not be as important because the rock being drilled will be far more competent than soft sediments and will hold up to the impact of a seismic source. In addition, it is likely that it will absorb far less energy than in the sedimentary case, so it is possible that the data will be far better in terms of signal-to-noise ratio and therefore easier to process.

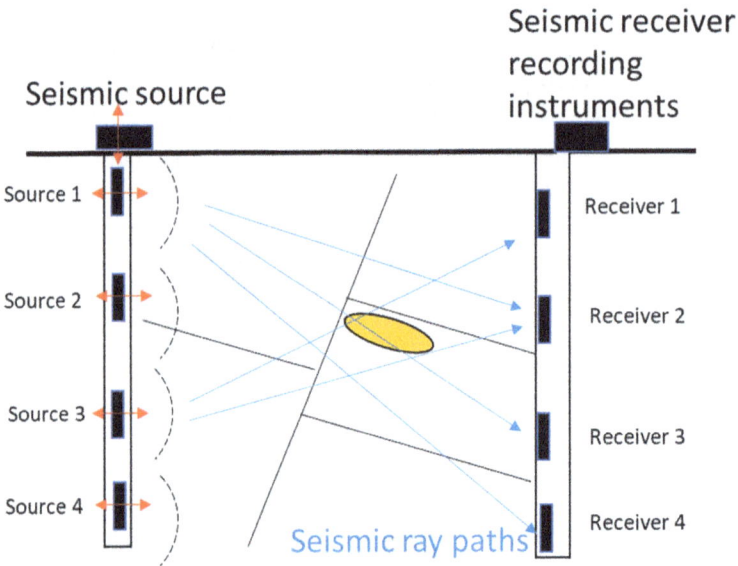

Figure 10.14: Crosswell tomography is performed by placing seismic sources down one borehole and receivers down another, with an area to be imaged between them, and then recording the seismic wave arrivals. Some ray-paths will be modified by energy passing through the missed gas zone (shown in yellow), and these ray-paths are in the zone of interest for seismic inversion.

When performing such tomographic surveys, we can place the source on the surface while the second borehole contains receivers that record the P- and S-wave arrivals. The difference in technology for hard rock fracture detection is that the receiver must record data in all three planes – x, y, and z – which is called *three-component data*. By using three-component (3C) geophones, we can compute the direction of arrival of both P- and S-waves, as well as their particle motion direction, which provides the fracture orientation. However, this method is not foolproof because it only displays the orientation of the last fracture the waves have passed through, which can easily cause confusion. The method is therefore non-unique, as interpreting a fracture that curves around can be inherently challenging.

Ideally, the downhole seismic source should be an S-wave generator, such as the orbital S-wave generator developed for such purposes [22]. However, if one is not available, then any surface or downhole seismic source must be used. To receive the P- and S-waves after they pass through a fracture, we need to use either downhole wall-locking 3C geophones or a downhole fiber optic cable. Recent developments in fiber optics have established distributed acoustic sensing (DAS) as a technology that enables real-time vibration measurements along the entire length of a fiber optic cable. Changes in the strain of the DAS fiber optic cable allow seismic vibrations to be recorded, and when suitably calibrated, these may be used to interpret the P-wave arrival, the later S-wave arrival, and the dominant S-wave particle motion direction.

The survey arrangements are as shown in Figure 10.15, with the well positioned between the boreholes.

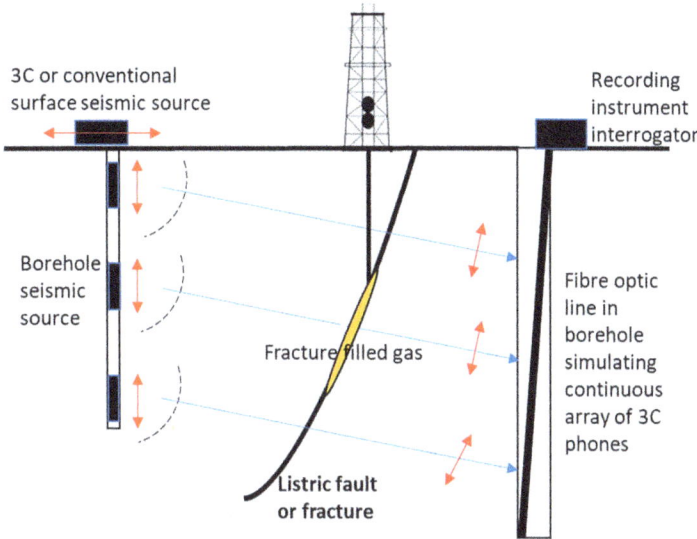

Figure 10.15: A surface or borehole seismic source transmits P- and S-waves through the fault or fracture,
which are received by the DAS cable. From this data, the fracture's presence and orientation are
computed, along with changes in velocity caused by the presence of fractures filled with H_2.

10.4 Conclusions

This chapter has covered the likely geophysical responses in the search for natural
hydrogen. However, prior to understanding the response, two typical models of the
geological locations of hydrogen have been proposed. The first model is found in shallow environments, where dolerites form the caprock, and it is well documented elsewhere. The second model involves deep sedimentary basins close to a crustal feature
or fractured basement, in which evaporites may trap the leaking hydrogen, typical of
a Neoproterozoic basin. It may also be possible that shallow dolerites form traps near
the surface in these basins.

The concept of searching for DHHIs in the deep basin model using established gas
exploration techniques has been discussed, with AVO/AVA methodologies being a
major part of that discussion. The joint interpretation of seismic data with magnetics
and gravity data was reviewed, highlighting the usefulness of gravity data in understanding the presence of evaporites and the role they play in the H_2 trapping process.
That section suggested that if the existing seismic data quality is good enough, applying AI techniques to such data may be more economical than deploying survey crews

to the field. It is often much easier and cheaper to obtain permits for conducting a soil sampling survey than for performing any form of geophysical surveys. If air-borne LIDAR instruments are developed in the near future (as suggested in one of these chapters), soil sampling surveys may even be replaced by the ability to map an area from the air.

The concept of using crosswell tomography techniques was then discussed. This methodology has been well established in the petroleum exploration industry for decades. However, in the sedimentary environment, it does suffer from issues related to the use of relatively weak energy sources downhole, whereas when recording data in a hard rock environment, it does not experience the same issues of energy loss unless affected by sedimentary layers. The sedimentary environment causes absorption of the seismic signal energy, whereas this is not the case with hard rock seismic profiling—instead, it is fracturing within the hard rock that breaks up signal energy (the imaging tool we want). Furthermore, the use of modern DAS recording systems with fiber optics makes field operations much simpler than ever, though the processing would need to compensate for that.

Normally, when exploring for natural hydrogen, a number of wells will need to be drilled. If the wells are not too far apart (no more than the depth of the wells to maintain a triangular or box-shape for easier processing), then a crosswell survey would be a cost-effective alternative to a surface seismic survey. If the data is of good quality, it will provide information on fracture geometry and width, which the 2D survey may not. The search for DHHIs in crosswell tomograms has not been performed to date, but an equivalent form of 'bright spot' H_2 detection may be possible if the data is of high quality.

The use of DHHIs provides a more scientific approach to interpreting the locations of gas traps than other subjective methods, and having a realistic geological model provides leads that might otherwise not be obvious. The DHHIs used in H_2/He exploration require seismic data to be reprocessed in order to perform an AVO analysis, whether they are bright spots or dim spots; however, we can readily see gas chimneys using conventional seismic processing methods. Nevertheless, DHHIs provide analytical support for the interpretation of a gas trap when there is a requirement to have high investor confidence before testing with a drill. A DHHI using crosswell tomographic methods is yet to be developed not because it is elusive but because the method has not been sufficiently tested, and there is little data available.

Here, this chapter outlines the potential for deep gas traps using existing open-file seismic, gravity, and magnetics DMIRS data, which would require reprocessing as well as confirmation that gravity and magnetic lines in the open-file data are correctly positioned before a truly accurate interpretation can be performed. The presence of two potential gas traps beneath shallow Table Hill dolerites, which also may be fed from a gas chimney, has been noted. However, it must be considered that the two lines reviewed here are in the deeper area of the northern Officer Basin (officially known as the Savory area), and it raises the question of whether the western edge of the northern Officer Basin of Australia may be shallower and, hence, more productive in terms of H_2/He. With the Monzon field expecting to produce from 3,500 m at 50c per kg (pers. Comment from Muro, 2023), this area may be more prospective and equally complex, but more remote from civilization—the bane of its past history.

Disclosures: The author confirms that he has no conflicts of interest and has received no institutional funding associated with the material presented in this chapter.

References

[1]   Prinzhofer A., Cisse C. S. T., Diallo A. B. Discovery of a large accumulation of natural hydrogen in Bourakebougou (Mali). International Journal of Hydrogen Energy. 2018, 43(42): 19315–19326. [Online]. Available http://dx.doi.org/10.1016/j.ijhydene.2018.08.193

[2]   Zgonnik V. The occurrence and geoscience of natural hydrogen: A comprehensive review. Earth-Science Reviews. 2020, 203: 103–140. https://doi.org/10.1016/j.earscirev 103140 (2020).

[3]   Briere D., Jerzykiewicz T. On generating a geological model for hydrogen gas in the southern Taoudeni Megabasin (Bourakebougou area, Mali), AAPG/SEG global meeting abstracts: 342–342. In: AAPG international conference and exhibition, Barcelona, Spain, 2016, 3–6. April 2016. http://doi.org/10.1190/ice2016-6312821.1

[4]   Muro I. Europe's first natural hydrogen project in Aragon, Spain: AAPG Int. Conf, 2023, Theme 6, Madrid.

[5]   Frery E., Langhi L., Maison M., Moretti I. Natural hydrogen seeps identified in the North Perth Basin, Western Australia. International Journal of Hydrogen Energy. 2021, 46: 31158–31173.

[6]   Windley B. Proterozoic collisional and accretionary orogens. Developments in Precambrian Geology. 1992, 10: 419–446. Chapter 11 Proterozoic Collisional and Accretionary Orogens – ScienceDirect.

[7]   Collins W. J. Hot orogens, tectonic switching, and creation of continental crust. Geology. 2002, 30(6): 535–538. Hot orogens, tectonic switching, and creation of continental crust | Geology | GeoScienceWorld.

[8]   Cawood P. A., Kroner A., Collins W., Kusky T. M., Mooney W. D., Windley B. F. Accretionary orogens through Earth history, 318, 1, Earth Accretionary Systems in Space and Time London: Geological Society Special Publications 2009.

[9]   Cawood P. A., Strachan R. A., Pisarevsky S. A., Gladkochub D. P., Murphy J. B. Linking collisional and accretionary orogens during Rodinia assembly and breakup: Implications for models of supercontinent cycles. Earth and Planetary Science Letters. 2016, 449(2016): 118–126. Linking collisional and accretionary orogens during Rodinia assembly and breakup: Implications for models of supercontinent cycles – ScienceDirect.

[10]   Ghienne J. F. Late Ordovician sedimentary environments, glacial cycles, and post-glacial transgression in the Taoudeni Basin, West Africa. Palaeogeography, Palaeoclimatology, Palaeoecology. 2003, 189: 3–4. 117–145 https://doi.org/10.1016/S0031-0182(2)00635-1

[11]   Walter M. R., Veevers J. J., Calvert C. R., Grey K. Neoproterozoic stratigraphy of the Centralian Superbasin, Australia. Precambrian Research. 1995, 73(1–4): 173–195. https://doi.org/10.1016/0301-9268(94)00077-5

[12]   Sheriff R. E. Reflectivity: Reflection Coefficient. Encyclopedic dictionary of exploration geophysics, Tulsa: Soc. of Expl. Geoph., P203, 1984. ISBN 0-931830-31-3.

[13]   Zoeppritz K. Erdbebenwellen. VIIIb. Über Reflexion und Durchgang seismischer Wellen durch Unstetigkeitsflächen. Nachrichten von der Königlichen Gesellschaft der Wissenschaften Zu Göttingen: Mathematisch-physikalische Klasse. 1919, 66–84.

[14]   Shuey R. T. A simplification of the Zoeppritz equations. Geophysics. 1985, 50(4): 609–614. doi: 10.1190/1.1441936

[15]   Castagna J. P., Swan H. W., Foster D. J. Framework for AVO gradient and intercept interpretation. Geophysics. 1998, 63: 948–956.

[16]   Avseth P., Mukerji T., Mavko G. Quantitative seismic interpretation: Applying rock physics tools to reduce interpretation risk, Cambridge university press, 2010.

[17]   Granli J. E., Arntsen B., Sollid A., Hilde E. Imaging through gas-filled sediments using marine shear-wave data. Geophysics. 1999, 64(3): 668–677.

[18]   Evans B. J., Rezaee R. Exploration for natural hydrogen in Officer Basin, Western Australia using open-file data. AAPG International Conference. 2023, Paper 3961078 Theme 6, Madrid.

[19]   Roden R., Chen C. W. Interpretation of DHI characteristics with machine learning. First Break. 2017, 35950. https://doi.org/10.3997/1365-2397.35.5.88069

[20]   Link C. A., McDonald J. A., Zhou H. W., Evans B. J. Cross-well tomography in a shallow clastic reservoir: Seventy-six West field, South Texas. SEG Technical Program Expanded Abstracts. 1991, 371–374. https://doi.org/10.1190/1.1888916

[21]   Dickens T. A. Diffraction tomography for crosswell imaging of nearly layered media. Geophysics. 1994, 59: 694–706. https://doi.org/10.1190/1.1443627

[22]   Daley T. M., Cox D. Orbital vibrator seismic source for simultaneous P- and S-wave crosswell acquisition. Geophysics. 2001, 66: 1471–1480. https://doi.org/10.1190/1.1487092

Charlie Ironside*, Mervyn Lynch, Jacob Martin, Mark Paskevicius, Mauricio Di Lorenzo, Craig E. Buckley, and Andrew Lockwood

Chapter 11
The development of an airborne, stand-off detection instrument for hydrogen gas

Abstract: The exploration of natural hydrogen as a carbon-free energy source has gained significant interest in recent years. However, the lack of efficient, large-scale detection methods has hindered its development. This study presents the development of an airborne, stand-off detection instrument for hydrogen gas, leveraging time-correlated single photon counting (TCSPC) Raman LIDAR technology. Unlike conventional fixed-point and handheld sensors, airborne Raman LIDAR offers remote, rapid, and highly specific detection of hydrogen seeps over large and often inaccessible areas.

The research outlines the advantages of airborne stand-off detection, including non-invasive data acquisition, real-time analysis, and cost-effective exploration, while also addressing the challenges posed by the weak Raman scattering signal of diatomic hydrogen molecules. A laboratory-based proof-of-concept instrument was developed to validate the technology, demonstrating a limit of detection of 9,000 ppm at 2 m. A numerical model was also designed to optimize system parameters for future airborne deployment, targeting an LOD of 1,000 ppm at 50 m with rapid integration times.

These findings confirm the feasibility of TCSPC Raman LIDAR for natural hydrogen exploration and fugitive emission monitoring, with potential applications in environmental assessments and energy resource mapping. Future work will focus on enhancing sensitivity and adapting the system for real-world airborne surveys.

Keywords: hydrogen, Raman, LIDAR, single photon detection, airborne survey

Acknowledgments: The initial work on this project was entirely reliant on generous support from Xcalibur https://xcaliburmp.com. We are also grateful to acknowledge support from the Australian government sponsored funding initiative Trailblazer: Resources Technology and Critical Minerals https://rtcm-trailblazer.au.

*Corresponding author: Charlie Ironside, Department of Physics and Astronomy, Curtin University, Perth, WA, Australia, e-mail: Charlie.Ironside@curtin.edu.au
Mervyn Lynch, Jacob Martin, Mark Paskevicius, Mauricio Di Lorenzo, Craig E. Buckley, Department of Physics and Astronomy, Curtin University, Perth, WA, Australia
Andrew Lockwood, Xcalibur Multiphysics, Perth, WA, Australia

https://doi.org/10.1515/9783111437040-011

11.1 Introduction

World-wide natural hydrogen gas seeps have been reported [1, 2] and there is growing commercial interest in exploiting these seeps as carbon-free sources of energy [3–6]. Hydrogen detection technology has played an important role in developing natural hydrogen utilization. Part of the reason that, until recently, this carbon-free energy resource was largely overlooked is the lack of routine use of H_2 detectors. To quote from [1] "It is difficult to estimate how many times hydrogen has not been identified in H_2 – rich samples because of the lack of a suitable detection technique to measure hydrogen concentrations."

Exploration for natural hydrogen has mostly relied on hand-held or fixed-point H_2 sensors. For example, a recent work [7] reported on a handheld electrochemical H_2 sensor (Geotech GA5000) that was thoroughly evaluated and used in a survey of a natural hydrogen seep in Western Australia.

The handheld or fixed-point H_2 gas sensors that are currently available [8–10] include catalytic combustion sensors, electrochemical sensors, semiconducting oxide sensors, thermal conductivity sensors, and a biological sensor [11]. These sensors have been mainly developed for safety applications in hydrogen processing plants. The US-based National Renewable Energy Laboratory (NREL) maintains a facility for assessing hydrogen sensors for H_2 safety applications [12, 13]. The NREL has reported [14] on using a network of fixed-point sensors for wide-area monitoring of H_2.

However, for natural hydrogen exploration, this chapter focuses on the detection of H_2 using indirect methods that do not require physical contact, referred to as "stand-off" methods, which offer the following potential advantages:

- **Remote and rapid detection**: Airborne stand-off H_2 detection enables the remote identification of hydrogen seeps over large and often inaccessible areas, significantly accelerating the exploration process compared to ground-based methods.
- **High specificity**: Airborne stand-off H_2 detection can be highly specific to H_2 and can differentiate it from other gases in the atmosphere. This specificity reduces false positives and helps ensure accurate identification of hydrogen seeps.
- **Vertical profiling**: Airborne stand-off H_2 detection can provide vertical profiles of hydrogen concentrations. This capability helps in understanding the dispersion and transport mechanisms of hydrogen in the atmosphere, which is crucial for locating the source of the seep.
- **Non-invasive method**: Airborne stand-off H_2 detection is non-invasive and does not require physical contact with the ground or probes, thereby preserving the integrity of the natural environment.
- **Integration with other sensors**: Airborne stand-off H_2 detection can be integrated with other remote sensing technologies, such as electromagnetic, gravitational field, and magnetic field sensors, to provide a comprehensive assessment of the exploration area and correlate hydrogen presence with other geological or environmental factors.

- **Real-time data acquisition**: The technology enables real-time data collection and processing, facilitating immediate analysis and decision-making during survey flights.
- **Reduced exploration costs**: By providing rapid, extensive, and accurate surveys, airborne stand-off H_2 detection can reduce the overall cost of hydrogen exploration by minimizing the need for extensive ground-based follow-up surveys.
- **Environmental monitoring**: Beyond exploration, airborne stand-off H_2 detection can continuously monitor fugitive hydrogen, aiding in environmental impact assessments and ensuring compliance with safety and environmental regulations.

So, in general, an airborne geophysical survey is a very useful method for investigating large areas quickly without harming the natural setting of the local environment [15].

A good candidate technology for airborne stand-off H_2 detection that could allow us to realize the above capabilities is Raman LIDAR, as it has been used extensively for monitoring the atmosphere [16–18] and for detecting H_2 fugitive emissions [19].

It should be noted that helium is also a valuable gas associated with hydrogen exploration [20], where the two gases have a radiological origin. Helium is a monatomic gas, and vibrational Raman scattering is not available. However, there has been some work [21] on using laser-induced breakdown spectroscopy for stand-off detection of helium in helium/hydrogen gas mixtures.

11.2 Stand-off detection of hydrogen

Light detection and ranging (LIDAR) for atmospheric studies is a remote sensing technique that uses laser pulses to measure properties of the atmosphere [22]. It operates by emitting pulses of laser light into the atmosphere and analyzing the backscattered signal. Typically, optical systems for stand-off detection of gases are based on LIDAR combined with rotational and vibrational spectroscopy.

Rotational and vibrational spectroscopy employs the interaction of molecules with light, causing the rotation and vibration of the atoms comprising the gas molecule [23]. Each molecule has its own unique set of rotational and vibrational energies and thus its unique spectrum of backscattered light.

Molecular rotational and vibrational modes with built-in dipole moments, such as those in CH_4, result in the absorption of infrared light. The absorption of infrared light by gases is fundamental to the development of differential absorption LIDAR (DIAL) systems used for stand-off detection in the hydrocarbon industry as part of their leak detection and repair programs [24, 25]. For example, using the absorption lines of methane combined with single photon LIDAR, methane leaks as low as 0.012 g/s and can be detected at distances over 90 m [26].

However, homo-nuclear diatomic gases such as N_2, O_2, and H_2 have a homogeneous distribution of charge and thereby lack a built-in electric dipole. They essentially do not have infrared absorption or have very weak infrared absorption [27, 28], which can be caused by the collision of molecules and electric quadrupole transitions [29]. Therefore, the DIAL systems used for stand-off detection in the hydrocarbon industry have not been utilized for H_2 detection, although a DIAL system for H_2 detection has been proposed but not demonstrated [30].

Thus far, stand-off detection of H_2 has relied on Raman scattering [31], which has been used to image an H_2 plume from liquid hydrogen [32] and an H_2 plume from an underground leak [33]. When Raman scattering is combined with LIDAR [19, 22, 34–37], it can be used to map an H_2 plume in 3D and can be utilized both for monitoring H_2 fugitive emissions and as an exploration tool for detecting natural hydrogen.

Airborne Raman LIDAR systems for mapping atmospheric gases have been developed previously [38–40]. In Figure 11.1, we illustrate the concept of using time of flight (TOF) Raman LIDAR to map H_2 on the surface of the Earth.

Airborne Raman LIDAR

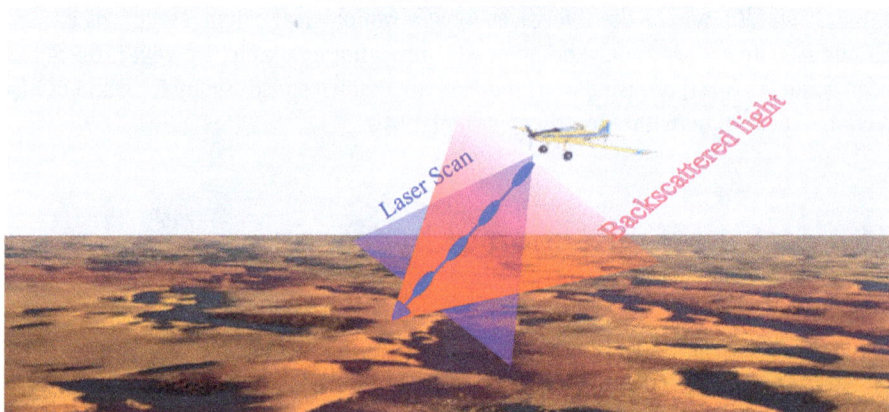

Figure 11.1: The airborne Raman LIDAR concept as an exploration tool for natural hydrogen; a pulsed laser mounted on the aircraft scans the terrain, and the backscattered light is collected by receiver optics also mounted on the aircraft. The backscattered light spectrum is monitored for signature H_2 Raman lines.

Specifically, to realize the advantages listed earlier, an airborne Raman LIDAR system for natural H_2 exploration needs to achieve a target performance that can be estimated from the existing data on H_2 seeps.

For example, Prinzhofer et al. [41] report on a particular geographical feature from Mali, a so-called "Fairy Circle," often associated with H_2 seeps. For this Mali seep, at the surface, the H_2 concentration is around 600 ppm (this value is 10^3 above the average background level of 530 ppb [42] for H_2). At another H_2 site, Chimaera, Turkey [43], there have been careful measurements of H_2 concentration, and a peak value of

120 ppm was observed. Thus, we could regard this as an approximate indication of the limit of detection (LOD) required for an exploration system to detect H_2 seeps. Another requirement of an exploration tool is the ability to map out H_2 surface concentration over an extensive terrain.

From the above considerations, target specifications for airborne detection of natural H_2, are as follows: a stand-off instrument is required to detect concentrations of H_2 in the range of 100–1,000 ppm at a distance of 50–100 m. The time taken for the instrument to acquire this level of H_2 detection will determine the spatial resolution of the mapping. For example, with the type of aircraft [44] currently used in airborne surveys, the typical speed is ≈120 mph or ≈60 m/s, so if the integration time required to reach an LOD of 600–1,000 ppm is around ≈1 s, that implies a lateral mapping resolution of ≈50 m.

11.2.1 Raman scattering

Figure 11.2 provides a simple conceptual illustration of light scattering by an H_2 molecule, including Rayleigh scattering and Raman scattering [45, 46]. There are two types of Raman scattering; the Stokes scattered photon, which has lower energy than the incident photon, and the anti-Stokes scattered photon, which has higher energy than the incident photon. At normal temperatures (≈300 K), Stokes scattering is the predominant form of Raman scattering and is used to identify molecules because each molecule has vibrational modes at a unique angular frequency (ω_v) and rotational modes at a unique angular frequency (ω_r).

The laser pump light has photon energy $\hbar\omega_p$ ($\hbar$ Planck's constant multiplied by 2π). The Raman Stokes light has lower photon energy, given by $\hbar\omega_{\text{vib}} = \hbar\omega_p - \hbar\omega_v$ and $\hbar\omega_{\text{rot}} = \hbar\omega_p - \hbar\omega_r$. By spectral analysis of the backscattered light, the presence of these lower photon energies can be determined and used to identify molecules.

Backscattered light also has a Rayleigh light component with the same photon energy as the laser pump light. However, the Rayleigh light cannot be used to identify molecules.

Tables 11.1 and 11.2 provide the relevant Raman scattering parameters for H_2 and atmospheric gases N_2 and O_2. These parameters are used in the LIDAR equation (11.1) to estimate the Raman LIDAR signal. The values of the Raman scattering parameters are derived from refs. [19, 34, 47, 48].

What we have described is called spontaneous Raman scattering (SRS) and is a weak (low probability) process. It is approximately 1×10^{-3} weaker than Rayleigh scattering. For example, typical of the type of scenarios discussed below, for H_2 concentration of 1% and a Raman pump laser operating at 355 nm with a pulse width of 1.1 ns, approximately ≈1×10^{-9} of the pump photons are converted into Raman photons. For a pulse energy of 1 μJ, there are ≈1.7×10^{12} pump photons, and so around ≈1.7×10^3

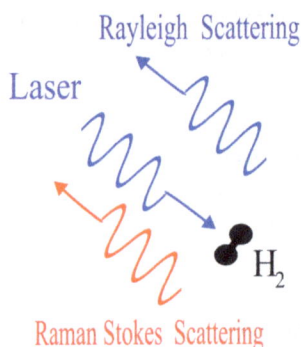

Figure 11.2: A schematic of the back scattering of light from an H_2 molecule. A small fraction of the laser light is scattered in all directions and consists of Rayleigh and Raman scattering. The Rayleigh scattering has the same photon energy as the laser light, $\hbar\omega_p$. The laser light interacts with the vibrational (energy $\hbar\omega_v$) modes and rotational (energy $\hbar\omega_r$) modes of the molecule, and the Raman Stokes photon energies are given by $\hbar\omega_{vib} = \hbar\omega_p - \hbar\omega_v$ v and $\hbar\omega_{rot} = \hbar\omega_p - \hbar\omega_r$ [47].

Raman photons. Although Raman scattering is a weak process, in multi-pass Raman pump configurations, it can measure H_2 concentrations with an LOD of 20 ppb [49].

It should also be noted that Stimulated Raman Scattering where a second laser beam at the Raman Stokes photon energy is also provided by a laser, can increase the Raman signal by orders of magnitude in techniques such as coherent anti-Stokes Raman scattering (CARS) [50]; however, these techniques are challenging to implement in Raman LIDAR intended for natural hydrogen exploration.

Table 11.1: Raman scattering Stokes shifts ($\lambda_0 = 355$ nm pump) for H_2, N_2, and O_2 vibrational (v) and rotational (r) modes.

Gas	λ_0	Vibrational Raman scattering		Rotational Raman scattering	
		Δv^v (cm^{-1})	λ^v (nm)	Δv^r (cm^{-1})	λ^r (nm)
H_2	354.7	4,160	416.1	587	362.2
N_2	354.7	2,330.7	386.67	75	355.65
O_2	354.7	1,103.3	369.15	60	355.46
H_2O	354.7	3,651.70	407.88	190	357

The backscattered light will also contain light from Rayleigh scattering at the Raman pump wavelength, approximately 10^3 times larger than the Raman lines.

H is the lightest atom, so H_2 has the largest rotational and vibrational Stokes shift of any gas. A Raman instrument with even modest spectral resolution, obtainable with thin-film optical filters, can nonetheless have high selectivity for H_2.

Table 11.2: Rayleigh and Raman cross-sectional values for scattering and Stokes wavelengths ($\lambda_0 = 355$ nm pump) for H_2 and the main atmospheric gases N_2 (0.8 of the atmosphere) and O_2 (0.2 of the atmosphere).

Gas	Rayleigh (cm^2 sr^{-1}]	Raman(Vib)(cm^2 sr^{-1})	Stokes (vib) λ (nm)
H_2	24.18×10^{-28}	7.07×10^{-30}	416.1
N_2	117.61×10^{-28}	2.28×10^{-30}	386.67
O_2	98.92×10^{-28}	2.68×10^{-30}	369.15

Another common atmospheric gas, water vapor, has a vibrational Stokes shift of 3,651 cm^{-1} close enough to H_2 that there is some overlap; however, this overlap can be accounted for [37].

11.2.2 Raman light detection and ranging (LIDAR)

Raman LIDAR is a specialized form of LIDAR that utilizes Raman scattering to map the composition and properties of the atmosphere. Unlike traditional LIDAR, which primarily measures distances and surface features, Raman LIDAR detects the molecular backscatter signals generated when laser light interacts with atmospheric constituents. This technology can acquire detailed information about atmospheric gases, humidity, and temperature profiles, which is crucial for studying air quality and meteorological phenomena with high accuracy and resolution. Raman LIDAR has been used extensively to map various atmospheric components. Much of the Raman LIDAR work has concentrated on measuring water vapor in the atmosphere [51]. For a more recent example, see Shi et al. [52] and references therein.

Here, we review research on TOF Raman LIDAR employed to map the spatial distribution of an H_2 cloud. So far, Raman LIDAR has not been employed in the search for natural hydrogen seeps. The technology has been developed for applications that include monitoring nuclear waste (nuclear waste generates hydrogen via water radiolysis) and detecting fugitive leaks to ensure safety in the burgeoning hydrogen economy.

Research groups that have implemented TOF Raman LIDAR for detecting H_2 include the following:
- Ball [53]
- Jeon et al. [54]
- Choi et al. [36, 55]
- Limery et al. [56]
- Ninomiya et al. [19, 34]
- Shiina [35]
- Stothard et al. [37]
- Voronina et al. [57]

Privalov and Shemanin [58] also conducted extensive work on modeling Raman LIDAR for H_2 detection.

To make a simple comparison between the groups, we adopt a generic scheme for Raman LIDAR, as illustrated in Figure 11.3. A Raman LIDAR consists of

– **Transmitter optics**: A pulsed source of light – usually a laser, but sometimes a light emitting diode, is directed at the target area.
– **Receiver optics**: A telescope for collecting light from the target area, some spectrum analysis that uses either an optical spectrometer or thin-film optical filters; there will also be a filter to reject Rayleigh scattering, residual laser light, and a detector. The detector can be either a photo multiplier tube, silicon photomultiplier detector (SiPM), or charge-coupled device (CCD) and operated in single photon or analog mode. There can be several Raman channels in the same receiver with optical filtering to select a Raman line of the target gas. The spectral resolution of the instrument determines its gas selectivity.
– **Signal processing**: To time-resolve (thereby spatially resolve), average, and improve the signal-to-noise ratio (and thereby gas sensitivity) by integrating the signal from the detector over many pulses.

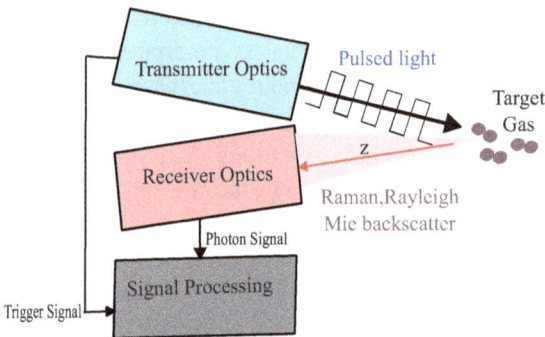

Figure 11.3: A generic Raman LIDAR system consisting of transmitter optics (usually a laser referred to as the Raman pump laser), receiver optics, and signal processing; the target gas is located at a distance Z.

Using the above scheme to characterize systems, in Tables 11.3–11.6 we compare the Raman LIDAR of various groups that have used Raman LIDAR systems for stand-off detection of H_2.

A variety of lasers have been used as Raman pump lasers. For Raman LIDAR, these are generally based on Nd:YAG lasers that are pulsed using a technique known as Q-switching, which produces nanosecond-long pulses at various pulse energies and repetition rates. The Raman cross section increases as the Raman pump laser wavelength, λ_p, becomes shorter; it increases as λ_p^{-4} and so the fundamental Nd:YAG laser wavelength of 1,064 nm is typically frequency-doubled to 532 nm, tripled to 355 nm, or quadrupled to 266 nm when used in Raman LIDAR applications.

Table 11.3: Comparison of Raman LIDAR transmitting optics.

Research Group	Raman Pump laser	λ_0(nm)	Pulse width (ns)	Rep rate (Hz)	Pulse energy (J)
Ball [53]	Q-switch laser	532 and 354.7	Not given	10	Up to 250×10^{-3}
Choi et al. [36]	Q-switch laser	354.7	1	10×10^3	3×10^{-6}
Jeon et al. [54]	LED	360	Not given	Not given	Not given
Limery et al. [56]	Q-switch laser	354.7	5.5	20	30×10^{-3}
Ninomiya et al. [34]	Q-switch laser	354.7	Not given	20	30×10^{-3}
Shiina [35]	LED	394	10	112×10^3	1×10^{-9}
Stothard et al. [37]	Q-switch laser	354.7	5	20	30×10^{-3}
Voronina et al. [57]	Q-switch laser	532	10	Not given	10×10^{-3}

The Q-switch laser is a Nd:YAG laser, which can be frequency-doubled (532 nm), tripled (355 nm), or quadrupled (266 nm).

The backscattered light collected by the receiver optics telescope requires spectral analysis so that the Raman Stokes lines from the various gases in the target area can be identified. For most gases of interest (see Table 11.1) there is a sufficient Stokes shift such that thin-film optical filters can provide enough spectral resolution to select Raman Stokes lines of interest. Usually, there is also a thin-film optical filter employed in the receiver optics to reject the Rayleigh scattering light, which is at the same wavelength as the Raman pump laser. Rayleigh light contains no information about the gas concentration, but it can contribute to noise.

Table 11.4: Comparison of Raman LIDAR receiver optics.

Research group	Collection optics (diameter (mm))	Spectrum analysis	Detector
Ball [53]	Telescope (diameter 101.6)	Grating spectrometer	CCD and PMT
Choi et al. [36]	Telescope (diameter 75)	Thin-film optical filters	PMT
Jeon et al. [54]	Beam expander	Spectrometer	Photodiode array
Limery et al. [56]	Telescope (diameter 152)	Thin-film optical filters	SiPM
Ninomiya et al. [34]	Telescope (diameter 212)	Thin-film optical filters	PMT
Shiina [35]	Telescope (diameter 127)	Thin-film optical filters	PMT
Stothard et al. [37]	Telescope (diameter 152)	Thin-film optical filters	SiPM
Voronina et al. [57]	Telescope (diameter 400)	Thin-film optical filters	PMT

The Raman Stokes light can be detected using a PMT or equivalent in single photon or analog mode. The detector signal is processed to measure the TOF, which is the time between the Raman pump laser trigger pulse and the arrival of the Raman Stokes signal. The TOF is generally in the range of ≈10–600 ns and is measured with ≈50 ps resolution. The TOF is then converted into a distance. Signal processing also includes an integration time over many pulses to increase the signal-to-noise ratio.

Table 11.5: Comparison of Raman LIDAR signal processing.

Research group	Detector	Signal processing
Ball [53]	CCD and PMT	Oscilloscope
Choi et al. [36]	PMT	Photon counter
Jeon et al. [54]	Photodiode array	Not given
Limery et al. [56]	SiPM	Photon counter
Ninomiya et al. [34]	PMT	Oscilloscope (5×10^9 sample/s)
Shiina [35]	PMT	Oscilloscope
Stothard et al. [37]	SiPM	Time-correlated single photon counting
Voronina et al. [57]	PMT	Oscilloscope

The performance of a Raman LIDAR system can be summarized in terms of the LOD for a target gas distance, Z (see Figure 11.3) and the integration time required to obtain a given LOD.

Table 11.6: Comparison of Raman LIDAR performance.

Research group	Limit of detection (LOD) %	Distance (m)	Integration time (s)
Ball [53]	1.79	6.58	15
Choi et al. [36]	0.2	10	30
Jeon et al. [54]	0.1	30	Not given
Limery et al. [56]	0.05	500	660
Ninomiya et al. [34]	Not given	50	Less than 30 for the entire scan
Shiina [35]	1	20	210
Stothard et al. [37]	0.05	45	30
Voronina et al. [57]	100	2	Not given

Note: All the works compared here employ spontaneous Raman scattering (SRS) [45]. There is also another approach using Coherent Anti-Stokes Raman Scattering (CARS), where a second Stokes beam interacts with the target gas. It is a considerably more complicated setup, but it has been proposed as a technique for finding H_2 leaks [50].

Overall, the results summarized above indicate that a Raman LIDAR is a promising candidate technology for airborne stand-off detection of natural hydrogen, which could realize the potential benefits outlined in the introduction.

11.3 Laboratory prototype of airborne Raman LIDAR for stand-off H_2 detection

In this section, we discuss the design of a lab-based prototype of an airborne Raman LIDAR for stand-off H_2 detection. The purpose of the design was to investigate Raman LIDAR as the basis for an airborne instrument for natural hydrogen exploration and to develop and validate numerical models of the instrument that would be used in the final design of the airborne instrument. The design of the lab-based prototype was mostly influenced by the works from Choi et al. [36] and Stothard et al. [37]. Their design offered the best prospects for the ultimate objective, which was to build an airborne instrument that is energy-efficient, lightweight, and robust.

In the final instrument, in addition to H_2, we plan to measure SRS from other atmospheric gases, such as N_2, to calibrate the concentration of H_2. Each gas has its own SRS wavelength, so each has its own Raman channel.

The design of the laboratory prototype for each Raman channel is described below.

Transmitter optics
- Laser: The Raman pump laser is an Nd:YAG diode-pumped solid-state laser that is Q-switched and frequency-tripled to 355 nm, delivering 3 µJ per pulse with a pulse width of 1.1 ns and a pulse repetition rate of 10 kHz (CryLas FTSS355-Q2).

Receiver optics
- Telescope: A 152 mm aperture Cassegrain reflector that focuses received light onto the detector.
- A thin-film optical bandpass filter with its center wavelength at the Stokes wavelength for the gas of interest, e.g., 417 nm for H_2 and 387 nm for N_2, and a bandpass of approximately 1–10 nm.
- A film optical filter that rejects Rayleigh scattering and residual laser light at 355 nm.
- Single photon detector (Hamamatsu MPPC module C13366-3050GD) – single photon avalanche detector (SPAD)
- Signal processing – TCSPC
- Time Tagger (Swabian Instruments Time Tagger Ultra, 42 ps time resolution) measures the time of arrival of a photon for each laser pulse after the trigger pulse from the laser and logs the TOF.
- A computer controls the Time Tagger and stores TOF data from the Time Tagger.

Figure 11.4 illustrates the layout of the lab-based prototype TCSPC Raman LIDAR system for a single Raman channel.

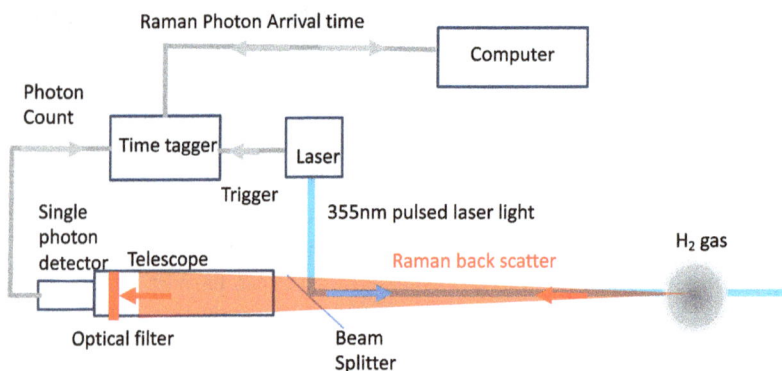

Figure 11.4: Raman LIDAR using time-correlated single photon counting (TCSPC). The pump laser is a Q-switched, diode-pumped, Nd:YAG laser, frequency-tripled to 355 nm – it produces 1.1 ns pulse at a 10 kHz repetition rate with ≈3 µJ per pulse.

The Raman light from the target is collected by a telescope and selected from the other backscattered light using an optical thin-film bandpass filter centered at the Raman line of interest, then focused on a single photon detector. The counts from the detector are recorded by the time tagger, which tags each photon with the time it is received relative to the trigger pulse from the laser. The time tagger records one photon count for each laser trigger pulse it receives. After each run of many laser pulses, a data file is created that logs the arrival time of the first photon received after the laser trigger pulse. From this file, a histogram of photon counts versus time of arrival (or time bin) can be generated.

In the laboratory, H_2 gas is produced at the cathode of an electrolysis cell at a rate of 20 mL/min.

The key features of TCSPC are as follows:

– a sync pulse from the laser initiates the time tagger at t_0.
– the laser beam is directed at the target area.
– the return light is spectrally filtered to reject Rayleigh light at 355 nm and to select light at the Raman Stokes line of interest.
– the single photon detector produces a pulse for return photons at time t_r.
– the time tagger records the time of arrival to determine the TOF, TOF $= t_r - t_0$, for only the first photon detected after t_0.
– the next laser trigger pulse initiates the next sequence, and each *TOF* is stored.
– the computer sorts TOF times into time bins of width t_b.
– the process continues for an integration time, T_i, with the total number of pulses for each run, $N = T_i \times$ repetition rate.

As mentioned previously, SRS is a weak process, and integration is required to increase the signal-to-noise ratio. The LOD is determined by the signal-to-noise ratio, and the trade-off is that a long integration time is required to reach a small LOD. However, for an airborne instrument, a longer integration time will result in poorer spatial resolution.

The laboratory prototype for each Raman channel is shown below (Figure 11.5).

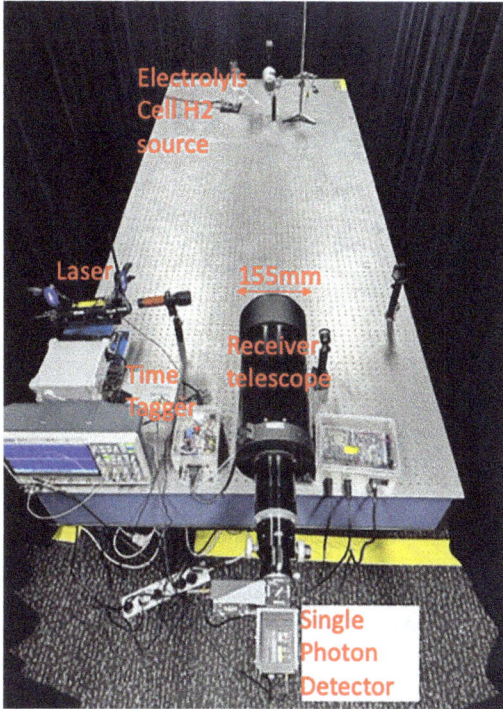

Figure 11.5: Photograph of the laboratory prototype front view: it shows the laser, the receiving optics, and the signal processing electronics in the foreground, while the electrolysis cell in the background produces H₂ at approximately 2 meters distance from the receiving optics.

11.3.1 Results from a TCSPC Raman LIDAR laboratory instrument

The raw data from the experiment were adjusted for background effects by subtracting the signal with no H_2 present (the electrolysis cell switched off) from the signal with H_2 present (the electrolysis cell switched on and producing 20 mL/min) – we label these plots no H_2 and H_2. The processed data (H_2 – no H_2) for the H_2 Raman channel are presented in Figure 11.6.

The system also has an N_2 channel (392 nm) that measures the concentration of N_2 gas. The data from the N_2 channel (392 nm) is normalizing for the different Raman scattering cross sections of H_2 ($(d\sigma/d\Omega) = 7.0710^{-30}$cm^2 sr^{-1}@ 355 nm) and N_2 ($(d\sigma/d\Omega) = 2.6810^{-30}$cm^2 sr^{-1}@ 355 nm).

If we assume that the Raman signal in each channel is given by eq. (11.1), the result provides the H_2 concentration as a function of distance, which is plotted in Figure 11.7.

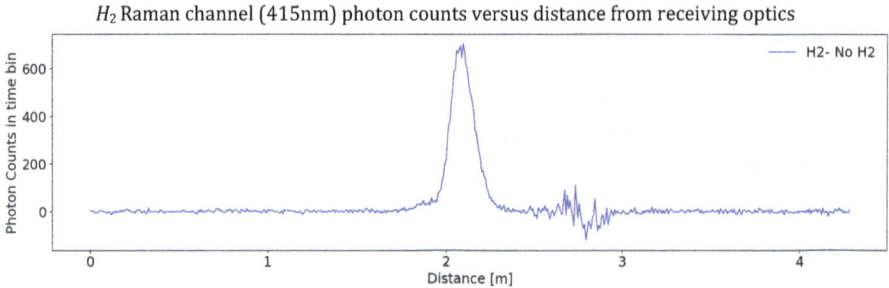

H_2 Raman channel (415nm) photon counts versus distance from receiving optics

Figure 11.6: The plot shows the H_2 Raman channel signal (electrolysis switched on) minus the No H_2 signal (electrolysis switched off) against the distance from the receiving optics.

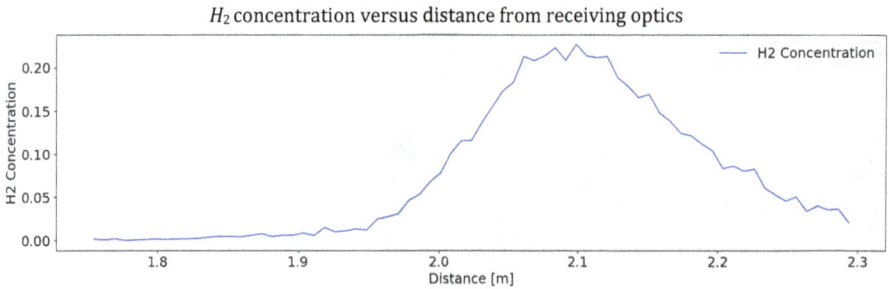

H_2 concentration versus distance from receiving optics

Figure 11.7: The figure shows H_2 concentration against distance from receiving optics – the H_2 concentration is derived from the data in Figure 11.6, divided by the N_2 signal, and normalized for the Raman scattering cross section.

Considering the data plotted in Figures 11.7 and 11.6, we estimate that, for a 60-second integration time, the LOD at 2 m is approximately 9,000 ppm.

11.3.2 Numerical model of time-correlated single photon counting (TCSPC) Raman LIDAR

In conjunction with the lab-based prototype TCSPC Raman LIDAR system, we also developed a numerical model of the TCSPC Raman LIDAR. The aim is to have a numerical model validated by the lab-based prototype and to use the numerical model to inform the design of the airborne system.

Here, we cover the TCSPC Raman LIDAR numerical model in broad outline.

11.3.2.1 Raman signal

We now implement a numerical model of TCSPC for Raman LIDAR and describe how it is validated. We begin by estimating the Raman signal energy, $E(\lambda_S, z)$, as outlined in [58]:

$$E(\lambda_S, z) = K_1 E_L \Delta z A_0 (d\sigma/d\Omega) N/z^2 \tag{11.1}$$

where $E(\lambda_S, z)$ is the Raman (Stokes) signal energy; λ_S is the Stokes wavelength; z is the distance to the region of interest; K_1 is the LIDAR constant; E_L is the pump laser pulse energy; Δz is the spatial extent of the sampled region – this is related to the time bin size, T_B, of the TCSPC system; $\Delta z = T_B \times c$; A_0 is the cross-sectional area of the collection optics aperture; $(d\sigma/d\Omega)$ is the Raman scattering cross section; and N is the molecular concentration.

For the numerical model of TCSPC, we need to convert the energy in the Stokes signal, $E(\lambda_S, z)$, into the number of Stokes photons, P_S, by dividing by the energy of the Stokes photons:

$$P_S(\lambda_S, z) = \frac{K_1 E_L \Delta z A_0 (d\sigma/d\Omega) N \lambda_S}{(z^2 hc)} \tag{11.2}$$

where h is Planck's constant and c is the speed of light.

The important parameters [19] for eq. (11.2) are given in Tables 11.1 and 11.2.

The TCSPC system converts the TOF into distance according to the formula $z = c(t_r - t_0)/2$, where $TOF = t_r - t_0$, t_0 is the time of the trigger pulse, and t_r is the time of the returning photon detected by the single photon detector.

Using the photon version of the Raman LIDAR equation (11.2), $P_S(\lambda_S, z)$, the average number of photons in a Raman channel centered on wavelength λ_S from a distance z, returned in a time bin of size T_B, is calculated. The TCSPC Raman LIDAR model uses Poisson statistics and converts this average number into a probability of zero photons detected, P_0; the probability of a photon being detected is $P_D = 1 - P_0$. From these probabilities, a weighted choice algorithm determines if a photon count is recorded.

11.3.2.2 Noise

The photon count in the H_2 Raman channel will also include noise from the following sources:
- Cross-talk from other channels including Rayleigh scattering.
- Dark counts from the detector (dark counts are counts produced by the single photon detector in the absence of light – dark counts are usually specified by the manufacturer of the single photon detector)

- Background light: Mostly solar light if the Raman LIDAR is operated outdoors during daylight [52].
- Fluorescence: There are many materials in the environment that produce fluorescence when exposed to ultraviolet light. For example, in organic material, most proteins produce fluorescence. In particular, for Raman LIDAR, laser-induced fluorescence (LIF) is an issue since it can be impossible to distinguish a Raman signal from the usually much larger LIF if they are both at the same wavelength. Thus, LIF can mask a Raman signal.

For a given Raman channel, the optical power, $P_B(\lambda_R)$, due to solar background at the Raman line wavelength, λ_R, is given by [59]

$$P_B(\lambda_R) = \frac{\pi^2}{16} E_\lambda \beta^2 A_0 T_R T_F \Delta_\lambda e^{\sigma z} \cos(\theta) \qquad (11.3)$$

where E_λ is the solar spectral radiance at λ_R [60], β is the field of view, T_R is the transmission coefficient of the receiving optic, T_F is the optical filter transmission coefficient at the Raman line wavelength, Δ_λ is the Raman filter's bandpass, σ is the atmospheric extinction coefficient at the Raman line wavelength, and θ is the angle between the receiver optic's optical axis and the normal axis of the target area.

One strategy for optimizing Raman LIDAR that emerges from this analysis of P_B (λ_R) is to operate with a Raman pump wavelength such that the Raman line of interest, λ_R, is in a part of the solar spectrum, where E_λ is at a minimum, and from [60], that is generally at wavelengths shorter than 280 nm, where there is almost zero solar background at the Earth's surface.

Further, as noted in [52], it is possible to operate Raman LIDAR in such a way that the Raman lines fall within the portion of the UV spectrum with little or no LIF.

11.3.3 TCSPC Raman LIDAR numerical model comparison with results

In Figure 11.8, experimental results from the laboratory-based TCSPC Raman LIDAR setup are compared with the numerical model calculations. As can be seen from Figure 11.8, and given the limitations of the model, there is broadly a good fit. The limitations of the model include the following:

- The dispersion of the H_2 gas plume is not included.
- The overlap factor [22] between the optimal collection efficiency and the range, z, is not considered.

Further, although noise from dark counts is included, it was unnecessary to include noise from background light, as the experiment was conducted indoors.

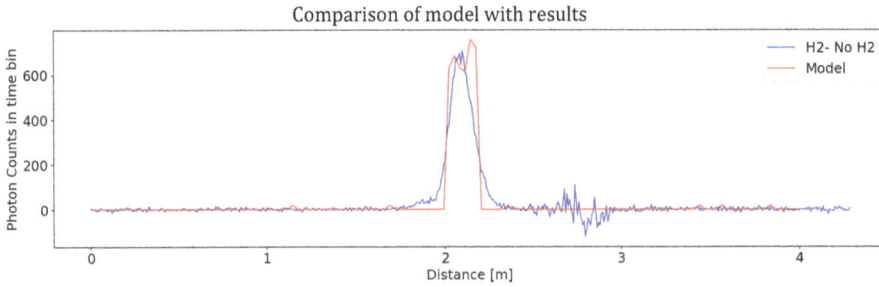

Figure 11.8: The Time Tagger data (see Figure 11.6) compared with the TCSPC Raman LIDAR model. The data (blue) is derived from two Time Tagger data files; the data taken without H_2 is subtracted from a data file taken with H_2 present. The TCSPC Raman LIDAR model fit (red) includes some adjustable parameters.

11.4 Conclusions

This chapter provides the background for developing a stand-off H_2 detection instrument as an exploration tool for identifying natural hydrogen seeps that could potentially be exploited later, if economically viable, as a carbon-free energy source.

A range of current state-of-the-art fixed-point hydrogen sensors has been reviewed. While hand-held sensors have been employed in characterizing existing hydrogen seeps, they are not suitable for surveying large areas as part of a systematic search for hydrogen seeps.

An extensive review of the literature on stand-off detection of hydrogen resulted in the selection of TCSPC Raman LIDAR for the proof-of-concept laboratory-based instrument.

The work carried out at Curtin University on a laboratory-based, proof-of-concept, stand-off detection of hydrogen instrument was based on TCSPC Raman LIDAR. By taking the ratio of the H_2 Raman signal to the atmospheric N_2 Raman signal, the instrument can map H_2 concentration as a function of distance. For a 60-sintegration time, the LOD has been estimated at 9,000 ppm H_2 at 2 m. A numerical model of TCSPC Raman LIDAR has been validated by fitting the experimental results and can be used to develop the design of an airborne instrument in the next phase of the project.

From the numerical model, we have gained insights into how the system must be improved to achieve the performance required of an airborne exploration tool for natural hydrogen detection and mapping. The next phase targets include an LOD of 1,000 ppm H_2 at 50 m with a 0.1-sintegration time in an outdoor environment. The key steps are to increase the Raman pump pulse energy and transition to a shorter wavelength to ensure more Raman photons in the H_2 channel and avoid any solar background.

References

[1] Zgonnik V. The occurrence and geoscience of natural hydrogen: A comprehensive review. Earth-Science Reviews. 2020, 203: 103140. [Online]. Available http://dx.doi.org/10.1016/j.earscirev.2020.103140

[2] Hand E. Hidden hydrogen. Science (New York, NY). 2023, 379(6633): 630–636. [Online]. Available https://www.science.org/content/article/hidden-hydrogen-earth-may-hold-vast-stores-renewable-carbon-free-fuel

[3] Stalker L., Talukder A., Strand J., Josh M., Faiz M. Gold (hydrogen) rush: Risks and uncertainties in exploring for naturally occurring hydrogen. The APPEA Journal. 2022, 62(1): 361–380. [Online]. Available https://www.publish.csiro.au/AJ/pdf/AJ21130

[4] GoldHydrogen, "Exploring for naturally occurring hydrogen resources," 2023. [Online]. Available: https://www.goldhydrogen.com.au

[5] Rezaee R. Assessment of natural hydrogen systems in Western Australia. International Journal of HydrogenEnergy. 2021, 46(66): 33068–33077. [Online]. Available http:/dx.doi.org/10.1016/ j.ijhydene.2021.07.149

[6] Blay-Roger R., Bach W., Bobadilla L. F., Reina T. R., Odriozola J. A., Amils R., Blay V. Natural hydrogen in the energy transition: Fundamentals, promise, and enigmas. Renewable Sustainable Energy Reviews. 2024, 189: 9. [Online]. Available http://dx.doi.org/10.1016/j.rser.2023.113888

[7] Mainson M., Heath C., Pejcic B., Frery E. Sensing hydrogen seeps in the subsurface for natural hydrogen exploration. Applied Sciences-Basel. 2022, 12(13): 12. [Online]. Available https://doi.org/10.3390/app12136383

[8] Hubert T., Boon-Brett L., Black G., Banach U. Hydrogen sensors – A review. Sensors and Actuators B-Chemical. 2011, 157(2): 329–352. [Online]. Available http://dx.doi.org/10.1016/j.snb.2011.04.070

[9] Chauhan P. S., Bhattacharya S. Hydrogen gas sensing methods, materials, and approach to achieve parts per billion level detection: A review. International Journal of Hydrogen Energy. 2019, 44(47): 26076–26099. [Online]. Available https://doi.org/10.1016/j.ijhydene.2019.08.052

[10] Swager T. M., Pioch T. N., Feng H. S., Bergman H. M., Luo S. X. L., Valenza I., J J. Critical sensing modalities for hydrogen: Technical needs and status of the field to support a changing energy landscape. Acs Sensors. 2024, 9(5): 2205–2227. [Online]. Available http://dx.doi.org/10.1021/acssensors.4c00251

[11] Opel F., Itzenhäuser M. A., Wehner I., Lupacchini S., Lauterbach L., Lenz O., Klähn S. Toward a synthetic hydrogen sensor in cyanobacteria: Functional production of an oxygen-tolerant regulatory hydrogenase in Synechocystis sp. PCC 6803. Frontiers in Microbiology. 2023, 14: 12. [Online]. Available https://doi.org/10.3389/fmicb.2023.1122078

[12] NREL, "Safety sensor testing laboratory," 2024. [Online]. Available: https://www.nrel.gov/hydrogen/sensor-laboratory.html

[13] Post M., Buttner W., Hartmann K., P. D., Thorson J., Wischmeyer T., "The NREL sensor laboratory: Status and future directions for hydrogen detection," 2021. [Online]. Available: https://research-hub.nrel.gov/en/publications/the-nrel-sensor-laboratory-status-and-future-directions-for-hydro

[14] Buttner W., Hall J., Coldrick S., Hooker P., Wischmeyer T. Hydrogen wide area monitoring of LH2 releases. International Journal of Hydrogen Energy. 2021, 46(23): 12497–12510. [Online]. Available https://doi.org/10.1016/j.ijhydene.2020.08.266

[15] Baranwal V. C., Rønning J. S. Airborne geophysical surveys and their integrated interpretation. Cham: Springer International Publishing, 2020, 377–400. [Online]. Available https://doi.org/10.1007/978-3-030-28909-614

[16] Newsom R., Chand D., Bambha R., "Raman lidar (rl) instrument handbook," ARM, Report, 2022. [Online]. Available: https://www.arm.gov/capabilities/instruments/rl

[17] TROPOS. Raman lidar. [Online]. Available: https://www.tropos.de/en/research/projects-
     infrastructures-technology/technology-at-tropos/remote-sensing/raman-lidar
[18] Yang F., Sua Y. M., Louridas A., Lamer K., Zhu Z., Luke E., Huang Y. P., Kollias P., Vogelmann A. M.,
     McComiskey A. A time-gated, time-correlated single-photon-counting lidar to observe atmospheric
     clouds at submeter resolution. Remote Sensing. 2023, 15(6): 11. [Online]. Available https://doi.org/
     10.3390/rs15061500
[19] Ninomiya H. Industrial applications of laser remote sensing Chapter 4: Gas sensing using Raman
     scattering. Bentham Science. 2012, [Online]. Available http://dx.doi.org/10.2174/
     97816080534071120101
[20] Joseph E. (2023) Hyterra delivers maiden independent prospective hydrogen and helium resource
     estimate in kansas. [Online]. Available: https://www.proactiveinvestors.com/companies/news/
     1035992/hyterra-delivers-maiden-independent-prospective-hydrogen-and-helium-resource-estimate-
     in-kansas-1035992.html
[21] Eseller K. E., Yueh F. Y., Singh J. P., Melikechi N. Helium detection in gas mixtures by laser-induced
     breakdown spectroscopy. Applied Optics. 2012, 51(7): B171–B175. [Online]. Available http://dx.doi.
     org/10.1364/ao.51.00b171
[22] Measures R. M. Laser remote sensing fundamentals and applications. Krieger Publishing
     Company, 1992.
[23] Banwell C. N., McCash E. M. Fundamentals of molecular spectroscopy. Indian Edition. 2017.
[24] Fox T. A., Barchyn T. E., Risk D., Ravikumar A. P., Hugenholtz C. H. A review of close-range and
     screening technologies for mitigating fugitive methane emissions in upstream oil and gas.
     Environmental Research Letters. 2019, 14(5). [Online]. Available https://doi.org/10.1088/1748-9326/
     ab0cc3
[25] NPL. (2024) Environmental monitoring. [Online]. Available: https://www.npl.co.uk/products-services/
     environmental
[26] Titchener J., Millington-Smith D., Goldsack C., Harrison G., Dunning A., Ai X., Reed M. Single photon
     lidar gas imagers for practical and widespread continuous methane monitoring. Applied Energy.
     2022, 306: 11. [Online]. Available https://doi.org/10.1016/j.apenergy.2021.118086
[27] Avetisov V., Bjoroey O., Wang J. Y., Geiser P., Paulsen K. G. Hydrogen sensor based on tunable diode
     laser absorption spectroscopy. Sensors. 2019, 19(23): 13. [Online]. Available https://doi.org/10.3390/
     s19235313
[28] Xiafukaiti A., Lagrosas N., Ogita M., Oi N., Ichikawa Y., Sugimoto S., Asahi I., Yamaguchi S., Shiina
     T. Optimization for hydrogen gas quantitative measurement using tunable diode laser absorption
     spectroscopy. Optics and Laser Technology. 2025, 180: 7. [Online]. Available http://dx.doi.org/10.
     1016/j.optlastec.2024.111587
[29] Mondelain D., De Casson L. B., Fleurbaey H., Kassi S., Campargue A. Accurate absolute frequency
     measurement of the S(2) transition in the fundamental band of H_2 near 2.03 μm. Physical Chemistry
     Chemical Physics. 2023, 7.[Online]. Available https://doi.org/10.1039/d3cp03187j
[30] Privalov V. E., Shemanin V. G. The concentration measurement of hydrogen molecules in the
     atmosphere: Lidar equation computer simulation for the differential absorption and scattering.
     Measurement Techniques. 2023, 7. [Online]. Available https://doi.org/10.1007/s11018-023-02157-1
[31] Weber A. Raman spectroscopy of gases and liquids. Berlin, Heidelberg: Springer, 1979. [Online].
     Available https://doi.org/10.1007/978-3-642-81279-8
[32] Hecht E. S., Panda P. P. Mixing and warming of cryogenic hydrogen releases. International Journal
     of Hydrogen Energy. 2019, 44(17): 8960–8970. [Online]. Available https://doi.org/10.1016/j.ijhydene.
     2018.07.058
[33] Sugimoto S., Ichikawa Y., Ogita M., Asahia I., Kamiji Y., Terada A., Hino R. Measurement of diffusion
     behavior of hydrogen leaked from buried pipe using raman imaging. In: Conference on electro-

optical remote sensing XIII, ser. Proceedings of SPIE, vol. 11160. Conference Proceedings, 2019. [Online]. Available https://doi.org/10.1117/12.2533032

[34] Ninomiya H., Yaeshima S., Ichikawa K., Fukuchi T. Raman lidar system for hydrogen gas detection. Optical Engineering. 2007, 46(9): 5. [Online]. Available http://dx.doi.org/10.1117/1.2784757

[35] Shiina T. Hydrogen gas detection by mini-Raman lidar. IntechOpen, 2018. [Online]. Available http://dx.doi.org/10.5772/intechopen.74630

[36] Choi I. Y., Baik S. H., Choi Y. S. Remotely measuring the hydrogen gas by using portable Raman lidar system. Optica Applicata. 2021, 51(1). [Online]. Available http://dx.doi.org/10.37190/oa210103

[37] Stothard D. J., Warden M. S., Spesyvtsev R., Kelly E., Leck J., Allen A., Squire J., Hepworth S., Malone S., Tunney J., et al. Long-range, range-resolved detection of H_2 using single-photon 'quantum' Raman: A condition monitoring tool for long-term storage of nuclear materials. In: Chemical, Biological, Radiological, Nuclear, and Explosives (CBRNE) sensing XXIII. SPIE, 2022, PC1211608. [Online]. Available https://doi.org/10.1117/12.2621423

[38] Heaps W. S., Burris J. Airborne Raman lidar. Applied Optics. 1996, 35(36): 7128–7135. [Online]. Available https://doi.org/10.1364/ao.35.007128

[39] Whiteman D. N., Rush K., Rabenhorst S., Welch W., Cadirola M., McIntire G., Russo F., Adam M., Venable D., Connell R., Veselovskii I., Forno R., Mielke B., Stein B., Leblanc T., McDermid S., Vomel H. Airborne and ground-based measurements using a high-performance Raman lidar. Journal of Atmospheric and Oceanic Technology. 2010, 27(11): 1781–1801. [Online]. Available https://doi.org/10.1175/2010jtecha1391.1

[40] Wu D. C., Wang Z. E., Wechsler P., Mahon N., Deng M., Glover B., Burkhart M., Kuestner W., Heesen B. Airborne compact rotational Raman lidar for temperature measurement. Optics Express. 2016, 24(18): A1210–A1223. [Online]. Available https://doi.org/10.1364/oe.24.0a1210

[41] Prinzhofer A., Cisse C. S. T., Diallo A. B. Discovery of a large accumulation of natural hydrogen in Bourakebougou (Mali). International Journal of Hydrogen Energy. 2018, 43(42): 19315–19326. [Online]. Available http://dx.doi.org/10.1016/j.ijhydene.2018.08.193

[42] Sand M., Skeie R. B., Sandstad M., Krishnan S., Myhre G., Bryant H., Derwent R., Hauglustaine D., Paulot F., Prather M., Stevenson D. A multi-model assessment of the global warming potential of hydrogen. Communications Earth Environment. 2023, 4(1): 12. [Online]. Available https://doi.org/10.1038/s43247-023-00857-8

[43] Etiope G. Massive release of natural hydrogen from a geological seep (Chimaera, Turkey): Gas advection as a proxy of subsurface gas migration and pressurised accumulations. International Journal of Hydrogen Energy. 2023, 48(25): 9172–9184. [Online]. Available http://dx.doi.org/10.1016/j.ijhydene.2022.12.025

[44] Tractor A. (2024) AT-502B. [Online]. Available: https://airtractor.com/aircraft/at-502b/

[45] Long D. A. The Raman effect: A unified treatment of the theory of raman scattering by molecules. England: John Wiley Sons Ltd, 2002.

[46] Popp J., Kiefer W. Raman scattering, fundamentals. Encyclopedia of Analytical Chemistry. 2006, [Online]. Available http://dx.doi.org/10.1002/9780470027318.a6405

[47] Compaan A., Wagoner A., Aydinli A. Rotational Raman-scattering in the instructional laboratory. American Journal of Physics. 1994, 62(7): 639–645. [Online]. Available https://doi.org/10.1119/1.17484

[48] Avila G., Tejeda G., Fernández J. M., Montero S. The rotational Raman spectra and cross sections of H_2O, D_2O, and HDO. Journal of Molecular Spectroscopy. 2003, 220(2): 259–275. [Online]. Available http://dx.doi.org/10.1016/s0022-2852(03)00123-1

[49] Singh J., Muller A. High-precision trace hydrogen sensing by multipass Raman scattering. Sensors. 2023, 23(11): 14. [Online]. Available https://doi.org/10.3390/s23115171

[50] Sugimoto S., Asahi I., Shiina T. A practical-use hydrogen gas leak detector using CARS. International Journal of Hydrogen Energy. 2021, 46(37): 19693–19703. [Online]. Available https://doi.org/10.1016/j.ijhydene.2021.03.101,

[51] Whiteman D. N., Melfi S. H., Ferrare R. A. Raman lidar system for the measurement of water-vapor and aerosols in the earths atmosphere. Applied Optics. 1992, 31(16): 3068–3082. [Online]. Available https://doi.org/10.1364/ao.31.003068

[52] Shi D. C., Hua D. X., Gao F., Chen T., Stanic S. Influence analysis of the detection accuracy of atmospheric water vapor using the solar-blind ultraviolet Raman lidar. Journal of Quantitative Spectroscopy Radiative Transfer. 2020, 251: 12. [Online]. Available https://doi.org/10.1016/j.jqsrt.2020.107032

[53] Ball A. J., "Investigation of gaseous hydrogen leak detection using Raman scattering and laser-induced breakdown spectroscopy," Thesis, 2005. [Online]. Available: https://ufdcimages.uflib.ufl.edu/UF/E0/01/05/45/00001/balla.pdf

[54] Jeon K. S., Sim J., Cho W. B., Park B. Research on long-range hydrogen gas measurement for development of Raman lidar sensors. International Journal of Hydrogen Energy. 2024, 67: 119–126. [Online]. Available http://dx.doi.org/10.1016/j.ijhydene.2024.04.126

[55] Choi I. Y., Baik S. H., Cha J. H., Kim J. H. Study of a method for measuring hydrogen gas concentration using a photon-counting Raman Lidar system. Korean Journal of Optics and Photonics. 2019, 30(3): 114–119. [Online]. Available https://doi.org/10.3807/kjop.2019.30.3.114

[56] Limery A., Cezard N., Fleury D., Goular D., Planchat C., Bertrand J., Hauchecorne A. Raman lidar for hydrogen gas concentration monitoring and future radioactive waste management. Optics Express. 2017, 25(24): 30636–30641. [Online]. Available https://doi.org/10.1364/oe.25.030636

[57] Voronina E. I., Privalov V. E., Shemanin V. G. Probing hydrogen molecules with a laboratory Raman Lidar. Technical Physics Letters. 2004, 30(3): 178–179. [Online]. Available https://doi.org/10.1134/1.1707159

[58] Privalov V. E., Shemanin V. G. Lidar measurement of the Raman differential cross section by hydrogen molecules. Optical Memory and Neural Networks. 2023, 32(1): 34–38. [Online]. Available http://dx.doi.org/10.3103/s1060992x23010034

[59] Manojlovic L. M., Barbaric Z. P. Optimization of optical receiver parameters for pulsed laser-tracking systems. IEEE Transactions on Instrumentation and Measurement. 2009, 58(3): 681–690. [Online]. Available http://dx.doi.org/10.1109/tim.2008.2005259

[60] Wald L. (2018) Basics in solar radiation at earth surface. [Online]. Available: http://dx.doi.org/10.13140/RG.2.2.36149.93920

Giuseppe Etiope* and Alexandra Orbán

Chapter 12
Surface gas geochemical exploration for natural hydrogen: uncertainties and holistic interpretation

Abstract: Surface gas geochemistry, including the analysis of surface fluid manifestations (seeps), gas in the soil and gas dissolved in shallow aquifers and springs, is a critical component of global exploration for natural hydrogen (H_2). As demonstrated in hydrocarbon exploration, this discipline can provide initial insights into the potential occurrence of gas resources underground prior to drilling and can motivate subsequent exploration activities ranging from geophysical surveys to drillhole surveys. However, the interpretation of the presence of H_2 in surface environments is not straightforward, and, as with hydrocarbon gas seepage, it is impossible to quantify the level of H_2 concentrations near the surface that could suggest a potentially economic resource. Here, we discuss the basic principles, advantages, and uncertainties of surface H_2 geochemistry, considering any potential background non-geological H_2 concentrations in soil, springs, and shallow aquifers. While determining a geological origin is easier for H_2 in gas seeps and hyperalkaline springs – particularly when the fluid manifestation exhibits distinct features of deep circulation (e.g., seeps and water with gases of unequivocal geogenic origin) – interpreting H_2 at ppmv levels in the soil or shallow aquifers requires prudence. Research conducted thus far indicates that near-surface biological activity can generate concentrations of H_2 in the soil in the order of 10^2–10^3 ppmv, up to 1% by volume. Consequently, it is important to exercise caution before hastily assigning a geological origin to H_2 in soil-gas at ppmv levels. The detection of H_2 in aquifers accessed by wells can also be equivocal due to the fact that, in addition to microbiological activity, some H_2 can result from well corrosion. A multiparametric, holistic approach that accurately evaluates the surface ecosystem and the origin of other gases associated with H_2 is therefore recommended. In particular, isotopic studies of the gases associated with H_2, especially radiocarbon analysis of CH_4 and CO_2, are strongly advised. An accurate evaluation of the geological context

*Corresponding author: Giuseppe Etiope, Istituto Nazionale di Geofisica e Vulcanologia, Sezione Roma 2, Rome, Italy; Faculty of Environmental Science and Engineering, Babes-Bolyai University, Cluj-Napoca, Romania, e-mail: giuseppe.etiope@ingv.it
Alexandra Orbán, Faculty of Environmental Science and Engineering, Babes-Bolyai University, Cluj-Napoca, Romania

https://doi.org/10.1515/9783111437040-012

and the system where H_2 is measured remains a fundamental component of any surface exploration strategy.

Keywords: natural hydrogen, gas seep, geochemical exploration, soil-gas, hydrogen isotopic signature

12.1 Introduction

Surface gas geochemistry is currently widely used in the initial steps of natural hydrogen (H_2) exploration, utilizing the fundamental principles of gas migration to the surface from source and reservoir rocks, which have been developed in petroleum geology and exploration. H_2-rich gas seeps and the presence of H_2 in the soil or shallow aquifers can, in fact, provide valuable insights into the potential existence of subterranean H_2 accumulations (e.g., [1–3]), just as hydrocarbon (gas and oil) seeps and microseepage have allowed petroleum reservoirs to be discovered (e.g., [4–7]). There are, however, fundamental differences between H_2 seeps and H_2 in soil-gas, shallow aquifers, or surface waters (lakes, rivers, ponds). H_2 in soil, surface, and shallow underground waters requires more challenging interpretations. In a gas seep, where H_2 may have concentrations at percentage levels (typically above 1% by volume) and the amount of gas flux expresses an advective migration driven by pressure gradients [2], there is little doubt that H_2 has an underground, geological origin. H_2 in soil-gas, where H_2 concentrations can be in the order of 10^1–10^3 ppmv, may instead have multiple origins – microbial, chemical, and geological (e.g., [2, 8], and references therein). Nevertheless, in many soil-gas studies, such as those related to "fairy circles," a paradigm is frequently assumed by which H_2 that is microbially generated in wet soils and aquifers can be rapidly consumed by bacteria and should not occur in the aerated vadose zone. The word "seep" is then cursorily used for H_2 concentrations in the order of tens to thousands of ppmv in the soil (e.g., [9–12]). As discussed below, this may be misleading.

   Here, the interpretations applicable to H_2 in seeps, soil-gas, springs, shallow aquifers, and surface waters are discussed, considering H_2 concentration ranges derived from our data inventory (updated to September 2024). Specifically, we will examine: (a) the potential of gas seeps as indicators of pressurized H_2 reservoirs; (b) the range of H_2 concentrations in hyperalkaline springs in serpentinized ultramafic rock systems; (c) the uncertainties in determining the origin of H_2 in near-surface environments, i.e., soil, shallow aquifers, and surface waters, taking into account the potential background (non-geological) H_2 concentrations; and (d) the need for a holistic approach to interpret any signal of H_2 at or near the surface, as a factor that may promote further exploration. This work does not address technical issues related to the reliability (errors of gas sampling procedures) or the accuracy of H_2 detecting instruments. The concepts reported in this chapter are a component of the road map for scientific research on natural hydrogen considered in the Technology Collaboration Program (TCP) task on "Natural hydrogen" launched in 2024 by the International Energy Agency [13].

12.2 The potential for gas seeps as indicators of pressurized H_2 accumulations

12.2.1 What is a "gas seep"?

The term "seep" is traditionally used in petroleum geology to define a surface manifestation of hydrocarbon-rich gas and oil related to a petroleum system (e.g., [4, 6, 7, 14]). A "seepage system" [6] is, in fact, a common, integral component of a petroleum system that includes source rocks, reservoirs, and traps. A seep is a fluid manifestation, either from dry ground (soil or rocks) or as bubble trains in water bodies (springs, rivers, lakes, ponds); therefore, it is generally visible with measurable gas flux into the atmosphere, often perturbing the terrain morphologically. The term "seep" is not used for geothermal, CO_2-rich manifestations, which are extensively referred to as "mofettes," "fumaroles," or more generally "geothermal fluid manifestations." However, "seep" has historically been used also for gas emissions that can have abiotic methane and hydrogen stemming from ultramafic rock systems, such as ophiolites and peridotite massifs (e.g., [7, 15, 16]). Thus, the use of the term "seep" for natural hydrogen is acceptable and is today widely used. Indeed, it is used too easily. In fact, it has also been associated with simple H_2 concentrations at ppmv levels observed in the soil-air. This aspect is discussed in Section 12.4.

12.2.2 The seep-reservoir connection

The petroleum industry emerged in the mid-1800s when it became evident that drilling in areas where gas and oil naturally seeped to the surface – namely in North America, Azerbaijan, Poland, and Romania, routinely resulted in the production of economically significant quantities of hydrocarbons. Surface seeps, whether in the form of bubbling pools, flames, mud volcanoes or dry gas vents, were in essence instrumental in the global discovery of gas-oil fields. Modern petroleum geology acknowledges that hydrocarbon seeps represent a migration of fluids from reservoirs [4, 5, 18], that the seepage system is an integral component of petroleum systems [6], and that over 80% of all existing hydrocarbon production has a seep association [19]. The accumulation feeding a seep, however, is not necessarily a major, deep, or producing reservoir of a petroleum field; in most cases, in fact, seeps stem from relatively shallow hydrocarbon accumulations, which may be part of a wider and productive

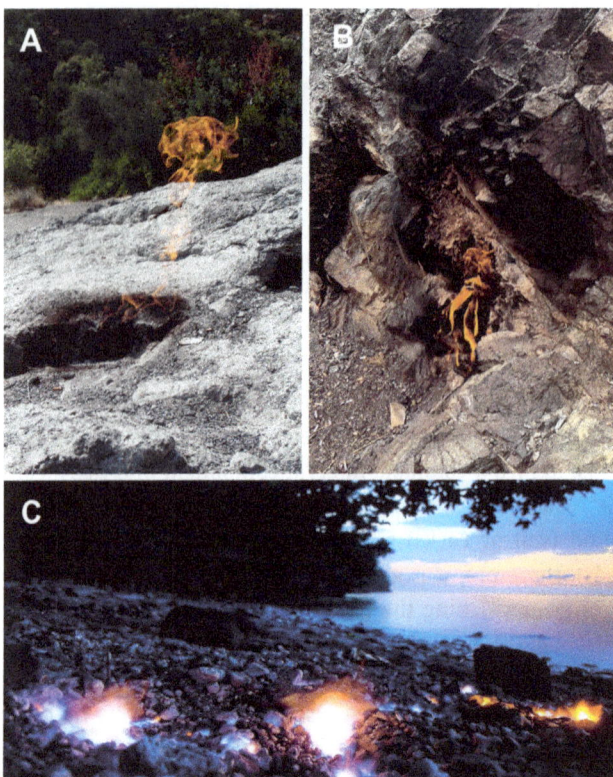

Figure 12.1: Examples of H_2-rich seeps. (A) Chimaera seep, Tekirova ophiolite, Turkey (photo by G. Etiope). (B) Kurtbagi seep, Kizildag ophiolite, Turkey (photo by W. D'Alessandro). C: Tanjung Api seep, East Sulawesi Ophiolite, Indonesia (photo by M. Levitin [17]).

petroleum system. The seep-reservoir connection is also demonstrated by the dependence of seep intensity on hydrocarbon exploitation activity, through pumping from reservoirs. Many seeps, in several countries, became inactive or weaker once hydrocarbon production began in nearby wells, as fluid extraction induced pressure reduction within the reservoir (e.g., [7], and references therein). The seep-reservoir coupling operates on the principles of fluid dynamics, whereby seeps are referred to as advective fluid migration (driven by pressure gradients following Darcy's law) and not diffusion (concentration gradients; Fick's law) [7, 20, 21]. Advection is confirmed by the lack of significant isotopic fractionation (which is instead a typical effect of diffusion) of gas observed at the surface: CH_4 in seeps and related reservoirs has, in fact, similar isotopic composition (e.g., [22]). Gas advection typically develops along permeable strata and tectonic discontinuities (faults and fracture systems) and it is caused by pressure gradients originating from pressurized accumulations. We do not enter here a detailed discussion on advection and diffusion in geological media, for which the reader may refer to several specific and review works (e.g., [2, 7, 20, 21, 23]). This concept, which is well-established in petroleum geology, can be transferred to abiotic gas seeps, including H_2 in non-sedimentary rocks. Accordingly, H_2-seeps, especially if intense, may disclose the existence of pressurized H_2-rich pools [2].

H_2-rich seeps, where the gas is released from dry ground (vents) or through bubbles in springs, are reported in several countries, mostly in ophiolite and peridotite massifs, with H_2 concentration values of up to 99 vol% (Huwayl Qufays, Oman) [24]. At least nine countries (Albania, Indonesia, Kosovo, New Caledonia, New Zealand, Oman, Philippines, Turkey, and the United States) have seeps with >10 vol% H_2.

It is assumed that H_2-seeps can also derive directly from the source rock (the "kitchen"), where H_2 is generated, or from reservoirs with a short (<100 years) residence time of gas, which are continuously refilled with H_2 (e.g., [25]). This would imply a fast and continuous generation of H_2, capable of sustaining the seep flux at a fixed position. In this respect, an examination of the known H_2 generation rates, either via serpentinization or radiolysis, suggests that H_2 source rocks can hardly sustain certain seeps [2, 26, 27], especially those with significant H_2 concentration and flux (Figure 12.1). Examples include the seeps of Chimaera, Tahtakopru, and Kurtbagi in Turkey, Tanjung Api in Indonesia [2, 28] and the intense bubbling seep in the Bulqizë ophiolite in Albania [27]. H_2 generation rates from radiolysis and low-temperature serpentinization (<150 °C) are lower than or comparable to those of biotic methane generation in shales [29]. Numerical simulations for the Chimaera and Bulqizë cases suggest the existence of pressurized reservoirs [26, 27]. Many H_2-seeps, then, generally contain other gases, such as CH_4, heavier hydrocarbons (ethane, propane, butane), and N_2, whose generation rates are not considered fast and certainly need to accumulate to sustain the surface gas manifestation. Models trying to explain a blend of "fresh" H_2 and old gas (e.g., [25]) cannot be applied universally and leave open questions. If H_2 is renewable over the human-life timescale, what is its relationship with the ^{14}C-free (fossil) abiotic CH_4 and C_{2+} hydrocarbons that are frequently observed in gas seeps in ultramafic rock systems?

12.3 H_2 dissolved in hyperalkaline water springs related to serpentinization

Hyperalkaline springs (pH > 9) related to continental serpentinization are widely studied in many countries. Some of them host gas bubbles that can be considered "seeps," as discussed in the previous section. Here, we focus on the amount of H_2 that is dissolved in the water released to the surface. Based on a literature survey, dissolved H_2 data (with H_2 concentrations typically expressed in µM or µg/L) are available from eight countries (Bosnia and Herzegovina, Canada, Greece, Italy, Oman, Philippines, Spain, the United States); the highest concentrations are around 600 µM, as reported in the United States, Spain, and Canada (681 µM at The Cedars in the Coast Range ophiolite [30]; 595 µM at the Ronda peridotite massif [31]; and 590 µM in the Tablelands ophiolite [32]). Figure 12.2 shows the histogram distribution of published dissolved H_2 data (inventory updated to September 2024).

There is little doubt that H_2 in these hyperalkaline springs is geological and derives from serpentinization (e.g., [30, 33, 34]). Therefore, they represent sites with H_2 generation, with deep water circulation interacting with the H_2 source rocks.

Understanding the hydrogeological circuit, in particular its maximum depth, is necessary to define the H_2-producing formations. Evaluating the H_2 amount discharged at the surface – that is, combining the water flow rate and the dissolved H_2 amount – will provide insights into the potential subsurface amounts and accumulation of H_2, as in the case of the seeps.

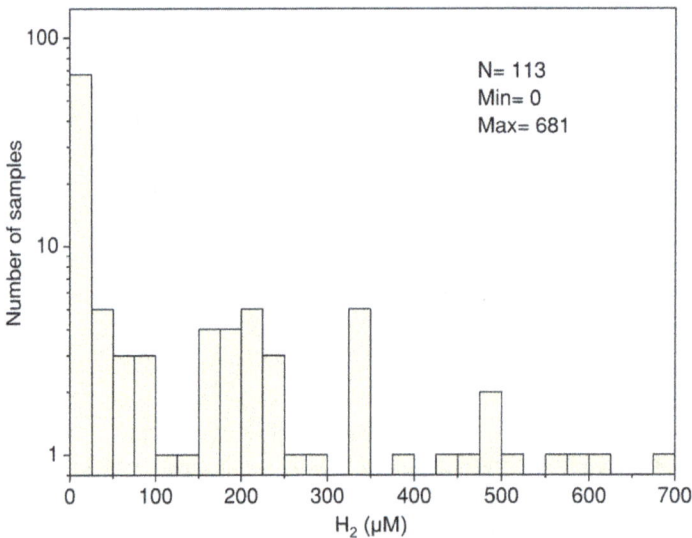

Figure 12.2: Histogram showing the distribution of published data on dissolved H_2 concentrations in hyperalkaline springs. Data inventory from 11 articles [30–32, 35–42].

12.4 Uncertainties in assessing H_2 origin in soil, shallow aquifers, and surface waters

While H_2-rich seeps and hyperalkaline waters, where H_2 concentration is often at percentage levels by volume and µM levels, are a typical manifestation of geological gas migration from depth (as discussed above), the presence of H_2 at low concentrations in soil (i.e., aerated soil, the upper portion of the vadose zone), shallow aquifers, and surface waters (lakes, rivers, and ponds) is not an unambiguous indication of geological gas. Multiple non-geological processes can, in fact, induce relatively high H_2 concentrations in the ground, groundwater, and wells, and interpreting the origin of H_2 is often challenging as, discussed in the following subsections.

12.4.1 H_2 in the aerated soil

Available soil-gas H_2 datasets are quite limited. Before the advent of soil-gas prospecting for natural hydrogen exploration, most geological gas studies reporting H_2 values in aerated soils were carried out for fault mapping and earthquake precursor research, without using specific analyses that may unambiguously demonstrate a geological H_2 origin (e.g., [43, 44]). Biological studies have largely addressed wetlands and the capacity of dry soil to act as an atmospheric H_2 sink; these studies are mainly based on laboratory tests and modeling (e.g., [45, 46]). Consequently, when natural hydrogen soil-gas exploration began, the potential concentrations of H_2 that could be produced biologically near the surface were poorly defined. Figure 12.3 shows the range of H_2 concentration in the soil reported within several countries, in the framework of natural hydrogen exploration (inventory updated to September 2024). Many natural hydrogen exploration studies have assumed that biological H_2 is virtually absent in aerated soils due to the notion of rapid microbial H_2 consumption (e.g., [47, 48]); therefore, any presence of H_2 in the soil has been considered a signal of seepage. Considering the atmospheric H_2 concentration of 0.5 ppmv, and adopting the notion that soil $H_2 > 1$ ppmv indicates seepage [12], most natural hydrogen exploration work (Figure 12.3), including studies of "fairy circles" or "subcircular depressions" (SCDs), interpreted soil-gas H_2 concentrations at the ppmv level as evidence of H_2 degassing from subsurface sources (e.g., [1, 9–12, 49–52]). It is known, however, that H_2 can be produced near the surface by multiple microbially mediated processes, including fermentation in wet soils or shallow aquifers, N_2 fixation, and cellulose decomposition by termites [53–56]. Additionally, H_2 can be generated by the oxidation or corrosion of ferrous minerals (e.g., [57]) and by the hydration of silicate radicals in basaltic soils [58].

A recent study reported H_2 concentrations in the range of 10^1–10^3 ppmv, with a peak of 1 vol%, in aerated soils along Alpine valleys in northern Italy [59]. Because the valleys were linked with tectonic faults, and H_2 was also associated with very high concentrations of CH_4 and CO_2 (up to 51 vol% and 27 vol%, respectively), seepage of geological gas was initially considered, in line with interpretations reported in the literature. Nonetheless, multiple isotopic analyses, in particular the radiocarbon (^{14}C) analysis of CH_4 and CO_2, revealed that H_2 was associated only with modern microbial gas ($F^{14}C$ of CH_4 and $CO_2 \sim 1$ [59]). Multiple lines of evidence, including (a) the absence of positive H_2 fluxes from the ground, (b) the absence of H_2 and other geological gases in springs, (c) the presence of H_2S in the highest H_2–CH_4–CO_2 soil sites, and (d) the presence of shallow aquifers, wet ground, and peat soil in the vicinity of the aerated soil sites with H_2, have suggested that H_2 stemmed from near-surface microbial processes (including fermentation). This case has provided a new reference for potential concentrations of non-geological H_2 in the soil and calls for prudence when cursorily assuming that ppmv levels of H_2 in soil represent seeps. Prudence is recommended, in particular, for the "fairy circles," SCDs of variable size that, in Russia, the United States, Australia, Brazil, Namibia, and Colombia, have been generally interpreted as expressions of geological H_2 emanations simply on the basis of ppmv detection of H_2 in soil and spatial correlation between H_2 and ground morphology [9–11, 49–51]. Many of these morphological features, especially those in the Carolina Bays [10], are wet zones with waterlogged soils and shallow aquifers; radiocarbon analysis of CH_4 observed at these sites may clarify whether the "circles" are really *loci* of geological degassing. Regarding the Carolina Bays, however, the general consensus is that they are simply the result of interaction between periglacial winds, groundwater levels, and an exposed water table (e.g., [60–62]).

Beyond the published data summarized in Figure 12.3, a wide number of unpublished cases exist reporting the serendipitous detection of H_2 in soil-gas, with concentrations in the order of 10^1–10^3 ppmv during studies aimed simply at testing H_2 sensors (Orbán, doctoral thesis in progress) or within the framework of prospecting focused on radon or carbon dioxide (Ciotoli, personal communication) in areas where natural hydrogen is not expected at all. Examples are found in urban gardens or sedimentary areas lacking biotic gas systems and with very deep igneous basement. This highlights the need to extend soil-gas H_2 measurements to any geological environment with varying soil conditions, regardless of the presence of potential geological H_2 sources.

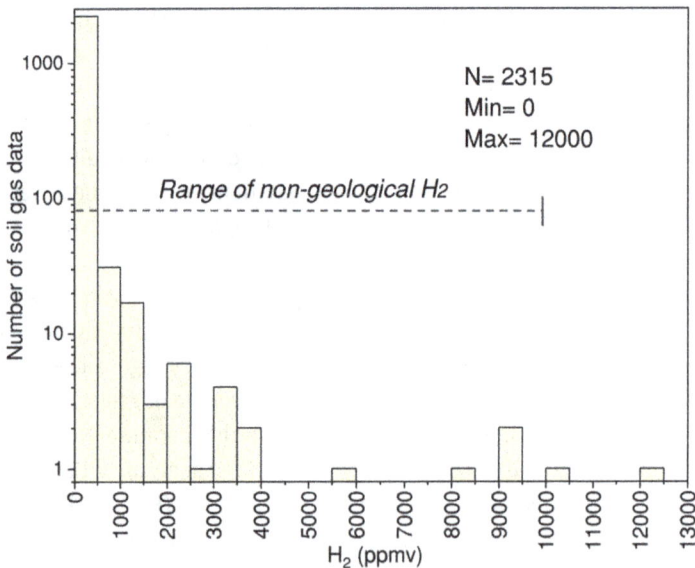

Figure 12.3: Histogram showing the distribution of H_2 soil-gas concentrations from the literature focused on natural hydrogen exploration. Data inventory from 20 articles [1, 9–12, 49–52, 59, 63–72].

12.4.2 H_2 in shallow aquifers and waterlogged soil

There is little available data on dissolved H_2 in shallow aquifers or water in soils, with most of the work having focused on microbial H_2 generation to study redox conditions in oxygen-depleted environments (e.g., [73]). H_2 dissolved in shallow (<100 m deep) groundwater can vary widely, depending on several factors such as microbial activity, redox conditions, the presence of other chemical species, and the geological environment. The concentration of dissolved H_2 in equilibrium with the atmosphere (air-saturated water, ASW) is 0.0004 µM (or 0.4 nM; atmosphere with 0.5 ppmv H_2; water at 20 °C). This is the background level of dissolved H_2 that can be observed in rainwater and surface waters in the absence of microbial or geological sources.

In aquifers and saturated soil and sediments (ponds, marshes, estuaries) with hypoxic or anoxic conditions and fermentation, dissolved H_2 concentration is typically on the order of 101–102 nM (Figure 12.4; e.g., [54, 74–76]). The highest dissolved biological H_2 concentration value that we could identify in the literature is 426 nM [58]. We may consider that values exceeding an order of magnitude of 10^3 nM are "extraordinarily high" from a biological point of view and can represent signals of subsoil geological degassing (if corrosion processes within boreholes, discussed below, can be excluded).

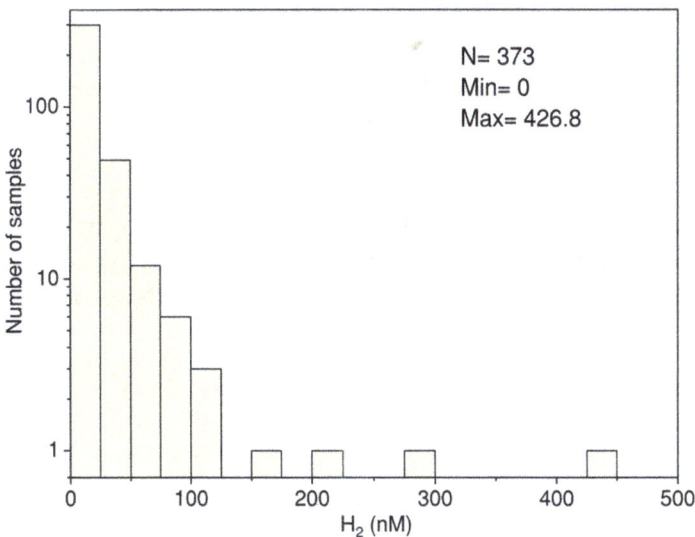

N= 373
Min= 0
Max= 426.8

Figure 12.4: Histogram showing the distribution of dissolved H_2 concentrations in aquifers and saturated sediments, not related to hyperalkaline waters or other geological H_2-bearing environments. Data inventory from 19 articles [54, 58, 73–75, 77–90].

12.4.3 The problem of the H₂ isotopic signature

Unfortunately, an isotopic analysis of H_2 that can be observed in the soil, aquifers, or springs may not help in assessing whether the gas is geological or biological, since the values of the $^2H/^1H$ isotopic ratio of H_2 (expressed as δ^2H) from both origins partially overlap [2]. For geological H_2, δ^2H can range from −100 to −1,000‰ [8]. The δ^2H value of biological H_2 within surface ecosystems generally ranges from −690 to −740‰ [91]. However, oxidized, residual biological H_2 can have δ^2H values up to −450‰ [46]. This biological range coincides with the isotopic signature of geologically produced H_2 over a wide range of temperatures [2], from continental serpentinization systems and Precambrian crystalline shields (δ^2H typically < −650‰), to geothermal-magmatic systems (δ^2H > −650‰). Geological H_2 could only be isotopically identified within the soil if it stems from very high-temperature systems, hydrothermal or magmatic sources with δ^2H > −450‰, and it is not re-equilibrated with shallow waters.

12.4.4 The problem of artificial H₂ caused by drilling or corrosion

Mechanical and chemical processes involving metals can generate artificial H_2 in the soil and aquifers. The soil-gas sampling methods typically use metallic probes that penetrate the ground by percussion. The mechanical friction between a metallic probe and hard stones, pebbles, or iron-rich granules occurring in the soil can, in fact, produce H_2. This process has been verified in several cases (e.g., H_2 up to ~10 vol% [64]). Artificial H_2 can also be produced by drilling and/or corrosion within any type of borehole with a metal casing. H_2 concentration build-up in a well casing of up to 9.2 vol% was observed in corroded geothermal boreholes [92], and up to 1.5 vol% was related to percussion drilling [93]. Accordingly, searching for natural hydrogen in aquifers through boreholes should consider the possibility that the observed H_2 is not natural. The corrosion conditions of the borehole casing should initially be carefully examined.

12.4.5 Can radon and helium help?

Radon (^{222}Rn) and helium (He) have been extensively used in surface geochemical exploration of seismic faults and hydrocarbons because of their typical origin from radioactive decay in underground rocks (or mantle origin in the case of ^3He-rich helium), enhanced migration along faults, and connection with oil-gas reservoirs, as documented in a wide body of literature (e.g., [94–101]). Accordingly, ^{222}Rn and He have been used to support the interpretation of the geological origin of H_2 observed in the ground (e.g., [1, 68]). The coexistence of crustal helium (^4He) anomalies and hydrogen in the soil has been suggested as a crucial element in determining the deep source of hydrogen, hence minimizing the possibility of misinterpreting surface biological hydrogen detection [68]. Nonetheless, this approach may not be conclusive if the radon and helium sources, that is, crystalline- or uranium-rich rocks or sediments, are near the surface; in these cases, the two gases observed in soil or shallow aquifers may not reflect gas migration from deep sources. In particular, He anomalies in the soil (concentrations above the atmospheric value of 5 ppmv) may simply reflect groundwater circulation in fractured igneous rocks at shallow depths (e.g., [102, 103]) and high concentrations of ^{222}Rn can be related to emanations from U-rich minerals near the surface [104–106]. In order to assess whether soil-gas radon is a tracer of gas migration from depth, it is necessary to evaluate the theoretical maximum values of ^{222}Rn produced in situ by its radium parent, based on the radium concentration in soil or near-surface rocks, the emanation power, the bulk density, and the porosity of the soil or near-surface rocks (e.g., [105, 106]). In parallel, it is important to demonstrate the presence of geological "carrier" gases (such as CO_2, CH_4, or N_2), which are essential for long-distance radon transport (e.g., [20, 23]). If relevant concentrations of CH_4 and/or CO_2 are found in the soil together with radon, their geological origin can only be proven through isotopic analyses, including radiocarbon (^{14}C), as discussed above [59].

12.4.6 Can soil-atmosphere flux measurements help?

When only biological H_2 occurs in the soil, its exhalation to the atmosphere is very limited or absent due to rapid H_2 consumption by micro-organisms and enzymes surrounding the H_2 generation zone (e.g., [53]). The presence of geological H_2 in the soil, instead, may be associated with pressurized gas pockets within soil pores or in the subsoil vadose zone, which can lead to positive H_2 fluxes at the soil-atmosphere interface [2]. A method based on H_2 flux measurement using a static closed-chamber was suggested to help evaluate whether H_2 in the soil is an expression of geological gas seepage or is related to biological activity [2]. The static closed-chamber is widely used to measure gas fluxes to the atmosphere, either for bio-ecosystem or geological applications (e.g., [107–110]). In the case of geological H_2, a positive flux, with clear H_2 concentration build-up within the chamber, can be recorded, and the flux may have a heterogeneous spatial distribution, typically correlated with faults. In the case of biological H_2, the flux is typically homogeneously extremely low, nil, or negative (soil sink [2]). However, since clayey or water-saturated layers may obstruct or limit gas outflows, the capability of the closed-chamber technique to detect advective exhalations depends on soil permeability. In some cases, methods that can record H_2 pulsations within the unsaturated vadose zone (e.g., [111]) may offer a viable solution. Nevertheless, the flux check may not be conclusive if used alone, and an integrated approach is always recommended, as discussed in the final section.

12.5 Holistic interpretation of H_2 in surface environments

According to the interpretation problems discussed above, when the H_2 concentration in surface environments (Figure 12.5) is not an expression of a clear gas seepage phenomenon (bubbling seeps, dry or burning vents, or gas transported by H_2-CH_4-rich hyperalkaline springs), the interpretation of the H_2 origin should be based on a multiparametric, integrated, and holistic approach. Simple geospatial or statistical analyses of soil-gas H_2 data, even if located in faulted zones, may be misleading. It is essential to investigate the origin of the gases that soil or aquifers show together with H_2 concentrations higher than those induced by interaction with the atmosphere or biological processes. Specifically, the following items should be considered:

(a) Isotopic and radiocarbon (^{14}C) analyses of CH_4 and CO_2 associated with H_2 are essential to determine the existence of geological degassing and, thus, support a deep H_2 origin.

(b) If helium and radon anomalies are observed, the depth of U-Ra-bearing rocks needs to be evaluated.

(c) The H_2 flux from the soil, for example, using the closed-chamber technique, may be tested as support.

(d) The soil-gas sampling should be validated by verifying that, in different types of soil (soft and hard), hammering or drilling does not produce artificial H_2 due to mechanical friction.

(e) The presence of ferrous materials (including natural minerals) in the soil, which can produce H_2 through oxidation or corrosion, should be verified.

(f) When H_2 is analyzed from boreholes, additional investigations should be carried out to verify corrosion processes within the well casing.

Anyway, all gas geochemical data must be interpreted, taking into account:

(g) The bio-ecosystem context (ground conditions), with particular attention to reducing environments that may produce H_2 and CH_4 (shallow aquifers, wet soil, peatlands, and wetlands).

(h) The geological context, that is, the existence and depth of potential H_2 sources (e.g., serpentinized peridotites, iron-rich rocks, and crystalline rocks). Figure 12.5 summarises the multiple H_2 sources in surface and near-surface environments that can be considered in gas geochemical H_2 exploration.

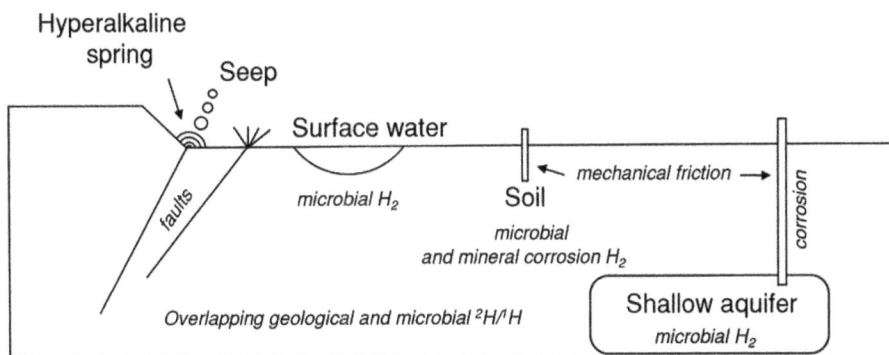

Figure 12.5: Schematic illustration of the systems considered in surface gas geochemistry and the potential multiple sources of H_2.

12.6 Concluding remarks

Similar to surface geochemical exploration of petroleum reservoirs, subsurface accumulations of natural hydrogen are mainly expected in areas hosting H_2-rich gas seeps, springs and soils. While gas seeps and H_2-rich hyperalkaline waters can provide an easy interpretation of the deep H_2 origin, the intensity of seepage may even suggest the existence of pressurized reservoirs. However, the occurrence of H_2 in the soil air or shallow aquifers is not a straightforward indication of geological gas. H_2 can be produced near the Earth's surface by multiple microbially mediated processes, and care should be taken to account for the possible production of artificial H_2 during soil-gas probing, well drilling, or by corrosion in boreholes. The interpretation of H_2 origin can be effectively supported by appropriate isotopic analyses of associated gases (we recommend, in particular, checking the radiocarbon of CH_4 and CO_2), with some prudence when using radon and helium, which can be reliable tracers of deep gas migration only if their radioactive source rocks are not shallow. In any case, the surface gas geochemical investigation of natural hydrogen should be supported by an integrated and holistic approach that considers, beyond the chemical parameters, the bio-ecosystem and geological context.

References

[1]   Lefeuvre N., Truche L., Donzé F. V., Ducoux M., Barré G., Fakoury R. A. Native H_2 Exploration in the Western Pyrenean Foothills. Geochemistry, Geophysics, Geosystems. 2021, 22: e2021GC009917.
[2]   Etiope G. Massive release of natural hydrogen from a geological seep (Chimaera, Turkey): Gas advection as a proxy of subsurface gas migration and pressurised accumulations. International Journal of Hydrogen Energy. 2023, 48(25): 9172–9184.

[3] Langhi L., Strand J. Exploring natural hydrogen hotspots: A review and soil-gas survey design for identifying seepage. Geoenergy. 2023, 1: 1.

[4] Link W. K. Significance of oil and gas seeps in world oil exploration. AAPG Bulletin. 1952, 36: 1505–1540.

[5] Macgregor D. S. Relationships between seepage, tectonics and subsurface petroleum reserves. Marine and Petroleum Geology. 1993, 10: 606–619.

[6] Abrams M. A. Significance of hydrocarbon seepage relative to petroleum generation and entrapment. Marine and Petroleum Geology. 2005, 22: 457–477.

[7] Etiope G. Natural Gas Seepage. The Earth's hydrocarbon degassing. Switzerland: Springer, 2015.

[8] Milkov A. V. Molecular hydrogen in surface and subsurface natural gases: Abundance, origins and ideas for deliberate exploration. Earth-Science Reviews. 2022, 230: 104063.

[9] Larin N., Zgonnik V., Rodina S., Deville E., Prinzhofer A., Larin V. N. Natural molecular hydrogen seepage associated with surficial, rounded depressions on the European Craton in Russia. Natural Resources Research. 2015, 24(3): 369–383.

[10] Zgonnik V., Beaumont V., Deville E., Larin N., Pillot D., Farrell K. M. Evidence for natural molecular hydrogen seepage associated with Carolina bays (surficial, ovoid depressions on the Atlantic Coastal Plain, Province of the USA). Progress in Earth and Planetary Science. 2015, 2.

[11] Frery E., Langhi L., Maison M., Moretti I. Natural hydrogen seeps identified in the North Perth Basin, Western Australia. International Journal of Hydrogen Energy. 2021, 46: 31158–31173.

[12] Davies K., Frery E., Giwelli A., Esteban L., Keshavarz A., Iglauer S. A natural hydrogen seep in Western Australia: Observed characteristics and controls. Science and Technology for Energy Transition. 2024, 79: 48.

[13] International Energy Agency, "Task 49: Natural hydrogen," 2020. (Accessed April 23, 2025, https://www.ieahydrogen.org/task/task-49-natural-hydrogen/.)

[14] Hunt J. M. Petroleum geochemistry and geology. New York: Freeman and Co, 1996.

[15] Abrajano T. A., Sturchio N. C., Kennedy B. M., Lyon G. L. Geochemistry of reduced gas related to serpentinization of the Zambales ophiolite, Philippines. Applied Geochemistry. 1990, 5: 625–630.

[16] Sano Y., Urabe A., Wakita H., Wushiki H. Origin of hydrogen-nitrogen gas seeps, Oman. Applied Geochemistry. 1993, 8: 1–8. 1993.

[17] Levitin M., Eternal fire on the beach: Tanjung Api, Sulawesi, 2022. (Accessed Sept 06, 2024,https://www.livetheworld.com/post/eternal-fire-on-the-beach-tanjung-api-sulawesi-ivid.)

[18] Ciotoli G., Procesi M., Etiope G., Fracassi U., Ventura G. Influence of tectonics on global scale distribution of geological methane emissions. Nature Communications. 2020, 11: 1–8.

[19] Aminzadeh F., Berge T. B., Connoly D. L. Hydrocarbon seepage: From source to surface, Society of Exploration Geophysicists and American Association of Petroleum Geologists, Tulsa, OK U.S.A. 2013.

[20] Etiope G., Martinelli G. Migration of carrier and trace gases in the geosphere: An overview. Physics of the Earth and Planetary Interiors. 2002, 129: 185–204.

[21] Rice G. K. Vertical migration in theory and in practice. Interpret. United Kingdom 2022, 10(1): SB17–SB26.

[22] Etiope G., Feyzullayev A., Baciu C. L. Terrestrial methane seeps and mud volcanoes : A global perspective of gas origin. Marine and Petroleum Geology. 2009, 26: 333–344.

[23] Malmqvist L., Kristiansson K. Experimental evidence for an ascending microflow of geogas in the ground. Earth And Planetary Science Letters. 1984, 70: 407–416.

[24] Neal C., Stanger G. Hydrogen generation from mantle source rocks in Oman. Earth and Planetary Science Letters. 1983, 66: 315–320.

[25] Prinzhofer A., Cacas-Stentz M. C. Natural hydrogen and blend gas: A dynamic model of accumulation. International Journal of Hydrogen Energy. 2023, 48: 21610–21623.

[26] Baciu C. L., Etiope G. A direct observation of a hydrogen-rich pressurized reservoir within an ophiolite (Tişoviţa, Romania). International Journal of Hydrogen Energy. 2024, 73: 402–406.

[27] Truche L., Donzé F. V., Goskolli E., Muceku B., Loisy C., Monnin C., Dutoit H., Cerepi A. A deep reservoir for hydrogen drives intense degassing in the Bulqizë ophiolite. Science. 2024, 383: 618–621.

[28] D'Alessandro W., Yüce G., Italiano F., Bellomo S., Gülbay A. H., Yasin D. U., Gagliano A. L. Large compositional differences in the gases released from the Kizildag ophiolitic body (Turkey): Evidences of prevailingly abiogenic origin. Marine and Petroleum Geology. 2018, 89: 174–184.

[29] Etiope G. Is natural hydrogen a renewable resource? A comparison between H_2 generation rates and surface seep fluxes. Poster Sess. Paris: H-NAT Conf, 2024b.

[30] Cook M. C., Blank J. G., Rietze A., Suzuki S., Nealson K. H., Morrill P. L. A Geochemical Comparison of Three Terrestrial Sites of Serpentinization: The Tablelands, the Cedars, and Aqua de Ney. Journal of Geophysical Research: Biogeosciences. 2021, 126: 1–20.

[31] Etiope G., Vadillo I., Whiticar M. J., Marques J. M., Carreira P. M., Tiago I., Benavente J., Jiménez P., Urresti B. Abiotic methane seepage in the Ronda peridotite massif, southern Spain. Applied Geochemistry. 2016, 66: 101–113.

[32] Szponar N., Brazelton W. J., Schrenk M. O., Bower D. M., Steele A., Morrill P. L. Geochemistry of a continental site of serpentinization, the Tablelands Ophiolite, Gros Morne National Park: A Mars analogue. Icarus. 2013, 224: 286–296.

[33] Schrenk M. O., Brazelton W. J., Lang S. Q. Serpentinization, Carbon, and Deep Life. Reviews in Mineralogy and Geochemistry. 2013, 75: 575–606.

[34] Vacquand C., Deville E., Beaumont V., Guyot F., Sissmann O., Pillot D., Arcilla C., Prinzhofer A. Reduced gas seepages in ophiolitic complexes: Evidences for multiple origins of the H2-CH4-N2 gas mixtures. Geochimica Et Cosmochimica Acta. 2018, 223: 437–461.

[35] Schwarzenbach E. M., Früh-Green G. L., Bernasconi S. The Ligurian ophiolite: An analogue to marine serpentinite-hosted hydrothermal systems. Mineralogical Magazine. 2011, 75: 1830.

[36] Boulart C., Chavagnac V., Monnin C., Delacour A., Ceuleneer G., Hoareau G. Differences in gas venting from ultramafic-hosted warm springs: The example of Oman and Voltri ophiolites. Ofioliti. 2013, 38(2): 143–156.

[37] Brazelton W. J., Thornton C. N., Hyer A., Twing K. I., Longino A. A., Lang S. Q., Lilley M. D., Früh-Green G. L., Schrenk M. O. Metagenomic identification of active methanogens and methanotrophs in serpentinite springs of the Voltri Massif. Italy: PeerJ, 2017.

[38] Cardace D., Meyer-Dombard D. R., Woycheese K. M., Arcilla C. A. Feasible metabolisms in high pH springs of the Philippines. Frontiers in Microbiology. 2015, 6: 1–17.

[39] D'Alessandro W., Daskalopoulou K., Calabrese S., Bellomo S. Water chemistry and abiogenic methane content of a hyperalkaline spring related to serpentinization in the Argolida ophiolite (Ermioni, Greece). Marine and Petroleum Geology. 2017, 89: 185–193.

[40] Etiope G., Tsikouras B., Kordella S., Ifandi E., Christodoulou D., Papatheodorou G. Methane flux and origin in the Othrys ophiolite hyperalkaline springs, Greece. Methane Flux and Origin in the Othrys Ophiolite Hyperalkaline Springs. 2013, 347: 161–174.

[41] Etiope G., Samardžić N., Grassa F., Hrvatović H., Miošić N., Skopljak F. Methane and hydrogen in hyperalkaline groundwaters of the serpentinized Dinaride ophiolite belt, Bosnia and Herzegovina. Applied Geochemistry. 2017, 84: 286–296.

[42] Ojeda L., Etiope G., Jiménez-Gavilán P., Martonos I. M., Röckmann T., Popa M. E., Sivan M., Castro-Gámez A. F., Benavente J., Vadillo I. Combining methane clumped and bulk isotopes, temporal variations in molecular and isotopic composition, and hydrochemical and geological proxies to understand methane's origin in the Ronda peridotite massifs (Spain). Chemical Geology. 2023, 642: 121799.

[43] Sugisaki R., Ido M., Takeda H., Isobe Y., Hayashi Y., Nakamura N., Satake H., Mizutani Y. Origin of hydrogen and carbon dioxide in fault gases and its relation to fault activity. Journal of Geology. 1983, 91(3): 239–258.

[44] Xiang Y., Sun X., Liu D., Yan L., Wang B., Gao X. Spatial Distribution of Rn, CO2, Hg, and H2 Concentrations in Soil Gas Across a Thrust Fault in Xinjiang, China. Frontiers in Earth Science. 2020, 8: 1–9.

[45] Conrad R. Soil microorganisms as controllers of atmospheric trace gases (H2, CO, CH4, OCS, N2O, and NO). Microbiological Reviews. 1996, 60: 609–640.

[46] Chen Q., Popa M. E., Batenburg A. M., Röckmann T. Isotopic signatures of production and uptake of H2 by soil. Atmospheric Chemistry and Physics. 2015, 15: 13003–13021.

[47] Rhee T. S., Brenninkmeijer C. A. M., Röckmann T. The overwhelming role of soils in the global atmospheric hydrogen cycle. Atmospheric Chemistry and Physics. 2006, 6: 1611–1625.

[48] Paulot F., Paynter D., Naik V., Malyshev S., Menzel R., Horowitz L. W. Global modeling of hydrogen using GFDL-AM4.1: Sensitivity of soil removal and radiative forcing. International Journal of Hydrogen Energy. 2021, 46(24): 13446–13460.

[49] Prinzhofer A., Moretti I., Françolin J., Pacheco C., D'Agostino A., Werly J., Rupin F. Natural hydrogen continuous emission from sedimentary basins: The example of a Brazilian H2 -emitting structure. International Journal of Hydrogen Energy. 2019, 44: 5676–5685.

[50] Moretti I., Geymond U., Pasquet G., Aimar L., Rabaute A. Natural hydrogen emanations in Namibia: Field acquisition and vegetation indexes from multispectral satellite image analysis. International Journal of Hydrogen Energy. 2022, 47: 35588–35607.

[51] Carrillo Ramirez A., Gonzalez Penagos F., Rodriguez G., Moretti I. Natural H2 emissions in Colombian ophiolites: First findings. Geosciences. 2023, 13: 358.

[52] Mathur Y., Awosiji V., Mukerji T., Scheirer A. H., Peters K. E. Soil geochemistry of hydrogen and other gases along the San Andreas fault. International Journal of Hydrogen Energy. 2024, 50: 411–419.

[53] Conrad R., Seiler W. Contribution of hydrogen production by biological nitrogen fixation to the global hydrogen budget. Journal of Geophysical Research. 1980, 85(C10): 5493–5498.

[54] Krämer H., Conrad R. Measurement of dissolved H_2 concentrations in methanogenic environments with a gas diffusion probe. FEMS Microbiology & Ecology. 1993, 12: 149–158.

[55] Sugimoto A., Fujita N. Hydrogen concentration and stable isotopic composition of methane in bubble gas observed in a natural wetland. Biogeochemistry. 2006, 81: 33–44.

[56] Pal D. S., Tripathee R., Reid M. C., Schäfer K. V. R., Jaffé P. R. Simultaneous measurements of dissolved CH_4 and H_2 in wetland soils. Environmental Monitoring and Assessment. 2018, 190: 176.

[57] Starkey R. L., Wigh K. M. Anaerobic corrosion of iron in soil. In American Gas Association, New York, 1945, p. 108.

[58] Dunham E. C., Dore J. E., Skidmore M. L., Roden E. E., Boyd E. S. Lithogenic hydrogen supports microbial primary production in subglacial and proglacial environments. Proceedings of the National Academy of Sciences of the United States of America. 2021, 118: 2.

[59] Etiope G., Ciotoli G., Benà E., Mazzoli C., Röckmann T., Sivan M., Squartini A., Laemmel T., Szidat S., Haghipour N., Sassi R. Surprising concentrations of hydrogen and non-geological methane and carbon dioxide in the soil. Science of the Total Environment. 2024, 948: 174890.

[60] Brooks M. J., Taylor B. E., Grant J. A. Carolina bay geoarchaeology and holocene landscape evolution on the upper coastal plain of South Carolina. Geoarchaeology - An International Journal. 1996, 11: 481–504.

[61] Rodriguez A. B., Waters M. N., Piehler M. F. Burning peat and reworking loess contribute to the formation and evolution of a large Carolina-bay basin. Quaternary Research. 2012, 77: 171–181. 2012.

[62] Moore C. R., Brooks M. J., Mallinson D. J., Parham P. R., Ivester A. H., Feathers J. K. The Quaternary evolution of Herndon Bay, a Carolina Bay on the coastal plain of North Carolina (USA): Implications for paleoclimate and oriented lake genesis. Southeastern Geology. 2016, 5(4): 145–171.

[63] Dugamin E., Truche L., Donzé F. V. Natural Hydrogen Exploration Guide. Geonum. 2019, 1: 16.

[64] Halas P., Dupuy A., Franceschi M., Bordmann V., Fleury J. M., Duclerc D. Hydrogen gas in circular depressions in South Gironde, France: Flux, stock, or artefact?. Applied Geochemistry. 2021, 127: 104928.

[65] Mainson M., Heath C., Pejcic B., Frery E. Sensing hydrogen seeps in the subsurface for natural hydrogen exploration. Applied Sciences. 2022, 12: 6383.

[66] Moretti I., Prinzhofer A., Françolin J., Pacheco C., Rosanne M., Rupin F., Mertens J. Long-term monitoring of natural hydrogen superficial emissions in a brazilian cratonic environment. Sporadic large pulses versus daily periodic emissions. International Journal of Hydrogen Energy. 2021, 46: 3615–3628.

[67] Prinzhofer A., Tahara Cissé C. S., Diallo A. B. Discovery of a large accumulation of natural hydrogen in Bourakebougou (Mali). International Journal of Hydrogen Energy. 2018, 43: 19315–19326.

[68] Prinzhofer A., Rigollet C., Lefeuvre N., Françolin J., Valadão de Miranda P. E. Maricá (Brazil), the new natural hydrogen play which changes the paradigm of hydrogen exploration. International Journal of Hydrogen Energy. 2024, 62: 91–98.

[69] Roche V., Geymond U., Boka-Mene M., Delcourt N., Portier E., Revillon S., Moretti I. A new continental hydrogen play in Damara Belt (Namibia). Scientific Reports. 2024, 14: 11655.

[70] Zgonnik V., Beaumont V., Larin N., Pillot D., Deville E. Diffused flow of molecular hydrogen through the Western Hajar mountains, Northern Oman. Arabian Journal of Geosciences. 2019, 12: 3.

[71] Zwaan F., Pilz P., Niedermann S., Zimmer M., Lefeuvre N., Petit V., Vieth-Hillebrand A., Gaucher E. C., Schmidt-Hattenberger C., Brune S. Soil and spring water gas data from an area with natural hydrogen (H$_2$) degassing. NW Pyrenean foreland, France: *GFZ Data Serv*, 2024.

[72] Frery E., Langhi L., Markov J. Natural hydrogen exploration in Australia – State of knowledge and presentation of a case study. APPEA J. 2022, 62(1): 223–234.

[73] Lovley D. R., Goodwin S. Hydrogen concentrations as an indicator of the predominant terminal electron-accepting reactions in aquatic sediments. Geochimica Et Cosmochimica Acta. 1988, 52: 2993–3003.

[74] Michener R. H., Scranton M. I., Novelli P. Hydrogen (H$_2$) distributions in the Carmans River estuary. Estuarine, Coastal and Shelf Science. 1988, 27: 223–235.

[75] Lovley D. R., Chapelle F. H., Woodward J. C. Use of dissolved H$_2$ concentrations to determine distribution of microbially catalyzed redox reactions in anoxic groundwater. Environmental Science & Technology. 1994, 28(7): 1205–1210.

[76] Chapelle F. H., Lovley D. R. Hydrogen concentrations in ground water as an indicator of bacterial processes in deep aquifer systems. In: *Proceedings of the First International Symposium on the Microbiology of the Deep Subsurface, Orlando FL, ed. CB Fliermans and TC Hazen*. 1990, p. 2–123.

[77] Conrad R., Phelps T. J., Zeikus J. G. Gas metabolism evidence in support of the juxtaposition of hydrogen-producing and methanogenic bacteria in sewage sludge and lake sediments. Applied and Environmental Microbiology. 1985, 50(3): 595–601.

[78] Hoehler T. M., Alperin M. J., Albert D. B., Martens C. S. Thermodynamic control on hydrogen concentrations in anoxic sediments. Geochimica Et Cosmochimica Acta. 1998, 62(10): 1745–1756.

[79] Jakobsen R., Albrechtsen H. J., Rasmussen M., Bay H., Bjerg P. L., Christensen T. H. H$_2$ concentrations in a landfill leachate plume (Grindsted, Denmark): In situ energetics of terminal electron acceptor processes. Environmental Science & Technology. 1998, 32: 2142–2148.

[80] Kuivila K. M., Murray J. W., Devol A. H., Novelli P. C. Methane production, sulfate reduction and competition for substrates in the sediments of Lake Washington. Geochimica Et Cosmochimica Acta. 1989, 53: 409–416.

[81] Kuivila K. M., Lovley D. R. Dissolved hydrogen concentrations in sulfate-reducing and methanogenic sediments. SIL Communications. 1996, 25: 55–62.

[82] McGuire J. T., Smith E. W., Long D. T., Hyndman D. W., Haack S. K., Klug M. J., Velbel M. A. Temporal variations in parameters reflecting terminal-electron-accepting processes in an aquifer contaminated with waste fuel and chlorinated solvents. Chemical Geology. 2000, 169: 471–485.

[83] Novelli P. C., Scranton M. I., Michener R. H. Hydrogen distributions in marine sediments. Limnol Oceanogr. 1987, 32(3): 565–576.

[84] Novelli P. C., Michelson A. R., Scranton M. I., Banta G. T., Hobbie J. E., Howarth R. W. Hydrogen and acetate cycling in two sulfate-reducing sediments: Buzzards Bay and Town Cove, Mass. Geochimica Et Cosmochimica Acta. 1988, 52: 2477–2486.

[85] Vroblesky D. A., Chapelle F. H. Temporal and spatial changes of terminal electron-accepting processes in a petroleum hydrocarbon-contaminated aquifer and the significance for contaminant biodegradation. Water Resources Research. 1994, 30(5): 1561–1570.

[86] Chapelle F. H., McMahon P. B., Dubrovsky N. M., Fujii R. F., Oaksford E. T., Vroblesky D. A. Deducing the distribution of terminal electron-accepting processes in hydrologically diverse groundwater systems. Water Resources Research. 1995, 31(2): 359–371.

[87] Chapelle F. H., Haack S. K., Adriaens P., Henry M. A., Bradley P. M. Comparison of Eh and H_2 measurements for delineating redox processes in a contaminated aquifer. Environmental Science & Technology. 1996, 30: 3565–3569.

[88] Chapelle F. H., Vroblesky D. A., Woodward J. C., Lovley D. R. Practical considerations for measuring hydrogen concentrations in groundwater. Environmental Science & Technology. 1997, 31: 2873–2877.

[89] Chapelle F. H., Lovley D. R. Competitive exclusion of sulfate reduction by Fe(III)-reducing bacteria: A mechanism for producing discrete zones of high-iron ground water. Groundwater. 1992, 30(1): 29–36.

[90] Chapelle F. H., McMahon P. B. Geochemistry of dissolved inorganic carbon in a Coastal Plain aquifer. 1. Sulfate from confining beds as an oxidant in microbial CO_2 production. Journal of Hydrology. 1991, 127: 85–108.

[91] Walter S., Laukenmann S., Stams A. J. M., Vollmer M. K., Gleixner G., Röckmann T. The stable isotopic signature of biologically produced molecular hydrogen (H_2). Biogeosciences. 2012, 9: 4115–4123.

[92] Feldbusch E., Wiersberg T., Zimmer M., Regenspurg S. Origin of gases from the geothermal reservoir Groß Schönebeck (North German Basin). Geothermics. 2018, 71: 357–368.

[93] Bjornstad B. N., McKinley J. P., Stevens T. O., Rawson S. H., Fredrickson J. K., Long P. L. Generation of hydrogen gas as a result of drilling within the saturated zone. Groundwater Monitoring & Remediation. 1994, 14(4): 140–147.

[94] Fleischer R. L., Mogro-Campero A. Mapping of integrated radon emanation for detection of long-distance migration of gases within the Earth: Techniques and principles. Journal of Geophysical Research: Solid Earth. 1978, 83(B7): 3539–3549.

[95] Philp R. P., Crisp P. T. Surface geochemical methods used for oil and gas prospecting – A review. Journal of Geochemical Exploration. 1982, 17: 1–34.

[96] Duddridge G. A., Grainger P., Durrance E. M. Fault detection using soil gas geochemistry. Quarterly Journal of Engineering Geology and Hydrogeology. 1991, 24: 427–435.

[97] Klusman R. W. Soil gas and related methods for natural resource exploration, John Wiley, New York, 1993, p. 483.

[98] McCarthy J. H., Reimer G. M. Advances in soil gas geochemical exploration for natural resources: Some current examples and practice. Journal of Geophysical Research. 1986, 91(B12): 12327–12338.

[99] Etiope G., Lombardi S. Evidence for radon transport by carrier gas through faulted clays in Italy. Journal of Radioanalytical and Nuclear Chemistry. 1995, 193(2): 291–300.

[100] Toutain J. P., Baubron J. C. Gas geochemistry and seismotectonics: A review. Tectonophysics. 1999, 304: 1–2,1–27.

[101] Fu C. C., Yang T. F., Du J., Walia V., Chen Y. G., Liu T. K., Chen C. H. Variations of helium and radon concentrations in soil gases from an active fault zone in southern Taiwan. Radiation Measurements. 2009, 43: S348–S352.

[102] Gregory R. G., Durrance E. M. Helium, radon, and hydrothermal circulation associated with the carnmenellis radiothermal granite of Southwest England. Journal of Geophysical Research. 1987, 92 (B12): 12567–12586.

[103] Gascoyne M., Wuschke D. M., Durrance E. M. Fracture detection and groundwater flow characterization using He and Rn in soil gases, Manitoba, Canada. Applied Geochemistry. 1993, 8: 223–233.

[104] Ball T. K., Cameron D. G., Colman T. B., Roberts P. D. Behaviour of radon in the geological environment: A review. Quarterly Journal of Engineering Geology and Hydrogeology. 1991, 24: 169–182.

[105] Akerblom G. Ground radon: Monitoring procedure in Sweden, paper presented at the "JAG" Disc. In Meeting on Radon Workshop, Geology, Environment, Technology 12, R. Astron. Soc, London, 1993.

[106] Guerra M., Etiope G. Effects of gas-water partitioning, stripping and channelling processes on radon and helium gas distribution in fault areas. Geochemical Journal. 1999, 33: 141–151.

[107] Livingston G. P., Hutchinson G. L. Enclosure-based measurement of trace gas exchange: Applications and sources of errors. In: Matson P. A., Harriss R. C. eds., Biogenic trace gases: Measuring emissions from soil and water, Oxford: Blackwell Science, 1995, 14e51.

[108] Norman J. M., Kucharik C. J., Gower S. T., Baldocchi D. D., Crill P. M., Rayment M., Savage K., Striegl R. G. A comparison of six methods for measuring soil-surface carbon dioxide fluxes. Journal of Geophysical Research: Atmospheres. 1997, 102(24): 28771–28777.

[109] Klusman R. W., Leopold M. E., LeRoy M. P. Seasonal variation in methane fluxes from sedimetrary basins to the atmosphere: Results from chamber measurements and modeling of transport from deep sources. Journal of Geophysical Research. 2000, 26: 24661–24670.

[110] Etiope G., Doezema L. A., Pacheco C. Emission of methane and heavier alkanes from the La Brea Tar Pits seepage area. Journal of Geophysical Research: Atmospheres. 2017, 122: 12008–12019.

[111] Cathles L., Prinzhofer A. What pulsating H_2 emissions suggest about the H_2 resource in the Sao Francisco Basin of Brazil. Geosciences. 2020, 10: 1–18.

Mahmoud Leila*, Fiammetta Mondino, Aya Yasser, Randy Hazlett

Chapter 13
Natural hydrogen favorability maps (NHFMs): a new concept for natural hydrogen exploration in different geological contexts

Abstract: Natural hydrogen has recently emerged as a cost-effective and environmentally friendly energy resource, naturally produced in various geological settings. These settings often contain specific rocks capable of generating natural hydrogen (natural H_2-generating rock types, or NH_2-GRTs) through processes such as serpentinization, radiolysis, and hydrothermal oxidation. The presence of natural hydrogen seeps, particularly in serpentinized ultramafic contexts, signifies active hydrogen generation systems. Additionally, soil hydrogen emissions, often associated with surface features like sub-circular depressions with sparse vegetation ("fairy circles"), serve as valuable exploration tools for identifying subsurface hydrogen systems.

The migration, accumulation, and entrapment of hydrogen resemble the dynamics of petroleum systems, requiring geological conditions conducive to storage and sealing. Most natural hydrogen generation systems involve interactions between NH_2-GRTs and geothermal water. To streamline exploration, we introduce the concept of natural hydrogen favorability maps (NHFMs), which link hydrogen-generating rock types with their associated generation mechanisms (e.g., serpentinization or radiolysis).

A comprehensive evaluation of the "hydrogen kitchen" involves understanding hydrogeology, hydrochemistry, and heat flow. Mapping the architecture of hydrogen-generating rock types alongside hydrogen kitchens provides a novel and effective tool to support decision-making in hydrogen exploration. The NHFMs can be further validated through surface indicators, such as fairy circles and hyperalkaline springs, which provide evidence of active water-rock interactions and hydrogen generation.

Keywords: natural hydrogen, water-rock interactions, natural hydrogen favorability maps (NHFMs), natural hydrogen exploration, subsurface hydrogen plays

*Corresponding author: Mahmoud Leila, School of Mining and Geosciences, Nazarbayev University, Astana, Kazakhstan, e-mail: Mahmoud.leila@nu.edu.kz
Fiammetta Mondino, Terra-A AG -Wuhrstrasse 14, 9490 Vaduz, Principality of Liechtenstein
Aya Yasser, Faculty of Science, Mansoura University, Mansoura, Egypt
Randy Hazlett, School of Mining and Geosciences, Nazarbayev University, Astana, Kazakhstan

https://doi.org/10.1515/9783111437040-013

13.1 Introduction

In order to align with the Paris Agreement's goal of limiting global warming to below 2 °C, both the global scientific community and political authorities must prioritize integrating alternative green energy sources into the energy mix and removing anthropogenic CO_2 from the atmosphere for safe storage. Achieving the net-zero emissions scenario outlined by the International Energy Agency (IEA) requires that hydrogen-based energy contribute at least 17,000 Terawatt hours (TWh) by 2050 [1]. Hydrogen (H_2) has recently emerged as an efficient, clean energy resource, as it can be used as fuel without emitting greenhouse gases. With a higher energy density than traditional hydrocarbons, H_2 also offers an efficient means for large-scale and long-term storage of electrical energy [2]. Currently, most H_2 energy is based on synthesis through different processes, including methane steam reforming (~50%), naphtha reforming (~30%), and coal gasification (~18%) – processes that emit substantial amounts of CO_2 [3]. Alternative methods, like bio-H_2 production from microalgae and cyanobacteria, require less energy [4]; however, these production routes vary in technology readiness level (TRL), leading to non-uniform costs for synthetic H_2. The conflict between rising fuel demand and the imperative to lower CO_2 emissions underscores the need for a global shift toward green, sustainable, and cost-effective energy resources. In this regard, scientific research is increasingly focused on naturally occurring H_2, which has been observed in various geological contexts worldwide (Figure 13.1). However, establishing a framework for subsurface natural H_2 exploration following the same concepts of the petroleum system is a critical missing piece.

13.2 Natural hydrogen -generating rock types "NH_2-GRTs"

The term "NH_2-GRTs" was introduced by [5] and refers to the type of rocks capable of generating hydrogen through one of the hydrogen generation mechanisms illustrated in Figure 13.1. Mafic and ultramafic rocks in ophiolites and greenstone belts are examples of H_2-GRTs that produce natural hydrogen via serpentinization and oxidation of $Fe^{(II)}$-rich minerals (e.g., olivine, pyroxenes) through interaction with thermal water as follows [6, 7]:

$$(6Fe_2SiO_4)_{olivine} + 7(H_2O)_{hydrothermal} \rightarrow (3Fe_3Si_2O_5(OH)_4)_{serpentine} + (Fe_3O_4)_{magnetite} + H_2$$

Natural H_2-rich gas seeps have been reported in serpentinized ophiolitic contexts around the globe, such as in the Balkans [8, 9], Italy (>4% [10]), Oman [11], the Philippines [12], Turkey [13], and the USA [14]. Under certain conditions (temperature >250 °C),

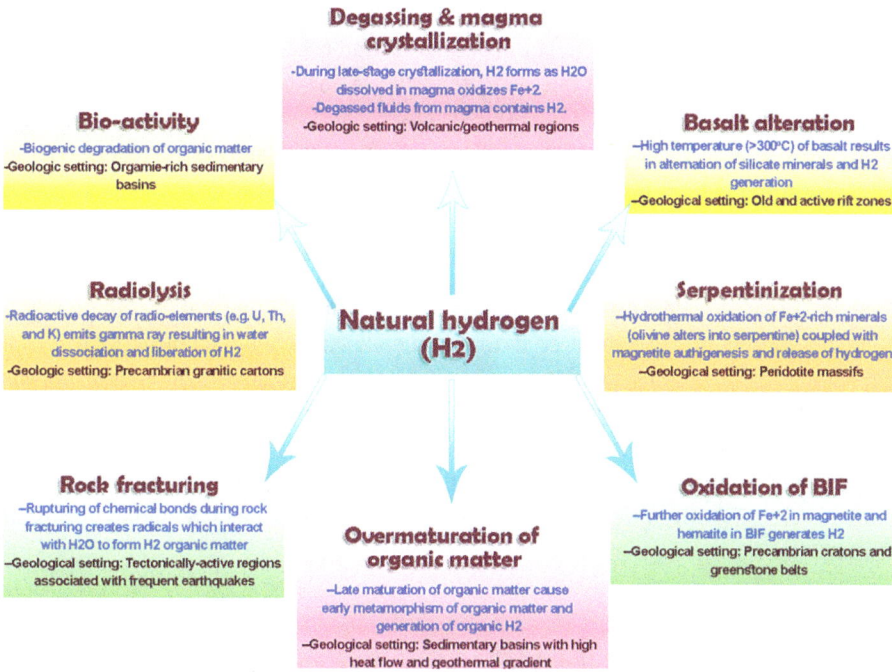

Degassing & magma crystallization
-During late-stage crystallization, H2 forms as H2O dissolved in magma oxidizes Fe+2.
-Degassed fluids from magma contains H2.
-Geologic setting: Volcanic/geothermal regions

Bio-activity
-Biogenic degradation of organic matter
-Geologic setting: Orgamie-rich sedimentary basins

Basalt alteration
--High temperature (>300°C) of basalt results in alternation of silicate minerals and H2 generation
--Geological setting: Old and active rift zones

Radiolysis
-Radioactive decay of radio-elements (e.g. U, Th, and K) emits gamma ray resulting in water dissociation and liberation of H2
-Geologic setting: Precambrian granitic cartons

Natural hydrogen (H2)

Serpentinization
--Hydrothermal oxidation of Fe+2-rich minerals (olivine alters into serpentine) coupled with magnetite authigenesis and release of hydrogen
--Geological setting: Peridotite massifs

Rock fracturing
--Rupturing of chemical bonds during rock fracturing creates radicals which interact with H2O to form H2 organic matter
--Geological setting: Tectonically-active regions associated with frequent earthquakes

Overmaturation of organic matter
--Late maturation of organic matter cause early metamorphism of organic matter and generation of organic H2
--Geological setting: Sedimentary basins with high heat flow and geothermal gradient

Oxidation of BIF
--Further oxidation of Fe+2 in magnetite and hematite in BIF generates H2
--Geological setting: Precambrian cratons and greenstone belts

Figure 13.1: An illustrative sketch summarizing the various mechanisms of natural H_2 generation.

serpentinization can occur very quickly, and the generation of H_2 may be continuous and renewable [6, 15].

U-, Th-, and ^{40}K-rich rocks in Precambrian cratons present another type of NH_2-GRTs. These rocks generate H_2 through water dissociation resulting from the ionized radiation emitted during the decay of these elements [16]. Radiolytic dissociation of water may occur in all subsurface environments and is independent of pressure and temperature ranges. The sole controlling factors on the amount of naturally generated hydrogen comprise the percentage of radioelements, the presence of a pore system (primary or secondary), as well as the residence time of water in contact with the radioelements [16, 17]. Examples of natural H_2 generation via radiolysis were reported in association with Neoproterozoic basins in Australia [18], USA [19], and South Africa [20].

Oxidation of $Fe^{(II)}$ incorporated in some minerals, such as biotite and amphibole, presents another source of natural H_2 [21, 22]. Namely, water interaction with mica-rich and peralkaline granites generates H_2 at high temperatures (>250 °C). This provides further opportunities for natural H_2 generation in olivine- and pyroxene-free rocks and sheds light on geothermal granitic aquifers as potential NH_2-GRTs. Similarly, oxidation of $Fe^{(II)}$ in mafic intrusive and extrusive rocks (e.g., olivine-basalt) generates H_2 [23, 24]. Interaction of these rocks with hydrothermal water at high temperatures (350–400 °C) results in the formation of iron-oxides (e.g., magnetite), silica, and H_2 [23]. Such conditions occur in both continental and oceanic settings, and H_2 generation through this mechanism has been reported in continental rift basins (e.g., Djibouti [24]). Banded iron formations (BIFs) in Archean cratons contain some mineral phases enriched in $Fe^{(II)}$; therefore, oxidation of these minerals would generate H_2 at temperatures less than 100 °C, as substantiated by [25]:

$$(Fe_3O_4)_{magnetite} + H_2O \rightarrow (Fe_2O_3)_{hematite} + H_2$$

Coal and organic matter-rich rocks can generate H_2 through thermal degradation and over-maturation of organic matter at approximately vitrinite reflectance (Ro) of ≥3.5% [26, 27]. Such conditions occur in very deep settings or in sedimentary basins with high heat flow [28]. The magnitude of organic H_2 is dependent on the initial total organic carbon (TOC) present, and approximately 25% of the TOC would be converted to H_2, which is two orders of magnitude greater than that generated by olivine in ultramafic rocks [29]. Organic H_2 has been reported in the deep Songliao Basin in China [28].

13.3 Evidence for natural H_2 generation

Generation of natural H_2 is sometimes associated with surface leakage and seeps in the form of dry or wet gas manifestations (e.g., Oman, Italy) [10, 11], or surficial subcircular depressions with scarce vegetation (fairy circles) [30-32]. Fairy circles have been reported in numerous regions around the globe and are commonly associated with H_2 emissions, such as in Australia [18, 33], Russia [34], the USA [35], and Namibia [36] (Figure 13.2). The fairy circles are distinguished based on their morphological characteristics, with a depth-to-equivalent diameter ratio of around 1% and a gentle slope with a maximum of around 3% [34].

Alkaline springs, which are associated with mafic and ultramafic contexts, provide evidence of water-rock interaction and the potential generation of natural H_2. These springs have been documented in numerous regions associated with ophiolite belts, including the Balkans, Italy, Oman, California, and the Philippines [8, 37-39]. The hyper-alkaline springs observed in these areas indicate active fluid-rock interactions, which involve the dissolution of rocks and the release of gases such as CH_4 and H_2 [39, 40]. Notably, hyper-alkaline springs develop in association with weathered, partially serpentinized peridotite massifs worldwide, such as those in Oman [37, 41], Canada [42], New Caledonia [43], the Philippines [44], New Zealand [45], Japan [46], and Turkey [47]. Most of these regions are characterized by H_2-rich gas seeps, which are mainly linked with the serpentinization of ultramafic massifs.

13.4 Reports on in-reservoir natural H₂ accumulations

Accumulations of natural H_2 in subsurface reservoirs have been reported in several regions around the globe (Figure 13.2). The most well-known case of in-reservoir H_2 accumulation has been reported in Bourakébougou, Mali, where carbonate reservoirs sealed by dolerite sills provide accumulation zones for natural H_2 [48, 49]. In Bouraké-bougou, the reservoir gas is almost pure H_2 (~98% volume) and is likely generated via deep serpentinization [49]. H_2 production from Bourakébougou carbonates commenced more than a decade ago without a notable drop in reservoir pressure, thereby suggesting a continuous recharge process.

The Monzon-1 well in the Ebro Basin, Spain, presents another example of in-reservoir natural H_2 accumulation. In the Monzon-1 well, natural H_2 has been reported in the folded Triassic sandstones of the Bunter Formation, which is capped by thick carbonates and evaporites of the Muschelkalk and Keuper formations [50]. The concentration of H_2 in the Monzon-1 well is up to 25% by volume and was likely formed by deep serpentinization of mantle rocks, migrating through deep-seated faults prior to accumulation in the Bunter sandstones. The presence of deep-seated conduits (faults) from the source (mantle rocks) to the trap highlights the Ebro Basin in Spain as a prospect for natural H_2 exploration.

Natural H_2 traps have also been reported in several basins in Australia. In the Amadeus Basin, H_2 concentrations of up to 11.46% volume were measured in the Mt Kitty-1 well [18]. Radiolysis is the primary generation mechanism of H_2, which accumulates in the fractured Precambrian granitic basement as well as in the Neoproterozoic Heavitree Quartzite, which is capped by the Gillen salt and evaporites. On Yorke Peninsula (South Australia), H_2 was detected in old petroleum wells, with concentrations of up to 84% in the Parara reefal limestone reservoir, which is capped by Permian mudstones [51]. In Western Australia, natural H_2 has also been reported in old petroleum wells drilled in the Canning Basin, with concentrations of 95.3% in the Meda-1 well, as well as in the Perth Basin in Drover-1 (H_2 ~ 18.5% volume) and Gingin-2 (H_2 ~ 26.5% volume). In most of these wells, the Devonian reef complex and Carboniferous clastics are the primary reservoir intervals [52].

Old petroleum wells in the Paris Basin, France, contain a natural H_2/CH_4 gas blend, which is often trapped in Jurassic carbonate reservoirs. The highest H_2 concentration (~52% volume) was reported in the Dogger carbonate of the Montreuil Aux Lions well [53]. The origin of natural H_2 in the Paris Basin is not clear; however, deep serpentinization of the Lizard-Rhenohercynian ophiolite and migration along the Bray fault system is the most likely source of H_2 in Dogger carbonates [53]. Another H_2/hydrocarbon gas blend was reported in the Vaux-en-Bugey field in the French Alps, which represents the first French methane production in the early twentieth century. In Vaux-en-Bugey, $H_2/He/CH_4$ was discovered in Jurassic and Triassic carbonates, where H_2 concentrations reached up to ~5% volume. The CH_4 is most likely thermogenic, but the source of H_2 and He is still unclear [54].

The recent H_2 (~84% volume) discovery in the Bulqizë chromite mine in Albania, with a flow rate of approximately 5 ± 1 L/s (at 25 °C and 1.031×10510^5 Pa), suggests its accumulation in a deep-rooted faulted trap in the Jurassic ophiolite [9]. In the USA, in-reservoir H_2 accumulations were observed in some old wells in the Nemaha uplift, Kansas, where H_2 occurs in concentrations of up to 91% mole in the DR1-A well and in variable concentrations ranging from 1.4% to 80% mole in other wells, such as the Scott and Heins wells. These accumulations are mainly hosted in the fractured basement aquifer, which is often sealed by thick Pennsylvanian mudstones [55]. The most likely source of H_2 in Nemaha High is the oxidation of Fe^{+2}-rich mafic rocks, where the deep-seated Humboldt fault acts as the main migration path for shallow H_2 accumulation [56].

Figure 13.2: Global distribution of fairy circles with soil H_2 seeps and reported H_2 in subsurface reservoir accumulations.

13.5 Defining subsurface natural H$_2$ plays

Subsurface natural H$_2$ reservoirs can be identified either by the presence of anomalous levels of H$_2$ in surface soil or bubbling H$_2$ in water springs [11, 30-32]. Studies conducted in several regions, such as the USA, Russia, Brazil, Australia, and Oman, have reported that soil H$_2$ concentrations on the order of 10^2–10^3 ppmv (parts per million by volume) are most likely linked to active subsurface H$_2$ generation and potential in-reservoir accumulation [11, 33-35]. However, as of now, it is impossible to correlate the levels of soil H$_2$ seepage with the actual volume of potentially generated or trapped H$_2$ in the subsurface. Moreover, the absence of gas seeps does not necessarily indicate the absence of subsurface H$_2$ accumulations. Determining the mechanisms of H$_2$ migration and exhalation into the soil and atmosphere may enable the identification of subsurface H$_2$ reservoirs. Essentially, H$_2$ advection through faults suggests the existence of a subsurface H$_2$ play, and distinguishing between diffusion and advection of soil H$_2$ is crucial for natural H$_2$ exploration [57]. However, to minimize exploration risks, a comprehensive understanding of H$_2$ generation, migration, and entrapment mechanisms in the subsurface is essential. To achieve this, integration of surface and subsurface geological data is required to define the favorable geological contexts for exploration and production.

13.6 Favorability maps concept

The concept and structure of favorability maps were initially framed in the hydrocarbon exploration sector in a process defined as play-based exploration (PBE). Major oil and gas companies developed this workflow in the early 1990s to optimize exploration activities. Herein, we adopt it for hydrogen using the same concept of petroleum PBE (Figure 13.3). This approach consists of multiple activity levels to investigate the exploration potential of a certain region, starting from the regional scale (basin and play analysis) and progressively moving to a local scale (lead and prospect) [58]. The basis for PBE is related to basin analysis and the critical subsurface elements, which, in combination, constitute a well-defined and successful petroleum system [59]. The ultimate goals of basin analysis are to (1) assess the basin type, fill, and structural evolution; (2) identify hydrocarbon generation, flux, and timing; and (3) define geologic boundaries where exploration potential is identified.

The PBE is based on the generation of conceptual maps, defined as "play-based maps," which are created through a specific framework in areas with favorable geological conditions for hydrocarbon accumulation [60]. The concept is to unlock the geological complexity of an exploration area into the main subsurface elements composing a specific play. A "play" groups together areas where the geologically critical components are comparable; these include all the subsurface elements required for the generation

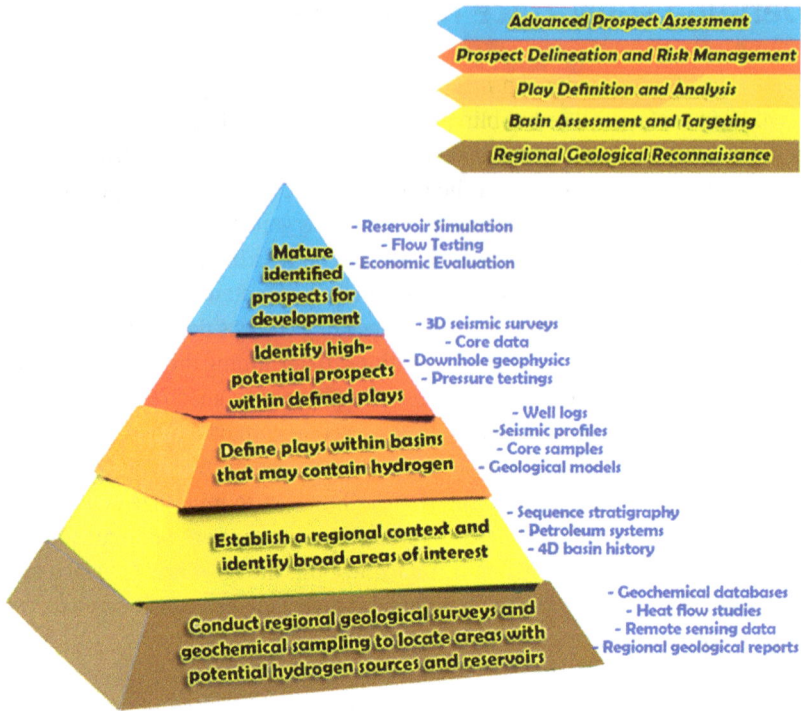

Figure 13.3: Play-based exploration workflow adopted for natural hydrogen systems, from regional/basin scale to prospect scale.

and accumulation of oil and gas, such as (1) reservoir, (2) seal, (3) source rock, (4) trap and trapping mechanisms, and (5) migration path presence and extension. All these elements are then compiled to identify where the most promising plays are located.

The ultimate goal of this exercise is to draw segment maps for each play element: each segment has similar probability characteristics, with boundaries reflecting potential changes (geology, confidence, probability, etc.). These maps will serve as the basis for generating a common risk segment mapping (CRS), a set of maps that include all geological and exploration elements generated [61]. CRS allows (1) evaluating play-scale risks and uncertainties; (2) identifying mitigation actions, (3) identifying, assessing, and screening potential opportunities; (4) assessing the Probability of Success (PoS) for potential areas; (5) improving decision-making; and (6) supporting a work program to move the identified opportunity from the play into the prospect level.

In geothermal resources, the exploration approach is also based on the definition of "plays," a geographically and stratigraphically delimited area where common geological factors exist, allowing resources to occur [62]. The same workflow is applied, including the integration of datasets that can constrain these variables either as direct or proxy evidence to identify areas with prospectivity [62, 63]. On the contrary, in the context of geothermal prospecting, geological data are not the only determining factors; surface data are critical and must be integrated into the workflow from the early stages. The exploration and development of a geothermal energy project require an integration of both subsurface and surface data, skills, and expertise. In geothermal exploration, the heat flow and geothermal gradient of the basin under exploration determine its potential for geothermal energy harvesting. Typically, in most sedimentary basins with geothermal gradients ranging between 28 and 35 °C/km, given the lower economic margins of geothermal energy, exploration and production capital expenditure benefit from smaller budgets compared to hydrocarbon projects. To identify the most favorable sites with geothermal potential, a GIS-based geoprocessing technique is employed. This technique involves compiling several GIS layers of spatial distribution of elements that favor the presence of a geothermal system. The datasets used in this analysis are generally grouped into two categories: surface elements (sink) and subsurface elements (source).

13.6.1 Natural hydrogen favorability maps (NHFMs)

Most natural H_2 generation mechanisms require interaction between NH_2-GRTs and geothermal water (Figure 13.1), while migration, accumulation, and entrapment necessitate a geological reservoir/seal similar to a petroleum system. Therefore, integration between a play-based concept and a geothermal exploration workflow is proposed herein as an efficient tool for natural H_2 exploration. NHFMs conceptually follow the segment maps of the oil and gas (O&G) exploration methods and the geothermal favorability maps, but the workflow to create them is adapted to the natural H_2 (NH) generation rock types and processes, which are quite different compared to O&G and geothermal. Particular attention is given to elements such as (1) the geodynamic setting of a region and its controls on the spatial distribution of potential hydrogen-generation rock types (e.g., peridotite lithotypes, radioelement-rich rocks, etc.), (2) the NH rock types (mineralogy, rheology, weathering intensity, petrophysical properties, etc.), (3) the relationship between rock types and hydrogen generation mechanisms (e.g., water-rock interactions), (4) the stratigraphic setting, (5) the reservoir characterization, (6) the tectonic setting and the link with the presence/absence of privileged migration paths, (7) the seal characterization, (8) the thermal properties of the region (heat-flow, thermal gradient, etc.) and the thermal window for NH generation, and (9) the hydrogeological setting (Figure 13.4).

In the case of NHFMs, the integration between surface and subsurface elements is also required. The surficial indicators for hydrogen generation include fairy circles, hyper-alkaline springs, and hydrogen surface measurements, which are all necessary to arrive at a proper NHFM. Moreover, the high costs of hydrogen transportation suggest that, as in the case of Geothermal Favorability Maps, a proper assessment of the surface, an understanding of the energy demand portfolio, and consumption distribution over time (current versus future customers, seasonal variations, business demands, etc.) are critical. The integration of all these play elements allows for the construction of NHFMs to assist in the identification and characterization of potential hydrogen prospects, the creation of a project funnel for prospect ranking, and support for decision-making.

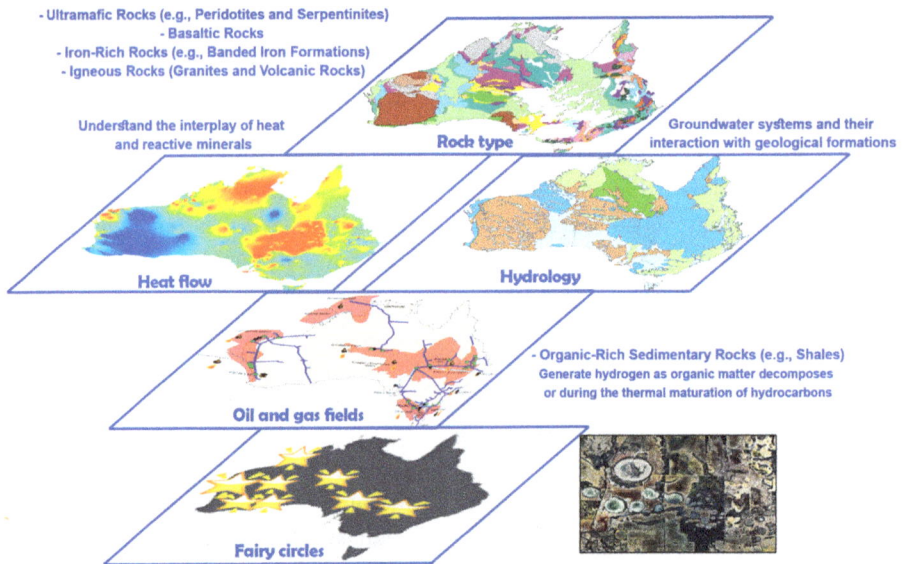

Figure 13.4: Example of NHFMs applied to Australia for locating potential prospects based on integrating geothermal and petroleum industry data with surficial indicators of H_2 emanations.

13.7 Application of NHFMs concept

While the exploitation of natural hydrogen is still in its early stages, the number of boreholes, identified potential deposits, and involved companies is growing rapidly. Although several research projects are investigating subsurface natural hydrogen systems around the globe, detailed data about specific plays have been kept confidential. However, numerous regions with a high probability for natural H_2 are well known (e.g., ophiolite contexts, underwater subduction zones, suture zones, etc.) [64, 65]. Identifying natural H_2 accumulations in the subsurface can be a complex task. One widely used method in petroleum exploration is seismic prospecting, which permits the identification of play elements (e.g., migration paths, reservoirs, structural, or stratigraphic entrapment). However, the utilization of seismic prospecting in natural H_2 exploration is still limited and should be combined with other tools (e.g., geochemical exploration). Therefore, locating the regions for starting a natural H_2 exploration campaign is the most critical step. This pre-exploration step depends on geological data collection and the identification of potential NH_2-GRTs. This will allow the identification of sweet spots for natural H_2 generation and thereby the prioritization of some regions over others for natural H_2 exploration (Figure 13.5).

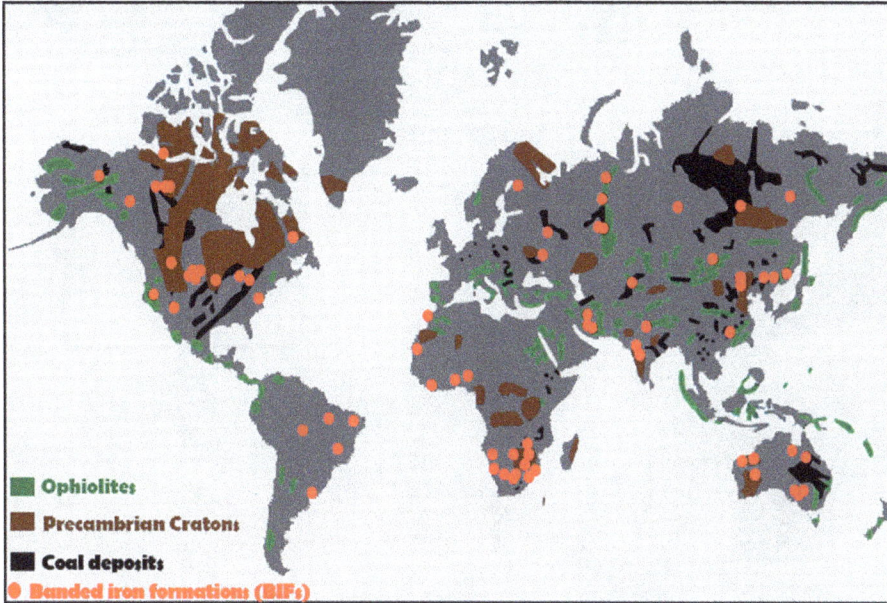

Figure 13.5: Global distribution of the potential NH_2-GRTs, illustrating the sweet spots for natural H_2 exploration.

Understanding the distribution of NH_2-GRTs forms the basis for constructing NHFMs, which help in identifying natural H_2 accumulation zones in the subsurface. This requires the presence of a natural H_2 generation system, such as the interaction between NH_2-GRTs and geothermal fluids to form H_2. Therefore, subsurface data should be available either in the form of seismic or borehole data to validate the existence of a natural H_2 generation system. Australia, Russia, South Africa, and the USA are among the few countries hosting all the NH_2-GRTs, thereby containing multiple sweet spots for natural H_2 exploration (Figure 13.5). Availability of subsurface data in Australia enabled us to apply the NHFMs concept to define the potential regions for natural H_2 accumulation in the subsurface (Figure 13.4). The Amadeus Basin is among the natural H_2 sweet spots in Australia. Natural H_2 has been reported in several wells in the southeastern part of the Amadeus Basin (e.g., Mt Kitty and Magee-1) [18]. The region is characterized by multiple fairy circles at the surface, which are correlated with gas chimneys rooted down to the Neoproterozoic basement, where natural H_2 is generated via radiolysis [18]. Integration of subsurface seismic, well data, surface

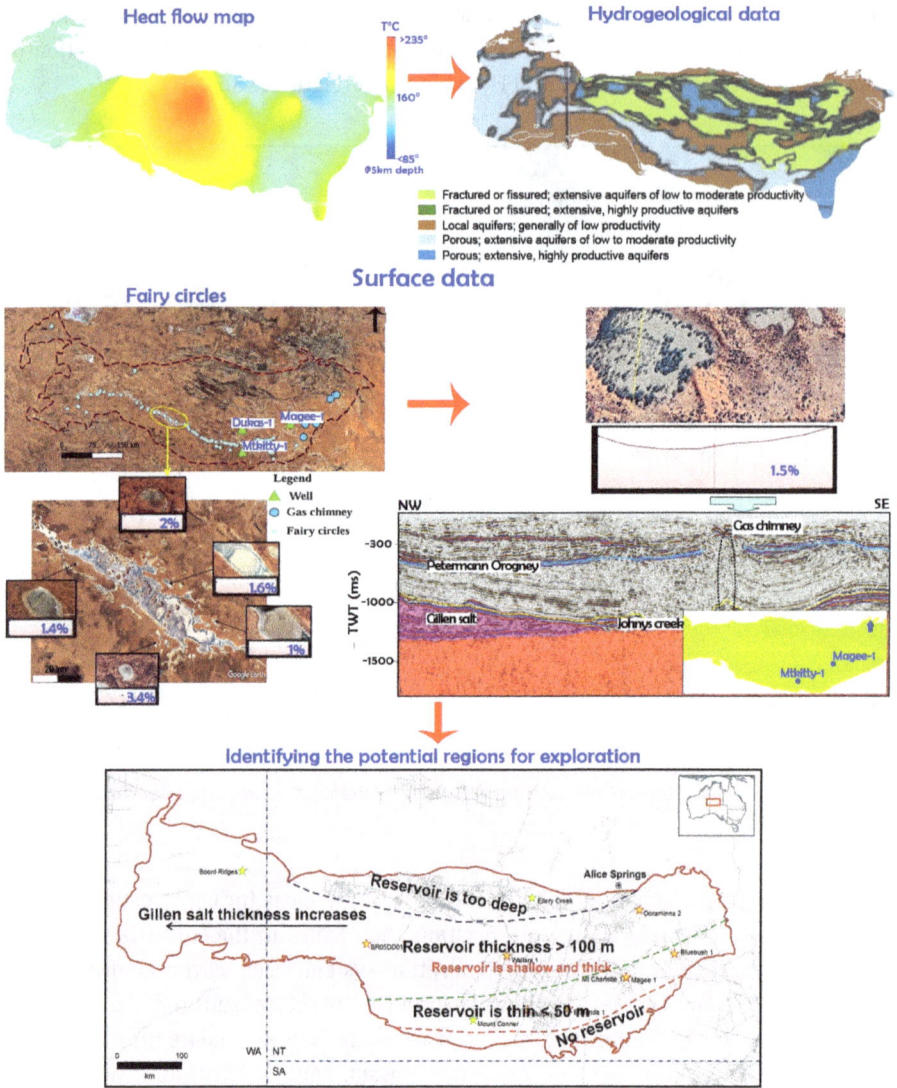

Figure 13.6: Applying the concept of NHFMs on the Amadeus Basin in Australia highlights the potential sweet spots for subsurface natural H_2 exploration.

data, and hydrogeological data facilitates the identification of potential regions for natural. By applying the NHFMs concept in the Amadeus Basin, potential regions containing all the natural H_2 system elements have been identified (Figure 13.6).

13.8 Conclusions

Considering the various mechanisms of natural H_2 generation that permit its existence in a wide range of geological contexts, it has emerged as a prominent, cost-effective energy resource that can play a paramount role in the global energy mix. However, establishing a framework for natural H_2 exploration is necessary for a more successful exploration campaign. Therefore, in this chapter, we have introduced the concept of natural H_2 favorability maps (NHFMs), which link petroleum system elements (source, reservoir, and seal) with geothermal and hydrogeological data (heat flow, geothermal aquifers). The NHFMs also account for the surficial observations of active H_2 systems, such as advective gas seeps and soil-H_2 (fairy circles). All these elements should exist in the vicinity of active natural hydrogen-generating rock types (NH$_2$-GRTs). Therefore, the first step for constructing NHFMs is to define the distribution of the NH$_2$-GRTs and link them with the mechanism of natural H_2 generation (e.g., serpentinization, radiolysis, etc.). Afterward integration with subsurface data (seismic, well data) is necessary to locate the subsurface natural H_2 accumulation zones. Therefore, the ultimate objective of NHFMs is to minimize risks associated with natural H_2 exploration and enable decision-makers to select favorable locations for exploration campaigns.

References

[1]   IEA. World Energy Outlook 2021. International Energy Agency. 2021. https://doi.org/10.1787/14fcb638-en

[2]   Gupta S. K., Kumari S., Reddy K., Bux F. Trends in biohydrogen production: Major challenges and state-of-the-art developments. Environ Technology. 2013, 34: 1653–1670.

[3]   Kalamaras C. M., Efstathiou A. M. Hydrogen production technologies: Current state and future developments. Conference Papers in Energy. 2013, 2013: 1–9.

[4]   Oey M., Sawyer A., Ross I., Hankamer B. Challenges and opportunities for hydrogen production from microalgae. Plant Biotechnology Journal. 2016, 14. 10.1111/pbi.12516

[5]   Lévy D., Roche V., Pasquet G., Combaudon V., Geymond U., Loiseau K., Moretti I. Natural H2 exploration: Tools and workflows to characterize a play. Science and Technology for Energy Transition. 2023, 78: 27. https://doi.org/10.2516/stet/2023021

[6]   Klein F., Bach W., Jöns N., McCollom T., Moskowitz B., Berquó T. Iron partitioning and hydrogen generation during serpentinization of abyssal peridotites from 15°N on the Mid-Atlantic Ridge. Geochimica Et Cosmochimica Acta. 2009, 73: 6868–6893.

[7]   Barbier S., Huang F., Andreani M., Tao R., Hao J., Eleish A., Prabhu A., Minhas O., Fontaine K., Fox P., Daniel I. A review of H2, CH4, and hydrocarbon formation in experimental serpentinization using network analysis. Frontiers in Earth Science. 2020, 8: 209.

[8]   Lévy D., Boka-Mene M., Meshi A., Fejza I., Guermont T., Hauville B., Pelissier N. Looking for natural hydrogen in Albania and Kosova. Frontiers in Earth Science. 2023, 11: 1167634.

[9]   Truche L., Donze F., Goskolli E., Muceku B., Loisy C., Monnin C., Dutoit H., Cerepi A. A deep reservoir for hydrogen drives intense degassing in the Bulqizë ophiolite. Science (New York, N Y). 2024, 383: 618–621.

[10] Leila M., Lévy D., Battani A., Piccardi L., Šegvić B., Badurina L., Pasquet G., Combaudon V., Moretti I. Origin of continuous hydrogen flux in gas manifestations at the Larderello geothermal field, Central Italy. Chemical Geology. 2021, 585: 120564. https://doi.org/10.1016/j.chemgeo.2021.120564

[11] Zgonnik V., Valérie B., Larin N., Daniel P., Deville E. Diffused flow of molecular hydrogen through the Western Hajar mountains, Northern Oman. Arabian Journal of Geosciences. 2019, 12: 71. https://doi.org/10.1007/s12517-019-4242-2

[12] Abrajano T. A., Sturchio N. C., Kennedy B. M., Lyon G. L., Muehlenbachs K., Bohlke J. K. Geochemistry of reduced gas related to serpentinization of the Zambales ophiolite, Philippines. Applied geochemistry. 1990, 5: 625–630. https://doi.org/10.1016/0883-2927(90)90060-I

[13] Hosgörmez H., Etiope G., Yalçin M. N. New evidence for a mixed inorganicand organic origin of the olympic chimaera fire (Turkey): A large onshore seepageof abiogenic gas. Geofluids. 2008, 8: 263–273.

[14] Morrill P., Kuenen J., Johnson O., Suzuki S., Rietze A., Sessions A., Fogel M., Nealson K. Geochemistry and geobiology of a present-day serpentinization site in California: the cedars. Geochimica Et Cosmochimica Acta. 2013, 109: 222–240.

[15] Worman S. L., Pratson L. F., Karson J. A., Schlesinger W. H.. Abiotic hydrogen (H_2) sources and sinks near the Mid-Ocean Ridge (MOR) with implications for the subseafloor biosphere. Proceedings of the National Academy of Sciences 2020, 117(24): 13283–13293.

[16] Lin L.-H., Hall J., Lippmann-Pipke J., Ward J., Lollar B., Deflaun M., Rothmel R., Moser D., Gihring T., Mislowack B., Onstott T. Radiolytic H_2 in continental crust: Nuclear power for deep subsurface microbial communities. Geochemistry, Geophysics, Geosystems. 2005, 6: Q07003.

[17] Dzaugis M. E., Spivack A. J., Dunlea A. G., Murray R. W., D'Hondt S. Radiolytic hydrogen production in thesubseafloor basaltic aquifer. Frontiers in Microbiology. 2016, 7: 76. https://doi.org/10.3389/fmicb.2016.00076

[18] Leila M., Loiseau K., Moretti I. Controls on generation and accumulation of blended gases (CH4/H2/He) in the Neoproterozoic Amadeus Basin, Australia. Marine and Petroleum Geology 2022, 140: 105643.

[19] Halford D. T., Karolytė R., Barry P. H., Whyte C. J., Darrah T. H., Cuzella J. J., Sonnenberg S. A., Ballentine C. J. High helium reservoirs in the Four Corners area of the Colorado Plateau, USA. Chemical Geology 2022, 596: 120790.

[20] Karolytė R., Warr O., Van Heerden E., Flude S., De Lange F., Webb S., Ballentine C. J., Sherwood Lollar B. The role of porosity in H2/He production ratios in fracture fluids from the Witwatersrand Basin, South Africa. Chemical Geology 2022, 595: 120788. https://doi.org/10.1016/j.chemgeo.2022.120788

[21] Murray J., Clément A., Fritz B., Schmittbuhl J., Bordmann V., Fleury J. M. Abiotic hydrogen generation from biotite-rich granite: A case study of the Soultz-sous-Forêts geothermal site, France, Appl. Geology Chemical. 2020, 119: 104631.

[22] Truche L., McCollom T., Martinez I. Hydrogen and abiotic hydrocarbons: molecules that change the world. Elements. 2020, 16: 13–18.

[23] Soule S. A., Fornari D. J., Perfit M. R., Ridley W. I., Reed M. H., Cann J. R. In corporation of seawater into mid-ocean ridge lava flows during emplacement. Earth and Planetary Science Letters. 2006, 252 (3–4): 289–307.

[24] Pasquet G., Houssein Hassan R., Sissmann O., Varet J., Moretti I. An attempt to study natural H_2 resources across an oceanic ridge penetrating a continent: The Asal-Ghoubbet Rift (Republic of Djibouti). Geosciences. 2021, 12(1): 16.

[25] Geymond U., Briolet T., Combaudon V., Sissmann O., Martinez I., Duttine M., Moretti I. Reassessing the role of magnetite during natural hydrogen generation. Frontiers in Earth Science. 2023, 11: 1169356.

[26] Allen P. A., Allen J. R. Basin analysis: Principles and application to petroleum play assessment, 3rd Ed. John Wiley & Sons, Chichester, West Sussex, 2013. 640.

[27] Mahlstedt N., Horsfield B., Weniger P., Misch D., Shi X., Noah M., Boreham C. Molecular hydrogen from organic sources in geological systems. Journal of Natural Gas Science and Engineering. 2022, 105: 104704.

[28] Horsfield B., Mahlstedt N., Weniger P., Misch D., Vranjes-Wessely S., Han S., Wang C. Molecular hydrogen from organic sources in the deep Songliao Basin, PR China. International Journal of Hydrogen Energy. 2022, 47(38): 16750–16774.

[29] Moretti I., Bouton N., Ammouial J., Carrillo Ramirez A. The H2 potential of the Colombian coals in natural conditions. International Journal of Hydrogen Energy. 2024, 77: 1443–1456.

[30] Myagkiy A., Moretti I., Brunet F. Space and time distribution of subsurface H_2 concentration in so-called "fairy circles": Insight from a conceptual 2-D transport model, BSGF – Earth Sci. Bull. 2020, 191: 13.

[31] Moretti I., Brouilly E., Loiseau K., Prinzhofer A., Deville E. Hydrogen emanations in intracratonic areas:. New Guide Lines for Early Exploration Basin Screening, Geosciences. 2021a, 11(3): 145.

[32] Moretti I., Prinzhofer A., Françolin J., Pacheco C., Rosanne M., Rupin F., Mertens J. Long-term monitoring of natural hydrogen superficial emissions in a Brazilian cratonic environment. Sporadic large pulses versus daily periodic emissions. International Journal of Hydrogen Energy. 2021b, 46(5): 3615–3628.

[33] Frery E., Langhi L., Maison M., Moretti I. Natural hydrogen seeps identified in the North Perth Basin, Western Australia. International Journal of Hydrogen Energy. 2021, 46(61): 31158–31173.

[34] Larin N., Zgonnik V., Rodina S., Deville E., Prinzhofer A., Larin V. N. Natural molecular hydrogen seepage associated with surficial, rounded depressions on the European Craton in Russia. Natural Resources Research. 2015, 24(3): 369–383.

[35] Zgonnik V., Valérie B., Deville E., Larin N., Daniel P., Farrell K. Evidence for natural molecular hydrogen seepage associated with Carolina bays (surficial, ovoid depressions on the Atlantic Coastal Plain, Province of the USA). Progress in Earth and Planetary Science:. 2015, 31: 1–15.

[36] Moretti I., Geymond U., Pasquet G., Aimar L., Rabaute A. Natural hydrogen emanations in Namibia: Field acquisition and vegetation indexes from multispectral satellite image analysis. International Journal of Hydrogen Energy. 2022, 47(84): 35588–35607.

[37] Chavagnac V., Monnin C., Ceuleneer G., Boulart C., Hoareau G. Characterization of hyperalkaline fluids produced by low-temperature serpentinization of mantle peridotites in the Oman and Ligurian ophiolites, Geochem. Geophys. Geosyst. 2013a, 14(7): 2496–2522.

[38] Chavagnac V., Ceuleneer G., Monnin C., Lansac B., Hoareau G., Boulart C. Mineralogical assemblages forming at hyper-alkaline warm springs hosted on ultramafic rocks: A case study of Oman and Ligurian ophiolites. Geochemistry, Geophysics, Geosystems. 2013b, 14: 2474–2495.

[39] Giampouras M., Garrido C. J., Zwicker J., Vadillo I., Smrzka D., Bach W., Peckmann J., Jiménez P., Benavente J., García-Ruiz J. M. Geochemistry and mineralogy of serpentinization-driven hyper-alkaline springs in the Ronda peridotites. Lithos. 2019, 350–351: 105215. https://doi.org/10.1016/j.lithos.2019.105215

[40] Arcilla C. A., Pascua C. S., Russell Alexander W. Hyper-alkaline groundwaters and tectonism in the Philippines: Significance to natural carbon capture and sequestration. Energy Procedia. 2011, 4: 5093–5101. https://doi.org/10.1016/j.egypro.2011.02.484

[41] Paukert Vankeuren A., Matter J., Kelemen P., Shock E., Havig J. Reaction path modeling of enhanced in situ CO2 mineralization for carbon sequestration in the peridotite of the Samail Ophiolite, Sultanate of Oman. Chemical Geology. 2012, s 330–331: 86–100.

[42] Szponar N., Brazelton W. J., Schrenk M. O., Bower D. M., Steele A., Morrill P. L. Geochemistry of a continental site of serpentinization, the Tablelands Ophiolite, Gros Morne national park: A Mars analogue. Icarus. 2013, 22: 286–296.

[43] Monnin C., Chavagnac V., Boulart C., Ménez B., Gérard M., Gérard E., Pisapia C., Quéméneur M., Erauso G., Postec A., Guentas-Dombrowski L., Payri C., Pelletier B. Fluid chemistry of the low temperature hyperalkaline hydrothermal system of Prony Bay (New Caledonia. Biogeosciences. 2014, 11: 5687–5706.

[44] Cardace D., Meyer-Dombard D. R., Woycheese K. M., Arcilla C. A. Feasible metabolisms in high pH springs of the Philippines. Frontiers in Microbiology. 2015, 6: 10. https://doi.org/10.3389/fmicb.2015.00010

[45] Pawson J. F., 2015. Abiotic Methane formation at the dun mountain Ophiolite, New Zealand. Master Thesis, 90p. http://dx.doi.org/10.26021/8976.

[46] Suda K., Ueno Y., Yoshizaki M., Nakamura H., Kurokawa K., Nishiyama E., Yoshino K., Hongoh Y., Kawachi K., Omori S., Yamada K., Yoshida N., Maruyama S. Origin of methane in serpentinite-hosted hydrothermal systems: The CH4-H2-H2O hydrogen isotope systematics of the Hakuba Happo hots spring. Earth and Planetary Science Letters. 2014, 386: 112–125.

[47] Yuce G., Italiano F., D'Alessandro W., Yalcin T. H., Yasin D. U., Gulbay A. H., et al., Origin and interactions of fluids circulating over the Amik Basin (Hatay, Turkey) and relationships with the hydrologic, geologic and tectonic settings. Chem Geol. 2014, 388: 23–39. https://doi.org/10.1016/j.chemgeo.2014.09.006

[48] Prinzhofer A., Tahara Cissé C. S., A.b D.. Discovery of a large accumulation of natural hydrogen in Bourakebougou (Mali). International Journal of Hydrogen Energy. 2018, 43(42): 19315–19326.

[49] Maiga O., Deville E., Laval J., Prinzhofer A., Diallo A. B.. Characterization of the spontaneously recharging natural hydrogen reservoirs of Bourakebougou in Mali. Scientific Reports. 2023, 13(1): 11876.

[50] Atkinson C., García-Curiel S., Matchette-Downes C., Munro I. Geological setting of natural "Gold" Hydrogen in the pyrenees and implications for exploration worldwide. A Presentation at SEAPEX SEC 2023, Singapore. 2023, March 8th, 23.

[51] Boreham C., Edwards D., Czado K., Rollet N., Wang L., Van Der Wielen S., Champion D., Blewett R., Feitz A., Henson A. Hydrogen in Australian natural gas: Occurrences, sources and resources. The APPEA Journal. 2021, 61: 163–191.

[52] Crostella A. A review of oil occurrences within the lennard shelf canning Basin western Australia. Geological Survey of Western Australia, Report. 1998, 56: 45.

[53] Lefeuvre N., Thomas E., Truche L., Donze F., Cros T., Dupuy J., Pinzon-Rincon L., Rigollet C. Characterizing natural Hydrogen occurrences in the Paris basin from historical drilling records. Geochemistry, Geophysics, Geosystems. 2024, 25(5): e2024GC011501.

[54] Deronzier J.-F., Giouse H. Vaux-en-Bugey (Ain, France): The first gas field produced in France, providing learning lessons for natural hydrogen in the sub-surface?. BSGF – Earth Sciences Bulletin. 2020, 191: 7.

[55] Coveney R. M. J., Goebel E. D., Zeller E. J., Dreschhoff G. A. M., Angino E. E. Serpentization and origin of hydrogen gas in Kansas. American Association of Petroleum Geologists Bulletin. 1987, 71: 39–48.

[56] Combaudon V., Sissmann O., Bernard S., Viennet J., Megevand V., Guillou C., Guelard J., Martinez I., Guyot F., Derluyn H., Deville E. Are the Fe-rich-clay veins in the igneous rock of the Kansas (USA) Precambrian crust of magmatic origin? Lithos. 2024, 474–475, 107583. https://doi.org/10.1016/j.lithos.2024.107583

[57] Etiope G. Massive release of natural hydrogen from a geological seep (Chimaera, Turkey): Gas advection as a proxy of subsurface gas migration and pressurised accumulations. International Journal of Hydrogen Energy. 2023, 48(25): 9172–9184.

[58] Royal Dutch Shell, 2014. Play Based Exploration, A Guide for AAPG's Imperial Barrel Award Participation.

[59] Allen P. A., Allen J. R. The Petroleum play basin analysis: principles and applications. Blackwell: Oxford, 2005.405–493.

[60] Lottaroli F., Craig J., Cozzi A. Evaluating a vintage play fairway exercise using subsequent exploration results: Did it work?. Petroleum Geoscience. 2017, 24(2). https://doi.org/10.1144/petgeo2016-150

[61] Grant S., Milton N., Thompson M. Play fairway analysis and risk mapping: An example using the middle jurassic brent group in the northern North Sea. In: Dore A. G., Sinding-Larsen R., eds., Quantification and prediction of petroleum resources, Norwegian petroleum society special publication, Elsevier, Amsterdam. 1996, 6: 167–181.

[62] Moeck I. Catalog of geothermal play types based on geologic controls. Renewable and Sustainable Energy Reviews. 2014, 37: 867–882.

[63] Phillips B., Ziagos J., Thorsteinsson H., Hass E. roadmap for strategic development of geothermal exploration technologies. In: Proceedings of 38th workshop on geothermal reservoir engineering. Stanford, CA: Stanford University, 2013. Feburary 11–13. 2013. SGP-TR-198.

[64] Liu Z., Perez-Gussinye M., García-Pintado J., Mezri L., Bach W. Mantle serpentinization and associated hydrogen flux at North Atlantic magma-poor rifted margins. Geology. 2023, 51(3): 284–289.

[65] Blay-Roger R., Bach W., Bobadilla L., Reina T., Odriozola J., Amils R., Blay V. Natural hydrogen in the energy transition: Fundamentals, promise, and enigmas. Renewable and Sustainable Energy Reviews. 2024, 189: 113888.

N. Ferrando, M.C. Cacas-Stentz*, F. Patacchini, F. Willien, and B. Braconnier

Chapter 14
Numerical simulation of hydrogen phase equilibrium and migration at basin scale

Abstract: This work presents research and development (R&D) efforts aimed at improving the modeling of natural hydrogen systems at the basin scale and illustrates how numerical basin modeling can contribute to the understanding of these systems.

The first development aims to accurately determine the single-phase and two-phase zones in natural hydrogen systems, which requires the use of rigorous phase equilibrium calculations. A full description of this type of calculation is given, emphasizing the role of the choice and parameterization of an appropriate equation of state.

New developments in numerical basin modeling are then considered. In addition to phase equilibria, numerical basin modeling must include the advection of dissolved species, diffusion and dispersion of gases, hydrogen source terms and consumption by microorganisms, as well as the consideration of faults and fractures.

Finally, the new basin modeling functionalities are illustrated through a simplified case study of a section of the Perth Basin in Australia, conducted using the ArcTem™ simulator. This case study demonstrates the appropriate conditions for the formation of free hydrogen accumulations and shows that a very low bacterial alteration rate is required to enable free gas accumulation, even under maximized hydrogen production hypotheses.

Keywords: numerical simulation, hydrogen phase equilibrium, hydrogen migration

14.1 Introduction

Climate change obliges society to develop clean energy sources and use decarbonized fuels. It motivates the recent interest in natural hydrogen, which is found leaking through the ground surface in many places around the world [1].

Intensive scientific research into its origins is underway, and it is now believed that most of it originates deep underground and migrates to these leak sites in a manner similar to hydrocarbons.

*Corresponding author: M.C. Cacas-Stentz, IFPEN, 1–4 av. de Bois-Préau, 92500 Rueil-Malmaison, France, e-mail: marie-christine.cacas-stentz@ifpen.fr
N. Ferrando, F. Patacchini, F. Willien, B. Braconnier, IFPEN, 1–4 av. de Bois-Préau, 92500 Rueil-Malmaison, France

https://doi.org/10.1515/9783111437040-014

In a similar manner to hydrocarbon systems, hydrogen systems are usually broken down into source, migration pathways, and subsurface accumulations. These "accumulations" imply that vapor-phase hydrogen can accumulate in some porous rocks, and they are the only economic targets of hydrogen explorers, at least for the time being. To predict whether and where such accumulations may exist, it is essential to understand the physical phenomena that control the source of hydrogen, those that control the migration pathways from the sources to these potential accumulations, and those that control accumulated hydrogen losses by leakage to the surface or by alteration.

14.1.1 The elements of a "hydrogen system"

A large number of papers have been published on the possible sources of natural hydrogen. Several processes have been identified as possible sources of natural hydrogen, including water reduction associated with iron oxidation [2, 3], water radiolysis [4], pyritization [5], primordial hydrogen from the mantle or core [1], and late cracking of organic matter during its metagenesis [6].

Less work has been done on migration [7–9] and trapping [10]. In contrast to hydrocarbons, which migrate predominantly in non-aqueous phases, hydrogen migrates both by advection as one of the species dissolved in water and as a component of a free gas phase. Advection is expected to be dominant in the deep subsurface because gas solubility increases with depth. At shallower depths, as thermodynamic conditions change, dissolved gas is exsolved to form a free gas phase, which can begin to migrate in addition to advection. Diffusion is another process that contributes to the migration of dissolved gases. Finally, adsorption and alteration by geochemical or biological processes can interact with migration by reducing the amount of migrating gas.

Few hydrogen-rich free gas accumulations have been discovered. The most famous may be the Bourakebougou accumulation in Mali [11], which contains 98% H_2, 1% N_2, 1% CH_4, and 500 ppm helium [12]. Like hydrocarbon reservoirs, hydrogen reservoirs exist as porous rocks with high gas saturation in their pores, capped by an impermeable "cap-rock" that prevents gas leakage. In contrast to hydrocarbon reservoirs, which are thought to be filled on geological timescales, some authors argue that hydrogen-rich gas reservoirs are dynamic, meaning that hydrogen has a much shorter residence time in the reservoir, which could be on the order of 100 to 1,000 years [7, 13].

14.1.2 Predicting migration paths and potential free gas accumulation

In recent years, many companies have begun to explore natural hydrogen. The exploration strategy so far has mainly consisted of searching for surface seeps along with favorable geological conditions for natural hydrogen production. Hydrogen exploration using quantitative approaches is still in its infancy [14, 15]. Prediction of hydrogen sources, migration pathways, and accumulations using phenomenological and quantitative approaches, such as those used in the oil and gas industry, must now be considered.

As with hydrocarbons, hydrogen exploration requires a thorough geological knowledge of the target zone, as well as a deep understanding of the physical phenomena that control its possible formation, migration, and accumulation. Predicting migration is all the more difficult because hydrogen-rich gas migration is controlled by advection, free gas flow, and gas exchange between the aqueous and free gas phases, whereas hydrocarbon migration occurs mainly as one or two fluid phases separated from water. Natural hydrogen-rich gases sampled in different hydrogen systems around the world are gas mixtures usually containing hydrogen, nitrogen, a small amount of helium, and occasionally methane [13]. Concentrations of hydrogen, nitrogen, and methane can vary from 1% to many tens of percent from one basin to another, whereas helium concentration can reach several percent.

Free gas appears at the bubble point of the gas-bearing water when the thermodynamic conditions for exsolution are reached [8]. The gas partitioning between the two phases is then controlled by its total amount and composition, water salinity, pressure, and temperature. When exsolution occurs, free gas migration can be initiated. From this stage, as pressure, temperature, and salinity evolve along the migration pathways, the distribution of gases in the two phases evolves as well.

Thus, an accurate fluid-phase equilibrium calculation is of primary importance for the quantitative modeling of natural hydrogen systems. Indeed, the relative amount and composition of the phases at thermodynamic equilibrium can directly control the development of a free gas accumulation, its composition, and its recharge rate. Many fluid properties, such as density and viscosity, also depend directly on these equilibrated compositions. Because it is essential for modeling hydrogen systems, we have focused on the phase equilibrium of brine, H_2, N_2, He, and CH_4 in Section 14.2.

When the thermodynamics of this brine-gas system are well understood, numerical basin modeling can be used to simulate the hydrogen system. Numerical basin modeling was developed for petroleum system exploration and can now be used to simulate hydrogen systems with a few improvements, including the modeling of exsolution and exchange between the brine and the free gas phase. This is the subject of Section 14.3.

14.2 Phase equilibrium modeling between H₂-rich gas and brine

14.2.1 Introduction

This section deals with the equilibrium calculation between two fluid phases. The first is a vapor phase containing hydrogen and three typical impurities found in natural hydrogen systems: nitrogen, methane, and helium. The proportions of these impurities vary according to the hydrogen production and consumption mechanisms that occur at each site. The second fluid phase is a brine, which is assumed to consist of water and dissolved sodium chloride. Except for the ions (Na^+ and Cl^-), all of these species can migrate from one phase to another. Figure 14.1 illustrates the phase equilibrium approach used in this study.

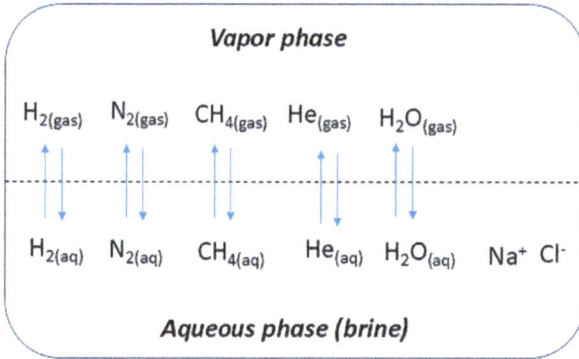

Figure 14.1: Principle of the phase equilibrium approach adopted in this study.

For basin or reservoir simulations, hydrogen migration must be modeled by considering hydrogen in its two forms: either in gaseous form in the vapor phase or in dissolved form in the aqueous phase. In this context, the aim of this section is to describe how a rigorous thermodynamic phase equilibrium is performed and what the recommended choices are for the thermodynamic model to be used for a natural hydrogen system. This rigorous first approach allows the precise determination of the partition coefficients of the different components as a function of temperature, pressure, and salinity.

14.2.2 The TP-flash algorithm

The resolution of a liquid-vapor thermodynamic equilibrium at an imposed temperature T and pressure P, commonly referred to as the *TP-flash calculation*, is obtained by solving the Rachford-Rice equation, which combines fugacity equality (thermodynamic equilibrium) with material balance [16]:

$$\Im = \sum_{i=1}^{n} \frac{(K_i - 1)z_i}{1 + (K_i - 1)\theta} = 0 \tag{14.1}$$

where z_i is the molar fraction of component i in the mixture:

$$z_i = \frac{n_i}{n^{tot}} \tag{14.2}$$

where n_i is the total mole number of component i and n^{tot} the total mole number in the mixture.

The vapor fraction θ is defined by

$$\theta = \frac{n^{vap}}{n^{tot}} \tag{14.3}$$

where n^{vap} is the total mole number of the vapor phase.

The partition coefficient K_i (also named K-value) of the component i is defined by the ratio of its molar fraction in the vapor phase x_i^{vap} and its molar fraction in the aqueous phase x_i^{aq}:

$$K_i = \frac{x_i^{vap}}{x_i^{aq}} \tag{14.4}$$

with

$$x_i^{vap} = \frac{n_i^{vap}}{n^{vap}} \tag{14.5}$$

and

$$x_i^{aq} = \frac{n_i^{aq}}{n^{aq}} \tag{14.6}$$

where n_i^{vap} and n_i^{aq} are the mole numbers of component i in the vapor and aqueous phases, respectively.

The partition coefficient, K_i, can be expressed in two different but equivalent ways, depending on the chosen thermodynamic approach.

14.2.2.1 Heterogeneous approach

In a heterogeneous approach, an activity model is generally used to model the fugacity in the aqueous phase, whereas an equation of state is used to model the fugacity in the vapor phase. In this approach, the partition coefficient is expressed as

$$K_i = \frac{x_i^{vap}}{x_i^{aq}} = \frac{H_i(T).\gamma_i^* \left(T, \text{salinity}, \overline{X^{aq}}\right)}{P.\varphi_{water}^{vap} \left(T, P, \overline{X^{vap}}\right)} \tag{14.7}$$

for a solute (dissolved gas, for example), and

$$K_{water} = \frac{x_{water}^{vap}}{x_{water}^{aq}} = \frac{P_{water}^{\sigma}(T).\gamma_{water} \left(T, \text{salinity}, \overline{X^{aq}}\right)}{P.\varphi_{water}^{vap} \left(T, P, \overline{X^{vap}}\right)} \tag{14.8}$$

for the solvent (water).

In these equations, H_i is the Henry constant of solute i in the solvent (water), γ_i^* its activity coefficient in this solvent, and φ_i^{vap} the fugacity coefficient of component i in the vapor phase. $\overline{X^{aq}}$ and $\overline{X^{vap}}$ stand for the composition vector in the aqueous phase and in the vapor phase, respectively. The asterisk (*) on the activity coefficient refers to the reference state adopted for solutes, which is a hypothetical state corresponding to an ideal solution at infinite dilution at the temperature and pressure of the mixture. With this convention, the activity coefficient tends toward unity when the solute concentration tends toward zero.

P_{water}^{σ} stands for the saturation pressure of water at the temperature T and γ_{water} its activity coefficient in the aqueous phase. For this solvent, the reference state is the pure component in the liquid state at the temperature and pressure of the mixture. In this convention, the activity coefficient tends toward unity when the solvent concentration tends toward unity.

14.2.2.2 Homogeneous approach

In a homogeneous thermodynamic approach, the same equation of state is used to model both fugacity in the aqueous phase and the vapor phase. In this case, the partition coefficient of any component i is expressed as

$$K_i = \frac{x_i^{vap}}{x_i^{aq}} = \frac{\varphi_i^{aq}\left(T, P, \overline{X^{aq}}\right)}{\varphi_i^{vap}\left(T, P, \overline{X^{vap}}\right)} \tag{14.9}$$

where φ_i^{aq} and φ_i^{vap} stand for the fugacity coefficient of the component i in the aqueous phase and the vapor phase, respectively.

14.2.3 The Søreide and Whitson equation of state

14.2.3.1 Description

Numerous studies on phase equilibrium modeling between light gases and brine have been reported in the literature. Most of them are related to methane, hydrocarbons, or

CO_2 solubility, either based on a heterogeneous approach using an activity model for the aqueous phase (e.g., [17–21]) or on a homogeneous approach using the same equation of state for all phases (e.g., [22–29]). Modeling works related to hydrogen solubility in brines are fewer in number and more recent [30–36]. Concerning nitrogen and helium solubility, only a very few modeling approaches have been published [37, 38].

Using a heterogeneous approach, especially with the Pitzer activity coefficient model [39], is particularly advantageous when the brine contains numerous different ionic species and when the ionic strength is significant. However, this remains a highly empirical model requiring numerous experimental data to adjust its parameters. It is also pressure-independent, necessitating the use of external corrections in fugacity calculations, such as employing a Poynting correction or an empirical correction in the Henry's constant formulation. In this study, the brine is considered to be composed only of dissolved halite, in a moderate concentration representative of those found in aquifers or sedimentary basins. Additionally, natural hydrogen systems are involved in subsurface areas where pressure can be significant, reaching up to several hundred bars. For these reasons, in this work, a homogeneous approach based on an equation of state for both phases is used. This equation of state will then be used to calculate the partition coefficient given in eq. (14.9).

The selected equation of state for this study is the one proposed by Søreide and Whitson [37] (hereafter referred to as the *SW model*). This choice is primarily motivated by its simplicity, particularly in how the salinity of the brine is accounted for. It has been extensively employed in the past to successfully model phase equilibria between light gases and brines and is available in various simulators [38]. The basis of this model is the Peng-Robinson cubic equation of state [40], with some modifications made specifically for water. The general form of the equation of state is as follows:

$$P = \frac{RT}{v-b} - \frac{a(T)}{v^2 + 2bv - b^2} \tag{14.10}$$

where P is the pressure, v is the molar volume, T is the temperature, R is the ideal gas constant, b is the repulsive parameter, and a is the attractive term.

The first term of this equation of state represents the Van der Waals repulsive interactions; it introduces the finite volume of molecules (the hard-sphere volume or co-volume, b, which is a measure of the volume of a molecule that other molecules cannot penetrate). The second term accounts for the attractive interactions between species, which are quantified by the attractive term, a:

$$a(T) = \sum_i \sum_j x_i x_j a_{ij}(T) \tag{14.11}$$

with

$$a_{ij}(T) = \sqrt{a_i a_j}\left(1 - k_{ij}\right) \tag{14.12}$$

where x, x_j are the molar fractions of the components i and j in the considered phase, a_i, a_j are the attractive energy terms for the components i and j, and k_{ij} is a binary interaction parameter. The attractive term a_i of a component i mainly depends on its critical properties and acentric factor, as summarized in Table 14.1:

$$a_i(T) = 0.457235 \frac{R^2 T_{c,i}^2}{P_{c,i}} \alpha_i(T)$$

(14.13)

where $T_{c,i}$ and $P_{c,i}$ are the critical temperature (K) and pressure (Pa), respectively, of component i. For any component except water, the *alpha function* is given by

$$\alpha_{water}(T) = \left[1 + 0.4530 \left(1 - \frac{T}{T_{c,water}} (1 - 0.0103.c_s^{1.1}) \right) + 0.0034 \left(\left(\frac{T}{T_{c,water}} \right)^{-3} - 1 \right) \right]^2$$

(14.14)

Finally, the general form for the binary interaction parameters k_{ij} between water (j) and another component i in eq. (14.12) is as follows:

$$k_{ij} = A_0 \left(1 + a_0.c_s^{\beta_0} \right) + A_1 \left(\frac{T}{T_{c,i}} \right) \left(1 + a_1.c_s^{\beta_1} \right) + A_2 \left(\frac{T}{T_{c,i}} \right)^2 \left(1 + a_2.c_s^{\beta_2} \right) + A_3 \exp \left(A_4 \left(\frac{T}{T_{c,i}} \right) \right)$$

(14.15)

All parameters of eq. (14.15), for both vapor and aqueous phases, are summarized in Section 14.2.3.2.

In this work, all TP-flash calculations using the SW model were performed using the thermodynamic library Carnot [41].

Table 14.1: Pure component parameters used in the SW model.

Compound	T_c (°C)	P_c (bar)	Acentric factor (–)
Hydrogen	−239.96	13.13	− 0.215993
Methane	−82.586	45.99	0.0115478
Nitrogen	−146.95	34 00	0.0377215
Helium	−267.95	2.275	−0.390032
Water	373.946	220.64	0.344861

14.2.3.2 Model parameterization

This section aims to present the binary interaction parameters used between the different gases considered in this work and the brine (eq. (14.15)), both in the aqueous phase. All parameters are summarized in Tables 14.1 and 14.3.

14.2.3.2.1 Hydrogen + brine

Lopez-Lazaro et al. [33] have recently collected and reviewed literature data related to hydrogen solubility in pure water and brines. This database has also been augmented using molecular simulation data at high temperatures, where no experimental data are available. A parameterization of the binary interaction parameter in the aqueous phase for the SW model has been proposed using this consistent database. This model reproduces the experimental data with an average deviation of less than 6%, for temperatures ranging from 0 to 300 °C and salinity ranging from 0 to 2 mol·kg^{-1}. The binary parameters in the vapor phase have been optimized to match experimental partial pressures up to 170 °C, with an average deviation of 7%.

More recent experimental data on hydrogen solubility [34], which were not used in the parameterization procedure, have been utilized to validate the interpolation capability of this model. As shown in Figure 14.2, this model can predict the data with an average deviation of 6%, including data with salinities higher than the original range (5 mol·kg^{-1}).

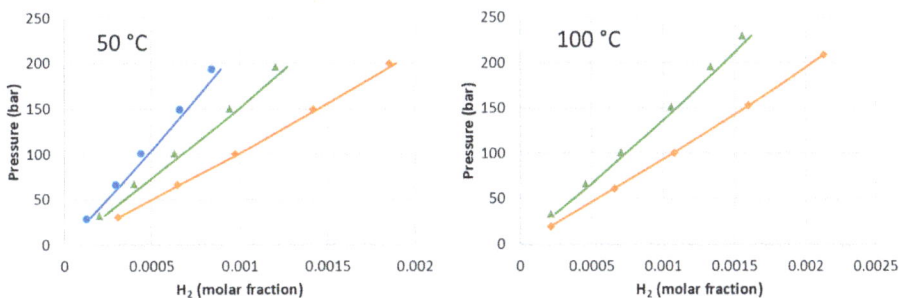

Figure 14.2: Hydrogen solubility in NaCl brine. Left: at 50 °C. Right: at 100 °C. Symbols: experimental data [34]: diamonds: molality = 1 mol·kg^{-1}; triangles: molality = 3 mol·kg^{-1}; circles: molality = 5 mol·kg^{-1}. Lines: this model.

14.2.3.2.2 Nitrogen + brine and methane + brine

The binary interaction parameters between N$_2$ and brine and between CH$_4$ and brine are directly taken from the original publication of SW [37]. As illustrated in Figures 14.3 and 14.4, this parametrization is evaluated using the experimental solubility data from [42] for temperatures ranging from 50 to 125 °C, pressures ranging from 1 to 500 bar, and salinities ranging from 0 to 4 mol·kg^{-1}. A good agreement is found between the experiments and the model predictions, with an average deviation of 4% for N$_2$ and 6% for CH$_4$.

14.2.3.2.3 Helium + brine

Regarding phase equilibrium between helium and brine, no existing parameterization of the SW model is available in the literature. Thus, a new parameter set is proposed

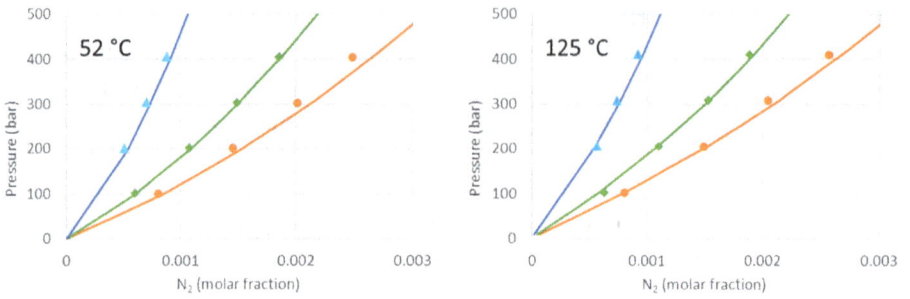

Figure 14.3: Nitrogen solubility in NaCl brine. Left: at 52 °C. Right: at 125 °C. Symbols represent experimental data [42]: circles indicate pure water; diamonds represent molality = 1 mol·kg^{-1}; triangles denote molality = 4 mol·kg^{-1}. Lines correspond to this model.

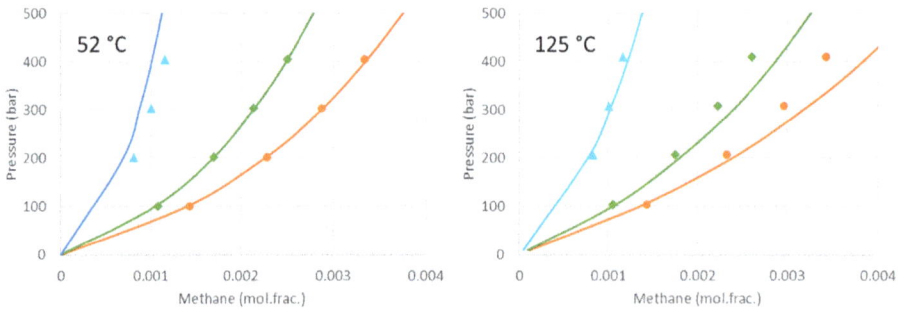

Figure 14.4: Methane solubility in NaCl brine. Left: at 50 °C. Right: at 125 °C. Symbols: experimental data [42]: diamonds: molality = 1 mol·kg^{-1}; triangles: molality = 3 mol·kg^{-1}; circles: molality = 5 mol·kg^{-1}. Lines: this model.

in this work. As summarized in Table 14.2, the experimental solubility data of helium in NaCl brine are provided in various forms, including Henry constants [43–46], Kuenen coefficients [44], and solubilities at given temperature and pressure ("TPx" data) [47–49]. To ensure data consistency for adjusting the parameters of eq. (14.14), all these data are converted into Henry constants according to the conversion rules recommended by the NIST [50]. The optimization of the equilibrium model based on pressure-independent Henry's constants is logical in the context of a natural hydrogen system. Indeed, due to the particularly low solubility of helium, it follows Henry's law even at high pressures (linear behavior between solubility and partial pressure of helium). For example, Figure 14.6 demonstrates that Henry's law is respected up to at least 300 bars. For the study of a natural hydrogen system, it is not necessary to calibrate the model for extreme temperatures and salinities. The experimental database used to adjust the model parameters is therefore restricted to temperatures up to 200 °C. Within this temperature range, the experimental salinities vary from 0 (pure water) to 3 mol·kg^{-1}.

In eq. (14.15), parameters A_1 to A_4 do not depend on salinity. For the binary inter-action in the aqueous phase, these parameters are first adjusted to match solubility data in pure water. Secondly, parameters α_0 to α_2 and β_0 to β_2 are adjusted to match solubility data in NaCl brine. Table 14.3 provides the optimized parameter set. Regard-ing the binary interaction parameter in the vapor phase, no experimental data are available to fit a parameter set. It is therefore assumed to be equal to 0.5, which is the order of magnitude of such binary parameters for water and gas in a vapor phase, as described by Søreide and Whitson [37].

With these optimized parameters, the Søreide and Whitson model is able to re-produce the Henry's constant of helium in pure water and brines with an average deviation of 4%, as illustrated in Figure 14.5.

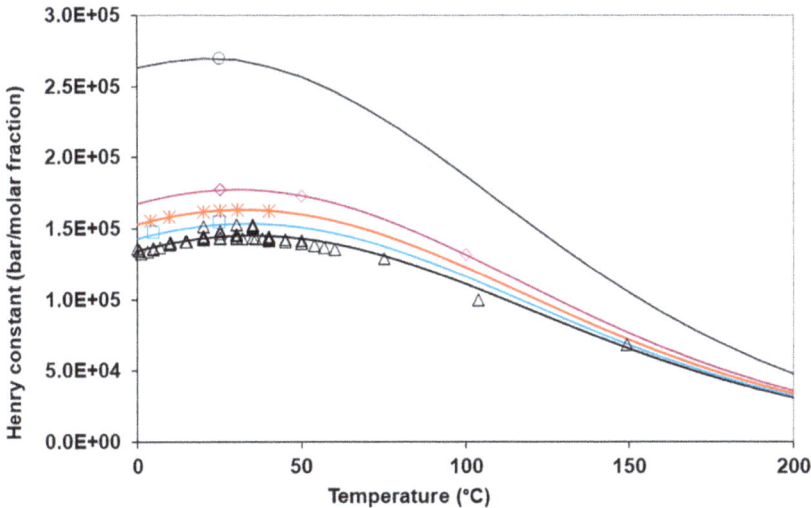

Figure 14.5: Henry's constant of helium in pure water and brines. Symbols represent experimental data, while lines represent this model. Black triangles indicate pure water; blue squares represent 0.3 mol·kg^{-1}; red crosses denote 0.6 mol·kg^{-1}; purple diamonds signify 1 mol·kg^{-1}; and green circles correspond to 3 mol·kg^{-1}.

This model is also directly evaluated using experimental solubility data under pres-sure [48, 49] as illustrated in Figure 14.6. An average relative deviation of 8% is ob-tained between model predictions and experiments.

14.2.3.2.4 Parameter summary
Pure component parameters for hydrogen, methane, nitrogen, helium, and water are reported in Table 14.1. The binary interaction parameters between brine and these compounds, recommended for the SW model, are provided in Table 14.3 for both the aqueous and vapor phases.

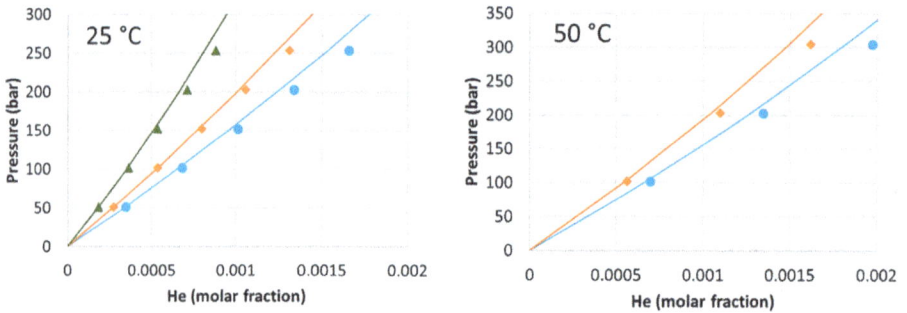

Figure 14.6: Helium solubility in NaCl brine. Left: at 25 °C. Right: at 50 °C. Symbols represent experimental data [48, 49]: circles indicate solubility in pure water; diamonds correspond to molality = 1 mol·kg^{-1}; triangles represent molality = 3 mol·kg^{-1}. Lines depict results from this model.

Table 14.2: Experimental database of helium solubility in pure water and in NaCl brine for the SW model parameterization.

Salinity	Solubility data type	Number of points	References
0 mol kg^{-1} (pure water)	Bunsen coefficient	41	[45, 46]
	Kuenen coefficient	6	[51]
	Henry constant	34	[43, 44]
	TPx	7	[47, 48]
0.3 mol kg^{-1}	Bunsen coefficient	2	[46]
0.6 mol kg^{-1}	Bunsen coefficient	7	[46]
1 mol kg^{-1}	TPx	4	[48, 49]
3 mol kg^{-1}	TPx	1	[49]

14.3 Numerical modeling of hydrogen systems

Because too many physical phenomena interact in hydrogen systems, human expertise is insufficient to quantitatively predict generation rates, migration paths, whether or not gas accumulations can form, where they can form, their composition, free gas supply rate, and leak rate. To meet the challenge of dealing with complex geological systems, oil and gas explorationists have developed numerical simulators that quantitatively model sedimentary basin history, petroleum generation, migration, and trapping. Therefore, it was particularly appropriate to test this type of tool for addressing hydrogen systems.

Table 14.3: Binary interaction parameters recommended for the SW model (eq. (14.15)) for the aqueous and vapor phases. For other component pairs, k_{ij} is assumed to be equal to 0.

		A_0	A_1	A_2	A_3	A_4	α_0	α_1	α_2	β_0	β_1	β_2	Ref.
H_2/H_2O	k_{ij}^{aq}	-2.513	0.181	-	-12.723	-0.499	6.8×10^{-4}	0.038	-	0.443	0.799	-	[33]
	k_{ij}^{vap}	2.500	-0.179	-	-	-	-	-	-	-	-	-	
CH_4/H_2O	k_{ij}^{aq}	-1.619	1.010	-0.169	-	-	6.5×10^{-21}	1.438×10^{-2}	2.1547×10^{-3}	1	1	1	[37]
	k_{ij}^{vap}	0.485	-	-	-	-	-	-	-	-	-	-	
N_2/H_2O	k_{ij}^{aq}	-1.702	0.443	-	-	-	0.0256	0.081	-	1	1	-	[37]
	k_{ij}^{vap}	0.4778											
He/H_2O	k_{ij}^{aq}	-5.103	0.117	-6.0×10^{-4}	-	-	-8.6×10^{-3}	-	-	1	-	-	This work
	k_{ij}^{vap}	0.500											

14.3.1 Basin modeling

Numerical basin modeling, often referred to as "basin modeling" [52], is a process-based numerical simulation of the geological evolution of a sedimentary basin and the many physical phenomena of interest that occur within it. For example, in petroleum exploration, these include heat transfer, compaction, fluid flow, organic matter maturation, expulsion, two- or three-phase migration, and hydrocarbon trapping. These numerical simulations provide 1D, 2D, or 3D maps, at different stages of basin evolution, of various quantities such as temperature, fluid pressure, flow rate of different phases, composition of hydrocarbon phases, rock porosity, and estimates of effective stress.

Some functionalities had to be added or extended to model natural hydrogen systems. These include the migration of dissolved species by advection, the calculation of phase equilibrium between the aqueous and free gas phases (see the previous section), diffusion and dispersion of dissolved gases, and gas sources and sinks.

We developed these new functionalities in the ArcTem™ calculator [53], which allows accurate modeling of flow associated with faults. This is particularly appropriate since fault flow is considered by many authors to be a major contributor to hydrogen migration at the basin scale [7, 9, 54].

These functionalities are detailed below.

14.3.2 Advection of dissolved species

Once hydrogen is dissolved in the aqueous phase w, its conservation of mass reads:

$$\frac{\partial}{\partial t}(z_w \rho_w s_w \phi) + \nabla \cdot (z_w F_w) = z_{w,\text{sup}} q_{w,\text{sup}} + z_{w,\text{ext}} q_{w,\text{ext}} \tag{14.16}$$

where ρ_w and s_w are the density and saturation of the aqueous phase w, respectively, and z_w is the mass fraction of H_2 in the phase (in kg of H_2 per kg of phase). Additionally, $q_{w,\text{sup}}$ and $q_{w,\text{ext}}$ are water sources corresponding to sedimentary deposits and to any external inflow with hydrogen compositions $z_{w,\text{sup}}$ and $z_{w,\text{ext}}$, respectively. For instance, $q_{w,\text{ext}}$ may represent an inflow of hydrothermal water at the domain boundary or a diagenetic process. The aqueous mass flux F_w is defined as $F_w = \rho_w U_w$, where is the Darcy velocity of the phase w. Combining the conservation of mass of the aqueous phase with the above equation yields the following conservation law for the mass of species i in w the phase:

$$\rho_w s_w \phi \frac{\partial z_w}{\partial t} + F_w \cdot \nabla z_w = (z_{w,\text{sup}} - z_w) q_{w,\text{sup}} + (z_{w,\text{ext}} - z_w) q_{w,\text{ext}} \tag{14.17}$$

The dissolution of H_2 into the aqueous phase w is achieved via a source term r_w added to the right-hand side of the above equation, and it takes the following form:

$$r_w = \frac{\partial}{\partial t}\left((z_{w,\text{diss}} - z_w)\rho_w\right) \tag{14.18}$$

where the quantity $z_{w,\text{diss}}\rho_w$ corresponds to the total mass of H_2 dissolved in the aqueous phase, calculated via the solubility law given later as a function of temperature, pressure, and molality. Accordingly, the combination of aqueous advection and dissolution is performed via a splitting approach, whereby flash equilibrium via the solubility law is computed first, followed by transport in the aqueous phase with the above source term.

14.3.3 Diffusion and dispersion

A divergence operator can be added to the righthand side of the advection equation to account for the diffusion and dispersion of the dissolved hydrogen. We thus obtain the following advective-conductive equation:

$$\rho_w s_w \phi \frac{\partial z_w}{\partial t} + F_w \cdot \nabla z_w = div(\Lambda_w \nabla z_w) + \left(z_{w,\text{sup}} - z_w\right)q_{w,\text{sup}} + \left(z_{w,\text{ext}} - z_w\right)q_{w,\text{ext}} \tag{14.19}$$

where Λ_w is the hydrogen conductivity in phase w. We assume it has the following expression:

$$\Lambda_w = \rho_w s_w \phi \left(D_{w,\text{diff}} \frac{a(T)}{\tau} + D_{w,\text{disp}}|U_w| \right) \tag{14.20}$$

where $D_{w,\text{diff}}$ is the diffusivity coefficient (measured in $m^2/\text{Myrs s}$) and $D_{w,\text{disp}}$ is the dispersion (measured in m). Moreover, τ is the rock tortuosity, which is dimensionless and greater than or equal to 1, and $a(T)$ is a temperature-dependent coefficient satisfying [55]:

$$a(T) = \frac{\mu_{w,25} T}{\mu_w(T) T_{25}} \tag{14.21}$$

where $\mu_w(T)$ and $\mu_{w,25}$ are the viscosities of the aqueous phase at the current temperature T and at 25 °C, respectively, and T_{25} is the conversion of 25 °C into kelvin units, that is, $T_{25} = 298.15$ K. We choose the Bingham formula for viscosity as a function of temperature:

$$\mu_w(T) = 21.5\left(T_c + \sqrt{8{,}078 + T_c^2} \right) - 1{,}200 \tag{14.22}$$

where $T_c = T - 264.75$ for the tortuosity, several estimations have been proposed in the literature; we use the following parameterized expression [56]:

$$\tau = \frac{1}{s_w \phi^{n-1}}$$

(14.23)

where n is an exponent is greater than or equal to 1, typical value 2.

14.3.4 Gas sources and sinks

As discussed above, many possible gas sources capable of producing natural hydrogen have already been identified, and many studies have been reported that aim to measure the production rates of the different generation processes [1]. While awaiting more mature gas generation models, gas sources are currently represented as user-defined source terms. These source terms are applied to predefined cells of the model and can be either (i) a hydrothermal inflow characterized by its flow rate, temperature, and composition, (ii) a free gas source characterized by its composition and production rate, or (iii) a dissolved gas concentration imposed on the aqueous phase.

Hydrogen is a highly reactive compound, and its alteration by geochemical or bacterial processes must be modeled [13]. However, since the evolution of basin modeling from oil and gas exploration to natural hydrogen exploration is ongoing, only bacterial alteration is considered in our simulator for the time being. Hydrogen alteration is modeled as a first-order degradation process [57], characterized by the degradation rate A (in years^{-1}):

$$\frac{dx_{H_2}}{dt} = -A\, x_{H_2}$$

(14.24)

where x_{H_2} is the hydrogen mass concentration in water.

The alteration rate, A, is modeled as a function of temperature and the alteration rate A_0 obtained at the optimal temperature for bacterial activity, as proposed by [58]:

$$A = A_0 * \left\{ \frac{T - T_{min}}{T_{opt} - T_{min}} \cdot \frac{\left(1 - e^{c(T-T_{max})}\right)}{\left(1 - e^{c(T_{opt}-T_{max})}\right)} \right\}^2 \quad \text{When } T \in [T_{min}, T_{max}]$$

$$A = 0 \qquad\qquad\qquad\qquad\qquad\qquad \text{When } T \notin [T_{min}, T_{max}]$$

(14.25)

where T is the temperature in K, T_{min}, T_{max}, and T_{opt} are the minimum, maximum, and optimal temperatures for bacterial activity in K, A_0 is the alteration rate at the optimal temperature (in year^{-1}) and c is a parameter calibrated so that A is maximum at $T = T_{opt}$.

14.3.5 Phase equilibrium

Rigorous calculation of phase equilibrium using an equation of state (see Section 14.2) is known to be accurate but too CPU-intensive to be implemented directly in basin simulators. To improve computational performance, the phase equilibrium calculation must be simplified at the risk of slightly reducing its accuracy. Although the gas composition partially controls the phase equilibrium, we essentially approximate the thermodynamic properties of the gas mixture by those of pure hydrogen. The phase equilibrium of pure hydrogen with brine has been studied by [34], who proposed a proxy to calculate the solubility threshold Sol(P,T,s) of hydrogen in brine as an analytical function of pressure P (bar), temperature T (K), and salt molality s. The validity ranges given by [34] are 50 °C < T < 100 °C, 0 < salinity < 5 mol·kg^{-1}, and 10 < pressure < 230 bar.

We have extended the validity range of this proxy to 15 °C < T < 180 °C, 0 < s < 5 mol·kg^{-1}, and to pressures up to 500 bar. The updated proxy approximates the solubility threshold calculated from rigorous flash calculations described in Section 14.2 to within 10%, which seems reasonable given the uncertainties in other basin model parameters. With the new calibration, hydrogen solubility is approximated by the following equation:

$$\text{Sol} = \left(0.0000001971PT + 0.02158\frac{P}{T} - 0.0001175P - 0.000000004468P^2 \right) e^{\left(0.004839s^2 - 0.1957s\right)}$$

$$(14.26)$$

To evaluate the error introduced by the assimilation of the gas mixture to pure hydrogen, we used the rigorous flash approach described in Section 14.2 to calculate the bubble point depth of water samples migrating upward and in equilibrium with different gas mixtures of He, H_2, CO_2 and CH_4, characterized by the same total molar amount but different compositions. We observed that varying the composition of the gas mixture can shift the bubble point depth by hundreds of meters, as shown in Figure 14.7. It is interesting to note that the gas mixture containing 98% hydrogen and 2% helium shows the deepest bubble point, which means that our approximation tends to overestimate the bubble point depth as soon as nitrogen or methane is introduced into the mixture.

Therefore, in principle, it is necessary to consider not only hydrogen but also the other gases commonly associated with H_2 in order to accurately predict the bubble point depth. However, at this stage of the development of our simulator, ArcTem™, we consider bubble point calculation only on the basis of an H_2 + brine system. Further development will take into account the other gases in the simulator.

14.3.6 Fault flow modeling

ArcTem™ can be run on 2D or 3D unstructured meshes, where faults can be represented as internal discontinuity surfaces with non-coincident meshes on both sides. Virtual planar cells can optionally be added to the fault walls to represent the damage

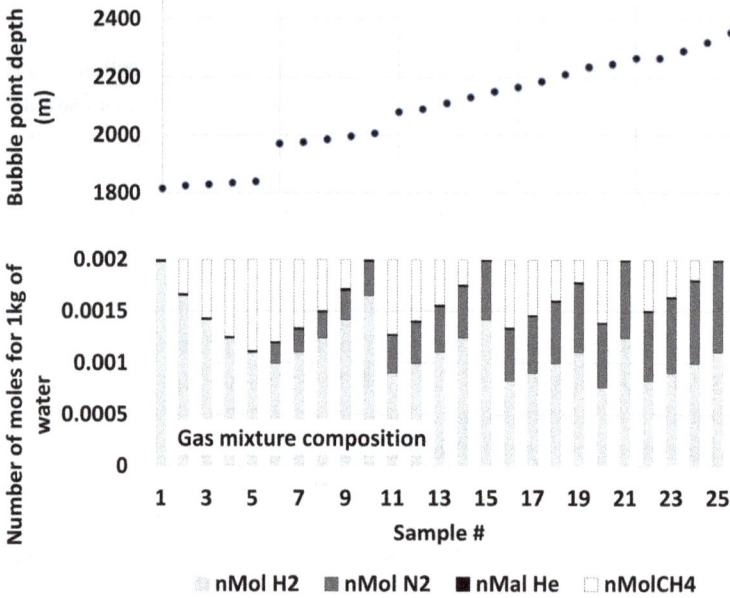

Figure 14.7: Analysis of the bubble point depth of gases mixed with pure water samples under a thermal gradient of 30 °C km^{-1} and a pressure gradient of 97 bars km^{-1}. The total molar amount of gas is constant, but the composition varies from one sample to another.

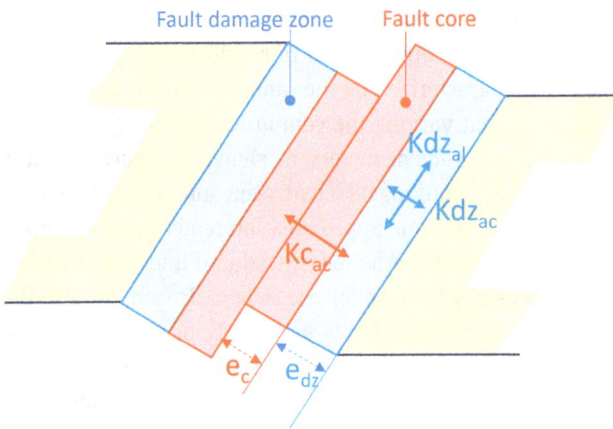

Figure 14.8: Fault discretization. *ec* and *edz* represent the fault core thickness and damage zone thickness, respectively. K_{cac} denotes the across-permeability of the core, while K_{dzal} and K_{dzac} represent the along-permeability and the across-permeability of the damage zone, respectively.

zone and core zone [53]. The damage zone cells (Figure 14.8) are characterized by their own along-fault and across-fault permeabilities and thickness, while the core zone is characterized by its thickness and across-fault permeability.

Using this fault flow modeling, faults can be represented as either drains, barriers, or neutral surfaces. More complex configurations can be modeled, such as a fault core acting as a barrier across the fault and a damage zone acting as a drain along the fault.

14.3.7 New outputs

New output variables must be considered for hydrogen exploration. These include the gas solubility threshold, gas concentration in the aqueous phase, hydrogen concentration in the free gas phase, and dissolved and free hydrogen fluxes. Indeed, if the hypothesis that dynamic accumulations can exist is confirmed, the recharge rate, also measured by the free gas flux at the accumulation inlet, becomes a key output variable along with free gas saturation and composition. Small free gas accumulations, but with high recharge rates and high hydrogen concentrations, could become new sweet spots, whereas accumulation volume and hydrocarbon composition have been the basic criteria for evaluating accumulations in O&G exploration.

14.4 Demonstration examples

In this section, we apply our basin simulator to a synthetic 2D section modified from a section of the Perth Basin (Australia) studied by [59]. This section is used here as a demonstrator, and the work presented below does not pertain to a proven hydrogen system. However, due to its complex structure, this section serves as an ideal playground to evaluate different hypotheses regarding hydrogen migration and accumulation.

Our model represents a 142-km-long and 7,000-m-deep section deposited over 150 Myrs, crosscut by a series of 23 normal faults (Figure 14.9). The heat flux at the basement/sediment interface is set to 50 mW·m^{-2}. The lithologic model includes 10 lithologies characterized by their own petrophysical properties and compaction curves. An impermeable layer was added to the original model to create a perfect seal above the fault termination in the western part of the section. The permeabilities are shown in Figure 14.9 at the present time. All expulsion saturations are set to less than 5%. The section was edited and restored [59] using the KronosFlow™ software to deliver a dynamic grid.

14.4.1 The different options

Starting from the dynamic grid, the basin simulator can be used to model an infinite range of hydrogen system scenarios, all based on different options and parameters concerning, among other factors, rock petrophysical properties, gas sources, and hydrogen alteration rates.

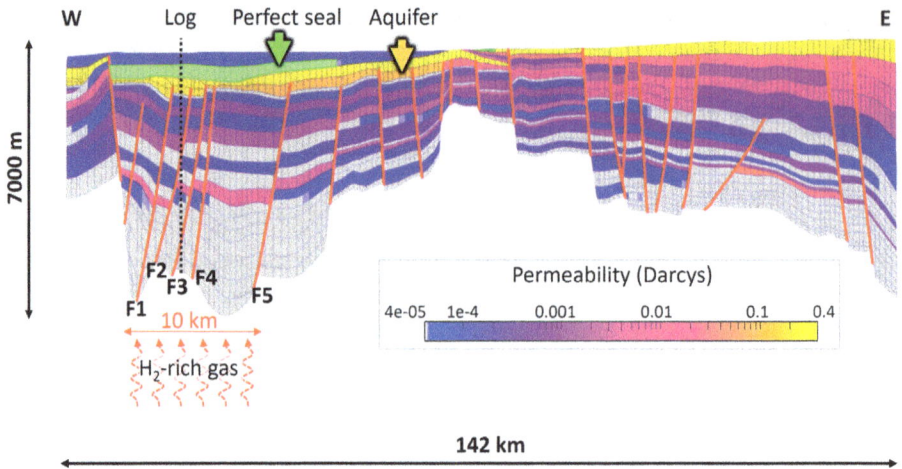

Figure 14.9: The cross section of interest.

In this work, we focus on hydrogen systems in which the gas is assumed to be delivered to the sedimentary cover through its interface with the basement, without assuming the nature of the process that generates it. We propose to analyze the sensitivity of gas migration and accumulation to the parameters of gas supply, fracture flow characteristics, and microbial degradation rate, all of which are usually very uncertain given the current state of knowledge of the deep subsurface. The different options considered in this work are listed below.

14.4.1.1 Gas supply options

Gas supply option #1, diffusive gas supply: In this option, we analyze a hydrogen system controlled by diffusion from a hydrogen "reservoir" located in the basement, just below the bottom of the sedimentary cover. We do not impose any fluid flow at the bottom of the sediments. Hydrogen delivery is assumed to start at −10 Myrs along a 10 km segment at the base of the western deep part of the model, below faults F1 to F5. It is modeled by an imposed gas concentration dissolved in water of 0.0007 kg·kg^{-1}, i.e., just below the saturation threshold, which is in the range of 0.0007 kg·kg^{-1} to 0.0009 kg·kg^{-1} at the base of the sediments in the relevant simulations. The diffusion coefficient at the reference temperature of 25 °C is set to 1.45×10^5 m^2.Myrs^{-1} according to [55], and n is set to 2 in eq. (14.23), meaning that tortuosity is assumed to be proportional to the inverse of porosity. In this option, only compaction flow and diffusion can migrate the dissolved gases.

Gas supply option #2, free gas supply: Here, we assume that free gas is supplied at the base of the model, as could occur if the gas source body consumed all the inflowing water and produced hydrogen, which was released as free gas to the sediment cover. Free gas is supplied uniformly along the 10 km segment at the base of the western deep part of the model, below faults F1–F5, as shown in Figures 14.10.6 and 14.11.6.

Gas supply option #3, hydrothermal supply enriched with dissolved gas: In this option, the gas is supplied by a hydrothermal in-flow at the base of the damage zones of faults F1–F5, as shown in Figures 14.10.1 to 14.10.5 and 14.11.1 to 14.11.5. This option represents a gas source body that would degrade only a small fraction of the in-flowing water, producing hydrogen that is dissolved and expelled to the sedimentary cover as dissolved gas. In the simulations, the hydrothermal flow is parameterized by its rate and dissolved gas concentration. Since it is often assumed that no free hydrogen migrates in the sediments at great depth, the dissolved gas concentration is set below the solubility threshold, which can change from one simulation to the next as the pressure at the bottom boundary increases with increasing injection rate. The input flow rate, although highly uncertain, is roughly constrained by the permeability of the fault, the maximum possible overpressure, and the maximum realistic flow velocity along the faults [7]. Here, two values of the hydrothermal flow rate averaged over the 10 km-long supply segment were tested, representing a low and a high hypothesis, respectively: 800 $m^3.year^{-1}.km^{-2}$ and 80,000 $m^3.year^{-1}.km^{-2}$.

14.4.1.2 Fault property options

Fault property option #1, neutral faults: according to this option, faults have no gouge and no damage zone. They are neither drains nor barriers, regardless of flow direction. They control flow only by the facies juxtaposition along the fault surface. In this option, the free gas source and the hydrothermal flow at the bottom boundary are uniformly distributed along a 10 km-long segment located at the root of faults F1–F5, as shown in Figures 14.10.4 and 14.11.4.

Fault property option #2, enhanced permeability faults: With this option, faults act as drains and do not compartmentalize the model. The fault damage zone width of all faults is set to 10 m, their along-fault permeability is set to the local facies permeability multiplied by 1.10^4, and their cross-fault permeability is set to that of the local facies. With this option, the free gas or the hydrothermal flow supplied at the bottom of the model is concentrated within the fault zones of faults F1 to F5, as shown in Figures 14.10.1 to 14.10.3, 14.10.5, 14.11.1 to 14.11.3, and 14.11.5.

14.4.1.3 Microbial alteration options

We tested two microbial degradation rates, $A_0 = 0.1\%$ $year^{-1}$ and $A_0 = 0.5\%$ $year^{-1}$. The minimum, optimum, and maximum temperatures of the thermal window of microbial activity were set at 5, 45, and 90 °C, respectively. Due to the simplifications made in the preliminary version of the calculator used in this paper, microbial degradation applies to the entire dissolved gas, not just hydrogen.

14.4.2 The studied scenarios

We ran several simulations combining a gas supply option, a fault property option, and different biodegradation rates to evaluate how these parameters control the occurrence of free gas accumulation and its economic potential.

The parameters of the simulations described in this paper are summarized in Table 14.4.

Table 14.4: Parameters used in the seven scenarios selected for this study. Average rates are calculated over the 10-km-long segment below the fault roots. *: averaged over a 10-km-long segment below faults F1–F5.

Simulation #	Averaged* gas supply (kg year^{-1} km^{-2})	Averaged* hydrothermal supply (kg year^{-1} km^{-2})	Gas supply mode	Fault flow properties	Bacterial alteration rate (% year^{-1})	Gas supply starting date
1	8.E + 00	8.E + 04	Hydrothermal	Enhanced permeability	No	−1 Myrs
2	8.E + 00	8.E + 04	Hydrothermal	Enhanced permeability	0.1	−1 Myrs
3	8.E + 00	8.E + 04	Hydrothermal	Enhanced permeability	0.5	−1 Myrs
4	8.E + 00	8.E + 04	Hydrothermal	Neutral	No	−1 Myrs
5	4.E – 02	8.E + 02	Hydrothermal	Enhanced permeability	No	−10 Myrs
6	4.E – 02	No	Free gas	Neutral	No	−10 Myrs
7	Not measured	No	Diffusive gas	Neutral	No	−30 Myrs

The calculated free gas saturation and free gas flow rate at present for Simulations 1 through 6 are shown in Figures 14.10 and 14.11, respectively. The results of Simulation #7 are not shown because no free gas appears in the simulation.

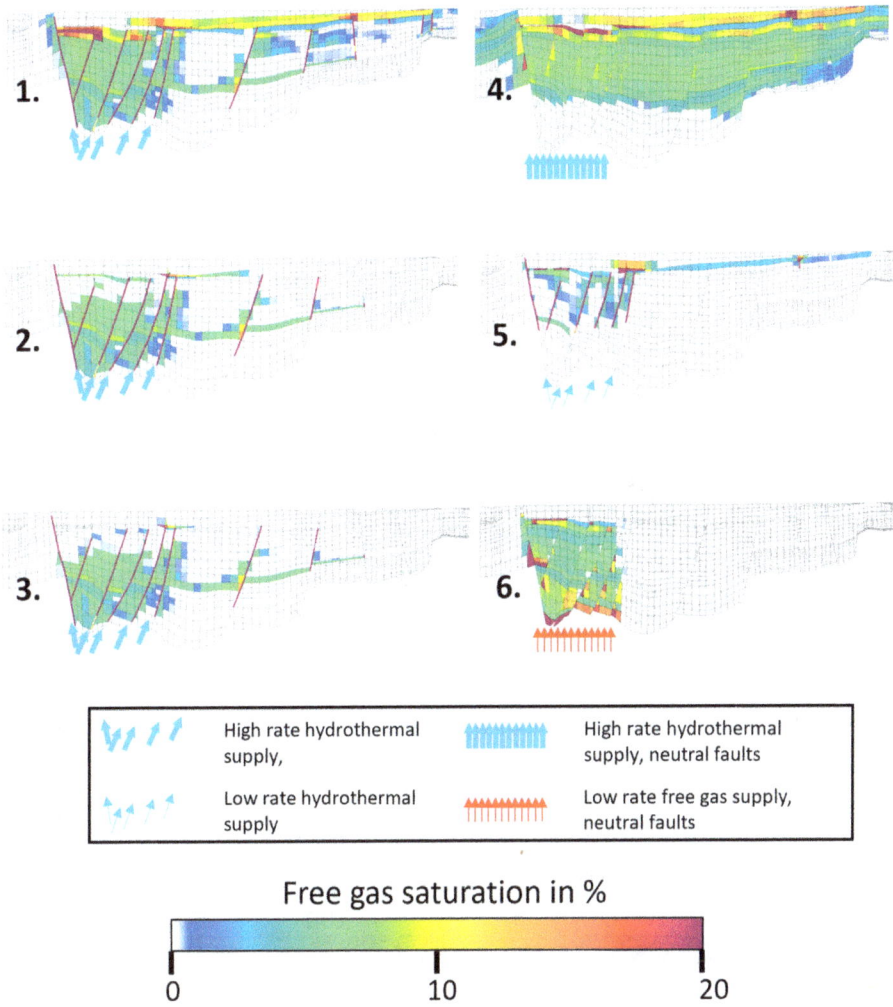

Figure 14.10: Free gas saturation calculated at present in scenarios #1 to #6 selected for this study.

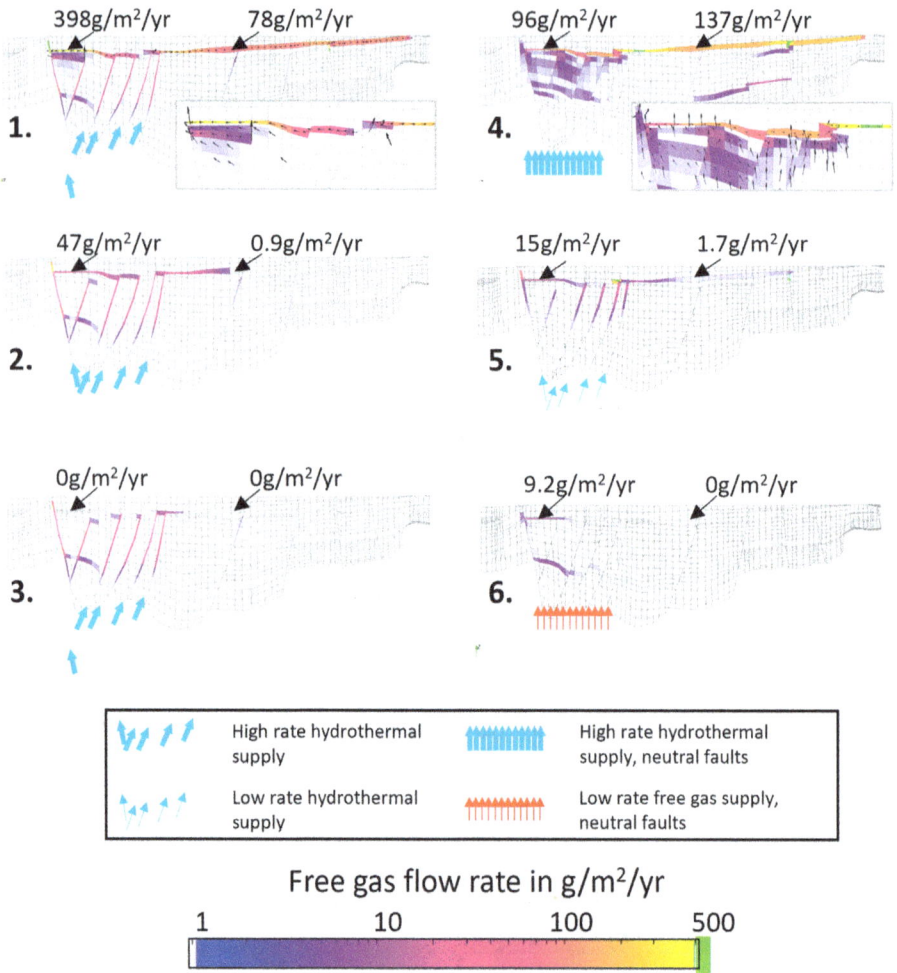

Figure 14.11: Free gas flow rate calculated at present in scenarios #1 to #6 selected for this study.

14.4.3 Simulation results

We made several observations and drew several conclusions from these results, which are listed below.

14.4.3.1 Sensitivity to supply modes

Diffusive supply: Diffusion has usually been neglected as a migration process for hydrocarbon gases such as methane [52], but it is sometimes considered a possible migration process for hydrogen because the diffusion coefficient of hydrogen is higher. For example, the diffusion coefficient of hydrogen is about three times that of methane at 1,000 m depth and 40 °C [13]. However, its solubility is about 20% lower, so the diffusive flux of dissolved hydrogen is not significantly higher than that of methane. This is confirmed by simulation #7 with neutral faults and diffusive gas injection, which shows no free gas 30 Myrs after the start of diffusive injection. The profile of the concentration is compared with that of the solubility threshold along the "Log" line (Figure 14.12). It shows that after 30 Myrs, the dissolved gas has reached the potential reservoir, but the dissolved gas concentration is still well below the saturation threshold, preventing the formation of a free gas phase.

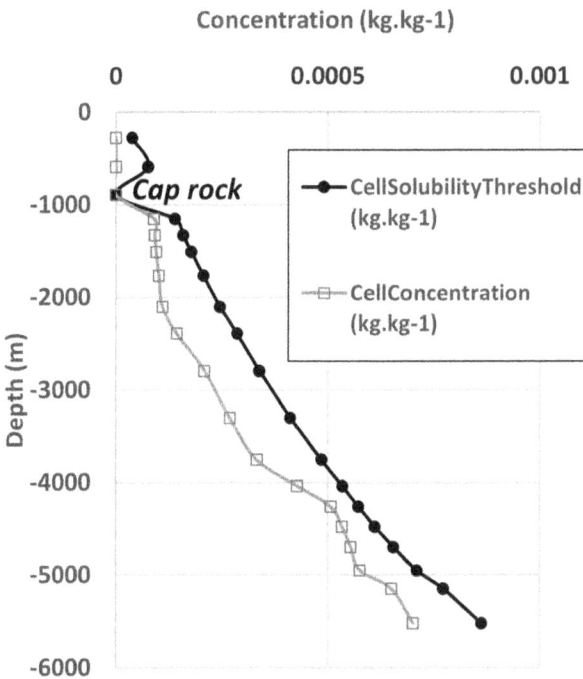

Figure 14.12: Comparison between the concentration profile and the saturation profile along the "Log" vertical line (see Fig. &&&), after 30 Myrs of diffusion from the basement.

Free gas supply versus hydrothermal supply: This comparison is addressed in Tests #5 and #6. In Figures 14.10.5 and 14.10.6, we can see that free gas occurs only in the shallow part of the section in the case of hydrothermal supply, whereas it also occurs at depth in the case of free gas supply. An analysis of the dynamics of these systems can explain this: in the case of free gas supply, immediately after the start of the supply, the free gas is dissolved upon being released into the sediments until the gas concentration reaches the solubility threshold. Gas bubbles appear and begin to rise, dissolving as they reach unsaturated water. In this process, a saturation front moves upward until it reaches the uppermost aquifers, whereas in the case of hydrothermal flow, the base of the section is short-circuited by the gas-enriched fluid. In terms of recharge potential, as measured by the free gas flow rate in the upper aquifers (Figures 14.11.5 and 14.11.6), the hydrothermal supply is also more favorable because the gas supply directly feeds the upper aquifers from the fault outlets instead of being dispersed throughout the section.

14.4.3.2 Sensitivity to hydrothermal supply rate: dynamic versus non-dynamic accumulations

This question is addressed by simulations #1 and #5. As intuition dictates, the high hydrothermal flow hypothesis produces more free gas than the low hypothesis (Figures 14.10.1 and 14.10.5). However, both simulations produce free hydrogen accumulations at the same locations. The difference between the two simulations is mainly in the free gas flow rate in the upper aquifer (Figures 14.11.1 and 14.11.5), which is on the order of 100 $g \cdot m^{-2} \cdot year^{-1}$ in the high supply case, whereas it is only a few $g \cdot m^{-2} \cdot year^{-1}$ in the low supply hypothesis. The high supply case shows a free gas recharge potential that is about 100 times larger than that of the low supply hypothesis, which is consistent with the ratio between the input gas rates of the two simulations.

14.4.3.3 Sensitivity to fault flow properties

This question is addressed in simulations #1 and #4, which have the same hydrothermal source but different fault characteristics. In the case of neutral faults (Figure 14.10.4), the overpressure induced by the inflow creates flow paths through the entire western part of the section, generating free gas bubbles in all cells above the bubble point depth. In the case of enhanced permeability faults (Figure 14.10.1), free gas is generated only near and within the hydrothermal faults and in the shallow part of the section above the hydrothermal fault terminations. The seepage velocity peaks at 100 $m \cdot year^{-1}$ at the fault roots in simulation #1, and the overpressure at the bottom of the model exceeds 500 bars in simulation #4. The enhanced permeability fault scenario produces quite different results from the neutral fault scenario in terms of recharge potential, as it generates an east-west free gas flow of over 500 $g \cdot m^{-2} \cdot year^{-1}$ along the accumulation to the west, whereas it is only a few tens of $g \cdot m^{-2} \cdot year^{-1}$ in the neutral fault case. Conversely, the free gas rate along the upper aquifer to the east of the fault outlets is significant in simulation #4, but the overpressure at the bottom of the model has reached lithostatic pressure, making this scenario quite unlikely.

14.4.3.4 Sensitivity to microbial alteration rate

Microbial alteration was applied to Scenario #1, which is most likely to generate free gas accumulations and recharge potential. At an alteration rate of 0.1% year^{-1}, free gas accumulations with limited saturation still appear below the cap rock, with a free gas flux reaching 40 g·m^{-2}·year^{-1} (Figure 14.11.2). At an alteration rate of 0.5%year^{-1}, only a small gas accumulation is generated at the outlet of Fault F4, with a free gas flux of 4 g·m^{-2}·year^{-1} (Figure 14.11.3).

14.4.4 Discussion of the demonstration example

Basin modeling is not yet a predictive tool for hydrogen systems because many parameters are highly uncertain, at least for the time being. The objective of this brief study was not only to explore the capabilities of basin modeling to simulate natural hydrogen systems but also to show how quantitative basin modeling can provide some insight into the necessary conditions for these parameters to enable hydrogen systems to produce economic gas accumulations. We have tentatively addressed this question in the context of a deep sedimentary basin supplied with hydrogen by basement hydrothermal faults.

At the beginning of our work, we designed an initial scenario in which hydrothermal flow is fed to the deep faults to the east of the section. Because the upper terminations of these faults are very close to the surface, we found that most of the gas leaks to the surface, so this scenario did not produce any accumulations. We then designed a second scenario where hydrothermal flow is fed to the faults to the west that terminate deeper, and we added a totally impermeable caprock over the fault terminations. This tends to redirect migration pathways from the fault terminations to the potential reservoir. The hydrothermal flow rate was calibrated to be very high at the bottom of the model, and the inlet hydrogen concentration was calibrated to be just below the saturation threshold, placing the bubble point depth at 6,000 m, which is deeper than usually thought. These choices make this second scenario as optimistic as possible in terms of hydrothermal gas supply rate and maximize the chances of generating a free gas accumulation. It follows that the daily gas rate imposed at fault roots is 22 g·m^{-2}·day^{-1}, which is comparable to the gas rate between 6.5 and 13.4 g·m^{-2}·day^{-1} proposed by [54] in the northern Pyrenees. The Darcy flow velocity along the faults is about 10 m·year^{-1}, which gives seepage velocities of a few tens of m·year^{-1} and a travel time along the faults of the order of a century, again consistent with residence times proposed by [7, 54]. Under these hypotheses, the model generates a dynamic accumulation with a free gas recharge rate of 400 g·m^{-2}·year^{-1} in the absence of microbial alteration. With a consumption rate of 0.1%·year^{-1}, some gas is retained in the reservoir, but the free gas flux is reduced to 47 g·m^{-2}·year^{-1}. No accumulation occurs at a degradation rate of 0.5 year^{-1}, which is about three orders of magnitude slower than the lowest consumption rate of 1%·day^{-1} proposed by [57]. However, in-situ hydrogen alteration by geochemical or microbial processes in a deep aquifer such as the one studied here remains understudied, and basin modeling in well-constrained basins could help to indirectly assess these degradation rates.

These results were obtained using a preliminary version of the functionalities of our basin simulator, ArcTem™ which is dedicated to natural hydrogen exploration. An important limitation is that gas mixture modeling is not yet available. The results shown in Figure 14.7 indicate that the bubble point depth is over-estimated in our simulations. Future developments will need to focus on more accurate modeling of fluid-rock interactions to enable quantitative predictions of hydrogen sources and sinks resulting from geochemical reactions. However, the main challenges remain in quantifying the parameters of the processes we simulate, as natural hydrogen migration at depth is still poorly understood.

14.5 Conclusion

In recent years, companies have begun to explore natural hydrogen, focusing on surface seeps and favorable geological conditions. Quantitative methods for predicting hydrogen sources and migration, similar to those used in oil and gas exploration, are still being developed. However, hydrogen migration is complex, involving advection, free gas flow, and phase exchange between water and gas. This work aims to provide some key elements for the development of numerical basin modeling for this purpose, along with an application example.

An important point is to be able to determine accurately the single-phase and two-phase zones in natural hydrogen systems, which requires the use of rigorous phase equilibrium calculations. A full description of this type of calculation is given in this work, emphasizing the role of the choice and parameterization of an appropriate equation of state. The Søreide and Whitson equation of state was chosen, as it is well-suited to modeling gas + brine phase equilibria, especially at high pressure. A complete parameterization of this equation of state is proposed in order to handle phase equilibria of (H_2 + CH_4 + N_2) + brine systems. This parametrization aims to accurately reproduce the solubilities of mixtures of these gases in brine over temperature, pressure, and salinity ranges representative of natural hydrogen systems.

The concepts to be considered in a basin simulation for natural H_2 systems are then described. In addition to phase equilibria, this simulation must include the advection of dissolved species, diffusion and dispersion of gases, hydrogen source terms, consumption by micro-organisms, and consideration of faults and fractures. The modeling of such a system is illustrated by a simplified case study of a section of the Perth Basin in Australia, realized with the ArcTem™ simulator. The simulations carried out allowed the study of different possible scenarios, depending on the hydrogen source term, its bacterial consumption rate, and the way fractures are represented. In particular, these scenarios were used to demonstrate the appropriate conditions for the formation of free hydrogen accumulations and to show that a very low bacterial alteration rate is required to allow free gas accumulation.

As a perspective, it is worth noting that, at this stage, basin modeling is not yet a predictive tool for hydrogen systems due to large uncertainties in key parameters. While the optimistic scenarios in this study show potential for free gas accumulation, significant challenges remain, including, among others, the need for more accurate modeling of fluid-rock interactions, hydrogen degradation rates, and gas mixture diffusion.

References

[1]   Zgonnik. The occurrence and geoscience of natural hydrogen: A comprehensive review. Earth-Science Reviews. 2020, 203: 103140.

[2]   Neal C., Stanger G. Hydrogen generation from mantle source rocks in Oman. Earth and Planetary Science Letters. 1983, 66: 315–320.

[3]   Marcaillou C., Munoz M., Vidal O., Parra T., Harfouche M. Mineralogical evidence for H2 degassing during serpentinization at 300 C/300 bar. Earth and Planetary Science Letters. 2011, 303(3–4): 281–290.

[4]   Truche L., Joubert J., Dargent M., Martz P., Cathelineau M., Rigaudier T., Quirte D. Clay minerals trap hydrogen in the Earth's crust: Evidence from the Cigar Lake uranium deposit, Athabasca. Earth and Planetary Science Letters. 2018, 493: 186–197.

[5]   Arrouvel C., Prinzhofer A. Genesis of natural hydrogen: New insights from thermodynamic simulations. International Journal of Hydrogen Energy. 2021, 46(36): 18780–18794.

[6]   Horsfield B., Mahlstedt N., Weniger P., Misch D., Vranjes-Wessely S., Han S., Wang C. Molecular hydrogen from organic sources in the deep Songliao Basin, PR China. International Journal of Hydrogen Energy. 2022, 47(38): 16750–16774.

[7]   Donzé F.-V., Truche L., Shekari Namin P., Lefeuvre N., Bazarkina E. F. Migration of natural hydrogen from deep-seated sources in the São Francisco Basin, Brazil. Geosciences. 2020, 10(9): 346.

[8]   Cheng A., Sherwood Lollar B., Gluyas J. G., Ballentine C. J. Primary N2–He gas field formation in intracratonic sedimentary basins. Nature. 2023, 615(7950): 94–99.

[9]   Lodhia B. H., Peeters L., Frery E. A review of the migration of hydrogen from the planetary to basin scale. Journal of Geophysical Research: Solid Earth. 2024, 129(6): e2024JB028715.

[10] Maiga O., Deville E., Laval J., Prinzhofer A., Diallo A. B. Trapping processes of large volumes of natural hydrogen in the subsurface: The emblematic case of the Bourakebougou H2 field in Mali. International Journal of Hydrogen Energy. 2024, 50: 640–647.

[11] Maiga O., Deville E., Laval J., Prinzhofer A., Diallo A. B. Characterization of the spontaneously recharging natural hydrogen reservoirs of Bourakebougou in Mali. Scientific Reports. 2023, 13(1): 11876.

[12] Prinzhofer A., Cissé C. S. T., Diallo A. B. Discovery of a large accumulation of natural hydrogen in Bourakebougou (Mali). International Journal of Hydrogen Energy. 2018, 43(42): 19315–19326.

[13] Prinzhofer A., Cacas-Stentz M. C. Natural hydrogen and blend gas: A dynamic model of accumulation. International Journal of Hydrogen Energy. 2023, 48(57): 21610–21623.

[14] Hidalgo J. C., Kauerauf A., Van Wijngaarden M. D. L., Torres C. S. Application of basin modeling technologies in the exploration of natural hydrogen systems. Proceedings of 85th EAGE Annual Conference & Exhibition. 2024, 2024(1): 1–5.

[15] Cacas-Stentz M., Braconnier B., Patacchini F. S., Lemgruber-Traby A., Willien F., Wolf S. Modeling basin-scale fault flow with unstructured meshes for hydrogen migration and lithium solute transport. Proceedings of 85th EAGE Annual Conference & Exhibition. 2024, 2024(1): 1–5.

[16] Mollerup J. M., Michelsen M. L. Calculation of thermodynamic equilibrium properties. Fluid Phase Equilibria. 1992, 74(1): 1–15.

[17] Duan Z. H., Moller N., Greenberg J., Weare J. H. The prediction of methane solubility in natural-waters to high ionic-strength from O-Degrees-C to 250-Degrees-C and from 0 to 1600 bar. Geochimica Et Cosmochimica Acta. 1992, 56(4): 1451–1460.

[18] Duan Z. H., Sun R. An improved model calculating CO2 solubility in pure water and aqueous NaCl solutions from 273 to 533 K and from 0 to 2000 bar. Chemical Geology. 2003, 193(3–4): 257–271.

[19] Duan Z. H., Mao S. D. A thermodynamic model for calculating methane solubility, density and gas phase composition of methane-bearing aqueous fluids from 273 to 523 K and from 1 to 2000 bar. Geochimica Et Cosmochimica Acta. 2006, 70(13): 3369–3386.

[20] Rumpf B., Xia J., Maurer G. Solubility of Carbon Dioxide in aqueous solutions containing acetic acid or sodium hydroxide in the temperature range from 313 to 433 K and at total pressures up to 10 MPa. Industrial & Engineering Chemistry Research. 1998, 37(5): 2012–2019.

[21] Xia J., Pérez-Salado Kamps Á., Rumpf B., Maurer G. Solubility of Hydrogen Sulfide in Aqueous Solutions of the single salts Sodium Sulfate, Ammonium Sulfate, Sodium Chloride, and Ammonium Chloride at Temperatures from 313 to 393 K and total pressures up to 10 MPa. Industrial & Engineering Chemistry Research. 2000, 39(4): 1064–1073.

[22] Zuo Y. X., Guo T. M. Extension of the Patel-Teja equation of state to the prediction of the solubility of natural gas in formation water. Chemical Engineering Science. 1991, 46(12): 3251–3258.

[23] Sorensen H., Pedersen K. S., Christensen P. L. Modeling of gas solubility in brines. Organic Geochemistry. 2002, 33: 635–642.

[24] Patel B. H., Paricaud P., Galindo A., Maitland G. C. Prediction of the salting-out effect of strong Electrolytes on Water + Alkane solutions. Industrial & Engineering Chemistry Research. 2003, 42(16): 3809–3823.

[25] Duan Z., Moller N., Weare J. H. Equations of State for the NaCl-H_2O-CH_4 system and the NaCl-H_2O-CO_2-CH_4 system: Phase Equilibria and volumetric properties above 573 K. Geochimica Et Cosmochimica Acta. 2003, 67: 671–680.

[26] Li J., Wei L., Li X. An improved cubic model for the mutual solubilities of CO_2–CH_4–H_2S–brine systems to high temperature, pressure and salinity. Applied Geochemistry. 2015, 54: 1–12.

[27] Courtial X., Ferrando N., de Hemptinne J.-C., Mougin P. Electrolyte CPA equation of state for very high temperature and pressure reservoir and basin applications. Geochimica Et Cosmochimica Acta. 2014, 142: 1–14.

[28] Ahmed S., Ferrando N., de Hemptinne J.-C., Simonin J.-P., Bernard O., Baudouin O. Modeling of mixed-solvent electrolyte systems. Fluid Phase Equilibria. 2018, 459: 138–157.

[29] Rozmus J., Hemptinne J. C. D., Galindo A., Dufal S., Mougin P. Modeling of strong electrolytes with ePPC-SAFT up to high temperatures. Industrial & Engineering Chemistry Research. 2013, 52: 9979–9994.

[30] Li D., Beyer C., Bauer S. A unified phase equilibrium model for hydrogen solubility and solution density. International Journal of Hydrogen Energy. 2018, 43(1): 512–529.

[31] Zhu Z., Cao Y., Zheng Z., Chen D. An accurate model for estimating H_2 solubility in pure water and aqueous NaCl solutions. Energies. 2022, 15(14).

[32] Torín-Ollarves G. A., Trusler J. M. Solubility of hydrogen in sodium chloride brine at high pressures. Fluid Phase Equilibria. 2021, 539: 113025.

[33] Lopez-Lazaro C., Bachaud P., Moretti I., Ferrando N. Predicting the phase behavior of hydrogen in NaCl brines by molecular simulation for geological applications. BSGF – Earth Sciences Bulletin. 2019, 190.

[34]   Chabab S., Théveneau P., Coquelet C., Corvisier J., Paricaud P. Measurements and predictive models of high-pressure H_2 solubility in brine (H_2O+ NaCl) for underground hydrogen storage application. International Journal of Hydrogen Energy. 2020, 45(56): 32206–32220.

[35]   Roa Pinto J. S., Bachaud P., Fargetton T., Ferrando N., Jeannin L., Louvet F. Modeling phase equilibrium of hydrogen and natural gas in brines: Application to storage in salt caverns. International Journal of Hydrogen Energy. 2021, 46(5): 4229–4240.

[36]   Kiemde A., Ferrando N., de Hemptinne J. C., Le Gallo Y., Réveillère A., Roa Pinto J. S. Hydrogen and air storage in salt caverns: A thermodynamic model for phase equilibrium calculations. Science and Technology for Energy Transition. 2023, 78(10): 1–16.

[37]   Soreide I., Whitson C. Peng-Robinson predictions for hydrocarbons, CO_2, N_2, and H_2S with pure water and NaCl brine. Fluid Phase Equilibria. 1992, 77: 217–240.

[38]   Chabab S., Ahmadi P., Théveneau P., Coquelet C., Chapoy A., Corvisier J., et al. Measurements and modeling of high-pressure O_2 and CO_2 solubility in Brine (H_2O + NaCl) between 303 and 373 K and pressures up to 36 MPa. Journal of Chemical & Engineering Data. 2021, 66(1): 609–620.

[39]   Pitzer K. S. Thermodynamics of Electrolytes I: Theoretical basis and general equations. Journal of Physical Chemistry. 1973, 77(2): 268–277.

[40]   Peng D. Y., Robinson D. B. A new two-constant equation of state: Industrial & engineering chemistry fundamentals. Industrial & Engineering Chemistry Fundamentals. 1976, 15(1): 59–64.

[41]   De Hemptinne J.-C., Ferrando N., Hajiw-Riberaud M., Lachet V., Maghsoodloo S., Mougin P., et al. Carnot: A thermodynamic library for energy industries. Science and Technology for Energy Transition. 2023, 78.

[42]   O'Sullivan T. D., Smith N. O. The solubility and partial molar volume of Nitrogen and methane in water and in aqueous sodium chloride from 50 to 125øC and 100 to 600 Atm. Journal of Physical Chemistry. 1970, 74(7): 1460–1466.

[43]   Krause D., Benson B. B. The solubility and isotopic fractionation of gases in dilute aqueous solution. IIa. solubilities of the noble gases. Journal of Solution Chemistry. 1989, 18(9): 823–873.

[44]   Potter R. W., Clynne M. A. The solubility of the noble gases He, Ne, Ar, Kr, and Xe in water up to the critical point. Journal of Solution Chemistry. 1978, 7(11): 837–844.

[45]   Benson B. B., Krause D. Empirical laws for dilute aqueous solutions of nonpolar gases. Journal of Chemical Physics. 1976, 64(2): 689–709.

[46]   Weiss R. F. Solubility of helium and neon in water and seawater. Journal of Chemical & Engineering Data. 1971, 16(2): 235–241.

[47]   Wiebe R., Gaddy V. L. The solubility of Helium in Water at 0, 25, 50 and 75° and at pressures to 1000 atmospheres1. Journal of the American Chemical Society. 1935, 57(5): 847–851.

[48]   Gardiner G. E., Smith N. O. Solubility and partial molar properties of helium in water and aqueous sodium chloride from 25 to 100.deg. and 100 to 600 atmospheres. Journal of Physical Chemistry. 1972, 76(8): 1195–1202.

[49]   Gerth W. A. Effects of dissolved electrolytes on the solubility and partial molar volume of helium in water from 50 to 400 atmospheres at 25 °C. Journal of Solution Chemistry. 1983, 12(9): 655–669.

[50]   Gamsjäger H., Lorimer J. W., Salomon M., Shaw D. G., Tomkins R. P. T. The IUPAC-NIST solubility data series: A guide to preparation and use of compilations and evaluations. Journal of Physical and Chemical Reference Data. 2010, 39(2).

[51]   Feillolay A., Lucas M. Solubility of helium and methane in aqueous tetrabutylammonium bromide solutions at 25 and 35.deg. Journal of Physical Chemistry. 1972, 76(21): 3068–3072.

[52]   Hantschel T., Kauerauf A. I. Fundamentals of basin and petroleum systems modeling. Springer Science & Business Media, 2009.

[53]   Faille I., Thibaut M., Cacas M. C., Havé P., Willien F., Wolf S., Agelas L., Pegaz-Fiornet S. Modeling fluid flow in faulted basins. Oil & Gas Science and Technology – Revue d'IFP Energies Nouvelles. 2014, 69(4): 529–553.

[54] Lefeuvre N., Truche L., Donzé F. V., Gal F., Tremosa J., Fakoury R. A., . . . Gaucher E. C. Natural hydrogen migration along thrust faults in foothill basins: The north Pyrenean frontal thrust case study. Applied Geochemistry. 2022, 145: 105396.

[55] Cussler E. L. Diffusion: Mass transfer in fluid systems. Cambridge university press, 2009.

[56] Lala A. M. A novel model for reservoir rock tortuosity estimation. Journal of Petroleum Science and Engineering. 2020, 192: 107321.

[57] Harris S. H., Smith R. L., Suflita J. M. In situ hydrogen consumption kinetics as an indicator of subsurface microbial activity. FEMS Microbiology Ecology. 2007, 60(2): 220–228.

[58] Rosso L., Lobry J. R., Flandrois J. P. An unexpected correlation between cardinal temperatures of microbial growth highlighted by a new model. Journal of Theoretical Biology. 1993, 162(4): 447–463.

[59] Frery E., Callies M., Giboreau R., Pagès A., Thomas C. Fault impact on hydrocarbon migration-2D complex modelling of the north Perth basin petroleum systems, Australia. Proceedings of 79th EAGE Conference and Exhibition. 2017, 2017(1): 1–5.

Part IV: **Global case studies and regional insights**

Corinne Arrouvel

Chapter 15
Natural hydrogen exploration in Brazil: from theory to fieldwork case studies

Abstract: Over the last few years, Brazil has become a promising country for natural hydrogen exploration, with seepage zones observed mainly in the São Francisco Basin and the state of Rio de Janeiro. While circular depressions have been associated with hydrogen leakage and studied in terms of long-term monitoring, other types of distinctive seepages are apparent as suggested by different private research and academic programs, using other methods. As a mineral-rich country, Brazil is today witnessing the discovery of an increasing number of sites with high hydrogen levels across several states, including Roraima, Tocantins, Ceará, Minas Gerais, Bahia, Goiás, Rio de Janeiro, Espírito Santo, São Paulo, Santa Catarina, Rio Grande do Sul, Maranhão, and Piauí. This chapter discusses several case studies and their potential for exploration of pilot programs, especially as new legal frameworks are being developed in Brazil to regulate hydrogen exploration under oil, gas, and mining licenses. For this purpose, a combination of theoretical techniques, data analyses from geological maps, and fieldwork measurements are suggested for the regions with the highest reported levels of hydrogen.

To date, the states of Rio de Janeiro and Espírito Santo have subsurface values exceeding 1% at depths of less than 1 m using a hydrogen detector. The regions with the highest hydrogen concentrations have often been reported near faults and fractures, with associated gamma radiation and magnetic anomalies. High iron content in rocks, particularly in cratonic areas and ophiolites, is believed to favor hydrogen generation, although other geological formations such as alkaline magmatic complexes have also shown high hydrogen concentrations. Using mineral identification and thermodynamic modeling, hydrogen generation can be simulated under varying pressure and temperature conditions. The formation of hydrogen in the Earth's crust is increasingly well understood, with theoretical models implicating redox reactions and radiolysis, both of which involve water. To identify potential reservoirs, other techniques must be implemented to complete the preliminary fieldwork such as long-term monitoring with semiconductor sensors, gravimetry, magnetometry, magnetotelluric, seismic studies, resistivity measurements, quantitative and qualitative mineral and soil substrate descriptions, as well as gas composition analysis including helium and isotope analyses.

Keywords: hydrogen exploration, Brazil, faults, hot spots, thermodynamic modeling

Corinne Arrouvel, UFRJ – IMQ/CM Campus Macaé, Rio de Janeiro, Brazil; Master Program PPGERE at LENEP-UENF, Macaé, Rio de Janeiro, Brazil; UFSCar – DFQM/CCTS Campus Sorocaba, São Paulo, Brazil, e-mail: corinne.arrouvel@imq.macae.ufrj.br

https://doi.org/10.1515/9783111437040-015

15.1 Introduction

With the global need to decarbonize the atmosphere, many countries are increasingly investing in hydrogen as a key energy source for integration into their energy grids. The transportation sector is also undergoing a transformation, incorporating hydrogen fuel into both private and public vehicles. However, for hydrogen gas to function as a truly clean primary energy source rather than merely an energy vector, it must exist in significant natural deposits.

Brazil's energy grid is predominantly powered by renewable sources such as hydroelectric, wind, and solar energy, with the latter two accounting for approximately 33% of the total grid capacity. Despite this green energy mix, 95% of the hydrogen currently produced in Brazil falls under the 'gray' category, which is derived from fossil fuels without carbon capture. To accelerate the transition to a low-carbon economy and capitalize on emissions reduction incentives, the oil industry is expanding investments in 'blue' hydrogen – produced via carbon capture and storage (CCS) – and 'turquoise' hydrogen, generated through methane pyrolysis. The latter is particularly promising due to its claimed zero-carbon emissions, as it does not release CO_2, and its residual solid carbon can be either stored in the soil or repurposed for other industrial applications.

The hydrogen market was projected to reach \$231 billion in 2024, with estimates rising to \$1.657 trillion by 2050 [1]. However, these projections primarily consider hydrogen produced via green and blue synthesis pathways, largely overlooking the potential of natural hydrogen. Recognizing this gap, researchers and industry stakeholders have intensified efforts to explore naturally occurring hydrogen reservoirs in Brazil.

Investigations into hydrogen seepages in Brazil began before 2019 when preliminary studies were conducted in four states – Ceará, Roraima, Tocantins, and Minas Gerais – by Prinzhofer et al. (GEO4U, private communication) and Engie [2]. The urgency of global crises, including the COVID-19 pandemic and the Ukraine-Russia conflict, prompted further research into hydrogen generation mechanisms within the continental crust. In 2019, Arrouvel and Prinzhofer [3] collaborated on thermodynamic simulations to better understand these processes. In 2022, the discovery of a hydrogen system in Maricá was announced, though the study's findings remained unpublished until a recent report [4]. More recently, in March 2024, Petrobrás formally launched its natural hydrogen research program, which commenced in 2023 in the state of Bahia. The company plans to drill its first exploratory well by 2029.

To assess Brazil's global potential for natural hydrogen production, new exploration approaches and reservoir optimization mechanisms are being developed. This chapter presents a combination of theoretical insights and experimental characterizations of potential hydrogen reservoirs in Brazil, incorporating ongoing fieldwork initiated in September 2023 across the states of Rio de Janeiro, São Paulo, and Espírito Santo. Preliminary data from these regions indicate hydrogen concentrations surpassing those previously recorded in subsurface studies (to a depth of 1 m), highlighting Brazil's potential as a significant player in the emerging natural hydrogen market.

15.2 Natural hydrogen exploration in Brazil

Most of the fundamental research in Brazil has been done by Prinzhofer and coworkers [2–5], mainly focused on the São Francisco Basin in the state of Minas Gerais. The article from Serratt et al. [6] reviewed the main available data in Minas Gerais, Ceará, Rio de Janeiro, Rio Grande do Sul, and Santa Catarina states. Those data are reported in Table 15.1. Other states are under investigation by companies and academics but no data are available for confidential reasons (e.g., Roraima, Bahia, Tocantins Maranhão, and Piauí) or in progress (e.g., Archipelago São Pedro and São Paulo in the state of Pernambuco). New investigations were conducted by the author in the states of São Paulo, Espírito Santo, and Rio de Janeiro (see Section 15.4).

Table 15.1: Main published H_2 occurrences in Brazil.

State	H_2 concentrations	Depth	References
Rio Grande do Sul	8.8%/1.5%	2.3 km/1.6 km	Serratt et al. [6]
Santa Catarina	1.04%	2.1 km	Serratt et al. [6]
Santa Catarina	2.43%	2.0 km	Serratt et al. [6]
Minas Gerais	1,150 ppm	80 cm	Prinzhofer et al. [2]
Maricá	2,777 ppm	80 cm	Prinzhofer et al. [4]
Minas Gerais	41.1%	860–912 m	de Freitas et al. [5]

It is important to note that, to date, no standard method exists for conducting consistent comparative studies in the literature, and some results may lead to misinterpretations if not carefully analyzed. For instance, it is widely accepted that hydrogen concentrations tend to increase with deeper drilling. Therefore, more reliable comparative studies can be achieved by using consistent field methodologies and diverse sampling approaches to identify optimal hydrogen sites.

Traditionally, hydrogen seepage zones were explored in cratonic areas, with satellite imagery aiding in locating 'fairy circles,' now more commonly referred to as circular depressions. Subsurface hydrogen data are usually measured down to 1 m.

Serratt et al. [6] reported data from wells at depths that cannot be directly compared to subsurface studies. The states of Santa Catarina and Rio Grande do Sul have the highest reported values, ranging from 1.04% to 8.8% at depths to 2 km in the Paraná intracratonic basin, though no subsurface measurements are reported for the wells. Various hypotheses have been proposed by the authors, including radiolysis from the granitic basement, biotite transformation, or kimberlite intrusions. Nevertheless, the association of layers of coal and an inverse trend in helium concentration versus dihydrogen concentration has not been explained by the authors [6]. The deep origin of hydrogen remains a possibility and its diffusion would be through faults. The highest recorded value in Brazil is by de Freitas et al. [5] with concentrations up to 41.1% at 860–912 m depth in the state of Minas Gerais. Those recent reports bring new hope in reactivating abandoned wells and in planning drilling test wells in productive regions.

An interesting observation from rare long-term monitoring studies is the consistent fluctuation of H_2 concentrations, with higher peaks at $x = 0.336$ (8 h), $x = 0.5$ (12 h), and $x = 1.002$ (24 h) [7] in a cratonic environment (x being the unit period per day). These pulses, with daily periodic emissions, have been observed by researchers [8, 9]. The graph in Figure 15.1 was simulated using meteorological data from the atmospheric pressure station in Montes Claros, Brazil, in 2018, near a studied fairy circle (Figure 15.2). Another peak occurs at 6 h, which could correspond to $x = 0.25$ in Moretti et al.'s graph [7], but due to background interference, this possibility cannot be confirmed. The main pulse is likely caused by the daily temperature variation, affecting the release of gases due to pressure changes, while the other peaks might represent resonance modes within the reservoir. These modes would be independent of the layer's thickness and with an analogy being sound wave propagation in a tube with resonant frequencies: the fundamental at $x = 1$, the first harmonic at $x = 1/2$, and subsequent harmonics at $x = 2/3$, $1/3$, and $1/4$. However, this phenomenon of resonance has not been observed by other authors [8] and might be contested.

Long-term monitoring data from other sources is not yet available due to the rarity of expertise and equipment worldwide; currently, only prototypes are used to detect hydrogen in the subsurface. Data collection would require at least 6 months of analysis, and possibly up to a year, to account for seasonal changes and to use a short

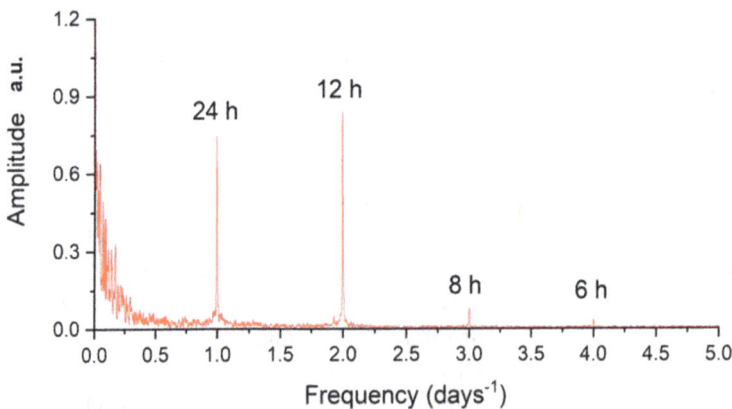

Figure 15.1: Fourier transform of pressure tides at Montes Claros city data collected in 2018 (courtesy of Luiz Fernando Roncaratti Junior (UnB, Brazil).

enough time step to collect accurate data. Temperature, humidity, rainfall, and tectonic activities (e.g., active faults and seismic events) directly affect the measurements and need to be correlated with H_2 emissions. This preliminary observation can be easily verified once a long-term monitoring program is launched in Brazil and other countries. Barometric changes are also known to be correlated with CO_2 fluctuations in the soil. Semi-daily and daily periods have been correlated to solar and lunar tides in relation to soil temperature, soil moisture, and atmospheric pressure [10, 11]. However, the factors influencing the proportions of the main gases – H_2, CO_2, CH_4, N_2, and He – are not yet fully understood. Various physical properties of the soil, such as particle size, permeability, and porosity, influence gas fluxes and their separation.

The link between faults and higher H_2 concentrations has been evidenced in Maricá [4] with gamma spectroscopy measurements. Gamma spectroscopy is becoming a powerful tool in search for H_2 reservoirs as it not only helps to localize faults but also helps to interpret the subsurface geology and estimate the radiogenic heat production in geothermal sites. The following section presents the methodology used in H_2 prospecting in which gamma spectroscopy of thorium peaks becomes a key proxy for the preliminary mapping of sites of interest.

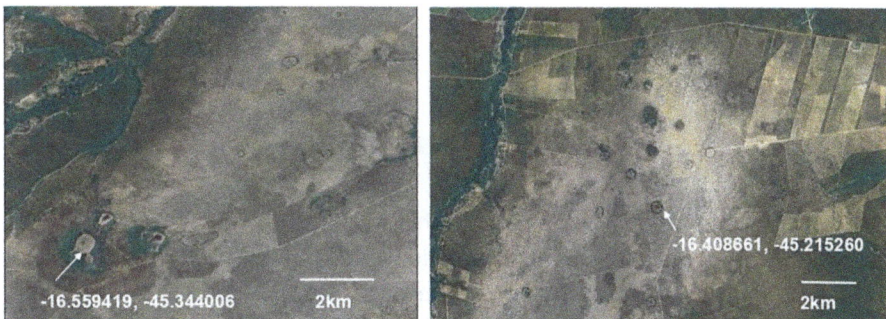

Figure 15.2: Satellite images of (a) fairy circles studied (2019) [2] at GPS – location 16.559419, −45.344006 and (b) similar structures nearby at −16.408661, −45.215260 (©2025 Google).

15.3 Methodology

To map the best areas for possible exploration, theoretical approaches are combined with fieldwork. Some traditional hints are first sought in localizing high levels of iron-based minerals (including oxides and sulfides) that can be associated with serpentinization and pyritization reactions. Some of the reactions have been theoretically modeled in a previous study [3] and in the field while identifying fairy circles (circular depressions associated with inhibited plant growth linked to H_2 emissions), such as the one studied in the state of Minas Gerais. The estimated average flux at this location is 385 m^3/day/km^2 [2, 7].

The search for proxies has been extended to other indices such as areas with higher radioactivity, magnetic anomalies, and faults. Four main regions have been prospected in Brazil within the FAPESP project: (1) Ipanema National Forest reserve (Floresta Nacional de Ipanema) – in the state of São Paulo; (2) Lakes region (Região dos Lagos) – in the state of Rio de Janeiro; (3) Vassouras – in the state of Rio de Janeiro; and (4) South of Espírito Santo state.

The theoretical approach is divided into two main tasks: (1) identifying proxies before the fieldwork and (2) proposing mechanisms for hydrogen formation and diffusion in the selected areas after the fieldwork. Another objective is to draw analogies between heterogeneous catalysis involving dehydrogenation reactions and the kinetics of geological hydrogen formation. It is well-known that iron, nickel, chromium, and manganese play a role in hydrogen formation. Recent discoveries have confirmed that a chromite mine in Albania is one of the largest hydrogen reservoirs found to date [12]. Unexpected sources of hydrogen are being discovered in nontraditional areas, often by chance, which underscores the need for a better understanding of the hydrogen system through a multidisciplinary approach.

15.3.1 Theoretical approach

15.3.1.1 Indices

Various proxies can be deduced from published studies to identify seepage (see [13]). The ongoing research program and exploration of natural hydrogen near the village of Bourakebougou in Mali, by Hydroma company in collaboration with GEO4U and IFP Energies Nouvelles, is a unique case study in which we can obtain important data that can be correlated to the present study (e.g., cratonic area and lithology). Maiga et al. [14] have shown that H$_2$ accumulation in a porous reservoir (e.g., carbonate and sandstone) and the necessity to have an impermeable layer of rocks (e.g., dolerite). From the Mali case, an increasing number of research programs are based on those data to prospect for new reservoirs around the world. Today, it is becoming increasingly accepted that other mechanisms and indices for hydrogen formation can be used to optimize H$_2$ exploration. For example, radiolysis, mantle rock alterations, obduction, and subductions are processes in vogue. Radiolytic processes imply a side interest as they indicate helium emission from rich uranium/thorium rocks (e.g., granite batholiths) and sediment deposits. Some uranium deposits are known in Brazil, in Minas Gerais state and south of Bahia state (in Lagoa Real-Caetite district), and it is the second country with the largest thorium reserves. Different types of deposits have been identified in many places, which could be good indicators for H$_2$ seepages. Localizing ophiolites is also a key indicator for the search for reservoirs and active subduction zones, as well as geothermal activities. While Brazil does not have known subduction zones, active faults and small earthquakes are occurrences of interest. Regarding geothermal activities, various locations have been summarized [15, 16], compiling data from the Geothermal Laboratory of the National Observatory, extending down to 3 km in depth. Paraná and Piauí are promising targets for geothermal exploration, and complementary techniques such as magnetometry, magnetotelluric, and gamma-ray spectrometry studies can help identify fractures and magma intrusions. Based on this complementary data, states of Minas Gerais and Pernambuco [17] are emerging as targets for geothermal and, consequently, hydrogen prospection. For gamma spectrometry surveying in geothermal zones, technical guidelines [18] can be applied to hydrogen surveying.

Another key proxy for hydrogen prospecting is to locate banded iron formation (BIF), which might be also referred to as itabirite in Brazil [19]. Itabira is a city in the state of Minas Gerais, known for exploring and exporting iron, and it has a high potential for H_2 formation and accumulation in sediment layers, localized at the southern São Francisco craton, in the Caue formation aged 2.1–2.6 Ga. Other BIF formations have been identified: the Caldeirão belt (aged 2.1–2.76 Ga), Carajás Mineral Province in Pará state (2.7 Ga, an Algoma oldest BIF type), and Jacadigo in Mato Grosso do Sul state (600 Ma, considered a young formation). The highest iron reserve is attributed to Cauê formation. We note that other BIFs might not have been related [19] in Brazil.

Table 15.2 summarizes the key indicators for hydrogen prospecting in Brazil. Additionally, a generalized geological map of South America (1:5,000,000 scale) is available through ArcGIS Online [20]. Future studies may integrate artificial intelligence (AI) to analyze spatial patterns of hydrogen seepages using satellite imagery and pair correlation functions [21].

Table 15.2: Key indicators for hydrogen prospecting in Brazil.

Category	Indicators
Formation	– Ophiolites, cratonic regions and banded iron formations (BIF)/itabirite – Basic/alkaline rocks, acidic areas, gabbro, and mafic/ultramafic rocks – Iron-rich minerals: magnetite, hematite, goethite, martite, biotite, ilmenite, limonite, amphibolite, chlorite, serpentine, pyroxene, pyrite, marcasite, arsenopyrite, siderite, and chromite – Thermal anomalies (hot spots) – Thorium-rich minerals (monazite and thorite)
Accumulation	– Porous reservoir rocks: limestone, dolomite, and sandstone – Sealing formations: schist, dolerite, salt, breccia, shale, laterite, and clays
Diffusion	– Circular ground depressions (*fairy circles*) – Active faults (e.g., shear zones) and fractures – Seismic zones

15.3.1.2 Thermodynamic simulations

The SUPCRTBL [22] program has been used to evaluate the direction of reactions under varying temperature and pressure conditions. The thermodynamic conditions are modeled for the gas phase of H_2 as it is released in the crust, considering its low solubility in water and accounting for both reductive and oxidative conditions. This method is the same as the one used before [3]. A list of complementary reactions is proposed, including the role of iron-based minerals in CO_2 sequestration forming hydrogen. Indeed, the coexistence of gases is still not well understood, and their proportions might vary as a function of the dynamic of the systems, the type of rocks, and biological activities. In the proposed reactions, helium and nitrogen will be passed over as the radiolysis and nitrogen database is limited. The first attempts to model nitrogen and hydrogen formations from ammonium in the fayalite-magnetite-quartz buffer and with the hematite-magnetite buffer have been done [23]. However, the lack of experimental data is a barrier to the improvement of the thermodynamic database. The presence of the other co-generated gases such as CH_4 and CO_2 will be linked to carbonate rock formation and decomposition through abiotic processes. Such reactions are also studied for H_2 storage and CCS. The main reactions are listed in Table 15.3.

Table 15.3: Reactions and thermodynamic data from SUPCRTBL, at 1 bar and 25 °C.

Reactions (RT ln $Q_r = 0$) 1 bar, 25 °C	$\Delta G_r°$ (kJ/ mol^{-1})	$\Delta H_r°$ (kJ/ mol^{-1})
$2\,Fe_3O_4(\text{magnetite}) + H_2O(l) \rightarrow 3Fe_2O_3(\text{hematite}) + H_2(g)$	29.252	37.907
$2\,Fe_3O_4(\text{magnetite}) + 4H_2O(l) \rightarrow 6\,FeOOH(\text{goethite}) + H_2(g)$	25.959	1.633
$2FeOOH(\text{goethite}) + 4H_2S(g) \rightarrow 2FeS_2(\text{pyrite}) + 4H_2O(l) + H_2(g)$	−154.575	−281.849
$Fe_2O_3(\text{hematite}) + 4H_2S(g) \rightarrow 2FeS_2 + 3H_2O(l) + H_2(g)$	−155.672	−293.940
$Fe_3O_4(\text{magnetite}) + 6H_2S(g) \rightarrow 3FeS_2 + 4H_2O(l) + 2H_2(g)$	−218.883	−421.956
$FeS(\text{pyrrhotite}) + H_2S(g) \rightarrow FeS_2 + H_2(g)$	−26.179	−52.309
$6Fe_2SiO_4(\text{fayalite}) + 7H_2O(l) \rightarrow Fe_3O_4(\text{magnetite}) +$ $3Fe_3Si_2O_9H_4(\text{greenalite}) + H_2(g)$	−83.858	−140.146
$3Fe_2SiO_4(\text{fayalite}) + 2H_2O(l) \rightarrow 2Fe_3O_4(\text{magnetite}) + 3SiO_2(\text{quartz}) + 2H_2(g)$	18.005	43.776
$2FeO(\text{ferropericlase}) + H_2O(l) \rightarrow Fe_3O_4 + H_2(g)$	−20.428	−12.762
$FeO(\text{ferropericlase}) + 2H_2S(g) \rightarrow FeS_2 + H_2(g) + H_2O(l)$	−79.770	−144.906
$3\,FeTiO_3(\text{ilmenite}) + H_2O(l) \rightarrow Fe_3O_4 + 3\,TiO_2 + H_2(g)$	20.402	29.508
$Fe_2Si_2O_6(\text{ferrosilite}) + H_2O(l) \rightarrow Fe_2O_3 + SiO_2 + H_2(g)$	15.282	27.508
$Fe_2Si_2O_6(\text{ferrosilite}) \rightarrow SiO_2(\text{quartz}) + Fe_2SiO_4(\text{fayalite})$	−0.470	0.280
$Fe_7Si_8O_{24}H_2(\text{grunerite}) + 4CO_2(g) \rightarrow 4FeCO_3(\text{siderite}) + Fe_3O_4 + 8SiO_2 + H_2(g)$	−86.673	−267.794
$3FeCO_3 + H_2O(l) \rightarrow Fe_3O_4 + 3CO_2(g) + H_2(g)$	107.877	277.453
$SO_2(aq) + 2H_2O(l) \rightarrow H_2SO_4(aq) + H_2(g)$	31.058	−14.947
$H_2S(g) + 2H_2O(l) \rightarrow SO_2(g) + 3H_2(g)$	206.299	268.995
$2H_2O(l) + CH_4(g) \rightarrow 4H_2(g) + CO_2(g)$	130.683	252.975
$2NH_3(g) \rightarrow 3\,H_2(g) + N_2(g)$	71.570	152.235

This set of reactions indicates the reactions that are favorable at the surface of the crust. Pyritization and serpentinization are exothermic reactions that occur more likely at shallow depths, while endergonic reactions are more expected with increasing depth.

15.3.2 Field methodology for hydrogen measurement and sampling

Once a zone has been selected, direct measurements are undertaken along unpaved paths preferentially to minimize anthropogenic influence. Since many zones of interest are on private properties, collaborations with ICMBio (Chico Mendes Institute for Biodiversity Conservation) have been prioritized to obtain authorization for conducting measurements within reserves. This collaboration also ensures the protection of the team while avoiding any conflicts of interest related to H_2 exploration, thereby emphasizing the importance of fundamental academic research and its dissemination.

15.3.2.1 Key instruments used in the field

The fieldwork relies on specialized gas detection instruments and sampling techniques to ensure accurate hydrogen measurements. The Variotec 460 Tracergas detector measures hydrogen concentrations from 0.1% to 100% by volume, with a precision of 0.1%. It uses a gas-sensitive semiconductor sensor for low concentrations and a thermal conductivity sensor for higher concentrations. Another key instrument is the GA5000 electrochemical sensor, which detects H_2 up to 0.1 vol% along with gases such as O_2, CH_4, and CO_2, with optional modules for CO and H_2S detection. The GA5000 has been used primarily for dry soils, with its application mainly in the first case study at Ipanema Forest, São Paulo.

To enable a comprehensive gas analysis, samples are collected in vacuum-sealed tubes at 10^{-5} torr using a turbomolecular pump for subsequent laboratory evaluation. Since the Variotec sensor is highly sensitive to humidity, measurements are conducted in relatively dry soils at depths of up to 1 m. In muddy or geothermal environments, modifications such as integrating a transfer pump connected to a possum belly, a settling reservoir for drilling fluids [24], would be necessary. Alternative sampling methods, including vials, funnels, and aliquots, as described in the literature [25], have also been considered, and emerging solutions from companies like Expro Co. offer new approaches for gas collection.

Hydrogen concentration measurements are performed systematically, with sampling occurring every 100 m in strategic areas. In locations where hydrogen levels exceed 1,000 ppm, repeat measurements are conducted nearby, followed by re-saturation tests. These tests involve pausing measurements and re-evaluating hydrogen levels in the same borehole after a short interval to assess gas replenishment rates. While there is currently no standardized methodology for hydrogen field detection, most studies in Brazil have followed a similar approach. Based on previous field experiences, drilling depths typically range from 20 cm to 1 m, based on the soil texture; GA5000 detectors are commonly used.

Measurement precision depends on hydrogen concentration levels and the sensitivity of the equipment. For values up to 100 ppm, the precision is ±1 ppm, whereas for concentrations between 100 and 1,000 ppm, the precision is at least ±10 ppm. In the field, the highest concentration peak is recorded immediately after drilling, with the Variotec detector providing a rapid response. The GA5000 sensor records a slightly delayed but broader peak. Because there is no universally established method for gas collection, detectors must be adapted based on field conditions. Several factors influence measurement accuracy, including the type of detector, whether a pump is used, and the operator's ability to handle syringes and tubes efficiently. Since leaks can occur during sample handling, hydrogen concentrations measured using gas chromatography (GC) are often lower than in situ readings obtained with portable gas detectors. To ensure consistency, a comparative study can be conducted by using two types of detectors and drilling equipment at the same time and location.

Preliminary field evaluations can provide valuable insights through simple observations and in-field techniques. Certain minerals, including magnetite, hematite, quartz, feldspar, mica (biotite and muscovite), orthogneiss, monazite, sandstone, limestone, and amphibole, can be identified visually. A magnet can be used to test for magnetite, while a Geiger counter, such as the FNIRSI GC-01, can detect radioactivity in minerals such as monazite. Structural indicators, including depressions, fractures, and shear zones, can also be identified through visual inspections, with geological maps serving as a valuable tool to confirm field observations.

While serpentinization and pyritization reactions do not fully explain some hydrogen emissions with high concentrations, other mechanisms of natural hydrogen formation are being explored. These hypotheses are based on observed field data, and further studies are aimed at refining our understanding of the processes responsible for hydrogen generation in geological settings.

15.4 Key case studies

Different motivations have influenced the selection of four strategic regions. The main attractive characteristics of Flona de Ipanema include high iron-based minerals and alkaline intrusions forming Morro de Araçoiaba, which is considered a hot spot formation (a thermal anomaly similar to the rare events that form hills such as Morro do Ferro in Poços de Caldas and Morro São João in Rio das Ostras) along with the adjacent ophiolite. In Vassouras, an unexplained magnetic anomaly crosses a major fault. The Lakes region is known for its alkaline and basic lakes, while in Espírito Santo, the coast is known for a high content of thorium-based minerals. Knowledge of the iron mine history and discoveries is a good indicator for hydrogen formation. If mining remains a controversial subject due to its negative impacts (e.g., Mariana disasters in 2015 and Brumadinho disasters in 2019, which continue to affect the ecosystem and drastically lower life expectancy), hydrogen exploration is increasingly linked to mining, as hydrogen production licenses are now being included within mining regulations in some countries (e.g., Spain) and under discussion in Brazil. The shift to natural hydrogen exploration may change the focus, positioning old mine workings as a future potential hot spot.

15.4.1 Ipanema National Forest – São Paulo

The Ipanema National Forest (Floresta Nacional de Ipanema) has been chosen due to the history of the steel industry in Brazil, indicating high levels of iron in the region, for the carbon-based minerals (limestone and dolomite deposits nearby, in Salto de Pirapora and Sorocaba), an ophiolite in Pirapora de Bom Jesus and some metavolcanic rocks. Some key characteristics of the region are now summarized and discussed further.

15.4.1.1 Historical facts

Iron exploration was crucial to the Industrial Revolution, and Afonso Sardinha pioneered iron forging in Brazil in 1589 at Morro de Araçoiaba, rich in magnetite. Sardinha patented his method in 1592. While slavery was rampant, iron forging thrived and Dom Pedro I established Brazil's first ironworks at Ipanema in 1682. Today, ICM-Bio manages the 5,180 ha Ipanema Forest, home to the Afonso Sardinha Trail, where magnetite and hydrogen emissions can be studied, extending to Lago do Cobra.

15.4.1.2 Geological formation

The region belongs to the lower Cretaceous. The Morro de Araçoiaba (also known as Morro de Ipanema) in the forest park is formed by alkaline magmatic intrusions aged 65–250 Ma, with a height of 968 m. The Paraná Basin, composed of sedimentary rocks, crosses the forest. A schematic topology [26] shows that Morro de Ipanema consists of amphibolite, pyroxene, and magnetite, which are favorable for H_2 formation. Apatite deposits, also present in and outside the forest park, may play a role in hydrogen generation, a topic further explored in subsequent sections. In the Sorocaba region, at

Serra de São Francisco, faults, shear zones, and rocky mounds are oriented NE-SW. Sorocaba city lies at the boundary of the Itararé Series (including the forest park) and the São Roque Series (featuring schist, granite, and limestone). The São Roque group crosses the Pirapora de Bom Jesus ophiolitic complex, approximately 80 km from the Ipanema Forest, with an age of about 620 Ma [27–29].

15.4.1.3 Mineralogy

The main minerals can be found in key studies by many authors [26, 30, 31], with other recent contributions [32, 33]. However, no clear information is available regarding the exact locations of these minerals, and there is no local data using gamma spectroscopy or magnetometry.

Table 15.4 provides the key iron-based minerals found in the region surrounding the Ipanema National Forest and São Roque series. Non-iron-based minerals are also listed in Table 15.4, although they may appear in some reactions.

Table 15.4: List of minerals reported surrounding the Ipanema National Forest.

Mineral name	Composition
Magnetite	Fe_3O_4
Hematite	Fe_2O_3
Goethite	$FeOOH$
Biotite	$K(Mg,Fe)_3AlSi_3O_{10}(F,OH)_2$
Olivine	$(Mg,Fe)_2SiO_4$
Amphibole hornblende	$(Na,Ca)_{2-3}Fe_5(Al, Si)_8O_{22}(OH)_2$
Ferrosilite	$Fe_2Si_2O_6$
Enstatite	$(Mg,Fe)_2Si_2O_6$
Amphibole actinolite	$Ca_2(Mg,Fe)_5Si_8O_{22}(OH)_2$
Cordierite	$(Mg,Fe)_2Al_4Si_5O_{18}$
Opaque	$Fe(Fe,Ti)_2O_4$
Pyrite, marcasite	FeS_2
Arsenopyrite	$FeAsS$
Mica biotite	$K(Mg,Fe)_3(AlSi_3O_{10})(F,OH)_2$
Alkali/potassic/calcium feldspar (e.g., shonkinite and labradorite)	$(K,Na,Ca)(Si,Al)_4O_8)$
Mica muscovite	$KAl_2(AlSi_3O_{10})(F,OH)_2$
Calcite	$CaCO_3$
Dolomite	$CaMg(CO_3)_2$
Wavelite	$Al_3(PO_4)_2(OH,F)_3.5H_2O$
Rutile	TiO_2
Chalcedony and quartz	SiO_2
Phlogopite	$KMg_3(AlSi_3O_{10})(F,OH)_2$

During some field inspections, some minerals and rocks can be easily identified, such as magnetite and sandstone. No X-ray diffraction (XRD)/X-ray fluorescence (XRF) analysis has been done in this fieldwork. Magnetite was easily recognized along the path due to its metallic appearance and attracting a magnet while sandstone is widespread in the region. Red layers of sedimentation are also often observable, indicating events rich in iron, oxidizing in hematite. A mine of calcite is located at the north of the reserve and was used in the construction of the main road crossing the reserve.

15.4.1.4 Hydrogen concentrations

A total of 178 measurements were conducted both inside and outside the forest on three separate occasions: 6–8 September 2023, 2–4 October 2023, and 16–21 December 2023. In the forest, the highest recorded hydrogen concentration was 5,000 ppm, detected along the trail near Acadebio, with surrounding areas within a 500 m radius exhibiting concentrations around 1,500 ppm. Outside the forest, the maximum concentration reached 9,000 ppm (GPS: −23.41244, −47.67307) in lateritic soil on the western side of the forest, where apatite deposits had previously been reported. This measurement was obtained using a drill bit at a depth of 1 m during the initial field campaign. Notably, hydrogen concentrations above 1,000 ppm were consistently detected in this region using the Variotec detector. Based on these findings, Bacaetava and Acadebio emerge as the most promising locations in the region for further investigation (Figure 15.3). Satellite imagery reveals the presence of fractures and depression-like structures in these areas, which may indicate potential hydrogen migration pathways.

In December 2023, additional measurements were taken at Bacaetava using a 50 cm drill bit compatible with the Dewalt drill hammer. The highest recorded value during this campaign was 1,400 ppm. However, it was not possible to replicate the previously recorded 9,000 ppm measurement at the same location (−23.41244, −47.67307). This discrepancy can be attributed to several factors, including the shallower drill depth and the softer soil texture caused by increased moisture following recent rainfall. These observations highlight the need for systematic research in hydrogen exploration to better understand how variations in equipment and environmental conditions influence recorded values. Long-term monitoring is essential for identifying the most promising sites for hydrogen accumulation. It is anticipated that higher concentrations will be detected with deeper drilling under drier conditions, as hydrogen levels are known to fluctuate throughout the day and in response to seasonal and weather variations.

The region has magmatic alkaline intrusions, crossing layers of sedimentary tillite (silt-clay matrix) from glacial deposits [34]. Hydrogen is therefore probably generated from alkaline intrusions, accumulating in porous rocks (e.g., limestone) capped by isolating layers and diffusing through local fractures, erosion, and depressions where the highest values have been reported.

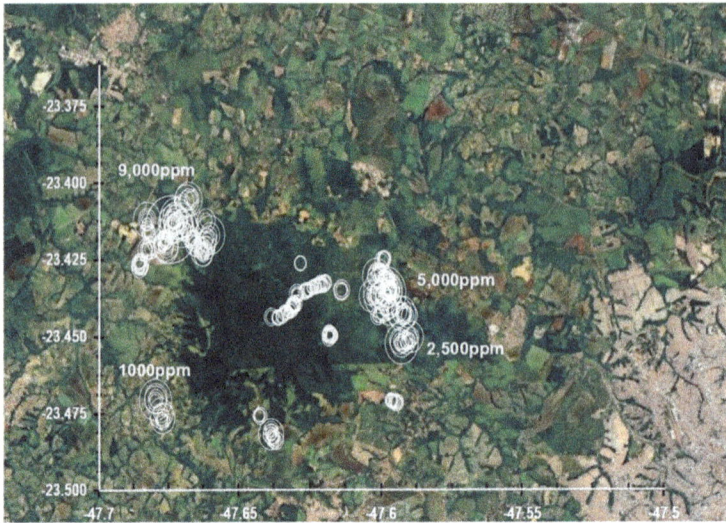

Figure 15.3: Satellite image of the main areas with the highest values at Ipanema National Forest and its surrounding area (white circles represent the GPS positions of H_2 measurements, not scaled to their concentrations) (©2025 Google).

15.4.2 Região dos Lagos – Rio de Janeiro

The region exhibits multiple geological proxies indicative of potential H_2 formation. One key factor is its alkalinity, which, combined with magnetic anomalies, suggests favorable conditions for hydrogen generation. Another notable characteristic is the presence of cratonic features. The Cabo Frio region has a cratonic origin and shares geological connections with the Congo Craton, which itself is linked to the São Francisco Craton – an area where hydrogen emissions have already been documented.

Additional indicators include the presence of a fault system extending between Saquarema and Macaé. The pre-field study strategy focused on identifying faults in proximity to surface magmatic and alkaline rocks to optimize the search for H_2 emissions. More than 400 measurements have been conducted in Região dos Lagos, covering the area between Saquarema and Macaé. Notably, the initial fieldwork and roadmap were developed independently, without prior knowledge of the findings from [4]. The gas analysis study in Maricá was published in April 2024; however, the authors did not disclose GPS coordinates for the measurement sites.

Another important observation from both this study and the existing literature, including CPRM (Companhia de Pesquisa de Recursos Minerais) reports [35], is that fault locations are not always precisely mapped. On-site visual inspections remain crucial for identifying local fractures and depressions. Moving forward, the integration of gamma spectrometry will be essential for correlating hydrogen emissions with geological structures. A significant correlation between elevated hydrogen emissions and increased radiation levels was previously identified in Maricá [4] using the MS1000 gamma spectrometer (Medusa), further reinforcing the importance of this method in future studies.

15.4.2.1 Saquarema-Ipitangas-Araruama

This area was selected due to the presence of magnetic anomalies and a fault system originating in Saquarema. In Ipitangas, hydrogen concentrations reach 3,500 ppm, while in Rio Seco, values as high as 6,500 ppm have been recorded. Along the path to Rio Pardo, H_2 concentrations reach 1% by volume. According to CPRM gamma spectroscopy data, the region contains alkaline minerals, particularly feldspar. Amphibole and mica are also present and may be associated with hydrogen formation, along with hematite as a by-product.

Araruama exhibits a unique geochemical profile, characterized by basic and hypersaline lakes. While sulfide minerals are known to contribute to hydrogen formation through the pyritization reaction, carbonaceous minerals and biominerals may also play a role in the H_2 system. The mineral dolomite $(Ca,Mg)CO_3$ is a strong candidate for hydrogen storage. Several measurements were conducted around the lake, but no significant hydrogen levels were detected. The soil texture in the area was not ideal for hydrogen detection – sandstones were highly porous, and a fine white clay known as 'Tabatinga' produced excessive dust during drilling, potentially interfering with the Variotec detector's functionality.

15.4.2.2 Rio das Ostras, Casimir Abreu

Rio das Ostras exhibits unique geological characteristics, including rare formations. One notable feature is the so-called Vulcão Billy, also known as Morro São João, which has a height of 806 m and serves as a popular tourist attraction. It has been classified as a magmatic alkaline intrusion associated with a thermal anomaly, making it a prime candidate for hydrogen generation. The site contains all the essential geological conditions for a 'hydrogen kitchen.' Syenite is a predominant rock type at Morro São João, which was formed between 65 and 135 million years ago during the Cretaceous-Tertiary period, similar to the Morro de Ipanema formation in São Paulo. Both hills are considered thermal anomalies, suggesting upwelling from the mantle.

Morro São João is covered with dense vegetation and has a highly humic and humid soil, which may have influenced the low H_2 levels detected along the path of the hill. However, measurements taken a few kilometers away yielded significantly higher values, with concentrations reaching up to 1.3% H_2 by volume in Casimiro de

Abreu, where a fault crosses the area. Additionally, a point located 3.2 km south of this site recorded hydrogen levels as high as 7,000 ppm (Figure 15.4).

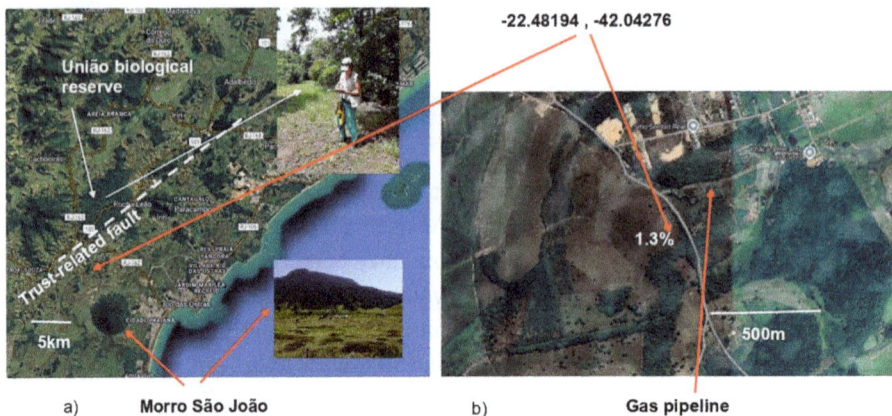

Figure 15.4: (a) Studied zone in the region of Rio das Ostras with 1.3 vol% of H_2 and (b) zoom at GPS −22.48194; −42.04276 with a gas pipeline nearby (©2025 Google; © Corinne Arrouvel).

A first fault was identified passing close to Rocha Leão, and for that reason, measurements have been carried out more systematically, thanks to access to the biological reserve belonging to ICMBio, Reserva Biológica da União. A total of 113 measurements have been done in the reserve and their GPS positions are indicated in Figure 15.5. We note that the highest measured value was at 7,500 ppm during a rainy day (at GPS location: −22.42499; −42.02059). It is known that rain and aquifers affect measurements, usually acting as a barrier to H_2 diffusion. On the other hand, a water cycle is necessary to enhance H_2 formation within the rocks. Interestingly, the hydrogeology and the location of the faults enable some correlations with zones of creosote leaks [36, 37]. Such leaks may possibly be correlated to hydrogen leaks and affect gas analyses. The highest reported hydrogen concentration shown in Figure 15.5 is at the cross section of two contractional faults or shear zones. From those preliminary results, cross sections of faults confirm the highest probability of locating hot spots of gas leakage. Further analyses and measurements will be necessary to identify the origin of hydrogen formation and the flux evolution during climate changes.

15.4.2.3 Macaé

The municipality is crossed by a main geological fault, and the hydrogen emission is expected around some strategic zones (see Figure 15.6). Two regions have been identified: at 6,000 and 8,500 ppm of H_2 (granite with orthopyroxene). Higher levels correspond to some faults reported more accurately [34]. It is expected that there are lo-

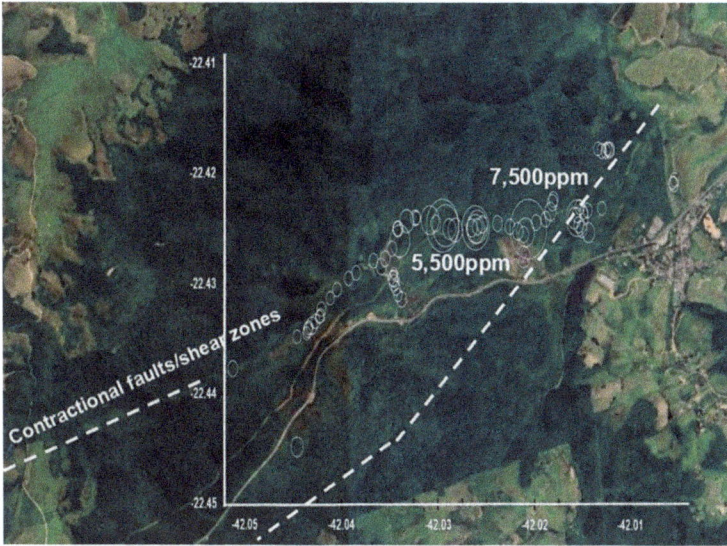

Figure 15.5: Satellite image indicating H_2 measurement zones at the União Biological Reserve-ICMBio (white circles represent the GPS positions of H_2 measurements, not scaled to their concentrations) and faults/shear zones (©2025 Google).

cally higher values than 8,500 ppm. Indeed, the first mapping was done under rainy conditions. Measurements have been done with some care to avoid equipment exposure to rain (the detector is sensitive to humidity and some measurements were aborted due to heavy rain).

Figure 15.6: Satellite image nearby Macaé city with the approximate positions of the highest H_2 concentrations near a fault (©2025 Google).

15.4.3 Vassouras – Rio de Janeiro

This region was chosen due to a magnetic anomaly and two main faults on CPRM maps [35]. Measurements were recorded along these faults on 26–30 December 2023, while it was raining. The first fault explored was crossing Vassouras and Paraiba do Sul on road BR393 and the second fault was parallel to the first one, below Nossa Senhora do Amparo. The highest concentration was close to Vassouras city with some fractures perpendicular to the main faults (see Figure 15.7).

a)   b)

Figure 15.7: (a) Satellite image of the zone of prospection around Vassouras (©2025 Google) and (b) measuring H_2 point along the trust-related fault on BR393 Road, Rio de Janeiro, Brazil. (photograph courtesy of Vanea de Souza Nogueira, December 2023) .

Another important site was found with high hydrogen concentrations of more than 1%, and interestingly with abnormal radiation (5.10 µSv/h using Geiger counter) in a private property. The landlords gave the authorization to enter their properties and carry out measurements. Due to the informal authorization, the location of their properties will not be revealed and the zone deserves to be further explored, including helium and radon detectors. In June 2024, we went back to the area to collect gases, called CC in Table 15.5, in another property with a high hydrogen level (1%) but no abnormal radiation had been observed during this second review and it was not possible to return to the same spot in the absence of the landlords. As expected, the hydrogen concentrations from GC analyses are lower than the measurements on-site with the Variotec detector and are not representative statistically due to the lack of resources. Other measurements were undertaken using a micro-GC Agilent, but results were unpublished.

A hypothesis would be linked to the formation of 'Pico do Cavalo Russo' in the Valença municipality, a zone with metamorphic rocks, and of Proterozoic age. It has not been possible to verify if the abnormal radiation comes from radioactive elements. A local study will be necessary for that purpose. However, in this complex, there have been reported outcrops of orthopyroxene with biotite and amphibole. Charnockite minerals are reported to be aged 2.1 Ba to 579 Ma by Zr-TIMS and monazite minerals of 563 Ma [38]. If confirmed, monazite could be one of the factors at the origin of hydrogen through radiolysis and that can explain some radioactive anomalies.

Table 15.5: Valença municipality (some private properties need formal authorization): collection of gases in June 2024.

	He (ppm)	Ne (ppm)	H$_2$ (ppm)	CH$_4$ (ppm)	CO$_2$ (%)	O$_2$ (%)	N$_2$ (%)
CC2	5.4	19.9	438.6	0.002	0.11	19.7	75.4
CC5	5.8	19.8	8,006.4	0.005	0.43	19	75.1
CC6	5.4	20.1	4,169	0.003	0.17	19.5	75.1
CC7	n.a.	19.5	3,509.8	0.002	0.24	19.8	74.6
Air	5.25	20.6	0.5	1.5	0.04	20.93	75.65

Analyses with a micro-GC by 45-8 Energy (n.a., not available)

15.4.4 South of the state of Espírito Santo

A road map was developed for several reasons. First, the road was chosen to allow measurement areas with an expected high level of radioactivity. For example, Guarapari is known for its high radioactive levels from monazite minerals. Indeed, the black sand tested at 'Areia Preta' (mixed with monazite, having a yellow color) exhibits radioactivity up to 35.20 µSv/h (the highest level recorded during the exploration). The sand has been collected and the black grains are magnetic with an aspect of magnetite. Contrary to expectations, the H$_2$ level was very low for three possible reasons: (1) the roads and building constructions prevent drilling in a natural environment; (2) the soil was too soft; or (3) the heavy rain affected the measurements. A second motivation for choosing this area was the geological formation of 'Pedra Azul' (blue stone). Such anomalies could imply a formation was near a granitic block. Several measurements had been done around the area, with some difficulties in drilling due to the presence of hard stones and low concentrations measured. The main explanation is that no hydrogen was escaped due to an isolating layer. Some sites were chosen by their proximity to apparent depressional structures but they turned out to be swamps. The heavy rains had an impact on the land, the depression-like structures were full of water, and H$_2$ measurements nearby were relatively low.

The highest level of H$_2$ was found at around 10 km from Pedra Azul (see Figure 15.8), in the direction of Guarapari, Urânia. The most probable explanation is that the fault was close by. Indeed, based on satellite images, the fault seems to cross Aracê and Cachoeira

Iracema. One feature of this location was high concentrations up to 23,000 ppm (2.3%), with a white and friable rock seeming to be linked with these high concentrations, as a substrate layer. A sample was collected and is to be analyzed by XRD/XRF. On the CPRM-ES maps [39], the highest concentration corresponds to a zone with different types of faults (transcurrent sinistral and contractional). Once more, faults become the more efficient proxy to prospect for hydrogen leaks.

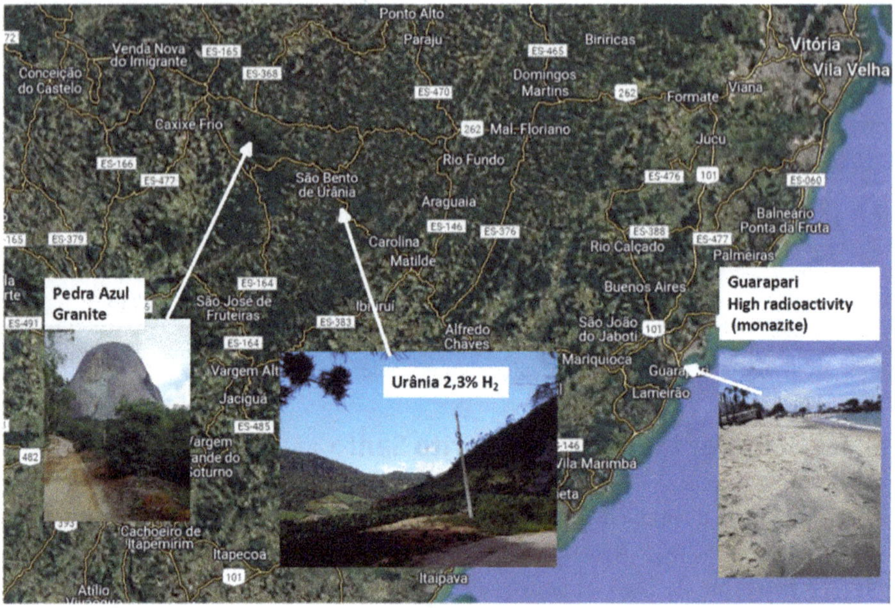

Figure 15.8: Map of the studied zone in the south of the state of Espírito Santo and with the highest H_2 concentration in Urânia (©2025 Google) (inset photographs © Corinne Arrouvel).

15.4.5 Possible redox reactions in the studied areas

A wide range of minerals has been listed previously and can participate in H_2 formation. Indeed, olivine, pyroxene, and amphibole are known to be part of the mechanism of serpentinization and can be transformed into micas. The presence of quartz also enables proposing other mechanisms (e.g., Bowen's reaction series and product from orthopyroxene). Basalt associated with magma degassing is therefore another possible mechanism for H_2 formation.

We intend to highlight the transformations involving iron and hydrogen under aqueous conditions and eventually with carbon dioxide that is involved with carbon-based minerals. Kimberlite has been identified in the south of Sorocaba (state of São Paulo) and there are some indices of possible coal layers. The reactions are then listed below and the thermodynamic data is deduced using SUPCRTBL program (negative

Gibbs energy ΔG_r means that the reaction is spontaneous at the surface of the soil and negative enthalpy ΔH means that the reaction is exothermic). Not all reactions involving listed minerals can be simulated due to the lack of experimental data.

These data express the feasibility of H_2 formation on the surface of the soil regarding the presence of minerals as products and reagents, reminding us that those mechanisms do not consider the kinetic aspects of the reactions. To evaluate the potential of the size of a reservoir from the reactions, other factors need to be considered through an estimation of the bacterial activities consuming hydrogen, a full gas analysis, long-term monitoring of gas emission, seismic studies, and modeling of reservoirs.

These reactions can also be applied to other regions in Brazil as some similarities will be appointed with a lithological description of formations in São Paulo state such as São Sebastião Island and in the state of Rio de Janeiro, involving the formation of alkaline magmatic intrusions, nepheline syenite such as in Poços de Caldas, Morro São João, and Catalão in Goiás.

We note that ferrosilite ($Fe_2Si_2O_6$) is more likely to be transformed to quartz and hydrogen at higher pressure from 300 bars in subduction zones and can also decompose to fayalite at ambient conditions. As subduction zones are unlikely in Brazil, this reaction of hydrogen formation should not be a process to consider. However, subduction zones are identified in South America with hydrogen seepage [40]. Then quartz and olivine are likely to coexist at the surface of the crust and the further transformation of olivine to serpentine and magnetite depends on hydrothermal fluxes.

In both cases, to predict the formation of H_2, it is necessary to have thermodynamic data of the transformed minerals (data that are not often accessible and need to be deduced from experiments in a laboratory under controlled pressure and temperature) and to include the effect of pH in fluid diffusion within connected porous rocks and fractures.

FeO (wüstite) is a mineral stable in the mantle under high pressure and temperature, and even if it has been identified in a furnace of the iron factory of Ipanema, for example, it is considered as an artifact probably obtained during steelmaking at high temperatures.

Pyroxene and amphibole are likely to play an important role in the H_2 formation, with the oxidation of iron to magnetite and hematite, reaction simplified as follows:

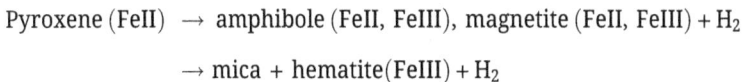

$$\text{Pyroxene (FeII)} \rightarrow \text{amphibole (FeII, FeIII), magnetite (FeII, FeIII)} + H_2$$

$$\rightarrow \text{mica} + \text{hematite(FeIII)} + H_2$$

Another reaction of interest is hydrogen formation when carbon dioxide reacts with amphibole. The reaction is spontaneous at the surface thermodynamically and the process has other interests in the field of CCS.

Indeed, the reaction $Fe_7Si_8O_{24}H_2(\text{grunerite}) + 4CO_2(g) \rightarrow 4FeCO_3(\text{siderite}) + Fe_3O_4(\text{magnetite}) + 8SiO_2(\text{quartz}) + H_2(g)$ is exergonic and exothermic under standard conditions. Figure 15.9 expresses the equilibrium limits under a range of extreme Q_r conditions, dominated by the ratio of gas partial pressure between CO_2 and H_2, using the same methodology [3]. By increasing the total pressure, the reaction will favor the formation of hydrogen (Le Chatelier's law in which $\Delta n(gas) < 0$). In addition, H_2 being volatile and grunerite decomposing into three types of minerals (siderite, magnetite, and quartz), the kitchen system is likely to present fractures. A degassing process will then be expected in the crust:

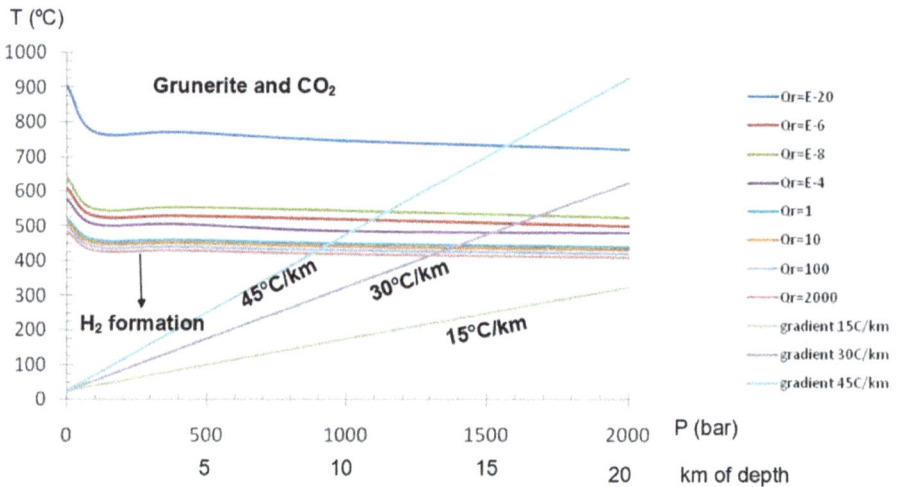

T (°C)

Grunerite and CO_2

45°C/km   30°C/km

H_2 formation   15°C/km

P (bar)
km of depth

Legend:
- Or=E-20
- Or=E-6
- Or=E-8
- Or=E-4
- Or=1
- Or=10
- Or=100
- Or=2000
- gradient 15C/km
- gradient 30C/km
- gradient 45C/km

Figure 15.9: On a pressure versus temperature plot, H_2 is formed below the temperature limit with the equation $Fe_7Si_8O_{24}H_2$ (grunerite) $+ 4CO_2(g) \rightarrow 4FeCO_3$ (siderite) $+ Fe_3O_4$ (magnetite) $+ 8SiO_2$ (quartz) $+ H_2(g)$.

$$\Delta G = \Delta G° + RT \ln Q_r$$

In the case of grunerite, $Q_r = \frac{P_{H_2}}{P_{CO_2}^4}$, where Q_r is the reaction quotient (without unit), combining the compound activities/fugacities as described in [3], neglecting the coefficients of activities/fugacities, and quantified by the partial pressures of the gases. R is the gas constant and T is the temperature.

The thermal gradient is often considered to be about 15–30 °C/km and the pressure gradient of 100 bar each km. Then we may consider 200 bar pressure at 20 km depth. We note that the pressure has a small influence on the direction of the reaction from 250 bar (2.5 km).

At a low gradient, 15 °C/km, the formation of hydrogen from amphibole is likely to occur in the crust, being then a good candidate for carbon storage. At the gradient 20 °C/km, the reaction is still favored down to 13 km considering $Q_r = 2,000$ in the extreme case of H_2 saturation.

The proposed reactions will be similar in other regions of Brazil with similar characteristics, such as magmatic inclusions in hot spots. Morro de Araçoiaba (São Paulo) and

Morro São João (Rio de Janeiro) have similarities in geochronology with alkaline intrusions. They have both high levels of hydrogen concentrations within a few kilometers in
radius from the hot spots, and the mechanism of formation is likely to be degassing from
the oxidation of iron-based minerals (e.g., olivine, amphibole, and pyroxene). The proposed mechanisms of formation are considered speculative as there is a lack of complementary data such as isotopic, gas composition, and mineral analyses. However, the thermodynamic results can be applied to other regions in Brazil and in the world. Indeed, hot
spots are reported in other regions of Brazil (e.g., Poços de Caldas in Minas Gerais with
hot springs, Itatiaia, Passa Quatro, and São Sebastião Island) and they might deserve
some exploration. Geothermal resources are also thermal anomalies that can be linked to
exothermic reactions involving hydrogen formation. Geothermal sites reported so far [15]
are therefore other regions to prospect and compare with other existing case studies (e.g.,
Soultz-sous-Forêts, France [41]). The identification of ophiolites is also another hint favoring hydrogen formation through the oxidation of iron-based minerals. Pirapora de Bom
Jesus ophiolitic complex in the state of São Paulo with chromitite ores is not a unique
case in Brazil. Such ophiolitic complexes have been identified in Quatipuru and Morro do
Agostinho (state of Pará) and Faixa Dom Feliciano (state of Rio Grande do Sul). The rich-
ness of mineral resources and geological characteristics in Brazil offer then many options
in the search for H_2 reservoirs in different states. Experimental laboratory studies under
controlled temperatures and pressure are necessary to complete the thermodynamic databases. The role of chromium-, nickel-, and titanium-based minerals, for instance, is not
measurable in the geological context as they can act as a catalyst or be transformed during magmatic evolution or lithosphere dynamics.

15.5 Conclusions and perspectives

While Brazil stays engaged with oil industry, exploration mainly conducted by Petrobrás and the National Agency of Petroleum, Natural Gas and Biofuels (ANP), an increasing number of Brazilian institutions and companies have started to dedicate
some research programs on natural hydrogen systems (e.g., UFSCar, ICMBio, UFRJ,
LENEP-UENF, GEO4U, and Quasis Energia) to accelerate the energy transition. From
the present acquired knowledge, different regions in Brazil deserve attention. Indeed,
while it has been claimed in 2024 that Maricá has a potential area for hydrogen exploration with concentrations up to 2,777 ppm at the subsurface (no exact location has
been revealed), other locations have recorded values higher than 1% at the subsurface
in the states of Espírito Santo and Rio de Janeiro. A combination of key proxies has
been at the origin of those sites with high hydrogen concentrations. First, Brazil has
similarities with the African continent (cratons), and has rich mineral resources, starting on the east side of Brazil (up to −58 longitude) with higher concentrations of iron
(based on mining exploration data). The choice of so-called hot spots (thermal anoma-

lies implying the geological formation of Morro de Araçoiaba (São Paulo) and Morro São João (Rio de Janeiro)) with alkaline intrusions give also high hydrogen concentrations near faults and fractures, where the gases are more likely to leak from its formation and/or accumulation. The locations of magnetic and thorium anomalies were also key indices to spot areas with high H_2 levels. Such indices are linked to an increase in magnetite concentration and the location of faults. From field observation and data collection, we deduce that the most important reactions are the oxidation of iron-based minerals from the mantle and magma degassing. Another important observation from thermodynamic and mineralogy results is that amphiboles participate in hydrogen formation through carbonic gas inclusion at the surface of the crust. This implies that the search for hydrogen reservoirs offers opportunities to include CCS and 'orange' hydrogen in research programs. An additional interest from hydrogen formation is the proximity of granitic formations (at the basement or extruded), implying helium formation through radiolytic processes.

Hydrogen accumulation might occur kilometers away from its formation and is more likely to accumulate in sediments and metamorphic rocks such as schist, limestone, sandstones, clays, and shales. Data on the location of potential reservoirs are scarce as Brazil is at the study-infancy in the universities and due to lack of release of confidential research results by private companies. Indeed, Petrobrás announced during a recent event co-organized by INMETRO and VAMAS ('Challenges of advanced materials and standardization needs for a greener world' at INMETRO on 25 September 2024) that the first pilot well should be drilled by 2029. No location of the pilot has been revealed but a program of exploration is under progress in the region of Bahia. Furthermore, long-term monitoring is also needed in order to launch such a pilot program, as the dynamics of hydrogen formation depend on its origin.

Even with promising results from preliminary mapping, it is expected to find higher concentrations at depth and by refining the location of faults using gamma spectroscopy. The analysis of the soil and minerals at the surface and at depth is also required in order to estimate the size of potential reservoirs. For that purpose, other methods are necessary, such as 2D and 3D seismic surveys, magnetotelluric, and gravimetry, which are methods that need to be adapted for the hydrogen system.

As Brazil is one of the leading countries on green hydrogen, it has significant mineral resources, high levels of hydrogen at the subsurface, considerable expertise in offshore oil exploration, and implements new laws to enable hydrogen exploration. The country is expected to fulfill its commitment to drastically reduce its carbon emissions by including exploration for hydrogen within governmental, academic, and private actions. In the short term, the production of hydrogen should benefit urban transportation locally as electric-hydrogen hybrid buses are projected for operation in the states of Rio de Janeiro and São Paulo and should benefit sustainable agriculture as hydrogen is the chemical feedstock with the highest demand for ammonia production.

Acknowledgments: Arrouvel thanks the São Paulo Research Foundation (FAPESP project number: 2022/12650-9) for funding and for enabling the start of fieldwork in Brazil. She is also grateful to the Chico Mendes Institute for Biodiversity Conservation (ICMBio) at the Ipanema National Forest in the state of São Paulo and the União Biological Reserve in the state of Rio de Janeiro (project numbers: Sisbio nos. 86576-1 and 92854-1) for supporting the research with information and infrastructure in the field. Arrouvel acknowledges the support in the field by Leonardo Silva de Oliveira, Rodinaldo Rodrigues Benedito (fireworker at ICMBio Ipanema National Forest), Vanea de Souza Nogueira, Maria Do Carmo Dos Anjos, José Carlos Vilar Amigo, and Mohammad Odeh Husein. The author is grateful to Henrique Fabrelli (CBPF, Brazil) for helping with the turbomolecular pump and Luiz Fernando Roncaratti Junior (UnB, Brazil) for sharing weather station data. The author sincerely thanks Dr. Reza Rezaee and Dr. Brian J. Evans for the valuable comments and corrections improving the manuscript. She is grateful to Marco Delfino for pointing out typographical and grammatical errors. Finally, Arrouvel is particularly grateful to Alain Prinzhofer for fieldwork training and fruitful discussions.

References

[1] https://www.astuteanalytica.com/industry-report/hydrogen-market, January 2025, report ID AA1221107.
[2] Prinzhofer A., Moretti I., Françolin J., Pacheco C., D'Agostino A., Werly J., Rupin F. Natural hydrogen continuous emission from sedimentary basins: The example of a Brazilian H2-emitting structure. International Journal of Hydrogen Energy. 2019, 44(12): 5676–5685.
[3] Arrouvel C., Prinzhofer A. Genesis of natural hydrogen: New insights from thermodynamic simulations. International Journal of Hydrogen Energy. 2021, 46(36): 18780–18794.
[4] Prinzhofer A. Rigollet, C., Lefeuvre, N., Françolin, J., Valadão de Miranda, P.E. Maricá (Brazil), the new natural hydrogen play which changes the paradigm of hydrogen exploration. International Journal of Hydrogen Energy. 2024, 62(1): 91–98.
[5] de Freitas V. A., Prinzhofer A., Françolin J. B., Ferreira F. J. F, Moretti I. Natural hydrogen system evaluation in the São Francisco Basin (Brazil), Science and Technology for Energy Transition. 2024, 79, 95.
[6] Serratt H., Cupertino J. A., Cruz M. F., Girelli T. J., Lehn I., Teixeira C. D., Oliveira H. O. S., Chemale Jr F. Southern Brazil hydrogen systems review. International Journal of Hydrogen Energy. 2024, 69, 347–357.
[7] Moretti I., Prinzhofer A., Françolin J., Pacheco C., Rosanne M., Rupin F., Mertens J. Long term monitoring of natural hydrogen superficial emissions in a Brazilian cratonic environment. Sporadic large pulses v. daily periodic emissions. International Journal of Hydrogen Energy. 2021, 46: 3615–3628.
[8] Simon J. B., Fulton P. M., Prinzhofer A., Cathles L. M. Earth tides and H2 venting in the São Francisco Basin, Brazil. Geosciences. 2020, 10: 414.
[9] Cathles L., Prinzhofer A. What pulsating H2 emissions suggest about the H2 resource in the Sao Francisco Basin of Brazil. Geosciences. 2020, 10: 149.

[10]   Moya M. R., Sánchez-Cañete E.P., Vargas R., López-Ballesteros A., Oyonarte C., Kowalski A.S., Serrano-Ortiz P., Domingo, F. CO2 dynamics are strongly influenced by low frequency atmospheric pressure changes in semiarid grasslands. Journal of Geophysical Research: Biogeosciences. 2019, v124: 902–917.

[11]   Siquiera I. P., Rodrigues L.A., Possa G., Roncaratti, L.F. Atmospheric tides over Brazil. Journal of the Royal Meteorological Society. 2023, 149: 2196–2205.

[12]   Truche L., Donzé F. V., Goskolli E., Muceku B., Loisy C., Monnin C., Dutoit H., Cerepi, A. A deep reservoir for hydrogen drives intense degassing in the Bulqizë ophiolite. Science. 2024, 08. Feb, 383: 618–621.

[13]   Langhi L., Strand J. Exploring natural hydrogen hotspots: A review and soil-gas survey design for identifying seepage. Geoenergy. 2023, 1(1), 1–15.

[14]   Maiga O., Deville E., Laval J., Prinzhofer A., Diallo A. B. Characterization of the spontaneously recharging natural hydrogen reservoirs of Bourakebougou in Mali. Scientific Reports. 2023, 13: 11876.

[15]   Hamza V. M., Eston S. M., Araujo R. L. C. Geothermal energy prospects in Brazil:. A Preliminary Analysis, Pure and Applied Geophysics. 1978, 117: 180–195.

[16]   Vieira F. P., Guimarães S. N., Hamza V.M. Updated assessment of geothermal resources in Brazil, Fourteenth International Congress of the Brazilian Geophysical Society, SBGf, Rio de Janeiro, Brazil. 2015, August 3–6.

[17]   Santos A. C. L., Padilha A. L., Pádua M. B., Vitorello I., Fuck R. A., Pires A.C. B. Magnetotelluric imaging of the southeastern, Borborema province. NE Brazil, Proceedings 13th International Congress of the Brazilian Geophysical Society, SBGf, Rio de Janeiro, Brazil. 2013, August 26–29.

[18]   MacCy A. T., Harley T. L., Younger P., Sanderson D. C., Cresswell A. J. Gamma-ray spectrometry in geothermal exploration: State of the art techniques. Energies. 2014, 7: 4757–4780.

[19]   Aftabi A., Atapour H., Mohseni S., Babaki, A. K. Geochemical discrimination among different types of banded iron formations (BIFs): A comparative review. Ore Geology Reviews. 2021, 136: 104244.

[20]   Tapias J. G., Schobbenhaus C., Ramírez N. E., Gutiérrez F. A., Zabala D. M. Mapping the geology of South America. Episodes. 2023, 46(4): 537–549.

[21]   Malvoisin B., Brunet F. Barren ground depressions, natural H2 and orogenic gold deposits: Spatial link and geochemical model. Science of the Total Environment. 2023, 856: 158969.

[22]   Zimmer K., Zhang Y., Lu P., Chen Y., Zhang G., Dalkilic M. M., Zhu C. SUPCRTBL: A revised and extended thermodynamic dataset and software package of SUPCRT92. Computers & Geosciences. 2016, 90: 97–111.

[23]   Jacquemet N. Prinzhofer A. The association of natural hydrogen and nitrogen: The ammonium clue? International Journal of Hydrogen Energy. 2024, 50: 161–174.

[24]   Davies K., Esteban L., Keshavarz A., Iglauer S. Advancing natural hydrogen exploration: Headspace gas analysis in water-logged environments. Energy & Fuels. 2024, 38: 2010–2017.

[25]   Lévy D., Roche V., Pasquet G., Combaudon V., Geymond U., Loiseau K., Moretti I. Natural H2 exploration: Tools and workflows to characterize a play, science and technology for energy. Transition. 2023, 78: 27.

[26]   Davino A. Geologia da Serra de Araçoiaba, Estado de São Paulo, Boletim IG. Instituto de Geociências, USP, 6, 129–144, 1975.

[27]   Santos E. O. Geomorfologia da região de Sorocaba e alguns de seus problemas. Boletim Paulista de Geografia. 2017, [S. l.], n. 12: 3–29.

[28]   Godoy A. M., Hackspacher P. C., Oliveira M. A., Araújo L. M. Evolução geológica dos batólitos granitóides neoproterozóicos do sudeste do estado de São Paulo, São Paulo. UNESP, Geociências. 2010, 29(2): 171–185.

[29]   Tassinari C. G., Munhá, J. M., Ribeiro, A., Correia C.T. Neoproterozoic oceans in the Ribeira Belt (southeastern Brazil): The Pirapora do Bom Jesus ophiolitic complex. Episodes. 2001, 24(4): 245–251.

[30]   Knecht T. As ocorrencias de minerios de ferro e pirita no estado Sao Paulo. Boletim IGG. 1939, 25.

[31] Felicíssimo J. História da siderurgia de São Paulo, seus personagens, seus efeitos. Boletim IGG. 1969, 49.

[32] Gomes C., Comin-Chiaramonti P., Azzone R. G., Ruberti E., Rojas G. E. Cretaceous carbonatites of the southeastern Brazilian platform: A review. Brazilian Journal of Geology. 2018, 48(2): 317–345.

[33] Perrotta M. M., Salvador E. D. Lopes R. D., D'agostino L. Z., Chieregati L. A., Peruffo N., Gomes S. D., Sachs L. L., Meira V. T., Garcia M. D., Filho J. V. Geologia e recursos minerais do estado de São Paulo: Sistema de Informações Geográficas - SIG. Rio de janeiro, Programa Geologia do Brasil: CPRM, 2006.

[34] Santos H. S., Gonçalves M. E., Gomes A. P. Características da Radioatividade Natural do Município de Macaé. Revista de Engenharia da Faculdade Salesiana. 2014, (1): 11–20.

[35] Heilbron M., Eirado, L. G., Almeida J. C. H. Geologia e recursos minerais do Estado do Rio de Janeiro, Belo Horizonte: CPRM, 2016.

[36] Vianna J. S., Ferreira M. I., Saraiva V. B., Machado P. V. Contaminação do solo por creosoto em uma Unidade de Conservação de Proteção Integral: o caso da Reserva Biológica União – RJ/Brasil, Boletim do Observatório Ambiental Alberto Ribeiro Lamego. Campos dos Goytacazes/RJ. jul./dez. 2016, 10(2): 131–153.

[37] ICMBio technical report, Plano de Manejo, Reserva Biológica União, Encarte 3 – Análise da Unidade de Conservação MMA/ICMBio, Rio de Janeiro, maio 2008.

[38] Heilbron M., Almeida J. C. H., Eirado L. G., Palermo N., Tupinambá M., Duarte B. P., Valladares C., Ramos R., Sanson M., Guedes E., Gontijo A., Nogueira J. R., Valeriano C., Ribeiro A., Célia Diana Ragatky C. D., Miranda A., Sanches L., Melo C. L., Roig H. L., Dios F. B., Fernández G., Neves A., Guimarães P., Dourado F., Lacerda V. G. Geologia da Folha Volta Redonda SF.23-Z-A-V: Escala, 1:100.000. Brasília; Rio de Janeiro: CPRM; UERJ, 2007.

[39] Vieira V. S., Silva M. A., Corrêa T.R., Lopes N.H. Mapa geológico do estado do Espírito, Santo: CPRM, 2018. https://rigeo.sgb.gov.br/handle/doc/15564

[40] Ramirez A. C., Gonzalez Penagos F., Rodriguez G., Moretti, I. Natural H_2 emissions in Colombian ophiolites: First findings. Geosciences. 2023, 13(12): 358.

[41] Murray J., Clément A., Fritz B., Schmittbuhl J., Bordmann V., Fleury, J.M. Abiotic hydrogen generation from biotite-rich granite: A case study of the Soultz-sous-Forêts geothermal site, France. Applied Geochemistry. 2020, 119: 104631.

Xueying Yin* and Bingchuan Yin

Chapter 16
Natural hydrogen in China: geological insights and exploration prospects

Abstract: Natural hydrogen is emerging as a promising clean energy source with significant potential to support China's carbon neutrality goals. This chapter provided a comprehensive review of natural hydrogen discoveries across China, focusing on distinctive geological settings, including geothermal regions, oil and gas basins, and non-oil and gas sedimentary basins. We explored diverse origins of natural hydrogen in China, including radiolysis, water-rock reactions, and mantle degassing, supported by geological and isotopic evidence. China's unique geological framework, characterized by small mafic and ultramafic rock belts and Mesozoic-Cenozoic basins, necessitated careful consideration for exploration. Effective exploration required a multidisciplinary approach, integrating geological, geochemical, and geophysical techniques to identify promising exploration areas, hydrogen sources, and reservoirs, alongside the development of a supportive regulatory framework to facilitate continued exploration.

Keywords: natural hydrogen, China, geothermal regions, radiolysis, water-rock reactions, mantle degassing

16.1 Introduction

Hydrogen is widely considered one of the most crucial energy sources of the twenty-first century [1]. China, as the world's largest producer and consumer of hydrogen, accounts for nearly one-third of global hydrogen production and consumption [2]. Currently, most of China's hydrogen is derived from coal, though there is a strong push toward cleaner alternatives [2, 3]. Significant investments have been directed towards electrolysis technologies, driven by China's 'dual carbon' strategy–its nationally declared goals to reach carbon peaking before 2030 and achieve carbon neutrality by 2060, announced in 2020 [3–5].

Despite these substantial investments, electrolysis has encountered numerous scaling challenges [3, 6]. In response, natural hydrogen has emerged as a promising alternative, offering the potential to supply large quantities of clean hydrogen more

*Corresponding author: Xueying Yin**, H2Terra Ltd Co., Beijing 102600, China,
e-mail: xyin@h2terra.com
Bingchuan Yin, H2Terra Ltd Co., Beijing 102600, China

https://doi.org/10.1515/9783111437040-016

economically and sustainably, thereby addressing some of the limitations associated with electrolysis [7].

In this chapter, we reviewed historical discoveries of natural hydrogen across China, providing an overview of the geological settings associated with these occurrences. We examined diverse origins of natural hydrogen in China supported by geological and isotopic evidence. We then discussed the unique geological conditions in China, which distinguished it from other regions globally and could influence exploration efforts. Finally, we outlined the essential steps required for successful exploration.

16.2 Summary of hydrogen discoveries in China

Reported natural hydrogen discoveries in China are summarized in Table 16.1, which are categorized into three primary types: geothermal hydrogen, hydrogen associated with oil and gas basins, and hydrogen associated with non-oil and gas basins. Unless otherwise noted, all hydrogen concentrations are reported by volume.

Table 16.1: Summary of hydrogen discoveries in China (all locations are marked in Figure 16.1).

Geological environments	Locations	Highest $c(H_2)$, % (site)	Samples collected	References
Geothermal environment	Tengchong	5.15 (Rehai)	Water	[8–11]
	Changbai Mountains	1.24 (Tianchi)	Water	[12, 13]
	Jimo, Jiaolai Basin	13	Water	[14]
Oil and gas basin	Zhanhua Sag, Bohai Bay Basin	19.23	Well gas	[15]
	Sebei, Qaidam Basin	99	Well gas	[16]
	Songliao Basin	26.89 (SK-2) 85.54 (Ershen-1)	Well gas	[17–20]
	Sanshui Basin	0.64	Soil gas	[21]
Non-oil and gas basin	Qianzhong Uplift, East Sichuan	65.42%	Shale-extracted gas	[22]
	Yanfeng Depression, Chuxiong Depression	43.79	Well gas	[23]
	Shangdu Basin	1.92	Logging mud gas	[24]
	Zhangbei Basin	0.38	Soil gas	Ting Kou (pers. comm.)

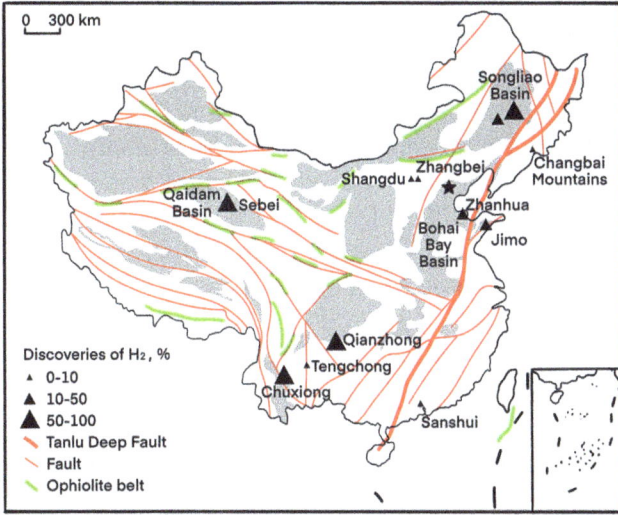

Figure 16.1: Natural hydrogen occurrences in China with only major basins and faults illustrated (see Table 16.1 for details).

16.2.1 Geothermal hydrogen discoveries in China

16.2.1.1 Geological context of geothermal hydrogen in China

In China, natural hydrogen seepage has been documented in several geothermal regions, including Tengchong county, Yunnan Province, and the Changbai Mountains, Jilin Province – two of the country's most prominent geothermal areas [8–13] (Figure 16.1). Research conducted as early as the 1990s revealed significant concentrations of natural hydrogen in geothermal gases from these regions, ranging from 0.04% to 5.15% in Tengchong [9] and approximately 2.56% in the Changbai Mountains [12]. The primary component of geothermal gas emissions in both regions was carbon dioxide (CO_2) [9, 12, 13]. For instance, in the Rehai geothermal area of Tengchong, CO_2 accounted for over 90% of the emissions, with nitrogen (N_2) comprising 0.8–7.85% [9]. Similarly, in the Julongquan geothermal area near Tianchi lake in the Changbai Mountains, CO_2 constituted 85.7–97.6% of the emissions, while N_2 contributed 2.4–7.9% [12].

More recently, the Jimo hot spring, located in the Jiaolai Basin, has been identified as a significant site for natural hydrogen, with dissolved hydrogen concentrations ranging from 2.4% to 12.5% [14] (Table 16.1 and Figure 16.1). Notably, the H_2/CH_4 ratio in this region was exceptionally high, reaching up to 46.5 [14]. Nitrogen was the primary dissolved gas at Jimo, with concentrations ranging from 83.8% to 89.4% [14].

Tengchong county hosts 63 hot spring areas, including 3 boiling springs, 25 hot springs, and 35 lukewarm springs [25, 26]. In the Changbai Mountains, hot springs are primarily distributed around Tianchi Lake, with prominent examples including the Changbai Julongquan hot springs, Tianchi Lakeside hot springs, and Jinjiang hot springs. The Julongquan hot spring area alone contains 165 individual springs, with temperatures reaching up to 86.5 °C and an average temperature of 59.64 °C [27]. By comparison, the Jimo geothermal field is relatively small, occupying an area of approximately 0.2 km^2, and is classified as a low-temperature geothermal system, unlike the higher-temperature volcanic systems of Tengchong and Changbai Mountains [14].

These geothermal systems were primarily influenced by fault structures, which determined their spatial distribution and geothermal characteristics [28–31]. In Tengchong, the hot springs were predominantly controlled by the NE-SW arc fault tectonic belts, which could be divided into three main zones: the Dayingjiang fault to the west, the Ruidian-Tengchong-Longchuan volcanic active fault belt at the center, and a series of eastern faults, including the Longchuanjiang, Gaoligongshan, and Nujiang faults [28–30]. In the Changbai Mountains region, geothermal activity was primarily governed by a combination of annular faults and NE- and NW-trending radial faults [31]. Similarly, the Jimo hot spring was predominantly influenced by NE- and NW-trending faults [32].

Geologically, the Tengchong Rehai geothermal field was capped by Miocene mudstone and volcanic rock, with an estimated thickness of approximately 300 m [30]. The underlying reservoir rock comprised the Precambrian Gaoligongshan Group and Late Cretaceous massive granite, characterized by extensive fracture networks [29, 33]. The Gaoligongshan Group consisted predominantly of quartzite, quartz schist, metamorphic granulite, plagioclase metamorphic granulite interbedded with hornblende-plagioclase metamorphic granulite, and hornblende schist [29, 30].

At the Changbai Mountains' hot springs, the caprock was primarily composed of Cenozoic basalt and andesite, which overlaid Cretaceous strata. The basement rocks included the Lower Proterozoic Ji'an Group and Laoling Group, composed mainly of marble, shallow granulite, metamorphic granulite, and plagioclase amphibolite [27, 31].

The Jimo hot spring was situated along the northeastern margin of the contact zone between the Laoshan granite body and a Cretaceous intrusion. The stratigraphy was dominated by the first section of the Qingshan Formation (K_1q^1) of the Lower Cretaceous, which consisted primarily of tuffaceous clastic rocks and volcanic rocks, with a cumulative thickness of approximately 1,500 m [32].

16.2.1.2 Possible origins of geothermal hydrogen in China

Isotopic studies indicated that the δD values of hydrogen (H_2) escaping from geothermal areas in China varied significantly, reflecting the influence of multiple geological processes and potential contributions from mantle-derived sources [11, 12, 14, 34]. These δD variations, when combined with other geochemical and isotopic markers, provided insights into the origin and migration pathways of geothermal H_2.

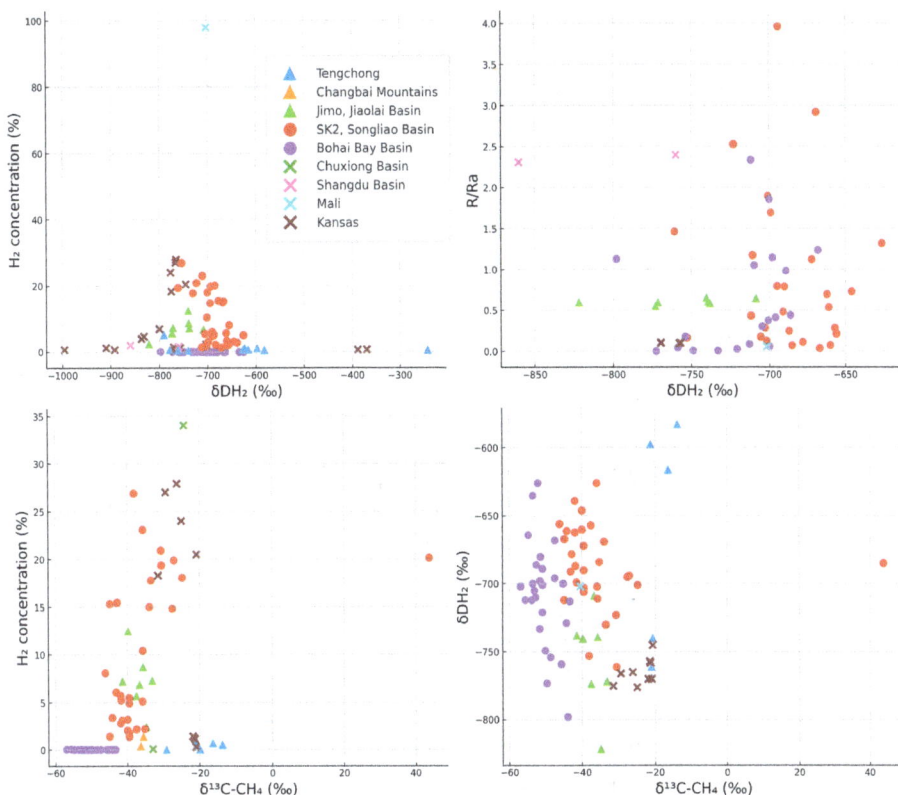

Figure 16.2: Isotopic plots of natural hydrogen occurrences in China, based on the available data. Note that most previous studies lacked systematically collected isotopic data. Data from Mali and Kansas are also included for comparison [35, 36].

In Tengchong, δD values of H_2 ranged from −791‰ to −583‰ [34], aligning with isotopic signatures typically associated with water–rock reactions (δD ≤ −650‰) (Figure 16.2) [37]. These reactions likely involved the oxidation of ferrous iron (Fe^{2+}) in deep-seated rocks, such as the metamorphic units of the Gaoligongshan Group, during hydrothermal circulation [10]. In addition, elevated H_2 concentrations (up to 5.15%) [10] and enriched δD values (e.g., −242.3‰) [34] were recorded during periods of increased seismic activity, indicating a mantle-derived hydrogen component [38, 39]. Prior to the hydrothermal explosion on September 20, 1999, hydrogen concentrations in mid-depth springs ranged from 0.38% to 5.15%, with an average of 1.18% [10]. Further evidence of deep magmatic involvement came from exceptionally high H_2 concentrations, exceeding 15% at Xiaogunguo hot spring in 2003 [11], suggesting that magmatic processes may have enhanced hydrogen release.

Stable isotope data for coexisting gases further supported these interpretations. In Tengchong, $\delta^{13}C$ (PDB) values ranged from −5.8‰ to −2.9‰ for CO_2 and −13.8‰ to

−21.2‰ for CH_4 [34]. In combination with $^3He/^4He$ ratios ranging from 1.19 to 4.49 R/ Ra, these values pointed to a significant mantle contribution [10].

In the Changbai Mountains, δD values of H_2 spanned a wide range, from −826‰ to −310‰ [12]. This broad isotopic spectrum overlapped with values typical of both radiolytic hydrogen (−550‰ to −350‰) and water–rock reactions (Figure 16.2) [37, 39, 40]. The more depleted δD values (as low as −826‰) were consistent with hydrogen generation through radiolysis of water in uranium- or thorium-rich rocks, while mid-range values indicated ongoing water–rock interaction. Some enrichment toward higher δD values suggested minor mantle-derived input, likely introduced via tectonic structures and magmatic pathways. This interpretation was supported by $\delta^{13}C$ values of CO_2 averaging −4.7‰ [12], a signature typically associated with magmatic degassing. Furthermore, mantle helium signals were observed in both Tengchong and Changbai Mountains, reinforcing the role of deep-seated faults and magmatic systems in channeling mantle gases into shallow geothermal reservoirs [41, 42].

The presence of natural H_2 in these regions was likely influenced by volcanic or deep magmatic processes. In Tengchong, 68 volcanoes, primarily composed of olivine basalt, andesite, dacite, and tuff, have contributed to the development of extensive basaltic layers and fracture networks through a complex multiphase eruptive history [43, 44]. Volcanic activity was dated from 4.3 million years ago to as recently as 0.07 million years ago [43–47]. The extensive basaltic eruptions, which blanketed 610 km^2 with flows up to 100 m thick [43, 44], also fractured the overlying strata, providing pathways for mantle gases to rise toward the surface. Similarly, in the Changbai Mountains, the Tianchi volcano provided a direct link between deep magma chambers and surface geothermal systems. Its alkaline basaltic lava and significant eruptions created conditions conducive to the storage and escape of mantle-derived H_2 alongside geothermal fluids [48–50].

Extensive magma chambers were identified beneath the Tengchong and Changbai geothermal areas using a range of geophysical methods [51–67]. In Tengchong, seismic and magnetotelluric (MT) surveys found a low-resistance body beneath Heikong Mountain, Dakong Mountain, and Xiaokong Mountain at depths of 12–30 km, spanning approximately 25 km from east to west. This feature was hypothesized to represent a magma chamber [52–55]. Low-resistance anomalies were also detected beneath the Mazhan-Tengchong-Ma'anshan and Wuhe-Longjiang-Tuantian stations, where the Earth's crust was thin, and Poisson's ratio was elevated [51]. These observations, coupled with a high geothermal gradient, strongly supported the presence of magma chambers in the region [56–58]. Further investigations using three-dimensional seismic imaging and inversion methods confirmed the existence of three magma chambers in the Tengchong geothermal area [55, 58–61]. Additionally, Seismic body-wave tomography and receiver function analyses revealed a low-velocity zone beneath the Tengchong geothermal area, extending through the mantle transition zone to depths of approximately 300 km, 410 km, and as deep as 660 km, indicating the presence of a deep-seated magmatic source [41, 60, 61].

In the Tianchi volcanic system, geophysical investigations identified multiple magma-related features spanning the upper mantle and lower crust. An S-wave velocity model (RG5.5) revealed a prominent low-velocity anomaly—referred to as a "low-velocity pocket"—in the upper mantle at depths of 38–65 km, with a lateral extent of 100–200 km and a 2.5% velocity reduction relative to surrounding material [62]. Complementary analyses, including seismic profiling, CT imaging, magnetotelluric (MT) surveys, and crustal tomography, identified a second magma body beneath the Tianchi crater at shallower depths. This crustal magma pocket, located at 9–15 km depth, extended approximately 80–90 km along the profile and 30–40 km north–south [63–66].

Seismic CT imaging also confirmed the existence of an upper mantle low-velocity zone beneath the volcano, consistent with the previously identified deep anomaly, and interpreted it as an upper mantle magma chamber [66]. Beneath the Tianchi crater, the crustal structure exhibited an inverted triangle-shaped low-velocity body and a concave Moho discontinuity with an amplitude of 2–6 km, forming what was interpreted as a crustal root of the volcanic system [67]. Together, these features suggested a vertically connected magmatic plumbing system, linking deep mantle reservoirs to shallower crustal magma storage zones, thereby facilitating the transfer of heat and mantle-derived gases and increasing the potential for seismic and geothermal activity.

Significant seismic and geothermal activity further supported the continued migration of H_2. In the Tengchong Rehai area, intense geothermal activity between 1993 and 2003 resulted in more than 20 large-scale explosive eruptions [43, 68]. In the Changbai Mountains, the frequency of volcanic earthquakes increased from an average of 7 per month before 2002 to over 70 per month between 2002 and 2005 [69]. Earthquakes with magnitudes of 4.4 and 4.0 were recorded at the Wangtian'e volcano, approximately 30 km from the Tianchi crater, between 2004 and 2005 [71].

Although geologically distinct from these volcanic systems, Jimo hot spring provided an additional example of hydrogen generation. Here, δD values of H_2 ranged from −822‰ to −709‰, while $\delta^{13}C$ values of CH_4 varied from −427‰ to −220‰ (VSMOW) [14]. These isotopic signatures strongly indicated hydrogen production through low-temperature water–rock reactions [14, 37]. The absence of a magmatic component made Jimo a valuable contrast to volcanic settings, underscoring the diversity of geological environments capable of generating natural hydrogen (Figure 16.2).

16.2.2 Hydrogen discoveries in oil and gas basins in China

16.2.2.1 Hydrogen discoveries in the Songliao Basin

Natural hydrogen was identified in four oil and gas basins in China, with the majority of research focused on the Songliao Basin (Table 16.1 and Figure 16.1). In the Songliao Basin, particularly in the Xujiaweizi Depression in eastern China, 17 hydrogen-enriched wells were identified, with H_2 concentrations reaching up to 85.54% [20]. Hydrogen (H_2) was present as an associated component in natural gas, with accompanying gases primarily including CH_4, CO_2, CO, and N_2 [19]. Most hydrogen-rich wells were located along the periphery or outside the boundaries of natural gas fields. While six wells reported industrial gas flow, their production capacities remained unevaluated [20].

The concentration of natural hydrogen generally increased with depth, reaching its highest levels within the Yingcheng Formation and the igneous reservoir in the basement. In contrast, the proportion of alkane gases, particularly CH_4, decreased with depth. Among non-hydrocarbon gases, these hydrogen-rich natural gas samples typically exhibited high N_2 content (generally exceeding 20%) and relatively low CO_2 levels, averaging less than 5%. Additionally, CO was present in about 39% of the analyzed samples, with an average concentration of 10.78%. The concentrations of N_2 demonstrated a depth-dependent variation consistent with that of H_2 [20].

The Xujiaweizi Fault Depression, located in the northern Songliao Basin, was a large-scale, north-south trending fault depression spanning 90 km in length, 55 km at its widest point, and covering an area of 5,350 km^2 [72]. Six layers of natural gas reservoirs were identified at depth within the depression, including the weathered crust of the basement rocks, igneous rocks of the Huoshiling Formation, conglomeratres of the Shahezi Formation, igneous rocks and conglomerates of the first member of the Yingcheng Formation, igneous rocks of the third member of the Yingcheng Formation, and gravels of the fourth member of the Yingcheng Formation [72]. The primary source of hydrocarbons in this area was the dark mudstones and coal-bearing strata of the Shahezi Formation. These hydrocarbons migrated vertically along faults or laterally through unconformities into reservoirs of the Yingcheng Formation and laterally into the basement rocks and igneous rocks of the Huoshiling Formation. The overlying dark mudstones of the second member of the Denglouku Formation and the first and second members of the Quantou Formation served as stable, region-wide caprocks [72–75].

The primary reservoirs for H_2 and He were situated within the upper Denglouku Formation, Yingcheng Formation, Huoshiling Formation, and Permian volcanic formations. The volcanic rocks in these strata were formed during episodes of intraplate magmatic activity within an extensional tectonic regime, likely linked to lithospheric thinning and asthenospheric upwelling in eastern China, driven by the subduction of the Pacific Plate during the late Mesozoic [76].

The Xujiaweizi Fault Depression exhibited a distinct dustpan-shaped morphology, transitioning from a fault-controlled structure in the west to an overrun structure in the east. Its structural configuration was primarily governed by three near-NNW-trending faults: the Xuxi, Xuzhong, and Xudong faults. The Xuxi Fault bifurcated into northern and southern branches due to the Xuzhong strike-slip fault, which was centrally positioned within the depression. These faults played a critical role in controlling natural gas accumulation [77]. Structurally, the depression demonstrated a dual-layer system: the lower faults formed a characteristic 'flower-shaped' structure, while the upper part exhibited a typical graben-horst configuration with limited fault connectivity between the two layers [77].

Tectonically, the Xujiaweizi Fault Depression underwent three major evolutionary stages, resulting in the development of three structural layers: fault depression, sag, and inversion [78]. During the early Huoshiling period (Late Jurassic to Early Cretaceous), intensified fault activity likely facilitated the formation of H_2 source rocks. In the Yingcheng period (Early Cretaceous), as fault activity diminished, the depression transitioned into a sag phase characterized by the formation of two volcanic rock sequences [79, 80]. By the Denglouku period (Late Cretaceous), faulting had ceased, and the region entered a post-rift thermal cooling and subsidence phase [81].

16.2.2.2 Possible origins of hydrogen in the Songliao Basin

Hydrogen in the Songliao Basin likely originated from multiple sources, as evidenced by geological and isotopic analyses [19]. Isotopic data from samples collected at the SK-2 well showed δD values for H_2 ranging from −685‰ to −667‰ in the Denglouku Formation, −730‰ to −695‰ in the Huoshiling Formation and basement strata, and −703‰ to −656‰ in the intermediate layers. These values were consistent with isotopic signatures typical of water-rock reactions and showed lighter δD values with increasing depth, indicating a deeper origin for some hydrogen. The multi-sourced origin of H_2 aligned with the characteristics of other associated gases. For example, δN values ranged from 0.8‰ to 7.2‰, suggesting that N_2 was primarily mantle-derived. In contrast, R/Ra ratios of helium, varying between 0.129 and 1.894, indicated contributions from both mantle and crustal sources (Figure 16.2) [19].

Radiolysis likely contributed to hydrogen production, as supported by the presence of uranium. In the Yingcheng Formation of the SK-2 well, two radioactive anomalies were identified at depths of 3,096.8–3,102.8 m and 3,168.3–3,170.9 m, with uranium concentrations of 20.5–29.3 ppm and 5.9–11.0 ppm, respectively. The latter anomaly also exhibited elevated thorium content (22.4–37.3 ppm) [82]. Uranium enrichment in the upper strata likely resulted from fault-volcanic activity, basin uplift, and erosion, which facilitated the migration of uranium-bearing groundwater, oil, and gas. Epigenetic reduction of hydrocarbons might lead to further uranium concentration. In contrast, the lower radioactive anomalies, primarily in agglomerate lava and tuff, were attributed to uranium adsorption by rhyolitic components with high thorium and clay mineral content. These U- and Th-rich strata likely generated hydrogen (H_2) and helium (He), which accumulated within the Yingcheng Formation or migrated along faults to reservoirs in the Denglouku and Qingshankou Formations [82, 83].

Hydrogen production through water-rock reactions was also plausible. The Yingcheng Formation in the Xujiaweizi area consisted of lava, pyroclastic rocks, and conglomerates, with interbedded pyroclastic and sedimentary rocks in transitional zones. Sedimentary zones included conglomerates, sandstones, siltstones, and mudstones [84]. The third member of the Yingcheng Formation in the Shengping Gas Field contained multiple volcanic rock bodies from different eruptive periods, forming lithologic-tectonic gas reservoirs with edge and bottom water [85]. In the Xushen Gas Field, gas-water layers were identified between depths of 3,948.8–3,996.4 m and 4,066.4–4,108.8 m. H_2-rich layers in wells such as Xushen 1, Xushen 18, and Xushen

6-106 were located above the gas-water layers, suggesting water-rock reaction processes. Formation water in the Yingcheng Formation, primarily of $NaHCO_3$ type with high mineralization, bicarbonate, and sodium content, provided favorable conditions for water-rock reactions [86].

The Huoshiling Formation, predominantly composed of the deep-water coarse debris and volcanic rock, also offered favorable conditions for water-rock reactions. Its lithology included volcanic rock, black mudstone, sandstone, intermediate-acid tuff breccia, and tuff lava [87]. The SK-2 core revealed that the Huoshiling Formation was 179.65 m thick, containing early Cretaceous basaltic andesite (141.6 ± 1.4 Ma) with Fe_2O_3T levels of 6.72–9.65%, indicating high iron content conducive to reactions producing hydrogen [87, 88].

Mantle degassing likely served as another source of hydrogen. Seismic surveys in the Xujiaweizi area revealed a heat flow diapir, interpreted as a 'mushroom cloud' structure on the Moho surface, likely representing a magma chamber or mantle plume. This structure corresponded to the top boundary of a detachment zone in the middle to upper crust at depths of 16–20 km [89]. Gravity and seismic data also identified a 3-km-thick, low-density layer at 18–20 km depth, accompanied by a low-velocity zone and high-conductivity layers (3–8 Ω/m), suggesting the presence of magma linked to the mantle [90, 91]. These findings aligned with isotopic evidence from CO_2 and He, confirming mantle contributions [19].

Further isotope and gas composition analyses supported these conclusions. For instance, CH_4 in the Denglouku Formation was thermogenic, while the Shahezi Formation contained mixed mantle and thermogenic CH_4 [19]. H_2 in the basement was predominantly mantle-derived, with its content influenced by mantle degassing and water-rock reactions involving Fe^{2+}-bearing minerals (e.g., pyroxene and hornblende). In deeper formations, N_2 and CO_2 transitioned from crustal to mantle origin with increasing depth, shifting from mixed organic-inorganic to predominantly inorganic sources. These findings highlighted the multifaceted origins of hydrogen in the Songliao Basin, encompassing radiolysis, water-rock reactions, and mantle degassing [20].

16.2.2.3 Hydrogen discoveries in other oil and gas basins

In the Bohai Bay Basin, a prominent oil and gas basin on the east coast of China, H_2 collected from pyrolysis of frozen, sealed cores from the Zhanhua Sag and Jiyang Sag was measured at concentrations of 19.23% and 4.54%, respectively (Table 16.1 and Figure 16.1). However, no further research had been conducted on the origins of this H_2 [15].

In the Qaidam Basin, located in Qinghai Province in western China, natural H_2 was identified in the Sebei gas reservoir, where H_2 adsorbed in clay layers reached concentrations as high as 99% [16] (Table 16.1 and Figure 16.1). Based on sedimentary, structural, and trace component analyses (e.g., Ar and He), Shuai et al. [16] proposed that this H_2 was primarily a product of microbial degradation of organic matter. However, δD values of H_2, ranging from −820‰ to −700‰, suggested a deeper origin linked to water-rock reactions.

In the Sanshui Basin, located in Guangdong Province, small oil, gas, and CO_2 reservoirs were identified in the northern region, including the Gaogang North Fault Block of the Baoyue Structure and the Tangbian North Fault Block of the Dalan Uplift [21]. Natural hydrogen (H_2) was detected in limestone fractures and sandstone pores, with concentrations reaching up to 4% and an average helium (He) content of 3.5% [92]. Soil gas analyses revealed multiple locations with elevated H_2 concentrations, with peaks reaching 6,948 ppm (Table 16.1 and Figure 16.1). δ^2H values below −650‰ indicated a crustal origin or deep serpentinization processes, whereas elevated $^3He/^4He$ ratios (1.14–4.56 R/Ra) and $\delta^{13}C$ values of CO_2 (−9.8‰) suggested a potential contribution from mantle-derived fluids [21].

16.2.3 Hydrogen discoveries in non-oil and gas basins in China

16.2.3.1 Geological context of hydrogen discoveries in non-oil and gas basins in China

In addition to geothermal fields and hydrocarbon basins, significant discoveries of H_2 have been reported in non-hydrocarbon basins (Table 16.1 and Figure 16.1) [22–24]. For instance, H_2 was detected during a shale gas exploration attempt in the eastern Qianzhong Uplift, with concentrations as high as 15.24% [22]. An even higher concentration of 43.79% was identified in the 2,411.3–2,425 m interval of the Wulong-1 well in the Chuxiong Basin, Yunnan Province [23].

In the Shangdu Basin, Autonomous Region of Inner Mongolia, the exploration of CO_2 gas reservoirs yielded H_2 concentrations of 1.92% and 1.55% in mud gas samples collected at depths of 371 m and 351 m, respectively, during the drilling of ST1 and ST2. Both boreholes were underlain by black basalt [93]. In the Shangdu Basin and its adjacent Zhangbei Basin, which shares the same tectonic framework, the extended outcrops of intermediate-basic rock series belonging to the Lower Chongli Group ($Ar_{2-3}C_1$) of the Neoarchean Guzuizi Formation ($Ar_{2-3}g$), interbedded with ultramafic rocks, were particularly notable. The Paleoproterozoic Hongqiyingzi Group (Pt_1h), comprising the Dongjingzi Group (Pt_1^2d) and the underlying Taipingzhuang Group (Pt_1^2t), was composed of metamorphic rocks with a foliation-parallel contact relationship between the two groups [93, 94].

Mesozoic strata in the area were primarily represented by Lower to Middle Jurassic formations, including the clastic rocks of the Xiahuayuan Formation (J_1x) and the volcanic rocks of the Tiaojishan Formation (J_2t). Lower Cretaceous formations, such as the Qingshila Formation (K_1q) and the Zhangjiakou Formation, consisted of sandstone, conglomerate, and mudstone, while the Middle Nantianmen Formation (K_2n) included fluvial sedimentary deposits. Neogene strata were characterized by the gray-black basalt of the Hannuoba Formation (E_3N_1h) and the claystone of the Kaidifang Formation (E_3N_1k), with Quaternary Holocene deposits (Q_3, Q_4) overlying these units [94–96].

These two basins were defined by three primary fault systems: the Shangyi-Chongli-Chicheng deep fault, the Kangbao-Weichang fault, and the Zhangbei-Gaoshanbao fault. These fault systems, together with the Precambrian basement and uplifted Cenozoic basalts, formed grabens and basalt platforms, serving as critical pathways for H_2 migration from the mantle. Near-east-west and northeast-trending faults controlled these structures, emphasizing their tectonic significance [94, 96].

16.2.3.2 Possible origins and reservoirs of hydrogen in non-oil and gas basins in China

For the Chuxiong Basin, the δCO_2 value (−3.1‰) suggested a typical inorganic origin (Figure 16.2), while the $^3He/^4He$ ratio (1.76×10^{-6}) and the $^{40}Ar/^{36}Ar$ ratio (1446) reflected characteristics consistent with a mantle-derived or crust-mantle composite gas source [23].

In the Shangdu Basin, δD values of H_2 ranged from −860‰ to −760‰, while $\delta^{13}C$ values of CO_2 ranged from −6.5‰ to −5.8‰, both indicative of a mantle-derived origin [24] (Figure 16.2). The Shangdu-Zhangbei Basin was a particularly noteworthy case as their diverse lithologies and fault systems provided potential source rocks (ultramafic and basaltic rocks), migration pathways (faults and fractures), and reservoir rocks (sedimentary and volcanic formations) for natural hydrogen. A range of factors likely contributed to the generation of H_2 in this area, including the multistage tectonic activity of the Shangyi-Chongli-Chicheng fault, eruptions of the Hannuoba Formation basalt, and subsequent seismic events [93–96].

Increasing evidence supported the interpretation that the Shangyi-Chongli-Chicheng fault represented a Paleoproterozoic suture line, characterized by intense deformation across multiple tectonic stages [97–99]. Mylonites in the region recorded tectonic activities spanning from the Late Proterozoic to the Mesozoic, transitioning from compressional ductile deformation and reverse faulting during the Yanshanian period to extensional brittle-ductile deformation and normal faulting during the Himalayan period [95, 98, 100–104]. This fault zone not only facilitated magma flow, contributing to the formation of the Hannuoba basalt strata, but also remained tectonically active since the Eocene [105, 106]. Additionally, the Zhangbei-Gaoshanbao fault, which was active during the Himalayan period, was hypothesized to have served as a conduit for upward gas migration [107].

Basalt layers, dikes, and fault zones provided pathways for mantle-derived fluids, enabling hydrogen-rich fluids to ascend from deep sources into the basin [108]. The Hannuoba volcanic rocks, with cycles of tholeiitic and alkali basalt formations, reflected an intraplate extensional environment that was conducive to gas release. Eruptions from the Paleogene through the Quaternary facilitated multistage basaltic activity, creating additional conduits for hydrogen migration [109, 110].

Seismic activity likely played a role in facilitating hydrogen migration [107, 111]. Over the 15-year period between 1993 and 2008, the City of Zhangjiakou experienced 3,095 earthquakes with magnitudes ranging from ML 1.0 to ML 4.0, including 40 events with magnitudes of ML ≥4.0 [112]. Most earthquakes were concentrated at the intersection of the Shangyi-Chongli-Chicheng Fault and the Zhangbei-Gaoshanbao Fault. Pre-seismic activity was marked by significant increases in radon (Rn), hydrogen (H_2), and mercury (Hg) emissions, which subsided shortly after the earthquakes [105]. These seismic events likely facilitated the movement of mantle-derived fluids into the aquifer along the fault system, potentially contributing to the accumulation or continuous replenishment of H_2 reservoirs [113].

Water-rock reactions provided another mechanism for H_2 generation in the basins. Basaltic eruptions caused substantial fluid intrusion into the basement strata, accompanied by a significant rise in fluid and rock temperatures, which promoted water-rock reactions [114, 115]. The Shangyi-Chongli-Chicheng Fault Zone, a major deep fault, was associated with late Archean basic-ultrabasic rock assemblages. In the Chicheng area, metamorphic peridotites, predominantly serpentinized harzburgite, were distributed within the biotite-plagioclase gneiss of the Hongqiyingzi Group, occurring as lenses or masses [93, 116]. These metamorphic peridotites were interpreted as ultramafic intrusions [117] or possibly as ophiolite fragments [99]. Additionally Tong et al. [118] identified 'spike structures' in the metamorphic peridotites, attributed to high-pressure decomposition of antigorite, which indicated that some of these blocks had undergone subduction processes.

The Archean strata of the Sanggan Group (Ar_3s), Chongli Lower Group (Ar_3C^1), and Chongli Upper Group (Ar_3C^2) consisted primarily of high-grade metamorphic rocks, including gneiss, granulite, and metamorphic granulite, which were significant sources of metamorphic iron ore and magnetite quartzite [94]. The peridotites within these basement strata, along with their associated metamorphic minerals and iron-rich rocks, were highly conducive to water-rock reactions, providing a robust geological basis for hydrogen generation [115].

These basins also exhibited conditions favorable for hydrogen accumulation, with potential reservoirs in fractured basement strata and coarse sandstones, effectively sealed by Cretaceous mudstone and Cenozoic basalt layers [93]. Beneath 35 m, the Lower Cretaceous Qingshila Formation (K_1q) lay in unconformable contact with the Hongqiyingzi Group of the basement. This formation consisted of mudstone and sandstone, with basal layers of coarse sandstone and gravel-bearing sandstone reaching a maximum thickness of 90 m [93]. A stable conglomerate or gravel-bearing coarse sandstone layer, approximately 53 m thick and primarily composed of gneiss breccia, underlay the area, providing substantial reservoir space. Additionally, multistage tectonic activity along the Shangyi-Chongli-Chicheng Fault significantly fractured the basement strata, creating further reservoir potential within the Paleoproterozoic Hongqiyingzi Group [95].

Caprock conditions were provided by the Mesozoic Cretaceous mudstone layers and the Cenozoic basalt layers. Multiple mudstone layers below 253 m exhibited a cumulative thickness ranging from 312 to 385.7 m [93]. The Zhangbei basalt platform, covering approximately 1,700 km², had a thickness varying between 30 and 150 m and was primarily composed of lava with a wedge-shaped geometry [94]. Initially formed through central-fissure eruptions, the volcanic activity later transitioned into strong central eruptions. The basalt, characterized by dense blocky, vesicular, and amygdaloidal textures, contained closed pores or pores filled with calcite and other minerals. These features made the basalt an effective caprock, capable of sealing subsurface gases [24, 119, 120].

16.3 Discussions

China's geological framework presented unique characteristics that necessitated tailored methodologies for natural hydrogen exploration. These geological nuances dictated the strategies and technological frameworks for exploration. The primary geological features influencing natural hydrogen exploration in China include:

1. **Relatively small mafic and ultramafic rock belts**: Although China is home to numerous mafic and ultramafic rock belts across various orogenic zones (Figure 16.1), the scale of these belts is generally limited [121, 122]. The primary exception is the Himalayan orogenic belt, which, while significant in size and geological importance, is located far from the major hydrogen utilization centers [2, 3]. The limited spatial extent of these rock belts could potentially constrain the scale and distribution of natural hydrogen reservoirs, though the specific impacts on exploration and hydrogen accumulation are not yet fully understood.

2. **Mesozoic-Cenozoic basins**: China's geological landscape is predominantly defined by Mesozoic-Cenozoic basins, as Precambrian basins are relatively rare [23, 123]. These Mesozoic-Cenozoic basins are concentrated along the Tan-Lu Deep Fault in eastern China, a major crust-penetrating structure that extends approximately 2,000 km from Lujiang in Hefei, Anhui Province, to Luobei in Jiamusi, Heilongjiang Province [124] (Figure 16.1). The Tan-Lu Fault is associated with a wealth of mineral deposits and a number of oil and gas basins [125]. Despite these geological advantages, current data from geothermal systems along the eastern coast and within the Tan-Lu Fault region indicated lower hydrogen concentrations (<0.5%), with sources likely originating from the crust [14]. This highlighted the need for further research to evaluate the potential role of the Tan-Lu Deep Fault in hydrogen generation and migration and to understand the geological conditions that could enhance hydrogen accumulation in these areas.

3. **Craton blocks**: The North China and Yangtze Cratons, both of which feature a significant number of basins with high exploration potential, have emerged as key regions for natural hydrogen occurrences. Each of these cratons contains 16 and 17 basins, respectively, with areas exceeding 4,000 km² (Figure 16.3). Many of

these basins could exhibit favorable conditions for hydrogen generation, migration, and trapping. Notably, the North China Craton hosts several ancient banded iron formations (BIFs), which are rich in magnetite – a mineral known to facilitate hydrogen generation through water-rock reactions [126]. While these cratons offer substantial exploration potential, they are more accessible than the Tarim Craton, China's largest basin, which presents logistical challenges due to its remote desert environment. The North China and Yangtze Cratons' relative accessibility and favorable geological features make them particularly attractive for future natural hydrogen exploration.

Given these geological characteristics, a comprehensive and multidisciplinary approach is essential for effective exploration. The recommended methodological framework encompasses the following steps:

1. Exploration Area Selection
China's extensive geological mapping, covering over 7 million km^2, serves as a critical resource for identifying regions with significant potential for natural hydrogen accumulation [127]. Area selection should be based on a combination of geological criteria, such as the presence of suitable source rocks, fault systems conducive to fluid migration, and formations that can act as reliable reservoirs and caprocks [115]. While geological data provide a solid foundation, additional research is required to further refine the conditions conducive to hydrogen generation, migration, and accumulation [35, 115]. As natural hydrogen exploration in China is still in the early stages, ongoing efforts to enhance geological models and datasets and incorporate new findings are essential for identifying the most promising exploration zones.

2. Geochemical Surveys and Hydrogen Seepage Monitoring
High-density, grid-based gas geochemical surveys, complemented by oriented geochemical profiles, have proven effective in delineating fault structures and detecting near-surface gas anomalies [100, 128]. Although still undergoing methodological refinement, soil gas surveys offer valuable insights into the migration patterns of hydrogen at shallow depths [129]. Long-term monitoring of hydrogen seepage is essential for tracking gas flow and emission dynamics, facilitating the identification of prospective subsurface reservoirs and improving our understanding of the spatial and temporal variability of natural hydrogen systems [130, 131].

While fairy circles—circular barren patches potentially linked to subsurface gas seepage—have been reported in regions such as Namibia and the São Francisco Basin in Brazil [113, 130–132], no such surface expressions have yet been documented in China. As a result, geochemical surveys and continuous hydrogen seepage monitoring remain the primary tools for hydrogen exploration in Chinese settings. Moving forward, the identification of subtle spectral anomalies associated with seepage-affected ground may offer additional indicators for locating active hydrogen emissions.

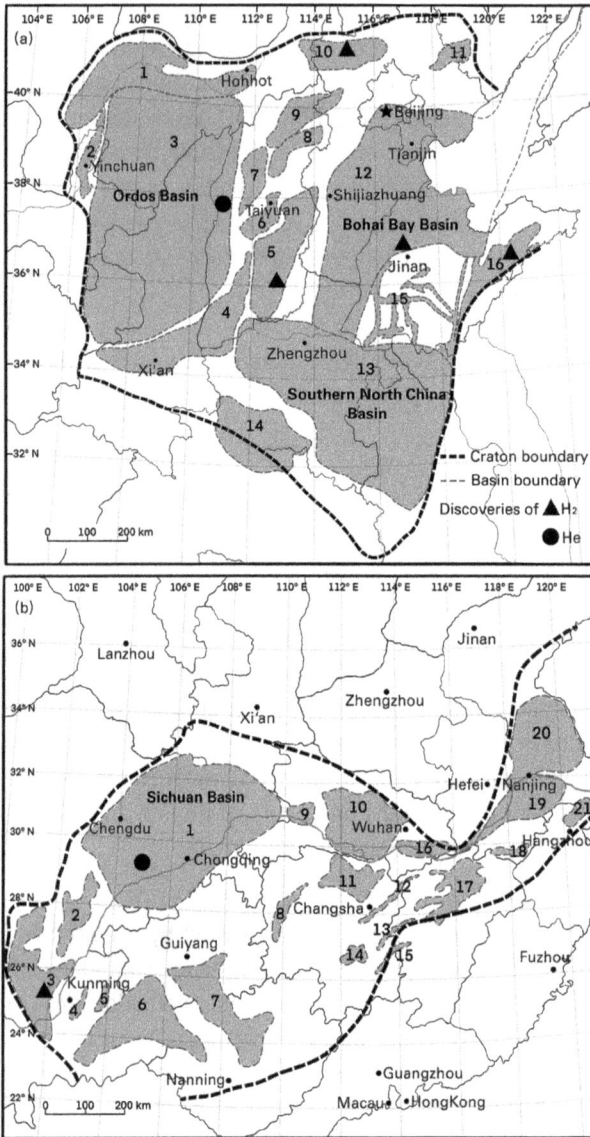

Figure 16.3: Natural hydrogen occurrences in the North China (a) and Yangtze Cratons (b), respectively. (a) North China Craton: 1, Hetao Basin; 2, Yinchuan Basin; 3, Ordos Basin; 4, Linwei Basin; 5, Qinshui Basin; 6, Taiyuan Basin; 7, Ningwu Basin; 8, Sanggan River Basin; 9, Datong Basin; 10, Zhangbei Basin; 11, Chengde Basin; 12, Bohai Bay Basin; 13, Southern North China Basin; 14, Nanxiang Basin; 15, Jining Basin; 16, Jiaolai Basin. (b) Yangtze Craton: 1, Sichuan Basin; 2, Xichang Basin; 3, Chuxiong Basin; 4, Kunming Basin; 5, Qvjing Basin; 6, Nanpan River Basin; 7, Guiyang-Liuzhou Basin; 8, Mayang-Huaihua Basin; 9, Zigui Basin; 10, Jianghan Basin; 11, Dongting Lake Basin; 12, Changsha-Pingjiang Basin; 13, Youxian Basin; 14, Hengyang Basins; 15, Chaling Basin; 16, Yangxin-Jinniu Basin; 17, Poyang Lake Basin; 18, Xiuning Basin; 19, Ningwu Basin; 20, Subei Basin; 21, Jiaxing Basin.

3. Identification of Hydrogen Sources and Generation Mechanisms

Ongoing research is needed to better characterize the relationship between hydrogen isotopic signatures and their geological sources, including whether the hydrogen is crustal or mantle-derived [14, 37, 39, 115]. Geophysical techniques, such as fluid geophysical exploration, electromagnetic surveys, and radioactive exploration methods, are crucial for identifying hydrogen kitchens [115, 132]. Additionally, studying iron-rich lithologies, such as BIFs, and radioactive formations, in combination with analyzing groundwater flow dynamics, can provide key insights into the geochemical processes that contribute to hydrogen generation, such as water-rock reactions and radiolysis [126]. A more refined understanding of these processes will enhance our ability to efficiently locate and exploit hydrogen resources.

4. Reservoir Characterization and Viability Assessment

To assess potential hydrogen reservoirs, advanced geophysical techniques, including two-dimensional and three-dimensional seismic surveys, are essential. These surveys enable detailed mapping of subsurface structures, including fault systems that may facilitate hydrogen migration and porous rock formations that may serve as potential storage reservoirs [132]. Seismic data also provide valuable information on the composition and location of caprock formations that prevent gas escape, which is essential for determining the long-term viability of a reservoir [133]. Understanding the structural and stratigraphic configurations of deep strata, as well as the gas content and permeability of potential reservoir rocks, is vital for evaluating hydrogen storage potential. Combined with geological modeling, these geophysical tools offer a comprehensive understanding of the subsurface architecture, facilitating the identification of high-potential hydrogen reservoirs [132].

5. Drilling and Production Feasibility Testing

Drilling is the definitive method for confirming the presence of natural hydrogen and evaluating its potential for large-scale production [35, 134]. Real-time logging during drilling operations is essential for monitoring gas flows, pressures, and gas composition, providing immediate insights into the characteristics of hydrogen reservoirs. This data is crucial for assessing reservoir properties such as gas saturation, pressure regimes, and potential for long-term production [35, 134]. Production tests, including well testing and production simulations, are necessary to determine flow rate, sustainability, and economic feasibility. These tests help optimize extraction strategies and provide critical information on the long-term viability of hydrogen production from a given reservoir [35].

In addition to advancing technological and geological understanding, the successful exploration of natural hydrogen in China will depend heavily on the establishment of a clear and robust regulatory framework [135]. As a novel mineral resource, natural hydrogen has not yet been incorporated into China's mineral rights registration system, which poses significant challenges to large-scale exploration and development. The absence of a formal legal framework complicates the classification, licensing and utilization of natural hydrogen, potentially deterring investment and hindering exploration efforts.

To address these challenges, policymakers should establish a dedicated regulatory framework for natural hydrogen, including provisions for exploration rights and licensing. Drawing on the experiences of countries like France and Australia, where natural hydrogen exploration has gained momentum, enabled by flexible, market-oriented policy frameworks and catalyzed by an active startup ecosystem, China can adopt best practices to integrate this new resource into its mineral resource management system. This would facilitate systematic exploration, encourage investment, and position China as a global leader in sustainable hydrogen energy development, in line with its broader goals for energy transition and carbon neutrality [7, 135].

The recent inclusion of hydrogen in China's new Energy Law is a promising step toward formalizing its status as a primary and secondary energy source [136]. This legal recognition enhances the appeal of natural hydrogen as an energy asset, increasing investor confidence and accelerating technological advancements, infrastructure development, and pilot projects. Effective implementation of this framework will require further measures, including the establishment of clear licensing procedures, market-driven incentives to foster innovation, and enhanced international collaboration. By addressing these regulatory challenges, China can unlock its vast natural hydrogen potential and play a pivotal role in the global hydrogen economy.

References

[1]   Johnston B., Mayo M. C., Khare A. Hydrogen: The energy source for the 21st century. Technovation. 2005, 25: 569–585.

[2]   China Hydrogen Industry Outlook. Shanghai, China: Boston consulting group, 2023. (Accessed on July 1, 2024, at https://web-assets.bcg.com/d4/2a/657fc0544c4e85a6c7533aed18fa/bcgxouyang-minggao-china-hydrogen-industry-outlook-en.pdf.)

[3]   China Hydrogen Industry Report 2024. Beijing, China: China EV100 low carbon institute, 2024. (Accessed on July 1, 2024, at http://www.wuhaneca.org/uploads/PDF/中国氢能产业发展报告2024.pdf.)

[4]   Global Hydrogen Review 2024. Paris, France: International energy agency, 2024. (Accessed on November 17, 2024, at https://iea.blob.core.windows.net/assets/89c1e382-dc59-46ca-aa47-9f7d41531ab5/GlobalHydrogenReview2024.pdf.)

[5]   Xiang P. P., He C. M., Chen S., Jiang W. Y., Liu J., Jiang K. J. Role of hydrogen in China's energy transition towards carbon neutrality target: IPAC analysis. Advances in Climate Change Research. 2023, 14: 43–48.

[6]   Oliveira A. M., Beswick R. R., Yan Y. A green hydrogen economy for a renewable energy society. Current Opinion in Chemical Engineering. 2021, 33: 100701.

[7]   Blay-Roger R., Bach W., Bobadilla L. F., et al. Natural hydrogen in the energy transition: Fundamentals, promise, and enigmas. Renewable and Sustainable Energy Reviews. 2024, 189: 113888.

[8]   Wang X., Xu S., Chen J., Sun M., Xue X., Wang W. Characteristics of hot spring gas composition and helium isotope composition in the Tengchong volcanic area. Chinese Science Bulletin. 1993, 38: 814–817.

[9] Shangguan Z. Variations of chemical compositions of mantle-derived magmatic gases and the origins during their ascent towards the surface. Geology Review. 1999, 45: 926–933.

[10] Shangguan Z., Bai C., Sun M. Mantle-derived magmatic gas releasing features at the Rehai area, Tengchong county, Yunnan Province, China. Science in China Series D-Earth Sciences. 2000, 43: 132–140.

[11] Shangguan Z., Zhao C., Li H., Gao Q., Sun M. Evolutionary characteristics of recent hydrothermal explosions in the Tengchong Rehai volcanic geothermal region. Bulletin of Mineralogy, Petrology and Geochemistry. 2004, 23: 124–128.

[12] Shangguan Z., Zheng Y., Dong J. Matter Source of released gas in Changbaishan Tianchi volcanic geothermal region. Science in China Series D-Earth Sciences. 1997, 27: 318–324.

[13] Shangguan Z., Sun M. Releasing characteristic of rare gas from mantle in Changbaishan Tianchi volcanic region. Chinese Science Bulletin. 1996, 41: 1695–1698.

[14] Hao Y., Pang Z., Tian J., et al. Origin and evolution of hydrogen-rich gas discharges from a hot spring in the eastern coastal area of China. Chemical Geology. 2020, 538: 119477.

[15] Li Z., Liu H., Liu P., et al. Characteristics and geological significance of escaping gas rich in natural hydrogen from pilot well BYP5 cores of lower sub-member of third member of Shahejie Formation and Zhanhua Sag, Bohai Bay Basin. Petroleum Geology & Experiment. 2024, 46: 979–988.

[16] Shuai Y. H., Zhang S. C., Su A. G., et al. Geochemical evidence for strong ongoing methanogenesis in Sanhu region of Qaidam Basin. Science in China Series D-Earth Sciences. 2010, 53: 84–90.

[17] Guo Z. Q., Liu J. F., Li G. S. A discussion on gas sources in deep gas fields Daqing oil field. Oil & Gas Geology. 2007, 28: 441–448.

[18] Dai J., Hu G., Ni Y., et al. Distribution characteristics of natural gas in Eastern China. Journal of Natural Gas Geoscience. 2009, 20: 471–487.

[19] Han S. B., Tang Z., Wang C., Horsfield B., Wang T., Mahlstedt N. Hydrogen-rich gas discovery in continental scientific drilling project of Songliao Basin, Northeast China: New insights into deep Earth exploration. Science Bulletin. 2022, 67: 1003–1006.

[20] Sun L., Feng Z., Jiang H., Zeng H. Geological survey and study of hydrogen-rich natural gas in Songliao Basin. Petroleum Geology & Oilfield Development in Daqing. 2024, 43: 7–16.

[21] Jin Z., Zhang P., Liu R., et al. Discovery of anomalous hydrogen leakage sites in the Sanshui Basin, South China. Science Bulletin. 2024, 69: 1217–1220.

[22] Qin C., Yu Q., Liu W., et al. Reservoir characteristics of organic-rich mudstone of Niutitang Formation in Northern Guizhou. Journal of Southwest University(Science & Technology. 2017, 39: 13–24.

[23] Li X., Liu Y., Wen J. Geochemical characteristics of the natural gas from well Wulong-1, Chuxiong Basin, and its geological significance. Natural Gas Industry. 2002, 22: 16–19.

[24] Li Y., Wei X., Lu J., Jiang T., Hang J. Origin of Cenozoic hydrogen in Shangdu Basin, Inner Mongolia Autonomous Region. Natural Gas Industry. 2007, 27: 28.

[25] Yunnan Second Geological Engineering Survey Institute. Geothermal resources survey and development evaluation report of Tengchong County, Baoshan city. Baoshan, Yunnan, China: Yunnan Second Geological Engineering Survey Institute, 2013.

[26] Long J. Distribution and formation characteristics of Tengchong geothermal field. Journal of Chengdu Institute of Geology. 1988, 15: 79–85. 1998.

[27] Zhang X., Guo L. Geological and geochemical characteristics of Changbai Mountain geothermal field. Jilin Geology. 2006, 25: 125–130.

[28] Shangguan Z., Sun M., Li H. Types of modern geothermal fluid activity in Tengchong area, Yunnan. Seismology, Geology. 1999, 2: 437–442.

[29] Fang N. Study on geological characteristics and formation mechanism of Tengchong Rehai geothermal field. Master's thesis. Yunnan, China: Kunming University of Science and Technology, 2013.

[30] Wang M. Hydrochemical and isotopic characteristics of underground hot water in Rehe area, 2020. PhD thesis. Beijing, China: China University of Geosciences (Beijing), 2020.

[31] Lin Y., Gao Q. Study on chemical characteristics of underground hydrothermal fluid in Tianchi volcanic area of Changbai Mountain. Geology Review. 1999, 45: 241–247.

[32] Luan G., Wang W., Liu D., Liu J. XRD determination division of Quaternary deposits depositional model of the warm spring geothermal geology. Acta Geoscientica Sinica. 2003, 24: 357–360.

[33] Liao Z., Yin Z., Jia X., Lv W. Conceptual model of Tengchong Rehe geothermal field. Journal – Geological Collections Group. 1997, 3: 2212–2220.

[34] Shangguan Z., Huo W. The δD value of H2 escaping from the hot area of Tengchong and its causes. Chinese Science Bulletin. 2002, 47: 148–150.

[35] Prinzhofer A., Cissé C. S., Diallo A. B. Discovery of a large accumulation of natural hydrogen in Bourakebougou (Mali). International Journal of Hydrogen Energy. 2018, 43: 19315–19326.

[36] Guelard J., Beaumont V., Rouchon V., et al. Natural H2 in Kansas: Deep or shallow origin. Geochemistry, Geophysics, Geosystems. 2017, 18: 1841–1865.

[37] Milkov A. V. Molecular hydrogen in surface and subsurface natural gases: Abundance, origins and ideas for deliberate exploration. Earth-Science Reviews. 2022, 230,104063.

[38] Moine B. N., Bolfan-Casanova N., Radu I. B., et al. Molecular hydrogen in minerals as a clue to interpret ∂D variations in the mantle. Nature Communications. 2020, 11: 3604.

[39] Boreham C. J., Edwards D. S., Czado K., et al. Hydrogen in Australian natural gas: Occurrences, sources and resources. APPEA Journal. 2021, 61: 163–191.

[40] Lin L. H., Slater G. F., Lollar B. S., Lacrampe-Couloume G., Onstott T. C. The yield and isotopic composition of radiolytic H_2, a potential energy source for the deep subsurface biosphere. Geochimica et Cosmochimica Acta. 2005, 69: 893–903.

[41] Xu M. J., Huang H., Huang Z. C., et al. Insight into the subducted Indian slab and origin of the Tengchong volcano in SE Tibet from receiver function analysis. Earth & Planetary Science Letters. 2018, 482: 567–579.

[42] Gao L., Shangguan Z., Wei H., Wucheng Z. Recent geochemical changes of hot-spring gases from Tianchi volcano area, Changbai Mountains, Northeast China. Earthquake Geology. 2006, 128: 358–366.

[43] Zhang L., Gao W., Su Y., Sun Z., Duan S., Zeng Q. Analysis of intense geothermal activity and seismic fluid changes in the Tengchong volcanic area. Plateau Earthquake. 2019, 31: 5–11.

[44] Li D. M., Li Q., Chen W. J. Volcanic activities in the Tengchong volcano area since Pliocene. Acta Petrologica Sinica. 2000, 16: 362–370.

[45] Li N., Zhang L. Y. A study on volcanic minerals and hosted melt inclusions in newly-erupted Tengchong volcanic rocks, Yunnan Province. Acta Petrologica Sinica. 2011, 27: 2842–2854.

[46] Liao Z. Tengchong volcanoes and geothermal energy. Geology Review. 1999, 45: 934–939.

[47] Mu Z., Tong W., Garniss H. C. The age of Tengchong volcanic activity and the source of magma. Chinese Journal of Geophysics. 1987, 30: 262–270.

[48] Liu X., Sui W., Xiang T., Wang X. Holocene volcanic activities and their features in Tianchi area, Changbaishan Mountains. Quaternary Science. 2004, 24: 638–644.

[49] Yang Q., Bo J. Status quo and prospects for research on Tianchi Volcano in Changbai Mountain. Journal of Catastrophology. 2007, 16: 133–139.

[50] Wei H., Liu R., Fan Q., Yang Q., Li N. Tianchi Volcano in Changbai Mountain – A polygenetic central volcano. Geology Review. 1999, 45: 257–262.

[51] Jiang M., Tan H., Zhang Y., et al. Geophysical model of the Mazhan-Gudong magma chamber in the Tengchong volcanic tectonic area, Yunnan. Acta Geoscientica Sinica. 2012, 33: 731–739.

[52] Bai D., Liao Z., Zhao G., Wang X. Inferring the magmatic heat source of the Tengchong Rehai hot field from MT detection results. Science Bulletin. 1994, 39: 344–347.

[53] Xu Y., Wang S., Zhang L., Yang X. Crust and upper mantle structure and its relations with magma activity of the Tengchong volcanic area. Progress in Geophysics. 2015, 30: 1034–1038.

[54] Li X., Xu Y., Wang S. Evidence of magma activity from S-wave velocity structure of the Tengchong volcanic area. Chinese Science Bulletin. 2017, 62: 2067–2077.

[55] Shen W. H., Liu S. L., Yang D. H., Wang W. S., Xu X. W., Yang S. X. The crustal and uppermost mantle dynamics of the Tengchong-Baoshan region revealed by P-wave velocity and azimuthal anisotropic tomography. Geophysical Journal International. 2022, 230: 1092–1105.

[56] Zhao C., Ran H., Chen K. Existing magma chambers in the Tengchong volcanic area inferred from relative geothermal gradients. Acta Petrologica Sinica. 2006, 22: 1517–1528.

[57] Yang H., Hu J., Hu Y., et al. Crustal structure in the Tengchong volcanic area and position of the magma chambers. Journal of Asian Earth Sciences. 2013, 73: 48–56.

[58] Zhang L., Hu Y. L., Qin M., et al. Study on crustal and lithosphere thicknesses of Tengchong volcanic area in Yunnan. Chinese Journal of Geophysics. 2015, 58: 1622–1633.

[59] Ye T., Huang Q., Chen X., et al. Magma chamber and crustal channel flow structures in the Tengchong volcano area from 3-D MT inversion at the intracontinental block boundary southeast of the Tibetan Plateau. Journal of Geophysical Research. Solid Earth. 2018, 123: 11112–11126.

[60] Zhang R., Wu Y., Gao Z., et al. Upper mantle discontinuity structure beneath eastern and southeastern Tibet: New constraints on the Tengchong intraplate volcano and signatures of detached lithosphere under the western Yangtze Craton. Journal of Geophysical Research. Solid Earth. 2017, 122: 1367–1380.

[61] Zhao C., Ran H., Wang Y. Present-day mantle-derived helium release in the Tengchong volcanic field, southwest China: Implications for tectonics and magmatism. Acta Petrologica Sinica. 2012, 28: 1189–1204.

[62] Guo L., Shizhuang M., Yushen Z. Application of seismic CT technology to study the magma chamber of Changbai Mountain volcano. CT Theory and Applications Research. 1996, 5: 47–52.

[63] Duan Y., Zhang X., Liu Z., et al. A study on crustal structures of Changbaishan-Jingpohu volcanic area using receiver function. Chinese Journal of Geophysics. 2005, 2: 352–358.

[64] Yang Z., Zhang X., Zhao J., Yang J., Duan Y., Wang S. Tomographic imaging of 3-D crustal structure beneath Changbaishan-Tianchi volcano region. Chinese Journal of Geophysics. 2005, 46: 352–358.

[65] Liu Z., Zhang X., Wang F., Duan Y., Lai X. Two-dimensional crustal Poisson's ratio from seismic travel time inversion in Changbaishan Tianchi volcanic region. Acta Seismologica Sinica. 2005, 27: 324–331.

[66] Zhang C., Zhang X., Zhao J., et al. Study on the crustal and upper mantle structure in the Tianchi volcanic region and its adjacent area of Changbaishan. Chinese Journal of Geophysics. 2002, 45: 812–820.

[67] Wang F., Zhang X., Yang Z. Two-dimensional crustal structure of Changbaishan-Tianchi volcanic region determined by seismic travel time inversion. Acta Seismologica Sinica. 2002, 24: 144–152.

[68] Li H., Yang C. Underground fluid study in Tengchong Rehai. Earthquake Research. 2000, 23: 231–238.

[69] Lv Z., Yang Q., Zhang H., Liu G., Gao J. Study on the relationship between Changbaishan-Tianchi volcanic seismic activity and deep earthquakes in the Northwest Pacific subduction zone. Seismology, Geology. 2007, 29: 470–479.

[70] Jiang C. Period division of volcanic activities in the Cenozoic Era of Tengchong. Earthquake Research. 1998, 21: 320.

[71] Gao Q. Volcanic hydrothermal activities and gas-releasing characteristics of Tianchi Lake Region, Changbai Mountains. Acta Geologica Sinica. 2004, 25: 345–350.

[72] Chen X., He W., Feng Z. Controlling effect of major faults on the gas reservoirs in the Xujiaweizi fault depression, Songliao Basin. Natural Gas Industry. 2012, 32: 53–58.

[73] Jiang C., Cang S., Wu J. Types and pooling patterns of deep gas reservoirs in Xujiaweizi fault depression. Natural Gas Industry. 2009, 29: 5–7.

[74] Li Y., Fu X., Zhang M. Fault deformation features and reservoir-controlling mechanisms of Xujiaweizi fault depression in Songliao basin. Journal of Natural Gas Geoscience. 2012, 23: 979–988.

[75] Ren Y., Zhu D., Wan C., Feng Z., Li J., Wang C. Natural gas accumulation rule of Xujiaweizi Depression in Songliao Basin and future exploration target. Petroleum Geology & Oilfield Development in Daqing. 2004, 23: 26–29.

[76] Meng F., Li J., Li M., et al. Geochemistry and tectonic implications of rhyolites from Yingcheng Formation in Xujiaweizi, Songliao Basin. Acta Petrologica Sinica. 2010, 26: 227–241.

[77] Hu M., Fu G., Lv Y., Fu X., Pang L. The fault activity period and its relationship to deep gas accumulation in the Xujiaweizi Depression, Songliao Basin. Geology Review. 2010, 56: 710–718.

[78] Fu X., Sha W., Wang L., Liu X. Distribution law of mantle-origin CO_2 gas reservoirs and its controlling factors in Songliao Basin. Journal of Jilin University(Earth Science Edition). 2009, 2: 253–263.

[79] Yan L., Hu Y., Ran Q., et al. Volcanic characteristics and eruption models of Yingcheng Formation No. 1 member in Xingcheng area, Xujiaweizi fault depression of Songliao Basin. Journal of Natural Gas Geoscience. 2008, 19: 821–825.

[80] Yin J., Liu H., Chi H. Evolution and gas-accumulation of Xujiaweizi Depression in Songliao Basin. Acta Petrologica Sinica. 2002, 23: 26–29.

[81] Hu W., Lv B., Zhang W., Mao Z., Leng J., Guan D. An approach to tectonic evolution and dynamics of the Songliao Basin. Chinese Journal of Geology. 2005, 40: 16–31.

[82] Zhang S., Zou C., Peng C., et al. Abnormally high natural radioactivity zones in the main borehole of the Continental Scientific Drilling Project of Cretaceous Songliao Basin: Geophysical log responses and genesis analysis. Chinese Journal of Geophysics. 2018, 61: 4712–4728.

[83] Hou H., Wang C., Zhang J. Deep continental scientific drilling engineering in Songliao Basin: Resource discovery and progress in earth science research. China Geology. 2018, 45: 641–657.

[84] Xiao J. Development characteristics of fluid inclusions in volcanic rocks of the Yingcheng Formation in the Xujiaweizi fault depression and their geological significance. Master's Thesis. Zhejiang, China: Zhejiang University, 2011.

[85] Wang Z. Accumulation analysis of volcanic gas reservoir in Yingcheng formation of Block Shengshen2-1 of Northern Part of Songliao Basin. Inner Mongolia Petrochemical Industry. 2013, 3: 135–140.

[86] Jin D. Research on the distribution relationship between gas reservoir and geochemical characteristics of formation water – An example from Xujiaweizi faulted depression of northern Songliao Basin. Inner Mongolia Petrochem Ind 2013, 137–139.

[87] Zhang Z., Huang F., Xu J. F., et al. Discovery and geological implications of the Early Cretaceous basaltic andesites in SK2 borehole. Acta Petrologica Sinica. 2022, 38: 1756–1770. 2022.

[88] Yin Y. Geochronology and petrology of the Triassic basement in the Songliao Basin and its tectonic significance. PhD Thesis. Jilin, China: Jilin University, 2022.

[89] Yun J., Jin Z., Yin J., et al. Reflection feature and geodynamic significance of deep seismic reflection in Xujiaweizi region of north Songliao Basin, China. Frontiers of Earth Science. 2008, 15: 307–314.

[90] Yang, B.J., Liu, C., Tang, J., et al. The basic characteristics of the structure of the Anda-Fengle nearly vertical seismic reflection profile of China. In: R. Carbonell, J. Gallart, & M. Tome (Eds.), Proceedings of the 8th International Symposium on Deep Seismic Profiling of the Continents and Their Margins. Barcelona, Spain: Consejo Superior de Investigaciones Científicas (CSIC) and University of Barcelona. 1998, 134.

[91] Zhang J., Wei P., Guo Y., et al. Discussion on the relationship between deep crustal structural characteristics and oil and gas fields in some petroliferous basins in China. Journal of Natural Gas Geoscience. 1998, 9: 28–36.

[92]   Zeng G., Tang Z. Origins of hydrocarbon and non-hydrocarbon gases in Sanshui Basin, Guangdong. Oil & Gas Geology. 1988, 4: 363–369.

[93]   Shi L., Feng Z., Shi X., et al. Detailed geological survey report on coal in Gonghui area, Zhangbei County, Hebei Province. Hebei, China: The Third Geological Brigade of Hebei Provincial Bureau of Geology and Mineral Resources, 2012.

[94]   Zhang Y., Zhang Z., Liu Z., Wei W., Zhang D. Regional geology of China (Hebei). Beijing, China: Geological Publishing House, 2017.

[95]   Li W. Study on metamorphic age and contact relationship of Hongqiyingzi Group in Zhangbei-Weichang stratigraphic community. Geology Hebei Province. 2022, 4: 46.

[96]   Ju Z., Li Q., Zhang Y., et al. Regional geology of Hebei Province, Beijing and Tianjin. Beijing, China: Geological Publishing House, 1989, 538–628.

[97]   Liu S., Lv Y., Feng Y., et al. Zircon and monazite geochronology of the Hongqiyingzi complex, northern Hebei. Geological Bulletin of China. 2007, 26: 1086–1100.

[98]   Ma C., Wang J., Zhang X., et al. Discussion on the formation and evolution history of faults in the Shangyi-uplift region of northern Hebei. China Geological Survey. 2020, 7(5): 88–94.

[99]   Liu H., Zhang H. Is the serpentinized peridotite in Chicheng, Northern Hebei, a fragment of Archaean ophiolite? Acta Petrologica Sinica. 2002, 38: 3760–3770.

[100] Qian H., Li L., Qian H., et al. The ductile searing belts of high-pressure granulite-garner-amphibolite in North China Platform. Acta Geoscientica Sinica. 1996, 17: 19–31.

[101] Wu Z. Deformation features of Chongli-Longhua-Fuxin structural zone and its control of mineralization. Journal of Geomechanics. 1997, 3: 73–81.

[102] Hu L., Song H., Yan D., et al. $^{40}Ar/^{39}Ar$ age records of mylonites in the Shangyi-Chicheng fault zone and geological significance. Science in China Series D-Earth Sciences. 2002, 23: 597–604.

[103] Wang H., Zhao F., Li H., et al. Zircon SHRIMPU-Pb age of the dioritic rocks from northern Hebei: The geological records of late Paleozoic magmatic arc. Acta Petrologica Sinica. 2007, 23: 597–604.

[104] Wang H., Chu H., Xiang Z., et al. The Hongqiyingzi Group in the Chongli-Chicheng area, northern margin of the North China Craton: A suite of late Paleozoic metamorphic complex. Frontiers of Earth Science. 2012, 23: 597–604.

[105] Yu J., Che Y., Zhang P., et al. Subsurface fluid anomalies before the Ms4.2 earthquake in Zhangjiakou, China. Earthquake. 1998, 18: 405–409.

[106] An S., Zhang G. Fault structure and earthquake zone distribution in Zhangjiakou-Xuanhua area. Journal of Xi'an Shiyou University, Natural Science Edition. 2016, 19: 83–87.

[107] Fang Z., Liu Y., Yang X., et al. Advance in the research on the source and migration mechanisms of gas in the seismic fault zone. Progress in Geophysics. 2012, 27: 483–495.

[108] Du L. Introduction to the new principles of earth science. Lanzhou, China: Lanzhou University Press, 2017.

[109] Wang Z., Liu S. Evolution of intraplate alkaline to tholeiitic basalts via interaction between carbonated melt and lithospheric mantle. Journal of Petrology. 2021, 62: egab025.

[110] Pasquet G., Idriss A. M., Ronjon-Magand L., et al. Natural hydrogen potential and basaltic alteration in the Asal–Ghoubbet rift, Republic of Djibouti. BSGF – Earth Sciences Bulletin. 2023, 194: 9.

[111] Satake H., Ohashi M., Hayashi Y. Discharge of H_2 from the Atotsugawa and Ushikubi faults, Japan, and its relation to earthquakes. Pure and Applied Geophysics. 1984, 122: 185–193.

[112] Liu Y., Li J., Wang Y., et al. The relationship between seismic activity and geological structure in Zhangjiakou area. Seismological and Geomagnetic Observation and Research. 2017, 38: 14–21.

[113] Donzé F. V., Truche L., Shekari Namin P., et al. Migration of natural hydrogen from deep-seated sources in the São Francisco Basin, Brazil. Geosciences. 2020, 10: 346.

[114] Zgonnik V. The occurrence and geoscience of natural hydrogen: A comprehensive review. Earth-Science Reviews. 2020, 203: 103140.

[115] Moretti I., Brouilly E., Loiseau K., et al. Hydrogen emanations in intracratonic areas: New guide lines for early exploration basin screening. Geosciences. 2021, 11: 145.

[116] Zhou X., Ni Z., Shi N. The geological and geochemical characteristics of Biotite Plagioclase Gneiss from Hongqiyingzi Group in Chongli Country, Northern Hebei Province. Journal of Southwest University Science & Technology. 2013, 28: 23–30.

[117] Bai W., Zhou M., Hu X., et al. Lithospheric tectonic evolution and mafic-ultramafic complex and mineralization characteristics of the North China Block. Beijing, China: Seismological Press, 1993.

[118] Tong Y., Ni Z., Wang R. Discovery of ultrabasic rocks with spinifex texture in northern Hebei, North China Craton. Mineralogy and Petrology. 2003, 23: 6–10.

[119] Cai H., Zhang S., Liu Y., et al. Detection and research of active faults in Zhangbei earthquake zone, Hebei Province. Shanxi Earthquake. 2003, 2: 23–28.

[120] Lu J., Wei X., Cao X., et al. Research on CO_2 gas pool-geological conditions in Shangdu area, Inner Mongolia. Northwest Geology. 2002, 35: 122–134.

[121] Zhang Q., Zhou G. Q., Wang Y. The distribution of time and space of Chinese ophiolites and their tectonic settings. Acta Petrologica Sinica. 2003, 19: 1–8.

[122] Li Z. P., Wu L., Yan L. L. Spatial and temporal distribution of ophiolites and regional tectonic evolution in Northwest China. Geological Bulletin of China. 2020, 39: 783–817.

[123] Xu X., Gao C., Huang Z., et al. Three stages of tectonic movements in formation of petroliferous basins in China. Oil & Gas Geology. 2005, 26: 155–162.

[124] Huang L., Liu C. Y., Kusky T. M. Cenozoic evolution of the Tan-Lu Fault Zone (East China) – Constraints from seismic data. Gondwana Research. 2015, 28: 1079–1095.

[125] He Y., Zhang Z., Mao J., Zhang R. The effect of Tan-Lu Fault on the formation of mineral deposits and oil-gas fields. Geotectonics and Metallogeny. 2002, 26: 10–15.

[126] Geymond U., Briolet T., Combaudon V., et al. Reassessing the role of magnetite during natural hydrogen generation. Frontiers of Earth Science. 2023, 11: 1169356.

[127] Xiao G., Mao X., Li M., et al. Progresses in China's basic geological survey and potential business orientation. China Geological Survey. 2014, 1: 1–14.

[128] Langhi L., Strand J. Exploring natural hydrogen hotspots: A review and soil-gas survey design for identifying seepage. Geoenergy. 2023, 1: 1–15.

[129] Patino C., Piedrahita D., Colorado E., Aristizabal K., Moretti I. Natural H_2 transfer in soil: Insights from soil gas measurements at varying depths. Geosciences. 2024, 14: 296.

[130] Cathles L., Prinzhofer A. What pulsating H_2 emissions suggest about the H_2 resource in the Sao Francisco Basin of Brazil. Geosciences. 2020, 10: 149.

[131] Moretti I., Prinzhofer A., Françolin J., et al. Long-term monitoring of natural hydrogen superficial emissions in a Brazilian cratonic environment. Sporadic large pulses versus daily periodic emissions. International Journal of Hydrogen Energy. 2021, 46: 3615–3628.

[132] De Freitas V. A., Prinzhofer A., Françolin J. B., Ferreira F. J., Moretti I. Natural hydrogen system evaluation in the São Francisco Basin (Brazil). Science and Technology for Energy Transition. 2024, 79: 1–25.

[133] Gao K., Creasy N. M., Huang L., Gross M. R. Underground hydrogen storage leakage detection and characterization based on machine learning of sparse seismic data. International Journal of Hydrogen Energy. 2024, 61: 137–161.

[134] 27 May 2024 ASX Announcement Ramsay Project Stage I – Interim exploration well testing update. Gold hydrogen. 2024. (Accessed January 20, 2025, at https://www.goldhydrogen.com.au/wp/wp-content/uploads/2024.05.27-ASX-Announcement-Exploration-Well-Testing-Interim-Update.pdf.)

[135] Gaucher E., Moretti I., Pélissier N., Burridge G., Gonthier N. The place of natural hydrogen in the energy transition: A position paper. European Geologist. 2023, 55: 5–9.

[136] Energy law of the people's republic of China. The state council of the people's republic of China, 2024. (Accessed on November 21, 2024 at https://www.gov.cn/yaowen/liebiao/202411/content_6985761.htm.)

Hyeong Soo Kim

Chapter 17
Potential occurrence and reservoirs of natural hydrogen based on the geological and tectonic setting of the Korean Peninsula

Abstract: Regions with high natural hydrogen (H_2) potential in the Korean Peninsula include the Hongseong-Imjingang Belt, formed in a Permian-Triassic subduction zone setting, and the Gyeongsang and Pohang Basins, developed under an extensional tectonic regime during the Cretaceous to Tertiary. Partially serpentinized ultramafic rocks (dunite, wehrlite, harzburgite, and lherzolite) in the Korean Peninsula are distributed sporadically within the Hongseong-Imjingang Belt and along faulted margins of the Gyeongsang Basin. Thermodynamic modeling in CaO-FeO-MgO-Al_2O_3-SiO_2-H_2O-Fe_2O_3 (CFMASHO) system suggests that H_2 can be generated under conditions of 100–370 °C and 0.5–6.0 kbar as olivine and pyroxene react with H_2O to form serpentine, brucite, and magnetite. Under these pressure-temperature conditions, as pressure decreases, rock density decreases (from 2.8 to 2.5 g/cm^3), while H_2 concentration increases. The maximum H_2 concentrations under a geothermal gradient of 20–40 °C/km are 300–700 ppm at a depth of approximately 3.5 km. In the Miocene Pohang Basin, H_2 concentrations identified in deep boreholes show a depth-dependent increase, reaching up to 1,000 ppm at a depth of 11–15 m. Previous studies on seismic velocity anomalies associated with subsurface faults and the mantle-derived helium signatures support the suggestion that H_2 emitted in the Pohang Basin is probably linked to hydrothermal activity in the subcontinental lithospheric mantle. Consequently, this study proposes two potential reservoirs and associated geological processes for H_2 in the Korean Peninsula: (1) H_2 generated through serpentinization in the shallow crust and (2) H_2 originating from the subcontinental lithospheric mantle beneath the Cenozoic extensional basin. Further research should refine the understanding of serpentinization kinetics and structurally controlled migration pathways for sustainable H_2 resource development.

Keywords: natural hydrogen, serpentinization, structurally controlled migration, Hongseong-Imjingang Belt, Pohang Basin

Acknowledgments: The authors declare no conflicts of interest relevant to this study. This work was supported by the National Research Foundation of Korea (NRF) grant funded by the Korean government (MSIT) (NRF-2022R1A2C1003840) and a Korea University Grant.

Hyeong Soo Kim, Department of Earth and Environmental Sciences, Korea University, Seoul 02841, Republic of Korea, e-mail: haskim2@korea.ac.kr

https://doi.org/10.1515/9783111437040-017

17.1 Introduction

Natural hydrogen, generated from various geological processes, represents a sustainable and economical source of carbon-free energy. Since the discovery of natural hydrogen within the Earth's interior, numerous mechanisms for its origin have been proposed ([1, 2] and references therein). A natural hydrogen (H_2) system encompasses the geological processes responsible for the formation, accumulation, migration, and seepage of natural hydrogen within the Earth's crust and mantle. The main mechanisms for H_2 generation in both continental and oceanic lithosphere are (1) rock-fluid interactions (e.g., serpentinization), (2) the radiolysis of water, and (3) degassing from the lower crust and mantle [3–12]. The estimated production of H_2 in the Precambrian lithosphere related to the serpentinization of ultramafic rocks and radiolysis ranges from 0.36 to 2.27×10^{11} mol/year with a maximum of approximately 23 Mt/year [4]. Recently, Ellis and Gelman [13] evaluated the potential of H_2 as an energy resource using a mass balance model. They estimated the total subsurface H_2 resource to range between 10^3 and 10^{10} million metric tons (Mt), with the most probable value being 5.6×10^6 Mt. Recovering just 2% (approximately 10^5 Mt) of the resource could meet global hydrogen demand for 200 years, providing twice the energy of all currently proven H_2 reserves.

The geology of the Korean Peninsula comprises a wide variety of rocks formed across the geological timescale, from the Precambrian basement to the Quaternary volcanic rocks. This study aims to discuss regions in the Korean Peninsula with the potential for H_2 generation and accumulation based on the geological and geophysical characteristics. An H_2 anomaly was recently identified in the Miocene sedimentary basin. The study examines the distribution and geological features of ultramafic rocks in the peninsula, which are attractive sources for H_2 generation [1–3], and presents the physicochemical conditions under which H_2 may be generated through the rock-water interaction of the ultramafic rocks. Consequently, the results on H_2 concentrations and thermodynamic modeling not only enhance understanding of the potential H_2 reservoirs based on the geological and tectonic characteristics but also provide foundational data for future H_2 exploration in the Korean Peninsula.

17.2 Geological and tectonic setting of potential natural hydrogen reservoirs in the Korean Peninsula

The basement of the Korean Peninsula primarily consists of the Paleoproterozoic and Neoproterozoic banded quartzo-feldspathic gneisses in the Gyeonggi and the Yeongnam massifs, separated by two Phanerozoic metamorphic belts: the Hongseong-Imjingang and the Ogcheon metamorphic belts (Figure 17.1). The Cretaceous Gyeongsang Basin, a terrestrial basin, formed under retro-arc and back-arc settings during the northward subduction of the Paleo-Pacific Plate (Figure 17.2a) [14, 15]. The basin contains a thick (~9 km) succession of siliciclastic sedimentary rocks and volcanic deposits (Figure 17.2a). Tertiary Pohang Basin, developed along the southeastern margin of the peninsula (Figure 17.2b), is a Miocene sedimentary basin associated with back-arc extension (e.g., pull-apart basin) driven by the opening of the East Sea [16]. The expansion of the East Sea is attributed to an extensional regime induced by the subduction of the Pacific Plate beneath the Eurasian Plate. The Pohang Basin comprises non-marine to deep-marine Miocene (~20 Ma) sedimentary strata, underlain basement rocks of Cretaceous sedimentary and volcanic rocks, and Late Paleozoic plutonic rocks (Figure 17.2b) [17].

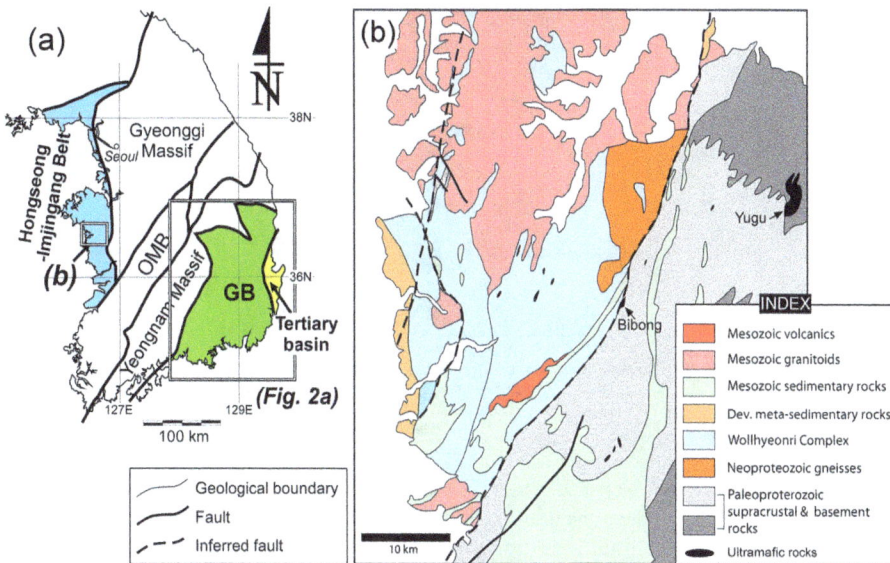

Figure 17.1: (a) Simplified tectonic map displays the locations of the Precambrian Gyeonggi Massif and Yeongnam Massif, the Phanerozoic Ogcheon Metamorphic Belt (OMB) and Hongseong-Imjingang Belt, the Cretaceous Gyeongsang Basin (GB), and the Tertiary Basin of the Korean Peninsula. (b) Geological map of the southern Hongseong-Imjingang Belt showing the distribution of ultramafic rocks with Bibong and Yugu peridotites. Modified from Kim et al. [18].

Ultramafic rocks on the southern Korean Peninsula occur sporadically within the Precambrian gneisses of the Hongseong-Imjingang Belt (Figure 17.1b) [19–21] and within the Cretaceous sedimentary rocks along the northern and eastern margins of the Gyeongsang Basin (Figure 17.2a). The Hongseong-Imjingang Belt experienced medium-to high-pressure regional metamorphism during a continental collisional orogeny during the Permian to Triassic [18, 22, 23], associated with the collision of the North and South China Blocks [24–26]. In addition, extensive magmatic activities occurred from the early Mesozoic to the Cretaceous [27, 28]. Ultramafic rocks at Andong region in the Gyeongsang Basin are in fault contact with Precambrian gneisses and Cretaceous sedimentary rocks along the northern boundary of the basin (Figure 17.2a) [29–31]. Along the eastern boundary, where the basin adjoins Tertiary sedimentary rocks, ultramafic rocks are located within the Ulsan active strike-slip fault system (Figure 17.2a) [32, 33]. The ultramafic rocks in the southern Korean Peninsula are linked to the Late Paleozoic to Early Mesozoic convergent plate boundary and the Cenozoic extensional tectonic regime.

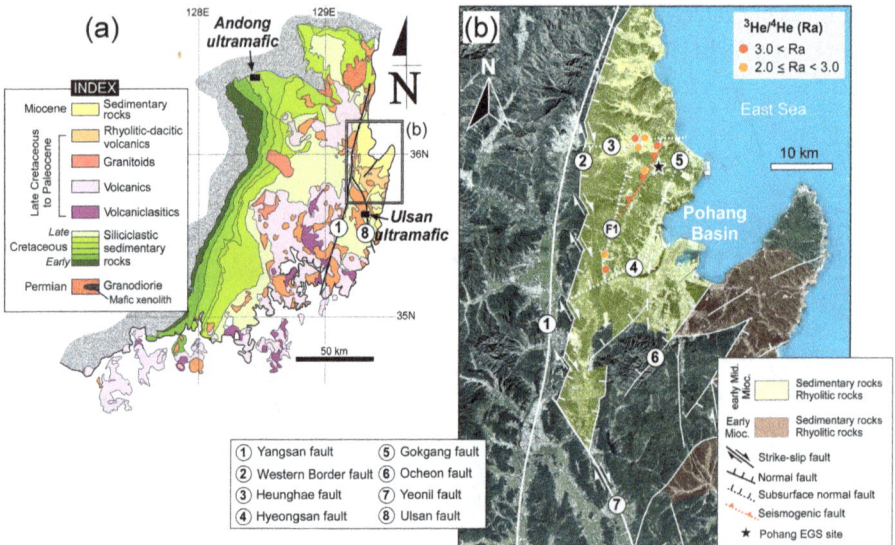

Figure 17.2: (a) Regional geological map of the Cretaceous-Tertiary depositional basins with locations of two main ultramafic rocks (Andong and Ulsan) at the southeastern margin of the Korean Peninsula (modified from Cheon et al. [15]). (b) Simplified geological map of the Miocene Pohang Basin showing main fault systems (modified from Son et al. [34]) with the Pohang enhanced geothermal system (EGS) site. Spots of helium (He) emission with $^3He/^4He$ (Ra) are marked (data from Kim et al. [35]).

Natural hydrogen anomalies in the peninsula have been identified in deep injection boreholes (PX1 and PX2) in the Pohang Basin (Figure 17.2b). These boreholes (~4.3 km length; Figure 17.3) were drilled between 2012 and 2016 at the Pohang Enhanced Geothermal System (EGS) site. Although the site was closed after the 2017 Pohang triggered earthquake [36], the two geothermal boreholes remain operational for monitoring groundwater levels and microseismicity. Logging data from two geothermal boreholes [36, 37] shows that the geology at the Pohang EGS site is dominated by Miocene (~20 Ma) sedimentary rocks to a depth of ~200 m, Cretaceous sedimentary and volcanic rocks to a depth of ~2,350 m, and Permian granodiorite with mafic xenoliths below ~2,350 m (Figure 17.3). The granodiorite, containing mafic xenoliths, was emplaced at 260–250 Ma [38]. Injection borehole PX1 is inclined and located on the hanging wall of the seismogenic fault, while the vertical borehole PX2 intersects the seismogenic fault, with its lowermost section situated on the fault's footwall [36]. Both boreholes are filled with groundwater at the lowermost open-hole section, with water levels measured at 15 m and 290 m in PX1 and PX2, respectively [36].

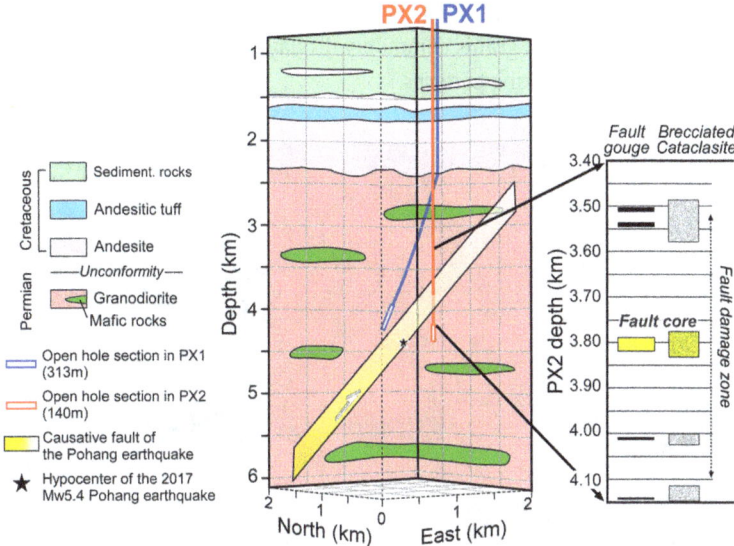

Figure 17.3: Geological column of the Pohang EGS site with two cased boreholes, PX1 and PX2 [33, 36]. The positions and widths of fault core and damage zone in PX2 were constructed based on logging data [37].

The Pohang Basin is bounded by the N-S-striking Western Border fault to the west and the NE-SW-striking Ocheon normal fault and NNW-SSE-striking Yeonil strike-slip fault to the south, all of which formed during the basin's extensional phase (Figure 17.2b) [34, 39]. The Pohang EGS site is surrounded by two NNE-SSW-striking normal faults (the Western Border and Gokgang faults) and two E-W- to ENE-WSW-striking normal faults (Heunghae and Hyeongsan faults; Figure 17.2b). The former is east-dipping listric or bookshelf-type faults, while the latter are conjugate normal faults (Figure 17.2b). The fault system in the Pohang Basin was formed during the Early Miocene during a back-arc extension [40].

17.3 Petrography of ultramafic rocks

Ultramafic rocks in the Hongseong-Imjingang Belt are identified as partially serpentinized dunite, harzburgite, and lherzolite based on the modal proportions of olivine (Ol), orthopyroxene (Opx), clinopyroxene (Cpx), and serpentine (Srp) [20, 21]. The Yugu lherzolite (sample YG) contains fine- to medium-grained Ol with Cpx and Opx porphyroclasts with magnetite (Mt) (Figure 17.4a), and exhibits partially developed foliation (Figure 17.4a). Srp and Mt mainly occur around the fine-grained Ol, with the amount of serpentine increasing in domains where foliation is well developed. Seo et al. [19] reported that serpentinized ultramafic rocks contain 10–38 vol% olivine, 3–17 vol% pyroxene, 30–90 vol% serpentine, and 3–4 vol% magnetite. The Bibong harzburgite (sample BB) consists of equigranular Ol, Opx, and Cpx, which are replaced by Srp, amphibole, and Mt during serpentinization [41]. The Srp around Ol grains commonly displays mesh/hour-glass textures and interpenetrating and/or interlocking patterns.

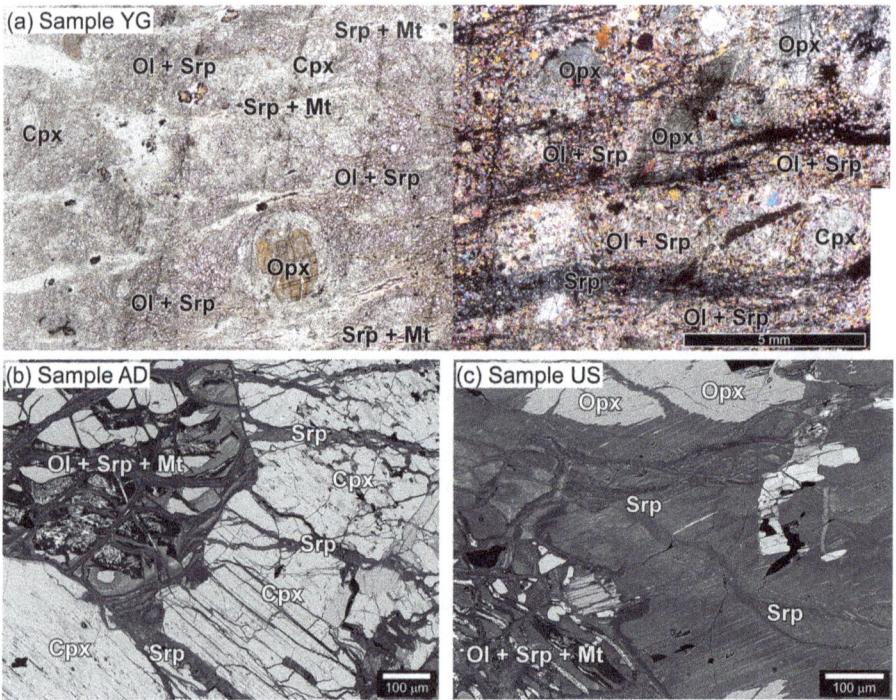

Figure 17.4: (a) Photomicrograph of partially serpentinized Yugu lherzolite (sample YG) in the Hongseong-Imjingang Belt showing olivine (Ol) + clinopyroxene (Cpx) + serpentine (Srp) + magnetite (Mt). The left and right parts are under plane-polarized light (PPL) and cross-polarized light (XPL), respectively. (b) Back-scattered electron (BSE) image of Andong wehrlite in the northern margin of the Gyeongsang Basin (see Figure 17.2a) showing mesh-forming Srp and fibrous Mt around relict Ol. (c) Back-scattered electron (BSE) image of Ulsan harzburgite (sample US) in the southeastern margin of the Gyeongsang Basin (see Figure 17.2a) showing bastite-forming Srp around relict Cpx.

Ultramafic rocks in the Andong and Ulsan regions include dunite, wehrlite, and harzburgite, along with minor pyroxenite, as classified by the modal proportions (% vol.) of Ol (30–90), Cpx (15–65), and Opx (2–50) [30–32, 41, 42, 43]. The ultramafic rocks in these areas are slightly serpentinized wehrlite (sample AD) and harzburgite (sample US). The mesh-forming Srp around relics of Ol grains displays oscillatory patterns in BSE images (Figure 17.4b) with or without Mt. Fibrous magnetite occurs only in dark gray-colored Srp (Figure 17.4b). Bastite textures are formed through the serpentinization of Cpx in the absence of Mt (Figure 17.4c).

17.4 Gas compositions in the Miocene Pohang Basin

Gas compositions at PX1 and PX2 of the Pohang EGS site were periodically measured using a portable gas analyzer from September 2023 to February 2024. The measured gas data are summarized in Table 17.1. The concentrations of O_2 and N_2 at PX1 and PX2 showed little variation, ranging from 18.6% to 22.2% (mean 21.0%) and 77.7% to 80.7% (mean 78.9%), respectively. However, the concentration of H_2 exhibited significant differences (Figures17.5a and b). At PX1, the H_2 concentration remained consistently below 25 ppm, regardless of the measurement depth (Figure 17.5a). In contrast, the H_2 concentration at PX2 ranged from 50 to 150 ppm at a depth of 1 m, and increased to a maximum of 1,000 ppm at depths between 5 and 15 m (Figure 17.5b and c). The mean H_2 concentrations measured at PX2 depths of 1, 3, and 5 m were 81 ± 23 ppm, 171 ± 45 ppm, and 424 ± 121 ppm, respectively (Figure 17.5c). In addition, the mean concentrations at depths of 8–10 and 11–15 m were 723 ± 134 ppm and 938 ± 46 ppm, respectively. H_2 concentrations increased with depth (between 1 and 15 m) in PX2 at an approximate rate of ~70 ppm/m (Figure 17.5c). The temporal variation of H_2 concentrations measured over a day at various depths exhibited a consistent trend in the PX2 borehole (Figure 17.5d). In February 2024, H_2 concentrations at PX2 were measured twice at depths of 3, 5, 10, and 11–13 m, with 5–10 min interval between measurements. The measured H_2 concentrations ranged from 78 to 171 ppm (mean 127 ppm) at 3 m, 136 to 276 ppm (mean 212 ppm) at 5 m, 628 to 735 ppm (mean 692 ppm) at 10 m, and 832 to 1,000 ppm (mean 935 ppm) at 11–13 m (Figure 5d).

Table 17.1: Summary of gas concentrations measured in boreholes PX1 and PX2 for 6 months.

Measurement date	Depth (m)	PX1							PX2						
		No. of meas.[1]	H2 (ppm) Ave.[2]	1σ	O2 (%) Ave.	1σ	N2 (%) Ave.	1σ	No. of meas.	H2 (ppm) Ave.	1σ	O2 (%) Ave.	1σ	N2 (%) Ave.	1σ
4 September 2004	1	13	15.1	3.7	19.9	0.2	80.0	0.2	15	76.9	13.1	18.6	0.2	81.4	0.2
11 September 2004	1	21	18.5	3.1	18.9	0.2	80.7	0.2	10	92.9	9.0	19.3	0.1	80.7	0.1
18 September 2004	1	20	14.7	1.5	19.7	0.1	79.9	0.1	20	82.9	10.2	19.7	0.2	80.3	0.2
25 September 2004	1	26	13.5	1.8	19.7	0.2	80.1	0.2	20	84.9	21.0	19.6	0.1	80.4	0.1
4 October 2024	1	25	11.7	1.3	19.8	0.1	80.0	0.1	15	61.5	4.6	19.6	0.1	80.4	0.1
9 October 2024	1	25	9.6	1.8	20.2	0.1	79.8	0.1	15	62.6	5.5	20.0	0.1	79.9	0.1
16 October 2024	1	25	9.2	1.5	20.3	0.1	79.7	0.1	25	63.3	4.2	19.9	0.2	80.1	0.2
23 October 2024	1	26	8.8	2.2	21.0	0.1	79.0	0.1	10	57.9	2.7	20.8	0.1	79.2	0.0
30 October 2024	1	25	13.3	2.3	20.9	0.2	79.0	0.2	20	82.3	6.1	21.1	0.2	78.9	0.2
06 November 2024	1	28	11.6	3.2	21.3	0.3	78.7	0.3	27	127.7	17.9	21.0	0.3	79.0	0.3
13 November 2024	1	25	9.0	3.3	21.4	0.1	78.5	0.1	25	86.4	11.1	21.1	0.1	78.8	0.1
20 November 2024	3	26	10.9	2.5	21.2	0.0	78.7	0.0	27	185.3	16.0	21.2	0.1	78.7	0.0
27 November 2024	3	25	7.9	2.0	22.6	0.3	77.3	0.3	25	97.8	7.4	22.4	0.2	77.5	0.2
4 December 2024	3	10	9.3	2.7	21.8	0.1	78.1	0.1	19	213.9	7.3	21.5	0.1	78.5	0.1
11 December 2024	3	25	5.1	2.8	21.7	0.1	78.2	0.1	25	198.2	16.1	21.3	0.2	78.7	0.2
21 December 2024	3	25	6.8	2.6	21.8	0.1	78.2	0.1	34	176.2	15.1	21.3	0.1	78.6	0.1
28 December 2024	10								24	761.4	20.2	21.1	0.2	78.9	0.2
4 January 2025	10								23	562.7	30.6	21.5	0.4	78.5	0.4
	15								16	935.3	25.1	21.7	0.5	78.2	0.5
11 January 2025	5								9	315.2	4.3	21.3	0.1	78.6	0.1
	10								18	635.7	19.7	21.0	0.2	78.9	0.2
	15								14	926.9	86.7	21.2	0.2	78.8	0.2

Date		meas¹							Ave²						
18 January 2025	5								27	522.7	32.6	21.2	0.5	78.8	0.5
	10	9	4.5	1.6	22.1	0.1	77.8	0.1	7	975.6	64.6	21.0	0.4	78.9	0.4
25 January 2025	5								34	446.4	30.6	21.5	0.5	78.4	0.5
1 February 2025	5								14	530.9	13.7	20.9	0.2	79.0	0.2
	8								7	795.7	19.1	21.1	0.1	78.8	0.1
	9								10	950.4	17.8	21.3	0.1	78.6	0.1
	10	18	8.2	1.6	21.9	0.3	78.0	0.3	9	168.3	1.8	22.0	0.0	77.9	0.0
5 February 2025	3	10	11.8	1.2	21.9	0.1	78.1	0.1	10	265.0	15.9	21.9	0.0	78.0	0.0
	5	10	12.8	0.6	22.1	0.0	77.8	0.0	11	672.5	15.3	21.7	0.0	78.2	0.0
	10	10	5.9	0.9	22.5	0.0	77.4	0.0	12	865.9	26.9	21.6	0.0	78.3	0.1
	11								10	898.5	16.4	21.5	0.1	78.4	0.0
	12								9	85.3	5.3	22.2	0.1	77.7	0.1
	3								10	159.5	8.9	22.0	0.0	77.9	0.0
	5								10	712.9	19.3	21.7	0.0	78.3	0.0
	10								11	961.5	24.6	21.5	0.1	78.4	0.0
	13														

Note: ¹meas., measurement; ²Ave., Average.

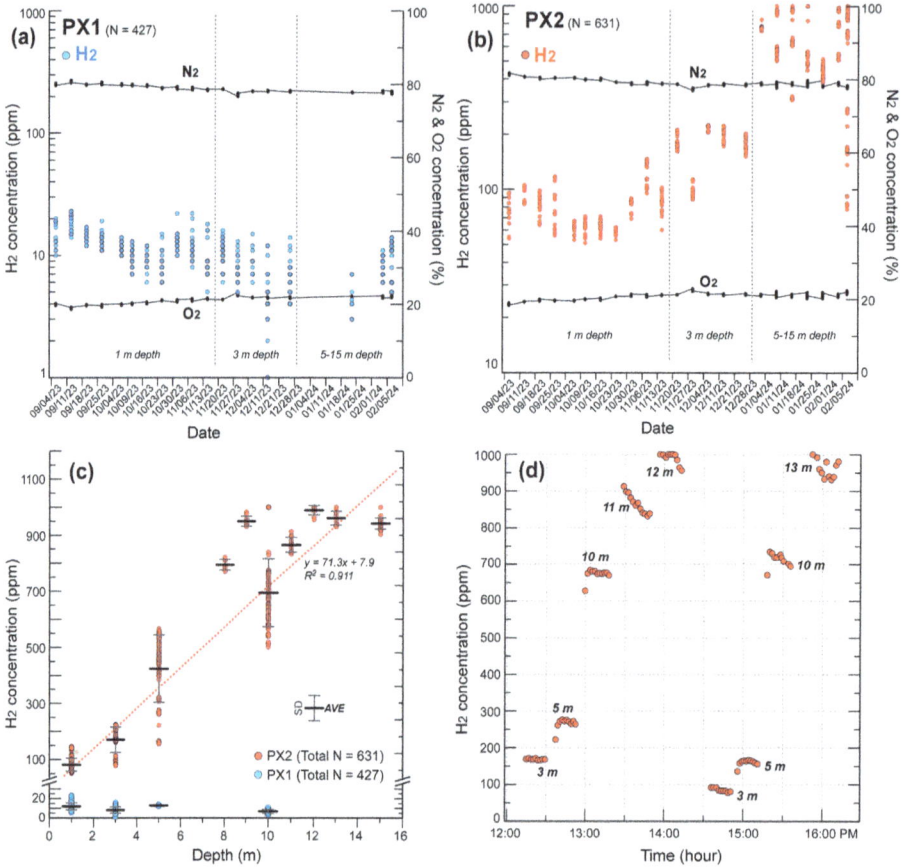

Figure 17.5: (a) H_2, O_2, and N_2 gas concentration by measured depths in PX1 for 6 months from 4 September 2023 to 5 February 2024. (b) H_2, O_2, and N_2 gas concentration by measured depths in PX2 for 6 months from 4 September 4 2023 to 5 February 2024. (c) H_2 concentration variation with depth in PX1 and PX2. (d) The change in H_2 concentrations measured according to time-depth over the course of a day in the PX2.

17.5 Natural hydrogen generation and potential reservoirs in the Korean Peninsula

17.5.1 Serpentinization of ultramafic rocks

To determine the phase relationships between ferromagnesian minerals and H_2 generation during serpentinization of the ultramafic rocks, pressure-temperature (P-T) phase equilibria modeling was conducted in the CaO-FeO-MgO-Al$_2$O$_3$-SiO$_2$-H$_2$O-Fe$_2$O$_3$ (CFMASHO) system using Perple_X (v. 6.8.6) [44] under conditions of 100–600 °C, 0.5–6 kbar, and H_2O saturation. The following solution models were used: Ol, Opx, Cpx, and chlorite (Chl) from Holland and Powell [45] and Holland and Powell [46], and Srp from Padrón-Navarta et al. [47]. Phases such as talc (Tlc), tremolite (Tr), brucite (Brc), and Mt were considered as ideal binary solid solutions. The bulk-rock compositions of the ultramafic rocks in the Hongseong-Imjingang Belt and the Gyeongsang Basin were measured using inductively coupled plasma-optical emission spectroscopy (ICP-OES) at Activation Laboratories, Ontario, Canada. FeO and Fe_2O_3 were analyzed using the traditional titration method. The bulk-rock compositions (mol%) used to calculate P-T pseudosections for samples AD, US, YG, and BB $CaO:FeO:MgO:Al_2O_3:SiO_2:Fe_2O_3$ are 4.32:8.10:47.15:0.68:38.99:0.77, 0.94:4.76:55.06:1.18:37.67:0.39, 1.67:4.69:52.95:0.88:39.34:0.47, and 0.41:4.18:57.41:0.66:36.94:0.26, respectively.

P-T pseudosections were calculated for partially serpentinized wehrlite (samples AD), harzburgite (samples US and BB), and lherzolite (sample YG) from the Hongseong-Imjingang Belt and the Gyeongsang Basin to constrain the variations of mineral phases and their association with H_2 generation during serpentinization. Srp in samples AD and US can be formed by the decomposition of Chl and Tlc at 480–590 °C and 0.5–6.0 kbar (Figure 17.6a and b). Ol decomposes at 320–360 °C and 0.5–6.0 kbar, resulting in H_2 generation (Figure 17.6a and b). It suggests that H_2 can be generated during serpentinization as the mineral assemblage of ultramafic rocks alterations to Cpx + Srp + Mt + Brc through the following reaction:

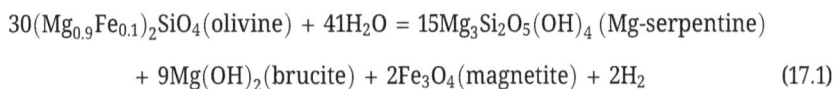

$$30(Mg_{0.9}Fe_{0.1})_2SiO_4(\text{olivine}) + 41H_2O = 15Mg_3Si_2O_5(OH)_4 \, (\text{Mg-serpentine})$$

$$+ \, 9Mg(OH)_2(\text{brucite}) + 2Fe_3O_4(\text{magnetite}) + 2H_2 \qquad (17.1)$$

The maximum modeled H_2 concentrations for the partially serpentinized wehrlite (sample AD) and harzburgite (sample US) are approx. 770 ppm and approx. 430 ppm, respectively, under pressure below 1 kbar, with H_2 concentration increasing as pressure decreases (Figure 17.6c and d). During serpentinization, the density of wehrlite and harzburgite sharply decreases as Srp forms at approx. 500 °C and brucite forms at approx. 400 °C with decreasing temperature (Figure 17.6c and d). Additionally, within the P-T range where H_2 is present, rock density decreases from 2.8 to 2.5 g/cm^3 as pressure decreases (Figure 17.6c and d). With a geothermal gradient of 20–40 °C/km, the H_2 concentration in sample AD can reach a maximum of ~700 ppm at ~120 °C and 1.2 kbar, and a minimum of ~570 ppm at ~350 °C and 4.5 kbar (Figure 17.6c). Similarly, the H_2 concentration generated in sample US can reach a maximum of ~360 ppm at ~120 °C and 1.2 kbar, and a minimum of ~220 ppm at ~350 °C and 4.5 kbar (Figure 17.6d).

H_2 generation, along with changes in mineral phase and reduction in rock density during the serpentinization of lherzolite (sample YG) and harzburgite (sample BB), follows patterns similar to those observed in ultramafic rocks in the Gyeongsang Basin. Srp and Brc form at 500–600 °C and 370–430 °C, respectively (Figure 17.7a and b), and the rock density shows an abrupt decrease within this temperature range (Figure 17.7c

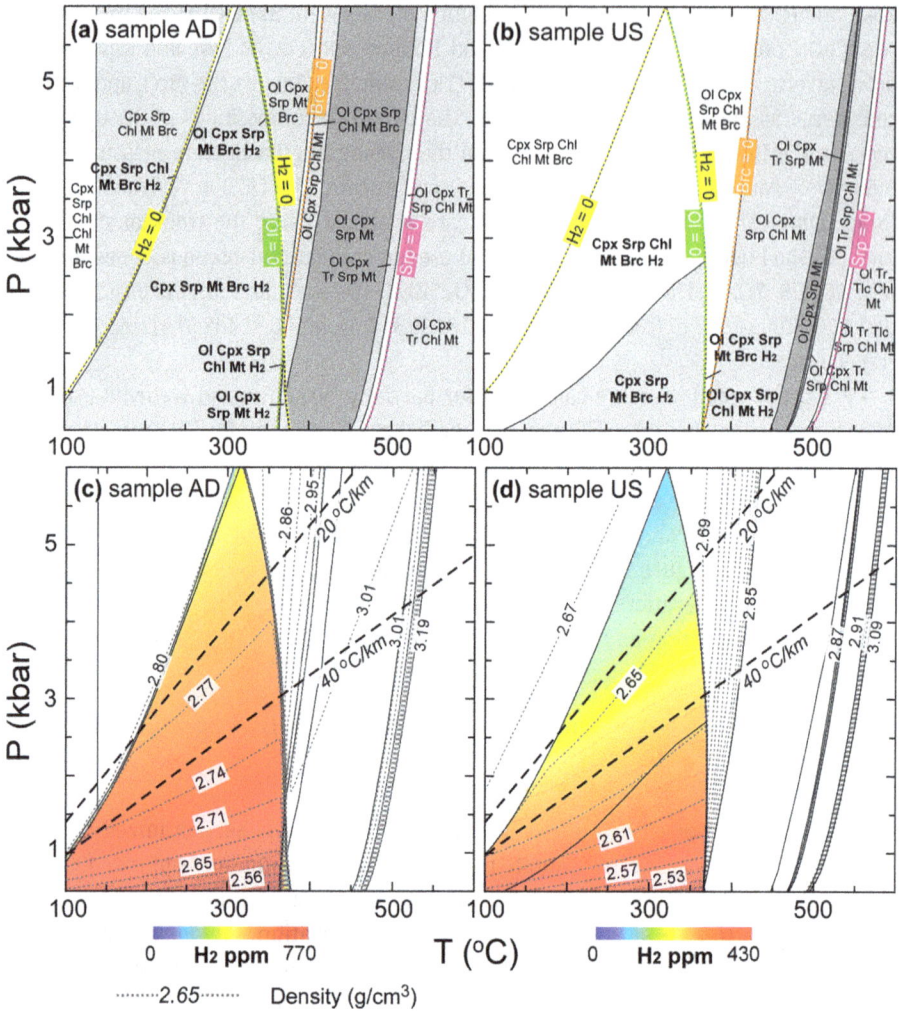

Figure 17.6: Thermodynamic modeling for H_2 generation. (a) and (b) P-T pseudosections of samples AD (wehrlite) and US (harzburgite) in CFMASHO system with H_2O in excess, respectively, the Gyeongsang Basin showing P-T conditions H_2-bearing phases. Highlighted dotted lines indicate zero-mode lines of Srp, Ol, Brc and H_2. (c) and (d) P–T pseudosections showing isodensity and concentration variations of H_2 generated from serpentinization of samples AD (wehrlite) and US (harzburgite), respectively. Dotted and dashed lines represent isodensity and geothermal gradients (20 and 40 °C/km).

and d). In these ultramafic rocks, H_2 is also generated through reaction (17.1), involving the decomposition of Ol and the formation of the mineral assemblage Cpx + Srp + Mt + Brc. H_2 concentrations increase to 420–560 ppm as pressure and density decrease (from 2.69 to 2.50 g/cm^3) (Figure 17.7c and d). When the geothermal gradient is 20–40 °C/km, the H_2 concentration generated in samples YG and BB can reach a maximum of 300–320 ppm at ~120 °C and 1.2 kbar, and a minimum of ~200 ppm at ~350 °C and 4.5 kbar (Figure 17.7c and d).

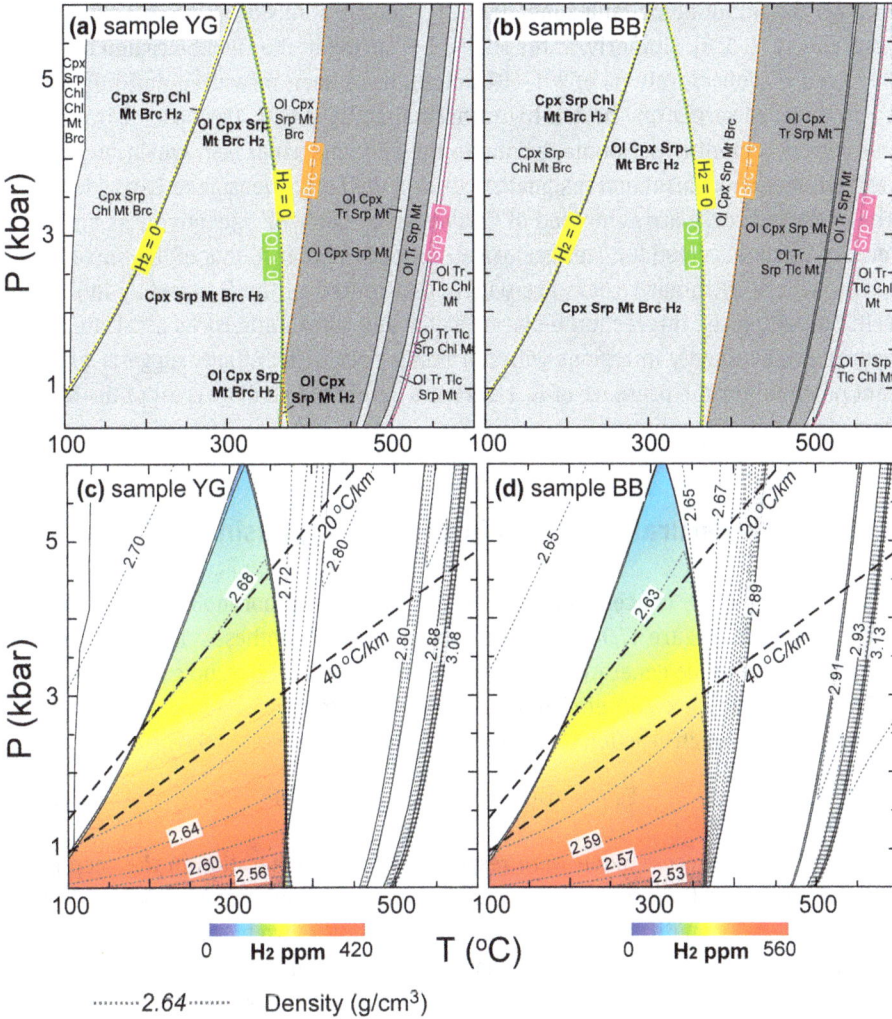

Figure 17.7: Thermodynamic modeling for H_2 generation. (a) and (b) P-T pseudosections of samples YG (lherzolite) and BB (harzburgite) in CFMASHO system with H_2O in excess, respectively, the Hongseong-Imjingang Belt showing P-T conditions in H_2-bearing phases. Highlighted dotted lines indicate zero-mode lines of Srp, Ol, Brc, and H_2. (c) and (d) P-T pseudosections showing isodensity and concentration variations of H_2 generated from serpentinization of samples YG (lherzolite) and BB (harzburgite), respectively. Dotted and dashed lines represent isodensity and geothermal gradients (20 and 40 °C/km).

H_2 generated by the serpentinization of ultramafic rocks has been reported in several countries. In the Chimaera region of Turkey and the western Hajar Mountains of Oman, H_2 produced by low-temperature (50–60 °C) serpentinization is released through the surrounding fault zones, with H_2 concentrations of 7.5–11.3% and 20–99%, respectively [1, 5, 8]. Similarly, at the Happo hot spring in the Hakuba region of Japan, dissolved H_2 concentrations of 201–664 µmol/L have been measured and interpreted as products of serpentine-hosted hydrothermal activity [48]. The basement rocks in this region, including peridotite, belong to the Hida marginal belt, which underwent regional metamorphism and magmatism during the Late Paleozoic to Early Mesozoic. The overlying rocks are composed of Cretaceous to Cenozoic igneous rocks and Quaternary volcanic rocks [49]. This geological setting is similar to that of the Korean Peninsula, where ultramafic rocks are widely distributed in the Hongseong-Imjingang Belt. Consequently, interaction between water and ultramafic rocks associated with hydrothermal activity in regions with ultramafic rock distributions suggests a significant potential for the presence of H_2 reservoirs within the shallow crust of the Korean Peninsula.

17.5.2 Deep natural hydrogen in the Pohang Basin

The variations in H_2 concentration measured in the deep boreholes PX1 and PX2 of the Pohang EGS (Figure 17.5) can be explained by two hypotheses. The first one suggests that H_2 could be generated by the corrosion of the steel boreholes at low temperatures [50]. In low-oxygen environments, Fe^{2+}-rich groundwater can facilitate Fe^{3+} precipitation through reaction (17.2), producing H_2 as a by-product [51]:

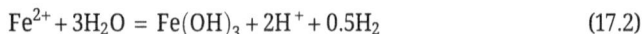

$$Fe^{2+} + 3H_2O = Fe(OH)_3 + 2H^+ + 0.5H_2 \qquad (17.2)$$

However, if this process alone were responsible for H_2 generation in the deep boreholes, H_2 concentrations in the two boreholes would be expected to be similar. In fact, the H_2 concentration differs by 50–100 times between the two boreholes.

The second hypothesis proposes that H_2 generated in the deep subsurface (mid-lower crust and uppermost mantle) of the Pohang Basin migrates upward and accumulates around subsurface fault zones beneath the basin, with some of the H_2 being released into the two boreholes. This hypothesis is supported by recent findings of seismic velocity anomalies and the emission of mantle-derived noble gases around the fault zones in the Pohang Basin. Lee et al. [52] reported that low-velocity anomalies in the crust and uppermost mantle of the Pohang Basin are associated with areas of high geothermal gradients and high permeability, suggesting that heat flow facilitates hydrothermal fluid circulation from the crust-mantle boundary. Additionally, Kim et al. [35] found that the composition of dissolved gases in groundwater near the Pohang EGS exhibits mixed characteristics of mantle and sedimentary origins. The helium (He) isotope ratio (^{3}He/^{4}He) reaches up to 3.83 Ra (Figure 17.2b), indicating a significant mantle contribution. The spatial distribution of mantle-derived He strongly correlates with the Heunghae and F1 fault zones (Figure 17.2b), suggesting that the subsurface fault zones in the Pohang Basin act as pathways for the migration of deep H_2- and He-bearing fluids. Thus, the large difference in H_2 concentration between boreholes PX1 and PX2 is attributed to the separation of groundwater in the hanging wall and footwall by the gouge layers (~3.8 km depth; Figure 17.3) of the seismogenic fault damage zone [37]. Specifically, H_2 generated in the deep subsurface migrates upward, where the fault gouge layer acts as a caprock, leading to the accumulation of H_2 in the footwall. As a result, deep groundwater with a high dissolved H_2 content flows into PX2, causing the observed high H_2 concentration in the uppermost section of PX2.

The Tertiary Pohang Basin on the Korean Peninsula, formed in a back-arc tectonic setting, have a relatively thin crust, causing the upper mantle, considered a source of deep H_2, to be relatively shallow [33, 52]. Also, active fault zones within the basin contain fault damage zones, characterized by low-permeability fault gouge layers and high-permeability brecciated cataclasite layers [37]. These fault-damaged zones can serve not only as conduits for the migration of gases or fluids originating from the deep subsurface but also as traps due to localized permeability anomalies [37, 53–56]. Notably, in many regions worldwide where ophiolites are present, H_2 is predominantly released along fault zones [1, 5, 8].

17.6 Implications and conclusion: natural hydrogen prospecting in the Korean Peninsula

Economically viable H_2 reservoirs, from an energy perspective, have been reported by some researchers [10, 57], but significant uncertainties remain. While H_2 holds the potential to represent a new frontier in the energy transition, further in-depth research is required to understand where and how this potentially valuable resource is generated and where it might accumulate. For instance, is it possible to discover large-scale H_2 accumulations similar to conventional petroleum or natural gas systems? The answer likely depends on the types and distribution of H_2 reservoirs in the shallow crust. H_2 reservoirs within the upper crust can take the form of two types (e.g., [58]): (1) pressurized reservoirs trapped beneath impermeable layers (e.g., salt, shale, fault gouge, and basalt), and (2) H_2 continuously generated within fractures and the matrix of basement rocks, gradually migrating, seeping, or accumulating in specific zones (e.g., beneath impermeable layers).

The two regions proposed in this study are currently in the early stages of research, making it impossible to assess their exploration or commercial potential. However, to facilitate future exploration of H_2 resources on the Korean Peninsula, it is necessary to establish quantitative criteria and evaluation systems for the following two indicators:

(1) Charge potential: This refers to the probability of H_2 being generated and accumulated within a specific basin or strata. Evaluating the charge potential associated with serpentinization requires geological characteristics [11, 12] such as (a) temperature, (b) total volume of ultramafic rocks, (c) the ratio of water to rock mass, (d) the $Fe^{3+}/\Sigma Fe$ ratio of ultramafic rocks, (e) the compositions and modal proportions of ferromagnesian minerals (e.g., olivine, clinopyroxene, orthopyroxene, and magnetite), and (f) the oxygen and SiO_2 fugacity (fO_2 and $fSiO_2$) of fluids. These factors collectively contribute to determining the charge potential of specific regions.

(2) Potential emplaced H_2: This represents the capacity for H_2 to accumulate in specific geological environments. A quantitative evaluation considers factors such as subsurface rock structures, the presence of cap rocks, sealing capacity, the permeability of subsurface rocks, and the extent of microbial activity that consumes hydrogen.

To accurately evaluate these two indicators, seismic, magnetic, and gravity surveys should be conducted to determine the spatial distribution of serpentinized ultramafic rocks in the Hongseong-Imjingang Belt, as well as the locations of subsurface faults in the Tertiary Pohang Basin. Furthermore, long-term monitoring of concentration changes in H_2 and other abiotic gases emitted from these regions is critical for accurate assessment.

References

[1]   Zgonnik V. The occurrence and geoscience of natural hydrogen: A comprehensive review. Earth Science Reviews. 2020, 203: 103140. https://doi.org/10.1016/j.earscirev.2020.103140

[2]   Milkov A. V. Molecular hydrogen in surface and subsurface natural gases: Abundance, origins and ideas for deliberate exploration. Earth Science Reviews. 2022, 230: 104063. https://doi.org/10.1016/j.earscirev.2022.104063

[3]   Sleep N. H., Meibom A., Fridriksson T., Coleman R. G., Bird D. K. H_2-rich fluids from serpentinization: Geochemical and biotic implications. Proceedings of the National Academy of Sciences. 2004, 101: 12818–12823. https://doi.org/10.1073/pnas.0405289101

[4]   Sherwood Lollar B., Onstott T. C., Lacrampe-Couloume G., Ballentine C. J. The contribution of the Precambrian continental lithosphere to global H_2 production. Nature. 2014, 516: 379–382. https://doi.org/10.1038/nature14017

[5]   Miller H. M., Matter J. M., Kelemen P., Ellison E. T., Conrad M. E., Fierer N., Ruchala T., Tominaga M., Templeton A. S. Modern water/rock reactions in Oman hyperalkaline peridotite aquifers and implications for microbial habitability. Geochimica et Cosmochimica Acta. 2016, 179: 217–241. http://dx.doi.org/10.1016/j.gca.2016.01.033

[6]   Sauvage J. F., Flinders A., Spivack A. J., Pockalny R., Dunlea A. G., Anderson C. H., Smith D. C., Murray R. W., D'Hondt S. The contribution of water radiolysis to marine sedimentary life. Nature Communications. 2021, 12: 1297. https://doi.org/10.1038/s41467-021-21218-z

[7]   Combaudon V., Moretti I., Kleine B. I., Stefánsson A. Hydrogen emissions from hydrothermal fields in Iceland and comparison with the Mid-Atlantic Ridge. International Journal of Hydrogen Energy. 2022, 47: 10217–10227. https://doi.org/10.1016/j

[8] Etiope G. Massive release of natural hydrogen from a geological seep (Chimaera, Turkey): Gas advection as a proxy of subsurface gas migration and pressurised accumulations. International Journal of Hydrogen Energy. 2023, 48: 9172–9184. https://doi.org/10.1016/j.ijhydene.2022.12.025

[9] Giuntoli F., Menegon L., Siron G., Cognigni F., Leroux H., Compagnoni R., Rossi M., Brovarone A. V. Methane-hydrogen-rich fluid migration may trigger seismic failure in subduction zones at forearc depths. Nature Communications. 2024, 15: 480. https://doi.org/10.1038/s41467-023-44641-w

[10] Truche L., Donzé F. V., Goskolli E., Muceku B., Loisy C., Monnin C., Dutoit H., Cerepi A. A deep reservoir for hydrogen drives intense degassing in the Bulqizë ophiolite. Science. 2024, 383: 618–621. science.org/doi/10.1126/science.adk9099

[11] Klein F., Bach W., McCollom T. M. Compositional controls on hydrogen generation during serpentinization of ultramafic rocks. Lithos. 2013, 2013, 178: 55–69. http://dx.doi.org/10.1016/j.lithos.2013.03.008

[12] Ely T. D., Leong J. M., Canovas P. A., Shock E. L. Huge variation in H_2 generation during seawater alteration of ultramafic rocks. Geochemistry, Geophysics, Geosystems. 2023, 24: e2022GC010658. https://doi.org/10.1029/2022GC010658

[13] Ellis G. S., Gelman S. E. Model predictions of global geologic hydrogen resources. Science Advances. 2024, 10: eado0955.

[14] Chough S. K., Sohn Y. K. Tectonic and sedimentary evolution of a Cretaceous continental arc–backarc system in the Korean peninsula: New view. Earth Science Reviews. 2010, 101: 225–249.

[15] Cheon Y., Ha S., Lee S., Son M. Tectonic evolution of the Cretaceous Gyeongsang Back-arc Basin, SE Korea: Transition from sinistral transtension to strike-slip kinematics. Gondwana Research. 2019, 83: 16–35. https://doi.org/10.1016/j.gr.2020.01.012

[16] Kim G. B., Yoon S. H., Chough S. K., Kwon Y. K., Ryu B. J. Seismic reflection study of acoustic basement in the South Korea Plateau, the Ulleung Interplain Gap, and the northern Ulleung Basin: Volcano-tectonic implications for Tertiary back-arc evolution in the southern East Sea. Tectonophysics. 2011, 504: 43–56.

[17] Sohn Y. K., Son M. Synrift stratigraphic geometry in a transfer zone coarse-grained delta complex, Miocene Pohang Basin, SE Korea. Sedimentology. 2004, 51: 1387–1408. doi: 10.1111/j.1365-3091.2004.00679.x

[18] Kim H. S., Kwon S., Kim S. W., Santosh M. Permo–Triassic high-pressure metamorphism in the central western Korean peninsula, and its link to Paleo-Tethyan Ocean closure: Key issues revisited. Geoscience Frontiers. 2018, 9: 1325–1335. https://doi.org/10.1016/j.gsf.2018.01.007

[19] Seo J., Choi S. G., Oh C. W., Kim S. W., Song S. H. Genetic implications of two different ultramafic rocks from Hongseong Area in the Southwestern Gyeonggi Massif, South Korea. Gondwana Research. 2005, 2005, 8: 539–552.

[20] Seo J., Oh C. W., Choi S. G., Rajesh V. J. Two ultramafic rock types in the Hongseong area, South Korea: Tectonic significance for northeast Asia. Lithos. 2013, 175–176: 30–39. http://dx.doi.org/10.1016/j.lithos.2013.04.014

[21] Park M., Jung H. Microstructural evolution of the Yugu peridotites in the Gyeonggi Massif, Korea: Implications for olivine fabric transition in mantle shear zones. Tectonophysics. 2017, 709: 55–68. http://dx.doi.org/10.1016/j.tecto.2017.04.017

[22] Oh C. W., Kim S. W., Choi S. G., Zhai M., Guo J., Sajeev K. First finding of eclogite facies metamorphic event in South Korea and its correlation with the Dabie–Sulu Collision Belt in China. Journal of Geology. 2005, 113: 226–232. https://doi.org/10.1086/427671

[23] Kim H. S., Ree J. H., Kang H. C., Yi K. Pressure–temperature–time–deformation (P–T–t–d) path for Devonian forearc deposits in the Imjingang Belt, South Korea: Implications for Permian–Triassic collisional orogenesis on the eastern margin of Eurasia. Journal of Metamorphic Geology. 2022, 40(3): 489–516. https://doi.org/10.1111/jmg.12636

[24] Chang K. H., Zhao X. North and South China suturing in the east end: What happened in Korean Peninsula? Gondwana Research. 2021, 22: 493–506.

[25] Dong Y. P., Sun S., Santish M., Zhao J., Sun J., He D., Shi X., Cheong C., Zhang G. Central China Orogenic Belt and amalgamation of East Asian continents. Gondwana Research. 2021, 100: 131–194. https://doi.org/10.1016/j.gr.2021.03.006

[26] Liu T., Hu Z., Zhang D., Li S., Cheong C., Zhou L., Wang G., Wang Z. The timing of the initial collision between the South and North China blocks constraining from the sediments in the eastern Sichuan Basin. Scientific Reports. 2023, 13: 22378. https://doi.org/10.1038/s41598-023-49498-z

[27] Cheong A. C. S., Jo H. J., Jeong Y. J., Li X. H. Magmatic response to the interplay of collisional and accretionary orogenies in the Korean Peninsula: Geochronological, geochemical, and O-Hf isotopic perspectives from Triassic plutons. Geological Society of America Bulletin. 2019, 131: 609–634.

[28] Cheong A. C. S., Jo H. J. Tectonomagmatic evolution of a Jurassic Cordilleran flare-up along the Korean Peninsula: Geochronological and geochemical constraints from granitoid rocks. Gondwana Research. 2020, 88: 21–44. https://doi.org/10.1016/j.gr.2020.06.025

[29] Choi P. Y., Lee S. R., Choi H.-I., Hwang J.-A., Kwon S.-K., Ko I.-S., An G.-O. Movement history of the Andong Fault System: Geometric and tectonic approaches. Geosciences Journal. 2002, 6: 91–102.

[30] Whattam S. A., Cho M., Smith I. E. M. Magmatic peridotites and pyroxenites, Andong Ultramafic Complex, Korea: Geochemical evidence for supra-subduction zone formation and extensive melt-rock interaction. Lithos. 2011, 127: 599–618.

[31] Kim N. K., Choi S. H. Petrogenesis of Late Triassic ultramafic rocks from the Andong Ultramafic Complex, South Korea. Lithos. 2016, 264: 28–40. http://dx.doi.org/10.1016/j.lithos.2016.07.042

[32] Seo J., Choi S. G., Kim J. W., Ryu I. C. Unique sodic–calcic skarn hosted by ultramafic rocks and albitite at the Ulsan skarn deposit, Gyeongsang Basin, South Korea. Ore Geology Reviews. 2019, 105: 537–550. https://doi.org/10.1016/j.oregeorev.2018.12.026

[33] Cheon Y., Shin Y. H., Park S., Choi J.-H., Kim D.-E., Ko K., Ryoo C.-R., Kim Y.-S., Son M. Structural architecture and late Cenozoic tectonic evolution of the Ulsan Fault Zone, SE Korea: New insights from integration of geological and geophysical data. Frontiers in Earth Science. 2023, 11: 1183329. 10.3389/feart.2023.1183329

[34] Son M., Song C. W., Kim M. C., Cheon Y., Cho H., Sohn Y. K. Miocene tectonic evolution of the basins and fault systems, SE Korea: Dextral, simple shear during the East Sea (Sea of Japan) opening. Journal of the Geological Society. 2015, 172: 664–680. doi: 10.1144/jgs2014-079

[35] Kim H., Lee H., Lee J., Lee H. A., Woo N. C., Lee Y. S., Kagoshima T., Takahata N., Sano Y. Mantle-derived helium emission near the Pohang EGS Site, South Korea: Implications for active fault distribution. Geofluids. 2020, 2359740: 14. https://doi.org/10.1155/2020/2359740

[36] Kim K. H., Ree J. H., Kim Y., Kim S., Kang S. Y., Seo W. Assessing whether the 2017 M w 5.4 Pohang earthquake in South Korea was an induced event. Science. 2018, 360(6392): 1007–1009. https://doi.org/10.1126/science.aat6081

[37] Korean Government Commission. Summary report of the Korean government commission on relations between the 2017 Pohang Earthquake and EGS Project. Geological Society of Korea, Seoul, South Korea. 2019. https://doi.org/10.22719/KETEP-20183010111860.

[38] Lee T. H., Yi K., Cheong C. S., Jeong Y. J., Kim N., Kim M. J. SHRIMP U-Pb zircon geochronology and geochemistry of drill cores from the Pohang Basin. Journal of the Petrological Society of Korea. 2014, 23: 167~185. http://dx.doi.org/10.7854/JPSK.2014.23.3.167

[39] Cheon Y., Son M., Song C. W., Kim J.-S., Sohn Y. K. Geometry and kinematics of the Ocheon Fault System along the boundary between the Miocene Pohang and Janggi basins, SE Korea, and its tectonic implications. Geosciences Journal. 2012, 16: 253–273.

[40] Son M., Kim I. S., Sohn Y. K. Evolution of the Miocene Waup Basin, SE Korea, in response to dextral shear along the southwestern margin of the East Sea. Journal of Asian Earth Sciences. 2005, 25: 529–544.

[41] Song S. H. Petrochemistry of the Peridotites within an Andong Ultramafic complex and characteristics of Asbestos occurrences. Journal of the Mineralogical Society of Korea. 2019, 32: 15–39.

[42] Kim K. H., Park J. K., Yang J. M., Satake H. A study on serpentinization of serpentinites from the Ulsan iron mine. Journal Korean Institute of Mining Geology. 1993, 26: 267–278.

[43] Koh S. M., Park C.-K., Soh W.-J. Preliminary study on the formation environment of serpentinite occurring in Ulsan area. Journal of the Mineralogical Society of Korea. 2006, 19: 325–336.

[44] Connolly J. A. D. The geodynamic equation of state: What and how. Geochemistry, Geophysics, Geosystems. 2009, 10: Q10014. https://doi.org/10.1029/2009GC002540

[45] Holland T. J. B., Powell R. Thermodynamics of order-disorder in minerals: II. Symmetric formalism applied to solid solutions. American Mineralogist. 1996, 81: 1425–1437. https://doi.org/10.2138/am-1996-11-1215

[46] Holland T. J. B., Powell R. An internally consistent thermodynamic dataset for phases of petrological interest. Journal of Metamorphic Geology. 1998, 16: 309–343. https://doi.org/10.1111/j.1525-1314.1998.00140.x

[47] Padrón-Navarta J. A., Sánchez-Vizcaíno V. L., Hermann J., Connolly J. A. D., Garrido C. J., Gómez-Pugnaire M. T., Marchesi C. Tschermak's substitution in antigorite and consequences for phase relations and water liberation in high-grade serpentinites. Lithos. 2013, 178: 186–196. http://dx.doi.org/10.1016/j.lithos.2013.02.001

[48] Suda K., Ueno Y., Yoshizaki M., Nakamura H., Kurokawa K., Nishiyama E., Yoshino K., Hongoh Y., Kawachi K., Omori S., Yamada K., Yoshida N., Maruyama S. Origin of methane in serpentinite-hosted hydrothermal systems: The CH_4–H_2–H_2O hydrogen isotope systematics of the Hakuba Happo hot spring. Earth and Planetary Science Letters. 2014, 386: 112–125. http://dx.doi.org/10.1016/j.epsl.2013.11.001

[49] Nakano S., Takeuchi M., Yoshikawa T., Nagamori H., Kariya Y., Okumura K., Taguchi Y. Geology of the Shiroumadake district. Quadrangle Series. 1: 50,000 Kanazawa (10). 2002. No. 25NJ-53-5-4, Geological Survey of Japan, AIST.

[50] Goebel E. D., Coveney R. M. J., Angino E. E., Zeller E. J., Dreschhoff G. A. M. Geology, composition, isotopes of naturally occurring H_2/N_2 rich gas from wells near junction city, Kansas. Oil & Gas Journal. 1984, 82: 215–222.

[51] Guélard J., Beaumont V., Rouchon V., Guyot F., Pillot D., Jézéquel D. Natural H_2 in Kansas: Deep or shallow origin? Geochemistry, Geophysics, Geosystems. 2017, 18: 1841–1865. https://doi.org/10.1002/2016GC006544

[52] Lee S., Song J. H., Heo D., Rhie J., Kang T. S., Choi E., Kim Y. H., Kim K. H., Ree J. H. Crustal and uppermost mantle structures imaged by teleseismic P-wave travel time tomography beneath the Southeastern Korean Peninsula: Implications for a hydrothermal system controlled by the thermally modified lithosphere. Geophysical Journal International. 2023, 235: 1639–1657. https://doi.org/10.1093/gji/ggad319

[53] Yeo I. W., Brown M. R. M., Ge S., Lee K. K. Causal mechanism of injection-induced earthquakes through the Mw 5.5 Pohang earthquake case study. Nature Communications. 2020, 11: 2614. https://doi.org/10.1038/s41467-020-16408-0

[54] Caine J. S., Evans J. P., Forster C. B. Fault zone architecture and permeability structure. Geology. 1996, 24: 1025–1028.

[55] Sibson R. H. Structural permeability of fluid-driven fault-fracture meshes. Journal of Structural Geology. 1996, 18: 1031–1042.

[56] Ferrill D. A., Morris A. P., Stamatakos J. A., Sims D. W. Crossing conjugate normal faults. AAPG Bulletin. 2000, 84(10): 1543–1559.

[57] McMahon C. J., Roberts J. J., Johnson G., Edlmann K., Flude S., Shipton Z. K. Natural hydrogen seeps as analogues to inform monitoring of engineered geological hydrogen storage. In: Miocic J. M., Heinemann N., Edlmann K., Alcalde J., Schultz R. A., eds., Enabling secure subsurface storage in future energy systems. Geological society, 528. London: Special Publications, 2024, 461–489. https://doi.org/10.1144/SP528-2022-59

[58] Jackson O., Lawrence S. R., Hutchinson I. P., Stock A. E., Barnicoat A. C., Powney M. Natural hydrogen: Sources, system and exploration plays. Geoenergy. https://doi.org/10.1144/geoenergy2024-002

Yunfeng Liang*, Wuge Cui, Arata Kioka, and Takeshi Tsuji*

Chapter 18
Natural hydrogen in Japan: general generation mechanisms, current work, and perspectives

Abstract: Natural hydrogen is widely distributed across the globe and has recently gained significant attention as a clean energy resource, with an increasing number of successful commercial developments. This chapter provides a concise review of the reaction mechanisms responsible for natural hydrogen generation through rock-water interactions, with a particular focus on mechanochemical reactions and the serpentinization of mafic and ultramafic rocks.

Drawing insights from natural hydrogen fields in Mali, we highlight the role of igneous rocks as caprocks, which serve to seal hydrogen reservoirs, alongside the conventional shale caprocks. Additionally, the sealing properties of porous breccia media for natural hydrogen retention are discussed.

Our current research utilizes first-principles molecular simulations to better understand chemical reaction mechanisms and identify the conditions under which they occur. Combined with literature studies, our findings suggest that surface reactions play a crucial role in both mechanochemical processes and serpentinization. Surface properties such as termination states, electron transfer and transition metal ion impurities are found to significantly influence chemical reactivity.

The knowledge gained from this study provides valuable insights for the exploration and development of natural hydrogen resources. Furthermore, it has implications for the advancement of orange hydrogen, a process that involves stimulating hydrogen generation by injecting water or CO_2 into reactive formations.

Finally, we present perspectives on the potential for natural hydrogen exploration in Japan, including the possibility of hydrogen reservoirs in the hydrate zone of the Nankai Trough and how they could enhance methane hydrate development. We also discuss hydrogen potential in subduction zones and the Hakuba field, emphasizing the vast yet unexplored opportunities for natural hydrogen in Japan.

*Corresponding author: Yunfeng Liang, Department of Systems Innovation, Graduate School of Engineering, The University of Tokyo, Tokyo 113-8656, Japan, e-mail: liang@sys.t.u-tokyo.ac.jp
*Corresponding author: Takeshi Tsuji, Department of Systems Innovation, Graduate School of Engineering, The University of Tokyo, Tokyo 113-8656, Japan, e-mail: tsuji@sys.t.u-tokyo.ac.jp
Wuge Cui, Arata Kioka, Department of Systems Innovation, Graduate School of Engineering, The University of Tokyo, Tokyo 113-8656, Japan

https://doi.org/10.1515/9783111437040-018

Keywords: natural hydrogen, ater-rock interactions, reaction mechanisms, surface reactions, serpentinization, first-principles molecular dynamics, methane hydrate, subduction zone, Hakuba field

18.1 Significance of natural hydrogen

Natural hydrogen is widely distributed in the world [1], and the investigation of its genesis is important across various domains, including the Earth's biosphere [2–4], lithosphere [2, 4], earthquake prediction [5], and outer space exploration [6]. Very recently, it has been considered as a potential clean energy resource [7]. In the biosphere, natural hydrogen potentially impacts the development and sustainability of life forms. Understanding the distribution and availability of natural hydrogen can provide insights into microbial ecosystem dynamics. Within the lithosphere, natural hydrogen can influence geological processes, including rock alteration, mineral transformations, and hydrothermal activity. It contributes to the overall geochemical balance and can provide valuable information for studying the Earth's geological history. Regarding earthquake prediction, the presence and behavior of natural hydrogen can serve as a potential indicator for seismogenic processes. Monitoring the hydrogen level and its interactions within fault zones may help predict and understand seismic activity. In outer space, exploring and harnessing natural hydrogen sources can have implications for organic history, habitability, and sustainability.

As a sustainable and clean energy source in the background of carbon neutrality, natural hydrogen in the subsurface has received significant attention [1, 7–9]. In contrast to blue and green hydrogen, natural hydrogen is known as gold hydrogen due to its potential economic benefits and already has a successful commercial development in Mali [8]. The number of natural hydrogen exploration and development companies is increasing. Helium co-existing in natural hydrogen reservoirs further increases the revenue, leading to great success (data from Golden Hydrogen Ltd., Australia). In addition, hydrogen generation can be stimulated by injecting water and CO_2 into reactive formations (i.e., through the reaction of water and reactive rocks), known as orange hydrogen [10, 11]. Natural hydrogen is generated in the subsurface, which is summarized in Figure 18.1 [4, 7]. Serpentinization [2, 12–14], mechanochemical reactions [15–19], degassing of magmas [20], radiolysis of water [14, 21], and thermogenic generation from a source rock [22] are considered possible sources of natural hydrogen. It was estimated that serpentinization of ultramafic rocks (such as peridotite and komatiite) in the upper crust at a depth of about 7 km may contribute 100 trillion tons of hydrogen, sufficient for 250,000 years at a rate of 400 million tons per year [10]. The contributions of other sources have also been estimated but only partially; all of them can be significant [14]. At the current stage, identifying the reaction conditions for rock-based hydrogen is essential for developing gold (white) and orange hydrogen as next-generation technologies and solutions for the energy transition in our society.

In this chapter, we will briefly explain the role of igneous rocks in generating hydrogen, focusing on those from mechanochemical reactions and serpentinization of mafic/ultramafic rocks. We will also briefly explain the role of igneous rocks as caprocks to seal the natural hydrogen. We show that first-principles molecular simulations can be employed to understand the chemical reaction mechanisms and identify the reaction conditions. Finally, we will briefly explain our perspectives on the potential and possible applications of natural hydrogen as an energy resource in Japan, including the possibility of finding hydrogen reservoirs in hydrate zone and how it can be profitable for developing methane hydrates as well; the potential of hydrogen in the subduction zone and how to find it; and the natural hydrogen in Hakuba field and our research plans. The knowledge gained in this study can aid in discovering and developing natural hydrogen.

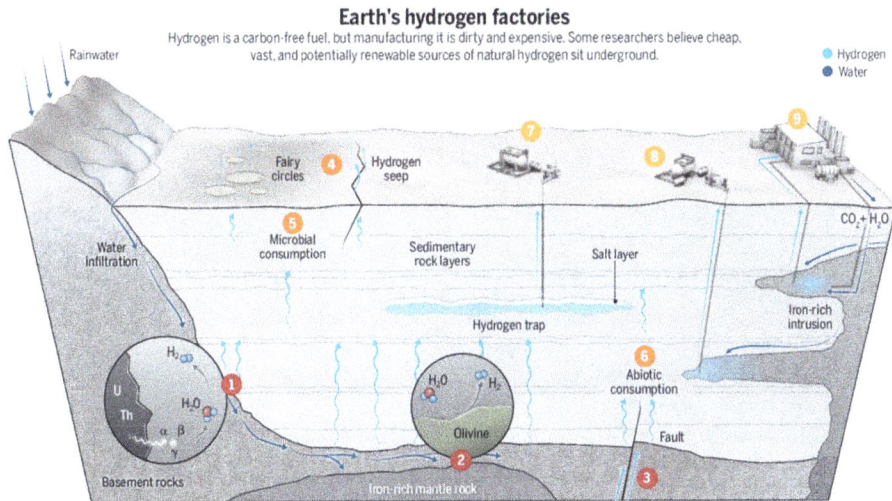

Figure 18.1: Earth's hydrogen factories: most of hydrogen generated from water and rock interactions. Reproduced with permission from Ref. [7]. Copyright 2023 The Authors, exclusive licensee American Association for the Advancement of Science, with free access.

18.2 Role of igneous rocks

18.2.1 Hydrogen generation by mechanochemical reaction

When we discuss rock and water interactions, it is important to note that the main rock referred to is igneous rock [23]. Although it is possible to generate hydrogen during high-speed friction experiments [19], the role of sedimentary rocks mainly serves as a container for natural hydrogen. Minerals (especially, the igneous minerals) occurring in the crust and shallow mantle have the potential to be one of the sources of natural hydrogen by the reaction of their fresh surface with water during fracturing [1, 3, 5, 15, 18, 19]. This phenomenon was first identified in the active Yamasaki fault in Japan [5], and hydrogen is one of the products of complex mechanochemical reactions that occur in fault-lubricated areas during earthquakes. In subsequent experimental studies, it has been confirmed that mechanochemical hydrogen generation is widespread in rock comminution and simulated faulting, not only in the tectonic fault regions [1, 3, 15, 18, 19]. H_2 generation is not only influenced by the type and surface area of the rocks but also correlated with the localized mineral surfaces [3, 18], which indicates that the direction and fracture location may affect the amount of hydrogen production. For example, calcite cannot produce hydrogen, by comparison with silicate minerals (Figure 18.2) [3].

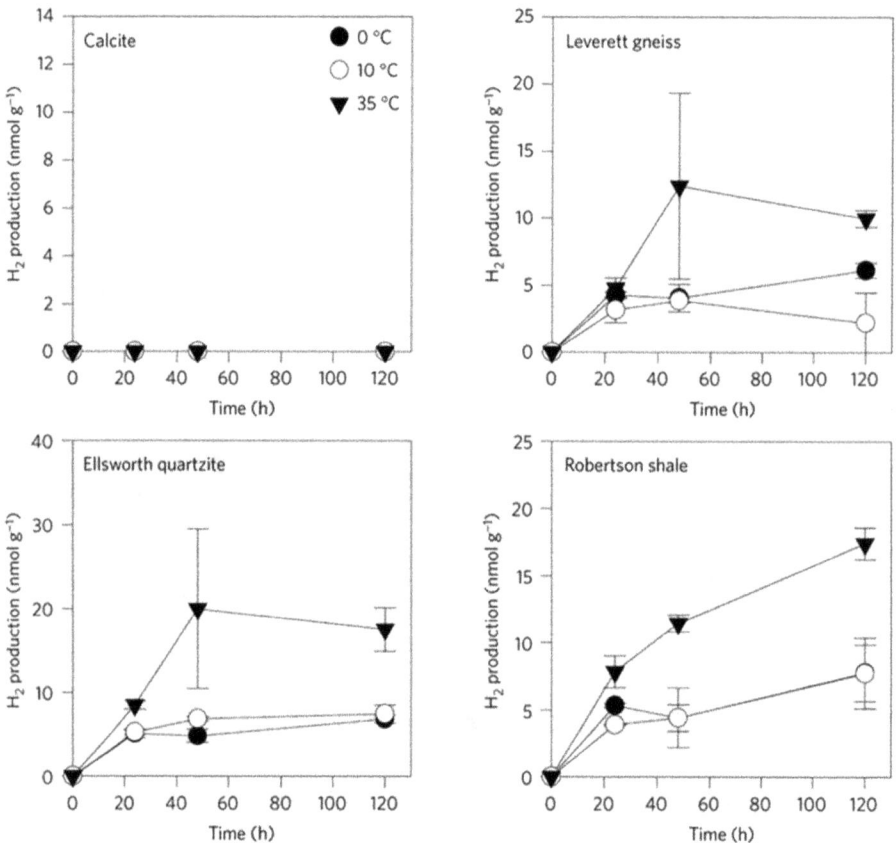

Figure 18.2: Hydrogen generation from felsic silicate rocks. Reproduced with permission from Ref. [3]. Copyright 2015 Springer Nature Limited.

For silicate and quartz minerals, there are two possible cleavage processes: one is homolytic cleavage and another is heterolytic cleavage through eq. (18.1) or (18.2), respectively:

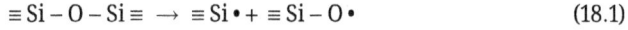

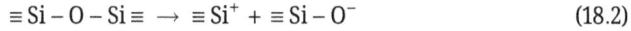

$$\equiv Si - O - Si \equiv \ \rightarrow \ \equiv Si \bullet + \equiv Si - O \bullet \tag{18.1}$$

$$\equiv Si - O - Si \equiv \ \rightarrow \ \equiv Si^+ + \equiv Si - O^- \tag{18.2}$$

The $\equiv Si\bullet$ radicals in silicate after rock comminution are thought to play a key role in producing hydrogen through the following equation [3, 15–19]:

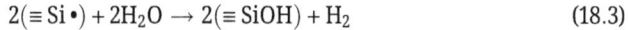

$$2(\equiv Si \bullet) + 2H_2O \rightarrow 2(\equiv SiOH) + H_2 \tag{18.3}$$

It is believed that calcite cannot generate hydrogen because it does not generate radicals during rock combination [3]. Silicon sites can be positively charged in heterolytic cleavage and interact with water through the following equation:

$$\equiv Si^+ + H_2O \rightarrow \equiv SiOH + H^+ \tag{18.4}$$

This chemical reaction was thought to generate acid conditions, not hydrogen gas (as the H^+ is charged). However, the chemical reaction is more complicated, as we will show in Section 18.3.1, the $\equiv Si\bullet$ radicals do play a role and the charge transfer as a transition state is also important. In any scenario, rock comminution causes a surface reaction between the newly cleaved surface and water, which may be linked to a low pH in the water. It supports field observations with a clear relationship between hydrogen concentration from active faults and spring water acidity [16]. As shown in Figure 18.3, the spring water was more acidic when the maximum hydrogen concentration was detected. A similar phenomenon was observed in laboratory experiments with quartz and water [17, 18].

In mechanochemical hydrogen generation, previous rock friction experiments have been detected in a wide variety of rocks [3, 19]. The right type of rock, a higher speed of friction, contains water at a certain "sweet" temperature, and the reducing nature of the environment, etc. are considered to be potential factors for improving

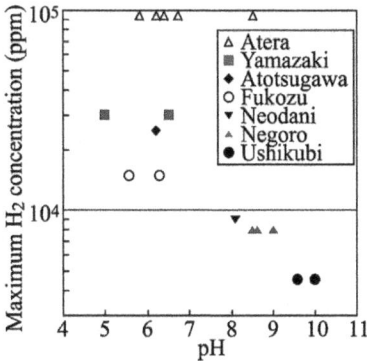

Figure 18.3: Relationship between hydrogen concentration from active faults and spring water acidity. Reproduced with permission from Ref. [16]. Copyright 2003 by the American Geophysical Union with free access for non-commercial purposes.

hydrogen production [15]. Furthermore, the water-rock ratio may play a certain role [17]. Mechanochemical generation of hydrogen not only makes people realize that fractures and earthquakes can generate hydrogen but also makes people realize the importance of faults in the search for natural hydrogen. Faults can serve as pathways for hydrogen migration and can also be the source of hydrogen generation.

18.2.2 Hydrogen generation by serpentinization of ultramafic rocks

Serpentinization of mafic/ultramafic rocks is believed to be the major source of natural hydrogen [4]. The serpentinization of ultramafic rocks can be expressed as follows [12]:

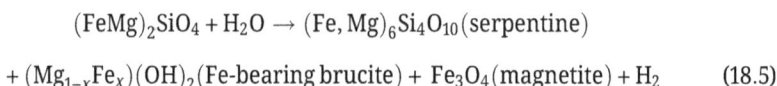

$$(FeMg)_2SiO_4 + H_2O \rightarrow (Fe, Mg)_6Si_4O_{10}(serpentine)$$

$$+ (Mg_{1-x}Fe_x)(OH)_2(Fe\text{-bearing brucite}) + Fe_3O_4(magnetite) + H_2 \qquad (18.5)$$

Numerical models based on chemical thermodynamics show that hydrogen generation is dependent on the temperature, giving a maximum at around 315 °C (Figure 18.4). Above 315 °C, the olivine is thermodynamically stable and only participates in hydrous alteration partially, so the hydrogen production decreases. Below 200 °C, the hydrogen production decreases significantly because of the partitioning of Fe^{2+} into brucite and the slow reaction kinetics. The water pH value increases with the temperature decreasing, as it is governed by the equilibrium of the aqueous phase and brucite through the following equation:

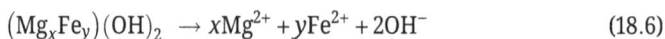

$$(Mg_xFe_y)(OH)_2 \rightarrow xMg^{2+} + yFe^{2+} + 2OH^- \qquad (18.6)$$

This means that water pH will be strongly alkaline as a consequence of the serpentinization process at low-temperature hydrothermal conditions, showing a pH of ~11 at 50 °C. It should be noted that the hydrogen generated can react with CO_2 or bicarbonate in the formation forming methane and other hydrocarbons [12].

The serpentinization process at low-temperature conditions (below 150 °C) has also been studied [13]. They have studied six different peridotite samples. It is generally believed that hydrogen production is related to the Fe^{2+} content; however, it is shown in Figure 18.5 that this is not the case. On the other hand, detailed analysis evidenced a catalysis role of spinel minerals such as magnetite, chromite, and gahnite. They hypothesized that the spinel surface may facilitate the charge transfer between water and Fe^{2+}. This finding is interesting, as the recently discovered natural hydrogen reservoir in Albania was underneath a chromite mine [9]. This hypothesis is worthy of further testing by first-principles simulations (as shown in Section 18.3.2).

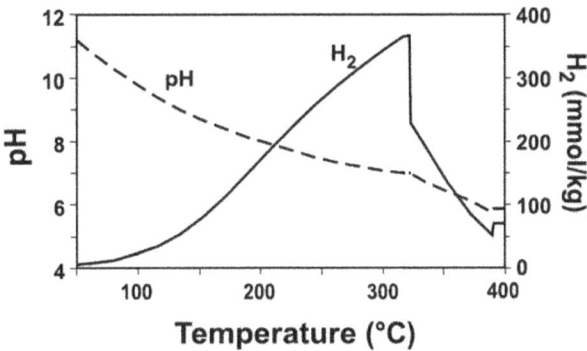

Figure 18.4: H_2 production and fluid pH as a function of temperature during serpentinization of ultramafic rocks. Reproduced with permission from Ref. [12]. Copyright 2008 Elsevier Ltd.

18.2.3 Role of igneous rock as a caprock

The discovery of a large accumulation of natural hydrogen in Mali was of great significance [7, 8]. It was reported that the price of natural hydrogen production can be 2–10 times less than that for green and blue hydrogen. There have been about 30 wells drilled, and the estimated volume in the field is about 60 billion m^3 of hydrogen [7]. Moreover, the volume continues to be filled with hydrogen influx. There are at least five stacked reservoir intervals with an area of about 8 km in diameter [8]. The dolerite sills are caprocks to seal the natural hydrogen (Figure 18.6). Dolerite is an igneous rock that can be a perfect seal for natural hydrogen if it is flat and widespread as a sill (i.e., parallel to the existing sediment structure). Furthermore, the traditional shale may play a role of seal in some wells (e.g., Bougou-13 Well) (figure 11 in Ref. [8].). This agrees with the work suggested by Rezaee (Curtin University), as presented in the Nanogeoscience 2024 Symposium held in Tokyo, 2024, and endorses the current efforts of geological hydrogen storage in our community. Interestingly, breccia filled with water may also serve as impermeable seals (e.g., Bougou-19 Well) (figure 11 in Ref. [8]). That means, even if a proper seal layer is not entirely effective, continuous H_2 generation from reactions such as water-rock interactions in ultramafic rocks can maintain a steady supply of hydrogen.

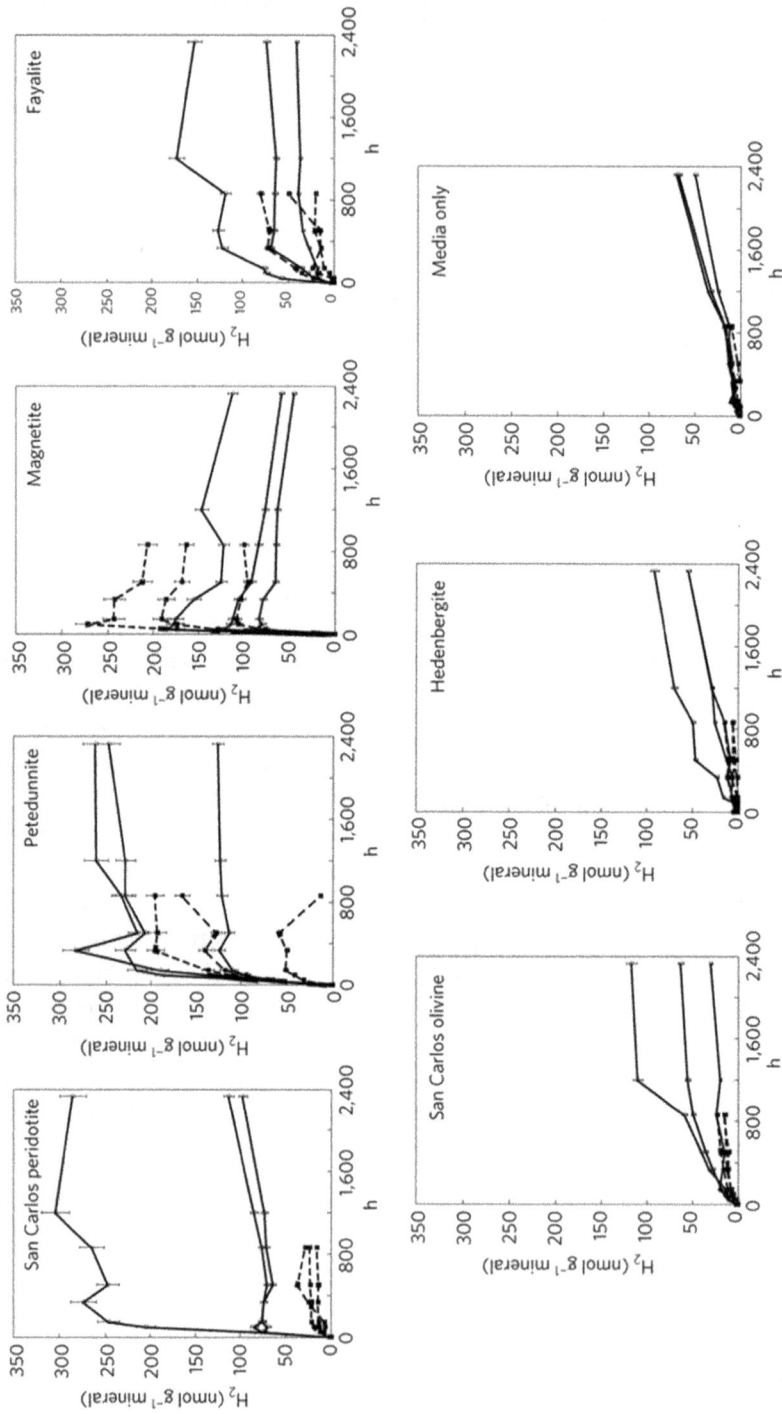

Figure 18.5: Hydrogen generation from mafic and ultramafic rocks. Reproduced with permission from Ref. [13]. Copyright 2013 Springer Nature Limited.

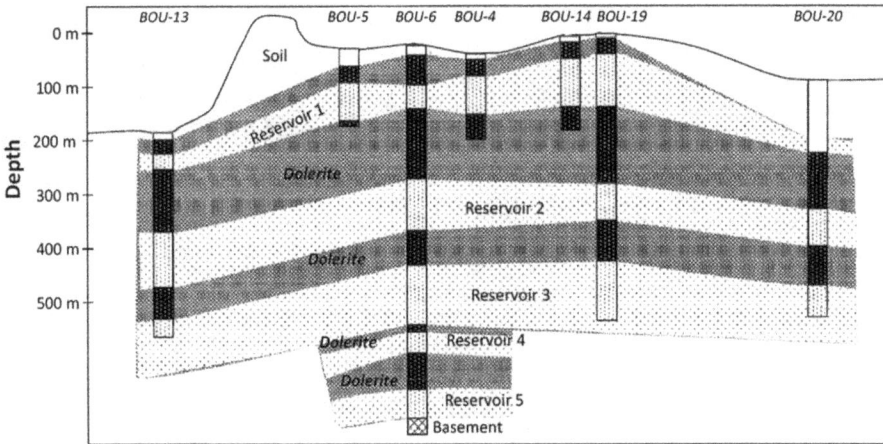

Figure 18.6: Dolerite sill as a caprock in Mali natural hydrogen field. Reproduced with permission from Ref. [8]. Copyright 2018 Hydrogen Energy Publications LLC (published by Elsevier Ltd.).

18.3 Reaction mechanisms revealed by first-principles simulations

18.3.1 Hydrogen generation from quartz and water

The study of natural hydrogen, particularly its generation through geological processes, holds vast potential for understanding both Earth's internal workings and possible extraterrestrial resources. One significant process contributing to natural hydrogen production involves the interaction of water and silicate minerals (e.g., by mechanochemical reaction). Quartz is often regarded as a model silicate mineral in the crust, which can generate hydrogen by interacting with water [17, 18]. In recent efforts, we have modeled the quartz-water system using first-principles molecular dynamics simulations to reveal the underlying reaction mechanisms [24]. Here, first-principles simulations are based on atomic interactions from electronic density functional theory [25]. Newton's law applies to atomic motions as those in classical molecular dynamics simulations; however, there is no need for classical potentials. This enables us to study chemical reactions. In the study, we focused on the Q3 (100) surface of quartz (Figure 18.7), which, according to surface energy considerations, is the most likely to expose $\equiv$Si• radicals – ideal candidates for hydrogen production [16–19]. Through this relationship, we observed a clear reaction pathway for the generation of molecular hydrogen, as depicted in Figures 18.8a and b. The random nature of rock fracturing means that mineral fragments produced during comminution often lack oxygen radicals, resulting in a reductive environment. This is crucial for hydrogen generation, particularly in small mo-

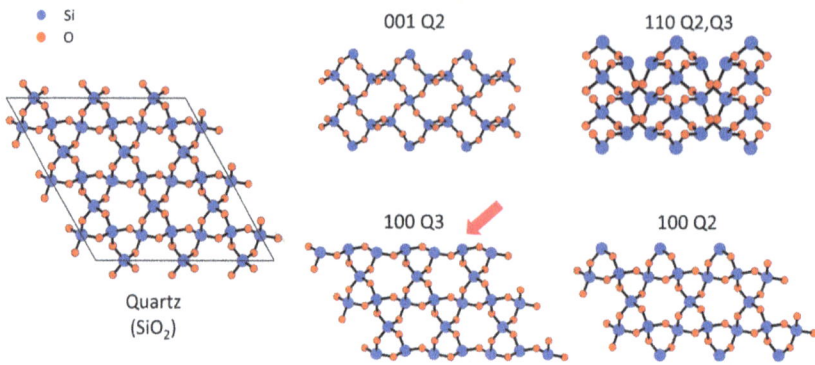

Figure 18.7: Quartz surfaces and the Q3 site. Reproduced with permission from Ref. [24]. Copyright 2024 under the Creative Commons CC BY 4.0 License.

lecular systems as employed in simulation studies. Without this reductive environment, any hydrogen atoms produced would immediately bond with oxygen forming silanol Si–OH groups.

Previous studies have suggested that there are two possible cleavage processes: one is homolytic cleavage and another is heterolytic cleavage [17, 26]. It is generally assumed that the first mechanism (i.e., homolytic cleavage) generates hydrogen (as the hydrogen molecule is zero-charged). Conversely, our simulations indicate that the hydrogen-generating reaction occurs via the second mechanism (i.e., heterolytic cleavage). There is no doubt that Si–OH will be formed by hydrolysis reaction. Si–H will be formed (instead of hydrogen molecules) without charge transfer. So, doing two simulations with charge transfer and without charge transfer allows us to justify the role of charge transfer and verify the hypothesis in the literature [24]. Charge analysis clearly shows the contrast between scenarios involving charge transfer and those that do not. In the absence of charge transfer, the silicon on the surface remains in a reduced state (+3.6 valence), diminishing its reducing properties and preventing hydrogen formation. However, when charge transfer is introduced, silicon achieves a + 4-valence state, facilitating the formation of $SiOH_2$ and subsequent hydrogen production (Figure 8c). Noteworthily, when positive charges were added to the system, hydrogen ions (H^+) were observed, supporting field observations [16] and previous experimental studies that also identified H^+ as a byproduct during similar reactions [17, 18].

In our simulations, we used quartz as a representative rock unit, given its widespread presence in the Earth's crust and its common occurrence in fault zones. To simulate the conditions of rock friction, we modeled fresh quartz surfaces, deliberately excluding any exposure to reactive oxygen species. This approach revealed that hydrogen could be generated under these conditions. Free radicals, which are often produced during high-energy events like friction, are inherently unstable subject to the environment. Our study suggests that while free radical reactions might be a component of the process, they cannot fully explain hydrogen generation in fault zones.

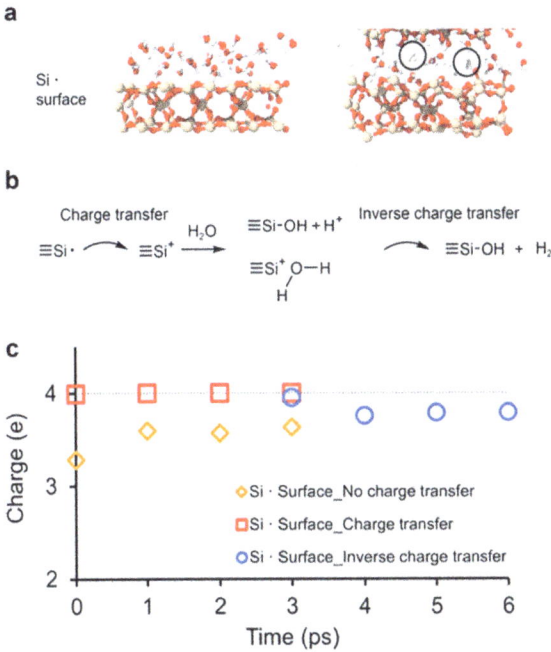

Figure 18.8: Reaction pathways of hydrogen generation from quartz and water. Reproduced with permission from Ref. [24]. Copyright 2024 under the Creative Commons CC BY 4.0 License.

The simulations highlighted the critical role of charge transfer in the process of natural hydrogen formation. In geological settings, such as during rock friction, the distribution of electrons across mineral surfaces becomes uneven. This electron redistribution can be modeled as charge transfer, a phenomenon observed in a variety of geological formations. During the process of rock friction, this uneven distribution of electrons could potentially lead to the creation of charge imbalances that favor the formation of hydrogen. In summary, the radicals are of significant importance; however, a charge transfer similar to the heterolytic cleavage of rock with charge transfer is considered to be one of the key steps [24]. This discovery offers a new perspective on the reaction pathways that contribute to natural hydrogen production.

In summary, our research has unveiled a detailed mechanism for hydrogen generation in fault zones and other geologically active environments. By combining molecular simulations with real-world geological observations, we have developed a clearer understanding of how friction, charge transfer and rock interactions contribute to the production of natural hydrogen. As we continue to explore these processes, it becomes evident that the conditions of deep geological environments, such as those found in subduction zones, may be particularly conducive to hydrogen generation. However, further evidence and more comprehensive studies are required to fully confirm the link between hydrogen production and seismic activity and to assess its po-

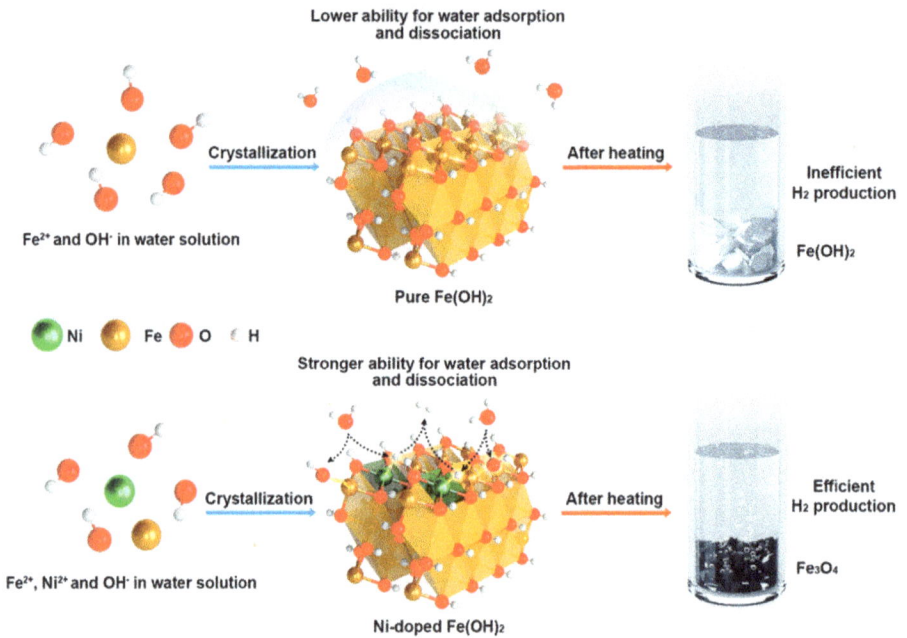

Figure 18.9: Experiment and simulation show that Ni^{2+} may catalyze water reduction on $Fe(OH)_2$. Reproduced with permission from Ref. [27]. Copyright 2021 Wiley-VCH GmbH.

tential as a sustainable resource. The study also emphasizes the importance of surface reactions, which should be important during the serpentinization process of mafic and ultramafic rocks. For example, it is noted that basalt generated hydrogen in the high-speed friction experiments [19].

18.3.2 Water reduction by $Fe(OH)_2$ oxidation

Serpentinization of mafic/ultramafic rocks is believed to be the major natural hydrogen process. First-principles simulation of serpentinization is still lacking, probably because the process is very complicated. Song et al. [27] have performed experiments on the hydrogen process from $Fe(OH)_2$ and studied the role of Ni^{2+} impurities (Figure 18.9). They further studied the adsorption and dissociation of water on the $Fe(OH)_2$ surface using first-principles simulations. Their experiments revealed that only 1% of Ni^{2+} impurities can enhance the hydrogen generation rate by 315 times. First-principles simulations showed that the water molecule can be adsorbed on the Ni^{2+}-doped $Fe(OH)_2$ surface much stronger than the pure $Fe(OH)_2$ surface. In contrast, the adsorption energies of OH^- on the two surfaces are almost the same. Furthermore, the energy at the transition state for water dissociation was also much reduced by the Ni^{2+} sites. The fundamental reason is that the Ni^{2+} sites have a higher electron density compared to the Fe^{2+} sites. Future studies should also explore the other impurities such as Cr^{3+} and the catalysis role of spinel minerals such as magnetite, chromite, and gahnite [13]. Knowledge of these studies can give us some guidance on the future exploration for natural hydrogen. It can be of great value for the development of natural hydrogen due to the additional profit of these valuable minerals.

18.4 Perspectives: natural hydrogen in Japan

18.4.1 Hydrogen reservoir in hydrate zone

The formation and distribution of natural hydrogen have become increasingly relevant in the context of both Earth's geochemical processes and its potential as a clean energy resource. In regions like Japan's Nankai area, substantial quantities of biogenic methane (CH_4) are stored as methane hydrate. The total amount of methane in methane hydrate was estimated to be about 1.1×10^{12} m^3 (40 trillion cubic feet), which is probably the largest methane hydrate deposit in the World [22]. In addition to methane hydrate, mud volcanoes are often observed in this area. Geogenic hydrogen is considered a potential contributor to this system [22, 28]. In the Nankai area, deep geological samples have revealed that hydrogen concentrations are three orders of magnitude higher at greater depths compared to shallow samples [28]. This discrepancy suggests that hydrogen generation may be occurring from deeper sources. The geological context of the Nankai subduction zone, with its continuous seismic faulting and fluid circulation, provides a possible explanation for this phenomenon (Figure 18.10). Notwithstanding, the thermogenic process in the subducted sediments can release hydrogen [22], and fault friction and water-rock interactions at the boundary of tectonic plates could play key roles in the production of hydrogen. Seismic imaging in the region has identified large faults extending from the deep plate interface, reinforcing the idea that hydrogen may be generated from these deeper plate boundaries [29].

It is our hope that we can discover hydrogen reservoirs in the methane hydrate zone. Furthermore, the recovered hydrogen could be used to enhance methane recovery from methane hydrates. Experiments show that a mixture of hydrogen and CO_2 can lead to a higher rate of methane recovery with different injection-production schemes across different laboratories [30, 31]. It was hypothesized that hydrogen could promote the exchange process and allow the exchange to readily occur. Molecular simulation showed that hydrogen molecules can diffuse freely in the methane hydrate solid phase, enter the occupied cages, and reduce the energy barrier of diffusion for the guest molecules that occupy the same cage with a hydrogen molecule [32]. As such, hydrogen alone may be used as an additive to enhance methane recovery. The produced gas is a mixture of hydrogen and methane; there is no need to separate, which provides us with cleaner en-

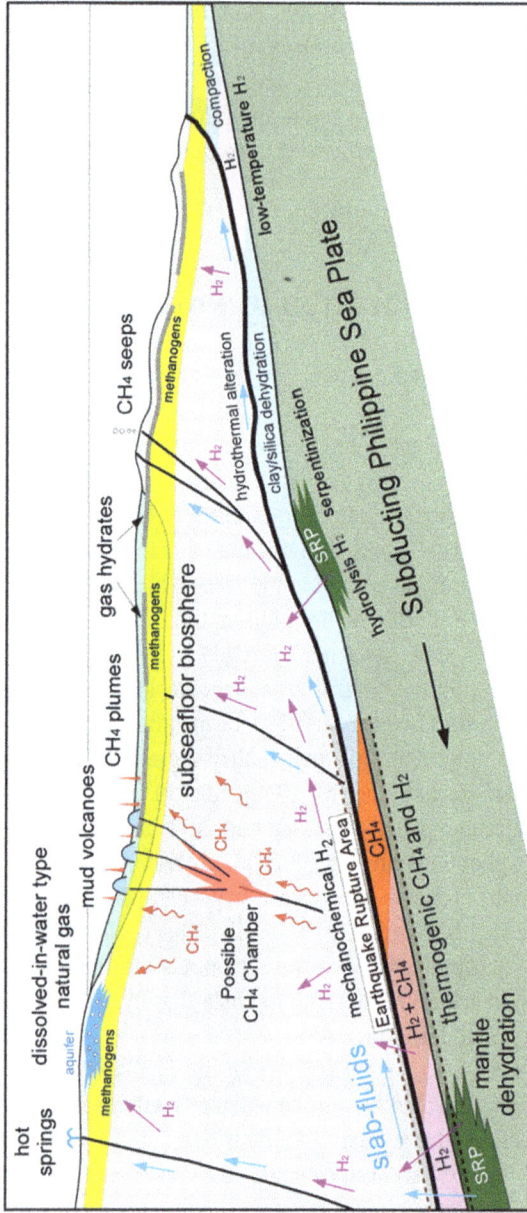

Figure 18.10: Hydrogen in the Nankai Trough. Methane in methane hydrate in Nankai Trough may partially result from biogenic production with hydrogen from the deep region. Seismic faulting and fluid circulation constantly occur, the H_2 may stem from fault friction and/or water-rock interactions. In addition, the thermogenic process in the subducted sediments can also be the source. Reproduced with permission from Ref. [22]. Copyright 2024 under the Creative Commons CC BY 4.0 license.

ergy. We believe that this can accelerate the commercialization process of methane hydrate and natural hydrogen as well. In the future, proactive considerations should be given to the possibility of trapped hydrogen in the hydrate zone.

18.4.2 Hydrogen in subduction zone

Serpentinization is the alteration process of mantle ultramafic rocks, such as peridotite and komatiite, to serpentine through interactions with water [12, 2, 33, 34]. This is one of the key processes to produce hydrogen gas associated with the subduction zone. Merdith et al. [33] proposed a theoretical framework and estimated the hydrogen production from the serpentinization in a subduction zone. In their work, they considered slab geometry, thermal profiles, and subduction histories of the world subduction zones and estimated the amount of residual abyssal peridotite (RAP) to serpentine during the subduction process over the last 5 million years. Their results suggested that the serpentinization process in the world's subduction zones could yield about 100,000 tons of hydrogen per year throughout the past five million years. If part of the generated hydrogen can be accumulated below a suitable caprock, it can be a significant energy source. As one can see, Japan has a wide area at the subduction zone, which is suitable for hydrogen production, showing a good yield of hydrogen [33]. Therefore, Japan could have huge potential for natural hydrogen. This has not included the hydrogen from other resources, such as those with a magmatic origin [20] and mechanochemical reaction during faulting [5, 16]. Given the widespread subduction zones in Japan, identifying potential systems for hydrogen commercial accumulation is of interest. Indeed, in an effort to investigate the influence of the igneous dome (in the over-riding plate) on the evolution of the accretionary prism in the Nankai Trough, an overpressured fluid or gas zone can be identified (Figure 18.11). There is no drilled well in this "reservoir," so we are not sure whether the over pressured fluid is hydrogen or methane gas. On the other hand, deep geological samples in this area have revealed the existence of hydrogen gas [28].

18.4.3 Natural hydrogen in the Hakuba field

Hakuba field is noteworthy for its hot springs, which contain natural hydrogen. The hot spring is strongly alkaline with a pH of over 11. Research by the Earth-Life Science Institute of the Tokyo Institute of Technology and others has revealed that the hot spring contains natural hydrogen, methane, and C_2–C_5 alkanes associated with the serpentinization process [35, 36]. Their research purpose is to unravel the mystery of the "origins of life." Very recently, it has attracted attention that there might be natural hydrogen reservoirs under the subsurface.

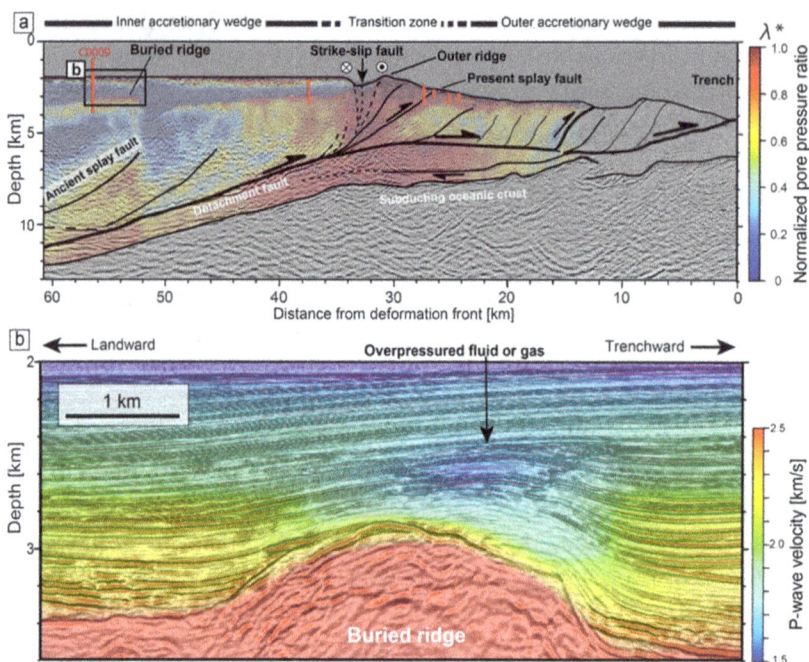

Figure 18.11: Seismic imaging of pathway and trapped gas in the Nankai Trough. Fracture or fault may promote the water-rock reaction generating H_2 as well as migration of H_2. The overpressured fluid or gas zone is likely methane, but there is a possibility of H_2. Note that a map showing the subduction profile worldwide and estimated H_2 production based on slab serpentinization can be found in Ref. [33]. Reproduced with permission from Ref. [29]. Copyright 2015 under the Creative Commons CC-BY license.

Hakuba, a village in Nagano, lies west of the Itoigawa-Shizuoka Tectonic Line (ISTL) (Figure 18.12), a major fault dividing Japan into western and eastern regions. The Shiroumadake area in Hakuba features an ultramafic complex spanning up to 8 km, comparable in size to the Bourakebougou natural hydrogen field in Mali [8]. Along the ISTL, hot springs with diverse water chemistry and gas compositions are distributed due to recent volcanic activity. Among these, Hakuba Happo is unique for its high concentrations of natural hydrogen gas, with water temperatures around 50 °C. Noble gas isotope analysis indicates that approximately 50% of helium originates from the mantle, while crustal contributions range from 17% to 35% [36]. In addition to ultramafic rock serpentinization, fracture systems may facilitate mechanochemical reactions.

A post-symposium field trip (Nanogeoscience 2024), organized by the University of Tokyo, investigated natural hydrogen degassing in the Hakuba Field. At Hakuba Happo, three hot spring wells were observed: Wells 1 and 3 are operational and located approximately 150 m apart, while Well 2 remains inactive due to scaling issues. Figure 18.13 presents a serpentinite sample collected near the Hakuba Happo hot spring wells, along the Matsu River, while Figure 18.14 displays data from Well 1, including hydrogen con-

(b) Volcanic chain

(a) Tectonic line

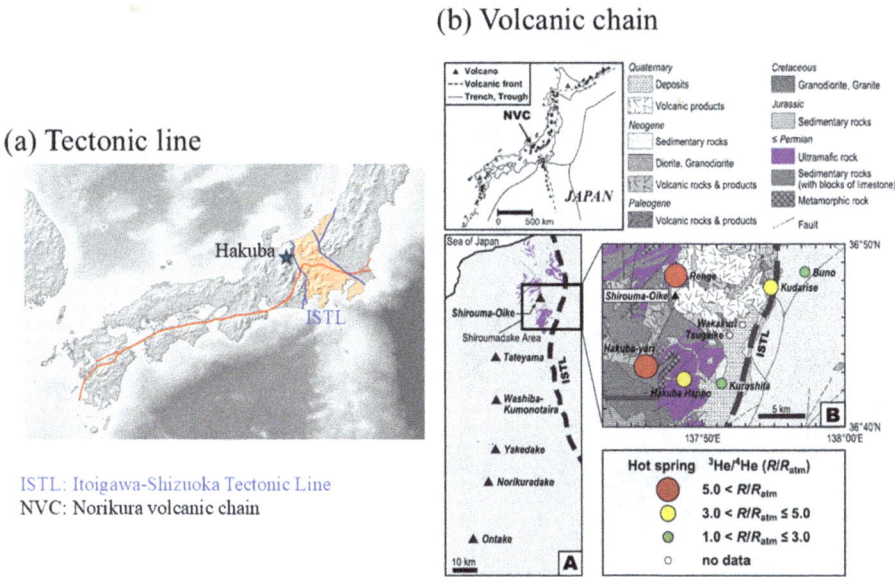

ISTL: Itoigawa-Shizuoka Tectonic Line
NVC: Norikura volcanic chain

Figure 18.12: Geological setting of Hakuba: (a) Itoigawa-Shizuoka Tectonic Line (ISTL) and (b) volcanic chain. The blue line on the left is the ISTL; the red line is the Median Tectonic Line. Panel (a) is reproduced with permission from the Japanese Active Fault database with the Legends added by authors. Panel (b) is reproduced with permission from Ref. [36]. Left panel: Copyright 2017 under the Creative Commons CC BY SA 3.0 License. Right panel: Copyright 2022 Elsevier B.V.

centration, water temperature, and pH. Hydrogen concentration was measured using a Portable Dissolved Hydrogen Meter ENH-2000 (TRUSTLEX, Japan). Water temperature was recorded with a PTFE-insulated thermistor TR-5,106 and Thermo Recorder TR42A (T&D, Japan). pH levels were determined using a glass electrode pH meter LAQUAtwin pH-33 (Horiba, Japan). The collected data strongly support the presence of natural hydrogen, generated through serpentinization. The water temperature was measured at 52.5 °C, with a hydrogen concentration of 1,480 ppb – close to the hydrogen solubility limit in water (~1,600 ppb) under ambient conditions (data from Molecular Hydrogen Institute). Considering temperature and the salting-out effect, the hydrogen content suggests saturation within the hot spring water. The pH of the water at ~50 °C was recorded at 10.78, while the room-temperature pH was approximately 11.465 which aligns well with thermodynamic predictions, which estimate a pH of ~11 at 50 °C [12].

Based on the discussions among the participants, the following themes were of interest:

1. An intensive geophysical survey in Hakuba (Shiroumadake area) will be planned next spring to have a better understanding of the Hakuba natural hydrogen system. A portable active sSeismic source (PASS) system [37, 38] can be installed near the hot spring well for long-term exploration. This system will be helpful in un-

Figure 18.13: Serpentinite near Hakuba hot spring. Photo taken during the Conference geological visit to Hakuba.

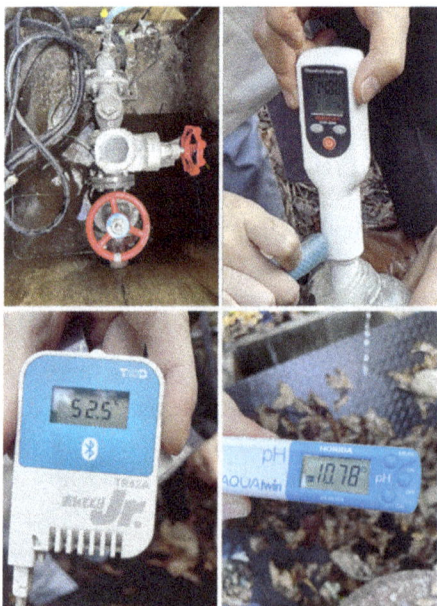

Figure 18.14: Hakuba hot spring well, hydrogen concentration, temperature, and pH were measured directly from Hakuba hot spring at Well 1. Photos were taken during the geological visit to Hakuba.

derstanding the fracture distribution in the hot spring reservoir and identifying potential seal structures nearby. A hydrogen sensor near the mixing tank will be installed to study the seasonal changes in hydrogen content.

2. Detailed analysis of the scale samples, fluid samples, mineral samples, and experimental data will help understand the detailed chemical reaction mechanisms, for example, whether the catalysts [13, 27] as documented in the literature exist and whether some other valuable minerals exist in the hydrothermal system.

3. Finally, active source seismic surveys may be conducted to find possible hydrogen reservoirs in a similar geological setting. In addition, possible fracture systems that may promote the hydrogen generation process, and the reservoirs sealed by shale or dolerite caprock, should be studied.

18.5 Conclusions

In this chapter, we first reviewed the significance of natural hydrogen. Then, we introduced the reaction mechanisms of natural hydrogen generation from rock-water interactions, focusing on mechanochemical reactions and serpentinization in mafic and ultramafic rocks. Learning from the Mali natural hydrogen field, we discussed the role of igneous rocks as cap rocks to trap natural hydrogen, in addition to conventional shale cap rocks. We also highlighted the uniqueness of the porous breccia media in trapping natural hydrogen. After that, we presented our current work using first-principles molecular simulations to understand the chemical reaction mechanisms and identify the reaction conditions. Together with literature studies, it is suggested that surface reactions play an important role in both the mechanochemical reactions and serpentinization processes. Surface states such as surface termination, electron transfer, and transition metal ion impurities are important for hydrogen generation. The knowledge we have gained is of great value for the discovery and development of natural hydrogen. It can also be used to develop orange hydrogen, which is generated by injecting water and CO_2 into reactive geological formations. Finally, we present our views on the potential of natural hydrogen as an energy resource in Japan, including the possibility of discovering hydrogen reservoirs in the hydrate zone of the Nankai Trough and its profitability for developing methane hydrate, the potential of hydrogen in subduction zones, and how it may be discovered. A combination of mechanochemical reactions in the fault systems and serpentinization of ultramafic rock offers a huge potential for natural hydrogen in Japan. We particularly introduced natural hydrogen in the Hakuba field. The field is noteworthy for its hot springs containing natural hydrogen. We have conducted a preliminary geological survey, which had interesting results. A more intensive geophysical survey is planned to better understand the natural hydrogen system in Hakuba as well as on Japan's national scale.

Acknowledgments: The authors acknowledge the Japan Society for the Promotion of Science (JSPS) for a Grant-in-Aid for Transformative Research Areas and Scientific Research (21H05202, 22H05108, 23K04647, and 24H00440), and are grateful to ENEOS Xplora Inc. for their financial support. We would like to express our sincere appreciation to Prof. Brian J. Evans (Curtin University) for his seminal talk entitled "Natural hydrogen, its geological regime and implications for Japan" at the University of Tokyo, for his insightful suggestions on future exploration of natural hydrogen in Japan during his visit, and his peer review of this manuscript. We wish to thank Prof. Zhijun Jin (Peking University), Prof. Reza Rezaee (Curtin University), and Prof. Yasuhiro Yamada (Kyushu University) for their invaluable contributions to the Nanogeoscience 2024 Symposium and suggestions during the geological survey in Hakuba. We also wish to thank Kazuyoshi Mizuno (Happoone Development Company) for his assistance in the geological survey.

Disclosure: Authors declare that they have no competing interests.

References

[1] Zgonnik V. The occurrence and geoscience of natural hydrogen: A comprehensive review. Earth-Science Reviews. 2020, 203: 103140.

[2] Kelley D. S., et al. A serpentinite-hosted ecosystem: The lost city hydrothermal field. Science. 2005, 307: 1428–1434.

[3] Telling J., et al. Rock comminution as a source of hydrogen for subglacial ecosystems. Nature Geoscience. 2015, 8: 851–855.

[4] Klein F., Tarnas D. J., Bach W. Abiotic sources of molecular hydrogen on earth. Elements. 2020, 16: 19–24.

[5] Wakita H., Kita I., Fujii N., Notsu K. Hydrogen release: New indicator of fault activity. Science. 1980, 210: 188–190.

[6] Waite J. H., et al. Cassini finds molecular hydrogen in the Enceladus plume: Evidence for hydrothermal processes. Science. 2017, 356: 155–159.

[7] Hand E. Hidden hydrogen. Science. 2023, 379: 630–636.

[8] Prinzhofer A., Cissé C. S. T., Diallo A. B. Discovery of a large accumulation of natural hydrogen in Bourakebougou (Mali). International Journal of Hydrogen Energy. 2018, 43: 19315–19326.

[9] Truche L., et al. A deep reservoir for hydrogen drives intense degassing in the Bulqizë ophiolite. Science. 2024, 383: 618–621.

[10] Osselin F., Soulaine C., Fauguerolles C., Gaucher E. C., Scaillet B., Pichavant M. Orange hydrogen is the new green. Nature Geoscience. 2022, 15: 765–769.

[11] Wang J., Watanabe N., Okamoto A., Nakamura K., Komai T. Pyroxene control of H_2 production and carbon storage during water-peridotite-CO_2 hydrothermal reactions. International Journal of Hydrogen Energy. 2019, 44: 26835–26847.

[12] McCollom T. M., Bach W. Thermodynamic constraints on hydrogen generation during serpentinization of ultramafic rocks. Geochimica et Cosmochimica Acta. 2009, 73: 856–875.

[13] Mayhew L. E., Ellison E. T., McCollom T. M., Trainor T. P., Templeton A. S. Hydrogen generation from low-temperature water–rock reactions. Nature Geoscience. 2013, 6: 478–484.

[14] Lollar B. S., Onstott T. C., Lacrape-Couloume G., Ballentine C. J. The contribution of the Precambian continental lithosphere to global H_2 production. Nature. 2014, 516: 379–382.

[15] Kita I., Matsuo S., Wakita H. H_2 generation by reaction between H_2O and crushed rock: An experimental study on H_2 degassing from the active fault zone. Journal of Geophysical Research. 1982, 87: 10789–10795.

[16] Kameda J., Saruwatari K., Tanaka H. H_2 generation in wet grinding of granite and single-crystal powders and implications for H_2 concentration on active faults. Geophysical Research Letters. 2003, 30: 2063.

[17] Saruwatari K., Kameda J., Tanaka H. Generation of hydrogen ions and hydrogen gas in quartz–water crushing experiments: An example of chemical processes in active faults. Physics and Chemistry of Minerals. 2004, 31: 176–182.

[18] Delogu F. Hydrogen generation by mechanochemical reaction of quartz powders in water. International Journal of Hydrogen Energy. 2011, 36: 15145–15152.

[19] Hirose T., Kawagucci S., Suzuki K. Mechanoradical H_2 generation during simulated faulting: Implications for an earthquake-driven subsurface biosphere. Geophysical Research Letters. 2011, 38: L17303.

[20] Wakita H., Sano Y. $^3He/^4He$ ratios in CH_4-rich natural gases suggest magmatic origin. Nature. 1983, 305: 792–794.

[21] Lin L. H., et al. Radiolytic H_2 in continental crust: Nuclear power for deep subsurface microbial communities. Geochemistry, Geophysics, Geosystems. 2005, 6: Q07003.

[22] Suzuki N., et al. Thermogenic methane and hydrogen generation in subducted sediments of the Nankai Trough. Communications Earth & Environment. 2024, 5: 97.

[23] Earle S. Physical Geology. First University of Saskatchewan Edition, 2019. https://openpress.usask.ca/physicalgeology/

[24] Cui W., Liang Y., Masuda Y., Hirose T., Tsuji T. Identifying general reaction conditions for mechanoradical natural hydrogen production. https://doi.org/10.21203/rs.3.rs-4001833/v1

[25] Marx D., Hutter J. Ab initio molecular dynamics: Basic theory and advanced methods. Cambridge University Press, 2009.

[26] Hochstrasser G., Antonini J. F. Surface states of pristine silica surfaces: I. ESR studies of Es' dangling bonds and of CO_2^- adsorbed radicals. Surface Science. 1972, 32: 644–664.

[27] Song H., Ou X., Han B., Deng H., Zhang W., Tian C., Cai C., Lu A., Lin Z., Chai L. An overlooked natural hydrogen evolution pathway: Ni^{2+} boosting H_2O reduction by $Fe(OH)_2$ oxidation during low-temperature serpentinization. Angewandte Chemie International Edition. 2021, 60: 24054–24058.

[28] Ijiri A., et al. Deep-biosphere methane production stimulated by geofluids in the Nankai accretionary complex. Science Advances. 2018, 4: eaao4631.

[29] Tsuji T., Ashi J., Strasser M., Kimura G. Identification of the static backstop and its influence on the evolution of the accretionary prism in the Nankai Trough. Earth & Planetary Science Letters. 2015, 431: 15–25.

[30] Ding Y. L., Xu C. G., Yu Y. S., Li X. S. Methane recovery from natural gas hydrate with simulated IGCC syngas. Energy. 2017, 120: 192–198.

[31] Sun Y. F., Wang Y. F., Zhong J. R., Li W. Z., Li R., Cao B. J., Kan J. Y., Sun C. Y., Chen G. J. Gas hydrate exploitation using CO_2/H_2 mixture gas by semi-continuous injection-production mode. Applied Energy. 2019, 240: 215–225.

[32] Waage M. H., Trinh T. T., van Erp T. S. Diffusion of gas mixtures in the sI hydrate structure. Journal of Chemical Physics. 2018, 148: 214701.

[33] Merdith A. S., Daniel I., Sverjensky D., Andreani M., Mather B., Williams S., Vitale Brovarone A. Global hydrogen production during high-pressure serpentinization of subducting slabs. Geochemistry, Geophysics, Geosystems. 2023, 24: e2023GC010947.

[34] Rupke L. H., Morgan J. P., Hort M., Connolly J. A. D. Serpentine and subduction zone water cycle. Earth & Planetary Science Letters. 2004, 223: 17–34.

[35] Suda K., et al. Origin of methane in serpentine-hosted hydrothermal systems: The CH_4-H_2-H_2O hydrogen isotope systematics of the Hakuba Happo hot spring. Earth & Planetary Science Letters. 2014, 386: 112–125.

[36] Suda K., et al. The origin of methane in serpentine-hosted hyperalkaline hot spring at Hakuba Happo, Japan: Radiocarbon, methane isotopologue and noble gas isotope approaches. Earth & Planetary Science Letters. 2022, 585: 117510.

[37] Tsuji T., Tsuji S., Kinoshita J., Ikeda T., Ahmad A. B. 4 cm Portable Active Seismic Source (PASS) for Meter- to Kilometer-scale imaging and monitoring of subsurface structures. Seismological Research Letters. 2022(94): 149–158.

[38] Tsuji T., Arakawa E., Tsukahara H., Murakami F., Aoki N., Abe S., Miura T. Signal propagation from portable active seismic source (PASS) to km-scale borehole DAS for continuous monitoring of CO_2 storage site. Greenhouse Gases: Science and Technology. 2024, 14: 4–10.

Manzar Fawad*, Scott Andrew Whattam, Abdullah Alqubalee,
and Ahmed Al-Yaseri

Chapter 19
Various elements of a potential hydrogen system in Saudi Arabia

Abstract: Natural or "white" hydrogen (H_2), a promising clean energy source with the potential to significantly reduce carbon emissions, has emerged. While exploration efforts have traditionally focused on conventional hydrocarbon reservoirs, reports of H_2 encounters while drilling for hydrocarbon in Australia and France and the recent discovery of natural H_2 accumulations in Mali point toward new possibilities for carbon-free energy production. With its vast geological diversity and unique tectonic setting, Saudi Arabia presents a compelling prospect for natural H_2 exploration. This study discusses the various elements comprising a potential H_2 System resulting in H_2 accumulations in Saudi Arabia, while leveraging existing geological and geophysical information. We identify key reservoir-seal pairs and regions exhibiting favorable H_2 generation and entrapment conditions by integrating relevant maps, cross sections, and subsurface temperature estimates. Our analysis highlights promising areas with ophiolitic source rocks, structural traps, and proximity to the potential kitchen areas. This assessment is a foundation for future exploration efforts, guiding targeted field surveys, geochemical sampling, and exploratory drilling campaigns. Identifying natural H_2 resources in Saudi Arabia could profoundly impact the nation's energy landscape, contributing significantly to its energy transition and global efforts to combat climate change.

Keywords: natural hydrogen system, Saudi Arabia, magnetic survey, fractures in ophiolites, play fairways

*Corresponding author: Manzar Fawad, Center for Integrative Petroleum Research, King Fahd University of Petroleum and Minerals, Dhahran 31261, Saudi Arabia,
e-mail: manzar.fawad@kfupm.edu.sa
Scott Andrew Whattam, Department of Geosciences, King Fahd University of Petroleum and Minerals, Dhahran 31261, Saudi Arabia
Abdullah Alqubalee, Ahmed Al-Yaseri, Center for Integrative Petroleum Research, King Fahd University of Petroleum and Minerals, Dhahran 31261, Saudi Arabia

https://doi.org/10.1515/9783111437040-019

19.1 Introduction

The global pursuit of clean and sustainable energy sources has led to a surge of interest in H_2 as a potential solution to mitigate carbon emissions and combat climate change. While industrial H_2 production methods have relied on fossil fuels directly or indirectly, the emergence of natural H_2, also known as "white H_2," offers a promising alternative with minimal environmental impact. Natural H_2 is generated through geological processes within the Earth's subsurface [1–4].

Evidence of H_2 seeps has been reported over the previous three decades. However, the 1998 discovery of a natural H_2 accumulation in Mali pointed toward an untapped potential. Recent studies of natural H_2 seeps in various geological settings, including Precambrian cratons [5–7], have fueled exploration efforts worldwide. These seeps highlight the potential of natural H_2 as a viable and abundant energy resource if trapped under suitable geologic conditions. Several countries, including Mali, Russia, the USA, Brazil, and Ukraine, have reported naturally occurring H_2, often associated with specific geological features such as circular depressions called fairy circles [5, 7–9].

H_2 exploration is similar to oil and gas exploration in that there is an analogous concept of source rock, reservoir, and trap. Three main types of source rocks for H_2 generation are identified, including (1) ultramafic/mafic/iron-rich rocks, (2) uranium-rich rocks, and (3) late-matured organic matter-rich rocks [1, 2, 5, 10, 11]. In the first case, the production of H_2 in serpentinites (hydrothermally altered ultramafic rocks) is attributed to the oxidation of Fe(II) by water (H_2O). In the second case, the production of H_2 is considered to be by the radiolysis of H_2O by natural radioactivity. Finally, H_2 may be produced during organic late maturation in the third case.

It is important to mention that a given sample's initial serpentinization in laboratory produced between 120 and over 300 mmol H_2 per kg of rock through water splitting methods, as indicated by mass balance and thermodynamic calculations (i.e., cases 1 and 2) [12]. However, the amount of produced H_2 is expected to be higher from organic-rich rocks (20 mg of H_2 from each g of total organic content (TOC)); yet the amount of H_2 still depends on the H_2 index and kerogen type [13].

According to Moretti et al. [5], there are three primary geological environments suitable for natural H_2 generation: (1) serpentinized ultramafic rocks at mid-ocean ridges; (2) compressional tectonic zones involving ophiolitic nappes; (3) Precambrian crystalline shields.

Saudi Arabia lies on the Arabian tectonic plate, which is subducting under the Eurasian plate, manifesting the foredeep of the Zagros collision zone [14–17] (Figure 19.1). In Saudi Arabia, vast ophiolitic ultramafic rocks, comprising serpentinites, have been reported in the western and northwestern segments of the exposed Precambrian Arabian Shield (e.g., as summarized by Johnson and Kattan [22]). These ophiolitic ultramafic rocks are accreted forearc fragments [18]. The Arabian Shield basement is overlain by a thick sedimentary cover, with thickness increasing toward the east and the northeast [16, 19, 20]. As Saudi Arabia possesses one of the world's most efficient and largest petroleum systems, it confirms that the reservoirs and caprocks are reliable. Therefore, we can deduce that all vital elements for a potential "H_2 system" are present in Saudi Arabia: (1) source rocks, (2) reservoirs, and (3) competent caprocks. Despite the growing interest in natural H_2 as a clean energy source, there remains a significant research gap in understanding the full potential of natural H_2 resources, especially in Saudi Arabia. While there is a promising geological setup and regions for natural H_2 exploration, there is a lack of subsurface data in the public domain (e.g., well logs, seismic, organic-rock geochemistry, and the produced hydrocarbon composition) to identify various factors influencing the distribution, abundance, and quality of natural resources in the country.

Additionally, further research is needed to optimize exploration and production techniques, assess the environmental and economic impacts of natural H_2 utilization, and develop appropriate policy and regulatory frameworks for the sustainable development of the natural H_2 sector. Addressing these research gaps is crucial for unlocking the full potential of natural H_2 as a viable and sustainable energy source in Saudi Arabia and beyond.

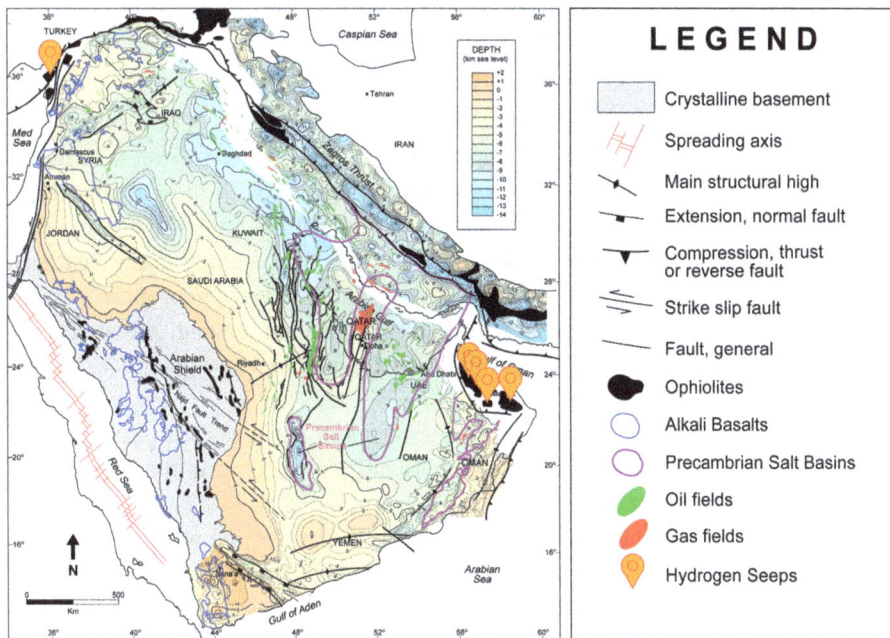

Figure 19.1: Major tectonic elements of the Arabian plate and its elevation under the sedimentary cover. The basin is shallow in the west and bounded by the exposed Precambrian Arabian Shield. It deepens gently in an easterly direction and reaches its maximum depth of several kilometers in the foredeep of the Zagros collision zone (modified after [15, 16, 22]). Serpentinization-related H_2 gas seeps are present in Oman and Kurtbagi, Turkey [9].

This study addresses this knowledge gap by preliminary screening for various elements of a potential H_2 system and identifying play fairways of natural H_2 accumulations in Saudi Arabia. By integrating existing geological and geophysical data, including maps, and cross sections, we seek to identify specific geological formations and regions that exhibit favorable conditions for H_2 generation and entrapment. Our analysis focuses on key geological criteria, such as the presence of source rocks, maturation/serpentinization possibilities, traps, and proximity to known H_2 seeps or occurrences [21]. Moreover, we develop a preliminary assessment methodology tailored to the unique geological context of Saudi Arabia, drawing upon insights from other regions with active H_2 exploration [11, 21]. This methodology will be a foundation for future exploration efforts, guiding targeted field surveys, geochemical sampling, and exploratory drilling campaigns. Ultimately, this research aims to pinpoint promising areas for further investigation and resource quantification, enhancing our understanding of Saudi Arabia's natural H_2 potential and its role in the global shift toward carbon-free energy solutions.

19.2 Tectonic setting of the Arabian plate and H_2 formation

The Arabian plate, a continental tectonic plate encompassing most of the Arabian Peninsula, is bordered by the African, Eurasian, Indian, and Anatolian Plates. Its northward movement at roughly 2 cm/year results in collision with the Eurasian Plate along the Zagros Mountains. The plate also undergoes rifting in the Red Sea and Gulf of Aden, subduction in the Makran Trench, and transform faulting in the Dead Sea Transform Fault. These tectonic processes have shaped diverse geological features that may influence natural H_2 generation and accumulation in Saudi Arabia. For instance, rifting has created extensional basins that could serve as potential sites for natural H_2 accumulation.

The Arabian Shield is exposed in the western and northwestern parts of Saudi Arabia. It comprises a series of intra-oceanic island arcs accreted in the late Proterozoic. These complexes are separated by sutures containing various levels of ophiolite sequences. The shield is overlain by sediments in the eastern and northeastern parts of Saudi Arabia, with the thickness increasing away from the shield toward the east-northeast (Figure 19.2). Sediments from Precambrian to Tertiary have been deposited in the basins.

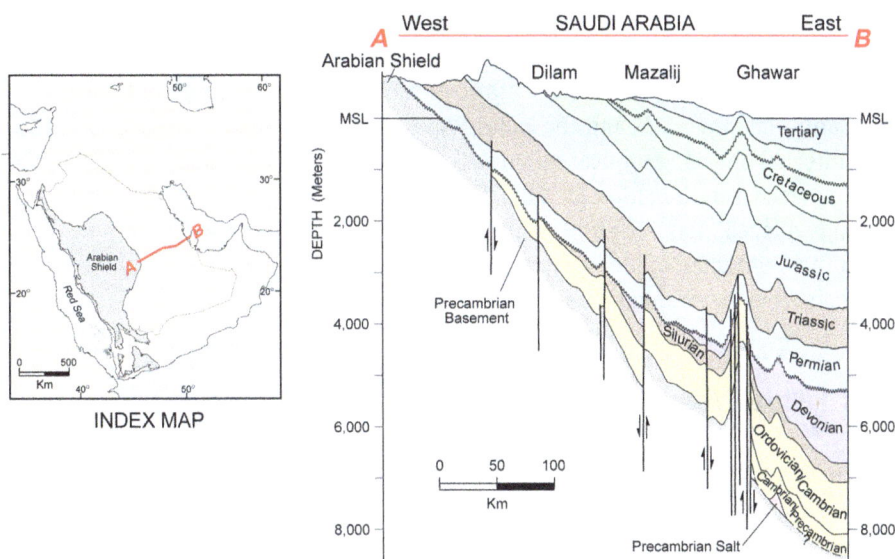

Figure 19.2: Geologic cross section (A–B) through Saudi Arabia toward Qatar with index map showing the profile location (modified after [16]).

19.3 Presence of H₂-generating ultramafic and mafic rocks (ophiolites)

Ultramafic and mafic rocks, such as those found in ophiolites, are crucial for H_2 generation through serpentinization. These rocks, remnants of ancient oceanic crust and uppermost mantle, are abundant in the western part of Saudi Arabia. Johnson and Kattan [22] highlight that ophiolites and related rocks within the Arabian-Nubian Shield represent remnants of oceanic lithosphere and associated supra-subduction zone (SSZ) assemblages. These ophiolites are often dismembered and tectonically interleaved with island arc volcanic and sedimentary rocks. Ophiolites were first recognized in the region (in the Nubian Shield) by Rittman [66], and starting with the founding work of Al-Shanti and Mitchell [65] and Bakor et al. [30], such rocks have been interpreted in the Arabian Shield as remnants of 890 Ma to 675 Ma oceanic crust and uppermost mantle (see review of Johnson and Kattan [22]). Tectonically, these ophiolites are indicators of arc-arc suturing and terrane amalgamation [23–27].

The Jabal Ess ophiolite (Figure 19.3) is the most complete ophiolite known in the Arabian Shield [28, 29]. It crops out in an area more than 30 km east-west by 5 km north-south as an assemblage of mantle peridotite, isotropic and layered gabbro, a dike complex, pillow basalt, and pelagic sediments up to 3 km thick. The ophiolite is located along the Yanbu suture and is bounded north and south by visible shear zones. The shear zones merge to the east, where the ophiolite ceases to be recognizable; to the west, the ophiolite is cut by the NW-trending sinistral fault system of the Da'ban and Durr shear zones. Mafic-ultramafic rocks continue to the south as the Sahluj mélange [24] and the Jabal Wask ophiolite [30–32]. Because of folding and internal shearing, typical ophiolite components are not always juxtaposed in the expected relationships at Jabal Ess. Peridotite dominates in the south as a strongly altered, massive serpentinite. Other (n ~ 20) similar but less complete serpentinite-dominated ophiolites and ultramafic massifs (Figure 19.3) of uncertain origin comprise the Arabian Shield.

19.3.1 Shield/basement features in magnetic survey data

Magnetic surveys may play a pivotal role in the search for source rocks for natural hydrogen. These surveys measure variations in the Earth's magnetic field caused by differences in the magnetic properties of underlying rocks and minerals. In natural hydrogen exploration, magnetic surveys can help identify areas with ultramafic and mafic rocks, which are often iron-rich and could be associated with hydrogen generation. Additionally, magnetic data can reveal geological structures like faults and fractures, which could act as pathways or traps for hydrogen accumulations. By interpreting magnetic anomalies, geologists can gain valuable insights into the subsurface geology and pinpoint promising areas for further investigation and potential drilling.

A magnetic survey map of Saudi Arabia shows two distinct sets of anomalies under relatively thin sediments (Figure 19.3). One EW trending feature lies in the northwest, whereas the second is NS trending just west of Riyadh. The magnetic anomalies in the sub-aerially exposed shield fairly well match the mapped ophiolite trends [22]. Since no significant basement structures are documented on these anomaly locations, these, therefore, likely represent paleo-sutures sandwiching magnetic ophiolitic rocks suitable for H_2 generation. In deeper parts of the sedimentary basins, the magnetic anomalies are either weak due to increase in depth, or the data is not available to the authors.

19.3.2 Fractures in ophiolites

Ophiolites are formed of oceanic crust and upper mantle that have been uplifted and exposed on land. While on the seafloor, these rocks can be subjected to tectonic forces that cause them to stretch and break, resulting in the formation of fractures. The subsequent emplacement process, where the oceanic crust and mantle are thrust onto the continental crust, can also cause significant deformation and fracturing of the ophiolite rocks. The ophiolites present in Arabian Shield also would have gone through deformation via uplift related to the Red Sea rifting and subsequent compres-

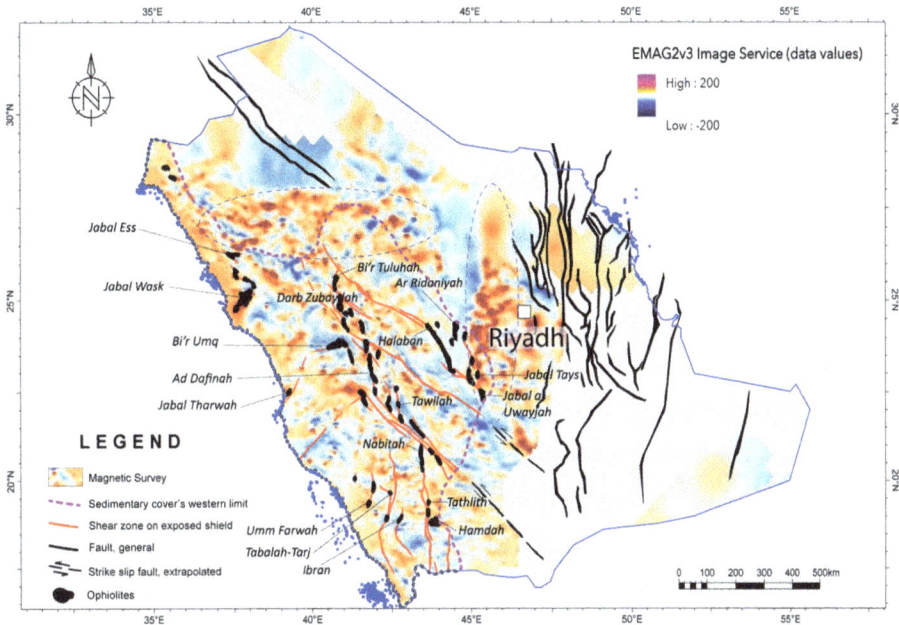

Figure 19.3: Magnetic survey map of Saudi Arabia with ophiolite bodies exposed in the Arabian Shield (modified from [33]), ophiolites and sheared zones in the exposed shield are taken from [22]; faults on the shield underlain by sedimentary cover are taken from [15, 16].

sion, particularly in the east, as a result of a collision with Eurasia, causing folding and thrusting in the Zagros zones [15, 16] (Figure 19.1).

Using Google satellite imagery and a geological map of Wadi Al Ay's, Saidy et al. [34] delineated fractures on four large serpentinite-dominated ophiolites in the Arabian Shield, namely, Jabal Ess, Jabal Wask, Jabal Tharwah, and Bi'r Tululah. The density of these fractures was mapped and classified into five categories: very high, high, moderate, low, and very low. The study found that areas with high to very high fracture density, represented by red and blue zones on the map (Figure 19.4), could potentially possess high (secondary) porosity and permeability. The permeability of the rock may allow water to percolate through and react with the minerals, leading to serpentinization and subsequent hydrogen production. Additionally, these interconnected fractures could provide conduits for the migration of generated hydrogen, potentially leading to its accumulation in suitable geological structures.

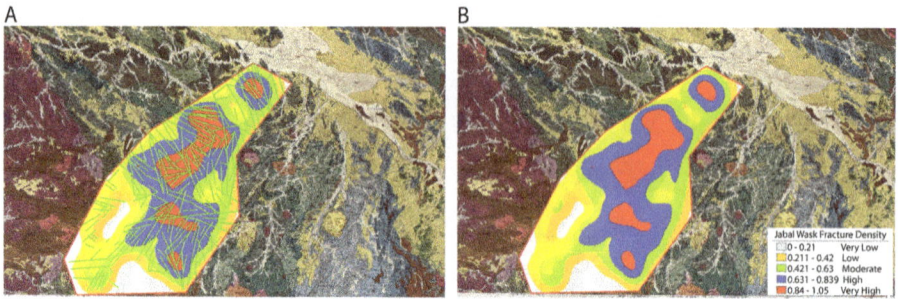

Figure 19.4: (A) Jabal Wask fracture lineaments and (B) Jabal Wask fracture density map (modified from [34]).

19.4 Hydrogeological features

Saudi Arabia has various aquifers and deep groundwater systems that could contribute to H_2 generation and migration (Figure 19.5). Groundwater interaction with specific rock types, particularly ultramafics, can lead to serpentinization, producing H_2. Additionally, deep groundwater systems could act as conduits for H_2 migration and accumulation in suitable geological traps.

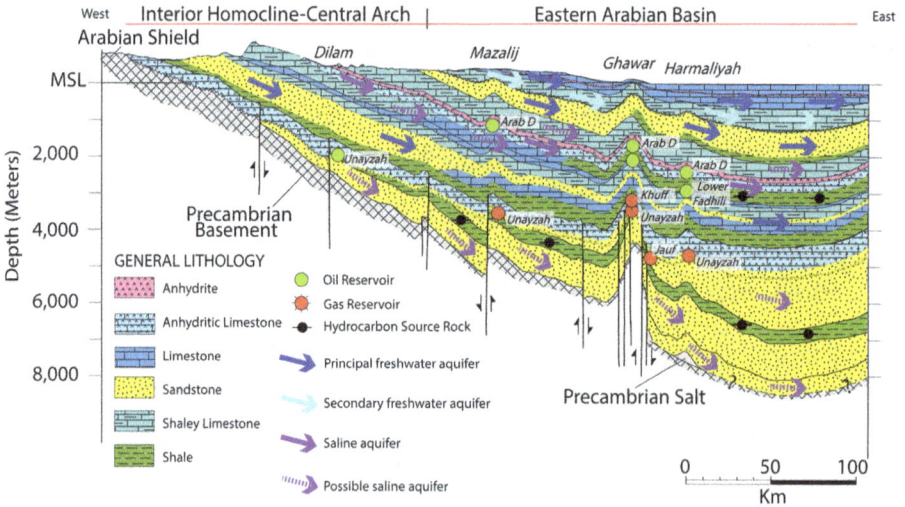

Figure 19.5: Geological cross section across central and eastern Saudi Arabia displaying the lithology and structure of the Interior Homocline-Central Arch and the Eastern Arabian Basin, highlighting principal and secondary groundwater aquifers, major regional aquitards/aquicludes, saline aquifers, water salinity, hydrocarbon reservoirs and source rocks (modified after [20]).

19.5 Subsurface temperature

The geological process of serpentinization, which can produce natural H_2 from water-rock interaction, relies heavily on suitable subsurface temperatures. Serpentinization, a hydrothermal reaction between water and ultramafic rocks, typically occurs at temperatures >200 °C, owing to the oxidation of reduced iron in the minerals [35].

The subsurface temperature depends on that area's temperature gradient and the measurement depth. The Arabian Shield's basement becomes deeper away from the section exposed in the western and northwestern parts of Saudi Arabia. Since the temperature data from the oil and gas wells is not publicly available, we used an average temperature gradient of 25 °C/km from exploration wells in Oman [36]. Assuming a mean annual temperature of 26 °C, we calculated the surface temperature of the Arabian Shield, taking depths from Konert et al. [16] (Figure 19.1). At least five regions can be identified where the top shield temperature exceeds 200 °C comprising the Wadi Sirhan Graben, Widyan Basin, Eastern Arabian Basin, Wajid Graben, and Rub'Al-Khali Basin (Figure 19.6). It suggests that the subsurface conditions in these regions of Saudi Arabia could be conducive to natural H_2 production through serpentinization. However, a detailed temperature gradient map will further refine temperatures at different depths and locations on top of the Arabian Shield to fully assess the potential for natural H_2 generation.

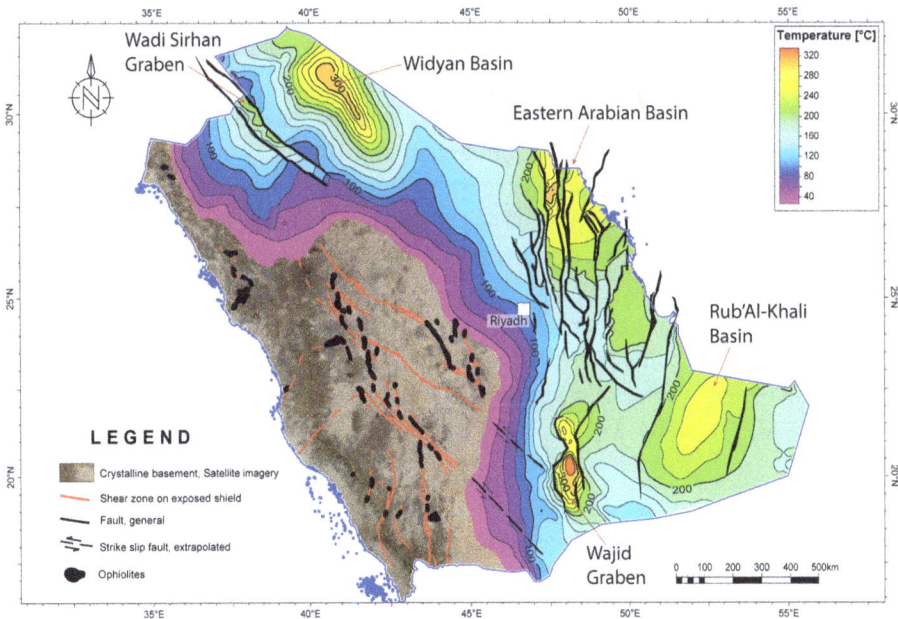

Figure 19.6: Temperature (°C) at the top of the Arabian Shield basement in Saudi Arabia estimated using a temperature gradient from Rolandone et al. [36], (ophiolites and sheared zones in the exposed shield are taken from [22]; faults on the shield underlain by sedimentary cover are taken from [15, 16]).

19.6 Stratigraphy and reservoir-seal pairs

Saudi Arabia's stratigraphy is characterized by Precambrian basement rocks overlain by Phanerozoic sedimentary rocks. The basement rocks are primarily composed of granitoids, gneisses, and metavolcanics sandwiching mafic and ultramafic rocks, which could provide potential sources of natural H_2 generation mainly through serpentinization. The sedimentary rocks consist mainly of carbonates, sandstones, and shales and have been influenced by tectonic events like rifting, folding and faulting. Siliciclastic successions in the Paleozoic dominate the Phanerozoic stratigraphy of Saudi Arabia, transitioning to a carbonate-dominated system in the Mesozoic (Figure 19.7). This shift is attributed to the evolving tectonic and depositional settings associated with the Proto-Tethys, Paleo-Tethys, and Neo-Tethys oceans.

Two main petroleum systems occur in the central and eastern parts of Saudi Arabia, including the Paleozoic and Mesozoic petroleum systems. The former is associated with Proto-Tethys/Paleo-Tethys, while the latter is associated with the Neo-Tethys [37]. The Paleozoic petroleum system is mainly sourced by the "hot shale" of the Lower Silurian Qusaiba Formation [38–41], and is characterized by reservoirs primarily in the Devonian Jauf Formation, the Permo-Carboniferous Unayzah Group, and the Permo-Triassic Khuff carbonate reservoirs. The major seal for the Paleozoic petroleum system is the Lower Triassic Sudair Shale. Other seals are also recognized due to the reservoirs beneath them, including the Lower Silurian Qusaiba, the Shaley interval of the lower Devonian Jauf (known as the D3B marker), tight carbonates, shales, and evaporites of the basal Khuff Formation [16, 37].

The Mesozoic petroleum system is mainly sourced from the organic-rich carbonates of the Upper Jurassic Tuwaiq mountain and Hanifa Formations [37, 42]. The main reservoirs of the Mesozoic petroleum system include the prolific Upper Jurassic Arab Formation and the Early Cretaceous Safaniya and Khafji Formations. The Upper Jurassic Hith Formation serves as a major regional seal for the Mesozoic petroleum system [37].

Diverse lithologies, stratigraphy, and complex depositional history have created many reservoir-seal pairs, contributing to Saudi Arabia's significant hydrocarbon reserves. The region provides a range of potential reservoirs and seal rocks for H_2 accumulation.

19.7 Structural and stratigraphic traps

Geological structures like faults and anticlines can trap and accumulate H_2. Saudi Arabia, a country well-known for its vast oil reserves, also contains diverse geological structures that serve as traps for valuable hydrocarbons. These traps can be classified into structural, stratigraphic, and combination type.

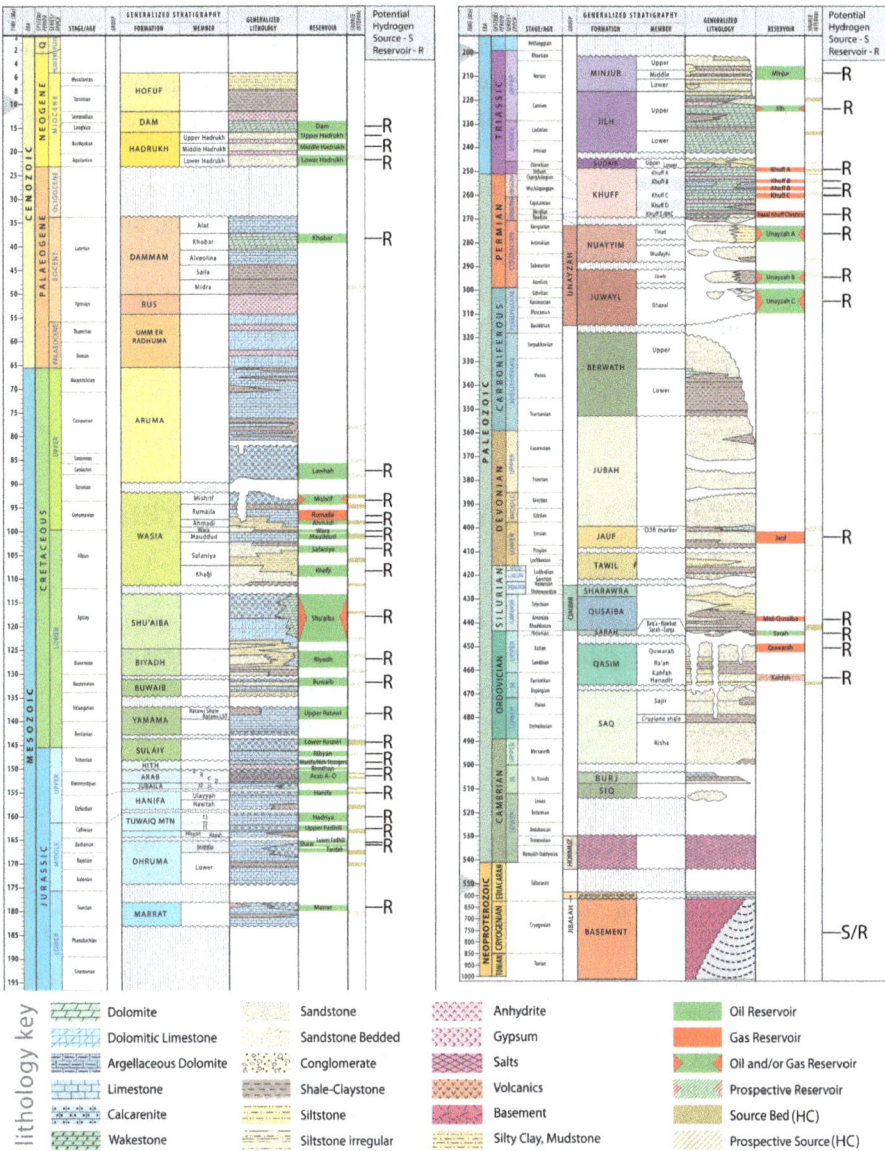

Figure 19.7: General stratigraphy, ages, and lithologies of sedimentary rocks in Saudi Arabia (modified after [37]).

Structural traps are formed by tectonic forces that deform rock layers, creating features like anticlines (upward folds) and fault traps (where rocks are displaced along a fault line). In the first case, examples are the Upper Ordovician structural traps of Dilam and AbuJifan in Central Saudi Arabia and Kahf and Jalamid in northern Saudi Arabia [16]. Stratigraphic traps result from changes in rock type or depositional patterns, such as pinch-outs (where a rock layer thins and disappears) or unconformities (erosional surfaces separating rock layers of different ages). For instance, the stratigraphic variability of the Permian Unayzah Formation, impacted by paleotopography and the continental depositional environment, contributed to the development of stratigraphic traps [16, 43].

As the name suggests, combination traps mix structural and stratigraphic elements. These traps play a crucial role in oil and gas accumulation, making them prime targets for exploration and production activities in Saudi Arabia. For instance, the Jauf gas field was found in a combination structural-stratigraphic trap along the flanks of the giant Ghawar structure, showcasing the success of exploration efforts in this complex geological setting, despite the poor seismic data quality for the pre-Khuff interval [44].

19.8 Proximity to known H_2 seeps or occurrences

The presence of known H_2 seeps or occurrences can indicate a region's potential for H_2 generation. So far, no natural H_2 seeps have been reported in Saudi Arabia; however, numerous H_2 seeps in neighboring Oman are related to much younger ophiolites. A seep in Turkey in the Kurtbagi area also represents a similar affinity (Figure 19.1), suggesting the presence of active H_2 generation processes in the region.

19.9 Other possible source rocks

19.9.1 Within-plate A-I-type granites

Though the known "highly" radioactive granitoid complexes within the Arabian-Nubian Shield (ANS) are limited in volume, the younger ANS granitoids record the major development and evolution of Neoproterozoic continental crust and extensive distribution of A-type (Anorogenic, Anhydrous, Alkaline) granitoids. Whereas 900–630 Ma I-type (Igneous protolith) granitoids dominate the early plutonic evolution of the ANS, within-plate A-type granitoids dominate 630–570 Ma plutons and provide a record of the final collision between eastern and western Gondwana. Some transitional A/I-type granitoids are highly mineralized in high-field-strength elements (HFSE) and rare-earth elements (REE) and are extremely radioactive. For example, the Al-Ghurayyah granitoid complex [45] represents a small, but highly mineralized, pluton exposed in the Midyan terrane of NW Saudi Arabia, the easternmost of a belt of similar bodies in Egypt [46]. In this complex, different types of pegmatites are documented: (1) northern pegmatites (type 1), emplaced into metavolcanics, exhibit complex mineral zonation and highly elevated $\sum$REE (436–13,115 µg/g) and Li (13–3,010 µg/g), but low concentration of U (2.82–94.3 µg/g) and Th (25.3–600 µg/g); (2) southwestern pegmatites (type 2) emplaced into microgranite exhibits simple mineral zonation with anomalous abundance of zircon (up to 10% of the mode), high $\sum$REEs (up to 6,000 µg/g), and extremely high radioactive element abundances (1,160 µg/g Th and 348 µg/g U) [45]. Such high radioactive element concentrations could conceivably provide enough energy to catalyze hydrolysis.

19.9.2 Overmatured organic-rich shales

Late pyrolysis of organic matter, particularly coal and shaly coal, has been observed to yield H_2, which may exist as a free gas phase [11, 13, 47]. As we noticed earlier, in the deep basins in Saudi Arabia, the estimated temperature exceeds 200 °C, thus making it possible for organic-rich shale to become overmatured. The likely places for this are the Wadi Sirhan Graben, Widyan Basin, Eastern Arabian Basin, Wajid Graben, and Rub'Al-Khali Basin (Figure 19.6). We discuss this further in the laboratory experiments section.

19.10 Discussion

19.10.1 Identification of potential H_2 play fairways

Our study identifies four potential H_2 play fairways (PF) in Saudi Arabia based on the combined geological criteria for H_2 generation and entrapment. These fairways are regions with high scores based on the presence of H_2-generating rocks (ultramafic and mafic), structural traps (faults, anticlines) and seals. The geological settings within each fairway are described as follows:

19.10.1.1 Play Fairway 1 (PF-1)

PF-1 is characterized by up-dip migration with various four-way closing traps. These fairways are in the up-dip of the Wadi Sirhan Graben, Widyan Basin, Eastern Arabian Basin, Wajid Graben, and Rub'Al-Khali Basin (Figure 19.8). Compressional stresses normally form these structures, providing a long-range lateral migration through a porous and permeable reservoir. An example of a fill-and-spill process trap is the Suman Platform in Saudi Arabia [48, 49]. Depending on the basin's geometry, such traps can be formed at very deep to shallow depths.

19.10.1.2 Play Fairway 2 (PF-2)

PF-2 is characterized by a flower-type trap with H_2 accumulating through vertical migration. The flower-type trap is formed by a combination of faults and folds, which can trap H_2 gas as it migrates vertically. It is located where the Najd Fault trend is buried under sediments in southern Saudi Arabia (Figure 19.8). The sedimentary cover containing potential reservoir-seal pairs can trap H_2 gas within a positive flower structure. Since the strike-slip faults are deep-seated in the basement, the possibility of migration of gas that was generated at deeper levels increases. The Bourabougou anticlinal closure in Mali that trapped H_2 is likely a flower structure.

19.10.1.3 Play Fairway 3 (PF-3)

PF-3 is characterized by a horst structure, an uplifted block of rock bounded by faults within the basement. These traps may be found in the East Arabian Basin (Figure 19.8), where the "Ghawar"-type structures deep at the basement level assume such a configuration. The horst structure can trap H_2 gas within a reservoir as it migrates upward along the bounding faults.

19.10.1.4 Play Fairway 4 (PF-4)

PF-4 is the deepest in Saudi Arabia (approximately >7,000 m), as this fairway is characterized by a basement horst structure capped by a salt seal. The salt was deposited in Precambrian times in the Wajid Graben, Rub'Al-Khali Basin, and some parts of the Eastern Arabian Basin (Figure 19.8). Fractures on the Arabian Shield especially within the ophiolitic bodies have been documented [34]. The salt seal is an effective caprock that prevents H_2 gas from escaping upward.

19.10.2 Comparison to global analogs

Comparing the potential H_2 systems in Saudi Arabia to known H_2 systems in other parts of the world shows that the hydro-geological settings in Saudi Arabia are similar to the basinal configuration in the Nehama Ridge, Kansas, in the US [51]. Helium (He), associated with H_2 in Nemaha ridge wells, however, indicates a rock with a radioactive source, as opposed to ophiolites in the Saudi case. The subsalt salt traps (PF-4) identified in Saudi Arabia could be similar to the prospects in Officer Basin, Australia, where a possible H_2 accumulation is manifested by bright spots in seismic below the salt layer overlying the basement (Personal communication with Rezae and Evans, 2023). Overall, the comparison to global analogs suggests that Saudi Arabia has the

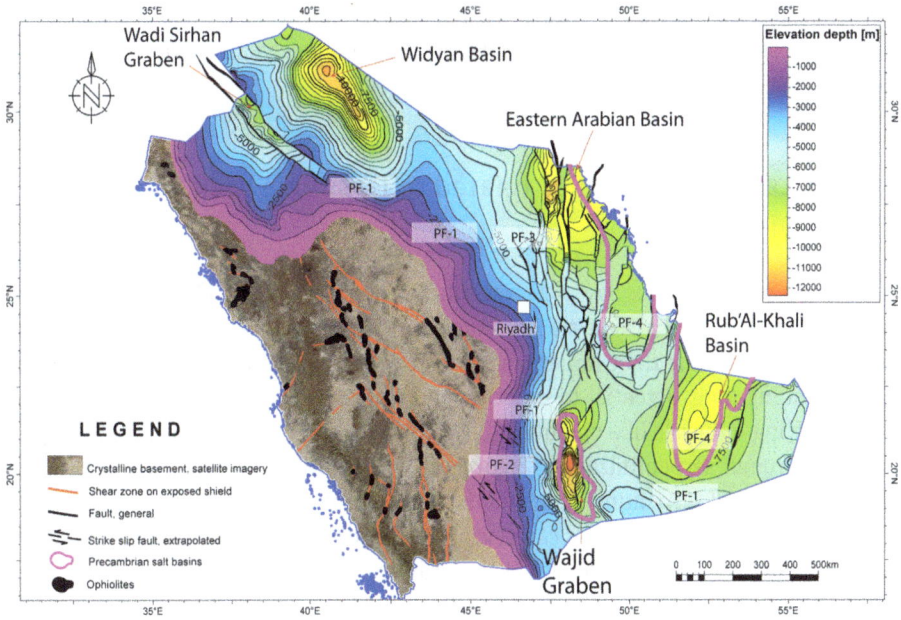

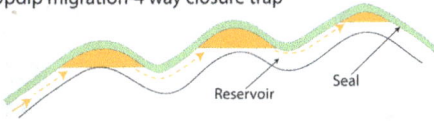

Play Fairway 1 (PF-1):
Updip migration-4 way closure trap

Play Fairway 3 (PF-3):
Vertical migration-Horst type trap

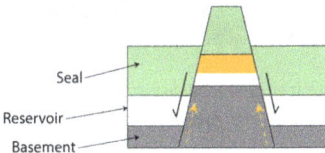

Play Fairway 2 (PF-2):
Vertical migration-Flower type trap

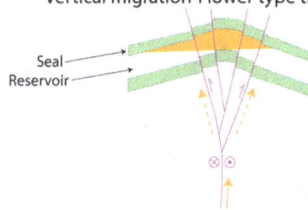

Play Fairway 4 (PF-4):
Vertical migration- Subsalt trap

Figure 19.8: Potential H_2 play fairways identified on an Arabian Shield basement elevation map (ophiolites and sheared zones in the exposed shield are taken from [22]; faults on the shield underlain by sedimentary cover are taken from [15, 16]). Play Fairway 1 (PF-1) is an up-dip migration with various four-way closing traps (modified from [49]), Play Fairway 2 is a Flower-type trap with H_2 accumulating through vertical migration (modified from [50], Play Fairway 3 (PF-3) is a horst structure (modified from [16]) and Play Fairway 4 (PF-4) is a basement horst structure capped by a salt seal. The orange colored arrows represent the H_2 migration directions.

potential to host significant natural H_2 resources. However, further exploration is needed to confirm this potential and to assess the economic viability of producing natural H_2 in Saudi Arabia.

19.10.3 Laboratory studies related to H_2 source rock and seal viability

Laboratory studies could play a crucial role in assessing the viability of potential H_2 source rocks and seals. These studies involve analyzing rock samples collected from the field to determine their mineralogical and petrophysical properties, essential for understanding their potential to generate and trap H_2. For source rocks, the focus is on identifying minerals like olivine and pyroxene, which can react with water to produce H_2 through serpentinization (Figure 19.9). Petrographic analysis helps to determine the extent of serpentinization and the availability of these reactive minerals. Additionally, geochemical analyses can reveal the presence of other elements or isotopes that might indicate the potential for H_2 generation.

Figure 19.9: (a) hand specimen and (b) micrograph of serpentinite from the Jabal Ess ophiolite (see Figure 19.3). Note the classic mesh texture in (a) and the olivine "islands" in (b) (above the Srp text). Abbreviations: Opx, orthopyroxene; Ol, olivine; Srp, serpentine.

The emphasis is on assessing seal rocks' integrity and ability to prevent H_2 leakage. This involves measuring their porosity, permeability and capillary properties, determining how easily fluids can flow through a rock core. By conducting these laboratory studies, researchers can gain valuable insights into the potential of different geological formations to serve as sources and seals for natural H_2, aiding in the exploration and development of this clean energy resource.

19.10.3.1 Potential of H_2 generation from organic-rich source rocks

Organic source rocks are a possible source of natural hydrogen production due to the presence of kerogen (mainly type I and type II). At high temperature, kerogen pyrolysis reactions can cause gas production, that is, of methane and hydrogen. To assess the feasibility and implications of utilizing organic-rich source rocks for hydrogen production, we conducted thorough analyses of source rock samples from different formations. Our experimental results reveal that high total organic carbon (TOC) samples (average 14.7% and ~800 H_2 index) with type I and II kerogens yield substantial hydrogen-rich gases, approximately ~40% H_2, and hydrocarbon gases (~50%) along with some H_2S depending upon total sulfur content, highlighting the influence of mineralogical and geochemical composition [52]. The study suggests that hydrocarbon source rocks with high organic matter content hold significant potential for maximizing energy gas recovery (>85%, including hydrogen and hydrocarbons) through the pyrolysis of their kerogen content. These findings demonstrate the viability of hydrogen gas production from unconventional sources such as organic-rich rocks, potentially impacting strategies for sustainable energy production and contributing significantly to global decarbonization efforts.

19.10.3.2 Evaluation of reactivity of Saudi Arabian sedimentary rocks to H_2

According to our experimental findings, the geochemical reactions between sandstone and carbonate (limestone and dolomite) rock minerals and hydrogen under subsurface conditions (1,500 psi and 75 °C) are limited, with very slight changes observed in surface morphology. As a result, minimal changes occurred in the pore properties of the sedimentary rocks, and the change in pore volume and porosity was minor in the presence of hydrogen, suggesting that alteration in large pores in sandstone and limestone rocks are expected to be limited [53–57]. However, a slow reaction was observed between H_2 and organic-rich source rock (XRD: 86.4% calcite, 4.8 quartz, 2.8 phosphate, 0.2 clay, 2.1 pyrite, 3.8 gypsum; and ~14 TOC, and 783 H_2 index) at similar conditions (1500 psi and 75 °C) after 80 days exposure time [58]. No hydrogen sulfide was observed from gas chromatography analysis at the end of the experiment, but traces of methane (0.018%) were detected. Pyrite is a common mineral found in organic-rich shales. However, it was found that pyrite can react with H_2 in the presence of calcite, producing H_2S gas. During this reaction, pyrite turns into pyrrhotite [59].

19.10.3.3 Evaluation of H_2 reactivity of cap rocks

Our experimental results indicate no notable changes in the mineralogy of halite and anhydrite with H_2 at subsurface conditions [60]. Yet, due to the dehydration process, gypsum displayed a complete transformation to bassanite when treated with H_2 at 75 °C and 1,450 psi [60]. However, when gypsum is treated with hydrogen at room temperature, no changes in mineral composition are reported. Thermogravimetric analysis shows that the tendency of mass reduction with temperature for all samples did not change, suggesting minor pore structure changes. Nitrogen adsorption/desorption isotherms show that the amount of adsorbed nitrogen after hydrogen treatment is similar to the initial conditions, implying that no significant alterations in pore geometry occurred, with no evidence of pore expansion or shrinkage. Consequently, negligible alterations are reported in specific surface area and total pore volume for all samples (not exceeding 5%), with stable pore size distribution profiles. Furthermore, we observed that the reactivity of one Saudi basaltic rock to H_2 is low. Thus, the potential of basalt to be a seal rock for hydrogen storage is also promising [61]. Structural trapping is the most critical mechanism utilized to prevent H_2 leakage from the subsurface due to the substantial gravity segregation of H_2. Thus, it is crucial to comprehend the wettability of caprocks under H_2 gas storage conditions. It has been observed by experimental works that most of the caprocks (containing mica, anhydrite, gypsum, calcite, quartz and halite) are strongly water-wet under subsurface conditions. As caprocks are strongly water-wet, this implies that H_2 has difficulties in displacing water from caprocks due to high capillary force; thus, caprocks should have a high sealing efficiency in the presence of H_2 [62, 63].

19.10.4 Uncertainties and limitations

This study acknowledges several uncertainties and limitations in assessing natural H_2 potential in Saudi Arabia. One of the main limitations is the lack of knowledge regarding H_2 migration and diffusion scenarios. When H_2 interacts with organic-rich shales, as when compacted and heated, the carbon molecules in organic-rich rocks may consume any available H_2 and form longer-chain hydrocarbons [10]. In the case where the H_2 migrates to a hydrocarbon reservoir, it might react to form more hydrocarbons.

Another limitation is the public domain's lack of seismic, well-log, and production data. Future studies will need to rely on existing geological and geophysical data, which may not be sufficient to assess the H_2 potential in all areas accurately. For example, the data may not resolve small-scale geological features that could be important for a given H_2 accumulation.

Finally, this study focuses on ophiolite-related H_2 generation. Other sources, such as hydrolysis, overmatured organic-rich rocks, and H_2 generation and migration from deep mantle levels, could contribute to Saudi Arabia's overall H_2 potential.

19.11 Conclusions

This study presents a preliminary assessment of natural H_2 potential in Saudi Arabia, highlighting key areas and geological formations that warrant further investigation. By integrating available geological and geophysical data, we have identified four promising H_2 play fairways with favorable conditions for H_2 generation, migration, and accumulation. These findings underscore Saudi Arabia's significant potential to contribute to the global transition toward cleaner energy sources.

The implications of this research are substantial for Saudi Arabia's energy landscape. Identifying potential natural H_2 resources aligns with the nation's strategic vision [64] to diversify its energy mix and reduce reliance on fossil fuels. Developing a natural H_2 industry could create new economic opportunities, foster technological innovation, and position Saudi Arabia as a leader in the emerging H_2 economy. Moreover, using natural H_2 as a clean energy source could significantly reduce greenhouse gas emissions, contributing to the country's environmental goals and global efforts to combat climate change.

19.12 Recommendations

We recommend five steps to fully realize the potential of natural H_2 in Saudi Arabia. By pursuing these recommendations, Saudi Arabia can utilize the full potential of its natural H_2 resources and pave the way for a cleaner, more sustainable and prosperous energy future.

19.12.1 Detailed field surveys

Conduct comprehensive field surveys, including geological mapping, geochemical gas anomaly/aquifer water sampling and geophysical exploration, to validate and refine the identified H_2 play fairways. These surveys should focus on characterizing the source rocks, H_2 generation, reservoir rocks, seal rocks, fracture systems, and structural traps in more detail.

19.12.2 Exploratory drilling and testing

Initiate exploratory drilling programs in the most promising areas to assess the quality and quantity of H_2 resources. Well-testing and analyzing the obtained gas samples will provide crucial information on such resources' production potential and commercial viability.

19.12.3 Development of a national H_2 exploration strategy

Formulate a national strategy for natural H_2 exploration and development, outlining clear goals, targets and timelines. This strategy should encompass all aspects of the H_2 value chain, from exploration and production to storage, transportation, and utilization.

19.12.4 Collaboration and knowledge sharing

Foster collaboration between academia, industry and government institutions to leverage expertise, share knowledge, and accelerate the development of the natural H_2 sector. More subsurface data (seismic, borehole logging, and hydrocarbon geochemistry) in the public domain will act as a catalyst for identifying and maturing an H_2 prospect. International collaboration with countries with an established natural H_2 presence could also provide valuable insights and lessons learned.

19.12.5 Investment in research and development

Invest in research and development to advance H_2 exploration, production, and utilization technologies. This includes developing innovative methods for resource assessment, optimizing production techniques, and exploring new applications for H_2 in applied energy sectors.

References

[1] Gaucher E. C. New perspectives in the industrial exploration for native hydrogen. Elements: An International Magazine of Mineralogy, Geochemistry, and Petrology. 2020, 16: 8–9.
[2] Milkov A. V. Molecular hydrogen in surface and subsurface natural gases: Abundance, origins and ideas for deliberate exploration. Earth-Science Reviews. 2022, 230: 104063.
[3] Moretti I. H 2: Energy vector or source?. Actualite Chimique. 2019, 442: 15–16.
[4] Smith N. J. P. It's time for explorationists to take hydrogen more seriously. First Break. 2002, 20: 246–253.

[5]   Moretti I., Brouilly E., Loiseau K., Prinzhofer A., Deville E. Hydrogen emanations in intracratonic areas: New guide lines for early exploration basin screening. Geosciences. 2021, 11: 145.

[6]   Rezaee R. Assessment of natural hydrogen systems in Western Australia. International Journal of Hydrogen Energy. 2021, 46: 33068–33077.

[7]   Zgonnik V., Beaumont V., Deville E., Larin N., Pillot D., Farrell K. M. Evidence for natural molecular hydrogen seepage associated with Carolina bays (surficial, ovoid depressions on the Atlantic Coastal Plain, Province of the USA). Progress in Earth and Planetary Science. 2015, 2: 1–15.

[8]   Larin N., Zgonnik V., Rodina S., Deville E., Prinzhofer A., Larin V. N. Natural molecular hydrogen seepage associated with surficial, rounded depressions on the European craton in Russia. Natural Resources Research. 2015, 24: 369–383.

[9]   Zgonnik V. The occurrence and geoscience of natural hydrogen: A comprehensive review. Earth-Science Reviews. 2020, 203: 103140.

[10]   Hand E. Hidden hydrogen. Science. 2023, 379: 630–636.

[11]   Lévy D., Roche V., Pasquet G., Combaudon V., Geymond U., Loiseau K., et al. Natural H 2 exploration: Tools and workflows to characterize a play. Science and Technology for Energy Transition. 2023, 78: 27.

[12]   Albers E., Bach W., Pérez-Gussinyé M., McCammon C., Frederichs T. Serpentinization-driven H2 production from continental break-up to mid-ocean ridge spreading: Unexpected high rates at the West Iberia margin. Frontiers in Earth Science. 2021, 9: 673063.

[13]   Mahlstedt N., Horsfield B., Weniger P., Misch D., Shi X., Noah M., et al. Molecular hydrogen from organic sources in geological systems. Journal of Natural Gas Science and Engineering. 2022, 105: 104704.

[14]   Johnson P. R. The Arabian–Nubian shield, an introduction: Historic overview, concepts, interpretations, and future issues. The Geology of the Arabian-Nubian Shield. 2021: 1–38.

[15]   Johnson P. R. Tectonic map of Saudi Arabia and adjacent areas. Deputy Ministry for Mineral Resources Technical Report USGS-TR-98-3 (IR-948). 1998.

[16]   Konert G., Afifi A. M., Al-Hajri S. A., Droste H. J. Paleozoic stratigraphy and hydrocarbon habitat of the Arabian Plate. GeoArabia. 2001, 6: 407–442.

[17]   Stern R. J., Moghadam H. S., Pirouz M., Mooney W. The geodynamic evolution of Iran. Annual Review of Earth and Planetary Sciences. 2021, 49: 9–36.

[18]   Cox G. M., Lewis C. J., Collins A. S., Halverson G. P., Jourdan F., Foden J., et al. Ediacaran terrane accretion within the Arabian–Nubian Shield. Gondwana Research. 2012, 21: 341–352.

[19]   Jaju M. M., Nader F. H., Roure F., Matenco L. Optimal aquifers and reservoirs for CCS and EOR in the Kingdom of Saudi Arabia: An overview. Arabian Journal of Geosciences. 2016, 9: 1–15.

[20]   Ye J., Afifi A., Rowaihy F., Baby G., De Santiago A., Tasianas A., et al. Evaluation of geological CO_2 storage potential in Saudi Arabian sedimentary basins. Earth-Science Reviews. 2023, 244: 104539.

[21]   Dugamin E., Truche L., Donze F. V. Natural hydrogen exploration guide. ISRN Geonum-NST. 2019, 1: 16.

[22]   Johnson P. R., Kattan F. H. The geology of the Saudi Arabian shield. Saudi Geological Survey. 2012: 177–208.

[23]   Genna A., Nehlig P., Le Goff E., Guerrot C., Shanti M. Proterozoic tectonism of the Arabian Shield. Precambrian Research. 2002, 117: 21–40.

[24]   Johnson P. R., Kattan F. H., Al-Saleh A. M. Neoproterozoic ophiolites in the Arabian Shield: Field relations and structure. Developments in Precambrian Geology. 2004, 13: 129–162.

[25]   Johnson P. R., Woldehaimanot B. Development of the Arabian-Nubian Shield: Perspectives on accretion and deformation in the northern East African Orogen and the assembly of Gondwana, vol. 206. London: Geological Society, Special Publications, 2003, 289–325.

[26]   Pallister J. S., Stacey J. S., Fischer L. B., Premo W. R. Precambrian ophiolites of Arabia: Geologic settings, U-Pb geochronology, Pb-isotope characteristics, and implications for continental accretion. Precambrian Research. 1988, 38: 1–54.

[27]   Stoeser D. B., Camp V. E. Pan-African microplate accretion of the Arabian Shield. Geological Society of America Bulletin. 1985, 96: 817–826.

[28]   Al-Shanti M. M. S. Geology and mineralization of the Ash Shizm-Jabal Ess area. Ph. D. thesis, King Abdulaziz University, Jeddah. 1982.

[29]   Shanti M., Roobol M. J. A late Proterozoic ophiolite complex at Jabal Ess in northern Saudi Arabia. Nature. 1979, 279: 488–491.

[30]   Bakor A., Gass I., Neary C. Jabal al Wask, northwest Saudi Arabia: An Eocambrian back-arc ophiolite. Earth and Planetary Science Letters. 1976, 30: 1–9.

[31]   Chevremont P., Johan Z. The Al Ays ophiolite complex. Deputy Ministry for Mineral Resources Open-File Report BRGM-OF-02–5. 1982: 65.

[32]   Ledru P., Auge T. The Al Ays ophiolitic complex; petrology and structural evolution. Saudi Arabian Deputy Ministry for Mineral Resources Open-File Report BRGM-OF-04–15. 1984: 57.

[33]   NOAA. EMAG2v3: Earth Magnetic Anomaly Grid (image service) 2016. https://hub.arcgis.com/data sets/b123cbf4686240798bcef8291deea77b/explore?location=24.082170%2C42.143086%2C7.32.

[34]   Saidy K., Fawad M., Whattam S. A., Al-Shuhail A. A., Al-Shuhail A. A., Campos M., et al. Unlocking the H2 potential in Saudi Arabia: Exploring serpentinites as a source of H2 production. International Journal of Hydrogen Energy. 2024, 89: 1482–1491.

[35]   McCollom T. M., Bach W. Thermodynamic constraints on hydrogen generation during serpentinization of ultramafic rocks. Geochimica Et Cosmochimica Acta. 2009, 73: 856–875.

[36]   Rolandone F., Lucazeau F., Leroy S., Mareschal J.-C., Jorand R., Goutorbe B., et al. New heat flow measurements in Oman and the thermal state of the Arabian Shield and Platform. Tectonophysics. 2013, 589: 77–89.

[37]   Cantrell D. L., Nicholson P. G., Hughes G. W., Miller M. A., Buhllar A. G., Abdelbagi S. T., et al. Tethyan petroleum systems of Saudi Arabia. In Marlow, L., Kendall, C., & Yose, L. (Eds.), Petroleum Systems of the Tethyan Region, AAPG Memoir, Tulsa, OK. 2014, 106: 613–639.

[38]   Abu-Ali M. A., Rudkiewicz J.-L.-L., McGillivray J. G., Behar F. Paleozoic petroleum system of central Saudi Arabia. GeoArabia. 1999, 4: 321–336.

[39]   Janjou D., Halawani M. A., Muallem M. S. A., Robelin C., Brosse J. M., Courbouleix S., et al. Explanatory notes to the geologic map of the Qalibah quadrangle. In Kingdom of Saudi Arabia, Ministry of Petroleum and Mineral Resources, Deputy Ministry for Mineral Resources, Directorate General of Mineral Resources (DGMR), Jiddah, Saudi Arabia. 1996: Sheet 28C.

[40]   Mahmoud M. D., Vaslet D., Husseini M. I. The lower Silurian Qalibah formation of Saudi Arabia: An important hydrocarbon source rock (1). AAPG Bulletin. 1992, 76: 1491–1506.

[41]   McGillivray J. G., Husseini M. I. The Paleozoic petroleum geology of central Arabia (1). AAPG Bulletin. 1992, 76: 1473–1490.

[42]   Cole G. A., Abu-Ali M. A., Aoudeh S. M., Carrigan W. J., Chen H. H., Colling E. L., et al. Organic geochemistry of the Paleozoic petroleum system of Saudi Arabia. Energy & Fuels. 1994, 8: 1425–1442.

[43]   Evans D. S., Bahabri B. H., Al-Otaibi A. M. Stratigraphic trap in the Permian Unayzah formation, Central Saudi Arabia. GeoArabia. 1997, 2: 259–278.

[44]   Wender L. E., Bryant J. W., Dickens M. F., Neville A. S., Al-Moqbel A. M. Paleozoic (pre-Khuff) hydrocarbon geology of the Ghawar area, eastern Saudi Arabia. GeoArabia. 1998, 3: 273–302.

[45]   Sulistyo H. F. A., Whattam S. A., Osman M., Khedr M. Z., Stern R. J., Chattopadhyay S., et al. Petrogenesis of the HFSE and REE mineralized Al-Ghurayyah granite pluton, NW Saudi Arabia. AGU Fall Meeting Abstracts. 2023, 2023: V41C-0136.

[46]   Stern R. J., Khedr M. Z., Whitehouse M. J., Romer R. L., Khashaba S. M. A., El-Shibiny N. H. Late Cryogenian and early Ediacaran rare-metal rich granites in the Eastern Desert of Egypt: Constraints from zircon ages and whole-rock Sr-and Nd-and feldspar Pb-isotopic compositions. Journal of the Geological Society. 2024, 181: jgs2023–068.

[47]   Horsfield B., Mahlstedt N., Weniger P., Misch D., Vranjes-Wessely S., Han S., et al. Molecular hydrogen from organic sources in the deep Songliao Basin, PR China. International Journal of Hydrogen Energy. 2022, 47: 16750–16774.

[48]   Arouri K., Panda S. K., Satti S. A., Yang Y. Migration tracers reveal long-range migration in the Summan exploration area. International Meeting of Organic Geochemistry, Interlaken, Switzerland, Abstract. 2011: 229.

[49]   Fustic M., Bennett B., Huang H., Larter S. Differential entrapment of charged oil – new insights on McMurray Formation oil trapping mechanisms. Marine and Petroleum Geology. 2012, 36: 50–69.

[50]   Huang L., Liu C. Three types of flower structures in a divergent-wrench fault zone. Journal of Geophysical Research: Solid Earth. 2017, 122: 10478–10497.

[51]   McIntyre A. Understanding the resource potential of natural subsurface hydrogen. AAPG Academy webinar, Tulsa, OK. 2023.

[52]   Gaduwang A. K., Tawabini B., Abu-Mahfouz I. S., Al-Yaseri A. Assessing hydrogen production potential from carbonate and shale source rocks: A geochemical investigation. Fuel. 2024, 378: 132923.

[53]   Al-Yaseri A., Fatah A., Alsaif B., Sakthivel S., Amao A., Al-Qasim A. S., et al. Subsurface hydrogen storage in limestone rocks: Evaluation of geochemical reactions and gas generation potential. Energy & Fuels. 2024, 38: 9923–9932.

[54]   Al-Yaseri A., Hussaini S. R., Fatah A., Al-Qasim A. S., Patil P. D. Computerized tomography analysis of potential geochemical reactions of carbonate rocks during underground hydrogen storage. Fuel. 2024, 361: 130680.

[55]   Al-Yaseri A., Fatah A., Adebayo A. R., Al-Qasim A. S., Patil P. D. Pore structure analysis of storage rocks during geological hydrogen storage: Investigation of geochemical interactions. Fuel. 2024, 361: 130683.

[56]   Al-Yaseri A., Yekeen N., Al-Mukainah H., Hassanpouryouzband A. Geochemical interactions in geological hydrogen storage: The role of sandstone clay content. Fuel. 2024, 361: 130728.

[57]   Al-Yaseri A., Fatah A. Impact of H2-CH4 mixture on pore structure of sandstone and limestone formations relevant to subsurface hydrogen storage. Fuel. 2024, 358: 130192.

[58]   Al-Yaseri A., Abu-Mahfouz I. S., Yekeen N., Wolff-Boenisch D. Organic-rich source rock/H2/brine interactions: Implications for underground hydrogen storage and methane production. Journal of Energy Storage. 2023, 63: 106986.

[59]   Truche L., Berger G., Destrigneville C., Guillaume D., Giffaut E. Kinetics of pyrite to pyrrhotite reduction by hydrogen in calcite buffered solutions between 90 and 180 C: Implications for nuclear waste disposal. Geochimica Et Cosmochimica Acta. 2010, 74: 2894–2914.

[60]   Fatah A., Al-Yaseri A., Theravalappil R., Radwan O. A., Amao A., Al-Qasim A. S. Geochemical reactions and pore structure analysis of anhydrite/gypsum/halite bearing reservoirs relevant to subsurface hydrogen storage in salt caverns. Fuel. 2024, 371: 131857.

[61]   Al-Yaseri A., Fatah A., Amao A., Wolff-Boenisch D. Basalt–hydrogen–water interactions at geo-storage conditions. Energy & Fuels. 2023, 37: 15138–15152.

[62]   Abdel-Azeim S., Al-Yaseri A., Norrman K., Patil P., Qasim A., Yousef A. Wettability of Caprock–H2–water: Insights from molecular dynamic simulations and sessile-drop experiment. Energy & Fuels. 2023, 37: 19348–19356.

[63]   Al-Yaseri A., Abdel-Azeim S., Al-Hamad J. Wettability of water-H2-quartz and water-H2-calcite experiment and molecular dynamics simulations: Critical assessment. International Journal of Hydrogen Energy. 2023, 48: 34897–34905.

[64]   Saudi Vision 2030. https://www.vision2030.gov.sa/en.

[65]   Al-Shanti A. M., Mitchell A. Late Precambrian subduction and collision in the Al Amar – Idsas region, Arabian Shield, Kingdom of Saudi Arabia. Tectonophysics. 1976, 30: T41–7.

[66]   Rittmann A. Geosynclinal volcanism, ophiolites and Barramiya rocks. The Egyptian Journal of Geology. 1958, 2: 61–66.

Part V: **Hydrogen storage, transportation, and environmental and technological challenges**

Quan Xie*, Adnan Aftab, Mohammad Sarmadivaleh, Lingping Zeng,
and Alireza Safari

Chapter 20
Underground hydrogen storage lessons for natural hydrogen systems

Abstract: Natural hydrogen presents significant potential to transform the global energy system and accelerate the transition to a low-carbon future. Despite this promise, research on developing natural hydrogen systems remains limited. This study builds on the extensive work on underground hydrogen storage (UHS) in depleted gas reservoirs to leverage existing knowledge and expertise for advancing natural hydrogen systems.

Both abiotic and biotic geochemical reactions, which occur during UHS in depleted gas reservoirs, are also relevant to natural hydrogen systems. These reactions cause hydrogen conversion and contamination over geological timescales, leading to variations in hydrogen concentrations across hydrogen-detected wells in diverse geological settings, including orebodies, volcanic gases, hydrothermal systems, and oil and gas wells. These fluctuations in hydrogen composition underscore the need for quantitative research to establish the hydrogen-based equation of state. While the reactions in natural hydrogen systems differ from those in natural gas reservoirs, the underlying physics and governing equations for multicomponent flow remain consistent. Thus, modeling codes tested for UHS may be adapted to manage and predict the behavior of natural hydrogen reservoirs.

Keywords: natural hydrogen, underground hydrogen storage, abiotic and biotic geochemical processes, reservoir performance, reservoir integrity

Acknowledgments: Q. Xie acknowledges the funding support from the Australian International Hydrogen Fellowship funded by the Australian Government and Future Energy Exports CRC through projects 23. RP2.0176 and 21.RP2.0091. A. Aftab acknowledges the RTP Scholarship supported by the Australian Government and Curtin University during his PhD research.

*Corresponding author: Quan Xie, Discipline of Energy Engineering, Curtin University,
26 Dick Perry Avenue, 6151 Kensington, Australia, e-mail: quan.xie@curtin.edu.au
Adnan Aftab, Mohammad Sarmadivaleh, Discipline of Energy Engineering, Curtin University,
26 Dick Perry Avenue, 6151 Kensington, Australia
Lingping Zeng, CSIRO Energy, Melbourne, VIC 3168, Australia
Alireza Safari, Discipline of Energy Engineering, Curtin University, 26 Dick Perry Avenue, 6151
Kensington, Australia; Department of Earth Resources Engineering, Graduate School of Engineering,
Kyushu University, 744 Motooka, Nishi Ward, Fukuoka, Japan

https://doi.org/10.1515/9783111437040-020

20.1 Introduction

20.1.1 Potential of natural hydrogen systems in future energy landscape

Natural hydrogen could revolutionize our low-carbon future, creating the biggest disruption to the global energy system in the coming decades. Natural hydrogen could also help energy companies advance their Climate Transition Action Plan in the future given its tremendous potential: (1) significant potential reserves – for example, stochastic model results predict that the potential hydrogen resource in place ranges from 10^3 to 10^{10} million metric tons (Mt) with the most probable value of 5.6×10^6 Mt [1]. If 2% of the predicted hydrogen can be recovered, it will contain more energy (1.59×10^{16} MJ) than all proven natural gas reserves on the Earth (8.4×10^{15} MJ); (2) *cost-effective production cost – for example,* the production cost from geological hydrogen is from 0.5 to $1.0 per kg, which can compete with grey hydrogen costs $0.9–3.2 per kg, blue $1.5–2.9 per kg, and green $3.0–7.5 per kg [2].

While natural hydrogen is in its infancy stage in knowledge, it has been found in various geological settings – orebodies, volcanic gases, and hydrothermal systems, even in oil and gas wells. For example, based on a comprehensive literature review [3], hydrogen was found in over 28 wells in iron ore mines with concentrations from 8.3% to 98%. Hydrogen was also found in volcanic gases in 17 locations with concentrations from 10% to 93%. Besides, hydrogen was found in 16 oil and gas wells with concentrations from 6% to 95.2%. Likewise, hydrogen was also found in sedimentary rocks, for example, 26 locations with hydrogen concentration from 4.2% to 98%.

In late 2023, Gold Hydrogen Limited reported up to 86% hydrogen and 17.5% helium in The Ramsay Project in SA. The latest publication [4] shows significant H_2 seepage on the Yilgarn Craton in WA with seasonal fluctuations – higher emissions after dry summers and reduced emissions following rainfall. Given that the major source of ultramafic rocks is located below the Yilgarn Craton, this may point to the direction of the serpentinization process for hydrogen generation [5].

Clearly, natural hydrogen resources have a rising momentum in revolutionizing the future energy landscape. However, limited research has been conducted on developing natural hydrogen systems that may be discovered in the future. Therefore, in this study, we examine how insights from current underground hydrogen storage (UHS) research can inform the development of future natural hydrogen systems.

20.1.2 Some examples of underground hydrogen storage research initiatives

The UHS community has been generating significant knowledge across several key areas, with several large consortia and research groups paving the way. Figure 20.1

provides an overview of some key projects related to UHS assessment and demonstration, spanning efforts by academia, industry, and regulatory bodies in different regions:

- US: Texas-Salt Moss Bluff Spindletop Clemens Dome [6]; Utah Delta, ACES Caverns, and SHASTA [7–9].
- UK: Tesside salt [10], HystorPor [11], Centrica's Rough Storage [12, 13], Lined Rock Shaft [14].
- EU: Hypster [15], HyStock [16], Hystories [17], HyStorage [18], HyUSPRe [19], Underground Sun Storage [20], EUH2STARS [21], H2eart [22], Hybrit-Lined Rock Cavern [23], Geogas H$_2$ [24].
- China: Hubei Underground Cavern Hydrogen Storage Facility [25], Large-Scale Deep Salt Cavern Storage in Henan [25], UHS in Salt Cavern, Pingdingshan [26].
- New Zealand: UHS in Taranaki [27], Hydrogen Geo-Storage in Aotearoa, University of Canterbury [28].
- Australia: RISC [29], CO2CRC [30], H2RESTORE [31], Future Energy Exports (FEnEx) CRC-Kantnook and Perth Basin, and Pilbara [32].
- **Argentina:** Hychico depleted field [33].

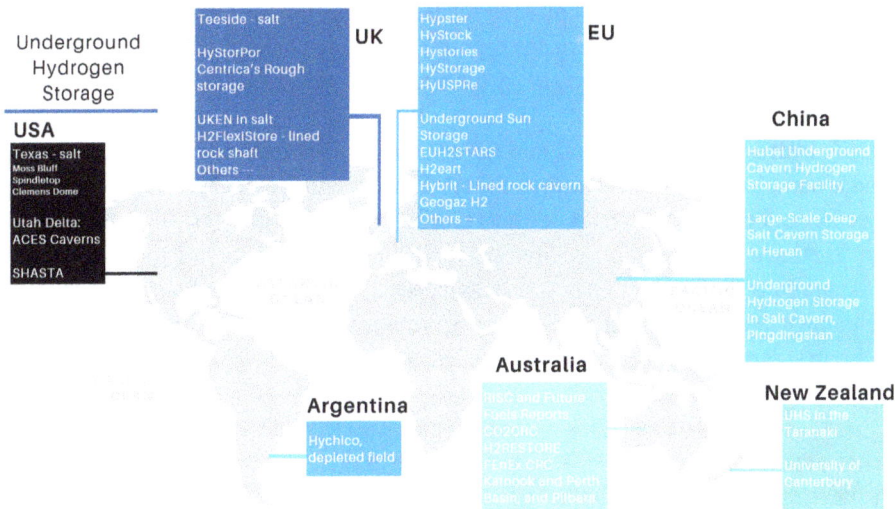

Figure 20.1: Some of the UHS projects worldwide (world map template created using Canva (https://www.canva.com)).

On the international front, the IEA-Hydrogen-TCP, Task 42 UHS, coordinated by TNO, the Netherlands, published its state-of-the-art review on UHS in 2023, namely, Technical Monitor Report [34], with contributions from experts across 57 organizations globally. In this report, six major themes are included, that is, (1) geochemical and microbial impacts, (2) storage (mechanical) integrity, (3) storage performance and screening, (4) facilities and wells, (5) economics and system integration, and (6) societal embedding. In this report, the state-of-the-art knowledge and the remaining technological gaps were classified. To build confidence in UHS across all themes, a final report, namely, Building Confidence in Underground Hydrogen Storage [35], was published in March 2025, where no major technical showstoppers were identified for commercial UHS development and upscaling.

In Europe, RAG Austria [36], with its partners, is at the forefront, leading efforts to develop the first-of-its-kind UHS solution in porous media, thus supporting a decarbonized European economy. Their focus lies in converting existing depleted gas reservoirs into hydrogen storage sites, aiming to demonstrate UHS in depleted natural gas reservoirs at technical readiness level (TRL) 8 by the end of the decade.

In the United States, several prominent consortia and groups are advancing hydrogen storage technologies. As a leading one, the Subsurface Hydrogen Assessment, Storage, and Technology Acceleration (SHASTA) [9] project was a comprehensive research initiative designed to advance UHS technologies and support the transition to zero-emission energy systems. SHASTA aimed to unlock hydrogen's potential as a versatile and sustainable fuel for transportation and power applications by focusing on critical areas such as reservoir performance, well component compatibility, and risk quantification. The project has already made significant strides, including delivering a comprehensive State of Knowledge report, developing advanced simulation capabilities, and creating a national inventory for hydrogen storage capacity assessment [9].

In Argentina, Hychico's UHS [33] project is an innovative initiative focused on storing hydrogen in depleted oil and gas reservoirs near their Hydrogen Plant. Initiated around 2010, the project aims to evaluate the potential of UHS through systematic injection and production cycles, testing reservoir capacity, tightness, and behavior. By constructing a specialized 2.3 km hydrogen pipeline and participating in international collaborations like the HyUnder Consortium, Hychico has advanced hydrogen storage technology [33]. The project underwent rigorous environmental impact assessments and represents a significant technological advancement in developing sustainable hydrogen energy infrastructure.

In Australia, collaborations among CO_2CRC, CSIRO, and Geoscience Australia are driving technical advancements in geological hydrogen storage and working toward the development of optimized methodologies for hydrogen storage site selection and facility demonstrations [37]. H2RESTORE [31] led by Lochard Energy is working toward developing innovative energy storage solutions. The Future Energy Exports CRC (FEnEx CRC) [38] has developed technical screening tools and a risk assessment matrix for UHS. Through laboratory-based research and techno-economic analyses, the team aims to deliver cost-effective UHS solutions in collaboration with Curtin University, Beach Energy, The University of Western Australia in Katnook, South Australia, and Perth Basin in Western Australia. Additionally, the University of Adelaide [39] leads a consortium to establish a national facility for integrated hydrogen flow research in geological formations, working with universities and industry partners.

20.2 Hydrogen conversion and contamination processes

20.2.1 Geochemical reactions: abiotic processes

To understand how natural hydrogen systems evolve in terms of hydrogen purity over geological time, it is crucial to identify the controlling geochemical reactions that may occur within these systems. Given the latest research work carried out in the UHS communities, four categories of geochemical reactions likely occur during the geological time of natural hydrogen systems. Those geochemical reactions trigger hydrogen conversion and contamination, which are summarized below.

Carbonate minerals: carbonate minerals consist of calcite ($CaCO_3$), dolomite [CaMg $(CO_3)_2$], magnesite ($MgCO_3$), siderite ($FeCO_3$), dawsonite [$NaAlCO_3(OH)_2$], and so on, where CO_3^{2-} can be reduced to CH_4 with the by-products of water and OH^-:

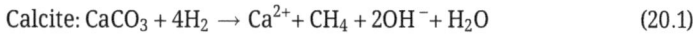

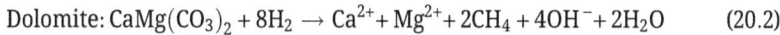

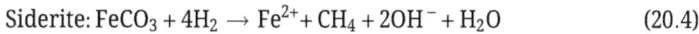

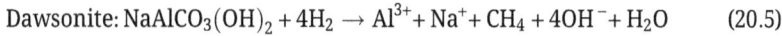

$$\text{Calcite: } CaCO_3 + 4H_2 \rightarrow Ca^{2+} + CH_4 + 2OH^- + H_2O \tag{20.1}$$

$$\text{Dolomite: } CaMg(CO_3)_2 + 8H_2 \rightarrow Ca^{2+} + Mg^{2+} + 2CH_4 + 4OH^- + 2H_2O \tag{20.2}$$

$$\text{Magnesite: } MgCO_3 + 4H_2 \rightarrow Mg^{2+} + CH_4 + 2OH^- + H_2O \tag{20.3}$$

$$\text{Siderite: } FeCO_3 + 4H_2 \rightarrow Fe^{2+} + CH_4 + 2OH^- + H_2O \tag{20.4}$$

$$\text{Dawsonite: } NaAlCO_3(OH)_2 + 4H_2 \rightarrow Al^{3+} + Na^+ + CH_4 + 4OH^- + H_2O \tag{20.5}$$

When natural hydrogen systems are rich in carbonate cementation, the redox reactions between these minerals and aqueous hydrogen lead to carbonate reductive dissolution, triggering hydrogen conversion and contamination. From the above-listed reactions, it can be seen that 1 mol CO_3^{2-} can consume 4 mol of hydrogen and generate 1 mol of methane.

Direct evidence from experimental work has demonstrated the redox reactions between carbonate minerals and hydrogen. For example, Bensing et al. [40] saturated claystone caprock samples containing calcite with hydrogen and 10 wt% NaCl solution at room temperature and hydrogen partial pressure of 150 bar. Petrographic analyses of SEM images on the samples before and after treatments were conducted to characterize the surface variation. Compared to the other scenarios, such as untreated samples or samples treated with dry hydrogen or only with NaCl brine, the sample treated with hydrogen and NaCl solution showed noticeable etching and dissolution of calcite grains. Flesh et al. [41] conducted various experimental analyses including microscopic, petrophysical (porosity and permeability) and CT imaging on sandstone samples containing carbonate minerals ageing with hydrogen and brine at different pressures, temperatures, and salinity conditions. They observed the dissolution of calcite, magnesite, and rhodochrosite (manganese carbonate, $MnCO_3$), implying that hydrogen did react with carbonates. Similar experimental observations were reported by Henkel et al. [42] and Pudlo et al. [43], and from numerical simulation by Hemme and Van Berk [44], Pichler [45], Bo et al. [46], Hassannayebi et al. [47], Zeng et al. [48], and Amid et al. [49]. However, further quantitative work is needed to calibrate the geochemical modeling rates and kinetics of H_2-brine-carbonate reactions using results from confined laboratory experiments.

Sulfate minerals: From sulfate minerals such as anhydrite ($CaSO_4$), gypsum ($CaSO_4$:$2H_2O$), anglesite ($PbSO_4$), barite ($BaSO_4$), and celestite ($SrSO_4$), the sulfate ion SO_4^{2-} can be reduced by hydrogen and generate H_2S (in gas/aqueous phase or further dissociate into HS^-):

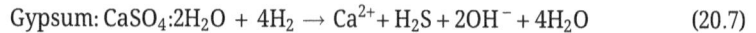

$$\text{Anhydrite: } CaSO_4 + 4H_2 \rightarrow Ca^{2+} + H_2S + 2OH^- + 2H_2O \tag{20.6}$$

$$\text{Gypsum: } CaSO_4{:}2H_2O + 4H_2 \rightarrow Ca^{2+} + H_2S + 2OH^- + 4H_2O \tag{20.7}$$

$$\text{Anglesite: } PbSO_4 + 4H_2 \rightarrow Pb^{2+} + H_2S + 2OH^- + 2H_2O \tag{20.8}$$

$$\text{Barite: } BaSO_4 + 4H_2 \rightarrow Ba^{2+} + H_2S + 2OH^- + 2H_2O \tag{20.9}$$

$$\text{Celestite: } SrSO_4 + 4H_2 \rightarrow Sr^{2+} + H_2S + 2OH^- + 2H_2O \tag{20.10}$$

The formed hydrogen sulfide compromises natural hydrogen purity. From an experimental perspective, Pudle et al. [43] treated sandstone samples with detectable fractions of sulfates (anhydrite and barite) with hydrogen at various temperatures (40–120 °C), pressures (4–20 MPa), and salinities (16–350 g/L). They reported that at high temperature, pressure, and saline conditions, anhydrite and barite were partial to completely dissolved after the aging process, supported by the evidence from tomographic analysis and increasing porosity. In the H2STORE project, Henkel et al. [42] also observed the dissolution of anhydrite as pore-filling material in sandstones after treatment with hydrogen with brine. Other results regarding sulfate redox reactions with hydrogen can be found in studies from Flesch et al. [41], Hemme and Van Berk [44], Lassin et al. [50], and Truche et al. [51].

Sulfide minerals: Pyrite (FeS_2) is one of the most common sulfide minerals. In the presence of hydrogen, it can be reduced to pyrrhotite mackinawite or troilite (FeS) [47, 52, 53]:

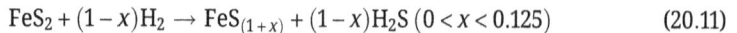

$$FeS_2 + (1-x)H_2 \rightarrow FeS_{(1+x)} + (1-x)H_2S \ (0 < x < 0.125) \tag{20.11}$$

The experimental observations and potential mechanisms of H_2-induced pyrite reduction are well discussed by Wiltowski et al. [54], Lambert et al. [55], Didier et al. [56], Moslemi et al. [57], and Truche et al. [52, 58]. Compared to other inorganic mineral reductive dissolutions such as carbonates or sulfates, the pyrite reduction can take place much quicker at reservoir conditions (temperature >90 °C and P(H_2) pressure >10 bar) [58]. The rate of pyrite reduction and H_2S generation would be further accelerated at more alkaline conditions [58]. Therefore, we suggest that natural hydrogen systems at shallow reservoir depth with lower reservoir temperature and pressure likely reduce the abiotic hydrogen conversion caused by the pyrite reduction.

Ferric iron-associated minerals: Ferric iron-associated minerals consist of oxides such as goethite [FeO(OH)] and hematite (Fe_2O_3), and some clay minerals containing Fe^{3+} such as smectite [59, 60] and kaolinite [61, 62]. The reductive dissolution of goethite and hematite can be described as follows:

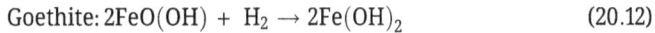

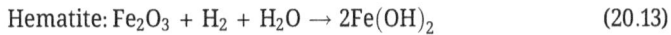

$$\text{Goethite: } 2FeO(OH) + H_2 \rightarrow 2Fe(OH)_2 \tag{20.12}$$

$$\text{Hematite: } Fe_2O_3 + H_2 + H_2O \rightarrow 2Fe(OH)_2 \tag{20.13}$$

From a geochemical point of view, abiotic iron reduction usually occurs at high temperature conditions 300–700 °C [63–67]. There is still a lack of experimental data regarding the performance of Fe^{3+} reduction in the presence of hydrogen and formation brine at reservoir temperatures and pressures. Nevertheless, Fe^{3+} reduction consumes much less hydrogen compared to CO_3^{2-} or SO_4^{2-}, for example, 2 mol Fe^{3+} consume 1 mol H_2 to reduce to Fe^{2+} whereas 1 mol CO_3^{2-} or SO_4^{2-} consumes 4 mol H_2 to reduce to CH_4 or H_2S. Therefore, ferric iron associated minerals may play less important roles in hydrogen conversion compared to carbonate and sulfate minerals.

20.2.2 Microbial reactions: biotic processes

Apart from abiotic geochemical reactions, there is a growing consensus that methanogenesis, acetogenesis, sulfate/sulfur reduction ($SO_4^{2-}/S \rightarrow H_2S/HS^-/S^{2-}$), and ferric iron reduction ($Fe^{3+} \rightarrow Fe^{2+}$) likely occur in the natural hydrogen systems the same as hydrogen storage in depleted gas fields [68, 69]. This is due to the ubiquitousness of bacteria communities/families to enable these biotic reactions in the subsurface [70–74]. In methanogenesis, the pre-existing CO_2 is reduced to CH_4 by H_2 where methanogenesis bacteria play the role of catalyst through the reaction [44, 45, 75, 76]:

$$CO_2 + 4H_2 \rightarrow CH_4 + 2H_2O \tag{20.14}$$

It is worth noting that subsurface micro-organisms can only use aqueous or dissolved hydrogen in metabolism [68]. Therefore, the presence of the formation brine is highly important for biotic (also abiotic geochemical reactions) hydrogen conversion. Given this reason, the process of methanogenesis can also be described as in the following two reactions:

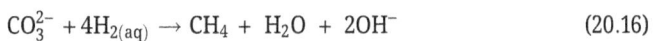

$$HCO_3^- + 4H_{2(aq)} \rightarrow CH_4 + 2H_2O + OH^- \tag{20.15}$$

$$CO_3^{2-} + 4H_{2(aq)} \rightarrow CH_4 + H_2O + 2OH^- \tag{20.16}$$

Methanogenesis has been observed in a Lobodice sandstone reservoir in the Czech Republic [77], Ketzin sandstone reservoir in Germany [78], and Lehen sandstone reservoir in Austria [79], where a portion of stored hydrogen was found to convert to methane in a produced gas mixture.

In acetogenesis, CO_2 is converted into CH_3COOH [68, 75, 76, 80]:

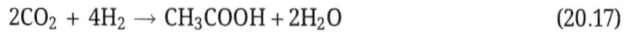

$$2CO_2 + 4H_2 \rightarrow CH_3COOH + 2H_2O \tag{20.17}$$

Or

$$2HCO_3^- + 4H_{2(aq)} \rightarrow CH_3COOH + 2H_2O + 2OH^- \tag{20.18}$$

$$2CO_3^{2-} + 4H_{2(aq)} \rightarrow CH_3COOH + 4OH^- \tag{20.19}$$

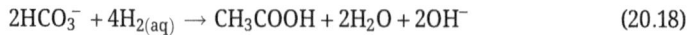

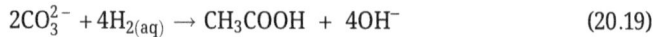

Sulfate/sulfur reduction is the most dominant microbial activity in geological subsurface sites including depleted gas reservoirs because of the faster kinetics compared to other types of biotic hydrogen oxidation processes [68, 80, 81]. The transformation from $SO_4{}^{2-}$ or S into hydrogen sulfide gas was reported in several geological hydrogen storage sites [77–79]. In fact, a field survey in France shows that the sulfate/sulfur reducing bacteria (for simplicity sulfate and sulfur reducers are collectively referred to as SRB) dominates the microbial community in seven natural gas storage sites [68, 82]. SRB can consume stored hydrogen and accelerate the reduction of $SO_4{}^{2-}$ or S into H_2S (in gas or the aqueous phase)/HS^-/S^{2-}:

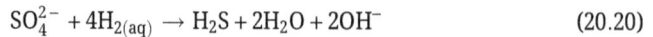

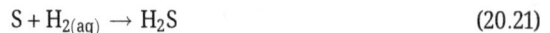

$$SO_4^{2-} + 4H_{2(aq)} \rightarrow H_2S + 2H_2O + 2OH^- \tag{20.20}$$

$$S + H_{2(aq)} \rightarrow H_2S \tag{20.21}$$

The SRB family is usually anaerobic and reproductive on the surface of sulfur particles in the form of biofilm [75, 83]. The generated H_2S can change the compositions of the gas mixture and compromise the purity and withdrawal efficiency of stored hydrogen. The microbially formed H_2S presented on a wellbore surface can also lead to casing/steel corrosion, triggering potential cracking and failure of the packer [84–87].

Ferric iron reduction triggered by iron reducing bacteria (IRB) can reduce Fe^{3+} into ferrous iron (Fe^{2+}), where the source of Fe^{3+} is mainly from the in situ minerals in the form of iron oxides or iron-bound minerals such as goethite [$FeO(OH)$] [88], hematite (Fe_2O_3) [89, 90], Fe-illite [91], Fe-smectite [92, 93], and Fe-chlorite [94]. The ferric iron reduction can consume extra portions of stored hydrogen that impairs storage efficiency. Accumulation and reproduction of IRB on the surface of a wellbore casing can lead to microbial corrosion along with SRB [86, 95, 96]. However, as Wiegel et al. [97] suggested, SRB is frequently scarce in oil/gas reservoirs. Besides, Fe^{3+} oxides are merely present in carbon-rich fields since they have been transformed into Fe^{2+} minerals over a long geological time span and hardly reform due to the anaerobic environment [68, 98]. Therefore, we suggest that the impact of SRB on hydrogen conversion in natural hydrogen systems is less than methanogenesis, acetogenesis, and sulfate/sulfur reduction in depleted gas reservoirs.

20.3 Natural hydrogen system performance

20.3.1 Impact of hydrogen conversion and contamination on natural hydrogen composition

Abiotic and biotic geochemical reactions have a significant importance while planning UHS technology. Simultaneously, these reactions affect the purity of natural hydrogen reservoirs due to hydrogen conversion and contamination. These processes primarily result in the generation of H_2S and CH_4, impacting the overall composition of natural hydrogen reservoirs. The interplay between these processes and the surrounding geological environment can vary depending on specific geological and reservoir conditions.

The percentage of natural hydrogen generation from the wells can be influenced by different factors, such as mineralogy of source rock, mineralogy of reservoir rock, flow properties, depositional environment, and microbial impact. For example, the microscopic fluid inclusions in quartz in Precambrian uranium deposits showed up to 100% of H_2 in Oklo, Gabon [89]. On the other hand, quartz and calcite showed 4% and 42.6% of H_2, respectively in Kurusaisk, Uzbekistan [3]. This indicates the in situ carbonated geochemical reactions may have occurred, thus triggering localized hydrogen conversion and contamination. However, there is a pressing need to examine the effect of geochemical reactions with and without microbial on the generation and flow properties of the natural hydrogen in the porous media.

In a more advanced case, the Bourakebougou field provides significant detail on natural hydrogen generation and the role of natural hydrogen reservoir rock in hydrogen generation and development [100]. The highest content of H_2 was observed in the shallowest main reservoir, which is dolomitic carbonate (Neoproterozoic cap carbonate). This is perceived as a negligible impact of abiotic and biotic reactions on hydrogen conversion and contamination. For example, from the Gas Chromatography analysis of gases produced, the upper layer (dolomitic carbonate) gives the highest H_2 concentration of 98% [100] with only 1% of each CH_4 and N_2.

It is worth noting that abiotic and/or biotic geochemical reactions may have occurred given the reactions from Section 20.2. However, the distribution of the gas composition in the reservoirs is dominated by hydrogen molecular diffusion and gravity segregation, resulting in a high concentration of H_2 gas on the top of structure. This account for the observation from the gas logging data, which reveals that the ratio of H_2/CH_4 increases with the elevation of the Reservoir 1's structural profile (*see figure 9 in* [101]) of Well BOU-19 in comparison to Well BOU-13 (*see figure 11 in* [101]). This also partially explains why CH_4 was detected in the upper carbonate reservoir.

Carbonate minerals might have reacted with H_2 in its environment of pressure and temperature due to the abiotic and biotic reactions. Although it is a very slow kinetic process given its high activation energy, in the geological time, 4 mol of hydrogen may be consumed and converted to 1 mol of CH_4 (from reactions (20.1)–(20.5)).

Moreover, if there are methanogen microbes in place, the geochemical reaction can be accelerated, leading to an even lower H_2 concentration with a higher CH_4 concentration. These localized abiotic and biotic geochemical reactions may also trigger localized heterogeneity in the formation. This may partially account for the high degree of heterogeneous porosities, such as 0.21–14.32% in the Bourakebougou field, which is mainly composed of a karstified geological process in which soluble rocks, such as limestone, dolomite, or gypsum, have undergone chemical weathering and dissolution to form distinctive features.

Natural hydrogen preserved in sulfate-rich reservoirs may only trigger negligible impurity of natural hydrogen in the system, given that this process does not produce any new gases. However, it is worth noting that this process (reactions (20.12) and (20.13)) triggers mineral transformation, for example, from goethite and hematite to amakinite (Fe(OH)2). In this geochemical process, regardless of the possible facilitation by microbes, the density of the minerals decreases from 4.28 g/cm^3 (goethite) and 4.2–5.3 g/cm^3 (hematite) to 2.925–2.98 g/cm^3 (amakinite). This process likely decreases pore volumes and reduces the permeability of the reservoirs, which may contribute to the low porosity of the reservoir [100].

Microorganisms can produce natural hydrogen in vitro. However, these bacteria always co-exist with hydrogen-utilizing bacteria [102]. Hence, biologically produced hydrogen is rapidly converted to other compounds. Reactions (20.14) and (20.17) may be related to the case of natural hydrogen conversion. Moreover, in closing, the drilling, logging and reservoir data of UHS wells could provide essential information for understanding a natural hydrogen reservoir system. Nevertheless, we believe that the impact of abiotic and biotic geochemical reactions differs from classical oil and gas reservoirs, given their unique geochemical and microbial activities. This is because the natural hydrogen reservoir possesses a dynamic system that is progressively recharged in H_2-rich gas at the production timescale [100].

20.3.2 Impact of hydrogen-rock physics on natural hydrogen development

Hydrogen-rock physics regulates the development of natural hydrogen systems. Figure 20.2 shows the main factors relating to hydrogen-rock physics are in situ wettability, interfacial tension, capillary pressure [103], hydrogen relative permeability [104], and hydrogen dispersion in natural hydrogen systems.

The state-of-the-art knowledge on wettability from both experimental tests (particularly the contact angle measurement) and numerical simulations (e.g., surface complexation modeling and molecular dynamics modeling) suggests that the wettability of almost every mineral is from strongly water-wet to weakly water-wet with hydrogen. This indicates that robust water management during natural hydrogen pro-

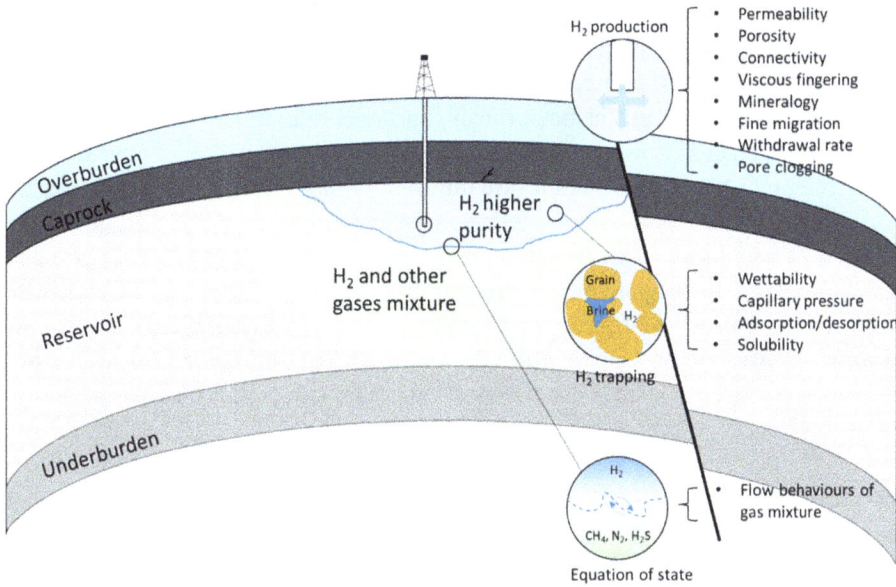

Figure 20.2: Hydrogen-rock physics description for natural hydrogen reservoir development (adapted from [105] with slight modifications).

duction plays a significant role in improving natural hydrogen recovery efficiency the same as the gas field's development.

20.3.2.1 Initialization of natural hydrogen systems

Given the abiotic and biotic geochemical reactions likely to occur in the natural hydrogen systems as revealed in Section 20.2, hydrogen likely exists with other gases, for example, CH_4 and H_2S, which are well reported from field observations [3]. Hydrogen dispersion and gravity characteristics will play an important role in hydrogen concentration distribution at the initial reservoir conditions before the depletion process. To predict the purity of hydrogen during the production process, it is important to understand the hydrogen concentration distribution over the depth of the reservoir. Therefore, it is important to understand the governing physics of hydrogen distribution within any reservoir.

Clearly, hydrogen's gravity forces, molecular diffusion, and mechanical dispersion, named as hydrodynamic dispersion, dominate the initialization process. Hydrogen molecular diffusion likely plays an important role in the initialization process of natural hydrogen reservoir. This is because the hydrogen molecular diffusion coefficient is 2–3 times higher than CH_4 and CO_2 depending on temperature and pressure. For example, in pure water at 298 K and standard pressure, the diffusion coefficient of hydrogen is around 5.13×10^{-9} m^2/s [106], with only 1.6×10^{-9} and 1.85×10^{-9} m^2/s for CO_2 [107] and CH_4 [108]. Increasing temperature and decreasing pressure can increase hydrogen diffusivity. The impact of hydrogen diffusivity can be classified into three categories: (1) upward diffusion through caprock, (2) interface diffusion in other in situ fluids, and (3) molecular diffusion into the same gas phase with impurities. The latest research [109] shows that upward diffusion through caprock leads to negligible loss of hydrogen through both numerical modeling and experiments. Also, given the much lower solubility of hydrogen in brine, the cross-phase diffusion of hydrogen likely results in a negligible impact on natural hydrogen development. Given hydrogen's higher molecular diffusion which is much less in gravity compared to CH_4, hydrogen's physical properties will affect the hydrogen concentration profile over reservoir depth, which can be modeled using existing codes (see Section 20.3.3).

20.3.3 Impac of hydrogen dynamics on natural hydrogen development

The impact of hydrogen dynamics on natural hydrogen development is attributed to two areas, that is, characterization of the equation of state (EOS) and multicomponent flow in porous media. Given that natural hydrogen systems likely comprise multiple gases, H_2, CH_4, CO_2, H_2S, and so on, to predict the performance of the development, the EOS needs to be better characterized. EOS is a thermodynamic equation relating state variables, which describe the state of matter under a given set of physical conditions, such as pressure, volume, temperature, or internal energy [110].

The specific model with the ability to simulate natural hydrogen systems, which can account for thermal, multiphase, and multi-component flow, remains scarce [111]. DuMU^{x}, ECLIPSE, TOUGH, OpenGeoSys-ECIPSE, and COMSOL are currently used for modeling of UHS or CO_2 geosequestration with consideration of other gas phases [12, 112–116]. Nevertheless, these simulators were initially developed for other applications rather than hydrogen-based fluid properties. To the best of our knowledge, none of them has been verified for UHS simulation against field observations and operations, including the solubility of hydrogen and gas mixtures in water/saline solutions and thermodynamic models for estimating key flow parameters such as the density and viscosity of hydrogen and gas mixtures [111]. Therefore, the characterization of EOS for hydrogen-rich gas mixtures requires further investigation.

To predict the hydrogen flow in natural hydrogen systems during the natural hydrogen development, it is important to model hydrogen mixing, which is associated with hydrogen molecular diffusion and mechanical dispersion, gravity, and heterogeneity of the reservoirs. While little dynamic modeling work has been conducted in natural hydrogen systems, existing rock flow dynamics codes likely provide a good quantitative agreement in predicting the performance of natural hydrogen development. Okoroafor et al. conducted an intercomparison of numerical simulation models for hydrogen storage in porous media using different codes [117]. Molecular diffusion, advection, gravity, radial flow, heterogeneity, and capillarity were modeled in four different scenarios at different scales, that is, one-, two- to three-dimension numerical modeling [117]. Five simulators were used for this intercomparison, for example, ECLIPSE 100, ECLIPSE 300, GEM, ADGPRS (Stanford University, in-house), and OPM Flow (open source, GNU General Public License). Their results show substantial agreement between results predicted from the different simulators. On average, the mean absolute error estimated at different times and spaces was less than 4% based on the numerical modeling [117]. This builds confidence that existing modeling codes may be effectively adapted for natural hydrogen system modeling with an improved EOS.

20.4 Natural hydrogen reservoir integrity

Key aspects influencing the storage integrity of natural hydrogen systems include the caprock's sealing capacity, the integrity of the caprock and reservoir, wells and the stability of faults. Published work shows that the same caprock that can seal to methane appears to have low risks of sealing to hydrogen [118]. This is because the hydrogen-brine-caprock has a higher capillary pressure compared to the methane-brine-caprock systems. More detailed documentation for the risks and uncertainties related to the storage integrity of UHS in porous rock can be seen in chapter 3 in Hydrogen TCP-Task 42 Underground Hydrogen Storage: Technical Monitor Report [34]. Here, we only present potential or perceived risks of reactivation of existing faults and fractures during natural hydrogen system development.

20.4.1 Faults: tensile damage, shear damage, and shear slip

Depletion of natural hydrogen systems results in the decrease of pore pressure, which, in turn, increases the effective stress on the rock matrix. This variation can lead to shear failure when the rock's shear strength is exceeded. This process may also cause tensile damage to reservoir and caprock, leading to the activation of existing fractures and generating tensile fractures [119, 120].The impact of tensile damage on caprock and fractures could be much more severe than shear damage in gas reservoirs with shallower burial depths [119, 121]. The development of tensile fractures can affect both permeability and permeability anisotropy [122] and accelerate hydrogen upward migration. Therefore, an accurate evaluation of tensile strength and trap ground stress is necessary to de-risk the tensile damage within the caprock and faults. A leak-off test [123, 124] and the AE Kaiser effect experiment [125, 126] are the two widely applied methods to test the in situ stress of reservoir and caprock.

For natural hydrogen reservoirs located in greater depth, the risk of shear damage on caprock and shear fractures could be more serious [127, 128]. The shear damage is caused by the rock mechanical heterogeneity induced by the triaxial principle stress difference and in situ stress change due to pressure variations [119, 129]. Shear damage can trigger sliding deformation along a mechanically weak plane, compromising caprock stability and aggravating hydrogen upward migration. Triaxial compression tests [130, 131] on core-scale samples combined with larger-scale 3D geomechanical modeling [132, 133] are commonly used to analyze the risk of shear damage on caprock and associated shear fracture propagation.

20.4.2 Friction coefficient and internal friction angle

Fractures and faults can slip when shear stress exceeds shear strength, which is a function of effective normal stress, friction coefficient, and internal friction angle. While the effective normal stress (total stress minus pore water stress [134]) and shear stress are affected by the stress regime change induced by hydrogen cycling, the friction coefficient and internal friction angle can be affected by the surrounding physicochemical conditions. For friction coefficient (the typical value is 0.6 but can increase to 0.85 [135, 136]), there is a greater value with a higher fraction of quartz and a less fraction of clays [137]. The friction coefficient decreases when water is added to the system, as it acts as a lubricating agent to reduce the cohesive force of the crystal particles and adjacent fracture planes [138]. While there is no evidence of friction coefficient change caused by the physical injection of gas, the geochemical reactions between H_2 and reactive minerals such as carbonates may still affect the friction coefficient [139]. Meanwhile, the adsorption of H_2 on clays and organic matter can decrease the internal friction angle, and thus reduce the shear strength (assuming other parameters are constant) and increase the risk of a fault's shear slip.

Moreover, abiotic geochemical reactions and microbial activities may change the in situ geochemical environment, such as pH and salinity, affecting rock surface species concentrations, surface potential and surface energy, and thus leading to subcritical crack growth [140–146] (see Figure 20.3). Subcritical crack growth corresponds to the slow fracture propagation at stress below the threshold of dynamic rupture [147–149]. When the energy release rate G (J/m^2) at the tip of a crack is greater than the rock surface energy γ_s (J/m^2), the propagation of an existing crack likely occurs [150]. The risk of subcritical crack growth on fracture extension cannot be omitted during the natural hydrogen development. However, there is a clear lack of understanding of the impact of geochemical, physiochemical together, and cyclic loading on subcritical crack growth and thus the integrity of natural hydrogen systems. Moreover, the literature lacks experimental data from grain-scale to core-scale to address this uncertainty. We see this as a perceived risk – *hydrogen penetration into faults through microfractures (edge-charged minerals) initiates chemically controlled disruptions of the*

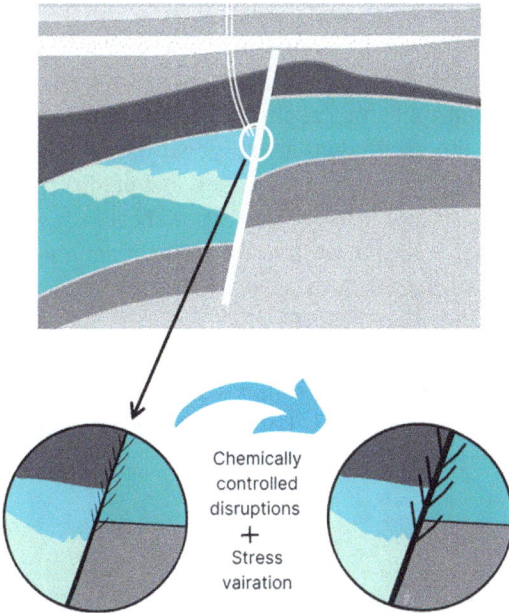

Figure 20.3: Schematic of microfracture propagation (subcritical crack growth) along main fractures and fault planes due to the surface energy change.

mineral structure, leading to localized environmental cracking also known as stress corrosion cracking. Figure 20.3 shows the hypothesis in a schematic diagram, which may need to be tested in the near future to understand the impact of the depletion process of natural hydrogen reservoirs on chemically controlled micro-fracture mechanics.

20.5 Conclusions

Natural hydrogen, also known as geological hydrogen, holds immense potential to revolutionize our low-carbon future, marking one of the most transformative shifts in the global energy system. However, limited research has been conducted on how to develop natural hydrogen systems. Building on extensive studies of UHS in depleted gas reservoirs [34], we aim to leverage this knowledge and expertise to drive the advancement of natural hydrogen systems.

Abiotic and biotic geochemical reactions, which occur during UHS in depleted gas reservoirs, also take place in natural hydrogen systems. These processes drive hydrogen conversion and contamination over geological timescales, leading to varying hydrogen concentrations across hydrogen-detected wells in diverse geological settings – such as orebodies, volcanic gases, hydrothermal systems, and even oil and gas wells.

As a result, the composition of hydrogen mixtures in natural hydrogen systems can fluctuate, highlighting the need for quantitative research to characterize the hydrogen-based EOS. While the abiotic and biotic reactions in natural hydrogen systems differ from those in natural gas reservoirs, the fundamental physics and governing equations of multicomponent flow remain the same. Consequently, existing modeling codes, which have been tested for UHS, may be adapted to manage and predict natural hydrogen reservoir behavior. Hydrogen-rock interactions trigger chemically controlled micro-fracture mechanics behavior, which may differ from that of natural gas reservoirs. However, conventional rock mechanics remains relevant to mitigate natural hydrogen reservoir integrity risks at the reservoir scale.

Declaration of competing interest: The authors declare that they have no known competing financial interests or personal relationships that could have appeared to influence the work reported in this chapter.

References

[1] Ellis G. S., Gelman S. E. Model predictions of global geologic hydrogen resources. Science Advances. 2024, 10(50): eado0955.
[2] Philip J., Ball K. C. Natural hydrogen: The new frontier. Geoscientist, 2022.
[3] Zgonnik V. The occurrence and geoscience of natural hydrogen: A comprehensive review. Earth-Science Reviews. 2020, 203: 103140.
[4] Krista Davies E. F., Giwelli A., Esteban L., Keshavarz A., Iglauer S. A natural hydrogen seep in Western Australia: Observed characteristics and controls. Science and Technology for Energy Transition, , 79,48 (2024).
[5] Frery E., et al. Natural hydrogen seeps identified in the north perth basin, western Australia. International Journal of Hydrogen Energy. 2021, 46(61): 31158–31173.
[6] Maraggi L. M. R., Moscardelli L. G. Hydrogen storage potential of salt domes in the Gulf Coast of the United States. Journal of Energy Storage. 2024, 82: 110585.
[7] Sorkhabi R., et al. A glimpse of the energy transition: Utah's new energy corridor. First Break. 2024, 42(10): 89–94.
[8] WSP. ACES Delta I Hydrogen Production and Storage. 2024 [cited 2025 18th January]; Available from: https://www.wsp.com/en-au/projects/aces-delta-i-hydrogen-production-and-storage.
[9] SHASHTA. DOE National Laboratories Investigate Subsurface Hydrogen Storage. 2024 [cited 2024 18th January]; Available from: https://edx.netl.doe.gov/sites/shasta/.
[10] Jahanbakhsh A., et al. Underground hydrogen storage: A UK perspective. Renewable and Sustainable Energy Reviews. 2024, 189: 114001.
[11] HyStorPor: Hydrogen Storage in Porous Media. 2023. p. https://blogs.ed.ac.uk/hystorpor/.
[12] Wallace R. L., et al. Utility-scale subsurface hydrogen storage: UK perspectives and technology. International Journal of Hydrogen Energy. 2021, 46(49): 25137–25159.
[13] Wood. *Wood to optimise hydrogen storage for Centrica's Rough field.* 2024 [cited 2025 24th January]; Available from: https://www.woodplc.com/news/latest-press-releases/2024/wood-to-optimise-hydrogen-storage-for-centricas-rough-field.
[14] Masoudi M., et al. Lined rock caverns: A hydrogen storage solution. Journal of Energy Storage. 2024, 84: 110927.

[15] HyPSTER, 1st demonstrator for green hydrogen storage. 2021. [cited 2025 18th January]; Available from: https://hypster-project.eu/.

[16] Gasunie. HyStock hydrogen storage. 2024 [cited 2025 18th January]; Available from: https://www.gasunie.nl/en/projects/hystock-hydrogen-storage.

[17] Hystories. Hydrogen Storage in European Subsurface. 2025 [cited 2025 18th January]; Available from: https://hystories.eu/.

[18] Storage, U.E. Hydrogen in Porous Rock Storage. 2025 [cited 2024 18th January]; Available from: https://www.uniper.energy/hystorage.

[19] HyUSPRe, Hydrogen Underground Storage in Porous Reservoirs. 2020. [cited 2025 18th January]; Available from: https://www.hyuspre.eu/.

[20] RAG. Underground Sun Storage 2030. 2024 [cited 2025 18th January]; Available from: https://www.rag-austria.at/en/research-innovation/underground-sun-storage-2030.html.

[21] EUH2STARS. 2024. p. https://www.euh2stars.eu/en/.

[22] H2eart – Creating the heart of the European Hydrogen Economy. 2024. p. https://h2eart.eu/.

[23] HYBRIT. HYBRIT: A unique, underground, fossil-free hydrogen gas storage facility is being inaugurated in Luleå. 2022 [cited 2025 18th January]; Available from: https://www.hybritdevelopment.se/en/hybrit-a-unique-underground-fossil-free-hydrogen-gas-storage-facility-is-being-inaugurated-in-lulea/.

[24] Capenergies. H2 GEOGAS STORAGE. 2022 [cited 2025 18th January]; Available from: https://www.capenergies.fr/en/project/h2-geogas-storage/.

[25] Bulletin C. H. China started construction of its first deep underground cavern hydrogen storage facility. 2024 [cited 2025 18th January]; Available from: https://chinahydrogen.substack.com/p/china-started-construction-of-its.

[26] Huang L., et al. A preliminary site selection system for underground hydrogen storage in salt caverns and its application in Pingdingshan, China. Deep Underground Science and Engineering. 2024, 3(1): 117–128.

[27] Higgs K. E., et al. Prospectivity analysis for underground hydrogen storage, Taranaki basin, Aotearoa New Zealand: A multi-criteria decision-making approach. International Journal of Hydrogen Energy. 2024, 71: 1468–1485.

[28] Yates E., et al. Hydrogen geo-storage in Aotearoa— New Zealand. New Plymouth, Aotearoa/New Zealand: Elemental Group, Ltd., 2021.

[29] RISC. *Hydrogen Storage Potential Of Depleted Oil And Gas Fields In Western Australia*. 2024 [cited 2025 18th January]; Available from: https://riscadvisory.com/hydrogen-storage-potential-of-depleted-oil-and-gas-fields-in-western-australia/.

[30] CO2CRC. Hydrogen Storage. 2025 [cited 2025 18th January]; Available from: https://co2crc.com.au/research/hydrogen-storage/.

[31] LOCHARD. A deep underground hydrogen storage project to shift energy seasonally 2024 [cited 2025 18th January]; Available from: https://www.lochardenergy.com.au/our-projects/h2restore/.

[32] Underground hydrogen storage. 2024: Future energy exports CRC. [cited 2025 18th January]; Available from: https://www.fenex.org.au/case-study/underground-hydrogen-storage/.

[33] Hychico. Underground Hydrogen Storage. 2024 [cited 2025 18th January]; Available from: https://hychico.com.ar/en/underground-hydrogen-storage-3/.

[34] Underground hydrogen storage: Technology monitor report. 2023. p. 153 pages including appendices.

[35] Hydrogen TCP-task 42, Building confidence in underground hydrogen storage. 2025. p. 55 pages including appendices.

[36] H2 E. S. European Underground H2 Storage System. 2025 [cited 2025 18th January]; Available from: https://www.euh2stars.eu/en/.

[37] Hydrogen storage. 2021: CO2CRC. p. https://co2crc.com.au/research/hydrogen-storage/.

[38] Underground hydrogen storage. Future energy exports CRC. 2024. https://www.fenex.org.au/case-study/underground-hydrogen-storage/

[39] CSIRO. Integrated experimental facility for characterisation of hydrogen flow in porous media during underground hydrogen storage. 2023 [cited 2025 18th January]; Available from: https://research.csiro.au/hyresearch/integrated-experimental-facility-for-characterisation-of-hydrogen-flow-in-porous-media-during-underground-hydrogen-storage/.

[40] Bensing J. P., et al. Hydrogen-induced calcite dissolution in Amaltheenton Formation claystones: Implications for underground hydrogen storage caprock integrity. International Journal of Hydrogen Energy. 2022https://eartharxiv.org/repository/view/3285/

[41] Flesch S., et al. Hydrogen underground storage – Petrographic and petrophysical variations in reservoir sandstones from laboratory experiments under simulated reservoir conditions. International Journal of Hydrogen Energy. 2018, 43(45): 20822–20835.

[42] Henkel S., et al. Effects of H2 and CO2 underground storage in natural pore reservoirs-findings by SEM and AFM techniques. In: The third sustainable earth sciences conference and exhibition. European Association of Geoscientists & Engineers, 2015.

[43] Pudlo D., et al. The impact of hydrogen on potential underground energy reservoirs. Geophysical Research Abstracts Vol. 20, EGU2018-8606, 2018 EGUGeneral Assembly 2018 ©Author(s) 2018. CC Attribution 4.0 license.

[44] Hemme C., Van Berk W. Hydrogeochemical modeling to identify potential risks of underground hydrogen storage in depleted gas fields. Applied Sciences. 2018, 8(11): 2282.

[45] Pichler M. Assessment of hydrogen rock interaction during geological storage of CH4-H2 mixtures. In: Second EAGE Sustainable Earth Sciences (SES) conference and exhibition. European Association of Geoscientists & Engineers, 2013.

[46] Bo Z., et al. Geochemical reactions-induced hydrogen loss during underground hydrogen storage in sandstone reservoirs. International Journal of Hydrogen Energy Volume 47, Issue 59, 12 July 2022, Pages 24861–24870

[47] Hassannayebi N., et al. Underground hydrogen storage: Application of geochemical modelling in a case study in the Molasse Basin, Upper Austria. Environmental Earth Sciences. 2019, 78(5): 1–14.

[48] Zeng L., et al. Hydrogen storage in Majiagou carbonate reservoir in China: Geochemical modelling on carbonate dissolution and hydrogen loss. International Journal of Hydrogen Energy. 2022.

[49] Amid A., Mignard D., Wilkinson M. Seasonal storage of hydrogen in a depleted natural gas reservoir. International Journal of Hydrogen Energy. 2016, 41(12): 5549–5558.

[50] Lassin A., Dymitrowska M., Azaroual M. Hydrogen solubility in pore water of partially saturated argillites: Application to Callovo-Oxfordian clayrock in the context of a nuclear waste geological disposal. Physics and Chemistry of the Earth, Parts A/B/C. 2011, 36(17–18): 1721–1728.

[51] Truche L., et al. Experimental reduction of aqueous sulphate by hydrogen under hydrothermal conditions: Implication for the nuclear waste storage. Geochimica Et Cosmochimica Acta. 2009, 73(16): 4824–4835.

[52] Truche L., et al. Kinetics of pyrite to pyrrhotite reduction by hydrogen in calcite buffered solutions between 90 and 180 °C: Implications for nuclear waste disposal. Geochimica Et Cosmochimica Acta. 2010, 74(10): 2894–2914.

[53] Hall A. Pyrite-pyrrhotine redox reactions in nature. Mineralogical Magazine. 1986, 50(356): 223–229.

[54] Wiltowski T., et al. Kinetics and mechanisms of iron sulfide reductions in hydrogen and in carbon monoxide. Journal of Solid State Chemistry. 1987, 71(1): 95–102.

[55] Lambert J., Simkovich G., Walker P. The kinetics and mechanism of the pyrite-to-pyrrhotite transformation. Metallurgical and Materials Transactions B. 1998, 29(2): 385–396.

[56] Didier M., et al. Adsorption of hydrogen gas and redox processes in clays. Environmental Science & Technology. 2012, 46(6): 3574–3579.

[57]   Moslemi H., Shamsi P., Habashi F. Pyrite and pyrrhotite open circuit potentials study: Effects on flotation. Minerals Engineering. 2011, 24(10): 1038–1045.

[58]   Truche L., et al. Sulphide mineral reactions in clay-rich rock induced by high hydrogen pressure. Application to disturbed or natural settings up to 250 °C and 30 bar. Chemical Geology. 2013, 351: 217–228.

[59]   Andrieux P., Petit S. Hydrothermal synthesis of dioctahedral smectites: The Al–Fe3 + chemical series: Part I: Influence of experimental conditions. Applied Clay Science. 2010, 48(1–2): 5–17.

[60]   Drits V., Manceau A. A model for the mechanism of Fe3 + to Fe2 + reduction in dioctahedral smectites. Clays and Clay Minerals. 2000, 48(2): 185–195.

[61]   Tardy Y., Nahon D. Geochemistry of laterites, stability of Al-goethite, Al-hematite, and Fe3 +-kaolinite in bauxites and ferricretes: An approach to the mechanism of concretion formation. American Journal of Science. 1985, 285(10): 865–903.

[62]   Trolard F., Tardy Y. A model of Fe3 +-kaolinite, Al3 +-goethite, Al3 +-hematite equilibria in laterites. Clay Minerals. 1989, 24(1): 1–21.

[63]   Munteanu G., Ilieva L., Andreeva D. Kinetic parameters obtained from TPR data for α-Fe2O3 and Auα-Fe2O3 systems. Thermochimica Acta. 1997, 291(1–2): 171–177.

[64]   Lin H.-Y., Chen Y.-W., Li C. The mechanism of reduction of iron oxide by hydrogen. Thermochimica Acta. 2003, 400(1–2): 61–67.

[65]   Lebedeva O. E., Sachtler W. M. Enhanced reduction of Fe2O3 caused by migration of TM ions out of zeolite channels. Journal of Catalysis. 2000, 191(2): 364–372.

[66]   Turkdogan E., Vinters J. Gaseous reduction of iron oxides: Part I. Reduction of hematite in hydrogen. Metallurgical and Materials Transactions B. 1971, 2(11): 3175–3188.

[67]   Monazam E. R., Breault R. W., Siriwardane R. Kinetics of hematite to wustite by hydrogen for chemical looping combustion. Energy & Fuels. 2014, 28(8): 5406–5414.

[68]   Thaysen E. M., et al. Estimating microbial growth and hydrogen consumption in hydrogen storage in porous media. Renewable and Sustainable Energy Reviews. 2021, 151: 111481.

[69]   Zivar D., Kumar S., Foroozesh J. Underground hydrogen storage: A comprehensive review. International Journal of Hydrogen Energy Volume 46, Issue 45, 1 July 2021, Pages 23436–23462

[70]   Schuchmann K., Müller V. Energetics and application of heterotrophy in acetogenic bacteria. Applied and Environmental Microbiology. 2016, 82(14): 4056–4069.

[71]   Hoehler T. M., et al. Acetogenesis from CO2 in an anoxic marine sediment. Limnology and Oceanography. 1999, 44(3): 662–667.

[72]   Aüllo T., et al. Desulfotomaculum spp. and related gram-positive sulfate-reducing bacteria in deep subsurface environments. Frontiers in Microbiology. 2013, 4: 362.

[73]   Pedersen K. Microbial life in deep granitic rock. FEMS Microbiology Reviews. 1997, 20(3–4): 399–414.

[74]   Brock T. D., Gustafson J. Ferric iron reduction by sulfur-and iron-oxidizing bacteria. Applied and Environmental Microbiology. 1976, 32(4): 567–571.

[75]   Dopffel N., Jansen S., Gerritse J. Microbial side effects of underground hydrogen storage–knowledge gaps, risks and opportunities for successful implementation. International Journal of Hydrogen Energy. 2021, 46(12): 8594–8606.

[76]   Gregory S. P., et al. Subsurface microbial hydrogen cycling: Natural occurrence and implications for industry. Microorganisms. 2019, 7(2): 53.

[77]   Amigáň P., et al. Methanogenic bacteria as a key factor involved in changes of town gas stored in an underground reservoir. FEMS Microbiology Ecology. 1990, 6(3): 221–224.

[78]   Stolten D., Emonts B. Hydrogen science and engineering, 2 volume set: Materials, processes, systems, and technology, vol. 1. John Wiley & Sons, 2016.

[79]   Pichler M. Underground sun storage results and outlook. In: EAGE/DGMK joint workshop on underground storage of hydrogen. European Association of Geoscientists & Engineers, 2019. Volume 2019, p.1 – 4 DOI: https://doi.org/10.3997/2214-4609.201900257

[80] Bernardez L., et al. A kinetic study on bacterial sulfate reduction. Bioprocess and Biosystems Engineering. 2013, 36(12): 1861–1869.

[81] Appelo C. A. J., Postma D. Geochemistry, groundwater and pollution. CRC press, 2004.

[82] Ranchou-Peyruse M., et al. Geological gas-storage shapes deep life. Environmental Microbiology. 2019, 21(10): 3953–3964.

[83] Hedderich R., et al. Anaerobic respiration with elemental sulfur and with disulfides. FEMS Microbiology Reviews. 1998, 22(5): 353–381.

[84] Ugarte E. R., Salehi S. A review on well integrity issues for underground hydrogen storage. J. Energy Resour. Technol. Apr 2022, 144(4): 042001 (10 pages)

[85] Bai P., et al. Initiation and developmental stages of steel corrosion in wet H2S environments. Corrosion Science. 2015, 93: 109–119.

[86] Enning D., Garrelfs J. Corrosion of iron by sulfate-reducing bacteria: New views of an old problem. Applied and Environmental Microbiology. 2014, 80(4): 1226–1236.

[87] Skovhus T. L., Enning D., Lee J. S. Microbiologically influenced corrosion in the upstream oil and gas industry. CRC press, 2017.

[88] Hui-Juan L., Jing-Jing P., Hong-Bo L. Diversity and characterization of potential H2-dependent Fe (III)-reducing bacteria in paddy soils. Pedosphere. 2012, 22(5): 673–680.

[89] Pineau A., Kanari N., Gaballah I. Kinetics of reduction of iron oxides by H2: Part I: Low temperature reduction of hematite. Thermochimica Acta. 2006, 447(1): 89–100.

[90] Pang J.-M., et al. Influence of size of hematite powder on its reduction kinetics by H2 at low temperature. Journal of Iron and Steel Research, International. 2009, 16(5): 07–11.

[91] Murad E., Wagner U. The thermal behaviour of an Fe-rich illite. Clay Minerals. 1996, 31(1): 45–52.

[92] Kostka J. E., et al. Growth of iron (III)-reducing bacteria on clay minerals as the sole electron acceptor and comparison of growth yields on a variety of oxidized iron forms. Applied and Environmental Microbiology. 2002, 68(12): 6256–6262.

[93] Kim J.-W., et al. The effect of microbial Fe (III) reduction on smectite flocculation. Clays and Clay Minerals. 2005, 53(6): 572–579.

[94] Lempart M., et al. Dehydrogenation and dehydroxylation as drivers of the thermal decomposition of Fe-chlorites. American Mineralogist: Journal of Earth and Planetary Materials. 2018, 103(11): 1837–1850.

[95] Herrera L. K., Videla H. A. Role of iron-reducing bacteria in corrosion and protection of carbon steel. International Biodeterioration & Biodegradation. 2009, 63(7): 891–895.

[96] Updegraff D. M. Microbiological corrosion of iron and steel – A review. Corrosion. 1955, 11(10): 44–48.

[97] Wiegel J., Hanel J., Aygen K. Chemolithoautotrophic thermophilic iron(iii)-reducer. In: Ljungdahl L. G., et al., ed., Biochemistry and physiology of anaerobic bacteria. New York: Springer New York, 2003, 235–251.

[98] Pannekens M., et al. Oil reservoirs, an exceptional habitat for microorganisms. New Biotechnology. 2019, 49: 1–9.

[99] Dubessy J., et al. Radiolysis evidenced by H2-O2 and H2-bearing fluid inclusions in three uranium deposits. Geochimica Et Cosmochimica Acta. 1988, 52(5): 1155–1167.

[100] Maiga O., et al. Characterization of the spontaneously recharging natural hydrogen reservoirs of Bourakebougou in Mali. Scientific Reports. 2023, 13(1): 11876.

[101] Prinzhofer A., Cissé C. S. T., Diallo A. B. Discovery of a large accumulation of natural hydrogen in Bourakebougou (Mali). International Journal of Hydrogen Energy. 2018, 43(42): 19315–19326.

[102] Hoehler T. M. Biogeochemistry of dihydrogen (H2). Metal Ions in Biological Systems, Volume 43-Biogeochemical Cycles of Elements. 2005, 9–48.

[103] Thaysen E. M., et al. Pore-scale imaging of hydrogen displacement and trapping in porous media. International Journal of Hydrogen Energy. 2023, 48(8): 3091–3106.

[104] Rezaei A., et al. Relative permeability of hydrogen and aqueous brines in sandstones and carbonates at reservoir conditions. Geophysical Research Letters. 2022, 49(12): e2022GL099433.

[105] Zeng L., et al. Hydrogen storage performance during underground hydrogen storage in depleted gas reservoirs: A review. Engineering. 2024, 40: 211–225.

[106] Ferrell R., Himmelblau D. Diffusion coefficients of hydrogen and helium in water. AIChE Journal. 1967, 13(4): 702–708.

[107] Tamimi A., Rinker E. B., Sandall O. C. Diffusion coefficients for hydrogen sulfide, carbon dioxide, and nitrous oxide in water over the temperature range 293–368 K. Journal of Chemical and Engineering Data. 1994, 39(2): 330–332.

[108] Witherspoon P., Saraf D. Diffusion of Methane, Ethane, Propane, and n-Butane in water from 25 to 43. The Journal of Physical Chemistry. 1965, 69(11): 3752–3755.

[109] Carden P., Paterson L. Physical, chemical and energy aspects of underground hydrogen storage. International Journal of Hydrogen Energy. 1979, 4(6): 559–569.

[110] Perrot P. A to Z of thermodynamics. Oxford University Press on Demand, 1998.

[111] Cai Z., Zhang K., Guo C. Development of a novel simulator for modelling underground hydrogen and gas mixture storage. International Journal of Hydrogen Energy. 2022, 47(14): 8929–8942.

[112] Sáinz-García A., et al. Assessment of feasible strategies for seasonal underground hydrogen storage in a saline aquifer. International Journal of Hydrogen Energy. 2017, 42(26): 16657–16666.

[113] Pfeiffer W. T., Beyer C., Bauer S. Hydrogen storage in a heterogeneous sandstone formation: Dimensioning and induced hydraulic effects. Petroleum Geoscience. 2017, 23(3): 315–326.

[114] Flemisch B., et al. DuMux: DUNE for multi-{phase, component, scale, physics, . . . } flow and transport in porous media. Advances in Water Resources. 2011, 34(9): 1102–1112.

[115] Yang L., et al. Numerical investigation of cycle performance in compressed air energy storage in aquifers. Applied Energy. 2020, 269: 115044.

[116] Pfeiffer W., Graupner B., Bauer S. The coupled non-isothermal, multiphase-multicomponent flow and reactive transport simulator OpenGeoSys–ECLIPSE for porous media gas storage. Environmental Earth Sciences. 2016, 75(20): 1–15.

[117] Okoroafor E. R., et al. Intercomparison of numerical simulation models for hydrogen storage in porous media using different codes. Energy Conversion and Management. 2023, 292: 117409.

[118] Hosseini M., et al. Capillary sealing efficiency analysis of caprocks: Implication for hydrogen geological storage. Energy & Fuels. 2022, 36(7): 4065–4075.

[119] Zheng D., et al. Key evaluation techniques in the process of gas reservoir being converted into underground gas storage. Petroleum Exploration and Development. 2017, 44(5): 840–849.

[120] McGrath A. G., Davison I. Damage zone geometry around fault tips. Journal of Structural Geology. 1995, 17(7): 1011–1024.

[121] Gudmundsson A. Active fault zones and groundwater flow. Geophysical Research Letters. 2000, 27(18): 2993–2996.

[122] Paul P. K., Zoback M. D., Hennings P. H. Fluid flow in a fractured reservoir using a geomechanically-constrained fault zone damage model for reservoir simulation.SPE Reservoir Evaluation & Engineering - Formation Evaluation, Aug. 2009, page 562-575.

[123] White A. J., Traugott M. O., Swarbrick R. E. The use of leak-off tests as means of predicting minimum in-situ stress. Petroleum Geoscience. 2002, 8(2): 189–193.

[124] Lin W., et al. Estimation of minimum principal stress from an extended leak-off test onboard the Chikyu drilling vessel and suggestions for future test procedures. Scientific Drilling. 2008, 6: 43–47.

[125] Michihiro K., Fujiwara T., Yoshioka H. Study on estimating geostresses by the Kaiser effect of AE. In: The 26th US Symposium on Rock Mechanics (USRMS). OnePetro, 1985.

[126] Holcomb D. J. General theory of the Kaiser effect. In: International journal of rock mechanics and mining sciences & geomechanics abstracts. Elsevier, 1993. Volume 30, Issue 7, December 1993, Pages 929–935

[127] Moeck I., Kwiatek G., Zimmermann G. Slip tendency analysis, fault reactivation potential and induced seismicity in a deep geothermal reservoir. Journal of Structural Geology. 2009, 31(10): 1174–1182.

[128] Chengyuan X., et al. Structural failure mechanism and strengthening method of fracture plugging zone for lost circulation control in deep naturally fractured reservoirs. Petroleum Exploration and Development. 2020, 47(2): 430–440.

[129] Kamali A., Ghassemi A. Analysis of injection-induced shear slip and fracture propagation in geothermal reservoir stimulation. Geothermics. 2018, 76: 93–105.

[130] Gong F., et al. Evaluation of shear strength parameters of rocks by preset angle shear, direct shear and triaxial compression tests. Rock Mechanics and Rock Engineering. 2020, 53(5): 2505–2519.

[131] Chang S.-H., Lee C.-I. Estimation of cracking and damage mechanisms in rock under triaxial compression by moment tensor analysis of acoustic emission. International Journal of Rock Mechanics and Mining Sciences. 2004, 41(7): 1069–1086.

[132] Fredrich J. T., et al. Geomechanical modeling of reservoir compaction, surface subsidence, and casing damage at the Belridge diatomite field. SPE Reservoir Evaluation & Engineering. 2000, 3(04): 348–359.

[133] Li L., et al. Numerical simulation of 3D hydraulic fracturing based on an improved flow-stress-damage model and a parallel FEM technique. Rock Mechanics and Rock Engineering. 2012, 45(5): 801–818.

[134] Simon A., Collison A. J. Pore-water pressure effects on the detachment of cohesive streambeds: Seepage forces and matric suction. Earth Surface Processes and Landforms. 2001, 26(13): 1421–1442.

[135] Suppe J. Absolute fault and crustal strength from wedge tapers. Geology. 2007, 35(12): 1127–1130.

[136] Sibson R. H. An assessment of field evidence for 'Byerlee' friction. Pure and Applied Geophysics. 1994, 142(3): 645–662.

[137] Samuelson J., Spiers C. J. Fault friction and slip stability not affected by CO_2 storage: Evidence from short-term laboratory experiments on North Sea reservoir sandstones and caprocks. International Journal of Greenhouse Gas Control. 2012, 11: S78–S90.

[138] Yao Q., et al. Mechanisms of failure in coal samples from underground water reservoir. Engineering Geology. 2020, 267: 105494.

[139] Bakker E., et al. Frictional behaviour and transport properties of simulated fault gouges derived from a natural CO_2 reservoir. International Journal of Greenhouse Gas Control. 2016, 54: 70–83.

[140] Zeng L., et al. Role of brine composition on rock surface energy and its implications for subcritical crack growth in calcite. Journal of Molecular Liquids. 2020, 303: 112638.

[141] Lu Y., et al. Analytical modelling of wettability alteration-induced micro-fractures during hydraulic fracturing in tight oil reservoirs. Fuel. 2019, 249: 434–440.

[142] Røyne A., Dalby K. N., Hassenkam T. Repulsive hydration forces between calcite surfaces and their effect on the brittle strength of calcite-bearing rocks. Geophysical Research Letters. 2015, 42(12): 4786–4794.

[143] Røyne A., Bisschop J., Dysthe D. K. Experimental investigation of surface energy and subcritical crack growth in calcite. Journal of Geophysical Research, VOL. 116, B04204, doi:10.1029/2010JB008033, 2011.

[144] Zeng L., et al. Effect of fluid-shale interactions on shales micromechanics: Nanoindentation experiments and interpretation from geochemical perspective. Journal of Natural Gas Science and Engineering. 2022, 101: 104545.

[145] Zeng L., et al. Effect of fluid saturation and salinity on sandstone rock weakening: Experimental investigations and interpretations from physicochemical perspective. Acta Geotechnica. 2022. Volume 18, pages 171 -186, (2023)

[146] Zeng L., et al. Interpreting micromechanics of fluid-shale interactions with geochemical modelling and disjoining pressure: Implications for calcite-rich and quartz-rich shales. Journal of Molecular Liquids Volume 319, 1 December 2020, 114117

[147] Atkinson B. K. Subcritical crack growth in geological materials. Journal of Geophysical Research: Solid Earth. 1984, 89(B6): 4077–4114.

[148] Rostom F., et al. Effect of fluid salinity on subcritical crack propagation in calcite. Tectonophysics. 2013, 583: 68–75.

[149] Xu M., Gupta A., Dehghanpour H. How significant are strain and stress induced by water imbibition in dry gas shales?. Journal of Petroleum Science and Engineering. 2019, 176: 428–443.

[150] Griffith A. A. VI. The phenomena of rupture and flow in solids. Philosophical Transactions of the Royal Society of London. Series A, Containing Papers of a Mathematical or Physical Character. 1921, 221(582–593): 163–198.

Mohammed Al Kindi*, Muhannad Al Hinai, Ahmed Al Abri,
and Zaid Al Siyabi

Chapter 21
Assessment of hydrogen storage in salt caverns in Oman

Abstract: We have undertaken a study to identify, screen, and assess potential salt caverns for underground hydrogen storage (UHS) in Oman. The project spanned two phases. The objectives of the first phase were to consolidate and interpret subsurface data across Oman, construct salt thickness maps, and evaluate prospective sites based on geological, operational and stakeholder considerations. During the first phase, 40 sites were selected across North, Central, and South Oman. These sites were broadly assessed for the depth to top salt, salt volume, thickness, terrain, and proximity to hydrogen sources. The second phase applied a more detailed evaluation, focusing on gross rock volume (GRV), salt purity, distribution, and stakeholder concerns, narrowing the selection to 10 sites and ultimately ranking four sites as the most promising: Qarat Al Kibrit, Al Noor, South Oman Salt Basin–6 (SOSB6) and Qarn Sahmah. Qarn Sahmah is highlighted for its strategic location, shallow depth, and high suitability for UHS. The other three top sites, Qarat Al Kibrit, Al Noor, and SOSB6, offer substantial storage capacity and align well with the geological and operational parameters. Each of the four top-ranked sites offers sufficient volume to store over 50,000 tons of hydrogen at depths shallower than 2,000 m. Calculations assume caverns that have a 75-m radius are 250 m in height, with each capable of storing approximately 35,000 tons of hydrogen under optimal conditions. To store 500,000 tons of hydrogen, an estimated 15 caverns per site would be required.

The four identified sites, and potentially many others, provide significant opportunities for hydrogen storage, bolstered by their geological attributes and strategic lo-

Acknowledgments: This study benefited from the discussions and support of Eng. Abdulaziz Al Shidhani (Hydrom Oman), Eng. Rumaitha Al Busaidi (Hydrom Oman), Eng. Maryam Al Farsi (Hydrom Oman), Mr. Husam Al Jabri (Hydrom Oman), Prof. Wilfried Baur (German University of Technology), Prof. Andreas Scharf (Sultan Qaboos University), Eng. Maram Al Balushi (Ministry of Energy and Minerals), Eng. Omer Al Mashaikhi (ESCC), Eng. Shifa Al Siyabi (ESCC) and Yousuf Al Darai (ESCC).

*Corresponding author: Mohammed Al Kindi, Earth Sciences Consultancy Centre (ESCC), Ghala Heights, Muscat, Oman, e-mail: malkindi@omanescc.com
Muhannad Al Hinai, Ministry of Energy and Minerals (MEM), Al-Khuwair, Ministry Streets, Muscat, Oman
Ahmed Al Abri, Hydrom Oman, Mina Al Fahal, Muscat, Oman
Zaid Al Siyabi, Vision Advanced Petroleum Solutions L.L.C (VAPS), Rusail Industrial City, Al Seeb, Muscat, Oman

https://doi.org/10.1515/9783111437040-021

cations. These findings position Oman as a key player in the emerging hydrogen economy, with further evaluations planned to advance the project to implementation.

Keywords: underground hydrogen storage, UHS, salt caverns, Oman, Fahud Salt Basin, Ghaba Salt Basin, South Oman Salt Basin, salt domes

21.1 Introduction

Hydrogen is emerging as a globally recognized sustainable energy source. However, its viability depends on the availability of efficient and scalable storage solutions to ensure consistent supply and integration into existing energy systems. Among the best identified locations for hydrogen storage are underground salt bodies [2, 9, 14]. These offer secure, large-scale, and cost-effective solutions. Large salt bodies can house significant volumes of hydrogen due to their vast size and thickness, low permeability, and high sealing capacity, preventing leakage and ensuring long-term containment. Moreover, the self-healing nature of salt enhances cavern integrity, providing a secure storage environment. Also, salt body plasticity allows them to absorb various stresses, reducing the risk of structural failures. Hydrogen stored in these caverns can be quickly withdrawn to stabilize energy grids and meet sudden demand, making them ideal for integrating intermittent renewable energy sources. Compared to other methods, salt caverns are cost-effective, offering long-term reliability with minimal maintenance.

This research evaluates the suitability of subsurface salt bodies across the salt basins in Oman for hydrogen storage by assessing them against different criteria and ranks them accordingly. The top-ranked salt bodies have been further evaluated for their storage capacity and potential arrangements of induced salt caverns. The research addresses the requirement for the potential emerging hydrogen economy for efficient, scalable, and secure storage solutions to support the energy transition in Oman and worldwide. Despite its extensive coverage and promising geological properties, the subsurface bodies of salt in the Ara Group lack proper screening and bridging of the gap between geological potential and operational implementation.

21.1.1 The Ara Salt in Oman

Oman is actively advancing its efforts to explore and harness the potential of hydrogen production as part of its commitment to sustainable energy development and global energy transition goals. Fortunately, the country is home to extensive salt deposits, known as the Ara Salt, that cover vast areas of North, Central, and South Oman (Figure 21.1). These deposits are integral to the country's subsurface geology and hold significant potential for various industries including petroleum, mining, and, today, underground hydrogen storage (UHS). The Ara Group dates to the Late Precambrian to Early Cambrian period (Figure 21.2). It is notable for its thickness and lateral continuity. The salt deposits of Ara can reach several hundred meters in thickness. They are predominantly found in three salt basins: the South Oman Salt Basin (SOSB), the Ghaba Salt Basin, and Fahud Salt Basin. These basins host numerous salt structures including salt domes, pillows, and other halokinetic features.

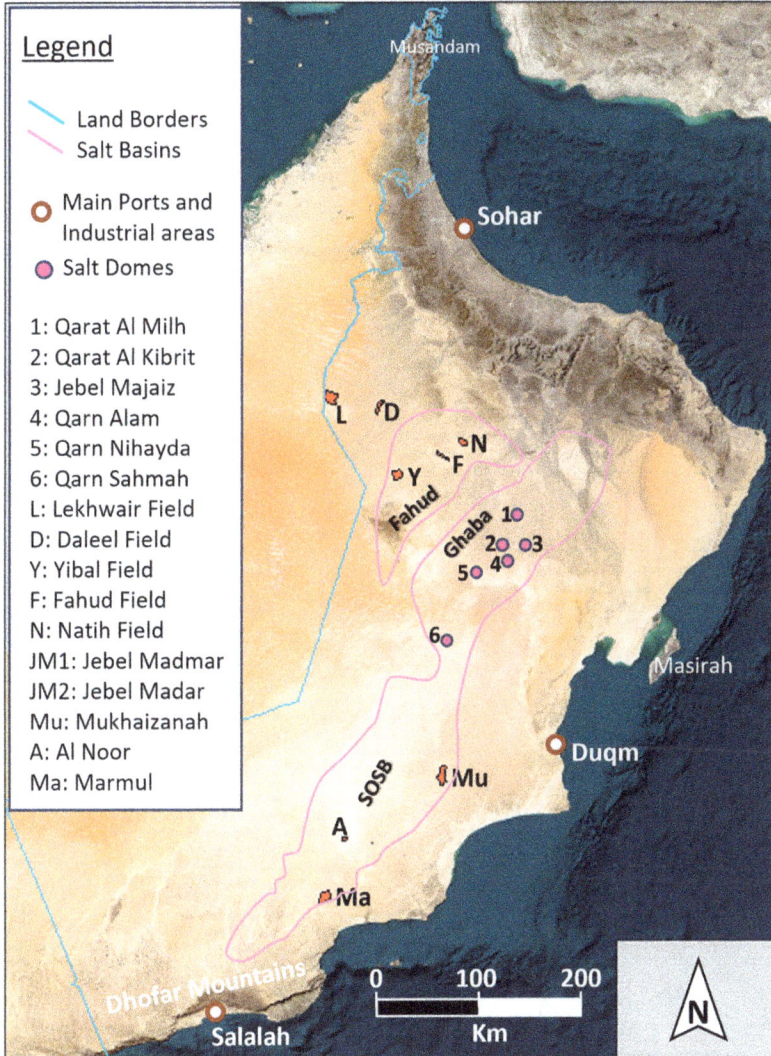

Figure 21.1: The salt basins (Fahud, Ghaba and SOSB) in Oman, with the locations of the surface-piercing salt domes and some of the main hydrocarbon. The map also shows the locations of six surface-piercing salt bodies in the Ghaba Salt Basin.

Among the most notable features of the Ara Group are the carbonate stringers, which play a crucial role in the region's hydrocarbon exploration and potential subsurface storage applications [5]. The carbonate stringers in the Ara Group are interbedded within the thick evaporitic layers of the Ara Salt (Figure 21.2). These stringers primarily consist of dolomite, limestone, and other carbonate minerals [7]. They were formed in a restricted marine environment, characterized by periodic flooding and evaporation, leading to the deposition of carbonates and subsequent encapsulation by evaporites. The carbonate stringers are typically found at various stratigraphic levels within the Ara Salt. They exhibit considerable lateral continuity and can vary in thickness from a few meters to several tens of meters. Structurally, these stringers are often associated with halokinetic movements, resulting in their significant folding and faulting [3]. In general, the Ara Salt deposits in Oman represent a significant geological resource with diverse industrial applications. Their extensive distribution, thickness, and favorable structural properties make them particularly suitable for UHS and hydrocarbon exploration.

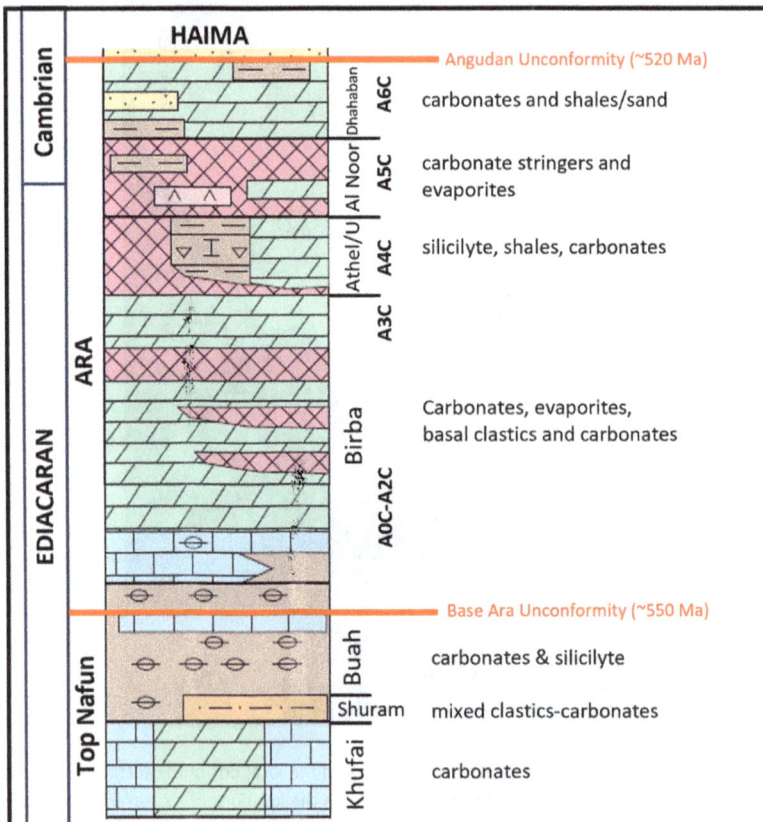

Figure 21.2: The stratigraphy of the Ara Group, with its salt and stringers (modified from [7, 10]).

21.1.2 The surface-piercing salt domes in Central Oman

The Ghaba Salt Basin of the Ara Group has six surface-piercing salt domes. These form significant geological features in Central Oman (Figure 21.1). These domes provide valuable geological insights and have significant potential for industrial applications. Their unique characteristics, such as thick salt layers, associated structural features, and carbonate stringers, make them valuable for hydrocarbon exploration and underground-storage considerations. This study explores the characteristics, geological significance, and potential use of the six major surface-piercing salt domes in Oman: Qarn Shamah, Qarn Alam, Qarat Kibrit, Jebel Majayiz, Qarat Al Milh, and Qarn Nihayda for UHS (Figure 21.3).

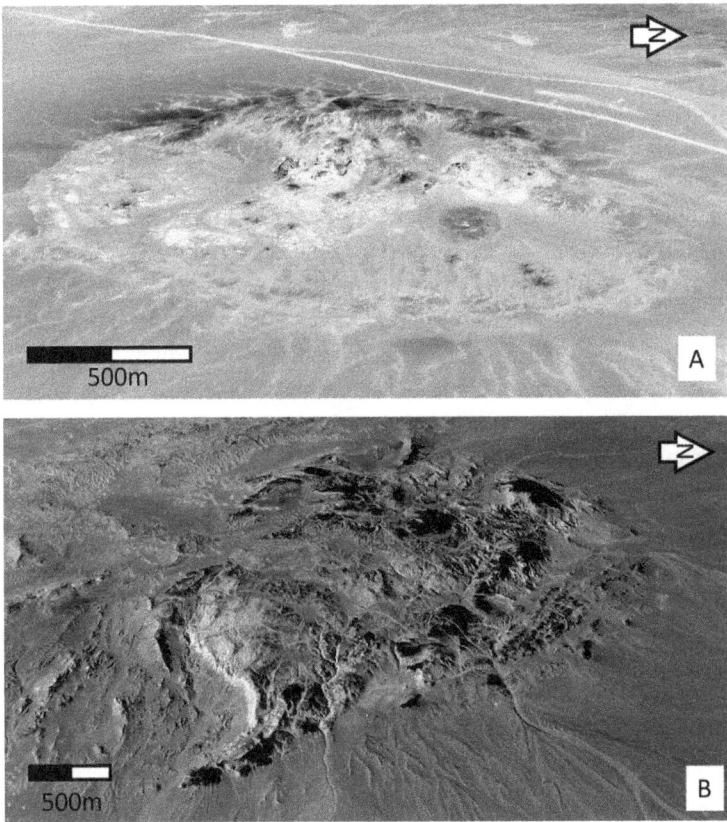

Figure 21.3: Oblique view of the surface expressions of (A) Qarat Al Kibrit and (B) Qarn Sahmah in the desert plain of Central Oman as shown on satellite images (Landsat). Qarat Al Kibrit has an elevated western edge with the middle part of the depression largely covered by evaporites. Qarn Sahmah salt dome has layered anhydrites which are stratigraphically interbedded with the stromatolitic dolostones and halite.

The surface-piercing salt domes in Central Oman have been studied by various authors [5, 10, 12, 13]. Large blocks of bedded Ara carbonates, called stringers or floaters,
are commonly exposed at the salt domes and form distinctive hills and ridges. These
exposed stringers consist mainly of carbonates, clastics (conglomerates, sandstones,
siltstones, and clays), volcanics, evaporites, and 'caprocks'. Halite is exposed in the
Qarn Sahmah, Qarat Kibrit, and Qarat Al Milh diapirs and is also present just below
the surface in shallow boreholes at Qarn Nihayda. Irregular, whitish anhydritic and
gypsiferous breccias, spongy residues and veins occur widely throughout the salt
domes, usually in close association with weathered evaporite bodies. Dissolution and
reprecipitation of evaporites and the dehydration of gypsum to anhydrite are interpreted as being common in and around the salt diapirs. The surface-piercing diapirs
are roughly circular to irregularly oval.

21.1.3 Hydrogen storage in salt bodies

Several studies have evaluated the potential of storing hydrogen in different parts of
the world. For example, Tarkowski and Czapowski [14] explored the feasibility of UHS
in Polish salt domes. These selections were based on geological and reservoir criteria,
such as the depth of the salt top (under 1 km), the minimal prior development of the
domes, and the thickness of salt bodies (preferably over 1,000 m). A comprehensive
analysis of 27 salt domes evaluated key geological parameters, including the penetration of overlying strata, cross-sectional dimensions, number of wells drilled, and estimated rock salt reserves. The study concluded that the undeveloped salt domes met
the criteria for leached salt caverns, providing advantages such as low permeability,
structural integrity and self-healing capabilities. While many salt domes have been
developed for mining, the study highlighted the potential for existing or planned
chambers to be repurposed for hydrogen storage. In a similar study, Liu et al. [9] investigated large-scale UHS sites in China. Salt caverns have significant advantages
over depleted reservoirs and aquifers for UHS. Key benefits include the salt's low permeability and porosity, engineering feasibility with large storage spaces (10–$100 \times$
10^4 m^3) at economically viable depths (600–2,000 m), and cost efficiency.

Cyran et al. [6] analyzed various examples of salt caverns used for UHS, focusing on
how geological factors influence their design and efficiency. The preferred cavern shape
is a vertical cylinder with a diameter of 50–80 m and a volume of 300,000–700,000 m^3.
Domal salt formations, such as those in Texas and Louisiana (USA) and Germany, provide the geotechnical simplicity necessary for constructing large caverns. In contrast,
bedded salt deposits, such as those in the Cheshire Basin (UK) and Jintan (China), require
tailored designs due to their thinner layers and interbedded impurities. These caverns
typically have volumes of 100,000–300,000 m^3 and utilize larger diameter-to-height ratios
to maximize storage capacity. Geological parameters like depth and salt layer thickness
significantly impact cavern stability and efficiency. Optimal depths for UHS caverns

range between 400 and 2,000 m. For instance, caverns in domal salt deposits like the Etzel salt dome (Germany) operate at depths of 900–1,200 m, while bedded salt caverns, such as those in Canada's Prairie Evaporite formation, reach depths of 1,800–1,900 m.

The studies by Aftab et al. [1] and Huang et al. [8] collectively provide a comprehensive understanding of the potential and challenges of UHS in geological formations. Aftab et al. [1] highlights the technical complexities of storing hydrogen in depleted reservoirs and salt caverns, emphasizing the need for robust trapping mechanisms and an understanding of hydrogen-mineral interactions. Meanwhile, Huang et al. [8] propose a systematic framework for selecting optimal UHS sites, using the Pingdingshan salt mine as a case study. These findings underscore the importance of further research to address the challenges of hydrogen leakage, microbial activity, and structural integrity.

21.2 Identifying the potential sites for UHS in the Ara Salt

The work presented here started by producing a reliable thickness map of the Ara Salt in the three salt basins in Oman: Fahud Salt Basin, Ghaba Salt Basin, and the SOSB. This was done by subtracting the seismic-interpreted top salt and base salt layers to identify the main salt bodies in these basins. The map in Figure 21.4 shows the thickness of the Ara Salt. The preliminary identification of the potential salt bodies for storage was based primarily on the availability of thick salt bodies and 3D seismic coverage for these salt bodies. Any salt body without 3D seismic is excluded from the first pass. Most of the identified prospects have existing petroleum wells either within the salt bodies or adjacent to them. Some of the prospects in South Oman do not have specific structural or field names. Therefore, they are identified as SOSB1 to SOSB8.

The main deliverables from the ranking are a shortlist of ranked salt bodies, an overview of identified prospects, and a preliminary assessment of the top four identified prospects. The screening included two main stages or passes:

- First-Pass Screening: A broad evaluation based on depth, volume, rock salt thickness, terrain, and proximity to hydrogen sources. In total 40 prospects are identified during the first past ranking across the basins of the Ara Salt (Figure 21.4 and Table 21.1).
- Second-Pass Screening: A detailed evaluation focusing on GRV, purity of salt, thickness, distribution of salt and stakeholders. Four sites were selected after the second phase of ranking.

Depth sections and salt thickness maps are used to analyze the prospects. An example of the cross sections is shown in Figure 21.5. Generally, the salt bodies in Central Oman are deep. Qarn Sahmah is the only shallow salt body in Central Oman that can

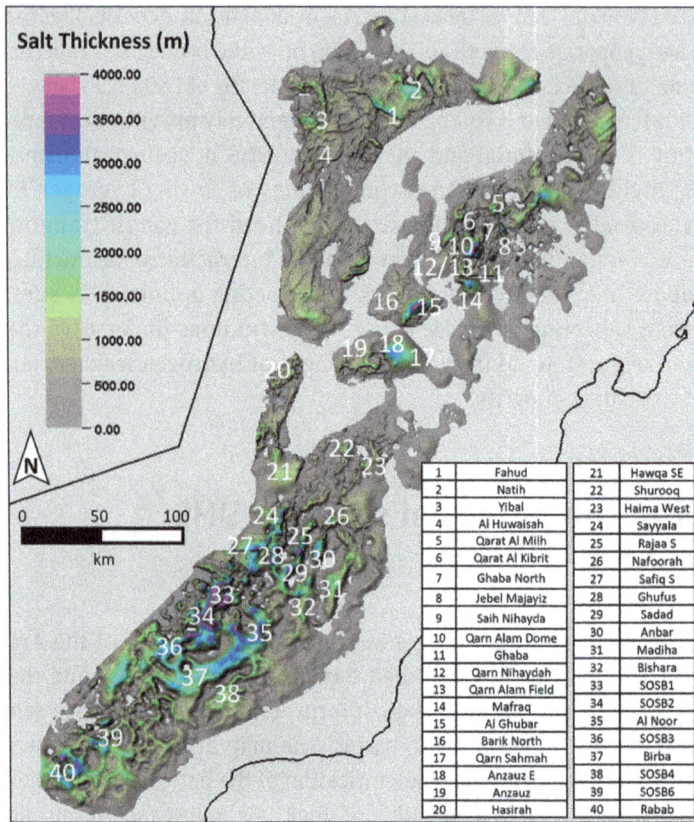

Figure 21.4: Salt thickness map with all the identified 40 prospects during the first-pass ranking.

be suitable for UHS. It is a surface-piercing salt dome which is only 150 km from Ad Duqum, see Figure 21.1, and a few kilometers from the current hydrogen concessions. The interpretation of the subsurface salt bodies is supported by the understanding of the salt tectonics and deformation styles in North, Central, and South Oman. Nearby drilled wells have been used in the past to constrain the top and base of Ara Group, as well as the top of Natih reflections. Most of the 40 screened salt bodies are deeper than 2,000 m.

Table 21.1: The 40 sites of salt bodies that were selected during the first pass of ranking and screening.

Ranking criteria definitions

Screening criteria	Element	Unsuitable (0)	Unfavorable (0.33)	Suitable (0.67)	Preferred (1)	Weight (%)
Capacity	Average depth of injection target	<300m or more than 2000m	300-500 or 1500-2000	500-800 or 1200-1500	800-1200 m	25
Capacity	Approximate GRV of salt body (m3)	<10⁷ m³	10⁷ - 10⁸ m³	10⁸ - 10⁹ m³	> 10¹⁰ m³	20
Containment	Rock Salt Thickness	<500m	500-750 m	500-1000 m	>1000 m	15
Containment	Distribution range of the target salt area (m2)	<5 km2	5-15 km2	15-35 km2	> 35 km2	15
Transport	(OR) Terrain	Wadi	wetlands/mountains/dunes/sab kha	forests / agricultural	not utilized	10
Transport	Distance to hydrogen source	>2,500 km (stepping)	Long distance shipping (<2,500 km)	Moderate distance pipelines (50-200 km)	Emitters co located in segment pipelines <50km	10
Stakeholders	Protection areas	Military protection areas (SSSI's)	> 80% nature protection areas (SSSI's)	80 – 30% nature protection areas (SSSI's)	<30% nature protecti on areas	5

Site scores

#	Site	Average depth (25)	Approx. GRV (20)	Rock Salt Thickness (15)	Distribution range (15)	Terrain (10)	Distance to H source (10)	Protection areas (5)	Total Score
1	Fahud	0.33	1	1	0.67	0.33	0.67	1	5
2	Natih	0							0
3	Yibal	0							0
4	Al Huwaisah	0							0
5	Qarat Al Milh	1	0.67	1	0.67	0.33	0.67	1	5.3
6	Qarat Al Kibrit	1	1	1	1	1	0.67	1	6.7
7	Ghaba North	0.33	0.67	1	0.67	1	0.67	1	5.7
8	Jebel Majayiz	1	1	1	1	1	0.67	1	6.7
9	Saih Nihayda	0							0
10	Qarn Alam Dome	1	1	1	1	1	1	1	7
11	Ghaba	0.67	1	1	1	1	1	1	6.7
12	Qarn Nihaydah	1	1	1	1	1	1	1	7
13	Qarn Alam Field	0							0
14	Mafraq	0							0
15	Al Ghubar	0							0
16	Barik North	0							0
17	Qarn Sahmah	1	1	1	1	1	1	1	7
18	Anzauz E	0							0
19	Anzauz	0							0
20	Hasirah	0							0
21	Hawqa SE	0							0
22	Shuroog	0							0
23	Halma West	0							0
24	Sawqah	0							0
25	Rajaa S	0							0
26	Nafoorah	0.33	1	1	1	1	1	1	6.3
27	Saih S	0							0
28	Ghufus	0							0
29	Sadad	0							0
30	Anbar	0.33	1	1	1	1	1	1	6.3
31	Madiha	0.33	1	1	1	1	1	1	6.3
32	Bishara	0							0
33	SOSB1	0							0
34	SOSB2	0							0
35	Al Noor	0.33	1	1	1	1	1	1	6.3
36	SOSB3	0.33	1	1	1	1	1	1	6.3
37	Birba	0							0
38	SOSB4	0							0
39	SOSB6	0.67	1	1	1	1	1	1	6.7
40	Rabab	0.33	1	1	1	1	1	1	6.3

Green is for the sites in North Oman, pink for the sites in South Oman, and blue for the sites in Central Oman. The site that is given zero value (red) is not further evaluated (grey cells).

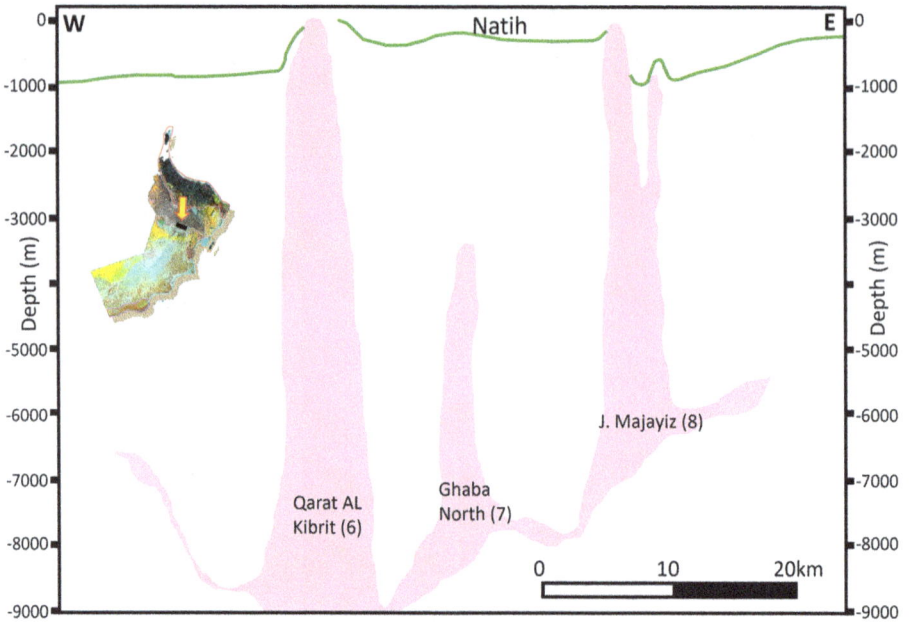

Figure 21.5: An example of a cross section, based on depth seismic data, passing through the salt bodies of Qarat Al Kibrit, Ghaba North, and Jebel Majayiz. The yellow arrow on the map points to the location of the section in Central Oman. The Late-Cretaceous reflection of the Natih Formation is shown here as a reference. Both Qarat Al Kibrit and Jebel Majayiz have a surface expression.

21.3 The ranking passes and their results

21.3.1 First-pass ranking

Table 21.1 summarizes the first-pass ranking for potential hydrogen storage sites. This ranking screened all of the identified 40 sites. The screening criteria are divided into various categories with specific elements that influence the ranking. Each element is evaluated based on predetermined conditions and the rankings are assigned accordingly. The elements assessed include the average depth of injection targets, approximate GRV of salt bodies, rock salt thickness, distribution range of the target salt area, distance to hydrogen sources and protection areas. The ranking is based on a general overview of the main criteria: capacity, containment, transport, and stakeholders. Each criterion includes in this ranking pass the main elements that help to rank the prospects and provide a good selection of the top selected ones.

For each element, there are four categories of ranking: unsuitable, unfavorable, suitable, and preferred. If the site receives an unsuitable ranking, it gets "0" value and therefore it turns red in the ranking list and will not be further screened in the following criteria. If the site is ranked unfavorable in a certain element, it will get "0.33," and it will get "0.67" if it is suitable and "1" if it is classified as preferred. Those ranking values will then be multiplied by the weight % of a certain ranking element to receive a weighted value for each element. The weight percentage of a certain element is based on its importance in ranking and select for a certain prospect. The more important the element is for ranking, the higher its weight percentage. For both first- and second-pass rankings, the highest weights (20–25%) are assigned to capacity and containment factors, particularly the depth, thickness, and lateral extent of the top salt body, as well as the proximity of fresh water aquifers or active faults. These elements are critical for distinguishing the most suitable target salt bodies for UHS and identifying any key showstoppers. Moderate weights (10–15%) are allocated to elements that could pose unforeseen containment risks or significantly increase hydrogen transportation costs, such as the presence of interlayers, seismic activity, and distance to the hydrogen source. Remaining elements are weighted at 5%, as they have minimal impact on the selection process or do not present major barriers to identifying optimal UHS candidates. The sum of all the weight percentages is 100%. The sum of the ranking, unweighted (without multiplications of the weight percentages) and weighted is shown at the bottom of the table.

The following list provided an overview of all the criteria, their elements and ranking categories:

Capacity – includes two main elements:
- – Average depth of injection target
- – Approximate GRV of salt body

Containment – includes two main elements:
- – Rock salt thickness
- – Distribution range of the target salt area (m^2)
- Transport – includes two main elements:
  - – Terrain
  - – Distance to a hydrogen source
- Stakeholders – includes one main element:
  - – Protection areas

The sites evaluated for hydrogen storage potential are ranked based on the above criteria. Each site has been assigned a score for each criterion, and these scores are used to determine their overall suitability, as described before. Generally, sites with thicker, shallower, and more extensive salt layers are preferred. Also, sites closer to hydrogen production facilities are more favorable due to reduced transportation costs and risks. The first-pass ranking is used for the following second-pass ranking. Out of 40 screened sites in the first-pass screening, 16 of them passed the screening, as shown in Table 21.2.

Table 21.2: List of the 16 sites that passed the first-pass screening, with their scores from the weighted values, as shown in the previous table.

No.	Prospect	Results	Ranking
10	Qarn Alam Dome	100	1
12	Qarn Nihaydah	100	2
17	Qarn Sahmah	100	3
6	Qarat Al Kibrit	96.7	4
8	Jebel Majayiz	96.7	5
11	Ghaba	91.75	6
39	SOSB6	91.75	7
30	Anbar	83.25	8
31	Madiha	83.25	9
35	Al Noor	83.25	10
36	SOSB3	83.25	11
40	Rabab	83.25	12
26	Nafoorah	83.25	13
5	Qarat Al Milh	78.45	14
7	Ghaba North (GN)	71.7	15
1	Fahud	5	16

21.3.2 Second-pass ranking

The second-pass ranking used the same main screening criteria: capacity (added to it solubility), containment, transport, and stakeholders. However, each criterion included more screening elements, compared to the first-pass ranking. Among the elements added in the second-pass ranking are:

- Capacity – the second-pass ranking included these extra elements:
    - The nature of the interlayer
    - Estimated geothermal gradient
    - Is salt present at the surface? (useful to sample the salt)
    - Average depth of injection target
- Containment – the second-pass ranking included these extra elements:
    - Fluid escape features and fissures
    - Lithology and thickness of roof
    - Lithology and thickness of floor
    - Faulting
    - Seismicity
- Transport – the second-pass ranking included these extra elements:
    - Reserved project areas
    - Distance to a power source
- Transport – the second-pass ranking included these extra elements:
    - Development/population

- Any existing petroleum operations in the structure
- Proximity to freshwater aquifers

The second-pass ranking uses the same approach for calculating the unweighted and weighted values to rank the prospects, as shown in Table 21.3. Following the second-pass ranking four sites were selected. These sites were: Qarat Al Kibrit, Al Noor, SOSB6, and Qarn Sahmah.

21.3.3 Preliminary evaluation of the identified four top prospects

The four selected prospects Qarat Al Kibrit, Al Noor, SOSB6, and Qarn Sahmah were further evaluated to assess the potential number of caverns that could be introduced within them. The volume of a single salt cavern can vary from 150,000 to 800,000 m^3 depending on the depth, pressure, geological features, and salt thickness. The energy density of salt caverns can vary between 214 kWh and 458 kWh/m^3 [4]. A salt cavern usually has a depth range of 600–2,000 m. This range is considered very suited and economical to store highly pressurized hydrogen [9]. According to Cyran [6], the target depth for salt caverns' location is between 400 and 500 m and depends on geological and mining conditions down to 2,000 m. The preferred cavern shape in the salt domes is a vertical cylinder several hundred meters high with a diameter of some 50–80 m. Such caverns have a volume of 300,000 m^3 to about 700,000 m^3 [6]. According to Plaat [11], caverns in bedded salt deposits usually have a volume in a range of 100,000 to 300,000 m^3. Van der Velde [15] argues that a typical hydrogen cave's size is 800,000 m^3 and is operated at pressures between 180 and 80 bar and contains 0.25 TWh working gas.

Given the parameters identified above, the following parameters are assumed for the sizes of salt caverns: the cavern will have a radius of 75 m and 250 m in height. Therefore, the volume of each cavern (cylindrically shaped) is approximately 4,400,000 m^3. The spacing distance between the boundaries of caverns is taken as 250 m. At 100 bar (10,000 kPa) and about 80 °C, a 5,000,000 m^3 salt cavern can store approximately about 30,000 tons of hydrogen. For example, if we need to store 500,000 tons of hydrogen per site, then we need around 15 salt caverns, assuming a vertical cylinder shape of 75 m and 250 m, spaced at 250 m between them. Looking back in more detail at the four selected salt domes Qarat Al Kibrit, Al Noor, SOSB6, and Qarn Sahmah to preliminarily assess their potential to accommodate at least 15 salt caverns, these sites have more than enough room to provide space and thickness to create such caverns at the desired depths. An example is given here for Qarat Al Kibrit (Figure 21.6).

Table 21.3: The 10 sites of salt bodies that were selected during the second pass of screening.

criteria	Element	Unsuitable (0)	Unfavorable (0.33)	Suitable (0.67)	Preferred (1)	Weight (%)	Q. Alam Dome (10)	Qarn Nihaydah (12)	Qarn Sahmah (17)	Qarat Al Kibrit (6)	Jebel Majayiz (8)	Ghaba (11)	SOSB6 (39)	Anbar (30)	Madiha (31)	Al Noor (35)
Capacity / Solubility	GRV of salt body (m3)	<10^7 m^3	10^7 - 10^8 m^3	10^8 - 10^9 m^3	> 10^9 m^3	20	0.67	0.67	0.67	0.67	1	0.67	1	1	1	1
	Ratio of interlayers (stringers) thickness vs. salt thickness	Stringers dominate the salt body (>40%)	Salt has thick/continuous stringers (20-40%)	Salt has thin stringers (10-20%)	Pure salt (<10%)	15	0.67	0.67	0.33	0.67	1	0.67	1	1	1	0.67
	The nature of the interlayer	Soluble content <5% or the porosity is large	The soluble content is from 5% to 10% and the porosity is small	The soluble content is from 10% to 15% and the porosity is small	The soluble content is >15% and the porosity is small	5	0.33	0.33	0.33	0.33	1	0.33	1	1	1	0.33
	Estimated Geothermal gradient	>40°C	>40°C	30-40°C	<30°C	5	0.67	0.67	0.67	0.67	1	0.67	1	1	1	0.67
	Is salt present at surface?	No	No	Only the stringers associated with salt	Yes	5	0.67	0.67	0.67	0.67	1	0.33	0.33	0.33	0.33	0.33
Containment	Average depth of injection target	<300m or deeper than 2000m	300-500 or 1500-2000	500-800 or 1200-1500	600-1200 m	25	1	1	1	1	–	1	0.67	1	1	1
	Rock Salt Thickness	<500m	500-750 m	750-1000 m	>1000 m	20	1	1	1	1	–	1	0.67	0.67	0.67	0.67
	Distribution range of the target salt area (km2)	<5 km2	5-15 km2	15-35 km2	> 35 km2	20	0.67	1	1	1	–	0.33	1	1	1	0.67
	Fluid escape features and fissures	Penetrating fissures through the salt developed	Fissures are developed but no penetrating fissures are formed	Few fissures are developed	Fissures are not developed	15	1	1	1	1	–	1	1	1	1	1
	Lithology and thickness of roof	-	Hard rock, thickness 30-50 m or tight rocks 50-100 m	Hard rock, thickness 50-100 m or tight rocks 100-200 m	Hard rock, thickness > 100 m or tight rocks > 200 m	5	1	1	1	1	–	1	1	1	1	1
	Lithology and thickness of floor	-	Hard rock, thickness 30-50 m or tight rocks 50-100 m	Hard rock, thickness 50-100 m or tight rocks 100-200 m	Hard rock, thickness > 100 m or tight rocks > 200 m	5	1	1	1	1	–	1	1	1	1	1
	Faulting	Recently active faults within 1 km around the area	There are major faulting within 1-5 km of the area	There are major faulting within 5-7 km of the area	There are no major faulting within 7 km of the area	20	0.67	0.67	0.67	0.67	0	0.67	1	1	1	1
	Seismicity	Frequent seismic events > 5 deg Richter scale	Infrequent seismic events < 5 deg Richter scale of deep origin	Rare seismic events of < 3 deg Richter scale	No recorded seismicity	15	1	1	1	1	–	1	0.67	0.67	0.67	1
Transport	(OR) Terrain	Wadi	wetlands / mountains/dunes/sabkha	forests / agricultural	not utilized	10	1	1	1	1	–	0.33	1	1	1	1
	Distance to hydrogen source (boundary of hydrogen concession)	>2,500 km (shipping)	>150 km (pipeline)/Long distance shipping (<2,500 km)	Moderate distance pipelines (50-200 km)	Emitters co located in segment / pipelines <50km	10	0.67	1	0.67	1	–	0.67	1	1	1	1
	Reserved Project Areas	Reserved Development Area	Proposed Development Area	-	No future plans in the area	10	1	1	1	1	–	1	1	1	1	1
	Distance to power source	>20km	>20km	5-20km	<5km	10	1	0.67	0.33	1	–	1	1	0.67	0.67	0.67
Stakeholders	Development / Population	>70% populated / built-up area	>50% of the area developed or planned	20-50% of the area developed or planned	Isolated installations / population centres	10	1	1	1	1	–	1	1	1	1	1
	Any existing petroleum Operations in the structure	Intensive operations across the salt body	Covers most of the salt body	Covers <50% of the salt body	None or they can be avoided	10	1	1	1	1	–	1	1	1	1	1
	Protection areas	Military	> 80% nature protection areas (SSSIs)	80-30% nature protection areas (SSSIs)	<30% nature protection areas	10	1	1	1	1	–	1	1	1	1	1
	Proximity to freshwater aquifers	In play segment	<5 km laterally or low-quality seal above	5-50 km laterally or medium quality seal above	>50km away or >100 m above shallowest store	20	1	1	1	1	–	0.67	1	1	1	1
					Total Score (unweighted)		18.3	18.3	18.3	20.0	0	16.6	19.3	19	19	19.6
					Total Score (weighted)	300	227	227	229	244	0	205	232	227	227	240

The extensive ranking shown here was developed during this study. The second-pass ranking is more thorough and includes many surface and subsurface aspects. The total score, if all criteria are 1, is 300. So, the values shown in the table below are out of 300. The total weighted score is the multiplication of the weight (%) of each criterion by the ranking (1, 0.67, and 0.3).

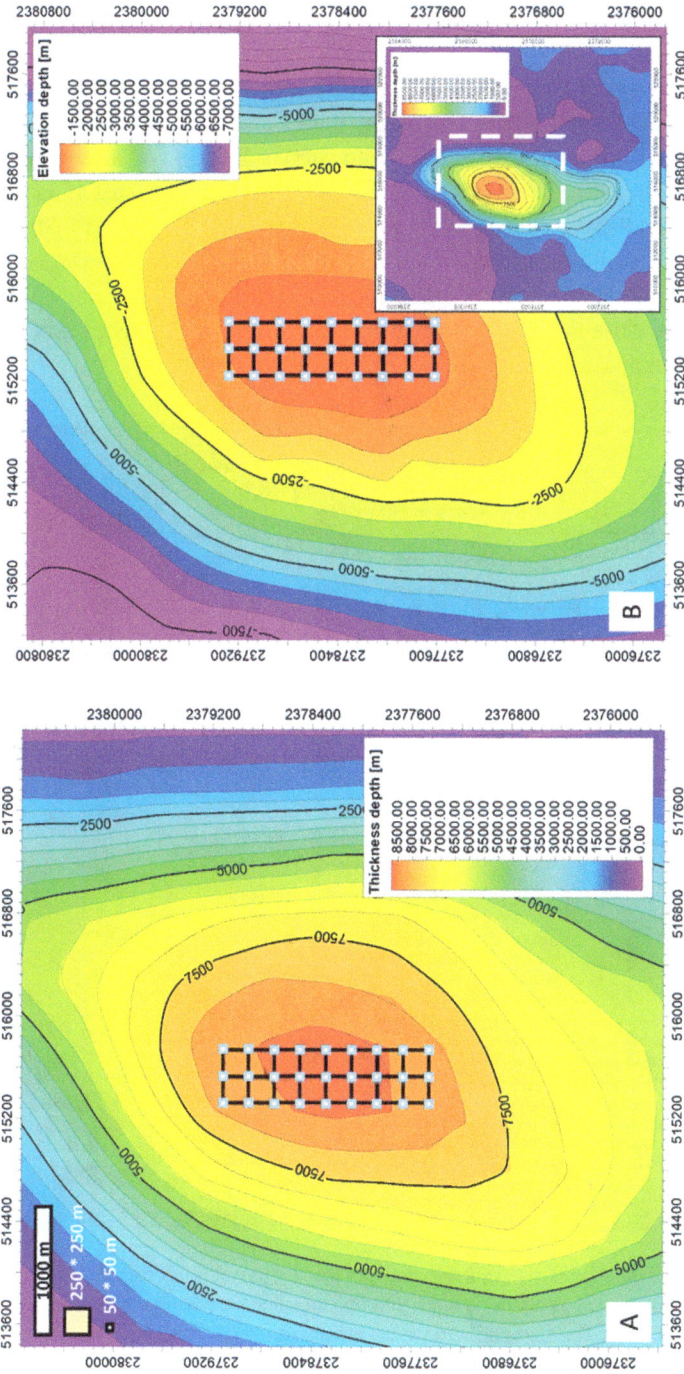

Figure 21.6: A potential grid arrangement of the salt caverns (with spacing of 250 m and radius of 75 m) greater than 2,000 m depth, on the (A) thickness map and (B) depth map of Qarat Al Kibrit UHS prospect. Note the scale in the coordinate grid numbers. The insert figure in (B) shows the full depth map of Qarat Al Kibrit and highlights the extent of the zoomed maps.

21.4 Final results and conclusions

Oman has a wide range of salt bodies spread across its three salt basins. In North and Central Oman, nearly all the salt bodies are deeper than 2 km, apart from those that pierce to the surface (six salt bodies). In South Oman, large salt bodies of several kilometers' width and length exist. However, only a few, mainly toward the east of South Oman Salt Basin, salt bodies are shallower than 2,000 m. The screening and rankings (first and second rankings) have resulted in the selection of four salt bodies (Qarat Al Kibrit, Al Noor, SOSB6, and Qarn Sahmah). The locations of salt bodies, as shown in Figure 21.7, seem to be close to the current hydrogen and renewable energy concessions defined across Oman.

#	Prospect	1st Pass Ranking
10	Qarn Alam Dome	1
12	Qarn Nihaydah	2
17	Qarn Sahmah	3
6	Qarat Al Kibrit	4
8	Jebel Majayiz	5
11	Ghaba	6
39	SOSB6	7
30	Anbar	8
31	Madiha	9
35	Al Noor	10

#	Prospect	2nd Pass Ranking
6	Qarat Al Kibrit	1
35	Al Noor	2
39	SOSB6	3
17	Qarn Sahmah	4

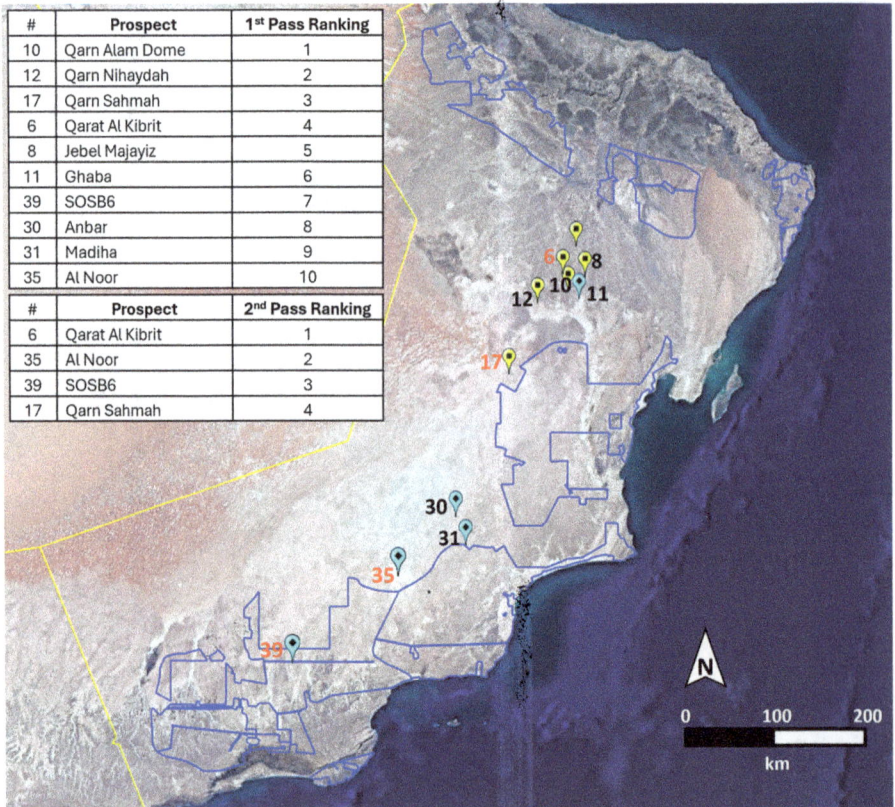

Figure 21.7: The locations of the 10 sites that were selected from the first-pass ranking, with the top four that were selected during the second-pass ranking highlighted as red numbers. The six surface-piercing salt domes are given yellow labels. The most northern salt dome was not selected from the first-pass ranking. The satellite map is from Landsat/Copernicus through GoogleEarth.

The evaluated salt bodies in Oman, including the top-ranked ones, demonstrate strong potential for UHS and compare favorably with global benchmarks. Situated at depths less than 2,000 m, these bodies offer excellent capacity, containment, and structural integrity, with massive storage volumes. Comparable to U.S. (e.g., Texas and Louisiana) and German (e.g., Etzel and Epe) sites, the identified sites in Oman provide similar depth ranges and storage capacities. The good purity of Ara Salt and minimal interlayers at the target storage parts of the salt bodies (top part of salt) give Oman a geological advantage over regions like China (Jintan Salt Formation), where interlayers challenge containment, and like the UK (Ceshire Basin), where the salt is thinly bedded. They also compare favorably to Poland's identified salt domes, which are less than 1,000 meters deep and feature similar geological conditions. Additionally, the proximity to existing and planned hydrogen concessions of the identified salt domes in Oman enhances economic feasibility, akin to some European examples.

With proper integration of advanced geological and engineering modeling and further economic and operational feasibility assessments, Oman's salt bodies could become globally competitive for large-scale hydrogen storage. However, the top-ranked salt bodies, as listed in this report, will require further evaluation to confirm their suitability for large-scale UHS. This future evaluation must include detailed geochemical analysis of the salt and its interlayers, utilizing the surface exposure of salt in the area, to identify any impurities that may impact on the containment reliability or may cause further complications in the process (e.g., presence of H_2S). Microbial and chemical lab analysis will need to test any possible contamination or microbial interaction. Moreover, advanced dynamic modeling techniques will be essential for simulating hydrogen injection, withdrawal, and storage over time. The long-term measurements and monitoring controls should be assessed by pilot tests. Such tests will validate the predicted sealing capacity and geomechanical integrity of the induced salt caverns, as forecasted by the subsurface models, in areas of interest. Moreover, given the potential for impacts on aquifers and other related local ecosystems, additional regulatory compliance in Oman and environmental impact assessments are critical.

Funding information: This research received funding from Hydrom Oman. The publishing of the paper is supported by Hydrom.

Conflict of interest: The authors declare that there are no conflicts of interest.

Data availability statement: The datasets used and/or analyzed during the current study are available from the corresponding author on reasonable request.

Authors' contribution: Study conception and design by Mohammed Al Kindi, Mohammed Al Hinai, Ahmed Al Abri, and Zaid Al Siyabi. Collection of data by Mohammed Al Kindi, Mohammed Al Hinai, and Ahmed Al Abri. Analysis, interpretation of data, and drafting of manuscript by Mohammed Al Kindi.

References

[1] Aftab A., Hassanpouryouzband A., Xie Q., Machuca L. L., Sarmadivaleh M. Toward a fundamental understanding of geological hydrogen storage. Industrial & Engineering Chemistry Research. 2022, 61(9): 3233–3253.

[2] Al-Kindi M. Overview of the carbon capture and storage opportunities in Oman. Geological Society, London, Special Publications. 2025, 550(1): SP550–2024.

[3] Al-Kindi M. H., Richard P. D. The main structural styles of the hydrocarbon reservoirs in Oman. Geological Society, London, Special Publications. 2014, 392(1): 409–445.

[4] Al-Rizeiqi N., Azzouz A., Liew P. Multi-criteria evaluation of large-scale hydrogen storage technologies in Oman using the analytic hierarchy process. Chemical Engineering Transactions. 2023, 106: 1117–1122.

[5] Al-Siyabi H. A. Exploration history of the Ara intrasalt carbonate stringers in the South Oman Salt Basin. GeoArabia. 2005, 10(4): 39–72.

[6] Cyran K. Insight into a shape of salt storage caverns. Archives of Mining Sciences. 2020, 65(2).

[7] Forbes G., Jansen H., Schreurs J. Lexicon of Oman subsurface stratigraphy. In: GeoArabia special publication 5. Bahrain: Gulf Petrolink, 2010, 371.

[8] Huang L., Fang Y., Hou Z., Xie Y., Wu L., Luo J., Wang Q., Guo Y., Sun W. A preliminary site selection system for underground hydrogen storage in salt caverns and its application in Pingdingshan, China. Deep Underground Science and Engineering. 2024, 3(1): 117–128.

[9] Liu H., Zhu Z., Yan Q., Yu S., He X., Chen Y., Zhang R., Ma L., Liu T., Li M., Lin R. A disordered rock salt anode for fast-charging lithium-ion batteries. Nature. 2020, 585(7823): 63–67.

[10] Peters J. M., Filbrandt J. B., Grotzinger J. P., Newall M. J., Shuster M. W., Al-Siyabi H. A. Surface-piercing salt domes of interior North Oman, and their significance for the Ara carbonate 'stringer' hydrocarbon play. GeoArabia. 2003, 8(2): 231–270.

[11] Plaat H. Underground gas storage: Why and how. Geological Society, London, Special Publications. 2009, 313(1): 25–37.

[12] Reuning L., Johannes S., Ansgar H., Urai J. L., Littke R., Kukla P. A., Rawahi Z. Constraints on the diagenesis, stratigraphy and internal dynamics of the surface-piercing salt domes in the Ghaba Salt Basin (Oman): A comparison to the Ara Group in the South Oman Salt Basin. GeoArabia. 2009, 14(3): 83–120.

[13] Schoenherr J., Urai J. L., Kukla P. A., Littke R., Schléder Z., Larroque J. M., Newall M. J., Al-Abry N., Al-Siyabi H. A., Rawahi Z. Limits to the sealing capacity of rock salt: A case study of the infra-Cambrian Ara Salt from the South Oman salt basin. AAPG Bulletin. 2007, 91(11): 1541–1557.

[14] Tarkowski R., Czapowski G. Salt domes in Poland–Potential sites for hydrogen storage in caverns. International Journal of Hydrogen Energy. 2018, 43(46): 21414–21427.

[15] Van der Velde F. Presentation: Large scale underground hydrogen storage. Presented in the Hydrogen Storage Seminar, by Casunie. Groningen, Netherlands, 2024.

Christopher Lagat

Chapter 22
Advancements and challenges in the transportation of natural hydrogen

Abstract: The transportation of natural hydrogen is pivotal in the global shift toward sustainable energy systems. This chapter examines the critical aspects of hydrogen transportation, focusing on the technical, economic, and regulatory dimensions that underpin its development. The discussion explores the main transportation pathways, high-pressure gaseous transport, low-temperature liquid transport, and material-based solid storage, and evaluates their respective benefits and challenges. Additionally, the chapter addresses the complexities of safety regulations and highlights the opportunities for innovation and international collaboration in scaling hydrogen transportation infrastructure. Through this analysis, the chapter aims to contribute to understanding how hydrogen can transition from a niche technology to a mainstream energy solution, fostering a global low-carbon economy.

Keywords: pipeline transportation, tube trailers, chemical carriers, liquid hydrogen, ammonia, policy frameworks, safety, policy, regulatory, sustainable energy

22.1 Introduction

One of the central challenges to integrating natural hydrogen into the energy market lies in its transportation. Unlike fossil fuels, which are relatively simple to transport due to their high energy density in liquid form, hydrogen presents significant logistical and technical barriers. Its low volumetric energy density in both gaseous and liquid states and its propensity for diffusion and leakage necessitate specialized infrastructure for safe and efficient transport [1, 2]. Effective hydrogen transportation infrastructure is critical for connecting production sites to end-use locations, whether for industrial processes, residential heating, or as a transportation fuel [3, 4]. The efficiency, cost, and safety of transportation methods are pivotal in determining hydrogen's competitiveness and feasibility as a global energy carrier [5, 6].

Hydrogen transportation relies on three main methods: high-pressure gaseous, low-temperature liquid, and material-based solid forms. Each presents distinct advantages and limitations that influence its suitability across applications. High-pressure

Christopher Lagat, WA School of Mines: Minerals, Energy and Chemical Engineering, Curtin University, Perth 6102, Australia, e-mail: christopher.lagat@curtin.edu.au

https://doi.org/10.1515/9783111437040-022

gaseous transport is ideal for short- to medium-range distribution due to its flexibility, but it demands costly compression systems and materials resistant to hydrogen embrittlement [1, 7]. Liquid hydrogen, which offers significantly higher energy density, requires cryogenic systems to maintain its boiling point of −253 °C, introducing complex engineering and economic challenges [5, 8]. Cryogenic cooling refers to the maintenance of hydrogen below its boiling point temperature allowing for more efficient storage and transport. Cryogenic systems are designed to maintain these low temperatures using advanced insulation, refrigeration, and vacuum technology to minimize heat transfer.

Material-based approaches, such as hydrogen carriers and metal hydrides, provide safer and more stable storage solutions but add complexities to hydrogen release and regeneration processes [4, 9].

Overcoming these challenges is essential to advancing hydrogen's role in the energy mix. Continued efforts in infrastructure innovation, materials science advancements, and the implementation of robust safety standards are required to enable efficient and secure hydrogen transportation across diverse contexts [2, 9]. Addressing these needs will determine the scalability of hydrogen transportation systems and their ability to support the growing demand for clean energy solutions.

To provide a comprehensive overview of hydrogen transportation, this chapter is organized into several key sections. Section 22.2 discusses the main distribution pathways and the current state of hydrogen transport infrastructure, including pipelines, tube trailers, and chemical carriers. Section 22.3 focuses on the technologies and materials required for safely compressing and transporting hydrogen in its gaseous form. This section will address the operational and safety concerns that accompany high-pressure hydrogen, as well as the economic considerations involved.

Section 22.4 delves into the cryogenic systems used to liquefy and transport hydrogen, covering the technical challenges and costs associated with this approach. Section 22.5 explores innovative storage methods that involve solid-state hydrogen carriers, such as metal hydrides and ammonia, which provide a safer, more stable means of transport but come with their unique limitations.

Section 22.6 provides an overview of the international standards and safety protocols necessary for hydrogen transport, highlighting historical incidents and outlining the regulatory landscape. This section will emphasize the importance of public perception in adopting hydrogen technologies, as safety remains a primary concern in its handling and transportation.

Finally, the chapter concludes with Section 22.7, which examines emerging technologies and potential policy measures that could facilitate safer, more efficient hydrogen transportation. This section will also address the need for global collaboration to standardize hydrogen transportation protocols and enhance cross-border distribution capabilities.

22.2 Hydrogen transportation mechanisms

This section will discuss the key methods for transporting hydrogen, focusing on pipelines, tube trailers, and chemical carriers. Each method will be analyzed in terms of its operational principles, advantages, limitations, and specific use cases within the hydrogen supply chain. The section will explore how pipelines enable large-scale and continuous hydrogen transport, the flexibility of tube trailers for smaller and decentralized demands, and the innovative use of chemical carriers for long-distance transportation. Additionally, the technical challenges, such as hydrogen embrittlement in pipelines, storage constraints in tube trailers, and energy-intensive processes for chemical carriers, will be addressed. This analysis aims to provide a comprehensive understanding of the critical role transportation mechanisms play in supporting the global hydrogen economy.

The transportation of hydrogen from production sites to consumption points is crucial for realizing its potential as a reliable and efficient energy source. Hydrogen distribution primarily relies on three key methods: pipelines, tube trailers, and chemical carriers. Each of these methods has distinct advantages and challenges, making them suitable for specific scenarios within the hydrogen supply chain [10].

Pipelines are the most efficient method for high-capacity hydrogen transport over established networks. Pipelines enable a continuous supply of hydrogen, making them ideal for large-scale industrial applications and long-term infrastructure. However, hydrogen pipelines require significant investment in infrastructure, including materials that resist embrittlement caused by hydrogen's interaction with metals under pressure. Embrittlement refers to the process by which a steel pipeline loses its ductility and becomes brittle, leading to a higher likelihood of cracking or failure under stress. In the case of hydrogen embrittlement, hydrogen atoms penetrate the metal's structure, diffusing into its lattice. This weakens the metal by reducing its ability to deform, thereby forming microcracks that can grow over time. The effect is particularly pronounced under high pressure or repeated stress, which makes it a significant challenge for hydrogen pipelines and infrastructure, as the materials used must resist this form of degradation. Additionally, constructing pipelines is geographically constrained to areas with high and stable hydrogen demand [1, 10].

Tube trailers, which transport compressed hydrogen in cylinders, offer flexibility for medium-distance transportation. They are particularly valuable for supplying smaller markets or areas without access to hydrogen pipelines. However, tube trailers are less efficient for large-scale transport due to their limited storage capacity and higher transportation costs. Their application is often limited to short-term or pilot-scale hydrogen projects [4, 11].

Chemical carriers, such as ammonia and liquid organic hydrogen carriers (LOHCs), represent a novel approach to hydrogen transport. These carriers chemically bond with hydrogen, enabling safe and efficient long-distance transportation. For example, ammonia, which can be easily liquefied and transported under mild conditions compared to liquid hydrogen, allows for hydrogen storage in a compact form. At the destination, hydrogen can be released from the carrier through a chemical process, although this requires additional energy and specialized infrastructure [10, 12].

Each transportation method comes with trade-offs in cost, efficiency, and scalability. Chemical carriers, while promising for long-distance and intercontinental transport, face technical challenges in optimizing hydrogen release and minimizing energy losses during processing [10].

As the global hydrogen economy continues to develop, research and innovation will play a vital role in enhancing the efficiency, cost-effectiveness, and scalability of these transportation methods. Establishing robust safety protocols and international standards will also be essential for secure hydrogen transportation across diverse applications and regions [11].

22.2.1 Pipeline transportation

Hydrogen pipelines, adapted from existing natural gas infrastructure, are an efficient and reliable means of transporting large volumes of hydrogen over long distances. Pipelines are particularly advantageous in established hydrogen economies, such as Europe and the United States, where networks of hydrogen pipelines are expanding to support industrial applications and facilitate the transition to low-carbon energy systems [13, 14].

The design and operation of hydrogen pipelines pose unique challenges due to the specific properties of hydrogen. Its small molecular size makes it prone to diffusion and leakage, which can result in hydrogen embrittlement. This issue is particularly critical for pipeline materials, as hydrogen embrittlement can weaken metals, leading to cracks and failures. To mitigate these risks, pipeline materials must possess high strength, toughness, and resistance to hydrogen-induced degradation. Commonly used materials include high-strength low-alloy steels, such as API 5L X52 or X60 grades, which offer a balance of durability and cost-effectiveness [15, 16].

Coatings and protective layers further enhance the safety and longevity of hydrogen pipelines. Internal coatings, such as those made from zinc-aluminum alloys, provide resistance to corrosion and minimize hydrogen permeability, ensuring the safe transport of hydrogen under high pressures. External coatings, typically composed of polyethylene or epoxy, protect the pipeline from external environmental factors such as soil corrosion [14, 17].

The flow rate of hydrogen through a pipeline can be modelled using the Darcy-Weisbach equation for compressible fluids:

$$Q = \frac{\Pi D^2}{4} . v$$

where Q is the volumetric flow rate of hydrogen (m³/s), D is the internal diameter of the pipeline (m), v is the flow velocity of hydrogen (m/s).

The efficiency of hydrogen pipelines depends on optimal operating conditions, including pressure and flow management. Hydrogen is typically transported at pressures ranging from 10 to 20 MPa, with compression stations placed at intervals to compensate for pressure drops caused by friction and thermal losses along the pipeline. Proper pipeline diameter and wall thickness are also crucial for minimizing energy losses and maintaining structural integrity over long distances [15]. Diameters are typically specified using nominal pipe size (NPS) in inches or diameter nominal (DN) in millimeters, with wall thicknesses categorized by schedules (e.g., Schedule 40 or 80). Higher schedules indicate thicker walls.

Hydrogen pipeline networks are evolving to support the growing hydrogen economy. Initiatives such as the European Hydrogen Backbone project aim to repurpose existing natural gas pipelines for hydrogen transport, reducing the need for new infrastructure and lowering costs. This approach highlights the potential for hydrogen pipelines to serve as a cornerstone of future energy systems by facilitating the efficient and scalable transport of hydrogen across regions [14, 17].

Existing pipelines may need retrofitting to handle hydrogen's higher diffusivity and pressure requirements, including thorough assessments of seals, welds and joints for potential leaks. Hydrogen's lower ignition energy and broader flammability range demand stricter safety protocols, along with advanced leak detection systems and enhanced safety management practices. Additionally, its lower energy density compared to natural gas necessitates higher flow rates (for the same energy) or the blending of hydrogen with natural gas to sustain energy supply levels, with blending currently being explored as an intermediate solution. According to the European Hydrogen Backbone study, the cost of repurposing existing natural gas pipelines for hydrogen use, including activities such as decommissioning from natural gas service, conducting water pressure tests, upgrading fittings, and removing old connections, is estimated to be approximately 10–15% of the cost required to construct entirely new hydrogen pipelines (Table 22.1).

However, operating hydrogen pipelines is anticipated to be more costly compared to natural gas pipelines. This is primarily because hydrogen requires about three times more compression power to achieve an equivalent energy flow rate as natural gas.

The study further highlights that the total capital expenditure per kilometer for repurposed hydrogen pipelines is roughly 33% of the cost associated with building new hydrogen infrastructure from scratch [18].

Hydrogen pipeline design is shaped by the material challenges posed by hydrogen's interaction with metals. Lower-strength steels are commonly used to address hydrogen embrittlement. Standards such as ASME B31.12 and AIGA guidelines specify stringent requirements for material selection, operating stresses, and fracture control to ensure safety. Surface treatments, coatings, and claddings are employed to minimize hydrogen penetration, although their effectiveness may vary depending on conditions. Design parameters often include reduced allowable stress levels compared to

Table 22.1: Investment costs for repurposing natural gas versus new hydrogen pipeline in 2020 [18].

Pipeline type	Pipeline diameter (inches)	Cost per km (USD)		
		Low	Average	High
Repurposing natural gas pipeline	<28	228,000	342,000	570,000
New hydrogen pipeline		1,596,000	1,710,000	2,052,000
Repurposing natural gas pipeline	28–37	228,000	456,000	570,000
New hydrogen pipeline		2,280,000	2,508,000	3,078,000
Repurposing natural gas pipeline	>37	342,000	570,000	684,000
New hydrogen pipeline		2,850,000	3,192,000	3,876,000

Note: It is assumed that the average exchange rate for EUR and USD was approximately 1 EUR = 1.14 USD in 2020.

natural gas pipelines and detailed evaluation of pre-existing defects to prevent crack growth. Additionally, hydrogen's effects on properties like fatigue crack growth and fracture toughness necessitate specialized testing and conservative design practices to maintain pipeline reliability and integrity.

22.2.2 Tube trailers

Tube trailers provide significant flexibility in hydrogen transportation, as they can be routed to meet demand without the need for permanent infrastructure. This characteristic makes them particularly useful for pilot hydrogen projects, remote industrial sites, and as a short-term solution in areas where hydrogen demand does not yet justify the investment in pipeline infrastructure [19]. By enabling on-demand delivery, tube trailers play an essential role in early-stage hydrogen market development, supporting diverse applications ranging from fueling stations to small-scale industrial operations [20].

However, tube trailers have limitations that constrain their use for large-scale or long-distance hydrogen transport. The most prominent challenge is their limited volume capacity. Even with advanced materials like type IV composite pressure vessels, the payload is restricted by trailer size and road weight limits. For example, current designs can transport approximately 640–1,000 kg of hydrogen, depending on the pressure and material used [20, 21]. High compression costs are another significant factor, as hydrogen must be stored at pressures of up to 250 bar (or higher) to optimize the payload while minimizing the trailer weight [15].

Safety is a critical consideration in the use of tube trailers for hydrogen transport. High-pressure hydrogen storage entails risks such as leaks, material fatigue, and pressure-related hazards during loading and unloading. To mitigate these risks, trailers are equipped with advanced safety features, including pressure relief devices and robust containment systems. Additionally, rigorous standards such as the International Organisation for Standardisation (ISO) guidelines and Department of Transportation (DOT) regulations govern the design, testing, and operation of tube trailers to ensure safe handling [20, 22].

Despite these challenges, tube trailers remain an indispensable option for hydrogen delivery, particularly in emerging markets and regions without fixed infrastructure. Continued advancements in materials and designs, such as lightweight composite tanks and higher-pressure capabilities, promise to enhance the efficiency and safety of tube-trailer operations in the hydrogen supply chain [21, 22].

22.2.3 Chemical carriers

Chemical carriers such as ammonia, methanol, and LOHCs present an innovative and efficient approach to hydrogen transportation. These carriers allow for hydrogen to be chemically bonded, transported safely, and released at the destination, overcoming challenges associated with hydrogen's low density and high reactivity [23].

The synthesis of ammonia as a hydrogen carrier follows the Haber-Bosch process:

$$N_2 + 3H_2 \xrightarrow{\text{yields}} 2NH_3$$

Ammonia is a well-established hydrogen carrier due to its high hydrogen content (17.6% by weight) and ease of liquefaction under moderate pressure and temperature conditions. Ammonia can be transported using existing liquefied natural gas (LNG) infrastructure, making it a viable option for large-scale and long-distance hydrogen transport. At the destination, ammonia can be catalytically decomposed to release hydrogen for use in industrial processes or as a fuel [24]. However, challenges such as ammonia toxicity and the energy requirements for decomposition remain critical considerations [25].

LOHCs, such as dibenzyltoluene and N-ethylcarbazole, represent another promising class of hydrogen carriers. LOHCs allow hydrogen to be stored in a chemically stable liquid form at ambient conditions, facilitating safe and efficient transport. The hydrogenation process binds hydrogen to the carrier compound, while the dehydrogenation process releases hydrogen for use (Figure 22.1). LOHC systems are particularly advantageous because they can utilize existing liquid fuel infrastructure, reducing the need for new investments in specialized facilities. Furthermore, the absence of hydrogen boil-off during storage enhances their suitability for long-term and international hydrogen trade [26, 27].

Methanol is another carrier with high versatility. It is liquid at ambient conditions and can be synthesized from hydrogen and carbon dioxide, providing an integrated pathway for carbon capture and utilization. Methanol's established infrastructure and its role as a precursor in various chemical industries further support its potential as a

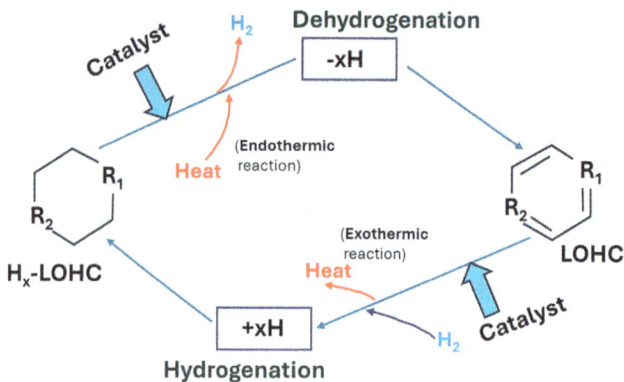

Figure 22.1: Liquid organic hydrogen carrier.

hydrogen carrier. However, the carbon emissions associated with methanol production and utilization need to be addressed to align with global decarbonization goals [28].

The adoption of chemical carriers in hydrogen transportation is gaining traction as countries invest in hydrogen economies. These carriers enable scalable and efficient hydrogen logistics, addressing key barriers such as storage, safety, and cost. However, ongoing research is required to optimize catalytic systems for hydrogen release, minimize energy losses, and mitigate environmental impacts associated with their production and use [1].

22.3 High-pressure gaseous hydrogen transportation

22.3.1 Compression technologies

High-pressure gaseous hydrogen transportation remains one of the most straightforward and widely adopted methods, particularly for short distances and moderate demand levels. Hydrogen is compressed into high-pressure tanks, pipelines or tube trailers, typically at pressures ranging from 200 to 700 bar [19]. This method's simplicity and flexibility have made it a critical component of early-stage hydrogen infrastructure. However, compression is energy-intensive, with estimates suggesting that compressing hydrogen to 700 bar can consume 10–15% of its total energy content [29].

The compressibility of hydrogen at high pressures is described by the real gas equation:

$$PV = ZnRT$$

where P is the pressure (Pa), V is the volume (m³), Z is the compressibility factor (varies with pressure and temperature), n is the number of moles of hydrogen, R is the gas constant (8.314 J/mol·K), and T is the temperature (K).

The real gas equation, also known as the Van der Waals equation, is an improved form of the ideal gas law that considers the behavior of real gases. Unlike ideal gases, real gases exhibit deviations due to the finite size of their molecules and the presence of intermolecular forces.

Modern compression technologies include reciprocating compressors, diaphragm compressors and emerging ionic compressors. Reciprocating compressors, which use pistons to compress hydrogen, are widely utilized due to their reliability and ability to achieve high pressures. Diaphragm compressors, on the other hand, rely on flexible diaphragms and are preferred in applications requiring ultra-high purity, as they minimize contamination risks [30]. Recently, ionic compressors, which use ionic liquids instead of traditional pistons, have shown promise in reducing mechanical wear and achieving higher compression efficiencies, making them a potential game-changer for hydrogen compression [31].

The choice of compression technology depends on various factors, including the required pressure, hydrogen volume, and application. These advancements in compression technologies are crucial for optimizing the cost and energy efficiency of high-pressure gaseous hydrogen transportation [19].

22.3.2 Design and structural requirements

Transporting hydrogen at high pressures necessitates specialized materials and structural designs to ensure safety and reliability. A significant challenge is hydrogen embrittlement, a phenomenon where hydrogen atoms diffuse into metals, causing cracks and compromising structural integrity. This risk is particularly high in traditional steel pipelines and tanks, making it necessary to use advanced materials such as high-strength stainless steel and composites [32].

Composite materials, particularly carbon-fiber-reinforced polymer (CFRP), are widely used in high-pressure hydrogen tanks due to their lightweight and superior strength-to-weight ratio. Type IV tanks, consisting of a polymer liner wrapped with carbon fiber, are capable of withstanding pressures up to 700 bar, making them ideal for transportation and storage applications [29]. Similarly, pipelines for high-pressure hydrogen transport are designed with specialized coatings and linings to mitigate embrittlement and corrosion. Valves and seals in high-pressure systems also require precise engineering to prevent leaks, which can lead to energy losses and safety risks [12].

The development of advanced materials and designs has significantly improved the feasibility and safety of high-pressure hydrogen transportation, making it a viable option for various industrial and commercial applications.

22.3.3 Challenges and limitations

While high-pressure gaseous hydrogen transportation offers flexibility and simplicity, it faces several challenges, including energy efficiency, economic cost, and safety concerns. Compressing hydrogen to high pressures consumes significant energy, which reduces the overall efficiency of the hydrogen supply chain. For instance, maintaining consistent pressure levels during storage and transport often requires multiple compression stages, further increasing energy use [29].

Economic costs are another significant barrier. The infrastructure required for high-pressure hydrogen transport, including compression facilities, pipelines, and storage tanks, is costly to install and maintain. Additionally, materials such as carbon fiber used in composite tanks are expensive, driving up the capital and operational costs of this method [19, 32]. Regular inspections and maintenance to ensure structural integrity and prevent leaks further add to the expense.

Safety remains a critical concern. Hydrogen's flammability and wide ignition range (4–75% in air) make it particularly susceptible to combustion in the event of a leak. The rapid release of compressed hydrogen can create jet flames, which are difficult to control and extinguish. As a result, high-pressure storage and transport systems require robust safety measures, including leak detection systems, automatic shut-off valves and explosion-proof equipment [53]. While these measures mitigate risks, they also introduce additional costs and complexities to the overall system.

22.3.4 Comparative analysis with other methods

Compared to other hydrogen transportation methods, high-pressure gaseous transport provides unmatched flexibility and responsiveness, particularly for short-distance and regional applications.

However, for long-distance or high-capacity transport, high-pressure gaseous hydrogen transport is less favorable compared to liquid hydrogen or chemical carriers. These alternatives offer higher energy densities and are more efficient for intercontinental trade and large-scale applications. While high-pressure transport remains an essential component of the hydrogen economy, its role will likely be complemented by other methods to address varying scales and distances [33].

22.4 Low-temperature liquid transportation

22.4.1 Liquefaction process

Transporting hydrogen in liquid form provides significant advantages in terms of energy density, enabling more hydrogen to be stored in a given volume compared to its gaseous counterpart [34]. However, the process of liquefying hydrogen is highly energy-intensive, requiring advanced cryogenic systems to cool hydrogen to −253 °C, the temperature at which it becomes a liquid. This extreme cooling process involves multiple stages, including compression, precooling, cryogenic cooling, and expansion, each demanding considerable energy input. Estimates indicate that hydrogen liquefaction can consume up to 30–40% of the hydrogen's total energy content, significantly impacting its overall efficiency [35].

The liquefaction process begins with compression, where hydrogen gas is pressurized, increasing its temperature. This is followed by precooling, which utilizes refrigerants such as nitrogen or helium to lower the hydrogen's temperature to the cryogenic range. Finally, cryogenic cooling is achieved using heat exchangers and Joule-Thomson expansion processes to bring the hydrogen to its liquefaction point [36]. Innovations in cryogenic technologies, such as magnetic refrigeration and mixed-refrigerant cycles, are being explored to improve efficiency, though they are still in the early commercial stages [31].

The energy required for liquefying hydrogen is given by:

$$Q = m.L_V$$

where Q is the energy required (J), m is the mass of hydrogen (kg), and L_V is the latent heat of vaporization for hydrogen (449 J/g).

Advances in thermal insulation and heat recovery systems are also being pursued to reduce the energy costs of liquefaction, potentially making liquid hydrogen a more viable solution for long-distance transportation [34].

22.4.2 Transportation vessels and insulation

Transporting liquid hydrogen requires specialized cryogenic vessels to maintain the ultra-low temperatures needed to prevent evaporation or "boil-off." Boil-off refers to the vaporization of hydrogen due to heat transfer, which can result in the loss of 1–3% of stored hydrogen per day during transit [35]. To mitigate this, advanced insulation systems, such as multilayer insulation and vacuum insulation, are used to minimize thermal conductivity.

Cryogenic storage tanks are typically double-walled and constructed from materials like stainless steel or aluminum, which are compatible with cryogenic temperatures and have low thermal conductivity. The space between the walls serves as a vacuum-insulated layer, reducing heat transfer. For long-distance transportation, liquid hydrogen is transported in ISO containers mounted on trucks or in specialized cryogenic tankers [37].

Intercontinental hydrogen transport, particularly via shipping, relies on vessels equipped with cryogenic tanks similar to those used for liquefied natural gas. These vessels require enhanced insulation due to hydrogen's lower liquefaction temperature. Boil-off management systems are often integrated into these vessels to handle vaporized hydrogen, either by safely venting it or recapturing and reliquefying the gas [38]. These systems ensure that liquid hydrogen maintains its energy content during transport over long distances.

22.4.3 Challenges of liquefaction

Despite its advantages, liquid hydrogen transportation faces several challenges. The energy-intensive nature of the liquefaction process is a primary drawback, consuming nearly one-third of the hydrogen's total energy content. This inefficiency affects the economic viability of liquid hydrogen, particularly in regions with high energy costs or limited renewable energy availability [36].

Boil-off losses during storage and transport present another challenge, especially for long-duration or intercontinental transport. Even with advanced insulation, heat transfer inevitably causes some hydrogen to vaporize. This issue is particularly pronounced in warmer climates, where boil-off can significantly reduce the hydrogen available for end use [37]. Current estimates indicate that liquefied hydrogen carriers experience boil-off losses ranging from 0.3% to 1.0% per day. Over a 10- to 12-day voyage, these losses can total approximately 10% of the cargo, particularly when operating in warmer climates. This is comparable to the boil-off rates observed in LNG transport systems (Kawasaki Heavy Industries, 2020; Hydrogen Council, 2021; IEA, 2022). These losses are primarily driven by hydrogen's extremely low boiling point, its low energy density, and the surrounding temperature conditions. Similar to LNG contracts, provisions for boil-off losses can be included in commercial agreements to account for the anticipated vaporization during transportation. While active cooling systems and advanced insulation materials mitigate these losses, they add to the cost and complexity of liquid hydrogen transport systems.

Safety concerns also play a critical role. Liquid hydrogen's extreme cryogenic temperature and rapid phase change upon contact with warmer environments pose risks such as frostbite and structural damage to containment systems. Moreover, hydrogen's flammability and wide ignition range (4–75% in air) make it highly susceptible to combustion in the event of a leak. To address these issues, liquid hydrogen transport systems are equipped with pressure relief valves, explosion-proof equipment, and monitoring systems to ensure safety during transit [38].

22.4.4 Comparative analysis

Liquid hydrogen transportation offers significant advantages in terms of volumetric efficiency, making it suitable for high-capacity, long-distance applications. Its high energy density allows for compact storage and makes it ideal for intercontinental trade and large-scale industrial use. By converting hydrogen to its liquid form, substantial volumes can be transported in a single shipment, reducing logistical complexity [31].

However, the high energy costs associated with liquefaction and boil-off losses make liquid hydrogen less competitive than other transport methods in certain scenarios. For shorter distances, gaseous hydrogen transport is often more economical due to lower operational costs.

While liquid hydrogen transport remains a key component of the hydrogen supply chain, its long-term viability will depend on technological advancements in cryogenic systems, boil-off management, and safety protocols. These innovations will enhance the efficiency and cost-effectiveness of liquid hydrogen transport, ensuring its role in the global hydrogen economy as demand for large-scale and intercontinental distribution grows.

22.5 Material-based solid transportation

22.5.1 Hydrogen storage materials

Material-based solid hydrogen transportation provides an innovative alternative to gaseous and liquid hydrogen storage by addressing key challenges related to safety, stability, and energy efficiency. This method involves storing hydrogen in solid materials, such as metal hydrides, chemical hydrides, and carbon-based materials. These materials absorb or chemically bond with hydrogen, releasing it on demand without the need for high-pressure or cryogenic systems (Figure 22.2). This approach enhances safety and cost-effectiveness, particularly for applications requiring portability and stability [39].

Metal hydrides are a prominent class of hydrogen storage materials. They store hydrogen by forming bonds with metals such as magnesium, palladium, or sodium, which absorb hydrogen under moderate pressures and temperatures and release it upon heating. Chemical hydrides, such as ammonia borane and sodium borohydride, chemically bind hydrogen, offering high storage densities but requiring catalytic or thermal processes for hydrogen release. Carbon-based materials, including metal-organic frameworks (MOFs) and activated carbons, use their high surface areas to adsorb hydrogen molecules, although their storage capacities are typically lower than those of hydrides [40].

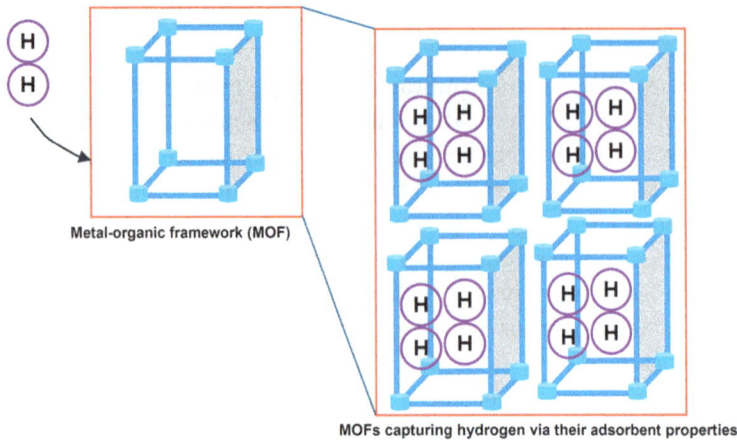

Figure 22.2: MOFs capturing hydrogen via their adsorbent properties.

22.5.2 Mechanisms of hydrogen absorption and release

The mechanisms for absorbing and releasing hydrogen depend on the material used. Metal hydrides absorb hydrogen through a bonding process with metal atoms, forming stable compounds. Hydrogen is released by heating the material, breaking these bonds and enabling controlled release. This process is advantageous for applications requiring consistent and precise hydrogen flow, such as fuel cells [39].

In chemical hydrides, hydrogen is stored in chemical bonds within the compound. Hydrogen release from these materials typically involves catalytic or thermal reactions, which can generate by-products requiring management or recycling. For example, ammonia borane releases hydrogen through thermal decomposition but produces by-products that must be handled to maintain system efficiency. Carbon-based materials, such as MOFs, adsorb hydrogen through physical processes, relying on high surface areas and pore structures. While these materials are effective at low pressures, their storage capacity remains a limitation, particularly for large-scale applications [41].

22.5.3 Advantages and challenges

Material-based hydrogen storage offers several advantages over traditional gaseous and liquid storage methods. By embedding hydrogen in solid materials, risks such as leakage, diffusion, and explosion are significantly reduced. This safety enhancement eliminates the need for high-pressure tanks or cryogenic systems, making material-based storage particularly suitable for applications prioritizing portability and safety [42].

However, there are notable challenges that limit the widespread adoption of this technology. The weight of many metal hydrides, such as magnesium hydride, impacts the overall energy density of the storage system, reducing its practicality for transportation applications where weight is a critical consideration. Furthermore, the absorption and release processes in metal hydrides require specific temperature and pressure conditions, limiting their flexibility in dynamic environments [43].

Chemical hydrides, while offering high storage densities, present challenges related to byproduct management and the complexity of hydrogen release processes. These factors increase operational costs, particularly in applications requiring rapid or continuous hydrogen supply. Similarly, carbon-based materials, though lightweight and versatile, currently have limited storage capacities, necessitating further advancements to achieve commercial viability [41, 44].

22.5.4 Applications and current research

Material-based hydrogen storage continues to attract significant research interest due to its potential to provide safe and stable hydrogen storage solutions. Metal hydrides are already utilized in stationary power applications, including backup power systems and microgrids, where weight and rapid hydrogen release are less critical. Their controlled hydrogen release capabilities make them well-suited for consistent energy demands [39].

Research efforts are focused on improving the hydrogen storage capacities, reducing the weight, and enhancing the release kinetics of these materials. For example, lightweight metal alloys, such as aluminum-based hydrides, are being investigated for their potential to reduce system weight without compromising storage density. Additionally, advancements in the thermodynamics of hydrogen release aim to lower the temperature and energy input required for desorption, making these materials more practical for broader applications [45].

Nanotechnology and material science advancements are also paving the way for next-generation carbon-based storage solutions, such as graphene-based materials and advanced MOFs. These materials offer high surface areas and tunable pore sizes, enabling improved hydrogen adsorption. However, achieving commercial-scale application of these technologies requires further research to address limitations in storage density and cost [46].

22.6 Safety regulations and transportation challenges

22.6.1 Current safety regulations

The transportation of hydrogen is governed by a robust framework of international, national, and industry-specific safety regulations aimed at mitigating the risks associated with hydrogen's unique physical and chemical properties. Hydrogen's high flammability, wide flammability range, and low ignition energy make it particularly prone to combustion and explosion hazards [47]. Regulatory bodies, including the International Organisation for Standardisation (ISO), the American Society of Mechanical Engineers (ASME), and the European Union, have established comprehensive safety standards to address these risks.

Standards such as ISO 16111 focus on the design and safety of hydrogen storage in metal hydride tanks, while ISO 19880-1 provides guidelines for hydrogen refueling stations, encompassing aspects like leak detection, safety distances, and emergency shut-off systems. ASME's Boiler and Pressure Vessel Code (BPVC) and ASME B31.12 specify the requirements for hydrogen pipelines and pressurized storage systems [48]. In the European Union, the ATEX directives ensure that hydrogen-handling equipment is explosion-proof and certified for use in hazardous environments [49].

Countries have also implemented national safety protocols tailored to their hydrogen markets. For instance, Japan's High-Pressure Gas Safety Act enforces rigorous safety measures for hydrogen facilities, while the United States Department of Transportation (DOT) regulates hydrogen transportation under the Code of Federal Regulations (CFR), covering hazardous materials transportation [50]. These measures collectively address the risks of hydrogen flammability, diffusion, and material embrittlement, creating a safer environment for hydrogen transport.

22.6.2 Risks associated with hydrogen transportation

Hydrogen transportation faces distinct challenges due to its chemical and physical properties, posing risks across all methods of transport – gaseous, liquid, and material-based. One significant risk is hydrogen embrittlement, where hydrogen atoms penetrate metallic structures, weakening their integrity and making them prone to fractures. This phenomenon is particularly critical in high-pressure pipelines and storage tanks, requiring the use of specialized embrittlement-resistant materials such as advanced alloys and composites, which increase costs [51].

Hydrogen's flammability and ease of forming explosive mixtures in air add another layer of risk. Its low ignition energy and high diffusivity mean that even minor leaks can lead to hazardous conditions, especially in confined spaces. For liquid hydrogen, the extreme cryogenic temperatures necessary for storage and transport introduce additional risks related to thermal insulation and boil-off losses [52]. Comprehensive safety systems, including leak detection, pressure relief valves, and ventilation systems, are integral to hydrogen transportation infrastructure to manage these risks.

The risks are heightened during loading, unloading, and transfer operations, where hydrogen is handled under varying pressure and temperature conditions. These processes require robust emergency shut-off mechanisms and real-time monitoring to prevent accidents. While these safety measures are effective, they add to operational complexity and costs, necessitating regular maintenance to ensure reliability [53].

22.6.3 Case studies of accidents and lessons learned

Several incidents have highlighted the importance of stringent safety protocols in hydrogen transportation. In 2004, a liquid hydrogen leak at a refueling station in Munich, Germany, led to an explosion, causing significant damage and minor injuries. Investigations revealed inadequate leak detection systems and faulty valve maintenance as key factors, emphasizing the need for reliable detection technologies and regular equipment checks [54].

In 2019, a high-pressure hydrogen storage system failure at a Santa Clara, California, hydrogen production and refueling facility resulted in an explosion. This incident disrupted regional hydrogen supplies and highlighted risks associated with high-pressure storage. Subsequent investigations prompted stricter regulations for valve design, pressure management systems, and regular inspections [55].

These incidents have spurred advancements in safety standards, including improved leak detection systems, better emergency response protocols, and enhanced training for personnel. Lessons learned underscore the importance of proactive maintenance, rigorous safety checks, and regulatory compliance to prevent similar accidents.

22.6.4 Addressing public concerns

Public perception remains a critical barrier to the widespread adoption of hydrogen technologies. Historical events such as the Hindenburg disaster have perpetuated concerns about hydrogen's safety. Addressing these misconceptions is essential to building public trust and fostering the acceptance of hydrogen projects [56].

Transparent communication about safety records, compliance with regulations, and risk mitigation strategies can help reassure the public. Educational campaigns highlighting modern safety measures, such as advanced leak detection and robust containment systems, can dispel outdated fears about hydrogen technology. Additionally, community engagement initiatives, including consultations and public forums, can address local concerns and promote awareness of hydrogen's safety [57].

Visible safety measures, such as real-time monitoring systems and clear signage at hydrogen facilities, also play a role in enhancing public confidence. These efforts, coupled with stringent regulatory oversight, demonstrate a commitment to safety and build trust in hydrogen infrastructure.

22.6.5 Regulatory and technological challenges

Despite advancements, significant challenges remain in implementing effective safety regulations and addressing technological limitations in hydrogen transportation. One key issue is the lack of uniform international standards, which complicates cross-border hydrogen trade and collaboration. Disparities in hydrogen purity specifications, storage pressures, and equipment standards create regulatory inconsistencies that hinder global integration [58].

Technological limitations further challenge the scalability of hydrogen transport. Current embrittlement-resistant materials are expensive, and leak detection systems require regular calibration to maintain accuracy. Continued research into affordable, durable materials and cost-effective safety technologies is essential to overcome these barriers [59].

There is also a need for improved data on hydrogen safety, particularly regarding long-term material performance under high-pressure and cryogenic conditions. Enhanced understanding of hydrogen embrittlement and other degradation mechanisms will inform safer infrastructure designs and operational protocols, contributing to more reliable transportation systems.

22.7 Future of hydrogen transportation

22.7.1 Technological innovations

The future of hydrogen transportation is closely tied to advancements in technology that address current challenges, including cost, efficiency, and safety. Emerging innovations in materials science, cryogenics, and chemical engineering are set to redefine hydrogen transportation methods. For example, the development of embrittlement-resistant materials, such as advanced alloys and composites, could extend the lifespan of pipelines and storage tanks while reducing costs. Similarly, lightweight composite materials, like carbon-fiber-reinforced tanks, are improving the strength-to-weight ratio, enhancing the feasibility of hydrogen storage in mobile applications [60].

Cryogenic technologies are also advancing, with ongoing research into efficient hydrogen liquefaction methods. Innovations such as magnetic refrigeration and advanced mixed-refrigerant cycles aim to reduce the energy costs associated with liquefaction, which currently consume 30–40% of hydrogen's total energy content. Additionally, developments in thermal insulation materials and boil-off management systems are expected to enhance the efficiency of liquid hydrogen transport, making it more suitable for long-distance applications [61].

In material-based hydrogen transportation, advancements in MOFs and LOHCs are enabling compact, scalable, and safe hydrogen storage. MOFs, with their high hydrogen adsorption capacities, offer potential energy densities superior to current methods. Meanwhile, LOHCs, which store hydrogen in a stable liquid form, eliminate the need for cryogenic or high-pressure systems, making them viable for international hydrogen trade. Enhanced catalytic processes are improving the efficiency of hydrogen extraction from LOHCs, further boosting their applicability [62].

22.7.2 Emerging distribution pathways

As the hydrogen economy expands, diverse distribution pathways are emerging to complement traditional methods like pipelines and tube trailers. One such pathway is the use of ammonia as a hydrogen carrier. Ammonia, which can be transported under moderate conditions, allows for long-distance and intercontinental hydrogen transport. At the destination, ammonia can be either used directly in fuel cells or converted back to hydrogen, offering flexibility in its application [63].

Decentralized hydrogen production and distribution are also gaining traction. Small-scale electrolysis systems powered by renewable energy sources enable local hydrogen production, reducing reliance on long-distance transport infrastructure. This approach is particularly beneficial in regions with abundant solar or wind energy, aligning with goals for energy decentralization and sustainability [42].

Additionally, digital technologies are enhancing hydrogen transportation through real-time monitoring and predictive maintenance. Internet of Things (IoT) devices, artificial intelligence (AI), and blockchain technologies enable detailed tracking of hydrogen flows, predictive analytics for leak prevention and secure documentation of supply chain activities. These innovations are expected to improve operational efficiency and safety while reducing costs [64].

22.7.3 Policy and economic factors

Supportive policy frameworks and economic incentives are critical for scaling hydrogen transportation infrastructure. Governments around the world are recognizing hydrogen's potential and implementing initiatives to drive its development. For example, the European Union's "European Clean Hydrogen Alliance" promotes investment in hydrogen infrastructure, including pipelines and refueling stations. Similarly, the U.S. Department of Energy is providing grants to accelerate research in hydrogen transportation technologies [60].

Economic tools such as subsidies, tax incentives, and carbon pricing mechanisms are leveling the playing field for hydrogen. Carbon pricing, in particular, increases the cost of fossil fuel-based alternatives, enhancing hydrogen's competitiveness. Public-private partnerships are also fostering collaboration across the hydrogen value chain, enabling the efficient development of infrastructure through shared resources and expertise. These coordinated efforts between governments and industry are essential for overcoming financial and logistical barriers to widespread hydrogen adoption [65].

22.7.4 Global collaboration and standards

Harmonizing global standards for hydrogen transportation is crucial for facilitating cross-border trade and establishing hydrogen as a globally traded commodity. Organizations such as the International Partnership for Hydrogen and Fuel Cells in the Economy (IPHE) and the International Energy Agency (IEA) are spearheading initiatives to standardize protocols related to hydrogen purity, storage conditions, and safety requirements. These efforts aim to minimize regulatory discrepancies, enhance safety measures, and streamline international hydrogen trade [66].

Standardization initiatives address critical technical and regulatory challenges associated with the hydrogen supply chain. For example, establishing uniform criteria for hydrogen purity ensures compatibility with fuel cells and other applications across borders. Similarly, consistent safety standards reduce risks during storage and transportation, fostering trust among trading partners. By aligning these standards, international organizations aim to create a cohesive framework that supports the scalability and reliability of hydrogen trade globally.

Global partnerships are playing a pivotal role in advancing the hydrogen economy. For instance, Japan's agreements with Australia and Saudi Arabia to import hydrogen underscore the potential for intercontinental supply chains. These partnerships demonstrate the technical and economic feasibility of large-scale hydrogen production, transportation, and distribution. Australia's investments in green hydrogen projects and Saudi Arabia's development of blue hydrogen infrastructure exemplify diverse approaches to hydrogen production that align with international demand. Such collaborations are not only laying the groundwork for a robust global hydrogen market but also promoting innovation in transportation technologies, such as liquefied hydrogen carriers and ammonia-based hydrogen transport [67].

The convergence of standardized protocols and global partnerships will be instrumental in building a sustainable and integrated hydrogen economy. Continued collaboration between governments, industry stakeholders, and international organizations is essential to overcoming technical barriers, reducing costs, and ensuring the long-term viability of hydrogen as a key energy carrier

22.7.5 Outlook for hydrogen transportation infrastructure

The outlook for hydrogen transportation is promising, with significant investments and technological advancements driving progress. Modular and scalable solutions, such as containerized hydrogen storage units and mobile refueling stations, are enhancing the adaptability of hydrogen infrastructure. These systems can be rapidly deployed in underserved regions, supporting the growth of emerging hydrogen markets [68].

The integration of hydrogen pipelines with existing natural gas networks is also being explored, allowing for efficient hydrogen transport without requiring entirely new infrastructure. Such hybrid systems could accelerate the adoption of hydrogen as a mainstream energy carrier. Additionally, advancements in cryogenic and pressurized hydrogen storage technologies are improving the efficiency and cost-effectiveness of long-distance hydrogen transportation. These innovations, coupled with the development of hydrogen-powered vehicles and vessels, highlight the transformative potential of hydrogen in the global energy landscape [69].

22.8 Discussion and conclusion

The transportation of natural hydrogen is a cornerstone in establishing its role as a clean and efficient energy carrier, vital to global decarbonization efforts. This chapter provides a detailed analysis of the primary transportation methods, including high-pressure gaseous transport, low-temperature liquid transport, and material-based solid storage. Each method offers distinct advantages while facing specific challenges related to energy efficiency, cost, safety, and scalability.

Pipelines emerge as the most effective option for large-scale, continuous transport in regions with existing infrastructure. They connect production facilities to end-use points, enabling economies of scale and reducing transportation costs per kilogram of hydrogen. Retrofitting existing natural gas pipelines for hydrogen transport can lower capital expenditures by 50–70%, depending on the condition and adaptability of the infrastructure [70]. However, building new pipelines in regions without an established hydrogen market requires significant investment. Moreover, hydrogen's small molecular size and tendency to cause material embrittlement necessitate the use of specialized materials and coatings, adding to overall costs [10].

Tube trailers provide a flexible solution for short- to medium-range hydrogen transport and are particularly valuable in emerging markets or during the early stages of infrastructure development. Their mobility makes them ideal for supplying fueling stations, remote industrial sites, and smaller-scale hydrogen projects. However, tube trailers have limited capacity, typically transporting 600–1,000 kg of hydrogen per trip depending on pressure and materials. The high energy requirements for compressing hydrogen to pressures exceeding 200 bar contribute to operational costs, making them less competitive for large-scale transport [71]. Safety considerations, including risks of high-pressure leaks and mechanical failures during loading and unloading, require strict regulatory compliance, further increasing costs [26].

Chemical carriers, such as ammonia and LOHCs, offer innovative solutions for long-distance and international hydrogen transport. Ammonia's high hydrogen content by weight (17.6%) and ability to liquefy under moderate conditions allow it to utilize existing liquefied natural gas infrastructure. However, reconverting ammonia back to hydrogen is energy-intensive, with current catalytic technologies requiring optimization to reduce costs and energy losses [70]. LOHCs, which enable hydrogen transport at ambient conditions, eliminate the need for high-pressure or cryogenic infrastructure. However, hydrogenation and dehydrogenation processes are both costly and energy-intensive, limiting their current commercial viability [19]. Despite these challenges, chemical carriers remain a promising option for regions where pipelines or tube trailers are impractical.

When comparing transportation costs, retrofitted pipelines are the most cost-effective solution for short and medium distances. Costs range from 0.1 to 1 USD/kg for distances of 0–50 km and regional transmission (51–500 km), increasing to 1–2 USD/kg for longer distances exceeding 1,000 km. New pipelines, while they involve higher capital investments, remain competitive, with similar cost ranges. For intercontinental transport over distances greater than 1,000 km, shipping liquefied hydrogen (LH_2) and ammonia (NH_3) becomes viable, though costs typically exceed 2 USD/kg. Trucking methods, including gaseous and liquefied hydrogen transport, are suitable for short distances (0–100 km), with costs in the range of 0.1–1 USD/kg, but become inefficient for long-distance or large-scale transport [72].

Safety regulations and standards are essential for the secure handling of hydrogen. International frameworks, coupled with advancements in leak detection, material technology, and emergency protocols, have significantly mitigated risks associated with hydrogen transportation. However, regulatory inconsistencies and technological limitations must still be addressed to enable hydrogen's role as a globally traded energy commodity.

Looking ahead, advancements in cryogenics, materials science, and digital technologies are expected to reduce costs and improve efficiency in hydrogen transport. Emerging pathways, such as ammonia carriers and decentralized production networks, offer scalable and flexible solutions to complement existing infrastructure. Policy support, economic incentives, and global collaboration will remain key drivers in the development of hydrogen transportation.

In conclusion, hydrogen transportation is pivotal to achieving a clean energy future. Continued innovation, investment, and international cooperation will ensure that hydrogen becomes a cornerstone of the global energy system, supporting decarbonization across industries and regions.

References

[1] Jaiswal-Nagar D., Dixit V., Devasahayam S.. Towards hydrogen infrastructure: Advances and challenges in preparing for the hydrogen economy, 2023.

[2] Salehi F., et al.. Overview of safety practices in sustainable hydrogen economy–An Australian perspective. International Journal of Hydrogen Energy. 2022, 47(81): 34689–34703.

[3] Pasman H. J., Rogers W. J.. Safety challenges in view of the upcoming hydrogen economy: An overview. Journal of Loss Prevention in the Process Industries. 2010, 23(6): 697–704.

[4]   Chen X., Zhang C., Li Y.. Research and development of hydrogen energy safety. Emergency Management Science and Technology. 2022, 2(1): 1–9.

[5]   Kotchourko A., Jordan T.. Hydrogen safety for energy applications: Engineering design, risk assessment, and codes and standards, Butterworth-Heinemann, 2022.

[6]   Moradi R., Groth K. M.. Hydrogen storage and delivery: Review of the state of the art technologies and risk and reliability analysis. International Journal of Hydrogen Energy. 2019, 44(23): 12254–12269.

[7]   Li H., et al.. Safety of hydrogen storage and transportation: An overview on mechanisms, techniques, and challenges. Energy Reports. 2022, 8, 6258–6269.

[8]   Mozakka M., Salimi M., Hosseinpour M.. Determining the challenges of transition to a hydrogen economy through developing a quantitative index. International Journal of Hydrogen Energy. 2024, 56, 1301–1308.

[9]   Ali M. S., et al.. Hydrogen energy storage and transportation challenges: A review of recent advances. Hydrogen Energy Conversion and Management. 2024, 255–287.

[10]   Faye O., Szpunar J., Eduok U.. A critical review on the current technologies for the generation, storage, and transportation of hydrogen. International Journal of Hydrogen Energy. 2022, 47(29): 13771–13802.

[11]   Gupta R., Basile A., Veziroglu T. N.. Compendium of hydrogen energy: Hydrogen storage, distribution and infrastructure, Woodhead Publishing, 2016.

[12]   Gerboni R., Salvador E.. Hydrogen transportation systems: Elements of risk analysis. Energy. 2009, 34(12): 2223–2229.

[13]   Witkowski A., et al.. Comprehensive analysis of hydrogen compression and pipeline transportation from thermodynamics and safety aspects. Energy. 2017, 141, 2508–2518.

[14]   Zhang C., et al.. Key technologies of pure hydrogen and hydrogen-mixed natural gas pipeline transportation. Acs Omega. 2023, 8(22), 19212–19222.

[15]   Tsiklios C., Hermesmann M., Müller T.. Hydrogen transport in large-scale transmission pipeline networks: Thermodynamic and environmental assessment of repurposed and new pipeline configurations. Applied Energy. 2022, 327, 120097.

[16]   Öney F., Veziro T., Dülger Z.. Evaluation of pipeline transportation of hydrogen and natural gas mixtures. International Journal of Hydrogen Energy. 1994, 19(10), 813–822.

[17]   Panfilov M.. Compedium of hydrogen energy–volume 2: hydrogen storage, transportation and infrastructure. Woodhead Publishing Series in Energy. 2016, str.

[18]   Regulators, E.U.A.f.t.C.o.E.. Transporting Pure Hydrogen by Repurposing Existing Gas Infrastructure: Overview of existing studies and reflections on the conditions for repurposing, 2021.

[19]   Reddi K., Elgowainy A., Sutherland E.. Hydrogen refueling station compression and storage optimization with tube-trailer deliveries. International Journal of Hydrogen Energy. 2014, 39(33): 19169–19181.

[20]   Bonner B.. Advanced hydrogen fueling station supply: Tube trailers. In: Air Products and Chemicals, Inc., Allentown: PA (United States), 2018.

[21]   Elgowainy A., et al.. Tube-trailer consolidation strategy for reducing hydrogen refueling station costs. International Journal of Hydrogen Energy. 2014, 39(35), 20197–20206.

[22]   Genovese M., et al.. Current standards and configurations for the permitting and operation of hydrogen refueling stations. International Journal of Hydrogen Energy. 2023, 48(51): 19357–19371.

[23]   Zaidman B., Wiener H., Sasson Y.. Formate salts as chemical carriers in hydrogen storage and transportation. International Journal of Hydrogen Energy. 1986, 11(5): 341–347.

[24]   He T., et al.. Hydrogen carriers. Nature Reviews Materials. 2016, 1(12): 1–17.

[25]   Bourane A., et al.. An overview of organic liquid phase hydrogen carriers. International Journal of Hydrogen Energy. 2016, 41(48): 23075–23091.

[26] Niermann M., et al.. Liquid Organic Hydrogen Carrier (LOHC) – assessment based on chemical and economic properties. International Journal of Hydrogen Energy. 2019, 44(13): 6631–6654.

[27] Kojima Y.. Hydrogen storage materials for hydrogen and energy carriers. International Journal of Hydrogen Energy. 2019, 44(33): 18179–18192.

[28] Sonthalia A., et al.. Moving ahead from hydrogen to methanol economy: Scope and challenges. Clean Technologies and Environmental Policy. 2021, 1–25.

[29] Dewangan S. K., et al.. A comprehensive review of the prospects for future hydrogen storage in materials-application and outstanding issues. International Journal of Energy Research. 2022, 46(12): 16150–16177.

[30] Becherif M., et al.. Hydrogen energy storage: New techno-economic emergence solution analysis. Energy Procedia. 2015, 74, 371–380.

[31] Zhang T., et al.. Hydrogen liquefaction and storage: Recent progress and perspectives. Renewable and Sustainable Energy Reviews. 2023, 176, 113204.

[32] Meda U. S., et al.. Challenges associated with hydrogen storage systems due to the hydrogen embrittlement of high strength steels. International Journal of Hydrogen Energy. 2023, 48(47): 17894–17913.

[33] Dawood F., Anda M., Shafiullah G.. Hydrogen production for energy: An overview. International Journal of Hydrogen Energy. 2020, 45(7): 3847–3869.

[34] Iulianelli A., Basile A.. Advances in hydrogen production, storage and distribution, Elsevier, 2014.

[35] Kim J., et al.. Key challenges in the development of an infrastructure for hydrogen production, delivery, storage and use. In: Advances in hydrogen production, storage and distribution, Elsevier, 2014, 3–31.

[36] Schlapbach L., Züttel A.. Hydrogen-storage materials for mobile applications. nature. 2001, 414(6861): 353–358.

[37] Aakko-Saksa P. T., et al.. Liquid organic hydrogen carriers for transportation and storing of renewable energy–review and discussion. Journal of Power Sources. 2018, 396, 803–823.

[38] Lee J.-S., et al.. Large-scale overseas transportation of hydrogen: Comparative techno-economic and environmental investigation. Renewable and Sustainable Energy Reviews. 2022, 165, 112556.

[39] Lavanya M., et al.. An overview of hydrogen storage technologies–key challenges and opportunities. Materials Chemistry and Physics. 2024, 129710.

[40] Osman A. I., et al.. Advances in hydrogen storage materials: Harnessing innovative technology, from machine learning to computational chemistry, for energy storage solutions. International Journal of Hydrogen Energy. 2024.

[41] Commission E.. A hydrogen strategy for a climate-neutral Europe, Melbourne, VIC, Australia: Global CCS Institute, 2020.

[42] McQueen S., et al.. Department of energy hydrogen program plan, Washington DC (United States): US Department of Energy (USDOE, 2020.

[43] Ousaleh H. A., et al.. An analytical review of recent advancements on solid-state hydrogen storage. International Journal of Hydrogen Energy. 2024, 52, 1182–1193.

[44] Rusman N., Dahari M.. A review on the current progress of metal hydrides material for solid-state hydrogen storage applications. International Journal of Hydrogen Energy. 2016, 41(28): 12108–12126.

[45] Shet S. P., et al.. A review on current trends in potential use of metal-organic framework for hydrogen storage. International Journal of Hydrogen Energy. 2021, 46(21): 11782–11803.

[46] Yu X., et al.. Recent advances and remaining challenges of nanostructured materials for hydrogen storage applications. Progress in Materials Science. 2017, 88, 1–48.

[47] Vivanco-Martín B., Iranzo A.. Analysis of the European strategy for hydrogen: A comprehensive review. Energies 2023. 2023, 16, 3866.

[48] Yang Y., et al.. Development of standards for hydrogen storage and transportation. In: E3S Web of Conferences, EDP Sciences, 2020.

[49]   Söderholm D.. Fire Safety in Hydrogen Processing Facilities-Design Considerations, 2023.
[50]   Raj P. K., Pritchard E. W.. Hazardous materials transportation on US railroads: Application of risk analysis methods to decision making in development of regulations. Transportation Research Record. 2000, 1707(1): 22–26.
[51]   Rolo I., Costa V. A., Brito F. P.. hydrogen-based energy systems: Current technology development status, opportunities and challenges. Energies. 2023, 17(1): 180.
[52]   Otto M., et al.. Optimal hydrogen carrier: Holistic evaluation of hydrogen storage and transportation concepts for power generation, aviation, and transportation. Journal of Energy Storage. 2022, 55, 105714.
[53]   Calabrese M., et al.. Hydrogen safety challenges: A comprehensive review on production, storage, transport, utilization, and CFD-based consequence and risk assessment. Energies. 2024, 17(6): 1350.
[54]   Patel P., et al.. A technical review on quantitative risk analysis for hydrogen infrastructure. Journal of Loss Prevention in the Process Industries. 2024, 105403.
[55]   Davies E., Ehrmann A., Schwenzfeier-Hellkamp E.. Safety of Hydrogen Storage Technologies. Processes. 2024, 12(10): 2182.
[56]   McCarthy M.. The hidden Hindenburg: The untold story of the tragedy, the Nazi secrets, and the Quest to rule the skies, Rowman & Littlefield, 2020.
[57]   Beasy K., Lodewyckx S., Mattila P.. Industry perceptions and community perspectives on advancing a hydrogen economy in Australia. International Journal of Hydrogen Energy. 2023, 48(23): 8386–8397.
[58]   Kumar S., et al.. Hydrogen safety/standards (national and international document standards on hydrogen energy and fuel cell). In: Towards Hydrogen Infrastructure, Elsevier, 2024, 315–346.
[59]   Patil R. R., et al.. Artificial intelligence-driven innovations in hydrogen safety. Hydrogen. 2024, 5(2): 312–326.
[60]   Vivanco-Martín B., Iranzo A.. analysis of the European strategy for hydrogen: A comprehensive review. Energies. 2023, 16(9): 3866.
[61]   Martín J. G.. The future of hydrogen: Seizing todays opportunities. Economía Industrial. 2022(424): 183–184.
[62]   Alves C., et al.. Hydrogen Technologies: Recent Advances, New Perspectives, and Applications, 2024.
[63]   Negro V., Noussan M., Chiaramonti D.. The potential role of ammonia for hydrogen storage and transport: A critical review of challenges and opportunities. Energies. 2023, 16(17): 6192.
[64]   Çelik D., Meral M. E., Waseem M.. Investigation and analysis of effective approaches, opportunities, bottlenecks and future potential capabilities for digitalization of energy systems and sustainable development goals. Electric Power Systems Research. 2022, 211, 108251.
[65]   Bade S. O., Tomomewo O. S.. A review of governance strategies, policy measures, and regulatory framework for hydrogen energy in the United States. International Journal of Hydrogen Energy. 2024, 78, 1363–1381.
[66]   Neef H.-J.. International overview of hydrogen and fuel cell research. Energy. 2009, 34(3), 327–333.
[67]   Lindner R.. Green hydrogen partnerships with the G lobal S outh. Advancing an energy justice perspective on "tomorrow's oil". Sustainable Development. 2023, 31(2), 1038–1053.
[68]   Sartory M., et al.. Modular concept of a cost-effective and efficient on-site hydrogen production solution, SAE Technical Paper, 2017.
[69]   Erdener B. C., et al.. A review of technical and regulatory limits for hydrogen blending in natural gas pipelines. International Journal of Hydrogen Energy. 2023, 48(14): 5595–5617.
[70]   Demir M. E., Dincer I.. Cost assessment and evaluation of various hydrogen delivery scenarios. International Journal of Hydrogen Energy. 2018, 43(22): 10420–10430.
[71]   Chatterjee S., Parsapur R. K., Huang K.-W.. Limitations of ammonia as a hydrogen energy carrier for the transportation sector. ACS Energy Letters. 2021, 6(12): 4390–4394.
[72]   Council H.. Hydrogen insights 2022, 2022.

David A. Wood* and Reza Rezaee

Chapter 23
Environmental impacts of hydrogen production and usage

Abstract: This chapter reviews the multifaceted and complex environmental impacts of hydrogen (H_2) from its human and subsurface influences, its supply chain including pipelines, underground storage, and the manner in which it is consumed as a combustion fuel and in fuel-cells. H_2 as an energy fuel has beneficial and adverse impacts on greenhouse gas (GHG) levels and air quality of the atmosphere compared to fossil fuels. These impacts depend on how it is generated, stored, transported, and consumed. When combusted H_2 tends to generate more oxides of nitrogen (NO_x) than fossil fuels, despite overall reductions in GHG. Green H_2, generated by electrolysis powered by off-peak renewable energy, without intermittent use of grid power supply, generates the least GHG. Blue H_2 captures some GHG by carbon capture and storage (CCS) but still contributes substantial GHG to the atmosphere. H_2 fuel cells offer the cleanest form of H_2 consumption, avoiding NO_x emissions but are too costly for mass-market uptake. Leakage of H_2 from above- and below-ground and H_2-supply infrastructure poses a major issue for H_2-fueled systems. Much uncertainty is associated with H_2-leakage from industrial-scale plants and small-scale H_2-fueled systems (vehicles and building heating systems). Robust life-cycle analysis, incorporating realistic leakage levels is essential to justify the building of H_2-supply chains and the potential exploitation of natural H_2 resources. Provisional analysis on a life-cycle basis suggests that exploitation of subsurface H_2 resources would only be environmentally beneficial from a GHG perspective from reservoirs containing >90% H_2 with minimal methane contamination.

Keywords: greenhouse gas impacts, oxides of nitrogen emissions, green versus blue H_2 generation, H_2 leakage, life-cycle analysis, biotic and abiotic natural H_2 sources, soil-based H_2 sink, Atmospheric H_2 to CH_4 conversion, role of contrails in H_2 aviation fuel impacts, fuel cell opportunities

*Corresponding author: **David A. Wood**, DWA Energy Limited, Lincoln, United Kingdom,
e-mail: dw@dwasolutions.com, orcid.org/0000-0003-3202-4069
Reza Rezaee, WA School of Mines: Minerals, Energy and Chemical Engineering, Curtin University, Perth 6102, Australia

https://doi.org/10.1515/9783111437040-023

23.1 Introduction

A substantial momentum has been achieved in recent years in the growth of hydrogen (H_2) production infrastructure in response to climate-change mitigation efforts to replace fossil fuels with viable clean alternative energies. The lower and higher heating values (LHV and HHV) at 24.85 C are 120 and 141.8 MJ/kg, respectively, providing it with a substantially higher energy density by weight than that of fossil fuels (e.g., gasoline 44 and 46 MJ/kg at 298 K). Liquid H_2 has an energy density (volume basis) about a factor of four lower than that of fossil fuels (e.g., 8 MJ/l versus 32 MJ/l for gasoline). This means that H_2 contains about two and a half times the energy density (weight basis) than natural (methane (CH_4)-rich) gas in a liquefied form (LNG). While H_2 gas has a higher energy density by weight, it has a lower energy density by volume, thereby requiring 2.8 times the volume to store H_2 than LNG [1]. These physical characteristics make H_2 a feasible alternative to fossil-fuel-driven energy supplies but poses some storage and transportation challenges [2] limiting its uptake in some commercial settings. At standard pressures, H_2 becomes liquefied at a temperature of − 253 °C, which makes it hazardous if leaked from storage, causing cold burns to animal and plant life as well as damage to soils and surrounding ecosystems. Moreover, leaked liquid H_2 poses possible explosions if mixed with air in confined spaces and exposed to an ignition source [2]. Worldwide consumption of H_2 was estimated in 2,023 at approximately 95 million metric tons (Mt)/year by the International Energy Agency (IEA) [3], and used mainly in industrial processes, not as an energy fuel.

The main attraction of adopting H_2 at scale as an energy fuel is that it offers the potential to reduce global CO_2 emissions substantially. However, some of the enthusiasm for H_2 uptake is based on the misguided expectation of many that its supply chain from production to consumption is not accompanied by GHG emissions or a range of other negative environmental impacts. Negative environmental impacts do occur but are complex and vary depending on the way in which H_2 is generated and/ or used. H_2 generates GHG impact indirectly as it reacts with various atmospheric components in the troposphere and stratosphere to generate other GHG, such as CH_4, ozone (O_3) and water vapor. H_2 indirectly modulates Earth's radiative balance in three key ways [4]: (1) by prolonging the longevity of CH_4 because it reduces the quantity of OH in the atmosphere, which serves to oxidize and destroy CH_4; (2) by generating HO_2 radicals, which produces tropospheric O_3 from reactions with NO_x (the Leighton relationship [5]; and (3) H_2 oxidizes in the stratosphere to generate water vapor. Such changes as these that contribute to atmospheric warming are referred to as positive radiative forcing elements, whereas the changes that contribute to atmospheric cooling are referred to as negative radiative forcing elements.

The mentioned complex indirect GHG impacts of H_2 have historically tended to be either overlooked or underestimated on the grounds that they are relatively short-lived [6]. However, H_2 is a highly significant reactive trace gas constituting a mixing ratio of about 530 parts per billion by volume (ppbv) of the Earth's atmosphere [7, 8].

CH_4 oxidation represents the largest H_2 source in the troposphere, whereas soil up-take (removal) functions as the main H_2 sink. H_2 has a short lifetime (~1.4 ± 0.2 years) in the troposphere, which is becoming shorter over time [9]as the planet warms. Thus, the entire atmosphere holds ~175 Tg of H_2 [10]. That H_2 component has accumulated partly through anthropogenic emissions, although a little more than half of it is generated by CH_4 other hydrocarbons (HC) undergoing photochemical oxidation in the atmosphere. However, H_2 atmospheric buildup is partly offset by H_2 removal from the atmosphere by rain entering the soil. H_2 is also destroyed by photochemical reactions within the atmosphere by reacting with the hydroxyl radical (OH) and by certain soil micro-organisms. Noyan [10] estimated that about 25% of atmospheric H_2 buildup is caused by human activities. However, that estimate may have underestimated the contribution of crustal H_2 seepage.

Generating H_2 from any carbon-based feedstock (fossil fuels, referred to as gray H_2), biomass, or fermentation techniques results in substantial GHG emissions. Combusting H_2 for energy generation results in substantially more atmospheric NO_x emissions than combusting methane. Generating H_2 by electrolysis (green hydrogen) only results in low GHG emissions if the energy source used for the electrolysis process is renewable (ideally from surplus wind and solar power facilities) [11]. If fossil fuels are used to generate grey H_2 substantial GHG emissions are generated in the production processes and a substantial amount of energy (from the fossil fuels) is consumed in those processes. Electrolysis is expensive due to its low efficiency and consumes substantial quantities of water making it unsuitable for many arid regions. The liquid waste generated by electrolysis processes is mainly associated with electrolyte replenishment leading to purging of moderately saline water, containing mainly sodium chloride and potassium chloride, which are deemed to be environmentally benign [12].

Despite the potential of green hydrogen, its uptake is hindered by the substantial inefficiency of the electrolysis methods (efficiency approx. 60–80%), which results in commercially uncompetitive production costs (about US$ 5/kg) [13]. Technological advancements are necessary to both improve efficiencies and reduce H_2 production costs under US$2/kg to make it commercially attractive. Once produced, H_2 still faces some costly infrastructure hurdles to overcome related to storage and transportation (above and below ground). H_2 production from biomass feedstocks can achieve similar yields to electrolysis at lower production costs and higher energy efficiency [14]. However, biomass gasification processes are difficult to scale up and typically involve land and energy-intensive feedstock supply chains. The use of supercritical water in the gasification process can improve process efficiency but is energy intensive, leads to tar as a waste product and increased corrosion issues for plant materials. Hence, further technology improvements are required to make bio-H_2 generation technologies commercially viable and sustainable on a large scale [14]. There are various other bio-H_2-generation processes [15], but all involve complexities that inhibit commercial uptake at scale [16]. These processes ferment biomass [17], conduct biological photolysis [18], or create electrolysis cells involving microbes [19]. By utilizing biowastes, such processes can be cost effective and reduce pollution. The latter has the potential to process biowastes, thereby potentially offering broader pollution-reduction benefits.

Historical data reveals that H_2 global emissions have increased by about 50% since the industrial revolution, with anthropogenic sources accounting for a little less than half of those emissions (Figure 23.1) [4]. Emissions peaked in the mid-1990s but have reduced since, due mainly to reductions in transportation equipment plant (improvements in engines and fuel consumption).

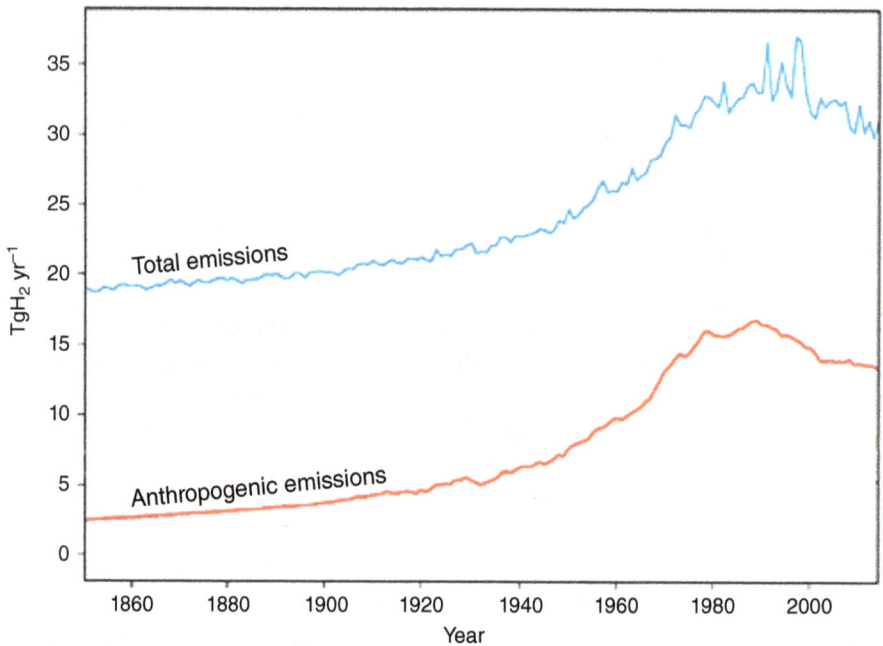

Figure 23.1: Historical hydrogen emissions. Total emissions peaked at 37.1 Tg/year in 1997. Modified from Paulot et al. [4].

All the H_2 generation processes available are associated with emissions of some GHG, including contributions associated with leakage of H_2 and CH_4 [20] (Figure 23.2), as well as NO_x and water vapor emissions from pure H_2 combustion.

If H_2 is to be used as a major component of a low-carbon energy supply chain, it needs to be stored cheaply and efficiently. The mobility of H_2 makes it vulnerable to leakage from storage and transportation facilities. Leakage can be exacerbated in the long-term by the metal embrittlement impacts of H_2 and leaked hydrogen contributes to GHG. Estimates for leakage in existing facilities suggest leakage rates in the range approximately 1% to >5% [20]. Underground hydrogen storage (UHS) is a potential means of storing hydrogen in bulk for short and long-term periods. However, H_2 stor-

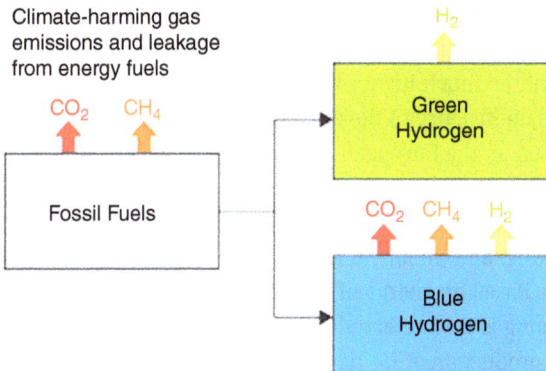

Figure 23.2: Primary climate-harming gases emitted and leaked from energy fuel supply chains. Modified from Ocko and Hamburg [6].

age in UHS involves substantial risks of leakage by penetrating the geological formations expected to seal such gas storage reservoirs. H_2 is highly reactive with many of the mineral components and organic materials present in common rock formations and soils, typically forming CH_4, water and carbon dioxide (CO_2) in various reactions. Moreover, some microbes present in soils and shallow rock consume H_2, others generate H_2 as part of their metabolic processes. In concentrated H_2-seepage conditions, high H_2 concentrations can degrade prevailing soil-based ecosystems and negatively impact agricultural yields.

This chapter addresses the various ways in which hydrogen supply chains can lead to adverse atmospheric emissions and potentially harm surface and subsurface ecosystems. It builds on the recent review of the challenges confronting the expansion of H_2-based energy systems [21].

23.2 Hydrogen production

H_2 is produced by various methods using different feedstocks and consuming different unit quantities of energy. Most commercial quantities of H_2 are produced by the steam reforming of CH_4, although other hydrocarbons (e.g., methanol) can also be used as feedstock for cracking processes. Steam methane reforming (SMR) processes lead to substantial direct GHG emissions. Electrolysis of water does not directly release GHG but indirectly may do so if fossil-fueled power generation is involved in the energy intensive process. According to EIGA [22], SMR consumes approximately 0.48 Nm^3 CH_4 to yield 1 Nm^3 H_2, but requires careful configuration to maximize exergy efficiency [23]: 1 Nm^3 refers to a normal cubic meter at standard conditions; temperature = 0 °C; absolute pressure = 1.01325 bar absolute. On the other hand, methanol cracking and electrolysis of water, respectively, consume 0.59 and 1.1 Nm^3 energy-equivalent CH_4 to yields 1 Nm^3 H_2. Clearly, if the electrolysis process is fueled by power generated by fossil fuels, then its indirect GHG emissions will be much higher than the direct GHG emissions of the reforming processes due to the inefficiency of the electrolysis process.

In terms of CO_2 emissions, based on the European Union average power generation fuel mix [22], steam reforming of CH_4 releases approximately 0.8 kg CO_2 associated with a yield of 1 Nm^3 H_2. This compares with about 1.2 kg 1 Nm^3 CO_2 associated with a yield of 1 Nm^3 H_2 from methanol cracking, and about 2.6 kg CO_2 associated with a yield of 1 Nm^3/H_2 generated by electrolysis. Based on energy efficiency and carbon footprint it is clear to see why CH_4 steam reforming is the preferred industrial process for generating H_2. To make the commercial scale production of H_2 via electrolysis environmentally sustainable and economical requires it to be powered by low-cost wind or solar energy. Global electrolyzer H_2 production rose from around 700 MW in 2022 to about 2GW in 2024 capacity, with most of the new build capacity constructed in China [24].

Based on IEA (2021) [25] projections for 2050 H_2 demand, Yedinak [26] considered two scenarios: (1) almost all H_2 demand continues is supplied by fossil fuels by the dominant and cheapest SMR process, which would generate approximately 6.6 Gt CO_2-eq/year by 2050 (~20% of current GHG emissions); (2) all new H_2 generation is provided by renewable-energy-powered electrolysis, equating to electrolysis plant increasing its share of the H_2 supply mix from about 0.1% (current) to about 88% in 2050. This would require the construction of around 3TW of new renewable power generation (mainly wind power based on cost) dedicated solely to powering electrolysis that equates to the total global current renewable energy capacity. The environmental consequences of scenario 1 are not acceptable in a net-zero energy system. On the other hand, the energy and H_2-plant infrastructure costs of scenario 2 represent a huge economic burden. Either a breakthrough in more efficient, GHG-emission free, H_2 generation processes is required or natural H_2 resources need to be exploited to meet the forecast H_2 demand. Nevertheless, substantial progress in building new H_2 production capacity was made in 2023/2024; for example, the sanctioning of seven major H_2 hubs in the United States of America (USA) with approximately US$1 billion of USA. Government grant awards to each Hub [27]. Major green hydrogen production initiatives also made progress in China, Saudi Arabia, and Australia in 2023/2024.

GHG emissions are not the only environmental impacts of concern related to H_2 production processes. CH_4 steam reforming uses nickel-based catalysts, whereas methanol cracking uses copper-based catalysts; both metals are toxic and can lead to soil contamination. Both of these processes involve the release to the atmosphere of some volatile organic compounds (VOCs) and discharge some contaminated water, which are potential pollutants unless the correct treatment steps are taken. H_2 generated by electrolysis initially contains some oxygen and water contamination, which need to be removed. De-oxidation is typically achieved by membranes or catalysts, and water is removed by drying; both processes consume additional energy. Oxygen's (O_2) flammability poses a safety hazard, requiring the oxygen produced by an electrolyzer to be safely collected and vented [12]. This can pose a problem for smaller scale electrolysis equipment located in buildings (e.g., to supply fuel-cell power units) when considered in conjunction with potential H_2 leakage and accumulation within buildings.

23.3 Hydrogen combustion

A common misconception is that pure H_2 consumed as fuel yields only H_2O in the process, which is only true when the H_2 is used to power fuel cells. Unfortunately, fuel cells remain costly to produce on a commercial scale, although technology breakthroughs are gradually reducing their costs [28]. Consequently, most H_2 currently consumed on an industrial scale to generate energy involves combustion in gas turbines or engines. Such combustion, particularly on the scale required in grid-connected thermal power plants, avoids CO_2 emissions but typically generates substantially more NO_x as an emission than natural gas combined-cycle-gas-turbine (CCGT) power generation [29]. A primary environmental challenge for H_2 combustion is, therefore, the thermal generation of NO_x [30]. Nitrogen dioxide (NO_2) is a potent air pollutant responsible for respiratory illnesses in animals and humans, as well as a GHG. Nevertheless, combusting H_2 produces minimal emissions of VOC, particulate matter (PM), and carbon monoxide (CO), which also degrade air quality and are responsible for respiratory illnesses, making it less damaging than fossil-fuel-derived combustion.

NO_x derived from systems that combust H_2 can be mitigated to an extent but to do so usually compromises their power-generation efficiency [31]. Treating flue-gas emissions to remove NO_x is expensive and process intensive. Standards for optimizing H_2-fueled combustion plants are currently lacking, meaning that users are not incentivized/penalized for failing to do so; a situation that requires carefully crafted regulation and monitoring if NO_x emissions are to be minimized. For internal combustion engines (ICEs), increasing the equivalence ratio (ER) at low engine speeds can increase energy performance and substantially reduce NO_x emissions [31]. The equivalence ratio measures the relative volume of O_2 needed for full combustion of H_2 injected into an engine cylinder; an ER <1 means there is excess O_2 to H_2 (fuel-lean) and ER >1 means there is excess H_2 to O_2 (fuel-rich) [32]. To minimize emissions of NO_x from most types of H_2 combustion plants but typically this comes at a price and reduces their energy efficiency. In such plant, energy output, combustion temperature, and NO_x emissions are highly sensitive to ER adjustments, as illustrated diagrammatically in Figure 23.3. Selecting the minimum NO_x configuration could make such plant, particularly domestic-scale H_2-fueled or hybrid H_2/CH_4-fueled gas boilers, unattractive from a cost alternative to other low-emissions heating options [30].

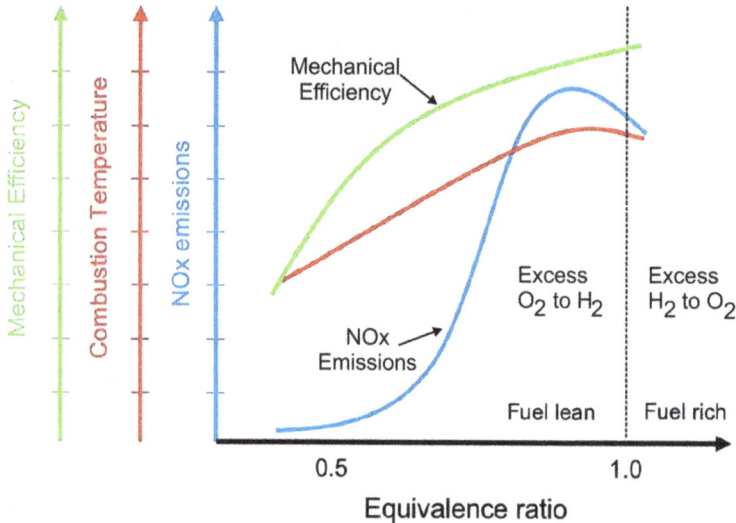

Figure 23.3: There is a trade-off in H_2 combustion plant: to minimize NO_x emissions it is typically necessary to sacrifice some energy-generation efficiency by tuning them to operate with leaner fuel (lower equivalence ratio). Modified from Lewis [30].

23.4 Blending hydrogen with hydrocarbon fuels

An alternative approach is to blend H_2 with hydrocarbon fuels to reduce GHG emissions and improve engine performance. The high burning velocity (flame speed) of H_2 makes it attractive for this purpose. For example, Tang et al. [33] revealed that an increase in the H_2 content of an H_2/hydrocarbon fuel blend decreased the minimum ignition energy but resulted in distinct changes in laminar flame speeds. Yasiry and Shahad [34] demonstrated the effectiveness of such blends based on experiments conducted with different fuel blends in attempts to increase their flame speeds and assess combustion pressure, and other attributes of blending liquid petroleum gas flames with a range of H_2 contents. It is apparent that different H_2/hydrocarbon combustion-fuel blends result in distinct GHG emissions, especially NO_x. Hence, tuning such engines for optimal energy production would likely not lead to minimum GHG impact.

Ozturk et al. [35] conducted experiments with different blends (0% to 30%) of H_2 and natural gas to fuel in a domestic gas-burning stove, recording emissions of CO, CO_2, NO_x, and energy efficiency of that combustion process. Over the blending range considered energy efficiency increased by about 5% as the H_2 content increased. This was associated with declining CO_2 and CO emissions but with fluctuating, though mainly declining, NO_x emissions (Figure 23.4). Compared to a 0% blending ratio, CO_2 emissions declined only by about 3.86% with a hydrogen ratio of 0.2 but declined by ~21% as the blending ratio reached 30%. However, based on a life-cycle analysis

(LCA), the global-warming potential (GWP) reduced only slightly (~1.8%) for a 30% H_2 blend calculated at 6.123 kg CO_2-eq./kg. However, uncertainty exists in the GWP values of different potential hydrogen production processes, storage facilities, and H_2 leakage assumptions and that study did not consider sensitivities relating to those uncertainties.

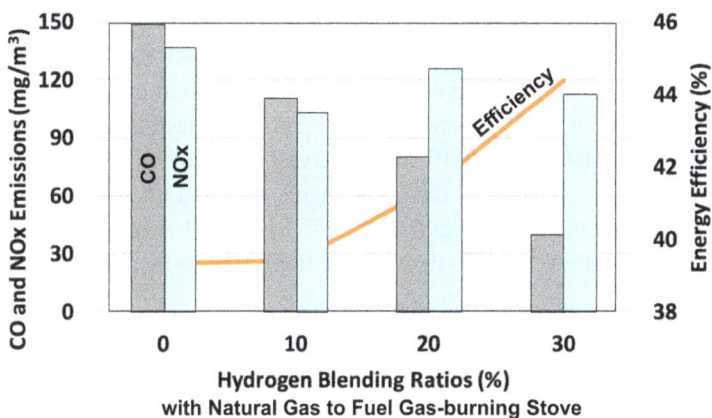

Figure 23.4: CO and NO_x emissions recorded and calculated energy efficiency associated with H_2: natural gas blending ratios of hybrid fuel compositions consumed in a small stove. Modified from Ozturk et al. [35].

23.5 Hydrogen as a road transport fuel

For use as a vehicle fuel for mass personal and light commercial vehicle transportation H_2 faces substantial environmental (from leakage) and infrastructure challenges in the storage and distribution sectors to be commercially viable. From the environmental perspective, H_2 fuel-cell vehicles (HFCV) can provide high energy efficiency, attractive power-to-weight trade-offs, and minimal GHG emissions. However, the commercial uptake of HFCV remains limited as they remain too costly to manufacture. Also, the low H_2 energy density (by volume) causes challenges for in-vehicle H_2-fuel storage, as do concerns regarding high-pressure safety and limited fuel-cell durability (approximately 120,000 km of driven distance), and paucity of existing H_2 refueling infrastructure [36].

In-vehicle H_2 storage systems remain under development, with energy density posing a key obstacle in improving the viability of all H_2-fueled vehicles [37]. H_2 compression of 4,900 psi achieves H_2-stored density = 23.32 kg/m^3, which is typically used in large vehicles. H_2 compression of 700 bar achieves H_2-stored density = 39.22 kg/m^3, a condition typically found in personal H_2-fuel-cell road vehicles [38]. All H_2-fueled vehicles struggle to provide effective H_2-fuel storage within the small confines of most vehicles.

H_2 blending with conventional gasoline for use in spark-ignition engines ICE acts to reduce unburned hydrocarbons and CO emissions with H_2 enrichment. H_2 blending in vehicles with compression-ignition (CI) engines, typically fueled by diesel, can diminish emissions of CO, CO_2, particulate matter (PM), and VOC but tend not to diminish emissions of NO_x [37]. Ghazal [39] assessed a range of diesel plus H_2 fuel blends (0.05% H_2 to 50% H_2; volume basis) with the CI engines operated at 1,000 rpm to 4,000 rpm. The experiments conducted also varied air: fuel ratios between 10 and 80. Engine performance increased by approximately 14% as the H_2 content of the fuel increased to about 40% when the CI engines were operated at high speed and at a high air-to-fuel ratio. The CI engine's thermal efficiency increased about 37% when operated at those conditions in comparison to the same engine operated on 100% diesel. Hamdan et al. [40] confirmed the increase in thermal efficiency of CI engines with H_2-enriched fuel blends, a reduction in particulate matter (PM), and an increase in NO_x emissions (Figure 23.5).

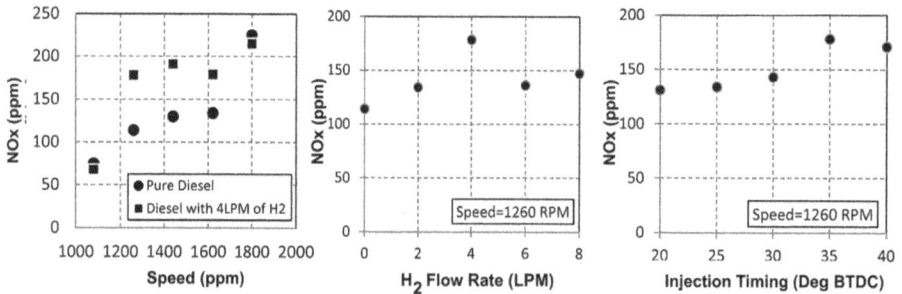

Figure 23.5: NO_x emissions generated by various H_2-diesel blends fueling a CI engine on trends. LPM = L/min. BTDC = before-top-dead-center piston position. Modified from Hamdan et al. [40].

Homogeneous-charge-compression-ignition (HCCI) engines incorporate beneficial attributes of ICE- and CI-powered vehicles. They do this by the compression of precise fuel: air blends promoting auto-ignition in the combustion chamber [41]. HCCI vehicles provide the benefit of reducing NO_x emissions without the need of a catalytic converter [42]. Nevertheless, HCCI engines suffer from engine knocking problems with certain fuels [43], which is the case when 100% H_2 fuel is used. HCCI-engine performance does improve when H_2 is blended either with diesel or natural hydrocarbon gases. With such fuel blends HCCI engines emit less CO, NO_x, and PM. Khaliq et al. [44] evaluated HCCI engines operated with a range of H_2-enriched fuel blends combusted in conditions ranging from 20% to 40% excess air. At operating temperatures between 13 and 41 °C, they achieved relatively high energy (44–48%) and exergy (32–36%) efficiencies.

The reality, based on current technologies and commercial factors, is that vehicles powered by H_2 combustion fuels (pure or hybrid) are unlikely to progress beyond limited market uptake. They have substantially more infrastructure requirements than electric vehicles. Battery-powered-hybrid-electric (BPHE) vehicles are not without their problems (e.g., high metal requirements to manufacture, driven-distance-range limitations without recharging and battery safety/ recycling concerns). However, such problems are more easily overcome by BPHE than those faced by H_2-fueled vehicles. Consequently, it seems unlikely that the environmental issues facing H_2-fueled road vehicles will need to be confronted and overcome in progress toward achieving current 2050 clean-energy target.

23.6 H_2 as a potential fuel for aircraft engines

In the aviation sector, CO_2 emissions increased by almost a factor of seven from 1960 to 2018 [45]. H_2-fueled aircraft engines release copious amounts of NO_x and H_2O (mainly in vapor form) into the high atmosphere leading to environmentally harmful contrails. Aircraft contrails account for a high proportion of aviation's total GHG emissions. Relatively high exhaust-gas temperatures of H_2 combusted in aircraft engines does, to an extent, impede contrails formation. On the other hand, the higher water vapor emissions of such engines do increase the likelihood of contrails forming at certain altitudes [46]. H_2-fueled combustion jet engines do not generate sulfate-PM-based aerosols, an outcome from the existing fleet of jet-kerosene-fueled aircraft. Such aerosols lead to crystals forming within contrails, causing them to enlarge and thereby reducing their negative GHG effects. The presence of nitrate aerosols emitted by H_2-fueled jet aircraft engines could also act to reduce their GHG impact [47]. H_2O and NO_x vapor emissions from H_2-powered jet aircraft are, however, quantitatively impacted by flying altitudes, jet-engine design, and operating configurations. Linear/cirrus contrails plus NO_x emissions are the main GWP impacts of modern jet aircraft. These emissions make a substantially greater contribution to global warming than aviation-related CO_2 emissions; approximately 3 times more according to Lee et al. [45]. H_2 escaping from the fuel storage tanks of potential H_2-powered aircraft could, in addition, be as much as 2% [48]. There is much uncertainty in that leakage figure as there have been few trials with H_2-powered aircraft to date [49].

The Resource to Climate Comparison Evaluator (RECCE) tool [50] attempts to estimate GHG contributions from aviation powered by different fuels and the costs of operating with each fuel type. The RECCE tool makes the speculative assumption that there would be no contrail-forcing impacts associated with H_2-fuel cell-powered aircraft. However, some have questioned the validity of this assumption and the failure of the RECCE models to consider potential GHG impacts from the substantial H2O emissions from H_2-fuel cell-powered aircraft [21, 46]. The high costs and emissions uncertainties of H2-powere aircraft make their commercial uptake unlikely in the next decade or so.

23.7 Environmental impacts of H₂ leakage from above-ground infrastructure

Uncertainties surrounding the H_2-leakage quantities that are likely to evolve from the expanded H_2 supply system envisaged for 2050 forming a cornerstone of the planned "net-zero" energy systems. Growing H_2 fuel and chemical feedstock supply chains (most of which do not currently exist) will collectively become the major sources of H_2 leakage, with an estimated contribution of 77% of industry-wide H_2 leakage [51]. New-technology-based H_2 supply networks and scaled-up infrastructure come with increased risk of H_2 leakage. A 2022 study assumed that leaking H_2 along supply chains could range from about 2.9% (low case) to about 5.6% [51]. By 2050, the IEA predicted that of 528 million tons [Mt] of H_2 would be consumed (approximately 5 times that consumed in 2020) [25]. Figure 23.6 illustrates the contributions of different supply-chain sectors to H_2 usage associated with that 2050 case. They concluded that "hydrogen leakage is expected to be a challenge for the hydrogen economy" but noted that there was substantial uncertainty in their leakage estimates, in both 2020 and 2050. The mentioned H_2-leakage assumptions [51] were re-used by the IEA their 2023 report [3].

IEA's Global Hydrogen Flow Forecast for 2050 (Net-zero Scenario of 2021)

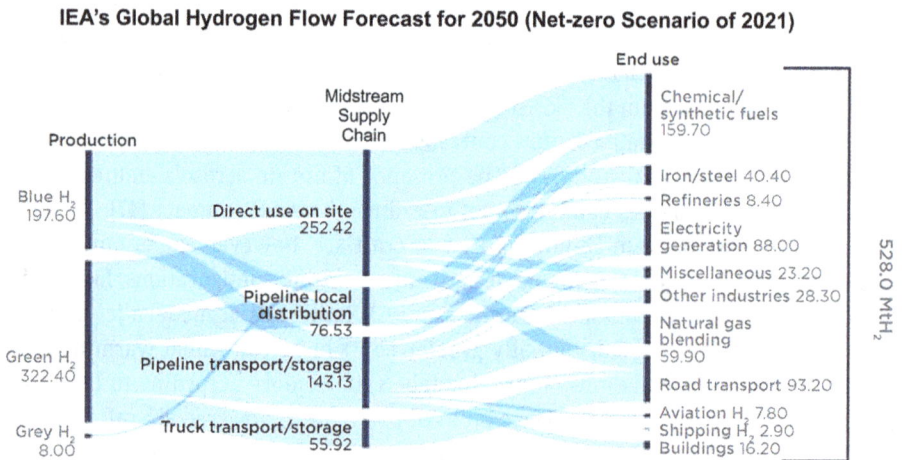

Figure 23.6: Forecast of global hydrogen flows (IEA 2050 projection to achieve their "net-zero" outcome). Modified from IEA [25].

Shu et al. [52] modeled the behavior of H_2 leaks from pipework in a confined space (e.g., commercial/residential buildings to better understand how the leaked H_2 would disperse and/or accumulate. H_2 leaks from pipework within buildings would tend to rise assuming constant-temperature conditions. In spaces that are thermally stratified, such as in space-heated buildings, the movement of such leaked gases tends oscillate and become focused at certain levels. Analysis applying the buoyant-jet model [52] revealed that variations in Froude number (Fr) influenced the heights within building at which leaked H_2 would likely accumulate (Figure 23.7). Fr considers the ratio of inertial force/gravitational force of various types of gas flow. Supercritical/rapid flow conditions (Fr > 1) exist when viscous force is greater than gravitational force. Subcritical/slow

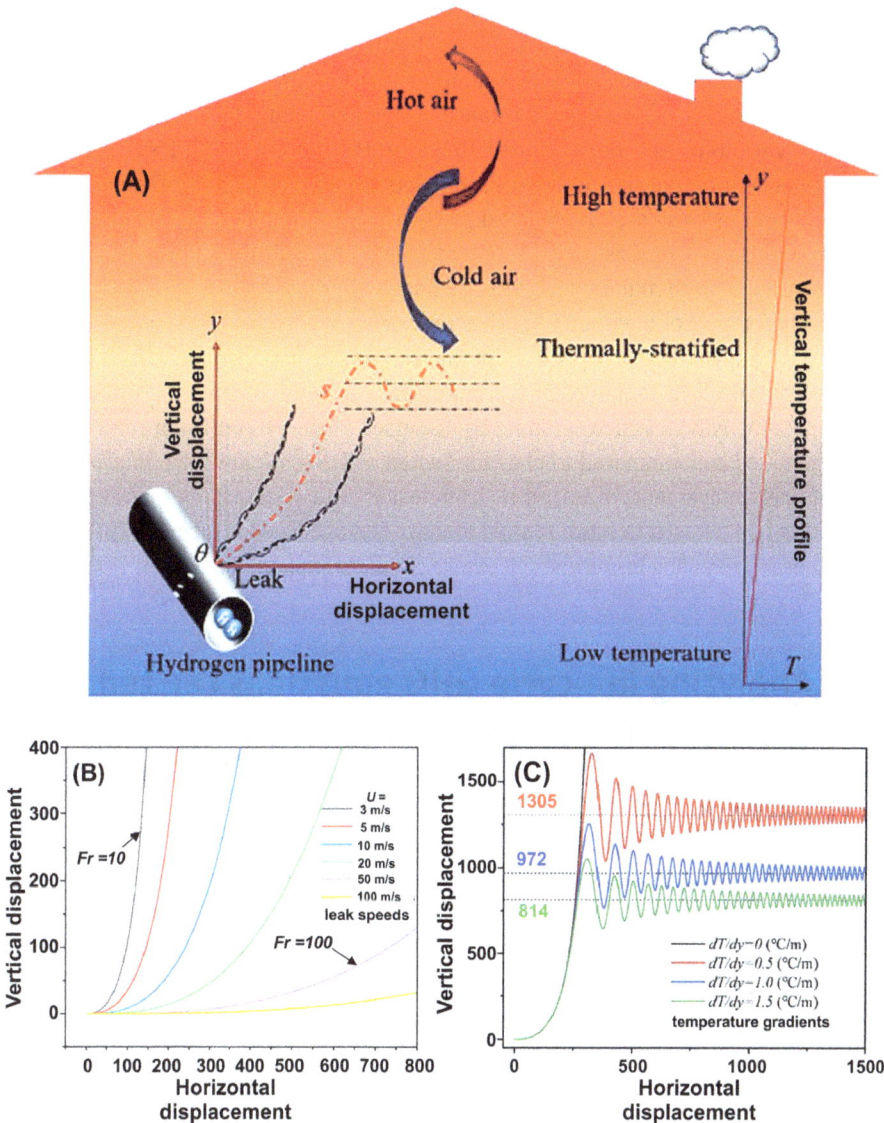

Figure 23.7: Prediction of H_2 building pipework leakage/diffusion characteristics in a space exhibiting thermal stratification. (A) flow trajectories versus temperature. (B) H_2 leakage velocity impacts on flow trajectories (Fr = Froude number). (C) H_2 leakage trajectory impacts for a range of building-space temperature gradients. Modified from Shu et al. [52].

flow conditions (Fr < 1) exist gravitational force is greater than viscous force. It is important to be able to model, predict, and understand the trajectory of H_2 leakage within confined spaces as such information can be used to inform building designs and ventilation to prevent H_2 accumulating and posing a significant safety hazard.

Zhang et al. [53] developed a H_2-leakage/diffusion model for a buried pipeline. Over time, such pipelines are prone to external corrosive reactions from related to the compositions of soil components and pore fluids. They are also prone internal H_2 embrittlement. Consequently, it is important to monitor such pipelines rigorously for signs of leakage. When H_2 leaks in such circumstances, it initially diffuses through the soil (overburden) at various rates and trajectories. However, H_2 diffusion through soil is complex and is dependent on the size of the leakage hole(s), soil type, pipeline pressure, and diameter. Simulation results show that leaked H_2 concentration increases as the duration of the leakage period increases. The distribution of the leaked H_2 tends to disperse over a symmetrical area surrounding points of leakage. Higher pressures lead to an increase in H_2-leakage rate. Longitudinal-diffusion movements dominate the spreading directions of the leak. As the size of the hole in the leaking pipeline increases, the H_2-leakage rate rises rapidly. The composition and pore space in soils surrounding leaking pipelines impact the way in which escaping H_2 disperses. Higher H_2-diffusion rates occur in sand-dominated soil compared to clay-dominated soil. Such knowledge improves estimates of likely movements of leaked H_2 from leaking buried pipelines, providing information that can be used to more rapidly identify leakage locations from H_2 seep monitoring, thereby reducing losses and the extent of environmental damage.

23.8 Analyzing life-cycle GHG emissions related to H_2 supply chains

The GHG effects of H_2 are short-lived relative to CO_2, H_2O, CH_4, and some other VOCs; they generate impacts for about 20 years. However, such impacts magnify if H_2 tropospheric and stratospheric emissions from all sources are considered. H_2 accumulates in these atmospheric layers non-uniformly. Globally, there is approximately 40 ppbv less H_2 in the polar areas of the northern hemisphere compared to similar latitudes of the southern hemisphere [54]. The influence of soil on the continental land masses explains this discrepancy, as it operates as a large-scale sink receiving some H_2 in the form of precipitation from the troposphere. This continuous soil removal effect of H_2 from the atmospheres [55] is substantially impacted by seasonal and multi-year climate conditions. The moisture content in soils, seasonal snow coverage, and the cumulative quantity of continental soil that prevails at any given time also influence the magnitude of this H_2 sink. Temperature-driven diffusion rates of H_2 in soils and the impacts of soil bacteria that consume H_2 in the shallowest soil layers also influence the effect of soil as a H_2 sink locally.

Data recorded 25 years ago [9] established the importance of soil as a sink for atmospheric H_2, identifying that it accounted for about 82% turnover of tropospheric H_2, with about 70% of that turnover occurring in the large continental land masses of the northern hemisphere. The VOC-photochemical-oxidation H_2 sink in the atmosphere is about 20% less effective than the soil H_2 sink. Deuterium/hydrogen isotopic ratio (δD) values and concentrations of H_2 in the troposphere reveal different H_2 cycles in the northern and southern hemispheres (Figure 23.8). The two H_2 sinks mentioned have quite distinctive δD values enabling the impacts of the H_2 sink related to soil to be quantitatively compared with the VOC-photochemical processes acting as an atmospheric H_2 sink. The distinctive northern hemisphere seasonal cycle matches fluctuations in snow coverage which can reach up to about 40% in mid-winter. Key effects of global warming are reducing the percentage the snow coverage throughout the world, leading to an average increase in soil temperatures, These changes are also reducing the areal extent of permafrost, and that loss is actually strengthening the soil H_2 sink, which acts to dampen climate change impacts in the northern hemisphere. However, these changes are ultimately leading to an increase in naturally (biotically) generated CH_4 emissions. However, soil plays a much less significant role as an H_2 sink in the southern hemispheres. These cycles are relevant to the type and scale of environmental impacts to be expected from of an increase in H_2 subsurface storage (UHS) and H_2 leakage from such facilities. H_2 leakage from UHS could upset the local balance between soil and atmosphere H_2 sinks. Figure 23.8 compares the H_2 seasonal cycles in the Earth's northern and southern hemispheres.

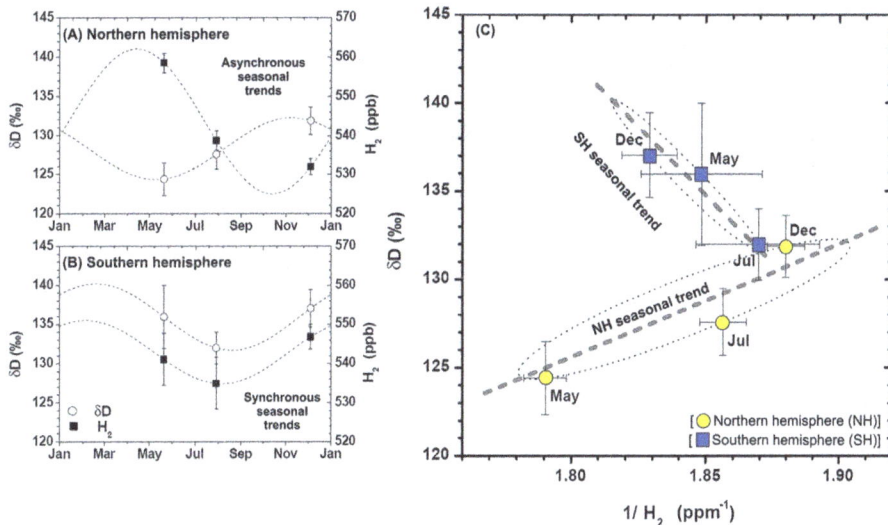

Figure 23.8: Seasonal relationships between atmospheric δD and H_2 variations in the Earth's hemispheres. Dots represent means and bars represent plus and minus standard-deviation errors. Modified from Rhee et al. [9].

Paulot et al. [4] demonstrated that the indirect GHG-emission impacts of H_2 in the troposphere and stratosphere are substantial. Radiative efficiency (RE; W/m^2 ppb) measures GHG strength of a particular atmospheric gas. The effects of H_2 leakage have been estimated to vary in terms of RE [8] to vary substantially depending on the assumption made (−0.25 W/m^2 if H_2 leakage is ignored; 0.5 W/m^2 high H_2 leakage included) leading to an average estimated RE = 0.18 W/m^2 ppb. That range of estimated H_2 RE was converted into mass [6] yielding a low-leakage case H_2 RE estimate of 3.64×10^{-13} W/m^2 kg and a high-leakage case H_2 RE estimate of 5.04×10^{-13} W/m^2 kg. Comparing these values with the CO_2 RE of 1.7×10^{-15} W/m^2 kg and CH_4 RE of 2.0×10^{-15} W/m^2 kg.highlights the substantial potential impact of H_2-leakage emissions on RE. On a mass basis, this implies that the actual GHG impact of H_2 is about 200× the GHG impact of CO_2 and about 2× the GHG impact of CH_4. Such effects tend to be underestimated when 100-year impacts of a single H_2-leakage event are considered. However, such analysis is deemed inappropriate in a scenario in which H_2 supply chains grow rapidly and leakage volumes continue to rise over coming decades [8]. Figure 23.9 illustrates how growing H_2-leakage emissions perpetuate the assumed short-term GHG impacts [6].

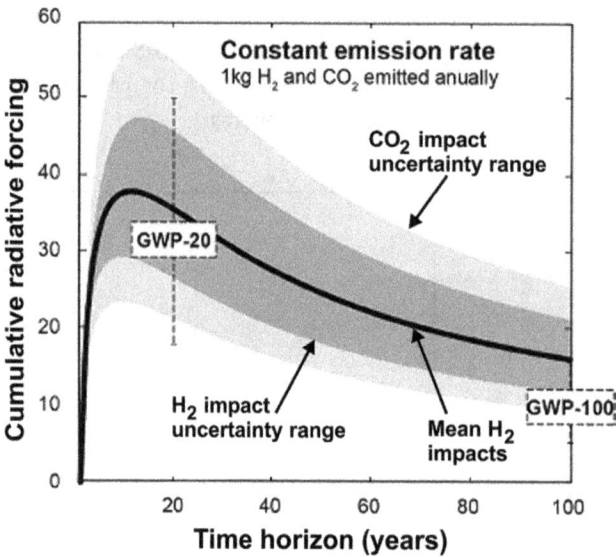

Figure 23.9: H_2 GWP atmospheric impacts compared to those of CO_2 in the atmosphere., assuming equal masses of emissions/gas/year. Modified from Ocko and Hamburg [6].

By modeling the impacts of 1Tg releases of CO_2 and H_2 to the atmosphere, it has been demonstrated [56] that the H_2 emission would have about 5X the GWP over the course of a century than that of CO_2. Such analysis highlights the significance of the H_2-leakage percentages from rapidly expanding H_2 supply chains. Unfortunately, to date, almost no data is available regarding H_2 emissions from commercial-scale infrastructure, leading to much uncertainty in the estimates of H_2-leakage percentages currently being considered [57]. This has led some to argue that rapid uptake of H_2 energy systems only makes sense at low H_2 leakage rates [21, 58]. Based on historic experience with natural gas supply chains, low emissions are only likely to materialize if robust leakage monitoring is enforced on supply-chain operators.

H_2 leakage influences the total GHG emissions resulting from green (renewable-energy-powered electrolysis) and blue (SMR facilities combined with adjunct CCS facilities) H_2 generation in different ways [6]. Leaking H_2 at a sustained rate of 1%, green H_2 production facilities are associated with low GHG emissions over both short- and long-term (about 100 years) horizons. Leaking H_2 at a sustained rate of 10%, green H_2 production facilities would incur much higher GHG impacts, particularly in their early years of operation. Nevertheless, such facilities would likely result in a substantial GWP reduction (about 80% lower), by long-term, sustained displacement fossil-fuel energy systems. If "blue" H_2-fueled energy supply chains are associated with substantial H_2/CH_4 emissions (including leakage), GWP impacts could actually worsen compared to existing fossil-fueled energy supply chains for several decades. On the other hand, by achieving low H_2/CH_4 emissions, blue H_2 supply chains under development could achieve beneficial GWP improvements compared to existing energy facilities when considered over 100+-year timeframes.

Based on the reported performances of CH_4-rich-gas producing facilities in North America, leakage rate assumptions in the range 1–3% have been used for blue H_2 infrastructure [59, 60]. However, some such sites have been recorded with emissions of about 5% [61]. Based on current uncertainties regarding the ability to contain H_2 at operating sites, it is not unreasonable to consider potential H_2 leakage rates between 1% and 10%. For blue H_2 facilities, for every kg of H_2 produced, 3 kg of CH_4 emissions are generated and need to be captured and sequestered to avoid leakage to the atmosphere.

A 1% H_2-CH_4 leakage scenario for blue H_2 supply [6] requires 1.01 times the quantity of H_2 delivered to be produced. Such a plant would also consume a quantity of CH_4 that would be 3.06 times the quantity of H_2 delivered. A 10% H_2-CH_4 leakage scenario for blue H_2 supply [6] requires 1.1 times the quantity of H_2 delivered to be produced and consume 3.43 times that quantity of CH_4. Hence, as the leakage rate increases the impact of potential CH_4 emissions grows substantially and only about two-thirds of that could be captured by existing carbon capturer technologies. LCA conducted for blue H_2 and green H_2 production supply chains considering such emissions scenarios [62] indicate a wide range of GHG outcomes compared to SMR-H_2 production facilities with no CCS capabilities, i.e., − 93%GHG (good) to + 46%GHG (bad) for blue H_2 compared to −66%GHG to −95%GHG for green H_2 production facilities. However, that study also confirmed that H_2-electrolysis production powered by electricity generated from coal combustion resulted in about 4 times the GHG impacts of SMR-H_2 production using current technologies.

Another LCA study [63], applying the Leiden-methane LCA method [64], considered six distinct H_2 production methods:

1. Unabated SMR
2. Mercury (Hg) electrolytic cell
3. Membrane electrolytic cell
4. Diaphragm electrolytic cell
5. Electrolysis powered by solar photovoltaic (PV)
6. Electrolysis powered by wind turbines

Five environmental consequences were assessed for each production method: (a) GWP (Figure 23.10A); (b) abiotic depletion factor (ADF; Figure 23.10B); (c) eutrophication/nitrification (E/N, Figure 10C); (d) nuclear radiation exposure; and (e) ozone layer depletion. SMR resulted in the worst environmental LCA outcomes, whereas electrolysis powered by wind energy resulted in the best environmental LCA outcomes. The poorer E/N outcomes for solar-PV-powered H_2-electrolysis production compared to wind-powered systems is due to the involvement of toxic metals in PV-cell production and the environmental impacts of extracting and recycling those metals.

The main construction/engineering activities and environmental impacts that are generic to many gas production and power plants (including H_2 production plants) that result in environmental impacts are:

- Constructing site (civil engineering) including transport access
- Installing process-plant equipment
- Installing pipework/H_2-storage facilities (above and/or below ground)
- Installing electricity-transmission/distribution lines, transformers and power storage facilities
- Drilling/completion of wellbores with monitoring capabilities (for CCS and UHS, as appropriate)
- End-of-life decommissioning of site and its processing plant
- LCA of ecosystems/biodiversity impacts
- Assessment of LCA air-quality impacts
- Assessment of LCA road-vehicle movements
- Assessment of LCA noise levels and local community disturbance impacts

LCA must take all the above-mentioned construction/engineering/environmental components into account when assessing environmental impacts.

Integrating LCA with net-energy analysis (NEA) can provide additional insight to H_2-production alternatives [66], as can commercial-scale simulations run with multiple sensitivity cases. NEA can establish if and how various energy-supply systems can deliver to consumers a net-energy surplus, taking into account all supply-chain energy losses [65]. Sensitivity LCA-NEA cases associated with a PV-powered H_2 production plant [66] identified substantial negative GWP and energy cost impacts associated with periods when grid-electricity back-up power supply was required.

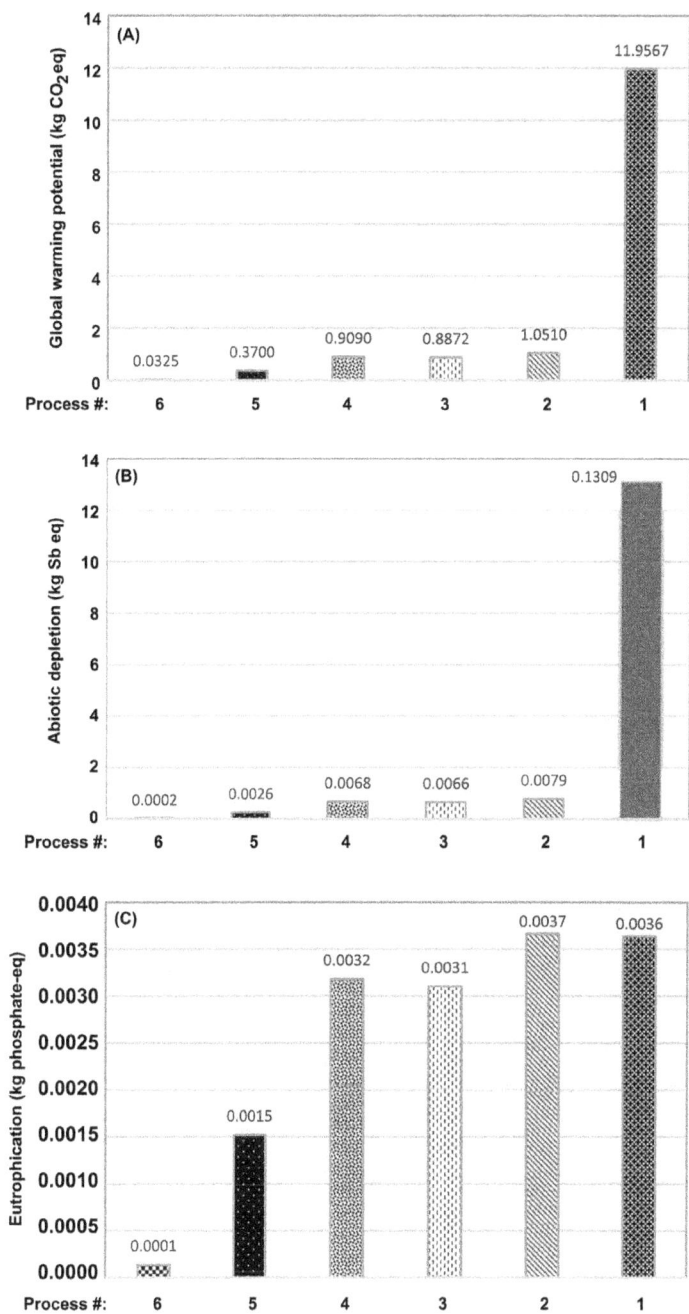

Figure 23.10: Three LCA impacts of six H$_2$ production methods: (A) GWP; (B) ADP; and (C) E/N. The numbers 6 to 1 on the horizontal axis refer to the six methods identified and numbered in the text. Modified from Suleman et al. [65].

A 2024 LCA study [67] compared the performances to 2050 of (a) coal-gasification-H_2 production, (b) SMR (with and without CCS), (c) biomass-gasification-H_2 production, and (d) renewable-energy-powered-electrolysis-H_2 production. The results of that study highlighted that SMR-CCS technologies (capturing only about 64% of their CH_4 emissions) dominated growing H_2 production up to 2050, such plants would be responsible for GHG emissions of about 1 Gt CO_2-eq. per year by 2050 (a substantial portion of total GHG emissions for net-zero energy scenarios). This implies that SMR-CCS technologies could become a long-term GHG burden, unless their GHG-capture efficiencies could be substantially improved. There is an increasing recognition that LCA needs to more comprehensively consider the GHG and broader environmental impacts of resource extraction, process-equipment manufacturing, operating and recycling, and waste-products generated [1] to provide holistic H_2-supply-chain impact assessments.

23.9 Underground hydrogen storage

Growth in demand for H_2 necessitates H_2-storage facilities with substantial capacities to balance supply-demand fluctuations [68]. UHS reservoirs offer this capability [69, 70]. Underground gas storage has been used around the world successfully for many decades to store natural (CH_4) gas in bulk [69], applying both short- and long-term (seasonal injection during off-peak season and recovery during periods of peak demand). Underground reservoirs are also now used extensively for storing captured CO_2 [70]. Four types of sub-surface reservoirs are considered potentially suitable for UHS: salt caverns (natural or artificial), former gas or oil producing reservoirs with known sealing formations, aquifers in geologically trapped/sealed structures, abandoned and repurposed mines (coal and other minerals).

Two main environmental risks are associated with UHS: (1) inadequate cap-rocks failing to trap H_2, and/or faulting and fractures through reservoir and seal formations, and (2) wellbore tubular/cement/casing failures causing H_2 leakage through UHS wellbores. Such risks also apply to the underground storage of other gases (e.g., CH_4 and CO_2), but the small molecular size, reactivity, and high mobility of H_2 make it more prone to leakage, with the potential for contamination of the overburden, soil, and/or groundwater by diffusion and leakage associated with near-surface reservoirs such as repurposed mine conduits [71]. H_2 solubility in reservoir formation water varies substantially depending on the salinity of the formation water and has substantial influence on how H_2 is stored and moves with UHS reservoirs [72].

High-diffusivity and low-viscosity characteristics of H_2 exacerbate its potential for leakage from UHS facilities [73]. Cap rock leakage is influenced by H_2's high diffusivity and buoyancy making micro-fractures, fractures, and faults easier to penetrate. High and sustained H_2-injection and reservoir pressure can negatively influence UHS leakage rate [74], as does the degree of tortuosity of the pore-size distributions of the cap rock formations [75]. Perforating H_2-injection wellbores close to the cap-rock can also increase the risk of leakage [76]. Extensive activity of sulfate-reducing bacteria and biologically induced methanogenesis of H_2 can also occur within some UHS reservoirs. Such processes generate water as a by-product, adversely increasing reservoir pressure over time and enhancing H_2 diffusivity into and through cap-rock formations [77]. UHS reservoir simulations also suggest that H_2 is more likely to leak through the caprock as reservoir depth increases [78]. Substantial reservoir-fluid hydrodynamic movements [79], geochemical reactions, and bio-degradation of H_2 within a UHS reservoir also influence the dispersal and composition of the fluids throughout a porous reservoir [80] and may influence leakage risks. Likewise, formation trap and cap-rock geometries that involve steep dips and high gas column thicknesses may increase UHS leakage risks [81].

Generally, depleted natural gas is relatively easy to repurpose as UHS facilities, because they have tried-and-tested traps, reservoirs, and seals. Hence, they pose less risk of cap-rock leakage or lateral formation dispersal than aquifers. A possibility exists to use the large number of depleted/inactive shale-gas wellbores located in fracture-stimulated formations in North America for UHS [82]. Such wells can potentially be re-purposed for UHS at relatively low cost. However, the quality of top and lateral seals is typically greatest in salt caverns, which should be preferred for UHS. However, salt caverns are not widely available and artificial salt caverns have limited capacities due to high leaching/construct costs. Assessing the suitability of any type of rock formation with a potential trapping mechanism for exploitation as a UHS facility from a leakage-risk perspective requires in-depth consideration of geomechanical factors that influence H_2 storage quantity and UHS productivity, with specific reference to H_2 physical/chemical characteristics [68, 83]. Storing H_2 as a liquid possible at surface but requires cryogenic temperatures (H_2 boiling point is approximately – 252.8 °C at atmospheric conditions). Such low temperatures make UHS in liquid form unfeasible.

Wellbore designs that ensure integrity under UHS conditions are essential to prevent H_2 leakage via cement/tubular/rock-formation interfaces (Figure 23.11) [84]. The risk of metal embrittlement caused by contact with H_2 requires the use of expensive steel alloys for UHS wellbore tubulars and multiple strings of cemented tubulars to serve as leakage barriers. The CH_4-rich gas storage industry has many years of experience operating thousands of wellbores, yet it still experiences periodic wellbore integrity failures, particularly in repurposed wells, some of which have serious environmental consequences. For example, the failure of the Aliso Canyon well (SS#25), with substantial gas leakage persisting for several months in 2015 and 2016, occurred in a repurposed well in a depleted gas/oil field in California (USA) [85]. Wellbore-integrity failures are therefore considered as a substantial risk for UHS facilities, and one of the most likely sources of major leakage from H_2 supply chains.

Fluid Leakage Routes Potentially Leading to Sub-surface Breaches of Well Integrity in Underground Hydrogen Storage Facilities

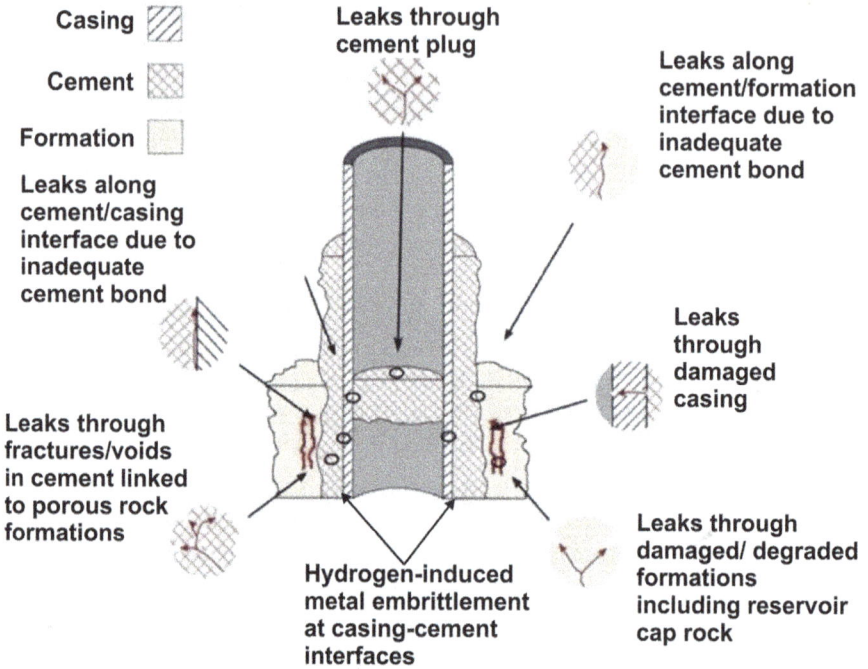

Casing

Cement

Formation

Leaks along cement/casing interface due to inadequate cement bond

Leaks through fractures/voids in cement linked to porous rock formations

Leaks through cement plug

Leaks along cement/formation interface due to inadequate cement bond

Leaks through damaged casing

Hydrogen-induced metal embrittlement at casing-cement interfaces

Leaks through damaged/ degraded formations including reservoir cap rock

Note: the small molecular size, reactivity and mobility of the hydrogen molecule means that it will more easily penetrate any defects that exist in well-bore casing and cement.

Figure 23.11: Potential UHS leakage points from wellbore tubulars and their causes. Modified from Wood [84] and Reinicke and Fichter [86].

23.10 Natural hydrogen resources and their environmental consequences

There are multiple potential sources of natural H_2 generated in the subsurface [87, 88]. These include biotic (H_2 generated within the soil and shallow sediments), abiotic processes (H_2 generated primarily in deeper crustal locations), and probably some traces of primordial H_2 located in the mantle originating from when the Earth originally formed within the solar system. The major abiotic sources of H_2 generated within the Earth's crust [89, 90] include:

(1) the complex oxidation-reduction processes involving Fe^{2+}-rich silicate minerals and water (e.g., serpentinization of certain Fe-rich mafic and ultramafic rocks, common in ophiolite formations) and Fe^{2+}-rich carbonates

(2) Oxidation of magnetite during the low-temperature metamorphism of Precambrian banded iron formations (BIF) converting magnetite (Fe_3O_4; containing iron as both Fe^{2+} and Fe^{3+}) into maghemite (γ-Fe_2O_3; containing iron only in Fe^{3+} form), in the presence of anoxic water present in aquifers and/or deep percolation, by the total oxidation reaction [91]:

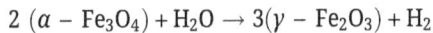

$$2\,(\alpha - Fe_3O_4) + H_2O \rightarrow 3(\gamma - Fe_2O_3) + H_2$$

(3) Radiolysis, the splitting of water and organic matter molecules with ionizing radiation [92]

(4) In the vicinity of volcanic activity, dominated by mid-ocean-ridge volcanism, and magma chambers, degassing of magma at low pressures involves the small quantities of dissolved water in the magma, oxidizing ferrous iron [90] by the reaction:

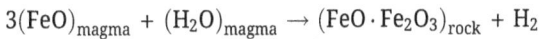

$$3(FeO)_{magma} + (H_2O)_{magma} \rightarrow (FeO \cdot Fe_2O_3)_{rock} + H_2$$

(5) Fault/fracture activity penetrating silica-rich formations (sandstones, siltstones, etc.) involve some rock crushing and mechanical fragmentation (mylonitization) forming free radicals that cause reactions in the presence of water resulting in H_2 generation [93] by the reaction:

$$Si + H_2O \rightarrow SiOH + H$$

H_2-bearing gases sampled from mid-ocean ridge volcanoes and serpentinites have recorded the highest average concentrations of H_2 (about 24% and around 21%, respectively), but samples associated with the other processes average only about 3.5% H_2 [94]. It is clearly essential to explore for and find much larger H_2-bearing reservoirs with much higher H_2 concentrations than have been found to data [95]. The various reactions described and others occurring within the Earth's crust indicate that a substantial amount of abiotic H_2 is being generated and circulated in a wide range of shallow and deep crustal environments in the presence of anoxic water. This is in addition to the biotic H_2 being generated in soils and shallow sediments by biotic processes.

Microbial activities exist in the soil and shallow formations that both generate and consume H_2 [96]. H_2-oxidizing bacteria exploit the H_2 molecule as an electron donor. They do this in different ways: aerobes combine H_2 with O_2 (as an electron acceptor) to form water; anaerobes (knell-gas bacteria) combine with sulfates or nitrogen dioxide. In the soil, some species can use H_2 to fix carbon in the presence of organic material, others fix nitrogen. H_2-oxidizing bacteria exist in many surface and subsurface environments including hot springs and deep-ocean hydrothermal vents. Hence, they consume vast quantities of hydrogen on a global basis. Micro-organisms also exist in subsurface environments (soils and relatively shallow rock formations) that can generate H_2 by several processes that do not require light energy. These processes typically involve the enzyme hydrogenase and include fermentation, N_2 fixation, anaerobic carbon monoxide oxidation, and phosphite oxidation [87]. Many of these processes have the potential to be adapted for commercial H_2-generation [97]. Both H_2-producing bacteria and methanogens are common in microbial communities in shallow carbon-rich formations including coal seams, where they contribute to coal-bed methane [98].

Because of this microbial activity, there is a substantial amount of near-surface microbial H_2 cycling activity. Approximately steady-state conditions currently exist between natural H_2 seeping from the continental land masses into the atmosphere and the prevailing H_2 concentrations in the atmosphere due to the destruction of much of that H_2 entering the atmosphere. There is, though, uncertainty regarding how much seeped/leaked H_2 the atmosphere and H_2-consuming microbial ecosystems in the soil/shallow rock formations can process into CH_4 and water vapor before it upsets that "equilibrium" and has more negative climate impacts. Boyd et al. [99] estimate that substantial quantities of H_2 generated by the abiotic process in the Earth's crust, more than 90% in some cases (e.g., Samail Ophiolite, Oman/UAE), are potentially consumed by subsurface microbial ecosystems. Some of the potential negative consequences of microbial activity on H_2 stored in UHS facilities are methanogenesis [100] causing loss of some stored H_2 due to conversion to CH_4, and the hydrogen sulfide (H_2S) production leading to iron sulfide precipitation and a degradation in porosity [75, 76]. H_2S traces in recovered UHS gas presents safety and corrosion hazards. The environmental impacts of H_2 microbial activity should not, therefore, be underestimated when considering H_2-resource exploitation or UHS projects.

In recent years, multiple studies have reported and described surface seeps of H_2-bearing gases, identified to be generated for the most part from crustal abiotic sources. These include sites in Russia [101], the USA [102]., Brazil [103], Oman [104], Australia [105, 106], Namibia [107], and Turkey [108]. These features tend to be characterized by multiple circular/elliptical surface anomalies, some with depressions, often water filled or marshy, surrounded by rings of degraded vegetation (referred to by some as "fairy circles"). In some regions, such features are aligned along fault trends. Most overlie Precambrian basement cratons or old ophiolite deposits with H_2 transported through faults connected to basement fault networks. The H_2 tends to be transported closer to the surface with advection related to percolating ground water into near-surface soils [109], in contrast to the soil diffusion processes associated with the dispersal of smaller concentrations of biotic H_2 [108]. H_2 concentrations are variable and not pure in such seeps, typically being accompanied by CH_4, CO_2, N_2, and helium(He) in a wide range of concentrations [110], which makes their exploitation problematic in terms of their potential GHG emissions, although He and N_2 (non-GHG) are not environmentally harmful. Indeed, He could be potentially extracted and exploited commercially in conjunction with H_2 [95]. On the other hand, too much N_2 entering the soil, groundwater and surface water does results in nitrification (E/N environmental impacts). However, of the four contaminating gases mentioned in natural H_2 accumulations, it is CH_4 and CO_2 that have the potential to substantially increase the GHG-emissions footprints of those gas mixtures.

Larin [101] studied the Borisoglebsk-Novokhopersk area (approximately 3,300 km^2) of central Russia. It is located where a Precambrian craton is relatively close to the surface (approximately 600 + m deep). Five hundred and sixty-two quasi-circular features were identified in this area as being caused by H_2 seepage. Most of the features had widths of <100 m but some reached 1 to 3 km in width. Only some of these features existed as water-filled depressions. The features were associated with soil bleaching and characterized by "a decrease of vegetation and microbial biomass." These described characteristics have environmental implications for sustained H_2 leakage, albeit at low percentages, from UHS facilities and future natural H_2 exploitation facilities involving subsurface reservoirs. Small but persistent amounts of H_2 leaking via capping/sealing formations, faults/fractures, and/or wellbore tubulars for sustained periods of time can lead to surface and sub-surface environmental damage. Potential negative influences on soil ecosystems and agricultural yields are poorly understood. Clearly, more research is required to quantify such effects. These potential risks should be addressed in robust environmental-impact assessments in advance of sanctioning the construction of such facilities.

Some early-stage research is being conducted to artificially stimulate H_2 production in the sub-surface by injecting water into Fe-rich formations and inducing oxidation reactions [111]. As well as determining whether such approaches are technically and commercially viable, it is essential that they consider the environmental implications, particularly the likelihood that at least some of the H_2 generated would be converted to CH_4 by subsurface processes and ultimately contribute to GHG emissions. The proposed technique was simulated to explain how it could potentially stimulate H_2 production in the sub-surface from suitable Fe-rich formations. The proposed method exploits formation weathering and oxidation reactions to generate H_2. This requires the drilling of injection and production wellbores and injecting water (and potentially CO_2) into the target formations. No pilot or commercial-scale testing has yet been conducted, although environmental concerns regarding the risk of poorly controlled potential CH_4, CO_2, and H_2 leakage from such projects [21, 95].

Brandt [112] conducted a provisional (early stage), partial LCA [113] concerning a potential natural-H_2 production site plus gas processing facility to identify the sensitivity of GHG emissions to produced gas compositions. The base case assumed a produced gas composition of H_2 (85%), N_2 (12%), and CH_4 (1.5%), assuming well productivity and depth ranges typical of natural gas reservoirs in the USA and generated site-boundary GHG emissions of ~0.4 kg CO_2-eq/kg H_2 produced. Fugitive emissions were identified as the main GHG emissions associated with the simulated site. However, the GHG emission results were highly sensitive to CH_4 concentrations in the produced gas. A case assuming a produced gas composition of H_2 (75%), N_2 (2.5%), and CH_4 (22.5%) generated site-boundary GHG emissions of about 1.5 kg CO_2-eq/kg H_2 produced. This analysis, as summarized in Figure 23.12, highlights that produced natural H_2 with less than about 90% H_2 concentration and/or high CH_4 components would result in relatively high GHG emissions and therefore, represent environmentally unattractive developments. Note that reservoired gases composed of only H_2 and He would plot below gases containing only H_2 and N_2 in Figure 23.12, making them attractive exploration targets [21].

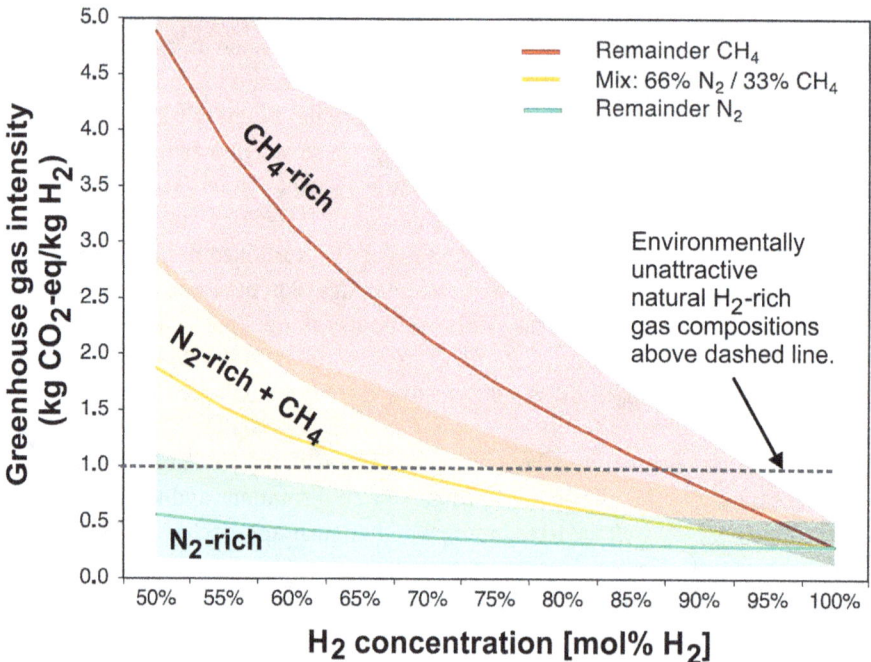

Figure 23.12: GHG intensity results of provisional life-cycle analysis of natural H_2 resources with different gas compositions including those rich in CH_4 and N_2, which are known to occur in the subsurface. Note that gases with substantial He contents would perform in a similar way to N_2-rich gas mixtures in this diagram. Modified from Brandt [112].

The need for hydrogen industry regulation to reduce its environmental footprint

Currently, due in part to the uncertainties regarding the technologies and future plants designed to potentially exploit natural H_2 or store green or blue H_2 production, H_2-focused legislation, regulation, or recommended best practices do not yet exist anywhere in the world. In general, the gas-processing and power-generation facility operators, and those regulating their activities, seem to be content to apply in-place natural-gas-related infrastructure and regulations, where possible. Do so leads to minimal changes, to stimulate investment, accelerate planning permissions and promptly implement new H_2-based infrastructure developments. However, it is clear from the characteristics of manufactured and naturally generated H_2 described in this chapter that H_2-specific safety, emissions/spill/leakage, environmental, ecosystem, and GHG emission risks require targeted regulation/legislation. Such focused regulations would ensure that potential adverse consequences would be more appropriately assessed, and specific mitigation steps taken to minimize their impacts prior to sanctioning H_2-focused facility developments. To justify and mandate such considerations as part of meaningful environmental impact and safety case studies, there is a need for H_2-explicit regulation and legislation [95]. Without such H_2-specific legislation there would likely be higher risks of accidents and spills (due to equipment failure/leakage), and environmental damage associated with new, H_2-based facilities developed under inappropriate existing legislation.

Several jurisdictions have prematurely formulated supportive policies to promote green/blue H_2 production facility construction, some providing support in the form of subsidy or grant or tax offset related to such facilities. However, these industry-supporting decisions have been made without H_2-specific legislation/regulation in place that considers the potential environmental consequences of substantially expanded H_2 supply chains [21, 109]. In respect of exploiting natural-H_2 resources, the existing CH_4-natural-gas-related legislation/regulation is not fit for purposes in respect of the GHG emissions, atmospheric, soil and ecosystem impacts, and leakage risks of H_2 supply chains.

23.11 Summary and conclusions

This review evaluates the complex interactions and environmental impacts of hydrogen molecules (H_2) of anthropogenic and natural origin within the atmosphere and soil. Understanding these impacts is important as global efforts are made to achieve near-carbon-free energy-supply systems involving substantial H_2 supply chains. Of the H_2 production processes available, only green H_2 (water-electrolysis processes powered only with renewable energy without fossil-fuel backup power supply) can substantially diminish the CO_2 emissions currently entering the atmosphere. However, existing electrolysis processes are energy inefficient and consume valuable water resources. The steam-methane reforming (SMR) process, the cheapest available and most widely used, releases the most GHG emissions of available H_2-generation processes. SMR emissions can be partly abated by adding carbon- capture-storage (CCS) facilities to generate blue H_2. However, the CCS is inefficient recovering up to about two-thirds of the GHG emissions and is expensive to install and operate. The depletion of natural hydrocarbon resources and the use of toxic metal catalysts are other negative environmental impacts of the SMR process.

When H_2 is combusted as a fuel for various energy generating, transport or petrochemical end use, no matter how the feedstock H_2 is produced, emissions of NO_x result that are in excess of those generated by fossil-fuel combustion. H_2 combustion is therefore associated with some negative impacts on air quality. Consuming H_2 in fuel cells does not involve its combustion and therefore avoids NO_x emissions, but fuel cells remain too expensive for many transport applications. To minimize adverse atmospheric effects of H_2 combustion requires reducing its NO_x emissions. To do so typically means that H_2 combustion equipment (turbines and engines) is tuned to provide sub-optimal energy outputs. Imposing such requirements on H_2-facility operators would likely prove difficult to regulate and enforce in practice.

From a cost perspective, in terms of infrastructure modification requirements, it is often effective to blend H_2 into plants/engines originally designed for hydrocarbon fuels. This can improve energy efficiency but typically only reduces GHG emissions by a relatively small percentage. H_2 can be used as a hybrid fuel with gasoline or diesel in internal combustion engines (ICE) and compression ignition (CI and HCCI) engines, respectively, acting to reduce GHG emissions. HCCI engines configured to consume hybrid H_2 with diesel or natural gas can also reduce NO_x emissions. However, the main challenges are related to the high-cost vehicle H_2-fuel-storage systems and the leakage/seepage of some H_2 from them. There are many environmental and cost challenges to overcome to make H_2 fuel use in aviation economically attractive and commercially viable. H_2-fuel combustion in aircraft engines would result in substantial emissions of NO_x and H_2O forming harmful contrails. The flying conditions (specifically altitude and jet-engine design) influence the scale of such emissions. Much uncertainty exists regarding the costs, environmental impacts, and risks of H_2-fuel leakage from aircraft, making it unlikely to become commercially adopted in coming decades.

Although it is widely recognized that H_2 leakage along the supply chain represents a substantial environmental issue, much uncertainty surrounds the likely magnitudes of H_2 leakage associated with expanding H_2 supply chains. Leakage estimates ranging between 1% and 10% have been evaluated. H_2 storage/transport infrastructure is most vulnerable to leakage. Leakage of H_2 in confined spaces such as in buildings using H_2-fueled heating/cooling systems poses safety and environmental hazards. Research conducted to date on the movements of H_2 leaked from pipelines, either buried or within buildings, reveal complexities that require designs specifically tailored to mitigate the likely impacts. Detailed LCA and net-energy analysis (NEA) applied to the entire H_2-supply chain are required to determine commercial feasibility as well as GHG emissions and other environmental impacts. To do this effectively, requires adequate consideration of H_2 sources/sinks in both atmosphere and soil, and their seasonal and spatial (northern versus southern hemisphere) variations. Although the GHG impacts of a single injection of H_2 into the atmosphere persist for more than about two decades (much shorter than CH_4/CO_2 emissions), sustained H_2 leakage, in growing quantities year-by-year, into the atmosphere generates substantial longer term GHG impacts.

To utilize H_2 in large quantities on a continuous basis requires large-scale H_2 storage, which can be most effectively achieved by utilizing underground H_2 storage reservoirs (UHS). UHS are prone to two environmental impacts: (1) as H_2 is a small and mobile molecule it poses a greater risk of leakage through the surrounding geological formations than CH_4 or CO_2; and (2) as H_2 is highly reactive it poses a greater risk of leakage through wellbore tubulars and annular cement than CH_4. UHS facilities are particularly vulnerable to H_2 leakage/seepage through cap rocks and wellbores. Such leakage could upset the balance of microbial H_2 generation/consumption within soils above a UHS facility's overburden, leading to soil degradation and reduced crop yields, as well as increased H_2 emissions to the atmosphere.

Complex and multiple biotic/abiotic H_2 generation processes are continuously at work in the Earth's crust and provide a sustained H_2 contribution to soils and the atmosphere. In recent years, surface seeps of gases rich in H_2 have been recorded and characterized in many countries. The majority of the known H_2 seeps are associated with soil bleaching and a decrease in vegetation and microbial biomass with negative consequences for agricultural yields. Much natural H_2 generated is relatively rapidly converted (on a geological timescale) into CH_4. The abiotic H_2-generation processes involving carbon-containing materials and biotic processes result in mixtures of H_2, CH_4, and CO_2 in varying proportions, with some of those gases becoming fixed/trapped in soils, and some ultimately being released to the atmosphere. Attempting to artificially stimulate H_2 production from underground Fe-rich formations or exploiting natural H_2 accumulations mixed with CH_4 and/or CO_2 is likely to lead to increased GHG releases, requiring GHG-emissions-mitigation steps to be taken. A provisional LCA applied to a hypothetical natural H_2 underground reservoir with gas contaminated with various amounts of CH_4 and N_2 suggests that gases with <90% H_2, especially those mixed mainly with CH_4 and/or CO_2, are likely to be unattractive from an environmental perspective due to their high GHG impacts on the atmosphere.

Nomenclature

ADF	Abiotic depletion factor
BPHE	Battery-powered hybrid electric (vehicles)
CCGT	Combined cycle gas turbines
CCS	Carbon capture and storage
CH_4	Methane
CI	Compression ignition
CO	Carbon monoxide
CO_2	Carbon dioxide
CO_{2-eq}	Carbon dioxide equivalent
DALY	Disability-adjusted life years
E/N	Eutrophication/nitrification
ER	Equivalence ratio
Fe^{2+}	Ferrous iron
Fe^{3+}	Ferric iron
Fr	Froude number

GHG	Greenhouse gas
GWP	Global warming potential
H_2	Hydrogen
HC	Hydrocarbons
HCCI	Homogeneous charge compression ignition
HFCV	Hydrogen fuel-cell vehicle
Hg	Mercury
ICE	Internal combustion engine
IEA	International Energy Agency
LCA	Life-cycle analysis
LNG	Liquefied natural gas
NEA	Net energy analysis
N_2	Nitrogen
NOx	Oxides of nitrogen
O_2	Oxygen
O_3	Ozone
PM	Particulate matter
PO_4	Phosphate
PV	Solar photovoltaic
RE	Radiative efficiency
RECCE	Resource to Climate Comparison Evaluator tool (Aviation Impact Accelerator)
rpm	Revolutions per minute
Sb	Antimony
SMR	Steam methane reforming
UHS	Underground hydrogen storage
VOC	Volatile organic compounds
δD	Deuterium/hydrogen (D/H) isotope ratio
$\gamma\text{-}Fe_2O_3$	Maghemite (containing iron only in Fe^{3+} form)

Disclosures: The authors confirm that they have no conflicts of interest and have received no institutional funding associated with the material presented in this chapter.

References

[1]   Osman A. I., Nasr M., Mohamed A. R., Al-Muhtaseb A. H., Ayati A., Fargali M., Al-Muhtaseb A. H., Al-Fateh A., Rooney D. W. Life cycle assessment of hydrogen production, storage, and utilization toward sustainability. WIREs Energy and Environment. 2024, 13: e526. https://doi.org/10.1002/wene.526

[2]   Atilhan S., Park S., El-Halwagi M. M., Atilhan M., Moore M., Nielsen R. B. Green hydrogen as an alternative fuel for the shipping industry. Current Opinion in Chemical Engineering. 2021, 31: 100668. https://doi.org/10.1016/j.coche.2020.100668

[3]   International Energy Agency (IEA). Towards hydrogen definitions based on their emissions intensity. International Energy Agency. 2023, 87pages. https://www.iea.org/reports/towards-hydrogen-definitions-based-on-their-emissions-intensity Accessed 1st October 2024

[4]   Paulot F., Paynter D., Naik V., Malyshev S., Menzel R., Horowitz L. W. Global modeling of hydrogen using GFDL-AM4.1: Sensitivity of soil removal and radiative forcing. International Journal of Hydrogen Energy. 2021, 46: 13446–13460. https://doi.org/10.1016/j.ijhydene.2021.01.088

[5]   Leighton P. Photochemistry of Air Pollution, Oxford: Elsevier Science, 1961. ISBN 978–0-323-15645-5

[6]   Ocko I. B., Hamburg S. P. Climate consequences of hydrogen emissions. Atmospheric Chemistry and Physics. 2022, 22: 9349–9368. https://doi.org/10.5194/acp-22-9349-2022

[7]   Novelli P. C., Lang P. M., Masarie K. A., Hurst D. F., Myers R., Elkins J. W. Molecular hydrogen in the troposphere: Global distribution and budget. Journal of Geophysical Research: Atmospheres. 1999, 104: 30427–30444. https://doi.org/10.1029/1999jd900788

[8]   Warwick N., Griffiths P., Keeble J., Archibald A., Pyle J., Shine K. Atmospheric implications of increased Hydrogen use, department for business. Energy and IndustrialStrategy. 2022, 75pages https://assets.publishing.service.gov.uk/media/624eca7fe90e0729f4400b99/atmospheric-implications-of-increased-hydrogen-use.pdf Accessed: 30 Sep 2024.

[9]   Rhee T. S., Brenninkmeijer C. A. M., Rockmann T. The overwhelming role of soils in the global atmospheric hydrogen. Atmospheric Chemistry and Physics. 2006, 6: 1611–1625.

[10] Noyan O. F. Some approach to possible atmospheric impacts of a hydrogen energy system in the light of the geological past and present-day. International Journal of Hydrogen Energy. 2011, 36(17): 11216–11228. https://doi.org/10.1016/j.ijhydene.2011.06.032

[11] Nowotny J., Veziroglu T. N. Impact of hydrogen on the environment. International Journal of Hydrogen Energy. 2011, 36: 13218–13224. 10.1016/j.ijhydene.2011.07.071

[12] EIGA. Environmental guidelines for permitting hydrogen plants producing less than2 tonnes per day Doc 220/19. 2019, 7 pages. https://www.eiga.eu/ct_documents/doc220-pdf/ [Accessed 1 October 2024]

[13] Sadeq A. M., Homod R. Z., Hussein A. K., Togun H., Mahmoodi A., Isleem H. F., Patil A. R., Moghaddam A. H. Hydrogen energy systems: Technologies, trends, and future prospects, Science of the Total Environment, 2024. vol. 939, 173622. https://doi.org/10.1016/j.scitotenv.2024.173622

[14] Amin M., Shah H. H., Fareed A. G., Khan W. U., Chung E., Zia A., Ur Z., Farooqi B., Lee C. Hydrogen production through renewable and non-renewable energy processes and their impact on climate change. International Journal of Hydrogen Energy. 2022, 47(77): 33112–33134. https://doi.org/10.1016/j.ijhydene.2022.07.172

[15] Osman A. I., Mehta N., Elgarahy A. M., Hefny M., Al-Hinai A., Al-Muhtaseb A. H., Rooney D. W. Hydrogen production, storage, utilisation and environmental impacts: A review. Environmental Chemistry Letters. 2022, 20(3): 2213–2213. https://doi.org/10.1007/s10311-022-01432-x

[16] Alagumala A., Devarajan B., Song H., Wongwises S., Ledesma-Amaro R., Mahian O., Sheremet M., Lichtfouse E. Machine learning in biohydrogen production: A review. Biofuel Research Journal. 2023, 10(2): 1844–1858. https://doi.org/10.18331/Brj2023.10.2.4

[17] Sarangi P. K., Nanda S. Biohydrogen production through dark fermentation. Chemical Engineering and Technology. 2020, 43(4): 601–612. https://doi.org/10.1002/ceat.201900452

[18] Javed M. A., Zafar A. M., Aly Hassan A., Zaidi A. A., Farooq M., El Badawy A., Lundquist T., Mohamed M. M. A., Al-Zuhair S. The role of oxygen regulation and algal growth parameters in hydrogen production via biophotolysis. Journal of Environmental Chemical Engineering. 2022, 10(1): 107003. https://doi.org/10.1016/j.jece.2021.107003

[19] Gautam R., Nayak J. K., Ress N. V., Steinberger-Wilckens R., Ghosh U. K. Bio-hydrogen production through microbial electrolysis cell: Structural components and influencing factors. Chemical Engineering Journal. 2023, 455: 140535. https://doi.org/10.1016/j.cej.2022.140535

[20] Derwent R., Simmonds P., O'Doherty S., Manning A., Collins W., Stevenson D. Global environmental impacts of the hydrogen economy. International Journal of Nuclear Hydrogen Production and Applications. 2006, 1(1): 57–67. https://doi.org/10.1504/IJNHPA.2006.009869

[21] Wood D. A. Critical review of development challenges for expanding hydrogen-fuelled energy. Fuel. 2025, 2025 287: 134394. https://doi.org/10.1016/j.fuel.2025.134394

[22] EIGA. Environmental impacts of hydrogen plants. Doc 122/18. 2018, 12pages. https://www.eiga.eu/uploads/documents/DOC122.pdf [Accessed 1 October 2024]

[23] Behroozsarand A., Wood D. A. Exergy losses for reformers involved in hydrogen and synthesis gas production compared. Chemical Engineering and Technology. 2019, 42(12): 2681–2690. https://doi.org/10.1002/ceat.201900078

[24] Al Mubarak F., Rezaee R., Wood D. A. Economic, societal and environmental impacts of available energy sources: A review. ENG. 2024, 5: 1232–1265. https://doi.org/10.3390/eng5030067

[25] International Energy Agency (IEA). Global Hydrogen Review October 21, 2021. 223 pages. https://doi.org/10.1787/39351842-en.

[26] Yedinak E. M. The curious case of geologic hydrogen: Assessing its potential as a near-term clean energy source. Joule. 2022, 6: 503–508. https://doi.org/10.1016/j.joule.2022.01.005

[27] Otillar S. R., Azar R., Rwejuna S., Bryant A. Hydrogen hub projects awarded $7 billion by us department of energy. Case & White 13th October 2023. https://www.whitecase.com/insight-alert/hydrogen-hub-projects-awarded-7-billion-us-department-energy [Accessed 14th October 2024]

[28] Mehmood A., Gong M., Jaouen F., Roy A., Zitolo A., Khan A., Sougrati M. T., Primbs M., Bonastre A. M., Fongalland D., Drazic G., Strasser P., Kucernak A. High loading of single atomic iron sites in Fe–NC oxygen reduction catalysts for proton exchange membrane fuel cells. Nature Catalysis. 2022, 5: 311–323. https://doi.org/10.1038/s41929-022-00772-9

[29] Cellek M. S., Pınarbaşı A. Investigations on performance and emission characteristics of an industrial low swirl burner while burning natural gas, methane, hydrogen-enriched natural gas and hydrogen as fuels. International Journal of Hydrogen Energy. 2018, 43: 1194–1207. https://doi.org/10.1016/j.ijhydene.2017.05.107

[30] Lewis A. C. Optimising air quality co-benefits in a hydrogen economy: A case for hydrogen-specific standards for NOx emissions. Environmental Science: Atmospheres. 2021, 1: 201–207. https://doi.org/10.1039/d1ea00037c

[31] Luo Q., Hu J. B., Sun B. G., Liu F., Wang X., Li C., Bao L. Z. Effect of equivalence ratios on the power, combustion stability and NOx controlling strategy for the turbocharged hydrogen engine at low engine speeds. International Journal of Hydrogen Energy. 2019, 44: 17095–17102. https://doi.org/10.1016/j.ijhydene.2019.03.245

[32] Luo Q., Sun B. Experiments on the effect of engine speed, load, equivalence ratio, spark timing and coolant temperature on the energy balance of a turbocharged hydrogen engine. Energy Conversion and Management. 2018, 162: 1–12. https://doi.org/10.1016/j.enconman.2017.12.051

[33] Tang C., Zhang Y., Huang Z. Progress in combustion investigations of hydrogen enriched hydrocarbons. Renewable and Sustainable Energy Reviews. 2014, 30,195–221. https://doi.org/10.1016/j.rser.2013.10.005

[34] Yasiry A. S., Shahad H. A. K. An experimental study of the effect of hydrogen blending on burning velocity of LPG at elevated pressure. International Journal of Hydraulic Engineering. 2016, 41: 19269–19277. http://dx.doi.org/10.1016/j.ijhydene.2016.08.097

[35] Ozturk M., Sorgulu F., Javani M., Dincer I. An experimental study on the environmental impact of hydrogen and natural gas blend burning. Chemosphere. 2023, 329: 138671. https://doi.org/10.1016/j.chemosphere.2023.138671

[36] Hassan Q., Azzawi I. D. J., Sameen A. Z., Salman H. M. Hydrogen fuel cell vehicles: Opportunities and challenges. Sustainability. 2023, 15: 11501. https://doi.org/10.3390/su151511501

[37] Sinigaglia T., Lewiski F., Santos Martins M. E., Mairesse Siluk J. C. Production, storage, fuel stations of hydrogen and its utilization in automotive applications-a review. International Journal of Hydrogen Energy. 2017, 42: 24597e611. https://doi.org/10.1016/j.ijhydene.2017.08.063

[38] Arsad A. Z., Hannan M. A., Al-Shetwi A. Q., Begum R. A., Hossain M. J., Jern Ker P., Indra Mahlia T. M. Hydrogen electrolyser technologies and their modelling for sustainable energy production: A comprehensive review and suggestions. International Journal of Hydrogen Energy. 2023, 48(72): 27841–27871. https://doi.org/10.1016/j.ijhydene.2023.04.014

[39] Ghazal O. H. Performance and combustion characteristic of CI engine fueled with hydrogen enriched diesel. International Journal of Hydrogen Energy. 2013, 38: 15469=76. https://doi.org/10.1016/j.ijhydene.2013.09.037

[40] Hamdan M. O., Selim M. Y. E., Al-Omari S. A. B., Elnajjar E. Hydrogen supplement co-combustion with diesel in compression ignition engine. Renew Energy. 2015, 82: 54–60. https://doi.org/10.1016/j.renene.2014.08.019

[41] Pachiannan T., Zhong W., Rajkumar S., He Z., Leng X., Wang Q. A literature review of fuel effects on performance and emission characteristics of low-temperature combustion strategies. Applied Energy. 2019, 251,113380. https://doi.org/10.1016/j.apenergy.2019.113380

[42] Sharma T. K., Rao G. A. P., Murthy K. M. Homogeneous Charge Compression Ignition (HCCI) engines: A review. Archives of Computational Methods in Engineering. 2016, 23: 623–657. https://doi.org/10.1007/s11831-015-9153-0

[43] Hairuddin A. A., Yusaf T., Wandel A. P. A review of hydrogen and natural gas addition in diesel HCCI engines. Renewable and Sustainable Energy Reviews. 2014, 32: 739–761. https://doi.org/10.1016/j.rser.2014.01.018

[44] Khaliq A., Khalid F., Sharma P. B., Dincer I. Energetic and exergetic analyses of a hydrogen-fuelled HCCI engine for environmentally benign operation. International Journal of Sustainable Energy. 2014, 33: 367–385. https://doi.org/10.1080/14786451.2012.744020

[45] Lee D. S., Fahey D. W., et al. The contribution of global aviation to anthropogenic climate forcing for 2000 to 2018. Atmospheric Environment. 2021, 244: 117834. https://doi.org/10.1016/j.atmosenv.2020.117834

[46] Forster P., Rap A., Rosen D. What are the likely non-CO2 impacts of hydrogen planes. Aviation Environment Federation, 2023, 5pages. https://www.aef.org.uk/2024/01/24/what-are-the-likely-non-co2-impacts-of-hydrogen-planes/ [Accessed 1st October 2024]

[47] Grobler C., Wolfe P. J., et al. Marginal climate and air quality costs of aviation emissions Environ. Environmental Research Letters. 2019, 14: 114031. https://doi.org/10.1088/1748-9326/ab4942

[48] Dray L., Schäfer A. W., Grobler C., et al. Cost and emissions pathways towards net-zero climate impacts in aviation. Nature Climate Change. 2022, 12: 956–962. https://doi.org/10.1038/s41558-022-01485-4

[49] Forster P., Storelvmo T., Armour K., Collins W., Dufresne J. L., Frame D., Lunt D. J., Mauritsen T., Palmer M. D., Watanabe M., Wild M., Zhang H. The Earth's energy budget, climate feedbacks, and climate sensitivity. Chapter 7. In: Climate change. the physical science basis, contribution of working group to the sixth assessment report of the intergovernmental panel on climate change, Cambridge University Press, United Kingdom 2021, 923–1054. https://www.ipcc.ch/report/ar6/wg1/ Accessed 30 Sep 2024.

[50] Aviation Impact Accelerator. Resource to Climate Comparison Evaluator tool (RECCE) https://recce.aiatools.org/ [Accessed 1st October 2024]

[51] Fan Z., Sheerazi H., et al. Hydrogen leakage: A potential risk for the hydrogen economy. Center on Global Energy Policy at Columbia University SIPA, 2022, 33pages. https://www.energypolicy.columbia.edu/wpcontent/uploads/2022/07/HydrogenLeakageRegulations_CGEP_Commentary_063022.pdf [Accessed 1st October 2024]

[52] Shu Z., Lei G., Liu Z., Liang W., Zheng X., Ma J., Lu F., Qian H. Motion trajectory prediction model of hydrogen leak and diffusion in a stable thermally stratified environment. International Journal of Hydrogen Energy. 2022, 47(3): 2040–2049. https://doi.org/10.1016/j.ijhydene.2021.10.103

[53] Zhang W., Zhao G. Leakage and diffusion characteristics of underground hydrogen pipeline. Petroleum. 2024, 10(2): 319–325. https://doi.org/10.1016/j.petlm.2023.06.002

[54] Ehhalt D. H., Rohrer F. The tropospheric cycle of H_2: A critical review. Tellus B: Chemical and Physical Meteorology. 2009, 61: 500–535. https://doi.org/10.1111/j.1600-0889.2009.00416.x

[55] Conrad R., Seiler W. Influence of temperature, moisture, and organic carbon on the flux of H_2 and CO between soil and atmosphere: Field studies in subtropical regions. Journal of Geophysical Research. 1985, 90,5699. https://doi.org/10.1029/jd090id03p05699

[56] Derwent R. G., Stevenson D. S., Utembe S. R., Jenkin M. E., Khan A. H., Shallcross D. E. Global modelling studies of hydrogen and its isotopomers using STOCHEM-CRI: Likely radiative forcing consequences of a future hydrogen economy. International Journal of Hydrogen Energy. 2020, 45(15): 9211–9221. https://doi.org/10.1016/j.ijhydene.2020.01.125

[57] Cooper J., Dubey L., Bakkaloglu S., Hawkes A. Hydrogen emissions from the hydrogen value chain-emissions profile and impact to global warming. Science of the Total Environment. 2022, 830: 154624. https://doi.org/10.1016/j.scitotenv.2022.154624

[58] Pearson J. K., Derwent R. Air pollution and climate change the basics, Routledge, United Kingdom, 2022. 187pages. ISBN 9781032275185

[59] Alvarez R. A., Pacala S. W., Winebrake J. J., Chameides W. L., Hamburg S. P. Greater focus needed on methane leakage from natural gas infrastructure. Proceedings of the National Academy of Sciences of the United States of America. 2012, 109: 6435–6440. https://doi.org/10.1073/pnas.1202407109

[60] Alvarez R. A., Zavala-Araiza D., Lyon D. R., Allen D. T., Barkley Z. R., Brandt A. R., Davis K. J., Herndon S. C., Jacob D. J., Karion A., Kort E. A., Lamb B. K., Lauvaux T., Maasakkers J. D., Marchese A. J., Omara M., Pacala S. W., Peischl J., Robinson A. L., Shepson P. B., Sweeney C., Townsend-Small A., Wofsy S. C., Hamburg S. P. Assessment of methane emissions from the U.S. oil and gas supply chain. Science. 2018, 361: 186–188. https://doi.org/10.1126/science.aar7204

[61] Wood D. A. Carbon-neutral LNG cargoes: A potentially valuable concept requiring improved transparency. Chapter 16. In: Wood D. A., Cai J., eds., Sustainable liquefied natural gas: Concepts and applications moving towards net-zero supply chains, Elsevier, United Kingdom, 2024, 445–475. https://doi.org/10.1016/B978-0-443-13420-3.00015-9

[62] Sun T., Shrestha E., Hamburg S. P., Kupers R., Ocko I. B. Climate impacts of hydrogen and methane emissions can considerably reduce the climate benefits across key hydrogen use cases and time scales. Environmental Science & Technology. 2024, 58: 5299–5309. https://doi.org/10.1021/acs.est.3c09030

[63] Suleman F., Dincer I., Agelin-Chaab M. Environmental impact assessment and comparison of some hydrogen production options. International Journal of Hydrogen Energy. 2015, 40(21): 6976–6987. https://doi.org/10.1016/j.ijhydene.2015.03.123

[64] Rigon M. R., Zortea R., Moraes C. A. M., Modolo R. C. E. Life cycle impact assessment methodology: Selection criteria for environmental impact categories. In: New frontiers on life cycle assessment, IntechOpen, 2019, http://dx.doi.org/10.5772/intechopen.83454

[65] Sgouridis S., Carbajales-Dale M., Csala D., Cheisa M., Bardi U. Comparative net energy analysis of renewable electricity and carbon capture and storage. Nature Energy. 2019, 4: 456–465. https://doi.org/10.1038/s41560-019-0365-7

[66] Palmer G., Roberts A., Hoadley A., Dargaville R., Honnery D. Life-cycle greenhouse gas emissions and net energy assessment of large-scale hydrogen production via electrolysis and solar PV. Energy & Environmental Science. 2021, 14: 5113–5131. https://doi.org/10.1039/D1EE01288F

[67] Wei S., Sacchi R., Tukker A., Suhd S., Steubing B. Future environmental impacts of global hydrogen production. Energy & Environmental Science. 2024, 17: 2157. https://doi.org/10.1039/D3EE03875K

[68] Davoodi S., Al-Shargabi M. A., Wood D. A., Longe P. O., Mehrad M., Rukavishnikov V. S. Underground hydrogen storage: A review of technological developments, challenges, and opportunities. Applied Energy. 2025, 381: 125172. https://doi.org/10.1016/j.apenergy.2024.125172

[69]   Al-Shafi M., Massarweh O., Abushaikha A. S., Bicer Y. A review on underground gas storage systems: Natural gas, hydrogen and carbon sequestration. Energy Reports. 2023, 9: 6251–6266. https://doi.org/10.1016/j.egyr.2023.05.236

[70]   Davoodi S., Al-Shargabi M. A., Wood D. A., Rukavishnikov V., Minaev K. Review of technological progress in carbon dioxide capture, storage, and utilization. Gas Science and Engineering. 2023, 117205070. https://doi.org/10.1016/j.jgsce.2023.205070

[71]   Liu W., Pei P. Evaluation of the influencing factors of using underground space of abandoned coal mines to store hydrogen based on the improved ANP method. Advance Material Science and Engineering. 2021, 2021: 9pages. https://doi.org/10.1155/2021/7506055

[72]   Longe P., Davoodi S., Mehrad M., Wood D. A. Combined deep learning and optimization for hydrogen-solubility prediction in aqueous systems appropriate for underground hydrogen storage reservoirs. Energy & Fuels. 2024, 38(22): 22031–22049. https://doi.org/10.1021/acs.energyfuels. 4c03376

[73]   Ghasemi M., Omrani S., Mahmoodpour S., Zhou T. Molecular dynamics simulation of hydrogen diffusion in water-saturated clay minerals; implications for Underground Hydrogen Storage (UHS). International Journal of Hydrogen Energy. 2022, 47(59): 24871–24885. https://doi.org/10.1016/j.ijhy dene.2022.05.246

[74]   Mahdi D. S., Al-Khdheeawi E. A., Yuan Y., Zhang Y. I. Hydrogen underground storage efficiency in a heterogeneous sandstone reservoir. Advances in Geo-Energy Research. 2021, 5(4): 437. https://doi. org/10.46690/ager.2021.04.08

[75]   Amid A., Mignard D., Wilkinson M. Seasonal storage of hydrogen in a depleted natural gas reservoir. International Journal of Hydrogen Energy. 2016, 41: 5549–5558. https://doi.org/10.1016/j. ijhydene.2016.02.036

[76]   Ershadnia R., Singh M., Mahmoodpour S., Meyal A., Moeini F., Hosseini S. A., Sturmer D. M., Rasoulzadeh M., Dai Z., Soltanian M. R. Impact of geological and operational conditions on underground hydrogen storage. International Journal of Hydrogen Energy. 2023, 48,1450–71. https://doi.org/10.1016/J.IJHYDENE.2022.09.208

[77]   Zivar D., Kumar S., Foroozesh J. Underground hydrogen storage: A comprehensive review. International Journal of Hydrogen Energy. 2021, 46(45): 23436–23462. https://doi.org/10.1016/j.ijhy dene.2020.08.138

[78]   Iglauer S. Optimum geological storage depths for structural H2 geo-storage. Journal of Petroleum Science & Engineering. 2022, 212: 109498.

[79]   Anikeev D. P., Zakirov E. S., Indrupskiy I. M., Anikeeva E. S. Estimation of diffusion losses of hydrogen during the creation of its effective storage in an aquifer. SPE Russian Petroleum Technology Conference. October 2021, https://doi.org/10.2118/206614-MS

[80]   Muhammed N. S., Haq M. B., Al Shehri D. A., Al-Ahmed A., Rahman M. M., Zaman E., Iglauer S. Hydrogen storage in depleted gas reservoirs: A comprehensive review. Fuel. 2023, 2023 337: 127032. https://doi.org/10.1016/j.fuel.2022.127032

[81]   Lysyy M., Fernø M., Ersl G. Seasonal hydrogen storage in a depleted oil and gas field. International Journal of Hydrogen Energy. 2021, 46: 25160–25174. https://doi.org/10.1016/j.ijhydene.2021.05.030

[82]   Singh H. Hydrogen storage in inactive horizontal shale gas wells: Techno-economic analysis for Haynesville shale. Applied Energy. 2022, 313: 118862. https://doi.org/10.1016/J.APENERGY.2022.118862

[83]   Sekar L. K., Kiran R., Okoroafor E. R., Wood D. A. Review of reservoir challenges associated with subsurface hydrogen storage and recovery in depleted oil and gas reservoirs. Journal of Energy Storage. 2023, 72(Part D): 108605. https://doi.org/10.1016/j.est.2023.108605

[84]   Wood D. A. Well integrity for underground gas storage relating to natural gas, carbon dioxide, and hydrogen. Chapter 19 In: Wood D. A., Cai J., eds., Sustainable natural gas drilling: Technologies and case studies for the energy transition, Elsevier, 2024, 551–576. https://doi.org/10.1016/B978-0-443-13422-7.00019-2

[85] Freifeld B., Oldenburg C., Jordan P., Pan L., Perfect S., Morris J., White J., Bauer S., Blankenship D., Roberts B., Bromhal G., Glosser D., Wyatt D., Rose K. Well integrity for natural gas storage in depleted reservoirs and aquifers. Sandia Report SAND2017-0599 December 2016. https://www.osti.gov/servlets/purl/1432270 [Accessed 14th October 2024]

[86] Reinicke K. M., Fichter C. Measurement strategies to evaluate the integrity of deep wells for CO_2 applications. Underground storage of CO_2 and Energy, Edited By Michael Z. Hou, Heping Xie, Jeoungseok Yoon CRC Press (Taylor Francis Group, Informa plc, United Kingdom

[87] Gregory S. P., Barnett M. J., Field L. P., Milodowski A. E. Subsurface microbial hydrogen cycling: Natural occurrence and implications for industry. Microorganisms. MDPI, Switzerland 2019, 7: 53. doi:10.3390/microorganisms7020053

[88] Klein F., Tarnas J., Bach W. Abiotic sources of molecular hydrogen on Earth. Elements. 2020, 16: 19–24. https://doi.org/10.2138/gselements.16.1.19

[89] Zgonnik V. The occurrence and geoscience of natural hydrogen: A comprehensive review. Earth-Science Reviews. 2020, 203: 103140.

[90] Wang L., Jin Z., Chen X., Su Y., Huang X. The origin and occurrence of natural hydrogen. Energies. 2023, 16: 2400. https://doi.org/10.3390/en16052400

[91] Geymond U., Briolet T., Combaudon V., Sissmann O., Martinez I., Duttines M., Moretti I. Reassessing the role of magnetite during natural hydrogen generation. Geochemistry - Frontiers in Earth Science. 2023, 11: 1169356. http://doi.org/10.3389/feart.2023.1169356

[92] Wang W., Liu C., Liu W., Wang X., Guo P., Wang J., Wang Z., Li Z., Zhang D. Dominant products and reactions during organic matter radiolysis: Implications for hydrocarbon generation of uranium-rich shales. Marine and Petroleum Geology. 2022, 137: 105497. https://doi.org/10.1016/j.marpetgeo.2021.105497

[93] Sato M., Sutton A. J., McGee K. A., Russell-Robinson S. Monitoring of hydrogen along the San Andreas and Calaveras faults in central California in 1980–1984. Journal of Geophysical Research: Solid Earth. 1986, 91: 12315–12326. https://doi.org/10.1029/JB091iB12p12315

[94] Milkov A. V. Molecular hydrogen in surface and subsurface natural gases: Abundance, origins and ideas for deliberate exploration. Earth-Science Reviews. 2022, 230: 104063. https://doi.org/10.1016/j.earscirev.2022.104063

[95] Wood D. A. Natural hydrogen resource exploitation must confront the issue that certain gas compositions are undesirable in terms of environmental sustainability. Advances in Geo-Energy Research. 2024, 15(3): 185–189. https://doi.org/10.46690/ager.2025.03.02

[96] Piché-Choquette S., Constant P. Molecular hydrogen, a neglected key driver of soil biogeochemical processes. Applied and Environmental Microbiology. 2019, 85(6): e02418–18. https://doi.org/10.1128/AEM.02418-18

[97] Khetkorn W., Rastogi R. P., Incharoensakdi A., Lindblad P., Madamwar D., Pandey A., Larroche C. Microalgal hydrogen production – A review. Bioresource Technology. 2017, 243: 1194–1206. https://doi.org/10.1016/j.biortech.2017.07.085

[98] Su X., Zhao W., Xia D. The diversity of hydrogen-producing bacteria and methanogens within an in-situ coal Seam. Biotechnology Biofuels. 2018, 11: 245. https://doi.org/10.1186/s13068-018-1237-2

[99] Boyd E. S., Colman D. R., Templeton A. S. Perspective: Microbial hydrogen metabolism in rock-hosted ecosystems. Frontiers in Energy Research. 2024, 12: 1340410. https://doi.org/10.3389/fenrg.2024.1340410

[100] Smigan P., Greksak M., Kozánková J., Buzek F., Onderka V., Wolf I. Methanogenic bacteria as a key factor involved in changes of town gas stored in an underground reservoir. FEMS Microbiol Letters. 1990, 73: 221–224.

[101] Larin N., Zgonnik V., Rodina S., Deville E., Prinzhofer A., Larin V. N. Natural molecular hydrogen seepage associated with surficial, rounded depressions on the European craton in Russia. Natural Resources Research. 2015, 24: 369–383. https://doi.org/10.1007/s11053-014-9257-5

[102] Zgonnik V., Beaumont V., Deville E., Larin N., Pillot D., Farrell K. M. Evidence for natural molecular hydrogen seepage associated with Carolina bays (surficial, ovoid depressions on the Atlantic Coastal Plain, Province of the USA). Progress in Earth and Planetary Science. 2015, 2: 31. https://doi.org/10.1186/s40645-015-0062-5

[103] Prinzhofer A., Moretti I., Francolin J., Pacheco C., d'Agostino A., Werly J., Rupin F. Natural hydrogen continuous emission from sedimentary basins: The example of a Brazilian H2-emitting structure. International Journal of Hydrogen Energy. 2019, 44: 5676–5685. https://doi.org/10.1016/j.ijhydene.2019.01.119

[104] Zgonnik V., Beaumont V., Larin N., Pillot D., Deville E. Diffused flow of molecular hydrogen through the Western Hajar mountains, Northern Oman. Arabian Journal of Geosciences. 2019, 12: 1–10. https://doi.org/10.1007/s12517-019-4242-2

[105] Frery E., Langhi L., Maison M., Moretti I. Natural hydrogen seeps identified in the north Perth Basin, Western Australia. International Journal of Hydrogen Energy. 2021, 46: 31158–31173. https://doi.org/10.1016/j.ijhydene.2021.07.023

[106] Aimar L., Frery E., Strand J., Heath C., Khan S., Moretti I., Ong C. Naturalhydrogen seeps or salt lakes: How to make a difference? grass patch example,Western Australia. Frontiers in Earth Science. 2023, 11: 1236673. https://doi.org/10.3389/feart.2023.1236673

[107] Moretti I., Geymond U., Pasquet G., Aimar L., Rabaute A. Natural hydrogen emanations in Namibia: Field acquisition and vegetation indexes from multispectral satellite image analysis. International Journal of Hydrogen Energy. 2022, 47: 35588–35607. https://doi.org/10.1016/j.ijhydene.2022.08.135

[108] Etiope G. Massive release of natural hydrogen from a geological seep (Chimaera, Turkey): Gas advection as a proxy of subsurface gas migration and pressurised accumulations. International Journal of Hydrogen Energy. 2023, 48: 9172–9184. https://doi.org/10.1016/j.ijhydene.2022.12.025

[109] Fitts C. R. Groundwater science, third edition. Academic Press, Elsevier, U.S.A. 2023. 661pages. https://doi.org/10.1016/C2016-0-00985-3

[110] McMahon C. J., Roberts J. J., Johnson G., Edlmann K., Flude S., Shipton Z. K. Natural hydrogen seeps as analogues to inform monitoring of engineered geological hydrogen storage. Geological Society of London, United Kingdom 2022, 528: 461–489. https://doi.org/10.6084/m9.figshare.c.6150485

[111] Osselin F., Soulaine C., Fauguerolles C., Gaucher E. C., Scaillet B., Pichavant M. Orange hydrogen is the new green. Nature Geoscience. 2022, 15: 765–769. https://doi.org/10.1038/s41561-022-01043-9

[112] Brandt A. R. Greenhouse gas intensity of natural hydrogen produced from subsurface geologic accumulations. Joule. 2023, 7(8): 1818–1831. https://doi.org/10.1016/j.joule.2023.07.001

[113] Bergerson J. A., Brandt A. R., Cresko J., Carbajales-Dale M., MacLean H. L., Matthews H. S., McCoy S., McManus M., Miller S. A., Morrow W. R., et al. Life cycle assessment of emerging technologies: Evaluation techniques at different stages of market and technical maturity. Journal of Industrial Ecology. 2020. 2020 24: 11–25. https://doi.org/10.1111/jiec.12954

[114] Ball P. J., Czado K. Natural hydrogen: The new frontier. Geoscientist 1 March 2022. https://geoscientist.online/sections/unearthed/natural-hydrogen-the-new-frontier/[Accessed 14 October 2024]

Md Mofazzal Hossain

Chapter 24
Drilling, construction, and completion of natural hydrogen exploration and production wells: emphasizing long-term well integrity

Abstract: As global energy demand increases, transitioning from fossil fuels to sustainable green energy sources is becoming more critical. Hydrogen, a clean energy source, is expected to play a key role in accelerating this transition. Natural hydrogen presents a significant opportunity in the energy sector as a clean alternative to traditional fossil fuels. However, developing natural hydrogen reservoirs poses unique challenges, particularly in subsurface exploration, drilling, well design, and construction for safe extraction and production operations. Hydrogen's distinct physical and chemical properties, such as low molecular weight, high diffusivity, and flammability, necessitate advanced techniques in drilling and well design. These techniques must ensure containment and safe production processes with the highest level of integrity.

Drilling and completing hydrogen wells require special attention. Unlike conventional hydrocarbon wells, hydrogen wells must account for hydrogen's chemical reactivity, its interaction with wellbore materials, completion hardware, and the surrounding formation (e.g., rock, cement). The impact of thermo-mechanical behavior, especially in dynamic situations during production operations with changing conditions (e.g., pressure, temperature), including the potential for leaks, must also be considered. Hydrogen can also cause metals to become brittle, posing challenges in selecting materials for wellbore construction and completion to maintain safe production operations. Innovations in drilling fluids, casing materials, and advanced completion techniques are essential for safely and economically extracting natural hydrogen from its reservoirs.

This chapter provides an overview of natural hydrogen, its reservoir characteristics, and the framework for drilling, constructing, and completing natural hydrogen exploration and production wells. It addresses the challenges of drilling exploration wells, and completing production wells, focusing on developing an engineering framework for designing and constructing these wells to ensure long-term well integrity. The chapter reviews and synthesizes wellbore integrity issues, detailing the causes of well integrity loss, the challenges encountered, and the associated risks. It proposes a

Md Mofazzal Hossain, Professor in Energy Engineering, School of WASM Minerals, Energy and Chemical Engineering, Curtin University, Perth, Australia, e-mail: md.hossain@curtin.edu.au

https://doi.org/10.1515/9783111437040-024

comprehensive engineering framework for designing, drilling, completing, and constructing natural hydrogen exploration and production wells. This framework offers guidance on drilling techniques, material selection, cementing practices, and real-time monitoring systems. Through a thorough review, this work aims to provide insights into harnessing natural hydrogen as a sustainable green energy resource.

Keywords: drilling, completion, well construction, natural hydrogen, underground hydrogen storage, cementing, elastomer, hydrogen embrittlement, hydrogen blistering, hydrogen-induced cracking, well integrity, wellbore casing materials, microbial-induced corrosion (MIC), sulfate-reducing bacteria (SRB), iron-reducing bacteria (IRB), hydrogenated nitrile butadiene rubber (HNBR), fluorocarbon elastomer (FKM), API casing

24.1 Introduction

Hydrogen has gained significant attention as one of the viable solutions in the transition to low-carbon energy systems. As a versatile and clean energy carrier, hydrogen is expected to play a key role in decarbonizing various sectors, including electricity generation, heating, transportation, and industrial processes [1]. Its use helps stabilize energy grids by compensating for the intermittent nature of renewable energy sources such as wind and solar power. This will contribute to a more resilient and flexible energy infrastructure, making hydrogen a viable energy solution. The International Energy Agency (IEA) projects a 47% rise in energy demand by 2050, with fossil fuels remaining the dominant source [2]. To reduce the reliance on fossil-based energy sources like oil, gas and coal, enhance the use of efficient and sustainable green energy, mitigate environmental impact, address climate change, and lower greenhouse gas emissions, hydrogen is considered a promising solution. When burned, hydrogen produces only water vapor, emitting no pollutants or greenhouse gases such as carbon dioxide.

Currently, hydrogen can be produced in three main ways: green hydrogen, produced by using renewable energy to electrolyze water; blue hydrogen, generated from natural gas steam conversion or coal-to-gas processes combined with carbon capture and storage; and gray hydrogen, derived from fossil fuels without carbon capture and storage [3]. Most of the hydrogen used today is gray hydrogen. In 2020, global hydrogen consumption was approximately 90 million tons, almost entirely produced from fossil fuels, resulting in nearly 900 million tons of CO_2 emissions. Although producing green hydrogen through water electrolysis is seen as the primary method for future hydrogen production, it is currently utterly energy intensive as well as expensive, and accounts for only 0.1% of total hydrogen production [4–6]. There are about 350 projects worldwide focused on hydrogen production via water electrolysis, and by 2030, the global supply of hydrogen from this method is expected to exceed

8 million tons [7]. However, this is still far below the 80 million tons demand projected by the International Energy Agency [8] and, consequently, natural hydrogen could potentially be a better source for providing sustainable green energy, meeting global energy demands, and ensuring energy security.

Natural hydrogen, formed through geological processes such as serpentinization and water-rock reactions, is found in subsurface reservoirs [7, 9]. Its potential as a scalable, low-carbon energy source has been demonstrated at sites like Bourakebougou, Mali, where shallow wells produce hydrogen with purities up to 98% [10].

Natural hydrogen reservoirs exhibit distinctive properties compared to conventional hydrocarbon reservoirs. Hydrogen is often found in shallow formations, such as dolomitic carbonates and porous sandstones, as observed in the Bourakebougou field in Mali. These reservoirs demonstrate a dual porosity system: primary porosity from the rock matrix and secondary porosity from karstic voids or fractures [10]. Hydrogen purity levels in these reservoirs are reported to be more than 98%, often accompanied by trace amounts of methane and nitrogen.

The unique geological formations associated with natural hydrogen include ophiolites and ultramafic rocks, where hydrogen generation occurs through serpentinization – a reaction between water and olivine-rich rocks that produces hydrogen as a byproduct [9]. Cratonic basins, ancient geological settings featuring iron-rich formations, contribute to hydrogen generation through water-rock interactions [10]. In sedimentary basins, hydrogen is found trapped beneath impermeable layers, such as dolerite sills, creating effective seals for accumulations [11].

Natural hydrogen reservoirs differ significantly from hydrocarbon reservoirs in several aspects, including physical properties, chemical reactivity, and hydrogen regeneration potential. Hydrogen's small molecular size and high diffusivity increase the risk of leakage compared to hydrocarbons, necessitating advanced sealing and containment strategies [9]. Additionally, hydrogen's reactivity with metals and other materials can lead to corrosion and embrittlement, presenting additional engineering challenges, especially in well design, construction, and completion for safe extraction [10]. Understanding these unique characteristics is essential for developing effective exploration, drilling, and production strategies for natural hydrogen extraction. Special considerations for drilling, completion, and construction of exploration and production wells include addressing the high diffusivity and reactivity of hydrogen to prevent leakage, degradation of wellbore materials, completion and production hardware, and the effects of hydrogen interaction with the surrounding wellbore (e.g., cement, formation) under dynamic operational conditions. Advanced drilling and well construction techniques must be employed to ensure the safe and efficient extraction of natural hydrogen, making it a viable and sustainable energy source for the future.

While natural hydrogen offers immense promise as a clean energy carrier, its extraction from geological formations presents unique challenges that must be addressed [12–14]. Table 24.1 summarizes the challenges and associated risks encountered in underground hydrogen storage (UHS) within depleted oil and gas reservoirs

[15, 16]. Similar risks and challenges can be encountered during the drilling of natural hydrogen exploration and production wells. These challenges are critical considerations for designing hydrogen exploration and production wells. The primary challenges associated with natural hydrogen extraction arise from its chemical reactivity, physical properties, and its potential to cause loss of well integrity, predominantly due to the degradation and failure of wellbore and completion materials, including the wellbore surroundings (e.g., cement and rocks), each of which poses unique risks to wellbore construction and long-term well integrity [17–19].

Table 24.1: Impact and potential risks associated with various activities/process.

Activities	Process	Impact	Potential risks
Geochemistry and Microbial activity	Chemical disequilibria Microbial growth	Biotic and abiotic gas production – Microbial clogging – Mineral dissolution and precipitation – Migration of fines – H_2 consumption and contamination	– Wellbore integrity issues such as casing, tubing, and completion hardware damages/failure, cement failure, near wellbore damages, H_2-induced corrosion and leakages – Loss of rock mechanical strength

Geomechanics	Stress changes	Mechanical strength of formation and strain	– Compaction
			 – Fault (re)activation
			 – Seismicity
			   – Subsidence
			   – Seal quality reduction
			 – Changes in reservoir's properties (porosity and permeability)

Hydrogeology and reservoir drive mechanism	Multiphase flow	Mixing and diffusion Unstable displacement Migration of fines Pressure build-up	Productivity impairment, reduction of hydrogen production

Hydrogen's chemical reactivity significantly impacts the materials used in wellbore construction. It reacts with minerals and dissolved solutes within the geological formation, altering chemical equilibrium and potentially leading to mineral dissolution or precipitation. This can degrade the cement sheath and compromise zonal isolation [20]. Furthermore, microbial metabolisms can be influenced by the presence of hydrogen, promoting microbial-induced corrosion (MIC) in metallic wellbore components and contributing to structural degradation over time. The reactivity extends to cement and elastomeric materials, altering their properties and increasing their susceptibility to micro-cracking or failure [15].

From a physical perspective, hydrogen's high diffusivity, low molecular weight, and low viscosity significantly increase its potential for leakage. Unlike CO_2 or natural gas, hydrogen can migrate more easily through completion hardware and sealing components, such as elastomers [21] as well as preferential pathways like micro-fractures in the cement sheath, casing cracks and diffusion through caprock [16, 20, 21]. In terms of medium- to long-term scenarios, these pathways can expand due to stress cycling, thermal fluctuations, or chemical interactions, further raising the risk of leakage into surrounding formations or aquifers. Hydrogen's ability to bypass sealing systems through diffusion poses additional challenges, making traditional containment strategies less effective compared to other gases.

Well integrity is a critical aspect of drilling and production operations, particularly for natural hydrogen wells. Ensuring well integrity involves maintaining the containment of the follow of hydrogen throughout the well's lifecycle, from drilling to decommissioning. Key challenges include managing the high diffusivity of hydrogen, which increases the risk of leaks, and its chemical reactivity, which can cause corrosion and embrittlement of well materials [22]. Effective well integrity management requires robust design and construction practices, including the use of advanced sealing technologies and corrosion-resistant materials [22]. During the planning and design phase, for instance, it is crucial to consider the worst-case scenarios including exposure to pressures and temperatures, erosion and corrosion mechanisms that could compromise well integrity [22]. The construction phase should involve the implementation of dual well barriers – one containing drilling fluid and the other being mechanical, using formation, cement, casing, and blowout preventers (BOPs) verified by pressure testing and logging [22]. Routine maintenance and testing of surface equipment and downhole barrier components are essential during production to avoid costly unplanned maintenance or downtime [22].

This chapter discusses the technical and engineering aspects of designing, constructing, and completing natural hydrogen exploration and production wells. It focuses on ensuring safe operation throughout the well's lifespan and mitigating any potential risks of well integrity loss during drilling, exploration, and production. It addresses key challenges related to well integrity, particularly from a medium- to long-term perspective. The main objective is to provide a comprehensive understanding of the mechanisms of well integrity loss and outline the necessary measures for designing, constructing, and maintaining natural hydrogen exploration and production wells. It covers principles of well design, material selection, construction techniques, crucial aspects of well integrity, and methods to minimize the risk of the loss of well integrity. Additionally, the chapter presents a practical framework for drilling, constructing, and completing natural hydrogen exploration and production wells, ensuring long-term integrity and containment.

24.2 Loss of well integrity

The term "well integrity" encompasses a multifaceted discipline that spans the entire lifecycle of a well, starting from drilling, well construction, and extraction to abandonment [23]. In the context of wells drilled for the exploration and production of natural hydrogen, well integrity is defined as the application of technical, operational, and organizational solutions to minimize the risk of failure resulting from uncontrolled fluid flow from underground formations [24]. For hydrogen exploration and production development projects, maintaining well integrity is critical due to hydrogen's unique properties, characteristics, and challenges outlined earlier.

Well integrity issues can be broadly classified into two categories: those related to well and completion design and those arising from operational challenges. Design-related challenges involve ensuring the integrity of the casing, cement, and completion and production hardware used to convert a drilled well into a producer. Beyond standard requirements such as pressure support and containment, the casing must also withstand any changes occurred from hydrogen's unique behavior under dynamic operational conditions, including pressure and temperature variations.

Cement sheaths play a crucial role in providing zonal isolation and sealing fluids behind the casing, including preventing the upward propagation of fluids through the interface between the casing and formation. However, hydrogen can diffuse through micro-cracks, creating potential paths for leakage and reducing well integrity over time. Hydrogen-induced chemical reactions, such as mineral dissolution or precipitation, can weaken the bond between the cement and surrounding rock [19], posing significant risks to well integrity. Advanced cement formulations are necessary to enhance bonding strength and maintain performance under dynamic conditions.

Completion hardware, such as packers, tubing, safety valves, slips, various sealing elements (e.g., O-rings, elastomers), and wellheads, can be affected by hydrogen diffusion, causing degradation of these materials and posing serious risks to well integrity. The selection of these components is vital for safe production operations while ensuring containment with the highest level of integrity. Materials used in these components must be carefully selected to be compatible with hydrogen's properties, such as high diffusivity, which can cause various issues related to material failures through mechanisms like hydrogen embrittlement, hydrogen blistering, and hydrogen stress cracking. Wellheads and various valves, including sealing elements, must be specifically engineered for hydrogen production to ensure long-term durability.

Operational challenges involve maintaining the integrity of tubing, annulus, wellheads, and trees including associated sealing elements during hydrogen production. Hydrogen's high reactivity can cause corrosion in tubing and annular materials, especially when exposed to water or corrosive agents, necessitating rigorous monitoring and regular integrity testing. Additionally, the wellhead and associated components are critical for maintaining control during operations. Hydrogen's low viscosity increases the risk of leaks, requiring advanced sealing technologies and real-time monitoring systems to detect and prevent failures [25, 26].

The potential leakage pathways for an abandoned well are illustrated in Figure 24.1 (a and b). As shown in Figure 24.1a, potential hydrogen leakage paths are most likely to develop between: (1) the wellbore if the casing material integrity is compromised, either through corrosion, embrittlement, or degradation caused by long-term exposure to hydrogen-rich conditions; (2) the casing-cement interface due to poor cementing practices or material incompatibility; (3) the cement matrix as a result of weak bonding or microcracks; (4) chemically induced pathways formed by interactions of hydrogen with brine, cement, or other geological materials; (5) induced fractures; (6) defects such as mud channels within the cement; and (7) flow at the cement-caprock interface [27]. For hydrogen production in older fields, particularly when recompleting an existing well (Figure 24.1b), leakage paths may develop between old and new cement, old cement and the formation, or casing and new cement. In such scenarios, assessing the integrity of both the casing and cement prior to hydrogen production is essential. Proper investigations should include laboratory experiments such as permeability tests, shear bonding strength analysis, and hydraulic bonding strength evaluations to assess the system's suitability. If potential leakage is detected due to compromised casing or cement integrity, remediation methods such as milling and replacing the casing, squeeze cementing, and/or re-cementing with new materials may be necessary.

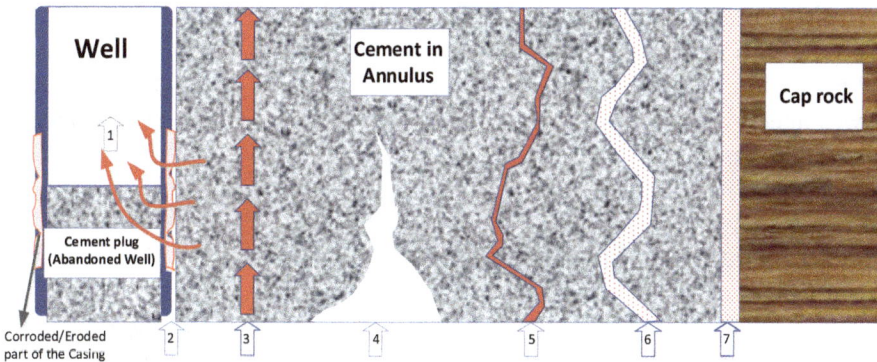

Figure 24.1: (a) Illustration of potential leakage pathways for H_2 injection and production well: (1) leaking through the casing damaged due to H_2 diffusion, microbial-induced corrosion, or other associated issues, (2) casing – cement interface, (3) Cement matrix, (4) bulk dissolution induced pathway, (5) fracture (induced), (6) cement defect/mud channel, and (7) cement-cap rock interface flow (adapted from [28]). (b) Schematic of potential leakage pathways for H_2 injection well [27].

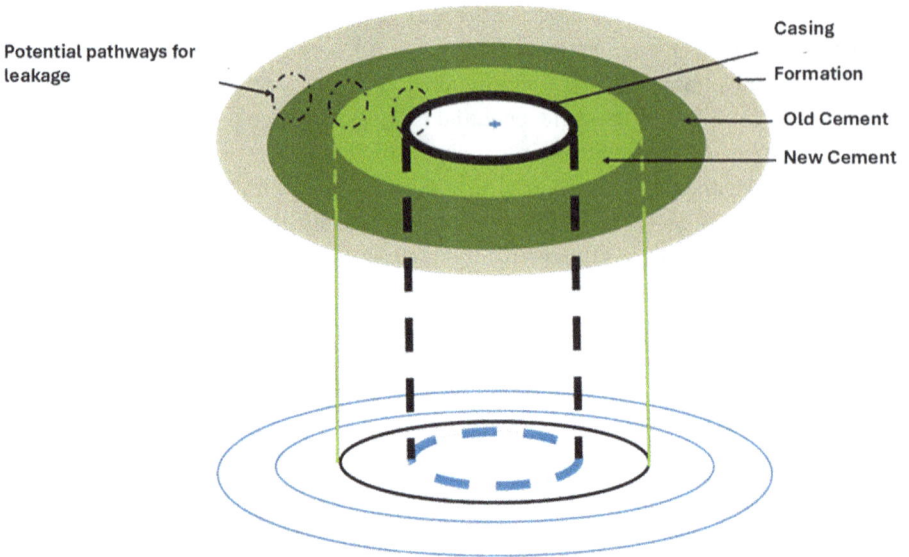

Figure 24.1 (Continued)

Identifying the root causes of hydrogen leakage is critical to preventing hazardous consequences and implementing effective mitigation measures. Understanding the mechanisms behind the formation of leakage pathways is paramount. These mechanisms are often related to corrosion, hydrogen embrittlement, chemical interactions; and mechanical, electrochemical, geochemical, and microbial factors. Further details on well integrity mechanisms associated with hydrogen exploration and production wells are discussed in the following section.

24.2.1 Mechanism of loss of well integrity

As illustrated in Figure 24.2, multiple mechanisms can cause loss of wellbore integrity. These include hydrogen embrittlement in casing and other metallic materials, hydrogen blistering, hydrogen-induced cracking (HIC), and failure of completion sealing materials (e.g., packers, elastomers), cement degradation, caprock sealing failure, and microbial-induced corrosion (MIC) caused by reactions on metal surfaces by sulfate-reducing bacteria (SRB). Hydrogen, as an active electron donor, participates in chemical and microbial reactions. Microbial survival in subsurface conditions depends on environmental characteristics, downhole conditions, especially pressure and temperature, and the chemical composition of various minerals and formation constituents. Reactions involving methanogens, acetogens, sulfate-reducing bacteria (SRB), and iron-reducing bacteria (IRB) can catalyze processes that compromise well integrity, especially when natural hydrogen comes into contact with hydrocarbons [29–31].

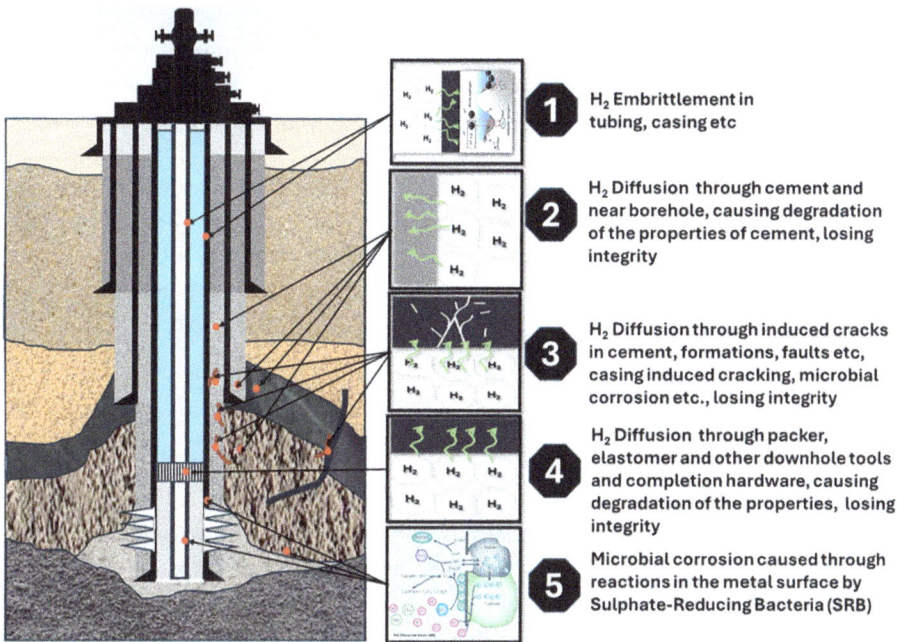

Figure 24.2: Mechanisms of the integrity loss of natural hydrogen exploration and production well.

24.2.1.1 Hydrogen embrittlement

Hydrogen's low molecular weight allows it to permeate through metallic and other materials (elastomers, sealing rings etc.), causing hydrogen embrittlement (HE). HE is a phenomenon where these materials become brittle and fracture due to the absorption of hydrogen. This process involves hydrogen atoms diffusing into the materials, reducing its ductility and tensile strength. HE can occur in various metals, including steels, nickel, titanium, and their alloys and various sealing components used as a part of well completion.

The mechanisms of HE include the formation of brittle hydrides, voids leading to high-pressure bubbles, enhanced decohesion at internal surfaces, and localized plasticity at crack tips [32]. Consequently, HE can lead to sudden and catastrophic failures in wellbore casing, tubing and other downhole completion hardware, and tools containing or exposed to hydrogen. All these essential components of the well can lose their ability to deform plastically due to the reduction of ductility and strength, making them more prone to cracking under stress. It can also lower the stress required for crack initiation and propagation, leading to premature failure of the well structures. Interestingly, HE becomes more severe at low pressure and temperature, where hydrogen diffusion is slower but more concentrated at crack tips. At higher temperatures and low pressure (below the threshold pressure), hydrogen may diffuse out of the metal, reducing chances of embrittlement [33]. While high pressure increases the amount of hydrogen absorbed into the metal increasing hydrogen embrittlement, there is often a threshold pressure beyond which embrittlement does not significantly increase.

Material embrittlement poses a significant challenge to long-term well integrity, particularly under stress cycling during natural hydrogen withdrawal operations. Hydrogen embrittlement can severely affect critical wellbore components such as casing, packers, and tubing, where prolonged exposure may weaken or compromise their structural integrity.

24.2.1.2 Hydrogen diffusion

Elastomers commonly used in injection and production wells for sealing components, such as packers, valves, gaskets, coatings, O-rings, and other critical elements, are susceptible to degradation due to hydrogen. This degradation becomes more pronounced under extended exposure to hydrogen or during repeated operational cycles, or under high-pressure and high-temperature conditions presenting a substantial risk to loss of well integrity by diffusion [21]. Failures in these components can lead to severe consequences, including costly workover operations, operational downtime, and, in extreme cases, catastrophic well failure. Such failures not only disrupt operations but also pose significant risks to health, safety, and the environment (HSE), with the potential for loss of human life.

Hydrogen diffusion into elastomers at high pressures can result in blistering or even explosions due to the rapid expansion of trapped gas when system pressure is quickly released, which can potentially cause explosive decompression or rapid decompression, and can compromise well integrity by creating pathways for hydrogen to vent through the annulus or causing uncontrollable casing pressure. Many elastomeric materials designed for oil and gas operations are unsuitable for hydrogen transport and injection, especially under conditions involving high pressure, temperature, or hydrogen purity.

Theiler et al. [34] conducted an experimental study to evaluate the impact of high-pressure hydrogen on cross-linked hydrogenated nitrile butadiene rubber (HNBR), commonly used as sealing elastomers in wellbores. Following the CSA/ANSI Standard, the material properties were tested before and after exposure to hydrogen at 100 MPa and 120 °C for 7 and 21 days. The study observed that the high-pressure hydrogen caused swelling, blistering, and microcracks due to rapid decompression, leading to decreased density and mechanical properties. The swelling behavior of HNBR elastomers shown in Figure 24.3 was observed during depressurization conditions demonstrating that the dissolved hydrogen expanded, causing the elastomer to swell and forming surface blisters (Figure 24.3c, d). These blisters appeared to be temporary and typically disappeared within 15 min, occasionally accompanied by a sound. While no significant surface damage was observed in the sample (Figure 24.3e), SEM images revealed microcracks on the surface (Figure 24.3g). This suggests that blister formation may induce microcracks, particularly under high-pressure operating conditions, thereby increasing the risk of failure through mechanisms such as extrusion or bending [35].

The risk of damage also rises significantly at elevated pressures and temperatures, as hydrogen has a lower threshold for decompression-related damage compared to gases like CO_2. Alhassan and Hossain [21] conducted numerical studies on hydrogen diffusion through fluorocarbon elastomer (FKM) used in wellbore packers, with an inner radius of 4.5 cm and an outer radius of 7.5 cm, at temperatures of 25 °C (L), 50 °C (M), and 100 °C (H). The simulated results, presented in Figure 24.4, demonstrate a strong dependence on temperature. As shown, hydrogen diffusion is negligible at low or ambient temperature (25 °C), increases to 1% mass fraction at medium temperature (50 °C), and further rises to 3% at high temperature (100 °C). This indicates a 4% increase in hydrogen diffusion from ambient to high temperature, highlighting the substantial impact of temperature on hydrogen diffusion through FKM.

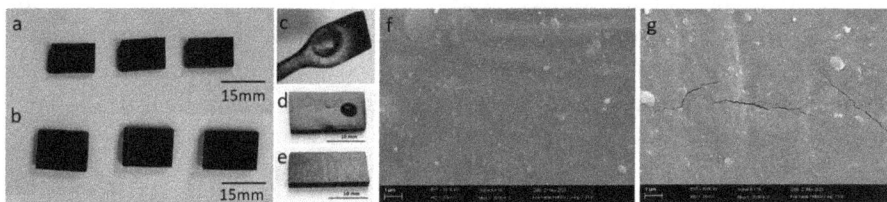

Figure 24.3: Swelling behavior of HNBR1 and HNBR2 samples: (a) HNBR1 before high-pressure hydrogen exposure, (b) immediately after exposure, showing swelling and blister formation upon rapid depressurization, (c) HNBR1 after 21 days exposure, (d) HNBR2 immediately after 7 days exposure, and (e) HNBR2 48 h post-exposure; SEM images: (f) HNBR2 before exposure and (g) HNBR2 after 7 days exposure to hydrogen at 100 MPa and 120 °C [34].

Swelling, hardening, or cracking are among the primary failure mechanisms of elastomers exposed to hydrogen. Three critical factors contribute to this degradation: the rate of decompression, the permeability of the elastomer to hydrogen, and the inherent strength of the elastomer material. Repeated compression and decompression cycles can further worsen damage, as the sealing material is subjected to cyclic stress. To ensure the reliability of sealing materials, all elastomers used in well completion or surface equipment that may encounter hydrogen should undergo rigorous testing to evaluate their resistance to hydrogen diffusivity and decompression effects. Advanced elastomer formulations tailored for hydrogen applications are essential in maintaining long-term well integrity and preventing operational failures.

The Effect of Temperature on Hydrogen Diffusion

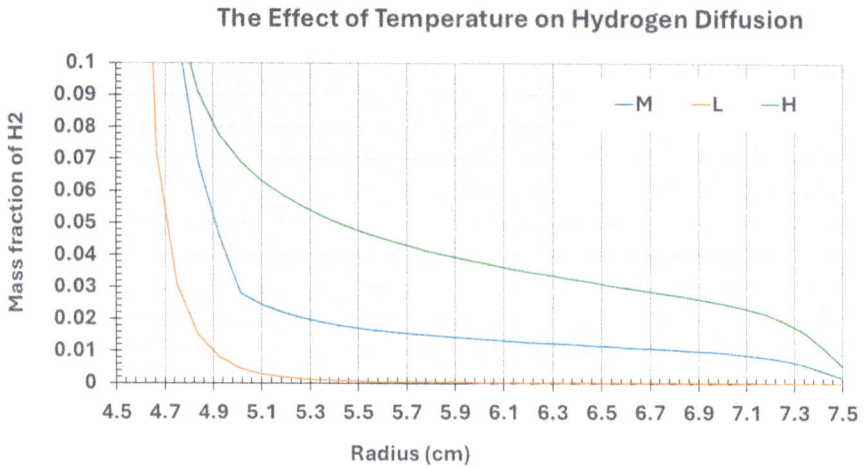

Figure 24.4: Temperature effect on hydrogen diffusion [21].

Medium- to long-term well integrity risks also include cumulative damage from thermal and pressure cycling during injection and withdrawal operations. These cycles create stress concentrations in the wellbore infrastructure, increasing the likelihood of micro-cracks in cement and casing materials. Over time, these micro-cracks can propagate, providing pathways for hydrogen to escape. Additionally, hydrogen-induced chemical reactions in the cement and at the cement-rock interface may weaken the bond strength, further reducing the long-term effectiveness of zonal isolation.

24.2.1.3 Hydrogen cement interaction

During natural hydrogen extraction, hydrogen interacts with the surrounding cement in the wellbore, potentially altering its physical properties. Previous studies have highlighted that hydrogen indirectly affects cement by initiating reactions that modify its characteristics [36]. According to API standards, industrial cements suitable for downhole conditions, considering depth, pressure, and temperature, are categorized into classes A to H, with Class G being the most commonly used [37]. These cements are primarily composed of lime (CaO), silica (SiO_2), alumina (Al_2O_3), iron oxide (Fe_2O_3), and gypsum ($CaSO_4 \cdot 2\,H_2O$). Hydration reactions transform these raw materials into the four main components of Portland cement: tricalcium aluminate (C_3A), tricalcium silicate (C_3S), tetracalcium aluminoferrite (C_4AF), and dicalcium silicate (C_2S) [38]. Hydrogen does not directly react with cement but must be converted to H^+ ions for interaction. Therefore, the availability of H^+ ions in the environment determines the extent of hydrogen-cement reactions [39]. Geochemical simulations suggest that hydrogen can reduce sulfates and iron in cement to sulfides and ferrous compounds, precipitating oxidized minerals like iron sulfides. This process may compromise the structural integrity of the cement. Equations (24.1–24.4) detail the thermodynamic reactions of cement minerals with H^+ ions [20, 40].

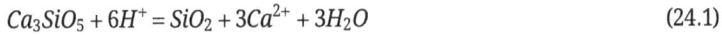

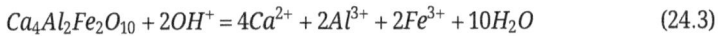

$$Ca_3SiO_5 + 6H^+ = SiO_2 + 3Ca^{2+} + 3H_2O \qquad (24.1)$$

$$CaO_3 + H^+ \leftrightarrow Ca^{2+} + HCO_3^- \qquad (24.2)$$

$$Ca_4Al_2Fe_2O_{10} + 2OH^+ = 4Ca^{2+} + 2Al^{3+} + 2Fe^{3+} + 10H_2O \qquad (24.3)$$

$$Fe_3O_3(s) + 4H^+ + H_2 = 2Fe^{2+} + 3H_2O \qquad (24.4)$$

In subsurface environments rich in hydrogen, single-celled organisms like sulfate-reducing bacteria (SRBs) consume hydrogen and produce H_2S in sulfur-rich conditions. This contributes to the chemical degradation of the cement through H_2S-induced corrosion [30, 41]. Sulfate reactions can form ettringite ($Ca_6(Al(OH)_6)_2 \cdot (SO_4)_3 \cdot 26\ H_2O$), whose expansion within the cement matrix causes internal stresses and cracking [41, 42].

Research on hydrogen-cement interactions is crucial to understanding hydrogen-induced degradation. Ugarte et al. [43] conducted experiments on Class H cement exposed to hydrogen for up to 168 days. Their findings indicated that prolonged hydrogen exposure increased the compressive strength of cement, reduced porosity and permeability by 6.1% and 4.2%, respectively, and enhanced Young's modulus and shear modulus by 4.7% and 5.0%, respectively. The study also observed a decrease in Poisson's ratio by 1.4% and an increase in potassium and calcium feldspar formation, which improved mechanical strength. Conversely, Maury et al. [44] found that hydrogen exposure under conditions of 10 MPa and 49 °C increased permeability by 175% and porosity by 1.9% after 14 days. The Young's modulus and Poisson's ratio decreased by 2.5% and 0.6%, respectively, suggesting compromised sealing capacity.

The cement can experience hydration, leading to degradation such as cracks, voids, and reduced mechanical strength. This issue is more pronounced at higher temperatures and pressures, typical of greater depths. Hydrogen interaction with cement can alter its primary minerals, further contributing to degradation. However, the exact nature of cement hydration with hydrogen under varying temperature and pressure conditions is still unclear [20]. Aftab et al. [38] investigated the effects of hydrogen on cement at 80 °C and 3000 psi, finding that chloride ions in saline environments accelerated hydration reactions. However, under these conditions, hydrogen caused no significant geochemical or structural alterations to the cement. The study noted that pulverizing the cement into fine particles may have limited insights into the broader implications for permeability, porosity, and elasticity.

In another study, Al-Yaseri et al. [36] examined hydrogen injection into Class G cement cores for 125 days under 1400 psi and 75 °C. They observed a slight increase in alite content and calcite growth, attributed to cement hydration processes. Following hydrogen injection, mass and density increased, while porosity decreased by 8.86%

and 8.43%, respectively. Minor reductions in permeability and slight increases in Poisson's ratio and dynamic Young's modulus were also noted.

Hussain et al. [45] assessed the impact of hydrogen on cured cement blocks and cement slurries at 1500 psi and 120 °F for 7 days. Results revealed that hydrogen entrapment in cement slurries reduced strength and increased viscosity. CT scans identified hydrogen-induced cracks, leading to decreased compressive strength in cured cement. Jacquemet et al. [46] conducted simulations that demonstrated reductive dissolution of cement minerals like ettringite and hematite, forming iron sulfides and oxides. These reactions had minimal impact on porosity due to the limited quantities of reacting materials.

Hydrogen diffusion through the cement sheath, a potential cause of gas leakage, was modeled by Dudun et al. [47]. Their simulations suggested complete penetration through a 35 cm cement sheath within ~7.5 days under anticipated well conditions. Hydrogen diffusion rates were found to increase with higher cement porosity and diffusion coefficients while decreasing with greater water saturation. Therefore, low-porosity cement offers better sealing capabilities.

The inconsistencies in hydrogen's impact on cement, as highlighted by experimental variations, underline the need for further studies. Current experiments, limited to durations up to 168 days, do not reflect the long operational lifespans of hydrogen wells. High-pressure and high-temperature conditions deep underground may intensify oxidative reactions, causing significant changes to cement properties. Expanding research to include longer exposure periods, diverse cement types, and realistic environmental conditions is imperative.

Ensuring cement integrity is critical for the safety of hydrogen wells. Key properties like permeability, thickening time, and compressive strength play a vital role. Chemical additives, such as retarders, accelerators, dispersants, and expanding agents, offer promising solutions for enhancing cement performance. While the effects of CO_2-containing fluids on cement integrity have been extensively studied, similar research for hydrogen interactions is lacking. Studies on CO_2 have demonstrated the effectiveness of additives like nanoclay, olive waste, graphite, and synthetic polypropylene fibers in mitigating cement degradation [48–50]. For natural hydrogen wells, a deeper understanding of hydrogen-induced degradation mechanisms could guide the development of specialized additives to enhance cement resilience and ensure operational safety.

Designing and constructing hydrogen exploration and production wells involves addressing unique challenges posed by hydrogen-cement interactions. The selection of cement formulations, particularly Class G and H cements, must consider potential hydrogen-induced changes in porosity, permeability, and strength. Incorporating chemical additives, such as retarders, dispersants, and nanomaterials, can enhance cement resilience, suppress harmful reactions, and minimize risks from microbial activity like sulfate-reducing bacteria. Reducing cement porosity and exploring impermeable coatings can mitigate hydrogen diffusion and gas leakage.

The design must also account for high-pressure, high-temperature (HPHT) conditions typical of deep reservoirs, which amplify oxidative and reductive reactions, necessitating materials tailored for such environments. Integrating monitoring systems, such as X-ray computed tomography and acoustic tools, into wells can detect early signs of degradation and inform proactive maintenance. Additionally, extending experimental studies to simulate long-term exposure is essential for addressing the operational lifespans of hydrogen wells effectively.

Collaboration with regulatory bodies to establish hydrogen-specific standards, alongside cross-industry learning from CO_2 storage, can ensure the integrity and safety of hydrogen wells. These measures will enable the efficient and secure extraction of hydrogen, contributing to the clean energy transition.

24.2.1.4 Microbial-induced corrosion

Microbial-induced corrosion (MIC): is caused by the metabolic activities of microorganisms, such as SRB, methanogens, and iron-reducing bacteria. These microorganisms produce corrosive substances like hydrogen sulfide, organic acids, and ammonia which accelerate the corrosion process [51, 52]. MIC often results in pitting and crevice corrosion, which can lead to significant material loss in localized areas [51], and can compromise the integrity of the wells leading to leaks and failures [52]. The temperature affects microbial activity, with most MIC-related bacteria thriving in moderate temperature ranges (20–60 °C). Higher temperatures can inhibit microbial growth and activity. The pressure impacts the solubility of gases like hydrogen sulfide in water, influencing the rate of microbial metabolism and corrosion [52].

24.2.1.5 Iron-reducing bacteria

Iron-reducing bacteria (IRB) also contribute to corrosion, albeit less significantly than SRB. Using hydrogen, IRB oxidizes metals and reduces iron (III) phases, relying on iron oxides, organic carbon, and minerals like smectite and chlorite. IRB and SRB can coexist with IRB utilizing acetate produced by SRB [53]. In iron oxide and organic carbon-rich environments, IRB may outcompete SRB due to its higher hydrogen affinity [54].

The discussion presented in this section highlights the significant technical challenges associated with hydrogen exploration and production wells, which require effective strategies to ensure long-term wellbore integrity and operational safety. Critical components like coated steels, high-chromium alloys, and hydrogen-resistant elastomers are necessary to minimize hydrogen-induced degradation. Cement formulations specifically designed for hydrogen wells, with additives to reduce permeability and improve bonding, are essential for maintaining well integrity over time.

Monitoring systems that provide real-time data on pressure, temperature, and micro-crack development are crucial for detecting early signs of degradation and enabling timely maintenance to avoid serious failures. These solutions need to be integrated into a wellbore design that expects both short-term and long-term risks, ensuring that the well can withstand the challenges posed by hydrogen.

The successful operation of natural hydrogen wells depends on several engineering and operational factors. Ensuring material compatibility is critical, as hydrogen embrittlement and degradation can weaken the physical and mechanical properties of key well components. Well integrity must be maintained throughout the drilling and production phases, requiring strict quality control during construction and completion, as well as regular integrity testing during operations.

Environmental and regulatory compliance is a critical consideration due to hydrogen's flammability and the significant safety and environmental risks posed by potential leaks. Many regulatory authorities are now establishing guidelines to ensure the safe operation of hydrogen production wells, aligning with sustainability goals. Preventing leaks, corrosion, and other forms of well degradation is fundamental to the successful exploration and production of natural hydrogen.

24.3 Fundamentals of drilling and well design for natural hydrogen exploration and production

Drilling wells for natural hydrogen exploration and production present significant differences compared to conventional hydrocarbon wells, largely due to the unique characteristics of natural hydrogen reservoirs. These reservoirs often exhibit high variability and fundamentally distinct features, including hydrogen's high diffusivity, chemical reactivity, and unique trapping mechanisms [9]. While hydrocarbon reservoirs are typically located in sedimentary basins, natural hydrogen reservoirs are frequently associated with dual porosity systems, fractures, vuggy or karstic voids, and basement geological features [10]. These complexities require specialized drilling and reservoir management strategies that differ from those used in conventional sedimentary hydrocarbon reservoirs. This section provides a comprehensive overview of the fundamental design principles and key considerations for the drilling, construction, and completion of natural hydrogen exploration and production wells.

24.3.1 Drilling considerations

While conventional drilling approaches, with careful selection of hydrogen-compatible materials for casing and cementing, can be employed for drilling natural hydrogen reservoirs, special considerations must be taken when reservoir characteristics vary significantly. For instance, reservoirs characterized by dual porosity systems, fractures, vuggy or karstic voids, and basement features necessitate specialized strategies. These strategies cover various aspects, including critical factors summarized in Table 24.2.

The presence of fractures and vuggy formations increases the risk of lost circulation, where drilling fluids escape into the formation rather than returning to the surface. To mitigate this, drilling fluids should be engineered with appropriate rheological properties and incorporate lost circulation materials (LCMs) tailored to the specific formation characteristics. Examples of LCMs include granular materials like nutshells, calcium carbonate, and corncobs; fibrous materials such as cedar fiber, sawdust, and shredded cane stalk; and flaky materials like mica and cellophane [55, 56]. Additionally, techniques such as managed pressure drilling and the use of aerated drilling fluids can help maintain wellbore stability and minimize circulation losses in such challenging environments [55].

Achieving effective zonal isolation in reservoirs with complex porosity and fracture networks requires careful selection of cement slurry properties. Lightweight cement slurries with enhanced flow properties can penetrate irregular voids, ensuring a competent seal. Moreover, incorporating additives that improve the cement's resilience to chemical interactions with hydrogen is crucial to maintain long-term well integrity [57]. Advanced cementing techniques, such as the use of expandable tubulars and casing while drilling, can also enhance cement placement and reduce the risk of formation damage [55].

The selection of casing and liner materials must account for the potential for hydrogen embrittlement and the mechanical stresses associated with fractured and karstic formations. High-chromium steels and nickel-based alloys offer superior resistance to hydrogen-induced degradation [9]. Additionally, implementing multilayer casing designs can provide redundant barriers, enhancing well integrity in complex geological settings. The use of expandable tubulars can also conform to irregular wellbore geometries, providing better zonal isolation and reducing the risk of casing deformation [55].

Employing real-time monitoring systems during drilling operations is essential to promptly detect and address issues such as fluid losses, wellbore instability, and gas influxes. Technologies like acoustic imaging and neutron tools can provide valuable insights into formation characteristics, enabling proactive adjustments to drilling parameters [10]. Predictive analytics, utilizing machine learning models, can also analyze operational data to forecast potential challenges and optimize drilling strategies [9].

By integrating these tailored approaches – customized drilling fluids, advanced cementing practices, appropriate casing material selection, and real-time monitoring – operators can effectively mitigate the challenges associated with drilling natural hydrogen reservoirs, ensuring safe and efficient resource extraction.

Table 24.2: Various aspects and critical factors to be considered for natural hydrogen reservoirs compared to conventional hydrocarbon reservoirs.

Aspect	Conventional hydrocarbon reservoir	Natural hydrogen reservoir	Critical factors for natural hydrogen
Reservoir characteristics	Sedimentary formations, typically sandstone or limestone	Dual porosity systems, fractures, vuggy or karstic voids, basement features	Geological complexity, risk of lost circulation, wellbore instability
Drilling fluid design	Standard drilling fluids with conventional LCMs	Customized drilling fluids with specific lost circulation materials (LCMs)	Mitigating lost circulation, maintaining wellbore stability
Cementing considerations	API standard cement slurries	API or other standard compatible with hydrogen cement with special additives. For instance, lightweight cement slurries with additives for hydrogen resilience	Ensuring effective cementing jobs, preventing formation of any leakage, induced micro-crack
Casing and liner materials	Typically conventional API standard steel casings	High-chromium steels, nickel-based alloys, multilayer casing designs, API casing with additional precautions like coatings, cathodic protection, or the use of non-API alloys	Resistance to hydrogen embrittlement, mechanical stresses, corrosion
Completion hardware	Standard completion hardware	Expandable tubulars, hydrogen-compatible materials	Ensuring containment, reducing the risk of casing deformation
Well integrity	Standard well integrity practices	Enhanced well integrity practices due to hydrogen's properties	Addressing hydrogen's high diffusivity, reactivity, and potential for embrittlement, blistering, rapid decompression, MIC, etc.
Operational challenges	Standard monitoring and maintenance	Real-time monitoring systems, predictive analytics	Detecting fluid losses, wellbore instability, gas influxes, leaks
Sealing technologies	Conventional sealing technologies	Advanced sealing technologies for hydrogen containment	Preventing hydrogen leaks, ensuring long-term durability

24.3.2 Material selection

The selection of materials for natural hydrogen wells is crucial for maintaining well integrity during hydrogen exploration and production. Materials, for instance, must be chosen for their resistance to hydrogen-induced cracking and corrosion. High-strength steels and certain nickel-based alloys are often preferred due to their superior resistance to hydrogen embrittlement [58]. Additionally, non-metallic materials such as polymers and composites are being explored for their potential to resist hydrogen permeation and chemical attack [17].

To elaborate further, the choice of high-strength steels and nickel-based alloys is driven by their ability to withstand the harsh conditions encountered in hydrogen wells, including high pressures and temperatures. These materials exhibit excellent mechanical properties and resistance to hydrogen embrittlement, which is a critical factor in preventing sudden material failure.

Non-metallic materials, such as advanced polymers and composites, offer additional benefits. These materials are being investigated for their ability to form effective barriers against hydrogen permeation, thereby reducing the risk of hydrogen leakage. Their chemical resistance also makes them suitable for use in environments where hydrogen may interact with other substances, potentially leading to degradation.

Incorporating these materials into well design requires a comprehensive understanding of their long-term performance under operational conditions. This includes evaluating their behavior in the presence of hydrogen, as well as their mechanical and chemical stability over time. By selecting materials that can withstand the unique challenges posed by hydrogen exploration and production, the integrity and safety of hydrogen wells can be ensured, supporting the efficient and secure extraction of hydrogen.

Casing materials: API standard steel casings are typically used for the design and construction of hydrocarbon (oil and gas) wells. They are not necessarily suitable and vulnerable to hydrogen embrittlement. High-chromium steels and nickel-based alloys are preferred for their superior resistance to hydrogen embrittlement and corrosion. Coated casings with epoxy or polymer-based barriers can further mitigate hydrogen diffusion and provide an additional layer of protection [36].

Cement materials: Cement serves as a barrier to ensure zonal isolation, but hydrogen can alter its chemical composition and microstructure over time. Class G cement, modified with silica or other additives, is commonly used for its low permeability and enhanced durability under well's zonal isolation and surrounding wellbore conditions [21]. Specialized cement formulations that resist hydrogen-induced microcracking and porosity changes are critical for maintaining long-term well integrity.

Elastomeric components: Completion hardware, such as packers and seals, relies on elastomeric materials for zonal isolation, sealing, and maintaining flow containment. To ensure well integrity, hydrogen-resistant elastomers (HRE) must be used to resist hydrogen degradation and damage. The selection of elastomeric components should consider several key factors, including chemical resistance, physical resistance, barrier properties, and tolerance to temperature and pressure variations.

Chemical resistance ensures that elastomers are chemically stable in the presence of hydrogen. This involves modifying the polymer structure to reduce reactivity with hydrogen, thereby preventing chemical degradation. For example, hydrogen-resistant elastomers such as FKM and HNBR are specifically designed to withstand hydrogen exposure without undergoing significant chemical changes. These materials demonstrate greater resistance to hydrogen permeation compared to traditional elastomers [21, 59].

Physical resistance is critical to prevent physical damage, such as blistering, rapid gas decompression (RGD) and microcracking. These issues often arise due to the rapid expansion of elastomers when pressure drops, which can cause the material to crack or blister [21, 59]. Elastomers must be engineered to withstand these mechanical stresses and maintain their structural integrity.

Barrier properties are essential for maintaining containment integrity. Elastomers used in packers, sealing rings, and other well components must have very low or no permeability to hydrogen to minimize penetration and reduce the risk of mechanical degradation. For example, elastomers with high fluorine content, such as FKM, typically exhibit lower hydrogen permeability. Moreover, elastomers must have the ability to retain their mechanical properties and maintain high integrity under dynamic temperature and pressure conditions. They must resist degradation even at elevated temperatures and pressures, ensuring reliable performance over the operational life of the wells [21, 60]. By carefully selecting elastomeric materials that meet these criteria, the integrity of completion hardware and overall well containment can be maintained effectively, even under challenging hydrogen production and operation conditions.

24.3.3 Containment integrity

Ensuring containment integrity is a critical aspect of natural hydrogen exploration and production wells to prevent leakage, and to maintain operational safety. At the well design and construction phase, the following key initiatives should be taken into consideration to address challenges associated with containment integrity.

Zonal isolation: Effective zonal isolation is crucial for preventing hydrogen leakage between storage zones and surrounding formations. High-quality cementing plays a pivotal role in achieving zonal isolation. Cement with low permeability and additives enhancing its sealing properties, such as silica flour and latex, has been identified as a key requirement for hydrogen wells [20, 36]. These formulations provide a robust barrier against hydrogen migration and help mitigate issues related to hydrogen diffusion and reactivity with cement. In addition, cement bond logs (CBLs) and ultrasonic tools are widely used to evaluate the quality of cement placement. These diagnostic tools allow operators to identify weak zones and perform remedial actions, such as cement squeezes, to ensure zonal isolation.

Pressure management: Pressure management is another critical factor in maintaining containment integrity. The drilling and well construction process must follow the principles of well design satisfying all the requirements associated with wellbore stability, pore and fracture pressure, and casing design principles, including meeting fracture pore and pressure requirements (i.e., safe drilling mud pressure window) for safe drilling operation and maintaining borehole stability. The pressure requirements should be determined utilizing geomechanical and earth modeling ensuring that the pressure during drilling and cementing operation remains below the fracture pressure and must not create induced fractures and/or microfracture, one of the primary causes of the creation of H_2 leakage pathways [27, 61]. It is also extremely important to monitor pressure gradients along the wellbore to avoid micro-cracking in the cement sheath. Such cracks can compromise zonal isolation and increase the risk of hydrogen leakage, especially during the production phase.

Temperature control: Temperature fluctuations during natural hydrogen withdrawal, caused by compression and expansion of hydrogen gas, can lead to thermal stresses. These stresses impact the mechanical integrity of the well, including casing and cement systems. Materials with high thermal tolerance, such as advanced cement formulations with thermal stabilizers, Tetra-polymer Nano-Composite (TPN) additives [62], and high-performance alloys for casings, are recommended for natural hydrogen wells. Research suggests that incorporating fiber-reinforced cements can further enhance the thermal resistance of the well structure [21].

Hydrogen diffusion control: The diffusion of hydrogen into steel casings, all other completions hardware, production surface equipment, and cementing materials leads to material degradation over time and poses significant risks to well integrity. To mitigate this, applying coatings with low or no hydrogen permeability, such as epoxy or ceramic-based barriers, has proven effective. These coatings serve as diffusion barriers, reducing the interaction between hydrogen and the base material [36].

Chemical compatibility: Hydrogen can chemically interact with minerals in cement and surrounding formations, potentially compromising structural integrity. For instance, hydrogen-induced sulfate reduction can lead to the formation of hydrogen sulfide (H_2S), which is corrosive to casing and cement [20]. The use of geochemically compatible cement additives and conducting site-specific geochemical assessments are essential measures to mitigate these risks.

Monitoring and testing: The successful implementation of hydrogen injection or production for underground hydrogen exploration and production projects depends on accurately identifying weak points and potential leakage pathways that may develop in wells under reservoir conditions. It is essential to have a clear and effective plan to mitigate these leakage risks. Monitoring and mitigation measures should be carefully considered to maintain well integrity in the event of H_2 leakage. Regular testing, such as ultrasonic testing and pressure decay analysis, is critical for detecting

Table 24.3: A summary of potential H$_2$ leakage sources, causes, effects, and mitigation plan.

Location/components	Functions	Challenges/failure risks	Causes	Effects	Mitigation
Wellhead	Flow control	H$_2$ leaking into the atmosphere	Corrosion, hydrogen embrittlement, hydrogen blistering (sealing elements, valve failure)	Uncontrolled flow	– Routine inspection – Use of H$_2$-compatible materials and seals, valves
Tubular packer	Isolation of tubing/casing, zonal isolation, and/or safety barrier	Failure in sealing elements (elastomers, O-ring, etc.)	H$_2$ diffusion causing embrittlement, blistering, swelling or cracking	Leakage inside the injection/production casing/liners	Install or replace (for existing well) packers with elastomers resistant to H$_2$ diffusion and embrittlement
Travel joint	Compensation of elongation due to thermal/mechanical Stresses	Failure in sealing elements	H$_2$ diffusion	Damaging injection/production casing, leaking through annular spaces that cause pressure build up at the wellhead	Use seal materials compatible with H$_2$ properties, and hydrogen diffusion-resistant materials in all conditions
Casing	Isolation from cements, and/or formations	Leak into formations	Corrosion, H$_2$ embrittlement, blistering, H$_2$-induced cracking	Uncontrolled flow	Batch repair or replace casing with H$_2$-compatible materials
Casing-cement interface	– Support casing – Ensure sealing against formation fluids, and injected H$_2$ – Zibal Isolation	– Loss of casing cement sealing against formation fluids and H$_2$ – Cement integrity failure	– Formation of leakage pathways, – Degradation of cement (loss of mechanical properties) – Loss of cement sheath integrity and overloading of the casing	– Changes in cement performance – Increase in cement permeability – Formation of potential H$_2$ leakage pathway	– Remediate with side-track well – Use of H$_2$-compatible cement plugs with enhanced properties

Cement strength	Ensure well integrity	Deterioration of cement mechanical properties	Chemical reaction leading to reduction in mechanical strength	Reduced stability and increased permeability (i.e., increased H_2 permeation and diffusion)	– Perform cement squeezing with improved formulations. – Consider well abandonment if needed
Fracture	Preferred pathway during injection/ production	Limited containment of H_2 in fractured formations	– Naturally occurring fractures in carbonate formations – Drilling-induced fractures in sandstones	Reduced reservoir contact between injected H_2 and formation	– Simulation should be run to identify potential pathways and establish equilibrium. – Special consideration for advanced cementing design, and use of special LCM to control potential loss circulation
Caprock	Provides essential sealing for the reservoir	Risk of uncontrolled flow	Unsuitability of caprock or seals for H_2 reservoir	Potential environmental damage, loss of projects, health and safety hazards	Pre-investigation of stability and integrity of cap rock is essential

early signs of damage. These methods help identify problems, such as micro-cracks in the cement sheath or casing issues before they become severe.

Real-time monitoring systems with advanced sensors provide continuous information on well conditions, such as pressure, temperature and gas composition. This real-time data supports proactive maintenance, reducing the chances of serious failures and ensuring the reliability of the system over time [20, 21]. To facilitate practical decision-making, Table 24.3 summarizes potential leakage pathways and corresponding mitigation strategies based on "what-if" scenarios. This structured approach offers a valuable framework for timely and effective interventions to safeguard well integrity and ensure safe drilling and production operations.

24.4 Framework of natural hydrogen well design and construction

Based on the compressive review and discussion presented in the previous section, it is obvious that designing natural hydrogen exploration and production wells is a complex, multidisciplinary engineering task. It requires a comprehensive understanding of the geological, mechanical, and chemical aspects of reservoir and well production systems. This process begins with discovering the field and involves detailed technical evaluations of the geological formation, including reservoir characteristics, rock mechanics, fluid properties, geomechanical and geochemical conditions, and the downhole environment (e.g., pressure and temperature), and evaluation of reservoir flow performance. Equally critical are meeting the operational safety requirements and developing strategies to ensure the highest level of well integrity throughout the well's lifecycle.

In summary, designing successful natural hydrogen wells requires a thorough understanding of hydrogen's unique physical and chemical properties and its interactions with well components, including casings, completion hardware, cement, and formation rocks. These interactions must be assessed for their short- and long-term impacts on well integrity. Key design issues include containment integrity, effective pressure and temperature management, control of hydrogen diffusion, and chemical compatibility among hydrogen, well materials, cements, and formation rocks. Addressing these challenges necessitates a systematic framework for well design and construction. To ensure safe and efficient hydrogen production, the following step-by-step engineering framework, as illustrated in Figure 24.5, is recommended for designing and constructing natural hydrogen exploration and production wells.

1. Reservoir Evaluation: The foundational step involves evaluating the reservoir system, including reservoir characteristics, geological features, properties, environmental impact assessments, and local considerations. Data from offset fields and reports

may be used as initial input for evaluating geological features such as porosity, permeability, pore pressure, fracture pressure, and geomechanical, geochemical characteristics and fluid characteristics. This step may also require community involvement to address impact assessment and community concerns. If the data from offset fields, available reports, and all analyses demonstrate the feasibility of the proposed field, the subsequent steps can proceed methodically, ensuring the design, construction, and deployment of the natural hydrogen exploration and production well align with safety and efficiency standards.

2. Preliminary Well Design and Material Selection: This step involves developing an initial well design tailored to the geological and operational requirements, addressing all challenges, issues, and concerns identified in step 1. It also includes a detailed design of the wellbore profile, drilling programs (including drilling fluid, cementing, and casing programs), and selection of appropriate casing (grade, types, materials), cements (grade, types, additives, mixing requirements, etc.), and other associated elements including equipment and infrastructure required to construct the well. The selected casing and cement materials must be compatible and resistant to hydrogen-induced degradation, considering high-performance API grade or non-API alloys for casing and hydrogen-resistant cement formulations (e.g., modified G-cement).

3. Drilling Operations: Employ advanced drilling techniques to ensure wellbore stability and compatibility with hydrogen, maintaining the highest level of integrity in workmanship to ensure safe drilling operations. Care should be taken in selecting the drilling mud and mud pressure to minimize formation damage, maintain wellbore stability, and avoid breakout and the creation of induced cracks or micro-cracks during drilling and cementing jobs.

4. Cementing and Cement Evaluation: Accurate cementing is crucial for maintaining the well's structural integrity and zonal isolation. The types of cement to be used require rigorous geochemical and cement-hydrogen interaction studies to ensure the selected cement is resistant to hydrogen. The quality of the cementing job, including curing strength, should be carefully evaluated to identify and fix weak zones if needed. CBL, VDL, and other advanced ultrasonic cement evaluation tools (e.g., Ultrasonic Pulse Echo Tool) and temperature logs may be considered to ensure quality and confirm the highest level of cement integrity.

5. Completion Design: This step involves equipping the drilled well with tubing and other necessary hardware, including packers, safety valves, flow mandrels, flow communicators (e.g., slips), and monitoring devices (e.g., pressure, and temperature gauges), to turn it into a safe producer. Particular attention should be given to ensuring that the materials selected for tools and components are compatible and resistant to hydrogen under all conditions, both short and long-term, ensuring the highest levels of integrity. It is also important to perform production performance tests of the completion design to ensure optimal operating conditions from the production perfor-

mance, dynamic reservoir behavior, and surface facility viewpoints. This process typically requires numerical simulation and is an iterative process, as illustrated in the engineering workflow in Figure 24.6.

6. Integrity Testing Prior to Injection or Production: Once the well is drilled and successfully completed, it must undergo rigorous integrity testing before operational deployment. This process involves performing pressure integrity tests to ensure well readiness and identifying and rectifying any weaknesses before operational deployment.

7. Operational Protocol Development: Clear protocols must be established for safe production cycles, ensuring operational parameters remain within safety margins. Develop emergency response procedures to handle unexpected conditions.

8. Monitoring System Development: Deploy real-time monitoring systems equipped with advanced sensors (e.g., fiber optics-enabled distributed sensors or distributed temperature sensors) to continuously track pressure, temperature, and gas composition, including identifying occurrences of leaks. These systems enable proactive maintenance by detecting early signs of well integrity compliance violations.

9. Maintenance and Repair Protocols: Implement a schedule for regular well integrity assessments by defining clear protocols for addressing identified issues promptly to maintain operational safety and mitigate health, safety, and environmental hazards.

The proposed framework is expected to serve as a preliminary guide for operators or engineers in the design, development, and construction of natural hydrogen exploration and production wells, addressing the unique challenges of hydrogen extractions and designing wells capable of long-term performance and reliability in a natural hydrogen exploration and production projects.

To facilitate the implementation of this framework, Table 24.4 summarizes the critical factors to be considered at each step (decision gate), along with recommended actions for further improvements and a guide to implement these recommendations.

24.5 Conclusions

The design and construction of natural hydrogen exploration and production wells requires a systematic and multidisciplinary approach to address the unique challenges posed by hydrogen's physical and chemical properties. Critical factors such as containment integrity, pressure and temperature management, material selection, and chemical compatibility must be carefully evaluated and incorporated into a comprehensive framework.

The proposed framework outlines a step-by-step process from reservoir evaluation to monitoring and maintenance, ensuring operational safety and long-term well

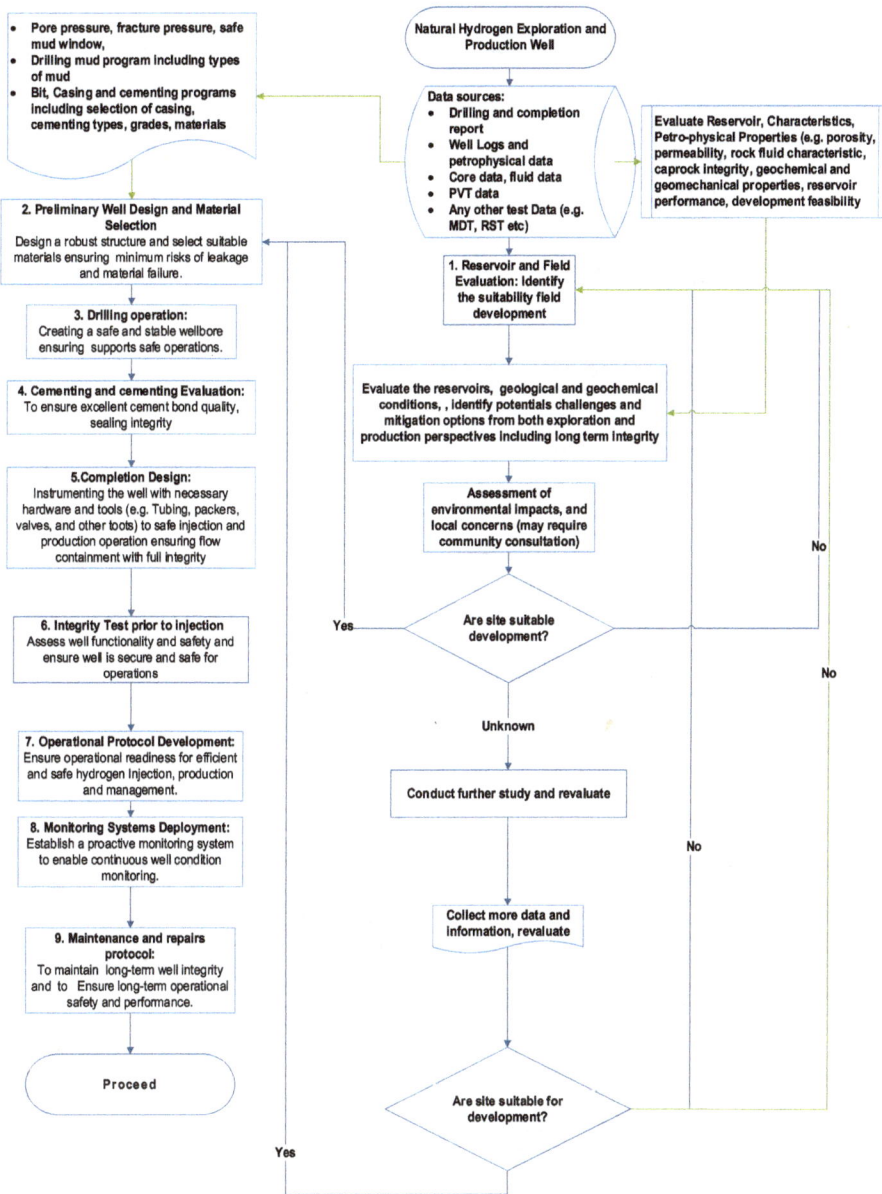

Figure 24.5: Engineering framework for the design, development and construction of natural hydrogen exploration and production well.

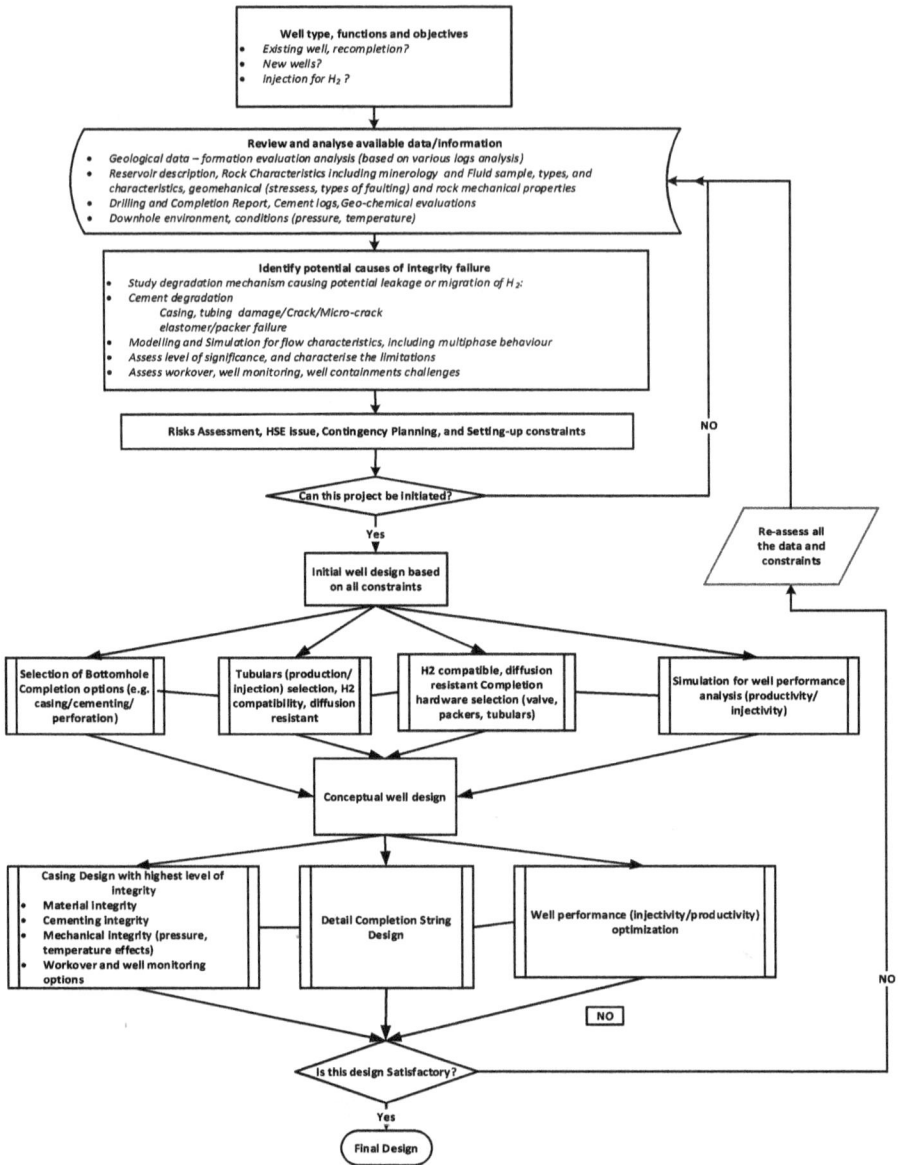

Figure 24.6: Engineering workflow for the design and completion of natural hydrogen exploration and production well (adapted from [27]).

Table 24.4: Step-by-step framework for designing and constructing hydrogen injection wells for underground hydrogen storage, including key decision points, further recommendations, and implementation guides.

Step	Key decision points/gate	Further recommendations	Guide to implement recommendations
1. Reservoir evaluation	– Feasibility of geological system (storage capacity, rock types, porosity, permeability, pore pressure, fracture pressure) – Environmental impact assessment – Community involvement and concerns	– Use data from offset fields and reports for initial evaluation – Engage with the community early to address concerns – Ensure thorough analysis of geomechanical and geochemical characteristics	– Collect and analyze geological data from offset fields and reports – Conduct environmental impact assessments – Organize community meetings and feedback sessions – Perform detailed geomechanical and geochemical studies
2. Preliminary well design and material selection	– Compatibility of materials with hydrogen – Selection of casing and cement materials	– Consider high-performance API grade or non-API alloys for casing – Use hydrogen-resistant cement formulations (e.g., Modified G-cement) – Tailor well design to geological and operational requirements	– Review material specifications and compatibility studies – Select appropriate casing and cement materials based on hydrogen resistance – Design wellbore profile and drilling programs according to site-specific data
3. Drilling operations	– Selection of drilling techniques – Choice of drilling mud and mud pressure	– Employ advanced drilling techniques for wellbore stability – Minimize formation damage and avoid induced cracks or micro-cracks	– Choose drilling techniques that ensure wellbore stability – Select drilling mud and adjust mud pressure to minimize formation damage – Monitor wellbore conditions continuously during drilling

(continued)

Table 24.4 (continued)

Step	Key decision points/gate	Further recommendations	Guide to implement recommendations
4. Cementing and cement evaluation	– Selection of cement types – Quality of cementing job	– Conduct rigorous geochemical and cement-hydrogen interaction studies – Use advanced ultrasonic cement evaluation tools (e.g., Ultrasonic Pulse Echo tool) – Evaluate curing strength and fix weak zones if needed	– Perform laboratory tests on cement formulations for hydrogen resistance – Use ultrasonic tools and temperature logs to assess cement quality – Identify and repair weak zones in the cement sheath
5. Completion design	– Compatibility of completion hardware with hydrogen – Performance of injection and production tests	– Ensure materials are resistant to hydrogen in all conditions – Perform numerical simulations and iterative testing – Optimize operating conditions for injection and production	– Select completion hardware based on hydrogen compatibility – Conduct numerical simulations to predict well performance – Test and refine completion design through iterative processes
6. Integrity testing prior to injection or production	– Results of pressure integrity tests – Identification of weaknesses	– Conduct rigorous integrity testing before operational deployment – Rectify any identified weaknesses promptly	– Perform pressure integrity tests on the completed well – Address any identified weaknesses before starting operations – Document and verify test results for compliance

7. Operational protocol development	– Establishment of safe operational parameters – Development of emergency response procedures	– Ensure protocols keep operational parameters within safety margins – Prepare for unexpected conditions with clear emergency procedures	– Develop and document operational protocols and safety margins – Create and train staff on emergency response procedures – Regularly review and update protocols as needed
8. Monitoring system development	– Deployment of real-time monitoring systems – Detection of leaks and integrity violations	– Use advanced sensors (e.g., fiber optics-enabled distributed sensors) – Enable proactive maintenance by detecting early signs of issues	– Install real-time monitoring systems with advanced sensors – Continuously track pressure, temperature, and gas composition – Set up alerts for early detection of integrity issues
9. Maintenance and repair protocols	– Schedule for regular well integrity assessments – Protocols for addressing identified issues	– Implement regular assessments to maintain operational safety – Define clear protocols for prompt issue resolution	– Establish a maintenance schedule for regular integrity checks – Develop protocols for quick and effective issue resolution – Train maintenance teams on protocols and procedures

integrity. Key recommendations include adopting advanced cement formulations for robust cement integrity, implementing hydrogen-resistant wellbores (e.g., casings), completion materials (downhole tools, packers, seal etc.), and integrating advanced real-time monitoring systems like optical fiber-based sensors (e.g., distributed temperature sensors, DTS – system) to continuously track well conditions.

Additionally, iterative testing during the completion design phase and proactive maintenance protocols are vital for optimizing performance and mitigating risks. These measures ensure that any potential issues are identified and addressed promptly, maintaining the well's integrity and operational efficiency.

While implementing this framework, it is crucial to pay careful attention to geological and local considerations. This includes compliance with regulatory requirements and adapting the framework to specific geological features of the storage site. Regulatory compliance ensures that all operations meet safety and environmental standards, which is essential for gaining approval and maintaining public trust. Geological considerations, such as the unique characteristics of the reservoir and surrounding formations, must be thoroughly understood and integrated into the design to prevent issues like leakage, structural or reservoir containment failure.

By following this structured approach and making necessary adjustments based on local and geological conditions, natural hydrogen wells can support safe, efficient, and sustainable development for the exploitation of natural hydrogen projects. This contributes significantly to the transition toward a low-carbon or carbon-free energy future, providing a reliable and clean energy solution. The successful implementation of this framework will not only enhance the performance and reliability of hydrogen exploration and systems but also play a crucial role in achieving global energy sustainability goals.

24.6 Recommendations for further advancement

For further advancements in drilling, construction, and completion of natural hydrogen exploration and production wells, research and development should focus on enhancing material performance, monitoring capabilities, and operational efficiency. Innovative materials such as advanced alloys, polymer composites, and self-healing cements can improve resistance to hydrogen-induced degradation, embrittlement, and chemical reactivity, ensuring longer well lifespans.

The integration of cutting-edge monitoring systems, such as distributed temperature sensors (DTS) [63, 64] and/or distributed acoustic sensing (DAS) [65, 66] and enhanced fiber-optic technologies, can provide high-resolution, real-time data for early detection of potential integrity issues. Moreover, advancements in machine learning and predictive analytics [67–69] can enable smarter decision-making by analyzing operational data to optimize injection and withdrawal processes and predict maintenance needs.

Collaborative field-scale pilot projects are essential for developing best practices for different geological and reservoir conditions. Policymakers and industry stakeholders should also support the establishment of standardized guidelines for hydrogen-specific well design and operational protocols to ensure safety and efficiency.

These advancements will not only enhance the reliability of natural hydrogen wells but also accelerate the adoption of clean and sustainable energy resources development.

References

[1]   Noussan M., et al. The role of green and blue hydrogen in the energy transition – A technological and geopolitical perspective. Sustainability. 2020, 13: 298.

[2]   U.S Energy Information Administration, International Energy Outlook 2021. Eia.gov., 2021.

[3]   Boreham C. J., et al. Hydrogen in Australian natural gas: Occurrences, sources and resources. The APPEA Journal. 2021, 61(1): 163–191.

[4]   IEA-Dubai, Global hydrogen review 2021. Public Report, 2021.

[5]   IEA, Global Hydrogen Review 2024, in Report. 2024, international Energy Agency.

[6]   IEA-Hydrogen-Review, Global Hydrogen Review 2021 Report extract Executive summary [EB/OL]. (2021-10-01) [2022-02-10], 2021.

[7]   Tian Q.-N., et al. Origin, discovery, exploration and development status and prospect of global natural hydrogen under the background of "carbon neutrality". China Geology. 2022, 5(4): 722–733.

[8]   Council H. Hydrogen decarbonization pathways: a life-cycle assessment. URL: https://hydrogencouncil.com/wp-content/uploads/2021/01/Hydrogen-Council-Report_Decarboniza tion-Pathways_Part-1-Lifecycle-Assessment.pdf, 2021.

[9]   Blay-Roger R., et al. Natural hydrogen in the energy transition: Fundamentals, promise, and enigmas. Renewable and Sustainable Energy Reviews. 2024, 189: 113888.

[10]   Maiga O., et al. Characterization of the spontaneously recharging natural hydrogen reservoirs of Bourakebougou in Mali. Scientific Reports. 2023, 13(1): 11876.

[11]   Czado K. Natural hydrogen: The race to discovery and concept demonstration. 2024.

[12]   Kruck O., et al. Overview on all known underground storage technologies for hydrogen. Project HyUnder–Assessment of the Potential, the Actors and Relevant Business Cases for Large Scale and Seasonal Storage of Renewable Electricity by Hydrogen Underground Storage in Europe, Report D. 2013, 3.

[13]   Matos C. R., Carneiro J. F., Silva P. P. Overview of large-scale underground energy storage technologies for integration of renewable energies and criteria for reservoir identification. Journal of Energy Storage. 2019, 21: 241–258.

[14]   Tarkowski R. Underground hydrogen storage: Characteristics and prospects. Renewable and Sustainable Energy Reviews. 2019, 105: 86–94.

[15]   Fernandez D. M., et al. A holistic review on wellbore integrity challenges associated with underground hydrogen storage. International Journal of Hydrogen Energy. 2024, 57: 240–262.

[16]   Heinemann N., et al. Enabling large-scale hydrogen storage in porous media–the scientific challenges. Energy & Environmental Science. 2021, 14(2): 853–864.

[17]   Lemieux A., Shkarupin A., Sharp K. Geologic feasibility of underground hydrogen storage in Canada. International Journal of Hydrogen Energy. 2020, 45(56): 32243–32259.

[18]   Luo X., et al. Underground hydrogen storage (UHS) in natural storage sites: A perspective of subsurface characterization and monitoring. Fuel. 2024, 364: 131038.

[19] Reitenbach V., et al. Influence of added hydrogen on underground gas storage: A review of key issues. Environmental Earth Sciences. 2015, 73: 6927–6937.

[20] Al Dandan E., Hossain M. M. Understanding of geochemical reactions in hydrogen-injected wells: Cement integrity for safe underground hydrogen storage. In: International petroleum technology conference. IPTC, 2024.

[21] Alhassan Y., Hossain M. M. Numerical modelling of hydrogen diffusion in elastomeric materials to evaluate wellbore integrity for hydrogen injected well. In: International petroleum technology conference. IPTC, 2024.

[22] AGR. Top 5 challenges associated with effective well integrity management, and how to overcome them. 2024.

[23] Aiken T. An introduction to the IEA GHG international research network on wellbore integrity. Energy Procedia. 2009, 1(1): 3539–3544.

[24] Norsok D. Well integrity in drilling and well operations. Norsok Standard D-010 Rev. 2013, 4.

[25] Sarker M. N., et al. Innovative materials and techniques for enhancing hydrogen storage: A comprehensive review of damage detection and preventive strategies. ASME Open Journal of Engineering. 2024, 3.

[26] Habib A. A., et al. Hydrogen-assisted aging applied to storage and sealing materials: A comprehensive review. Materials. 2023, 16(20): 6689.

[27] Hossain M. M., Amro M. Drilling and completion challenges and remedies of CO2 injected wells with emphasis to mitigate well integrity issues. In: SPE Asia pacific oil and gas conference and exhibition. SPE, 2010.

[28] Viswanathan H. S., et al. Development of a hybrid process and system model for the assessment of wellbore leakage at a geologic CO2 sequestration site. Environmental Science & Technology. 2008, 42(19): 7280–7286.

[29] Ugarte E. R., Salehi S. A review on well integrity issues for underground hydrogen storage. Journal of Energy Resources Technology. 2022, 144(4): 042001.

[30] Loto C. Microbiological corrosion: Mechanism, control and impact – A review. The International Journal of Advanced Manufacturing Technology. 2017, 92(9): 4241–4252.

[31] Muyzer G., Stams A. J. The ecology and biotechnology of sulphate-reducing bacteria. Nature Reviews Microbiology. 2008, 6(6): 441–454.

[32] Lee J. A., Woods S. Hydrogen embrittlement. 2016.

[33] Li Q., et al. Hydrogen impact: A review on diffusibility, embrittlement mechanisms, and characterization. Materials. 2024, 17(4): 965.

[34] Theiler G., et al. Effect of high-pressure hydrogen environment on the physical and mechanical properties of elastomers. International Journal of Hydrogen Energy. 2024, 58: 389–399.

[35] Koga A., et al. Evaluation on high-pressure hydrogen decompression failure of rubber O-ring using design of experiments. International Journal of Automotive Engineering. 2011, 2(4): 123–129.

[36] Al-Yaseri A., et al. On hydrogen-cement reaction: Investigation on well integrity during underground hydrogen storage. International Journal of Hydrogen Energy. 2023, 48(91): 35610–35623.

[37] Parrott L. Effect of changes in UK cements upon strength and recommended curing times. Concrete (London). 1985, 19(9).

[38] Aftab A., et al. Geochemical integrity of wellbore cements during geological hydrogen storage. Environmental Science & Technology Letters. 2023, 10(7): 551–556.

[39] Fatah A., Al Ramadan M., Al-Yaseri A. Hydrogen impact on cement integrity during underground hydrogen storage: A minireview and future outlook. Energy & Fuels. 2024, 38(3): 1713–1728.

[40] Szabó-Krausz Z., et al. Wellbore cement alteration during decades of abandonment and following CO2 attack–A geochemical modelling study in the area of potential CO2 reservoirs in the Pannonian Basin. Applied Geochemistry. 2020, 113: 104516.

[41] Kutchko B. G., et al. H2S–CO2 reaction with hydrated Class H well cement: Acid-gas injection and CO2 Co-sequestration. International Journal of Greenhouse Gas Control. 2011, 5(4): 880–888.

[42] Collepardi M. A state-of-the-art review on delayed ettringite attack on concrete. Cement and Concrete Composites. 2003, 25(4–5): 401–407.

[43] Ugarte E. R., Tetteh D., Salehi S. Experimental studies of well integrity in cementing during underground hydrogen storage. International Journal of Hydrogen Energy. 2024, 51: 473–488.

[44] Maury Fernandez D., et al. Effects of hydrogen on Class H well cement's properties under geological storage conditions. In: ARMA US rock mechanics/geomechanics symposium. ARMA, 2023.

[45] Hussain A., et al. Experimental investigation of wellbore integrity of depleted oil and gas reservoirs for underground hydrogen storage. In: Offshore technology conference. OTC, 2022.

[46] Jacquemet N., Chiquet P., Grauls A. Hydrogen reactivity with (1) a well cement-PHREEQC geochemical thermodynamics calculations. In: 1st geoscience & engineering in energy transition conference. European Association of Geoscientists & Engineers, 2020.

[47] Dudun A., Feng Y., Guo B. Numerical simulation of hydrogen diffusion in cement sheath of wells used for underground hydrogen storage. Sustainability. 2023, 15(14): 10844.

[48] Mahmoud A. A., Elkatatny S., Mahmoud M. Improving Class G cement carbonation resistance using nanoclay particles for geologic carbon sequestration applications. In: Abu Dhabi international petroleum exhibition and conference. SPE, 2018.

[49] Mahmoud A. A., Elkatatny S. Improved durability of Saudi Class G oil-well cement sheath in CO2 rich environments using olive waste. Construction and Building Materials. 2020, 262: 120623.

[50] Mahmoud A. A., et al. The use of graphite to improve the stability of Saudi class G oil-well cement against the carbonation process. ACS Omega. 2022, 7(7): 5764–5773.

[51] Rao P., Mulky L. Microbially influenced corrosion and its control measures: A critical review. Journal of Bio-and Tribo-Corrosion. 2023, 9(3): 57.

[52] Pal M. K., Lavanya M. Microbial influenced corrosion: Understanding bioadhesion and biofilm formation. Journal of Bio-and Tribo-Corrosion. 2022, 8(3): 76.

[53] Rickard D. Sedimentary iron biogeochemistry. In: Developments in sedimentology. Elsevier, 2012, 85–119.

[54] Lovley D. R., Phillips E. J. Organic matter mineralization with reduction of ferric iron in anaerobic sediments. Applied and Environmental Microbiology. 1986, 51(4): 683–689.

[55] Alkinani H. H., A.T.T.A.-H., Flori R. E., Dunn-Norman S., Hilgedick S. A., Amer A. S., Alsaba M. T. Drilling strategies to control lost circulation in Basra oil fields, Iraq. In: 2018 AADE fluids technical conference and exhibition. Hilton Houston North Hotel, Houston, Texas: American Association of Drilling Engineers, 2018, April 10–11, 2018.

[56] Alkinani H. H. A comprehensive analysis of lost circulation materials and treatments with applications in Basra's oil fields, Iraq: Guidelines and recommendations. Missouri University of Science and Technology, 2017.

[57] Gaurina-Međimurec N., et al. Drilling fluid and cement slurry design for naturally fractured reservoirs. Applied Sciences. 2021, 11(2): 767.

[58] Heinemann N., et al. Hydrogen storage in porous geological formations–onshore play opportunities in the midland valley (Scotland, UK). International Journal of Hydrogen Energy. 2018, 43(45): 20861–20874.

[59] EPS. Elastomers in Hydrogen Service, 5 Critical Things to Consider, Retrieved from Engineering Pipeline Solution. 2024 12/12/2024; Available from: https://engps.com.au/2024/05/31/elastomers-and-hydrogen-5-things-to-consider/

[60] Simmons K. L., et al. H-Mat hydrogen compatibility of NBR elastomers. 2021.

[61] Hossain M., Rahman M., Rahman S. Hydraulic fracture initiation and propagation: Roles of wellbore trajectory, perforation and stress regimes. Journal of Petroleum Science and Engineering. 2000, 27 (3–4): 129–149.

[62] Khan M. I., Al-Ghamdi S. G. Hydrogen economy for sustainable development in GCC countries: A SWOT analysis considering current situation, challenges, and prospects. International Journal of Hydrogen Energy. 2023, 48(28): 10315–10344.

[63] Ghafoori Y., et al. A review of measurement calibration and interpretation for seepage monitoring by optical fiber distributed temperature sensors. Sensors. 2020, 20(19): 5696.

[64] Ouyang L.-B., Belanger D. Flow profiling by Distributed Temperature Sensor (DTS) system – Expectation and reality. SPE Production & Operations. 2006, 21(02): 269–281.

[65] He Z., Liu Q. Optical fiber distributed acoustic sensors: A review. Journal of Lightwave Technology. 2021, 39(12): 3671–3686.

[66] Rossi M., et al. Assessment of Distributed Acoustic Sensing (DAS) performance for geotechnical applications. Engineering Geology. 2022, 306: 106729.

[67] Salem A. M., Yakoot M. S., Mahmoud O. A novel machine learning model for autonomous analysis and diagnosis of well integrity failures in artificial-lift production systems. Advances in Geo-Energy Research. 2022, 6(2): 123–142.

[68] Ragab A. M. S., Yakoot M. S. E., Mahmoud O. Application of machine learning algorithms for managing well integrity in gas lift wells. In: SPE Asia pacific oil and gas conference and exhibition. SPE, 2021.

[69] Savini M. B., Domec B. S. Big data and machine learning: Effective tools for ensuring well integrity. In: Offshore technology conference Asia. OTC, 2020.

Yongqiang Chen

Chapter 25
Effect of salt on rock wettability and gas interactions in natural hydrogen reservoirs

Abstract: Rock salts widely exist in geological settings, such as sandstones, carbonates, and shales. This study employs molecular dynamics (MD) simulations to explore the wettability and gas interactions between hydrogen (H_2), CO_2, CH_4, and clay impurities at the molecular scale on salt surfaces. Results reveal that water preferentially forms a sealing layer on rock salt surfaces. Compared with CO_2, CH_4 is less likely to mix with H_2, minimizing hydrogen contamination in the reservoirs. Kaolinite prefers adsorbing CO_2 impurities rather than H_2, which indicates less polluted natural hydrogen with clays in the reservoirs. These findings provide valuable insights into understanding natural hydrogen behaviors as a function of rock salt properties. While this study highlights the importance of wettability in rock salt environments, future work is recommended to explore the influence of dynamic environmental factors and validate the results experimentally.

Keywords: rock salts, natural hydrogen, gas impurities, clay impurities, organic matter, molecular dynamics simulations

25.1 Introduction

Salts, such as halite, sulfates, anhydrite, and gypsum, are generated from evaporation and weathering, which is part of rock structure and widely exists in geological settings. Although rock salts are widely spread, their role in the flow and transport of natural hydrogen has long been neglected. Especially, the hydrogen-fluid-rock interactions (also known as wettability) govern porous flow and accumulation capacity of hydrogen [1–3], but how the salts affect hydrogen wettability remains unclear.

Acknowledgments: The author acknowledges funding from the Permanent Carbon Locking Future Science Platform (OD-233488) and the Impossible Without You (IWY) program (OD-233197), CSIRO. The Australian Research Council is gratefully acknowledged for supporting author's DECRA fellowship (DE250100674).

Yongqiang Chen, WA School of Mines: Energy Research Unit/Permanent Carbon Locking FSP, CSIRO, Kensington, WA 6151, Australia, Minerals, Energy and Chemical Engineering, Curtin University, Kensington, WA 6151, Australia; e-mail: yongqiang.chen@csiro.au, chenyongqiang86@foxmail.com

https://doi.org/10.1515/9783111437040-025

In porous media, hydrogen flow is affected by multiple parameters, such as surface wettability, capillary force [4], interfacial tension [5], fluid saturation, and pore structure. Among these parameters, the hydrogen surface wettability refers to the relative adherence of hydrogen to brine on the rock surface, which can be evaluated by the contact angle at the hydrogen-brine-rock interface. Based on Young's equation, the surface wettability governs the capillary pressure [6]. Given that the capillary pressure is critical to trapping H_2 in the porous media, understanding the H_2-rock interaction provides insights into the state of H_2 in underground reservoirs.

Recent studies revealed the hydrogen wettability on sandstone and calcite surfaces. Iglauer et al. [1] measured hydrogen wettability on sandstone (quartz) surface. They found that the surface was more hydrogen-wet with increasing temperature, pressure, and concentration of organic matter. Al-Yaseri et al. [7] further measured the contact angle on various clays. They identified that the hydrogen wettability was a function of clay types. The kaolinite surface had the least hydrogen wetting compared to illite and montmorillonite under the same pressure and temperature. Vo Thanh et al. [8] comprehensively evaluated the hydrogen surface wettability of 513 data points collected from literature by various machine learning algorithms, including XGBoost, RF, LGRB, and Adaboost_DT. These algorithms were applied to assess the hydrogen surface interactions against salinity, pressure, temperature, and substrate minerals. Their machine learning studies proved that the substrate mineralogy is the most impactful variable among the identified four parameters. Although the experimental and relevant statistical analysis provided solid evidence, the underlying mechanisms remain insufficiently explored for H_2-salt rock interactions at the molecular scale.

Molecular dynamics (MD) simulation is a unique method that provides insights into hydrogen surface wettability at the molecular scale. MD utilizes high precision force fields to estimate the atom-atom interactions in a wide range of temperatures and pressures. In addition, MD calculates the accurate surface species density, radial distribution function ($g(r)$, RDF), and interphase energy. For example, Chen et al. [9] examined hydrogen wettability as a function of salinity, mineralogy, and organic matter. The MD method successfully identified the evolution of hydrogen wettability with increasing salinity: a more hydrogen wetting siloxane surface than aluminum surface. The amine ($-NH$) group is more hydrogen wetting than the carboxylic (-COOH) group. The density distribution and $g(r)$ plotting prove that MD is an effective method to capture the effective molecular groups responsible for hydrogen wettability. Abdel-Azeim et al. [10] explored the impact of gas impurities (CH_4 and CO_2) on H_2-brine interfacial tension and hydrogen wettability through the MD method. Their results demonstrated that the CO_2 can significantly reduce the interfacial tension (IFT) between gas and brine. Furthermore, in silica rocks, the CO_2 presence in the H_2 mixture showed lowered capillary pressure and thus can minimize snap-off and reduce H_2 loss in saline aquifers. In addition, the MD method can accurately predict the H_2 diffusion in porous media, which is key to estimating the H_2 loss in the caprock [11]. Taken together, the MD method has been proven to be a valuable tool to discover the hidden physics of H_2-rock interactions.

Although plenty of studies have revealed the importance of hydrogen wettability in sandstones, carbonates, and shale rocks, hydrogen wettability as a function of rock salts remains uncharted. Furthermore, salt is usually precipitated with organic matter, which decomposes under high temperature and high pressure. This thermal decomposition is a source of natural hydrogen. However, due to the complicated reservoir conditions, the properties of the co-existing salts-organic matter-gas mixture are challenging to explore via conventional experiments or regular reservoir simulations, which thus demands a novel method to decipher the H_2-salt interactions. In this chapter, the MD method was selected to investigate the key factors (mineralogy, organic matter, salinity, ion type, and salt rocks) underlying the H_2-salt rock interactions.

25.2 Methodology

25.2.1 Force field to simulate H_2, CO_2, and CH_4

To estimate the inter-atom forces for underground gases, a couple of force fields have been developed to evaluate the thermodynamic properties under in situ conditions. For a comprehensive review and the readers' reference, we have included the most used force fields for the gases in Table 25.1. In our model, we employed the Cygan force field [12] for CO_2 molecules, the Yang force field [13] for H_2, and the OPLSAA force field [14] for CH_4. In addition, SPC/E [15] and ClayFF [16] force fields were applied for water and kaolinite molecules, respectively. During the simulation, the kaolinite unit was treated to be rigid to guarantee a stable calculation. OPLSAA force field was selected for the decanoic acid and quinoline molecules. Halite was evaluated by the JC$_{LB}$ force field listed in Table 1 of Nezbeda et al. [17].

In Table 25.1, EPM and EMP2 force fields are from Harris and Yung [18]; SAFT-γ force field is from Avendaño et al. [19]; TraPPE force field is from Potoff and Siepmann [20]; TraPPE-flex force field is from Perez-Blanco and Maginn [21]; Higashi force field is from Higashi et al. [22]; Zhang force field is from Zhang and Duan [23]; Cygan force field is from Cygan et al. [12]. In Table 25.2, the TraPPE force field is from Martin and Siepmann [24]; the SAFT-γ force field is from Lafitte et al. [25]. In Table 25.3, the Yang force field is from Yang et al. [13]; the Buch force field is from Buch [26]; Marx force field is from Marx and Nielaba [27]; SAFT-γ Mie force field is from Nikolaidis et al. [28]. Among the force field parameters, the unit of ε_{ab}/k_B is kelvin, the unit of σ_{ab} is Å, the unit of q is e, the unit of r_0 is Å, the unit of θ_0 is degree, the unit of k_r is kJ/(mol* Å^2), the unit of k_θ is kJ/(mol*rad^2), m and n refer to exponents in the equation of Lennard-Jones potential.

Table 25.1: Frequently used force fields for CO_2 [29, 30].

		EPM	EPM2	SAFT-γ	TraPPE	TraPPE-flex	Higashi	Zhang	Cygan
ε_{ab}/k_B (K)	O-O	0.1648	80.507	–	79	79	–	82.656	80.378
	C-C	0.0576	28.129	361.69	27	27	236.1	28.845	28.144
σ_{ab} (Å)	C-C	2.785	2.757	3.72	2.8	2.8	3.72	2.7918	2.8
	O-O	3.064	3.033	–	3.05	3.05	–	3	3.028
q(e)	C	0.6645	0.6512	–	0.7	0.7	–	0.5888	0.6512
(Å)	C-O	1.161	1.149	–	1.16	1.16	–	1.163	1.162
(°)	C-O	180	180	–	180	180	–	180	180
k_r (kJ/(mol* Å²))	–	–	–	–	–	8610.7	–	–	8443
k_θ (kJ/(mol*rad²))		–	–	–	–	468.61	–	–	451.9
m		12	12	23	12	12	12	12	12
n		6	6	6.66	6	6	6	6	6

Table 25.2: Frequently used force fields for CH_4.

	CH_4			
	ε_{ab}/k_B	σ_{ab}	m	n
TraPPE	148	3.73	12	6
SAFT-γ	153.36	3.7412	12.65	6

Table 25.3: Frequently used force fields for H_2.

	H_2				
	ε_{ab}/k_B	σ_{ab}	q	m	n
Yang	10	2.72	0	12	6
Buch	34.2	2.96	0	12	6
Marx	36.7	2.958	+0.468/−0.936	12	6
SAFT- γ Mie	18.355	3.1586	0	7.813	6

25.2.2 Simulation configuration

LAMMPS (Version 03-Mar-2020) package has been employed to perform the MD simulations. During the simulation, the NaCl slab was erased with internal molecular forces, frozen at its initial positions, and the initial velocity set to 0. All the forces for each mobile group were calculated while the deformation of the NaCl slab was excluded to guarantee a stable simulation.

The system's energy was minimized at the beginning to reduce the extra forces for a smooth simulation. Then the mobile group was relaxed in a micro-canonical ensemble (NVE) for 250 ps. After the relaxation, the mobile group was simulated in a canonical ensemble for 25 ns to ensure reaching an equilibrium state. The time step was 1 fs for all the operations with a periodic boundary condition applied to the simulation cell in all three dimensions. The cut-off distance for long range forces (LJ potential and electrostatic force) was 10 Å. The pair potential between unlike atoms was calculated through the Lorentz-Berthelot mixing rule. The gravity effect was not considered at the molecular scale.

The simulation was performed in a cubic cell with a dimension of $56.4 \times 56.4 \times 62.04$ Å (Figure 25.1). The simulation box comprises a NaCl slab and a solution domain. To mimic the gas impurities and complicated sedimentary environment of rock salts, the solution domain contains various gases, clays, and organic matter, such as H_2, CH_4, CO_2, H_2O, kaolinite units, quinoline, and carboxylic acid. To be specific, the solute domain consists of 300 H_2, 50 CH_4, 20 CO_2, 30 H_2O, and 3 kaolinite units together with 5-decanoic acid and quinoline molecules to represent the gas mixture, clay impurities, with organic matter remaining in the rock salts.

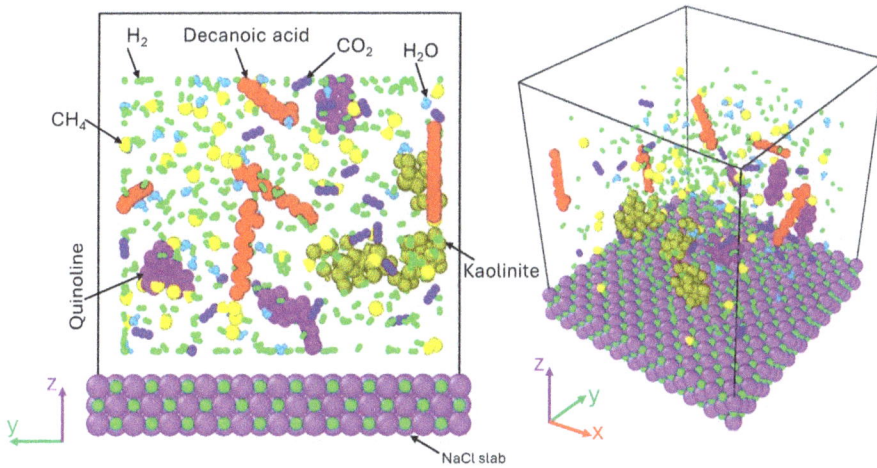

Figure 25.1: Simulation box at the initial stage. Left: orthogonal view; right: perspective view. Calculated by the Packmol and visualized by OVITO. In the solute domain, each molecular was colored as a group.

25.2.3 Data analysis and visualization

To characterize the spatial distribution of H_2, CH_4, CO_2, H_2O, and kaolinite between the NaCl slab, the density distribution was plotted for H_2, CH_4, CO_2, H_2O, and kaolinite along the vertical direction of the NaCl surface. Note: the center of mass (COM) was used for density distribution analysis. Furthermore, $g(r)$ was computed for the target molecule to evaluate the inter-atom interactions. $g(r)$ estimates the probability of one atom presenting itself around the central atom at a specific position. The visualization was achieved by the OVITO [31].

25.3 Results

25.3.1 Density distribution in the simulation box

To analyze species' spatial distribution, the density distribution is plotted (as shown in Figure 25.2). The density plotting reveals that most gas species accumulate near the NaCl slab. The first layer is H_2O, which appears at 7.83 Å and then follows CO_2, CH_4, and H_2. This distribution pattern suggests that H_2 is less likely to be in direct contact with the surface of rock salts while the water is closely coated on the surface. However, the density plotting shows that most H_2 tends to accumulate near the NaCl with no significant peaks observed except the two peaks near NaCl slabs. Furthermore, kaolinite and quinoline are more distant to the slab than gases, which implies that kao-

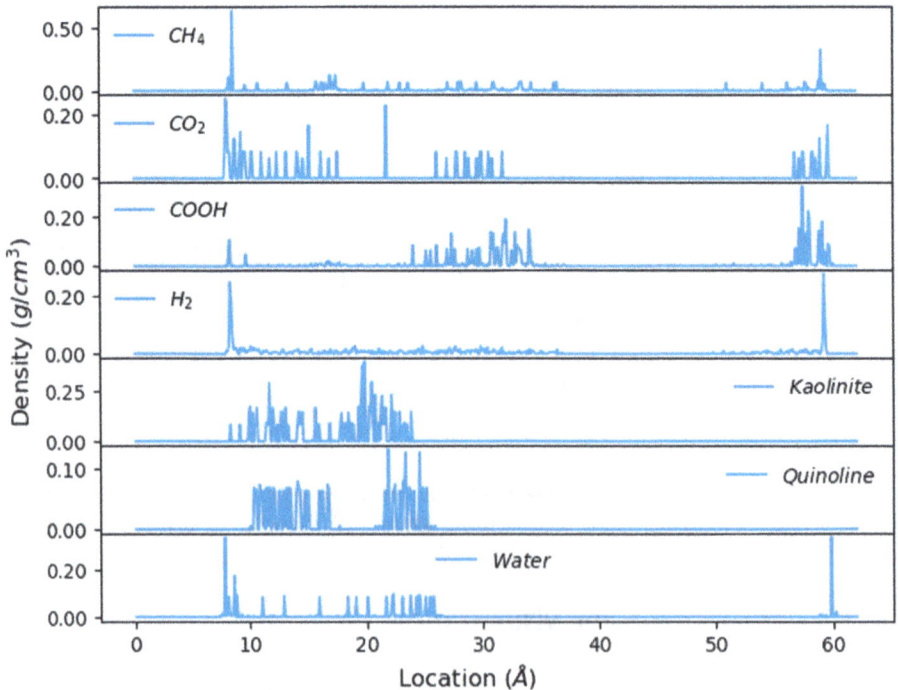

Figure 25.2: Density distribution along the direction vertical to NaCl slab.

linite and quinoline are less inclined to be absorbed onto rock salts' surface. However, the carboxylic acid group sticks to the surface of the rock salts. Taken together, the density plotting proves that water is strongly attached to rock salts while H_2 mostly distributes near the NaCl surface. This distribution pattern reveals that water vapors can be a sealing layer to significantly reduce H_2 leakage from the reservoirs.

25.3.2 Distribution of $g(r)$ near the NaCl slab

The $g(r)$ near the NaCl slab reveals the probability of atoms appearing near the NaCl slab, which indicates the adsorption capability of atoms. In Figure 25.3, the $g(r)$ around the Cl atom is characterized. The plotting shows that the H atom in the H_2O is most likely to be observed around Cl in the NaCl slab. Its first peak is at a position around 2 Å while the first peak near Cl is around 4 Å for H in CH_4, around 3.2 Å for H in H_2, around 3.2 Å for C in CH_4, around 3.2 Å for O in CO_2, and around 3.0 Å for O in H_2O, respectively. It clearly shows that the Cl atom in the NaCl slab is more likely to adsorb water molecules than other gases. Thus, the rock salt tends to be water wet, which indicates a wet water layer could be generated on the wall of salt caverns. This water layer can prevent direct contact between rock salt and the stored gases.

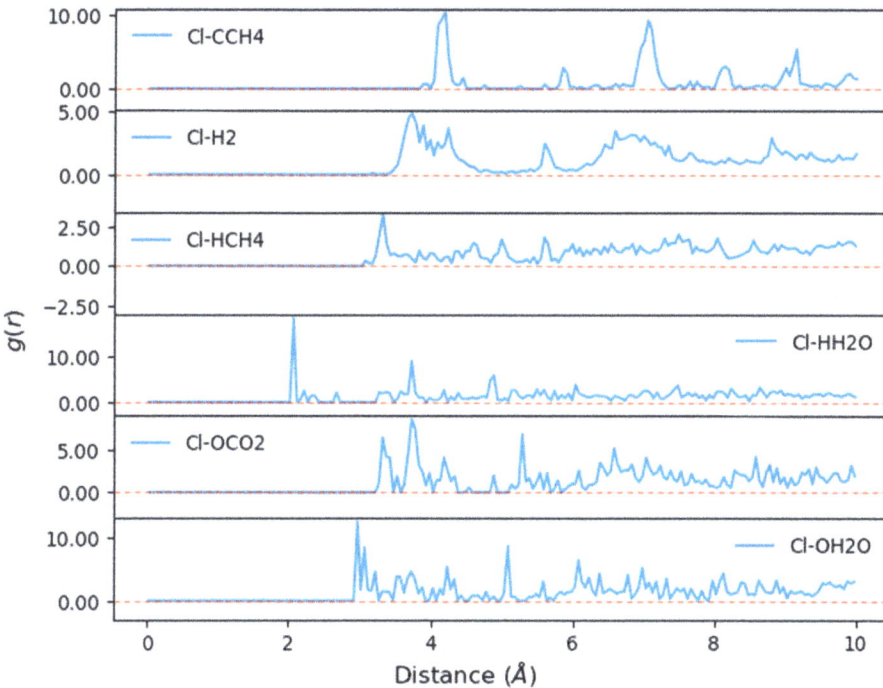

Figure 25.3: Distribution of $g(r)$ near the Cl in NaCl slab.

Figure 25.4 reveals the $g(r)$ distribution around the Na atom in the NaCl slab. The peak of $g(r)$ appears in a sequence of O in H_2O, H in H_2O, O in CO_2, O in CO_2, H_2, and CH_4. This distribution confirms that the water molecules are more likely to be adsorbed on the NaCl slab. According to the position of the first peak of $g(r)$, the CO_2 molecules are the second most absorbable gases in the rock salt. In addition, a close peak position was observed between the Na atom and atoms in H_2 and CH_4 molecules, which suggests similar adsorption capability of Na to these two gases. Furthermore, the $g(r)$ value distribution further proves the relative adsorption capability of Na to gases. The $g(r)$ value is over 20 for Na and O in H_2O, and over 50 for Na and O in CO_2, which are higher than the $g(r)$ value between Na and atoms in H_2 and CH_4.

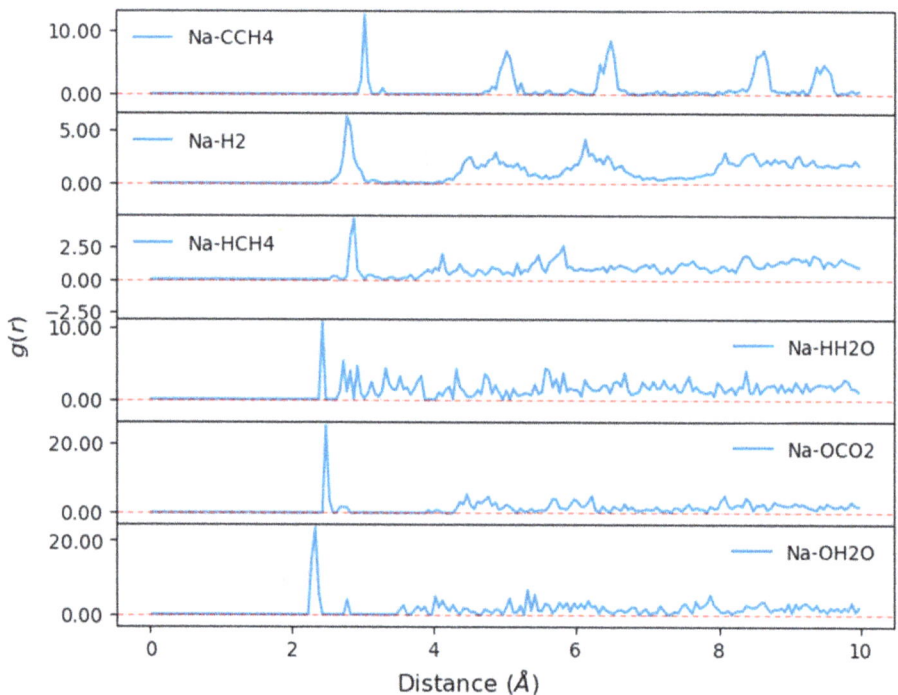

Figure 25.4: Distribution of $g(r)$ near the Na in NaCl slab.

To summarize, water is absorbed more readily to the NaCl slab while the CH_4 and CO_2 seem to be closer to the Cl atom than H_2. This distribution suggests that the H_2 is less likely to be adsorbed by rock salts. This $g(r)$ character demonstrates that the salt can lower H_2 loss caused by rock salt adsorption and a water layer can strengthen the integrity of reservoirs containing H_2.

25.3.3 Distribution of g(r) near H₂ molecules

The $g(r)$ between H_2 and gases reveals the probability of atoms appearing near H_2, which indicates the mixing trend between H_2 and gases (as shown in Figure 25.5). The first non-zero $g(r)$ is between H_2 and H in H_2O molecules, which positions at around 2 Å. Then the position of the first non-zero $g(r)$ point follows H in CH_4, C in CO_2, O in CO_2, C in CH_4, and O in H_2O. This $g(r)$ distribution suggests that H_2O molecules are more likely to be observed at a closer distance around H_2 molecules. In addition, the peak value of $g(r)$ is over 50 for O in CO_2 centered around H_2. Sorting this peak value in descending order is C in CH_4 centered around H_2 (over 25), C in CO_2 centered around H_2, H in CH_4 centered around H_2, O in H_2O centered around H_2, and H in H_2O centered around H_2. This sequence demonstrates the highest probability of observing CO_2 around H_2 molecules, which suggests a stronger mixing capacity between H_2 and CO_2 over other gas impurities. To summarize, the distribution of $g(r)$ near H_2 molecules reveals that H_2 molecules tend to mix with CO_2 while being less inclined to mix with CH_4. This pattern implies that CH_4 could be a better cushion gas than CO_2 to avoid pollution from cushion gas mixing.

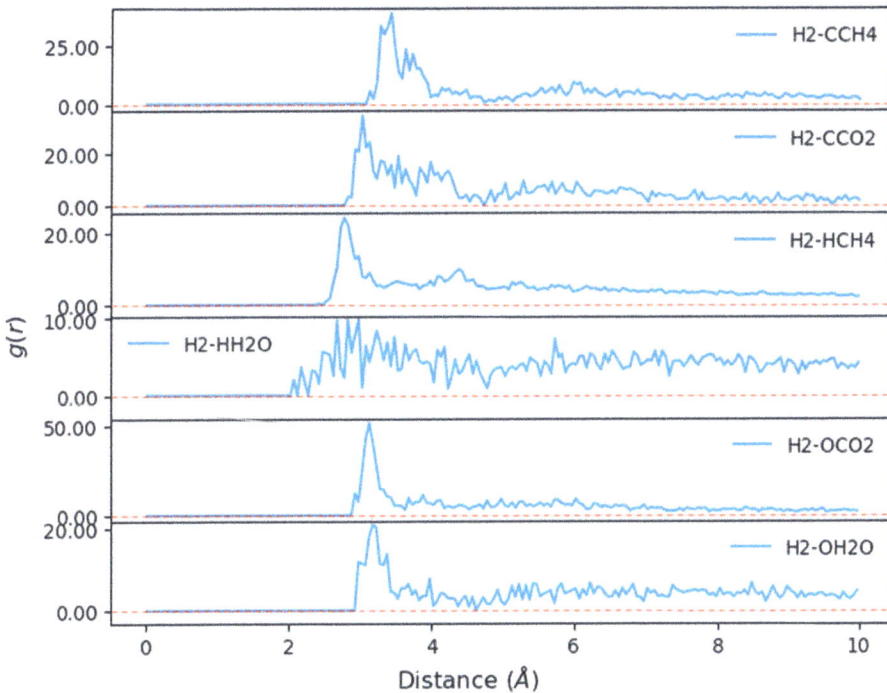

Figure 25.5: Distribution of $g(r)$ near H_2 molecules.

25.3.4 Distribution of g(r) near Al and Si in kaolinite molecules

The $g(r)$ near the Al reveals the probability of atoms appearing near the Al element in the aluminum layer, which indicates the adsorption tendency between gases and the aluminum layer (as shown in Figure 25.6). The first non-zero $g(r)$ is between Al and O in H_2O molecules. It follows O in CO_2 molecules, H in H_2O molecules, H in CH_4 molecules, C in CH_4 molecules, and H in H_2 molecules. This sequence suggests that Al in an aluminum layer has a preference to adsorb H_2O over CO_2 and CH_4. Furthermore, the value of $g(r)$ between Al and O in H_2O is over 250, which is bigger than any other $g(r)$ value. The second biggest $g(r)$ value is over 200 for between Al and H in H_2O molecules. However, the smallest $g(r)$ value is observed between Al and H in H_2 molecules. This $g(r)$ value interaction indicates that the Al in the aluminum layer has stronger adsorption to H_2O than gases, especially the H_2.

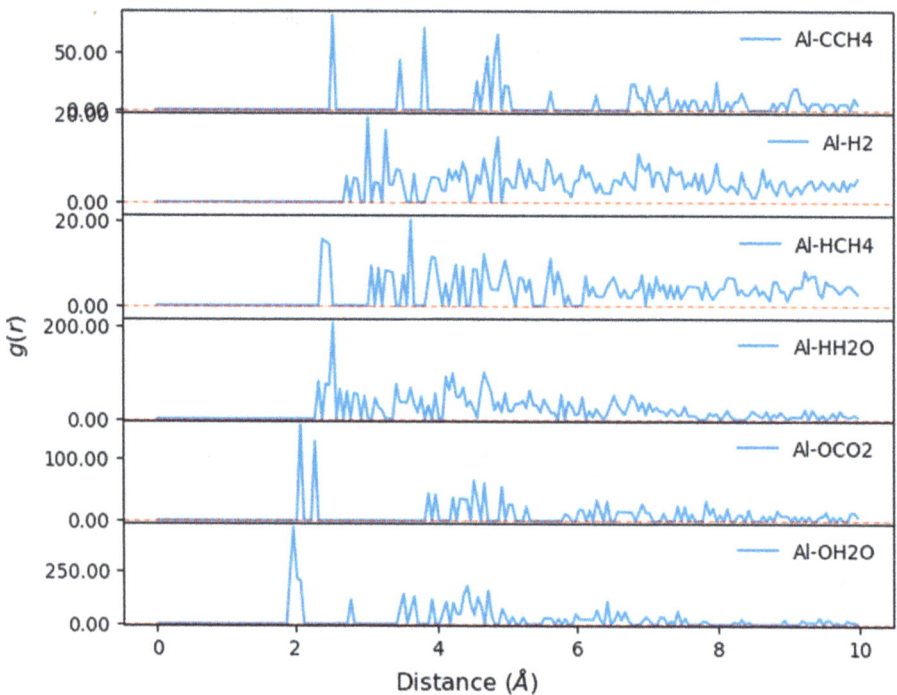

Figure 25.6: Distribution of g(r) near Al element in kaolinite molecules.

The $g(r)$ near the Si reveals the probability of atoms appearing near the Si element in the siloxane layer, which indicates the adsorption tendency between gases and the siloxane layer (as shown in Figure 25.7). The first non-zero $g(r)$ is between Si and O in H_2O molecules. It follows O in CO_2 molecules, H in H_2O molecules, H in CH_4 molecules, C in CH_4 molecules, and H in H_2 molecules. This sequence suggests that Si in the siloxane layer has a preference to adsorb H_2O over CO_2 and CH_4. This pattern is in line with the $g(r)$ around Al in aluminum. In addition, the value of $g(r)$ between Si and O in H_2O is over 1000, which is bigger than any other $g(r)$ value of Si. This value is bigger than the $g(r)$ of Al. Consistent with the $g(r)$ value of Al, the smallest $g(r)$ value is observed between Si and H in H_2 molecules. Taken together, the $g(r)$ plotting proves that the H_2 is not inclined to adsorb onto kaolinite while water is more likely to be adsorbed. This characteristic indicates that kaolinite is a favorable factor to adsorb water and other gas impurities and thus reduces H_2 loss in rock salts.

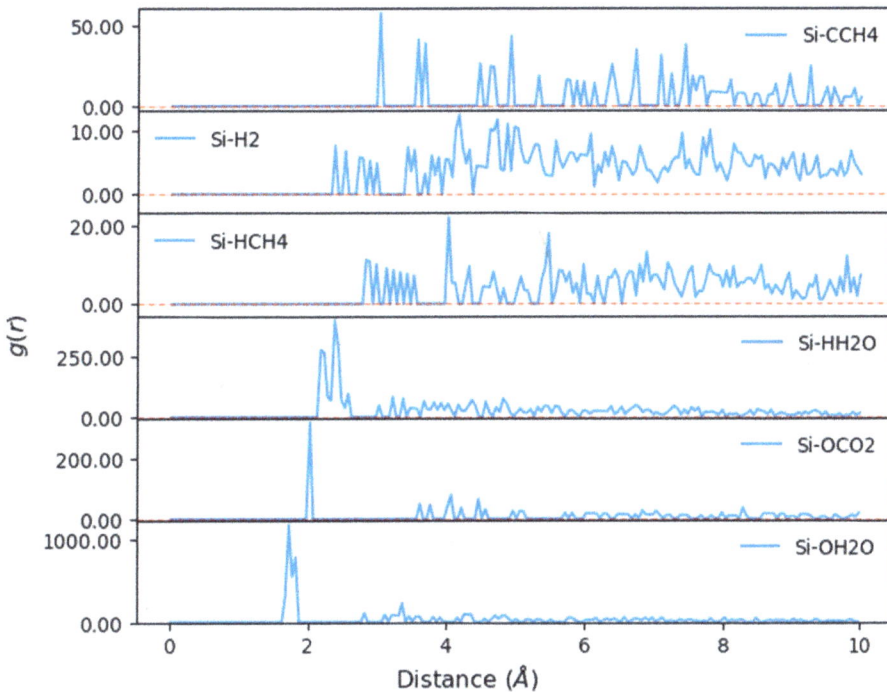

Figure 25.7: Distribution of $g(r)$ near Si element in kaolinite molecules.

25.3.5 Distribution of $g(r)$ near O element in a hydroxyl group of kaolinite molecules (indicating the hydrogen bonding formation)

The above $g(r)$ plotting shows that the hydroxyl group in kaolinite has a strong affinity to water molecules. A Hydroxyl group in kaolinite was reported to form hydrogen bonding with polar molecules [9, 32]. The $g(r)$ near the O in the hydroxyl group was thus analyzed to reveal the probability of atoms appearing near the hydroxyl group. The first non-zero $g(r)$ is between O in the hydroxyl group and H in H_2O molecules (as shown in Figure 25.8). It follows the first non-zero point of $g(r)$ between O in the hydroxyl group and H in H_2O molecules. In addition, the $g(r)$ value between O in the hydroxyl group and atoms in H_2O molecules reaches 100 and 200 for H and O atoms, respectively, which outnumbers any $g(r)$ values of other gas molecules.

Furthermore, compared to CO_2 and CH_4, it is expected that there is a low possibility to be observed of H_2 around the hydroxyl group of kaolinite. The $g(r)$ value is over 10 for H_2 while the $g(r)$ values are over 50 and 20 for CO_2 and CH_4, respectively (Figure 25.8). Taken together, all Figures 25.6–25.8 demonstrate that the existence of kaolinite is favored to absorb water and CO_2 impurities. Thus, rock salts with a certain amount of kaolinite can be an advantage in reducing impurities for natural hydrogen recovery.

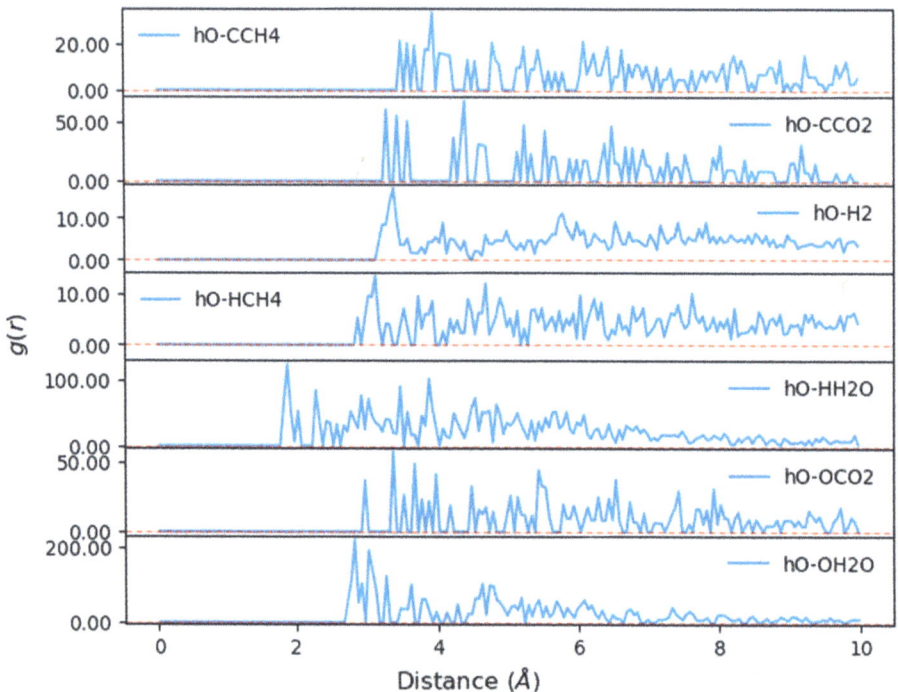

Figure 25.8: Distribution of $g(r)$ near O element in hydroxyl group of kaolinite molecules.

25.4 Discussion

Water seems to be favorable conditions to maintain a reservoir's integrity. Clay impurities reduce H_2 pollution. The density plotting reveals that water mainly accumulates near the NaCl slab. Among all the gases, water forms the most inner layer on the salt rock surface. This water layer prevents the direct contact of H_2 with the salt wall. Thus, the water molecules behave as a sealing layer to reduce H_2 leakage. This is in line with the literature. From the density plotting of Zhang et al. [33], their results confirmed that the clay surface prefers adsorbing CO_2 rather than H_2, which is in line with the $g(r)$ plotting in this study. The $g(r)$ plotting demonstrates that the $g(r)$ value between the H in H_2 and O in CO_2 outnumbers the $g(r)$ value for H_2 (Figures 25.7 and 25.8). Thus, a certain amount of clay in the salt rocks is a favorable condition to fix CO_2 impurities, which reduces H_2 adsorption and thus increases the efficiency of natural hydrogen recovery.

The $g(r)$ between H_2 and gas impurities reveals the mixing capacity between species and fills the knowledge gaps of previous studies. For example, Zhao et al. [34] found that H_2 is most diffusive in water under critical conditions while CO_2 is least diffusive in water. The diffusion coefficient of H_2 can reach 75×10^{-8} m^2/s while the diffusion coefficients of CH_4 and CO_2 are around 25 and 17×10^{-8} m^2/s, respectively. The literature study showed the highly diffusive character of H_2 in the water near critical points (600–670 K, 250 atm). However, its mixing property between gases has not been investigated. This study shows that although CO_2 is not diffusive as in the literature [34], the $g(r)$ between H_2 and CO_2 is higher than H_2 and CH_4, which demonstrates that H_2 is more likely to mix with CO_2 rather than CH_4. This feature indicates that CH_4 could be a better cushion gas than CO_2 to minimize the mixing and contamination of H_2 in salt caverns. Jia et al. [35] confirmed this with a pore network modeling. Their investigation uncovered that H_2 was more diffusive in the CO_2 than in the CH_4. Their finding is consistent with our MD results: CO_2 shows superior containment effects on H_2 compared to CH_4. However, in our simulations, the gravity force is not included and the gravity effect for mixing between gases was not evaluated, which could lead to a discrepancy between the molecular scale and meter scale.

25.5 Conclusion and implications

Salts widely exist in reservoir formations and caprocks, where they coexist with gas mixtures, clays, and organic matter. Due to this complicated geological environment, the properties of H_2-rock salt interaction are less understood at the molecular scale. We thus design and implement MD studies to reveal the H_2-rock salts interactions. We find that water and gas impurities can significantly affect the interfacial interactions and gas mixing properties.

The density distribution reveals a water-coated salt rock, which can prevent direct gas contact and thus behave as a sealing layer. This implies that a certain amount of water in the rock salts is favorable to preserving natural hydrogen.

The $g(r)$ value and distributions reveal the mixing properties between gases. We find that H_2 is most likely to mix with CO_2 rather than CH_4. Therefore, CH_4 is better than CO_2 to avoid natural hydrogen contamination.

Furthermore, the $g(r)$ between kaolinite and gases reveals that kaolinite is inclined to combine with CO_2 and water rather than H_2. This characteristic indicates that a certain amount of clays in the rock salts could immobilize the gas impurities and thus reduce gas pollution of the natural hydrogen reservoirs.

This MD study provides one of the first insights into hydrogen wettability on salt surfaces at the molecular scale. This study evaluates the role of each gas impurity, kaolinite mineral, organic matter, and water in H_2 adsorption and mixing. The results can help screen the optimal physiochemical conditions to explore potential natural hydrogen reservoirs.

25.6 Knowledge gaps and limitations

Although this study provides insights into the H_2-salt interactions, a couple of knowledge gaps have been identified as follows:

The NaCl shape was found to affect the water-salt rock interactions. Lanaro and Patey [36] identified the effect of a crystal shape in the dissolution process. Their study found that the cubic, spherical, tablet-like, and rod-like crystals followed different dissolution rate laws. However, given the rigid model is applied to NaCl in this study, the effects of crystal shape in H_2-rock salt interactions have not been included in current research.

Chen et al. [37] plotted the density distribution on the kaolinite. Their results showed that H_2 is closer to the kaolinite than H_2O, which was not in line with this study. This could be the salt effect. In this study, the salt was fixed as a rigid body while in the study of Chen et al. [37] fixed the kaolinite as a slab and treated salts as free electrolytes, which could be the reason for inconsistent results. A further study (either experimentally or mathematically) is needed to decipher the relative adsorption capacity between H_2 and H_2O in halite.

No experiments have been performed to decode the diffusion between H_2, CH_4, and CO_2 as well as their relative adsorption in the rock salts. The diffusion coefficients were not included in the MD simulations. Importantly, the H_2 diffusion coefficients in CH_4 and CO_2 mixtures have not been investigated with the variable percentage of water moisture. Furthermore, the relative adsorption of H_2, CH_4, and CO_2 in salt caverns has not been measured as a function of water percentage experimentally. Filling these knowledge gaps could be useful to minimize H_2 loss by adsorption and reduce H_2 pollution from gas mixing.

This study deciphers the mixing properties between H_2 and impurity gases at the molecular scale. The gravity is not coupled in the simulations. The gravity could play a critical role in the mixing between gases. It is thus recommended that the gravity effect be considered for the reservoir simulations.

Four force fields were listed in Table 25.2 for H_2. These available force fields rely on different physical assumptions. For example, the Max force field employs charged hydrogen atoms [27] while all other three force fields assume neutral-charged hydrogen atoms. However, the effect of hydrogen charge has not been evaluated. Given that Na and Cl are charged atoms with significant columbic force effects, the selection of hydrogen force fields needs to be fully investigated.

References

[1]   Iglauer S., Ali M., Keshavarz A. Hydrogen wettability of sandstone reservoirs: Implications for hydrogen geo-storage. Geophysical Research Letters. 2021, 48(3): e2020GL090814.

[2]   Aghaei H., et al. Host-rock and caprock wettability during hydrogen drainage: Implications of hydrogen subsurface storage. Fuel. 2023, 351: 129048.

[3]   Thaysen E. M., et al. Hydrogen wettability and capillary pressure in Clashach sandstone for underground hydrogen storage. Journal of Energy Storage. 2024, 97: 112916.

[4]   Hashemi L., Blunt M., Hajibeygi H. Pore-scale modelling and sensitivity analyses of hydrogen-brine multiphase flow in geological porous media. Scientific Reports. 2021, 11(1): 8348.

[5]   Doan Q. T., et al. Molecular dynamics simulation of interfacial tension of the CO2-CH4-water and H2-CH4-water systems at the temperature of 300 K and 323 K and pressure up to 70 MPa. Journal of Energy Storage. 2023, 66: 107470.

[6]   Young T. III. An essay on the cohesion of fluids. Philosophical Transactions of the Royal Society of London. 1805, 95: 65–87.

[7]   Al-Yaseri A., et al. Hydrogen wettability of clays: Implications for underground hydrogen storage. International Journal of Hydrogen Energy. 2021, 46(69): 34356–34361.

[8]   Vo Thanh H., et al. Predicting the wettability rocks/minerals-brine-hydrogen system for hydrogen storage: Re-evaluation approach by multi-machine learning scheme. Fuel. 2023, 345: 128183.

[9]   Chen Y., et al. Effect of salinity, mineralogy, and organic materials in hydrogen wetting and its implications for underground hydrogen storage (UHS). International Journal of Hydrogen Energy. 2023.

[10]   Alshammari S., et al. The influence of CH4 and CO2 on the interfacial tension of H2-Brine, Water–H2–rock wettability, and their implications on geological hydrogen storage. Energy & Fuels. 2024, 38(16): 15834–15847.

[11]   Hubao A., et al. H2 diffusion in cement nanopores and its implication for underground hydrogen storage. Journal of Energy Storage. 2024, 102: 113926.

[12]   Cygan R. T., Romanov V. N., Myshakin E. M. Molecular simulation of carbon dioxide capture by montmorillonite using an accurate and flexible force field. The Journal of Physical Chemistry C. 2012, 116(24): 13079–13091.

[13]   Yang X., et al. A molecular dynamics simulation study of PVT properties for H2O/H2/CO2 mixtures in near-critical and supercritical regions of water. International Journal of Hydrogen Energy. 2018, 43(24): 10980–10990.

[14]   Jorgensen W. L., Maxwell D. S., Tirado-Rives J. Development and testing of the OPLS all-atom force field on conformational energetics and properties of organic liquids. Journal of the American Chemical Society. 1996, 118(45): 11225–11236.

[15]   Berendsen H. J. C., Grigera J. R., Straatsma T. P. The missing term in effective pair potentials. The Journal of Physical Chemistry. 1987, 91(24): 6269–6271.

[16]   Cygan R. T., Liang -J.-J., Kalinichev A. G. Molecular models of hydroxide, oxyhydroxide, and clay phases and the development of a general force field. The Journal of Physical Chemistry B. 2004, 108(4): 1255–1266.

[17]   Moučka F., Nezbeda I., Smith W. R. Molecular force fields for aqueous electrolytes: SPC/E-compatible charged LJ sphere models and their limitations. The Journal of Chemical Physics. 2013, 138(15).

[18]   Harris J. G., Yung K. H. Carbon dioxide's liquid-vapor coexistence curve and critical properties as predicted by a simple molecular model. The Journal of Physical Chemistry. 1995, 99(31): 12021–12024.

[19]   Avendaño C., et al. SAFT-γ force field for the simulation of molecular fluids. 1. A single-site coarse grained model of carbon dioxide. The Journal of Physical Chemistry B. 2011, 115(38): 11154–11169.

[20]   Potoff J. J., Siepmann J. I. Vapor–liquid equilibria of mixtures containing alkanes, carbon dioxide, and nitrogen. AIChE Journal. 2001, 47(7): 1676–1682.

[21]   Perez-Blanco M. E., Maginn E. J. Molecular dynamics simulations of CO2 at an ionic liquid interface: Adsorption, ordering, and interfacial crossing. The Journal of Physical Chemistry B. 2010, 114(36): 11827–11837.

[22]   Higashi H., et al. Diffusion coefficients of aromatic compounds in supercritical carbon dioxide using molecular dynamics simulation. Kanazawa University, 1998.

[23]   Zhang Z., Duan Z. An optimized molecular potential for carbon dioxide. The Journal of Chemical Physics. 2005, 122(21).

[24]   Martin M. G., Siepmann J. I. Transferable potentials for phase equilibria. 1. United-atom description of n-alkanes. The Journal of Physical Chemistry B. 1998, 102(14): 2569–2577.

[25]   Lafitte T., et al. Accurate statistical associating fluid theory for chain molecules formed from Mie segments. The Journal of Chemical Physics. 2013, 139(15).

[26]   Buch V. Path integral simulations of mixed para-D2 and ortho-D2 clusters: The orientational effects. The Journal of Chemical Physics. 1994, 100(10): 7610–7629.

[27]   Marx D., Nielaba P. Path-integral Monte Carlo techniques for rotational motion in two dimensions: Quenched, annealed, and no-spin quantum-statistical averages. Physical Review A. 1992, 45(12): 8968–8971.

[28]   Nikolaidis I. K., et al. Modeling of physical properties and vapor – liquid equilibrium of ethylene and ethylene mixtures with equations of state. Fluid Phase Equilibria. 2018, 470: 149–163.

[29]   Aimoli C. G., Maginn E. J., Abreu C. R. A. Force field comparison and thermodynamic property calculation of supercritical CO2 and CH4 using molecular dynamics simulations. Fluid Phase Equilibria. 2014, 368: 80–90.

[30]   Chen L., Wang S., Tao W. A study on thermodynamic and transport properties of carbon dioxide using molecular dynamics simulation. Energy. 2019, 179: 1094–1102.

[31]   Stukowski A. Visualization and analysis of atomistic simulation data with OVITO–the Open Visualization Tool. Modelling and Simulation in Materials Science and Engineering. 2009, 18(1): 015012.

[32]   Chen Y., Xie Q., Niasar V. J. Insights into the nano-structure of oil-brine-kaolinite interfaces: Molecular dynamics and implications for enhanced oil recovery. Applied Clay Science. 2021, 211: 106203.

[33]   Zhang M., et al. Molecular simulation on H2 adsorption in nanopores and effects of cushion gas: Implications for underground hydrogen storage in shale reservoirs. Fuel. 2024, 361: 130621.

[34]   Zhao X., et al. Numerical study of H2, CH4, CO, O2 and CO2 diffusion in water near the critical point with molecular dynamics simulation. Computers & Mathematics with Applications. 2021, 81: 759–771.

[35] Jia Z., et al. Pore-scale binary diffusion behavior of Hydrogen-Cushion gas in saline aquifers for underground hydrogen storage: Optimization of cushion gas type. Fuel. 2025, 381: 133481.
[36] Lanaro G., Patey G. N. Molecular dynamics simulation of NaCl dissolution. The Journal of Physical Chemistry B. 2015, 119(11): 4275–4283.
[37] Chen Y., et al. Effect of salinity, mineralogy, and organic materials in hydrogen wetting and its implications for underground hydrogen storage (UHS). International Journal of Hydrogen Energy. 2023, 48(84): 32839–32848.

Index

https://doi.org/10.1515/9783111437040-026

www.ingramcontent.com/pod-product-compliance
Lightning Source LLC
Chambersburg PA
CBHW080340220326
41598CB00030B/4567